AF661245

Encyclopaedia of
SOLID GEOMETRY

Encyclopaedia of
SOLID GEOMETRY

Volume 2

R.K. Pandey

ANMOL PUBLICATIONS PVT. LTD.
NEW DELHI - 110 002 (INDIA)

ANMOL PUBLICATIONS PVT. LTD.
H.O.: 4374/4B, Ansari Road, Daryaganj,
New Delhi-110 002 (India)
Ph.: 23278000, 23261597
B.O.: No. 1015, Ist Main Road, BSK IIIrd Stage
IIIrd Phase, IIIrd Block,
Bangalore - 560 085 (India)
Visit us at: www.anmolpublications.com

Encyclopaedia of Solid Geometry
© Reserved
First Edition, 2009
ISBN 978-81-261-4010-7 (Set)

PRINTED IN INDIA

Printed at Mehra Offset Press, Delhi.

Contents

Preface

The Encyclopaedia of Solid Geometry is so presented that students will gain a knowledge of the Properties of Planes and Lines in space; Polyhedrons, Cones, Cylinders, and Spheres; further practice in the processes of logic which were begun in Plane Geometry; a realization that they must understand solid geometry if they wish to comprehend nature and art and the fascinating and diversified developments of modern means of Communication, Transportation, Construction, and Fabrication; and skills and attitudes that will be helpful as a part of the training essential in meeting his own personal problems of everyday living.

Author

List of Symbols

$\pm$	$\not\subset$	$\subset$	$\supset$	$\subseteq$	$\in$	$\notin$	∂
$\neq$	$\equiv$	$\approx$	$\angle$	∇	Σ	Π	Ω
ϕ	α	β	χ	ε	γ	η	λ
μ	ν	$\perp$	π	θ	ρ	σ	τ
$\Re$	ω	ξ	ψ	ζ	Υ	$\exists$	$\int$
$\iint$	$\iiint$	$\oint$	$\times$	$\sqrt{\ }$	$\therefore$	$\because$	Δ
$\geq$	$\leq$	$\Rightarrow$	$\Downarrow$	$\Leftrightarrow$	∞	$\propto$	$\mathbb{R}$
$\mathbb{Z}$	$\mathbb{N}$	$\mathbb{C}$	$\mathbb{Q}$				

Chapter 1

Plane and Spherical Trigonometry

It is assumed in this chapter that readers are familiar with the usual elementary formulas encountered in introductory trigonometry. We start the chapter with a brief review of the solution of a plane triangle. While most of this will be familiar to readers, it is suggested that it be not skipped over entirely, because the examples in it contain some cautionary notes concerning hidden pitfalls.

This is followed by a quick review of spherical coordinates and direction cosines in threedimensional geometry. The formulas for the velocity and acceleration components in twodimensional polar coordinates and three-dimensional spherical coordinates are developed. This is a fairly long section, and it will be essential reading for those who are contemplating making a start on celestial mechanics. The rotation of axes in two and three dimensions, including Eulerian angles and the rotation matrix of direction cosines.

A number of commonly encountered trigonometric formulas are gathered for reference.

PLANE TRIANGLES

This section is to serve as a brief reminder of how to solve a plane triangle. While there may be a temptation to pass rapidly over this section, it does contain a warning that will become even more pertinent in the section on spherical triangles.

Conventionally, a plane triangle is described by its three angles *A, B, C* and three sides *a, b, c,* with *a* being opposite to *A*, *b* opposite to *B*, and *c* opposite to *C*.

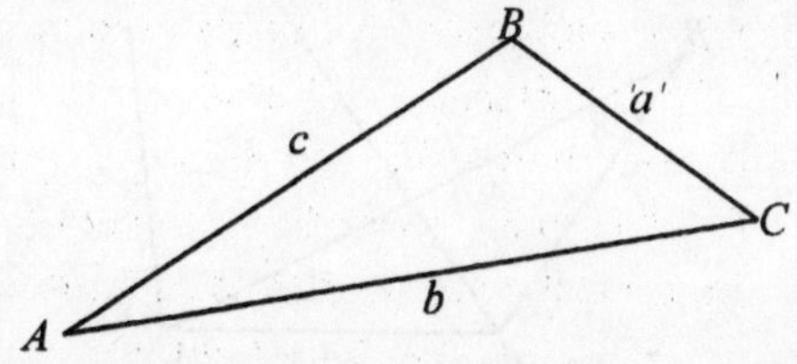

It is assumed that the reader is familiar with the sine and cosine formulas for the solution of the triangle:

$$\frac{a}{\sin A} = \frac{b}{\sin B} = \frac{c}{\sin C}$$

and $$a^2 = b^2 + c^2 - 2bc\cos A,$$

and understands that the art of solving a triangle involves recognition as to which formula is appropriate under which circumstances. Two quick examples - each with a warning - will suffice.

Example: A plane triangle has sides $a = 7$ inches, $b = 4$ inches and angle $B = 28°$. Find the angle A.

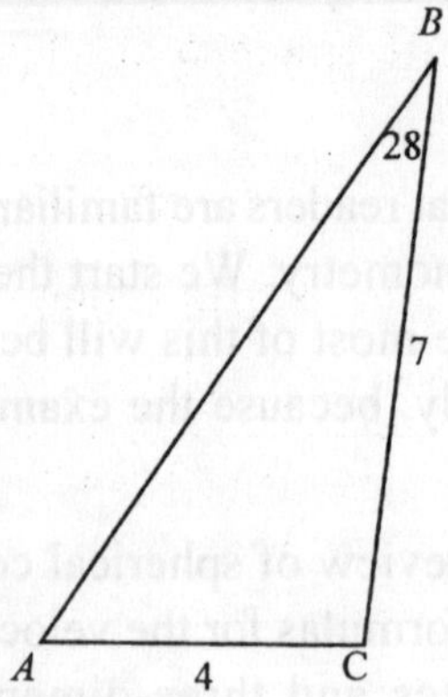

We use the sine formula, to obtain

$$\sin A \frac{7\sin 28°}{4} = 0.821575$$

$$A = 55°\ 14'.6$$

The pitfall is that there are two values of A between 0° and 180° that satisfy $\sin A = 0.821575$, namely 55° 14'.6 and 124° 45'.4. Figure shows that, given the original data, either of these is a valid solution.

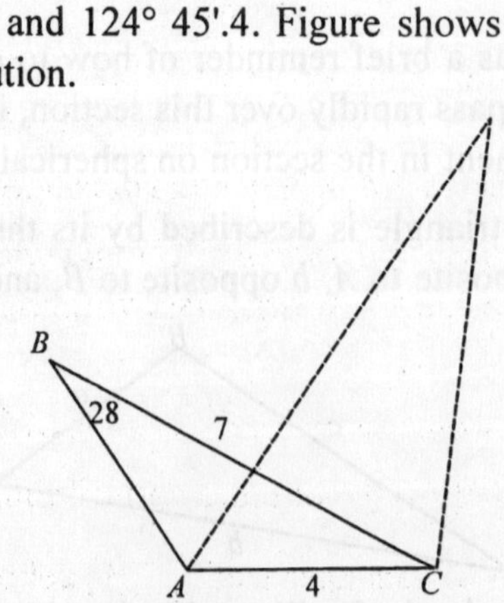

The lesson to be learned from this is that all inverse trigonometric functions ($\sin^{-1}$, $\cos^{-1}$, $\tan^{-1}$) have two solutions between 0° and 360°. The function $\sin^{-1}$ is particularly troublesome since, for positive arguments, it has two solutions between 0° and 180°. The reader must always be on guard for "quadrant problems" (i.e. determining which quadrant the desired solution belongs to) and is warned that, unless particular care is taken in programming calculators or computers, quadrant problems are among the most frequent problems in trigonometry, and especially in spherical astronomy.

Example: Find x in the triangle illustrated in figure.

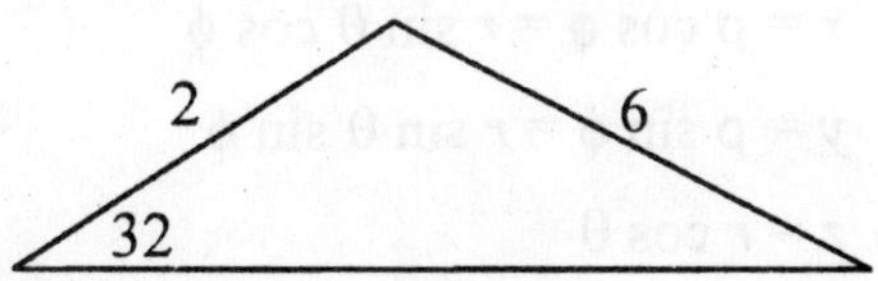

Application of the cosine rule results in

$$25 = x^2 + 64 - 16x \cos 32°$$

Solution of the quadratic equation yields

$$x = 4.133 \text{ or } 9.435$$

This illustrates that the problem of "two solutions" is not confined to angles alone. Figure is drawn to scale for one of the solutions; the reader should draw the second solution to see how it is that two solutions are possible.

The reader is now invited to try the following "guaranteed all different" problems by hand calculator. Some may have two real solutions. Some may have none.

The reader should draw the triangles accurately, especially those that have two solutions or no solutions. It is important to develop a clear geometric understanding of trigonometric problems, and not merely to rely on the automatic calculations of a machine.

Developing these critical skills now will pay dividends in the more complex real problems encountered in celestial mechanics and orbital computation.

The area of a plane triangle is $\frac{1}{2}$ × base × height, and it is easy to see from this that

$$\text{Area} = \frac{1}{2}\, bc \sin A = \frac{1}{2}\, ca \sin B = \frac{1}{2}\, ab\cos C$$

By making use $\sin^2 A = 1 - \cos^2 A$ and $\cos A = (b^2 + c^2 - a^2)/(2bc)$, we can express this entirely in terms of the lengths of the sides:

$$\text{Area} = \sqrt{s(s-a)(s-b)(s-c)}.$$

where s is the semi-perimeter $\frac{1}{2}(a + b + c)$.

CYLINDRICAL AND SPHERICAL COORDINATES

It is assumed that the reader is at least somewhat familiar with cylindrical coordinates (ρ, ϕ, z) and spherical coordinates (r, θ, ϕ) in three dimensions, and I offer only a brief summary here. Figure illustrates the following relations between them and the rectangular coordinates (x, y, z).

$$x = \rho \cos \phi = r \sin \theta \cos \phi$$

$$y = \rho \sin \phi = r \sin \theta \sin \phi$$

$$z = r \cos \theta$$

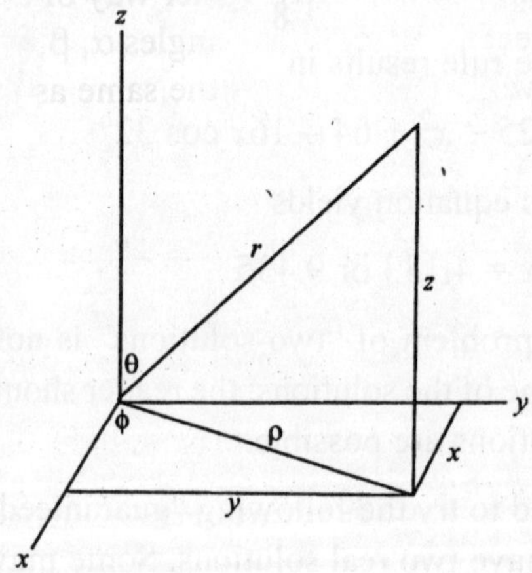

The inverse relations between spherical and rectangular coordinates are

$$r = \sqrt{x^2 + y^2 + z^2}$$

$$\theta = \cos^{-1} \frac{1}{\sqrt{x^2 + y^2 + z^2}}$$

$$\theta = \cos^{-1} \frac{1}{\sqrt{x^2 + y^2}} = \sin^{-1} \frac{y}{\sqrt{x^2 + y^2}}$$

The coordinates r, θ and ϕ are called, respectively, the "radial", "polar" or "meridional", and "azimuthal" coordinates respectively.

Note that r is essentially positive (the symbol " denotes the positive or absolute value of the square root). The angle θ is necessarily between 0° and 180° and therefore there is no quadrant ambiguity in the evaluation of θ. The angle ϕ, however, can be between 0° and 360°. Therefore, in order to determine ϕ uniquely, both of the above formulas for ϕ must be evaluated; or the signs of x and y must be inspected. It does not suffice to calculate ϕ from $\phi = \tan^{-1}(y/x)$ alone.

The reader, however, should be aware that some computer languages and some hand calculator functions will inspect the signs of x and y for you and will return ϕ in its correct quadrant. For example, in FORTRAN, the function ATAN2(X,Y) (or DATAN2(X,Y) in double precision) will return ϕ uniquely in its correct quadrant (though perhaps as a negative angle, in which case 360° should be added to the outputted angle) provided the arguments X and Y are inputted with their correct signs. This can save an immense amount of trouble in programming, and the reader should become familiar with this function.

Direction cosines

The direction to a point in three dimensional space relative to the origin can be described, as we have seen, by the two angles , and Æ. Another way of describing the direction to a point, or the orientation of a vector, is to give the angles α, β, γ that the vector makes with the x-, y- and z-axes, respectively. The angle γ is the same as the angle θ.

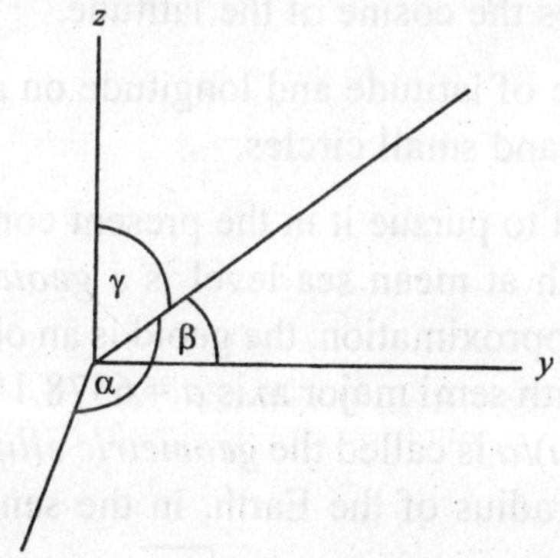

More commonly one quotes the cosines of these three angles. These are called the *direction cosines*, and are often denoted by (l, m, n). It should not take long for the reader to be convinced that the relation between the direction cosines and the angles θ and ϕ are

$$l = \cos\alpha = \sin\theta\cos\phi$$

$$m = \cos\beta = \sin\theta\sin\phi$$

$$n = \cos\gamma = \cos\theta$$

These are not independent, and are related by

$$l^2 + m^2 + n^2 = 1.$$

A set of numbers that are multiples of the direction cosines - i.e. are proportional to them - are called *direction ratios.*

Latitude and Longitude

The figure of the Earth is not perfectly spherical, for it is slightly flattened at the poles. For the present, however, our aim is to become familiar with spherical coordinates and with the geometry of the sphere, so we shall suppose the Earth to be spherical. In that case, the position of any town on Earth can be expressed by two coordinates, the latitude ϕ, measured north or south of the equator, and the longitude λ, measured eastwards or westwards from the meridian through Greenwich.

These symbols, ϕ for latitude and λ for longitude, are unfortunate, but are often used in this context. In terms of the symbols θ, ϕ for spherical coordinates that we have used hitherto, the east longitude would correspond to ϕ and the latitude to $90° - \theta$.

A plane that intersects a sphere does so in a circle. If that plane passes through the centre of the sphere (so that the centre of the circle is also the centre of the sphere), the circle is called a *great circle*. All the meridians (the circles of fixed longitude that pass through the north and south poles) including the one that passes through Greenwich, are great circles, and so is the equator. Planes that do not pass through the centre of the sphere (such as parallels of latitude) are *small circles*. The radius of a parallel of latitude is equal to the radius of the sphere times the cosine of the latitude.

We have used the example of latitude and longitude on a spherical Earth in order to illustrate the concepts of great and small circles.

Although it is not essential to pursue it in the present context, we mention in passing that the true figure of the Earth at mean sea level is a *geoid* - which merely means the shape of the Earth. To a good approximation, the geoid is an oblate spheroid (i.e. an ellipse rotated about its minor axis) with semi major axis $a = 6378.140$ km and semi minor axis $c = 6356.755$ km. The ratio $(c - a)/a$ is called the *geometric ellipticity* of the Earth and it has the value 1/298.3. The mean radius of the Earth, in the sense of the radius of a sphere having the same volume as the actual geoid, is $\sqrt[3]{a^2c} = 6371.00$ km.

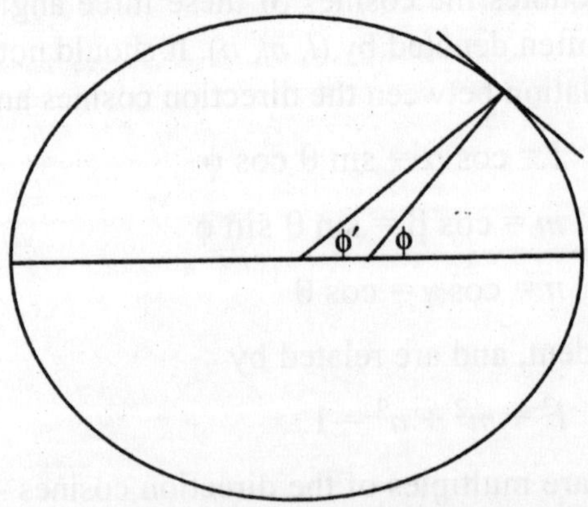

It is necessary in precise geodesy to distinguish between the geographic or geodetic latitude ϕ of a point on the Earth's surface and its geocentric latitude ϕ' Their definitions evident from figure. In this figure, the ellipticity of the Earth is greatly exaggerated; in reality it would scarcely be discernible. The angle ϕ is the angle between a plumb-bob and the equator. This differs from ϕ' partly because the gravitational field of a spheroid is not the same as that of an equal point mass at the centre, and partly because the plumb bob is pulled away from the Earth's rotation axis by centrifugal force.

The relationship between ϕ and ϕ' is

$$\phi - \phi' = 692'.74 \sin 2\phi - 1'.16 \sin 4\phi.$$

VELOCITY AND ACCELERATION COMPONENTS

i. Two-dimensional Polar Coordinates: Sometimes the symbols *r* and , are used for two-dimensional polar coordinates, but in this section I use (ρ, ϕ) for consistency with the (*r*, θ, ϕ) of three-dimensional spherical coordinates. In what follows I am setting vectors in boldface. If you make a print-out, you should be aware that some printers apparently do not print Greek letter symbols in boldface, even though they appear in boldface on screen. You should be on the look-out for this. Symbols with ^ above them are intended as *unit* vectors, so you will know that they should be in boldface even if your printer does not recognize this. If in doubt, look at what appears on the screen.

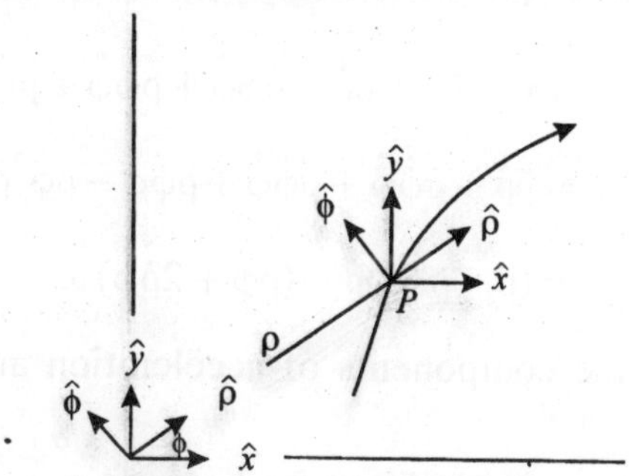

Figure shows a point *P* moving along a curve such that its polar coordinates are changing at rates ρ and ϕ. The drawing also shows fixed unit vectors $\hat{x}$ and $\hat{y}$ parallel to the *x*- and *y*-axes, as well as unit vectors $\hat{\rho}$ and $\hat{\phi}$ in the radial and transverse directions. We shall find expressions for the rate at which the unit radial and transverse vectors are changing with time. (Being unit vectors, their magnitudes do not change, but their directions do.)

We have $\quad \hat{\rho} = \cos\phi\hat{x} + \sin\phi\hat{y}$

and $\quad \hat{\phi} = -\sin\phi\hat{x} + \cos\phi\hat{y}$

$\therefore \quad \dot{\hat{\rho}} = -\sin\phi\dot{\phi}\hat{x} + \cos\phi\dot{\phi}\hat{y} = \dot{\phi}(-\sin\phi\hat{x} + \cos\phi\hat{y})$

$\therefore \quad \dot{\hat{\rho}} = \dot{\phi}\hat{\phi}.$

In a similar manner, by differentiating equation with respect to time and then making use of equation, we find

$$\dot{\hat{\phi}} = -\dot{\phi}\hat{\rho}$$

Equations give the rate of change of the radial and transverse unit vectors. It is worthwhile to think carefully about what these two equations mean.

The position vector of the point P can be represented by the expression $\rho = \rho\hat{\rho}$. The velocity of P is found by differentiating this with respect to time:

$$v = \dot{\rho} = \dot{\rho}\hat{\rho} + \rho\dot{\hat{\rho}} = \dot{\rho}\hat{\rho} + \rho\dot{\phi}\hat{\phi}$$

The radial and transverse components of velocity are therefore $\dot{\rho}$ and $\rho\,\dot{\phi}$ respectively.

The acceleration is found by differentiation of equation, and we have to differentiate the products of two and of three quantities that vary with time:

$$a = \dot{v} = \ddot{\rho}\hat{\rho} + \dot{\rho}\dot{\hat{\rho}} + \dot{\rho}\dot{\phi}\hat{\phi} + \rho\ddot{\phi}\hat{\phi} + \rho\dot{\phi}\dot{\hat{\phi}}$$

$$= \ddot{\rho}\hat{\rho} + \dot{\rho}\dot{\phi}\hat{\phi} + \dot{\rho}\dot{\phi}\hat{\phi} + \rho\ddot{\phi}\hat{\phi} - \rho\dot{\phi}^2\hat{\rho}$$

$$= (\ddot{\rho} - \rho\dot{\phi}^2)\hat{\rho} + (\rho\ddot{\phi} + 2\dot{\rho}\dot{\phi})\hat{\phi}.$$

The radial and transverse components of acceleration are therefore $(\ddot{\rho} - \rho\dot{\phi}^2)$ and $(\rho\ddot{\phi} + 2\dot{\rho}\dot{\phi})$ respectively.

ii. Three-dimensional Spherical Coordinates

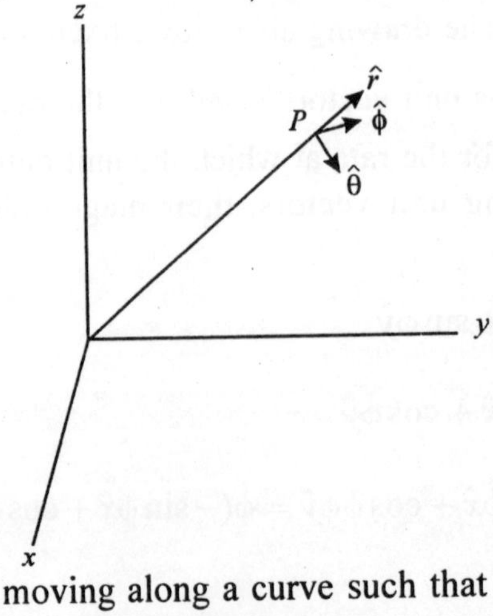

In figure, P is a point moving along a curve such that its spherical coordinates are changing at rates $\dot{r}, \dot{\theta}, \dot{\phi}$. We want to find out how fast the unit vectors $\hat{r}, \hat{\theta}, \hat{\phi}$ in the radial, meridional and azimuthal directions are changing.

We have $\hat{r} = \sin\theta\cos\phi\hat{x} + \sin\theta\sin\phi\hat{y} + \cos\theta\hat{z}$

$$\hat{\theta} = \cos\theta\cos\phi\hat{x} + \cos\theta\sin\phi\hat{y} - \sin\theta\hat{z}$$

$$\hat{\phi} = -\sin\phi\hat{x} + \cos\phi\hat{y}$$

$$\therefore \quad \dot{\hat{r}} = (\cos\theta\dot{\theta}\cos\phi - \sin\theta\sin\phi\dot{\phi})\hat{x} + (\cos\theta\dot{\theta}\sin\phi + \sin\theta\cos\phi\dot{\phi})\hat{y} - \sin\theta\dot{\theta}\hat{z}.$$

We see, by comparing this with equations that

$$\dot{\hat{r}} = \dot{\theta}\hat{\theta} + \sin\theta\dot{\phi}\hat{\phi}$$

By similar arguments we find that

$$\dot{\hat{\theta}} = \cos\theta\dot{\phi}\hat{\phi} - \dot{\theta}\hat{r}$$

and $\quad \dot{\hat{\theta}} = -\sin\theta\dot{\phi}\hat{r} - \cos\theta\dot{\phi}\hat{\theta}$

These are the rates of change of the unit radial, meridional and azimuthal vectors.

The position vector of the point P can be represented by the expression $r = r\hat{r}$. The velocity of P is found by differentiating this with respect to time:

$$v = \dot{r} = \dot{r}\hat{r} + r\dot{\hat{r}} = \dot{r}\hat{r} + r(\dot{\theta}\hat{\theta} + \sin\theta\dot{\phi}\hat{\phi})$$

$$= \dot{r}\hat{r} + r\dot{\theta}\hat{\theta} + r\sin\theta\dot{\phi}\hat{\phi}$$

The radial, meridional and azimuthal components of velocity are therefore $\dot{r}$, $r\dot{\theta}$, and $r \sin\,\theta\dot{\phi}$ respectively. The acceleration is found by differentiation of equation. [It might not be out of place here for a quick hint about differentiation. Most readers will know how to differentiate a product of two functions. If you want to differentiate a product of several functions, for example four functions, *a, b, c* and *d*, the procedure is *(abcd)' = a'bcd + ab'cd + abc'd + abcd'*. In the last term of equation, all four quantities vary with time, and we are about to differentiate the product.]

$$a = \dot{v} = \ddot{r}\hat{r} + \dot{r}(\dot{\theta}\hat{\theta} + \sin\theta\dot{\phi}\hat{\phi}) + \dot{r}\dot{\theta}\hat{\theta} + r\ddot{\theta}\hat{\theta} + r\dot{\theta}(\cos\theta\dot{\phi}\hat{\phi} - \dot{\theta}\hat{r})$$

$$+\dot{r}\sin\theta\dot{\phi}\hat{\phi} + r\cos\dot{\theta}\dot{\phi}\hat{\phi} + r\sin\ddot{\theta}\hat{\theta} + r\sin\theta(-\sin\theta\dot{\phi}\hat{r} - \cos\theta\dot{\phi}\hat{\theta})$$

On gathering together the coefficients of $\dot{r}, \hat{\theta}, \hat{\phi}$, we find that the components of acceleration are:

Radial: $\ddot{r} - r\dot{\theta}^2 - r\sin^2\theta\dot{\phi}^2$

Meridional: $r\ddot{\theta} + 2\dot{r}\dot{\theta} - r\sin\theta\cos\theta\dot{\phi}^2$

Azimuthal: $2\dot{r}\dot{\theta}\sin\theta + 2r\dot{\theta}\dot{\phi}\cos\theta + r\sin\theta\ddot{\phi}$

SPHERICAL TRIANGLES

As with plane triangles, we denote the three angles by *A, B, C* and the sides opposite to them by *a, b, c*. We are fortunate in that we have four formulas at our disposal for the solution of a spherical triangle, and, as with plane triangles, the art of solving a spherical triangle entails understanding which formula is appropriate under given circumstances. Each formula contains four elements (sides and angles), three of which, in a given problem, are assumed to be known, and the fourth is to be determined.

Three important points are to be noted before we write down the formulas.

1. The formulas are valid only for triangles in which the three sides are arcs of great circles. They will not do, for example, for a triangle in which one side is a parallel of latitude.
2. The sides of a spherical triangle, as well as the angles, are all expressed in angular measure (degrees and minutes) and not in linear measure (metres or kilometres). A side of 50° means that the side is an arc of a great circle subtending an angle of 50° at the centre of the sphere.
3. The sum of the three angles of a spherical triangle add up to more than 180°.

In this section are now given the four formulas without proof, the derivations being given in a later section. The four formulas may be referred to as the sine formula, the cosine formula, the polar cosine formula, and the cotangent formula. Beneath each formula is shown a spherical triangle in which the four elements contained in the formula are highlighted.

The sine formula:

$$\frac{\sin a}{\sin A} = \frac{\sin b}{\sin B}\left(= \frac{\sin c}{\sin C}\right)$$

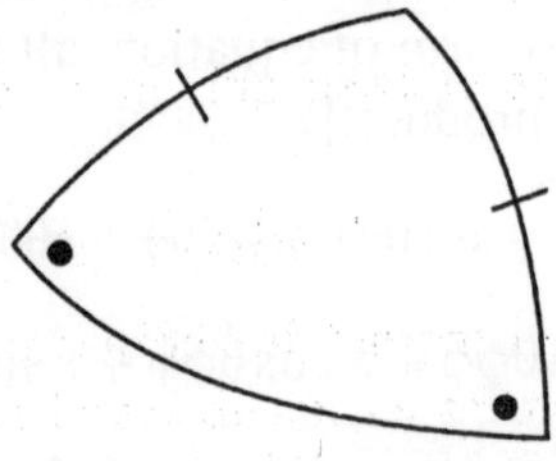

The cosine formula: $\cos a = \cos b \cos c + \sin b \sin c \cos A$

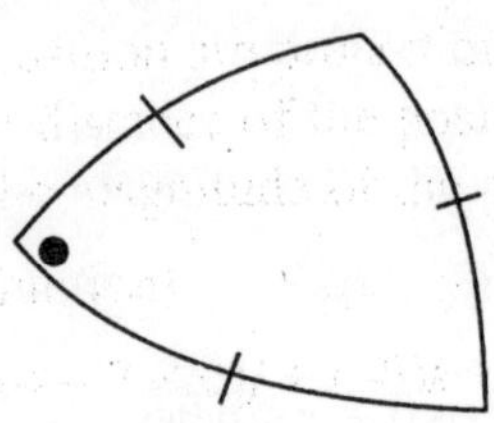

The polar cosine formula:

$$\cos A = -\cos B \cos C + \sin B \sin C \cos a$$

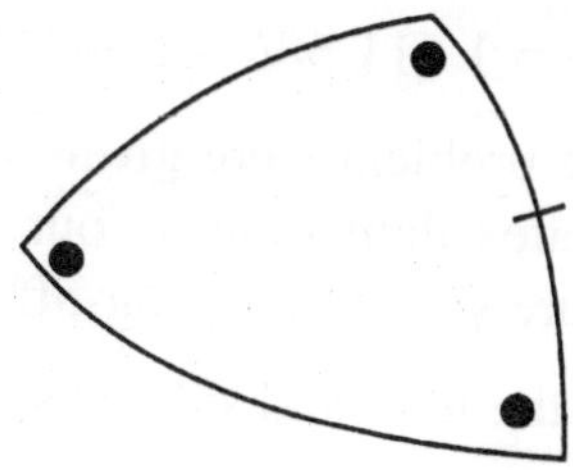

The cotangent formula:

$$\cos b \cos A = \sin b \cot c - \sin A \cot C$$

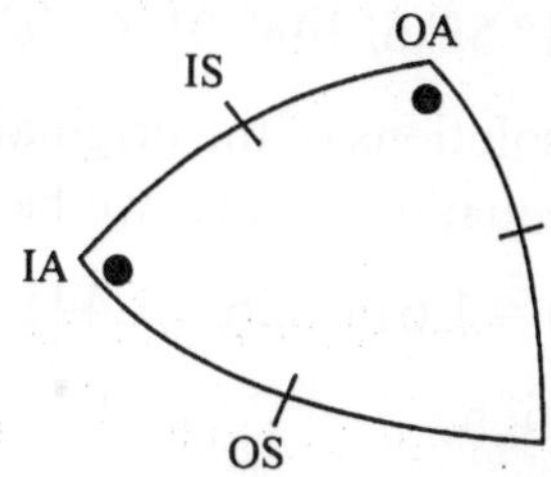

The cotangent formula is a particularly useful and frequently needed formula, and it is unfortunate that it is not only difficult to commit to memory but, even with the formula written out in front of one, it is often difficult to decide which is b, which is A and so on. However, it should be noted from the drawing that the four elements, side-angle-side-angle, lie adjacent to each other in the triangle, and they may be referred to as outer side (OS), inner angle (IA), inner side (IS) and outer angle (OA) respectively. Many people find that the formula is much easier to use when written in the form

$$\cos (\text{IS}) \cos (\text{IA}) = \sin (\text{IS}) \cot (\text{OS}) - \sin (\text{IA}) \cot (\text{OA})$$

The reader will shortly be offered a goodly number of examples in the use of these formulas. However, during the course of using the formulas, it will be found that there is frequent need to solve deceptively simple trigonometric equations of the type

$$4.737 \sin \theta + 3.286 \cos \theta = 5.296$$

After perhaps a brief pause, one of several methods may present themselves to the reader - but not all methods are equally satisfactory. I am going to suggest four possible ways of solving this equation. The first method is one that may occur very quickly to the reader as being perhaps rather obvious - but there is a cautionary tale attached to it. While the method may seem very obvious, a difficulty does arise, and the reader would be advised to prefer one of the less obvious methods. There are, incidentally, two solutions to the equation between 0° and 360°. They are 31° 58'.6 and 78° 31'.5.

Method i.

The obvious method is to isolate cos θ:

$$\cos\theta = 1.611\,686 - 1.441\,570 \sin\theta.$$

Although the constants in the problem were given to four significant figures, do not be tempted to round off intermediate calculations to four. It is a common fault to round off intermediate calculations prematurely. The rounding-off can be done at the end.

Square both sides, and write the left hand side, $\cos^2\theta$, as $1 - \sin^2\theta$. We now have a quadratic equation in sin θ:

$$3.078125 \sin^2\theta - 4.646\,717 \sin\theta + 1.597\,532 = 0.$$

The two solutions for sin θ are 0.529 579 and 0.908 014 and the four values of θ that satisfy these values of sin θ are 31° 58'.6, 148° 01'.4, 78° 31'.5 and 101° 28'.5.

Only two of these angles are solutions of the original equation. The fatal move was to square both sides of the original equation, so that we have found solutions not only to

$$\cos\theta = 1.611\,686 - 1.441\,570 \sin\theta$$

but also to the different equation

$$-\cos\theta = 1.611\,686 - 1.441\,570 \sin\theta.$$

This generation of extra solutions always occurs whenever we square an equation. For this reason, method (i), however tempting, should be avoided, particularly when programming a computer to carry out a computation automatically and uncritically.

Method ii.

This method makes use of the identities

$$\sin\theta = \frac{2t}{1+t^2}, \cos\theta = \frac{1-t^2}{1+t^2},$$

where $t = \tan\frac{1}{2}\theta$.

When applied to the original equation, this results in the quadratic equation in t:

$$8.582t^2 - 9.474t + 2.010 = 0$$

with solutions $t = 0.286528$ and $t = 0.817410$.

The only values of θ between 0° and 360° that satisfy these are the two correct solutions 31° 58'.6 and 78° 31'.5.

It is left as an exercise to show, using this method algebraically, that the solutions to the equation

$$a \cos \theta + b \cos \theta = c$$

are given by

$$\tan \frac{1}{2}\theta = \frac{a \pm \sqrt{a^2 + b^2 - c^2}}{b + c}$$

Method iii.

We divide the original equation

$$4.737 \sin \theta + 3.286 \cos \theta = 5.296$$

by the "hypotenuse" of 4.737 and 3.286; that is, by $\sqrt{(4.737^2 + 3.286^2)} = 5.765151$.

Thus $\quad 0.821\,661 \sin \theta + 0.569\,976 \cos \theta = 0.918\,623$

Now let $0.821\,661 = \cos \alpha$ and $0.569\,976 = \sin \alpha$ (which we can, since these numbers now satisfy $\sin^2 \alpha + \cos^2 \alpha = 1$) so that $\alpha = 34° \, 44'.91$.

We have $\quad \cos \alpha \sin \theta + \sin \alpha \cos \theta = 0.918\,623$

or $\quad \sin (\theta + \alpha) = 0.918623$

from which $\quad \theta + \alpha = 66° \, 43'.54$ or $113° \, 16'.46$

Therefore $\quad \theta = 31° \, 58'.6$ or $78° \, 31'.5$

Method iv.

Methods ii and iii give explicit solutions, so there is perhaps no need to use numerical methods. Nevertheless, the reader might like to solve, by Newton-Raphson iteration, the equation

$$f(\theta) = a \sin \theta + b \cos\theta - c = 0,$$

for which $f'(\theta) = a \cos \theta - b \sin \theta$.

Using the values of *a, b* and *c* from the example above and using the Newton-Raphson algorithm, we find with a first guess of 45° the following iterations, working in radians:

0.785 398

0.417 841

0.541 499

0.557 797

0.558 104

0.558 104 = 31°58'.6

The reader should verify this calculation, and, using a different first guess, show that Newton- Raphson iteration quickly leads to 78° 31'.5.

Having now cleared that small hurdle, the reader is invited to solve the spherical triangle problems below. Although these twelve problems look like pointless repetitive work, they are in fact all different. Some have two solutions between 0° and 360°; others have just one. After solving each problem, the reader should sketch each triangle - especially those that have two solutions - in order to see how the two-fold ambiguities arise. The reader should also write a computer program that will solve all twelve types of problem at the bidding of the user.

Answers should be given in degrees, minutes and tenths of a minute, and should be correct to that precision. For example, the answer to one of the problems is 47° 37'.3. An answer of 47° 37'.2 or 47° 37'.4 should be regarded as wrong. In celestial mechanics, there is no place for answers that are "nearly right".

An answer is either right or it is wrong. (This does not mean, of course, that an angle can be measured with no error at all; but the answer to a calculation given to a tenth of an arcminute should be correct to a tenth of an arcminute.)

DERIVATION OF THE FORMULAS

Before moving on to further problems and applications of the formulas, it is time to derive the four formulas which, until now, have just been given without proof. We start with the cosine formula.

There is no loss of generality in choosing rectangular axes such that the point A of the spherical triangle ABC is on the z-axis and the point B and hence the side c are in the zx-plane. The sphere is assumed to be of unit radius.

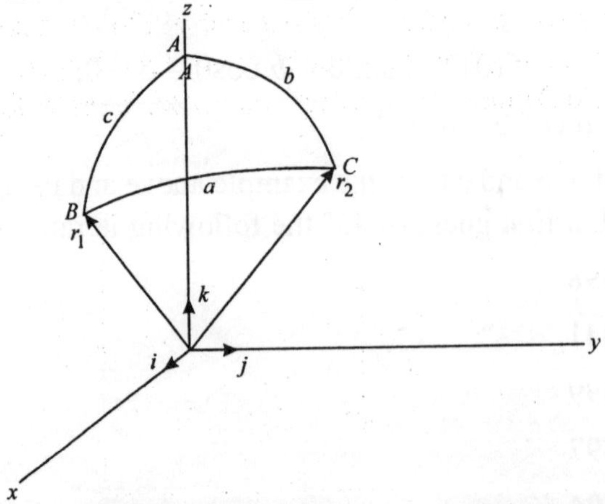

If i, j and k are unit vectors directed along the x–, y– and z–axes respectively, inspection of the figure will show that the position vectors of the points B and C with respect to the centre of the sphere are

$$r_1 = i \sin c + k \cos c$$

and $\quad r_2 = i \sin b \cos A + j \sin b \sin A + k \cos b$

respectively. The scalar product of these vectors (each of magnitude unity) is just the cosine of the angle between them, namely cos a, from which we obtain immediately

$$\cos a = \cos b \cos c + \sin b \sin c \cos A.$$

To obtain the sine formula, we isolate cos A from this equation, square both sides, and write $1 - \sin^2 A$ for $\cos^2 A$. Thus,

$$(\sin b \sin c \cos A)^2 - (\cos a - \cos b \cos c)^2,$$

and when we have carried out these operations we obtain

$$\sin^2 A = \frac{\sin^2 b \sin^2 c - \cos^2 a - \cos^2 b \cos^2 c + 2\cos a \cos b \cos c}{\sin^2 b \sin^2 c}.$$

In the numerator, write $1 - \cos^2 b$ for $\sin^2 b$ and $1 - \cos^2 c$ for $\sin^2 c$, and divide both sides by $\sin^2 a$.

This results in

$$\frac{\sin^2 A}{\sin^2 a} = \frac{1 - \cos^2 a - \cos^2 b - \cos^2 c + 2\cos a \cos b \cos c}{\sin^2 a \sin^2 b \sin^2 c}..$$

At this stage the reader may feel that we are becoming bogged down in heavier and heavier algebra and getting nowhere. But, after a careful look at equation, it may be noted with some delight that the next line is:

Therefore

$$\frac{\sin A}{\sin a} = \frac{\sin B}{\sin b} = \frac{\sin C}{\sin c}.$$

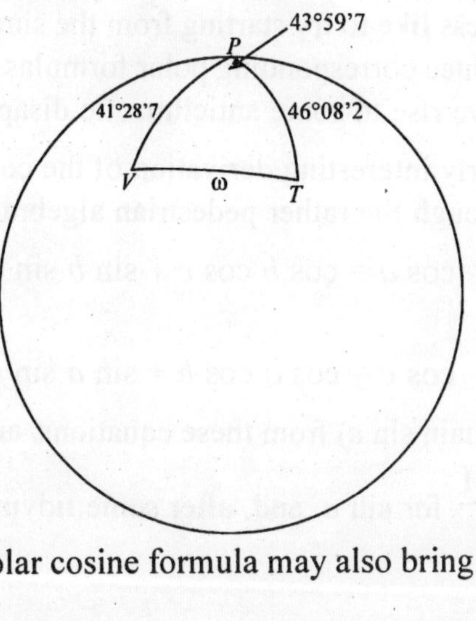

The derivation of the polar cosine formula may also bring a small moment of delight.

In figure, $A'B'C'$ is a spherical triangle. ABC is also a spherical triangle, called the polar triangle to $A'B'C'$. It is formed in the following way. The side BC is an arc of a great circle 90° from A'; that is, BC is part of the equator of which A' is pole.

Likewise CA is 90° from B' and AB is 90° from C'. In the drawing, the side $B'C'$ of the small triangle has been extended to meet the sides AB and CA of the large triangle. It will be evident from the drawing that the angle A of the large

triangle is equal to $x + a' + y$. Further, from the way in which the triangle ABC was formed, $x + a'$ and $a' + y$ are each equal to 90°. From these relations, we see that

$$A + A = [(x + a') + y] + [x + (a' + y)]$$

or

$$2A = 180° + x + y = 180° + A - a'$$

Therefore $A = 180° - a'$

In a similar manner, $B = 180° - b'$ and $C = 180° - c'$

Now, suppose $f(A', B', C', a', b', c') = 0$ is any relation between the sides and angles of the triangle $A'B'C'$. We may replace a' by $180° - A$, b' by $180° - B$, and so on, and this will result in a relation between A, B, C, a, b and c; that is, it will result in a relation between the sides and angles of the triangle ABC.

For example, the equation

$$\cos a' = \cos b' \cos c' + \sin b' \sin c' \cos A'$$

is valid for the triangle $A'B'C'$. By making these substitutions, we find the following formula valid for triangle ABC:

$$-\cos A = \cos B \cos C - \sin B \sin C \cos a,$$

which is the polar cosine formula.

The reader will doubtless like to try starting from the sine and cotangent formulas for the triangle $A'B'C'$ and deduce corresponding polar formulas for the triangle ABC, though this, unfortunately, may give rise to some anticlimactic disappointment.

I know of no particularly interesting derivation of the cotangent formula, and I leave it to the reader to work through the rather pedestrian algebra. Start from

$$\cos a = \cos b \cos c + \sin b \sin c \cos A$$

and

$$\cos c = \cos a \cos b + \sin a \sin b \cos C.$$

Eliminate $\cos c$ (but retain $\sin c$) from these equations, and write $1 - \sin^2 b$ for $\cos^2 b$. Finally substitute $\frac{\sin c \sin A}{\sin C}$ for $\sin a$, and, after some tidying up, the cotangent formula should result.

Examples: At this stage, we have had some practice in solving the four spherical triangle formulas, and we have derived them. In this section we encounter examples in which the problem is not merely to solve a triangle, but to gain some experience in setting up a problem and deciding which triangle has to be solved.

1. The coordinates of the Dominion Astrophysical Observatory, near Victoria, British Columbia, are

Latitude 48° 31'.3 N Longitude 123° 25'.0 W

and the coordinates of the David Dunlap Observatory, near Toronto, Ontario, are

Latitude 43° 51'.8 N Longitude 79° 25'.3 W

How far is Toronto from Victoria, and what is the azimuth of Toronto relative to Victoria?

The triangle to be drawn and solved is the triangle PVT, where P is the Earth's north pole, V is Victoria, and T is Toronto. On figure are marked the colatitudes of the two cities and the difference between their longitudes.

The great circle distance ω between the two observatories is easily given by the cosine formula:

$$\cos\omega = \cos 41°\ 28'.7 \cos 46°\ 08'.2 + \sin 41°\ 28'.7 \sin 46°\ 08'.2 \cos 43°\ 59'.7$$

From this, we find $\omega = 30^0\ 22'.7$ or 0.53021 radians. The radius of the Earth is 6371 km, so the distance between the observatories is 3378 km or 2099 miles.

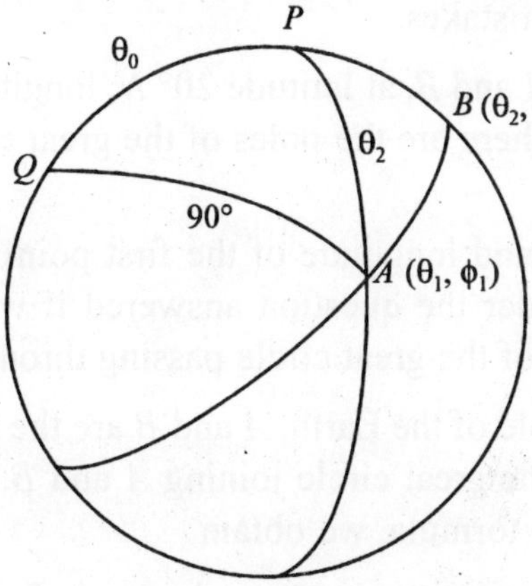

Now that we have found ω, we can find the azimuth, which is the angle V, from the sine formula:

$$\sin V = \frac{\sin 46°08'.2 \sin 43°59'.7}{\sin 30°22'.7} = 0.990\,275$$

and hence $V = 82°\ 00'.3$

But we should now remember that $\sin^{-1} 0.990\ 275$ has two values between 0° and 180°, namely 82° 00'.3 and 97° 59'.7.

Usually it is obvious from inspection of a drawing which of the two values of $\sin^{-1}$ is the required one. Unfortunately, in this case, both values are close to 90°, and it may not be immediately obvious which of the two values we require. However, it will be noticed that Toronto has a more southerly latitude than Victoria, and this should easily resolve the ambiguity.

We could, of course, have found the azimuth V by using the cotangent formula, without having to calculate ω first.

Thus

$$\cos 41°\ 28'.7 \cos 43°59'.7 = \sin 41°\ 28'.7 \cot 46°\ 08'.2 - \sin 43°\ 59'.7 \cot V$$

There is only one solution for V between 0° and 180°, and it is the correct one, namely 82° 00'.3. A good drawing will show the reader why the correct solution was the acute rather than the obtuse angle (in our drawing the angle was made to be close to 900 in order not to bias the reader one way or the other), but in any case all readers, especially those who were trapped into choosing the obtuse angle, *should take careful note of the difficulties that can be caused by the ambiguity of the* function $\sin^{-1}$.

Indeed it is the strong advice of the author never to use the sine formula, in spite of the ease of memorizing it.

The cotangent formula is more difficult to commit to memory, but it is far more useful and not so prone to quadrant mistakes.

2. Consider two points, A and B, at latitude 20° N, longitude 25° E, and latitude 72° N, longitude 44° E. Where are the poles of the great circle passing through these two points?

Let us call the colatitude and longitude of the first point (θ_1, ϕ_1) and of the second point (θ_1, ϕ_2). We shall consider the question answered if we can find the coordinates (θ_0, ϕ_0) of the poles Q and Q' of the great circle passing through the two points.

In figure, P is the north pole of the Earth, A and B are the two points in question, and Q is one of the two poles of the great circle joining A and B. The figure also shows the triangle PQA. From the cosine formula, we obtain

$$0 = \cos\theta_0 \cos\theta_1 + \sin\theta_0 \sin\theta_1 \cos(\phi_1 - \phi_0),$$

from which

$$\tan\theta_0 = \frac{\cos\theta_1}{\cos(\phi_1 - \phi_0)}.$$

Similarly from triangle PQB we would obtain

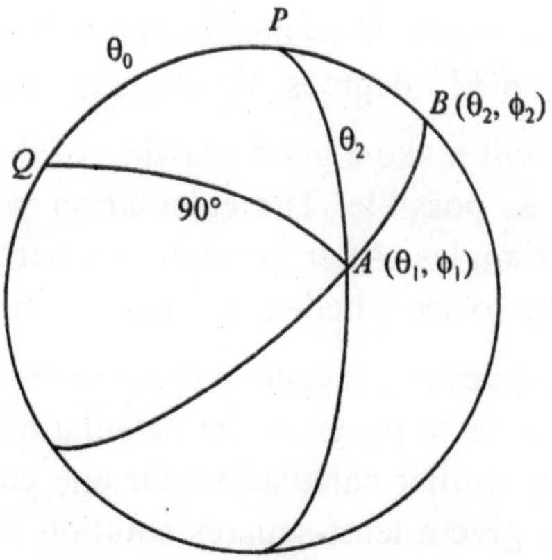

$$\tan\theta_0 = \frac{\cos\theta_1}{\cos(\phi_1 - \phi_0)}.$$

These are two equations in θ_0 and ϕ_0, so the problem is in principle solved. Equate the righthand sides of the two equations, expand the terms cos $(\theta_1 - \phi_0)$ and cos $(\theta_2 - \phi_0)$, gather the terms in sin ϕ_0 and cos ϕ_0, eventually to obtain

$$\tan\phi_0 = \frac{\cos\theta_2\cos\phi_1 - \cos\theta_1\cos\phi_2}{\cos\theta_1\sin\phi_2 - \cos\theta_2\sin\phi_1}.$$

If we substitute the angles given in the original problem, we obtain

$$\tan\phi_0 = \frac{\cos 18°\cos 25° - \cos 70°\cos 44°}{\cos 70°\sin 44° - \cos 18°\sin 25°} = -3.7476946$$

from which $\phi_0 = 284° \; 56'.4$ or $104° \; 56'.4$

Note that we get two values for ϕ_0 differing by 180°, as expected.

We then use either of the equations for tan θ_0 to obtain θ_0 (It is good practice to use both of them as a check on the arithmetic.) The north polar distance, or colatitude, must be between 0° and 180°, so there is no ambiguity of quadrant.

With $\phi_0 = 284° \; 56'.4$, we obtain $\theta_0 = 117° \; 03'.3$

and with $\phi_0 = 104° \; 56'.4$, we obtain $\theta_0 = 62° \; 56'.7$

and these are the coordinates of the two poles of the great circle passing through *A* and *B*. The reader is strongly urged actually to carry out these computations numerically in order to be quite sure that the quadrants are correct and unambiguous. Indeed, dealing with the quadrant problem may be regarded as the most important part of the exercise.

Here is a challenging exercise and an important one in meteor astronomy. Two shower meteors are seen, diverging from a common radiant. One starts at right ascension 6 hours, declination +65 degrees, and finishes at right ascension 1 hour, declination +75 degrees.

The second starts at right ascension 5 h, declination +35 degrees, and finishes at right ascension 3 hours, declination +15 degrees. Where is the radiant?

The assiduous student will make a good drawing of the celestial sphere, illustrating the situation as accurately as possible. The calculation will require some imaginative manipulation of spherical triangles. After arriving at what you believe to be the correct answer, look at your drawing to see whether it is reasonable.

The next step might be to develop a general trigonometrical expression for the answer in terms of the original data, or to program the calculation for a computer, so that it is henceforth available for any similar calculation. Or one can go yet further, and write a computer program that will give a least-squares solution for the radiant for many more than two meteors in the shower. I find for the answer to the above problem that the radiant is at right ascension 7.26 hours and declination +43.8 degrees.

ROTATION OF AXES, TWO DIMENSIONS

In this section we consider the following problem. Consider two sets of orthogonal axes, O*x*, O*y*, and O*x'*, O*y'*, such that one set makes an angle θ with respect to the other. A point *P* can be described either by its coordinates (*x*, *y*) with respect to one "basis set" O*x*, O*y*, or by its coordinates with respect to the other basis set O*x'*, O*y'*. The question is, what is the relation between the coordinates (*x*, *y*) and the coordinates (*x'*, *y'*)?

We see that OA = x, AP = y, ON = x', PN = y', OM = $x \cos\theta$, MN = $y \sin\theta$,

∴ $x' = x\cos\theta + y\sin\theta$.

Also MA = NB = $x\sin\theta$, PB = $y\cos\theta$,

∴ $y' = -x'\sin\theta + y\cos\theta$.

These two relations can be written in matrix form as

$$\begin{pmatrix} x' \\ y' \end{pmatrix} = \begin{pmatrix} \cos\theta & \sin\theta \\ -\sin\theta & \cos\theta \end{pmatrix} \begin{pmatrix} x \\ y \end{pmatrix}.$$

There are several ways of obtaining the converse relations; that is, equations for *x* and *y* in terms of *x'* and *y'*. One way would be to design drawings similar to (b) and (c) that show the converse relations clearly, and the reader is encouraged to do this. Another way is merely to solve the above two equations (which can be regarded as two simultaneous equations in *x* and *y*) for *x* and *y*. Less tedious is to interchange the primed and unprimed symbols and change the sign of θ. Perhaps the quickest of all is to recognize that the determinant of the matrix

$$\begin{pmatrix} \cos\theta & \sin\theta \\ -\sin\theta & \cos\theta \end{pmatrix}$$

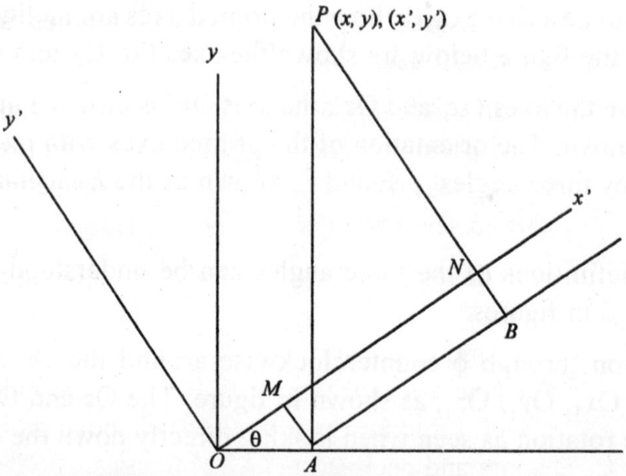

is unity and therefore the matrix is an orthogonal matrix. One important property of an orthogonal matrix M is that its reciprocal M^{-1} is equal to its transpose $\tilde{M}$ (formed by transposing the rows and columns). Therefore the converse relation that we seek is

$$\begin{pmatrix} x \\ y \end{pmatrix} = \begin{pmatrix} \cos\theta & \sin\theta \\ -\sin\theta & \cos\theta \end{pmatrix}^{-1} \begin{pmatrix} x' \\ y' \end{pmatrix} = \begin{pmatrix} \cos\theta & -\sin\theta \\ \sin\theta & \cos\theta \end{pmatrix} \begin{pmatrix} x' \\ y' \end{pmatrix}.$$

The reader might like to try all four methods to ensure that they all arrive at the same result.

ROTATION OF AXES, THREE DIMENSIONS. EULERIAN ANGLES

We now consider two sets of orthogonal axes O*x*, O*y*, O*z* and O*x* ', O*y* ', O*z* ' in three-dimensional space and inclined to each other. A point in space can be described by its coordinates (x, y, z) with respect to one basis set or (x', y', z') with respect to the other. What is the relation between the coordinates (x, y, z) and the coordinates (x', y', z')?

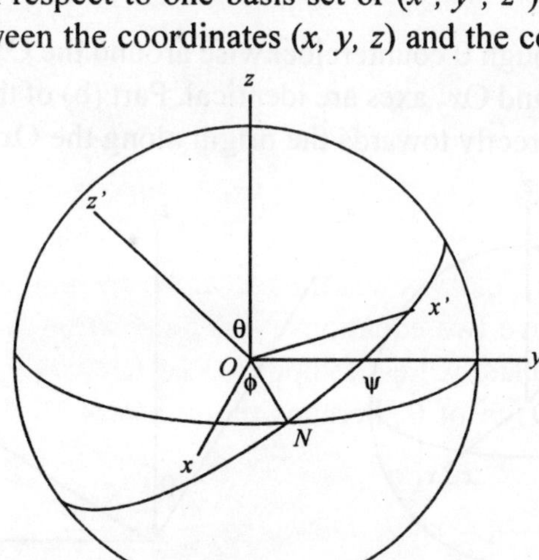

We first need to describe exactly how the primed axes are inclined with respect to the unprimed axes. In the figure below are shown the axes Ox; Oy and Oz.

Also shown are the axes Ox' and Oz'; the axis Oy' is directed behind the plane of the paper and is not drawn. The orientation of the primed axes with respect to the unprimed axes is described by three angles „ Æ and È, known as the *Eulerian angles*, and they are shown in figure.

The precise definitions of the three angles can be understood by three consecutive rotations, illustrated in figures.

First, a rotation through ϕ counterclockwise around the Oz axis to form a set of intermediate axes Ox_1, Oy_1, Oz_1, as shown in figure. The Oz and Oz_1 axes are identical. Part (b) shows the rotation as seen when looking directly down the Oz (or Oz_1) axis.

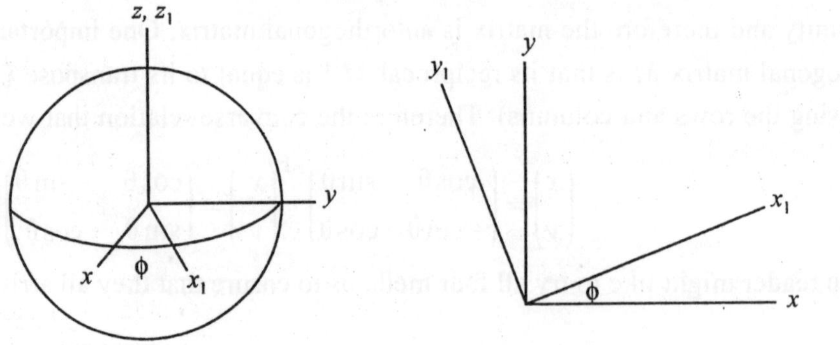

(a) (b)

The relation between the (x, y, z) and (x_1, y_1, z_1) coordinates is

$$\begin{pmatrix} x_1 \\ y_1 \end{pmatrix} = \begin{pmatrix} \cos\theta & \sin\theta \\ -\sin\theta & \cos\theta \end{pmatrix} \begin{pmatrix} x \\ y \end{pmatrix}$$

Next, a rotation through θ counterclockwise around the Ox_1 axis to form a set of axes Ox_2, Oy_2, Oz_2. The Ox_1 and Ox_2 axes are identical. Part (b) of the figure shows the rotation as seen when looking directly towards the origin along the Ox_1 (or Ox_2) axis.

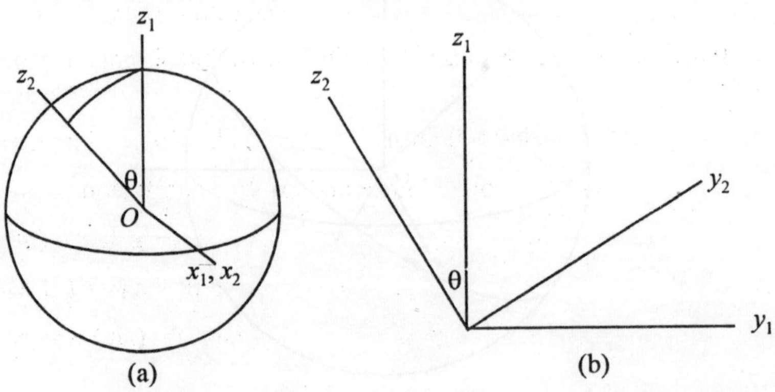

(a) (b)

The relation between the (x_1, y_1, z_1) and (x_2, y_2, z_2) coordinates is

$$\begin{pmatrix} y_2 \\ z_2 \end{pmatrix} = \begin{pmatrix} \cos\theta & \sin\theta \\ -\sin\theta & \cos\theta \end{pmatrix} \begin{pmatrix} y_1 \\ z_1 \end{pmatrix}.$$

Lastly, a rotation through ψ counterclockwise around the Oz_2 axis to form the set of axes *Ox', Oy' Oz'*. The Oz_2 and Oz' axes are identical. Part (b) of the figure shows the rotation as seen when looking directly down the Oz_2 (or *Oz'*) axis.

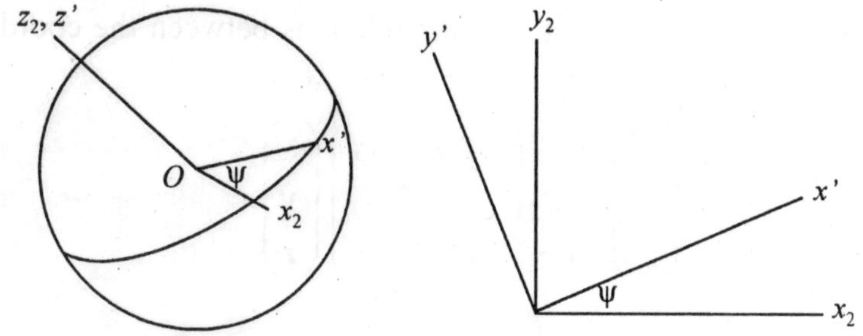

(a) (b)

The relation between the (x_2, y_2, z_2) and (x', y', z') coordinates is

$$\begin{pmatrix} x' \\ y' \\ z' \end{pmatrix} = \begin{pmatrix} \cos\psi & \sin\psi & 1 \\ -\sin\psi & \cos\psi & 1 \\ 0 & 0 & 1 \end{pmatrix} \begin{pmatrix} x_2 \\ y_2 \\ z_2 \end{pmatrix}$$

Thus we have for the relations between (x', y', z') and (x, y, z)

$$\begin{pmatrix} x' \\ y' \\ z' \end{pmatrix} = \begin{pmatrix} \cos\psi & \sin\psi & 1 \\ -\sin\psi & \cos\psi & 1 \\ 0 & 0 & 1 \end{pmatrix} \begin{pmatrix} 1 & 0 & 0 \\ 0 & \cos\theta & \sin \\ 0 & -\sin\theta & \cos\theta \end{pmatrix} \begin{pmatrix} \cos\phi & \sin\phi & 0 \\ -\sin\phi & \cos\phi & 0 \\ 0 & 0 & 1 \end{pmatrix} \begin{pmatrix} x \\ y \\ z \end{pmatrix}$$

On multiplication of these matrices, we obtain

$$\begin{pmatrix} x' \\ y' \\ z' \end{pmatrix} = \begin{pmatrix} \cos\psi\cos\phi - \sin\phi\sin\psi & \cos\psi\sin\phi + \cos\theta\cos\phi\sin\psi & \sin\psi\sin\theta \\ -\sin\psi\cos\phi - \cos\theta\sin\phi\cos\psi & -\sin\psi\sin\phi + \cos\theta\cos\phi\cos\psi & \cos\psi\sin\theta \\ \sin\theta\sin\phi & -\sin\theta\cos\phi & \cos\theta \end{pmatrix} \begin{pmatrix} x \\ y \\ z \end{pmatrix}$$

The inverse of this may be found, as in the two-dimensional case, either by solving these three equations for *x, y* and *z* (which would be rather tedious); or by interchanging the primed and unprimed quantities and reversing the order and signs of all operations (replace ψ with $-\phi$, θ with $-\theta$, and ϕ with $-\psi$) which is less tedious; or by recognizing that the determinant of the matrix is unity and therefore its reciprocal is its transpose, which is hardly tedious at all. The reader should verify that the determinant of the matrix is unity by multiplying it out and making use of trigonometric identities.

The reason that the determinant must be unity, however, and that the rotation matrix must be orthogonal, is that rotation of axes cannot change the magnitude of a vector.

Each element of the matrix is the cosine of the angle between an axis in one basis set and an axis in the other basis set. For example, the second element in the first row is the cosine of the angles between Ox' and O*y*.

The first element of the third row is the cosine of the angles between O*z*' and O*x*. The matrix can be referred to as the matrix of direction cosines between the axes of one basis set and the axes of the other basis set, and the relations between the coordinates can be written

$$\begin{pmatrix} x' \\ y' \\ z' \end{pmatrix} = \begin{pmatrix} c_{11} & c_{12} & c_{13} \\ c_{21} & c_{22} & c_{23} \\ c_{31} & c_{32} & c_{33} \end{pmatrix} \begin{pmatrix} x \\ y \\ z \end{pmatrix}$$

$$R' = CR.$$

You will note the similarity of the forms of the direction cosines to the cosine formula for the solution of a spherical triangle, and indeed the direction cosines can all be derived by drawing and solving the relevant spherical triangles. You might (or might not!)enjoy trying to do this.

The matrix C of the direction cosines is orthogonal, and the properties of an orthogonal matrix are as follows. The reader should verify this using the formulas for the direction cosines in terms of the Eulerian angles.

The properties also apply, of course, although more trivially, to the rotation matrix in two dimensions.

(a) det $C = \pm 1$

(det $C = -1$ implies that the two basis sets are of opposite chirality or "handedness"; that is, if one basis set is right-handed, the other is left-handed.)

(b) The sum of the squares of the elements in any row or any column is unity. This merely means that the magnitudes of unit orthogonal vectors are indeed unity.

(c) The sum of the products of corresponding elements in any two rows or any two columns is zero. This is merely a reflection of the fact that the scalar or dot product of any two unit orthogonal vectors is zero.

(d) Every element is equal to its own cofactor. This a reflection of the fact that the vector or cross product of any two unit orthogonal vectors in cyclic order is equal to the third.

(e) $C^{-1} = \tilde{C}$, or the reciprocal of an orthogonal matrix is equal to its transpose.

The first four properties above can be (and should be) used in a numerical case to verify that the matrix is indeed orthogonal, and they can be used for detecting and for correcting mistakes.

For example, the following matrix is supposed to be orthogonal, but there are, in fact, two mistakes in it. Using properties (b) and (c) above, locate and correct the mistakes. (It will become clear when you do this why verification of property (b) alone is not sufficient.) When you have corrected the matrix, see if you can find the Eulerian angles θ, ϕ and ψ without ambiguity of quadrant.

As a hint, start at the bottom right hand side of the matrix and note, from the way in which the Eulerian angles are set up, that θ must be between 0° and 180°, so that there is no ambiguity of quadrant.

The other two angles, however, can lie between 0^0 and 360^0 and must be determined by examining the signs of their sines and cosines. When you have calculated the Eulerian angles, a further useful exercise would be to prepare a drawing showing the orientation of the primed axes with respect to the unprimed axes.

$$\begin{pmatrix} +0.0752848827 & -0.5186744682 & +0.8516507396 \\ -0.5531104732 & -0.7323630008 & +0.3971312619 \\ -0.8296993375 & -0.4421589632 & -0.3420201433 \end{pmatrix}$$

Note, as a matter of good computational practice, that the numbers are written in groups of three separated by half spaces after the decimal point, all numbers, positive and negative, are signed, and leading zeroes are not omitted.

Trigonometric Formulas

I gather here merely for reference a set of commonly-used trigonometric formulas. It is a matter of personal preference whether to commit them to memory. It is probably fair to remark that anyone who is regularly engaged in problems in celestial mechanics or related disciplines will be familiar with most of them, at least from frequent use, whether or not any conscious effort was made to memorize them.

At the very least, the reader should be aware of their existence, even if he or she has to look to recall the exact formula.

$$\frac{\sin A}{\cos A} = \tan A$$

$$\sin^2 A + \cos^2 A = 1$$

$$1 + \cot^2 A = \csc^2 A$$

$$1 + \tan^2 A = \sec^2 \text{A}$$

$$\sec A \csc A = \tan A + \cot A$$

$$\sec^2 A \csc^2 A = \sec^2 A + \csc^2 A$$

$$\sin (A \pm B) = \sin A \cos B \pm \cos A \sin B$$

$$\cos (A \pm B) = \cos A \cos B \mp \sin A \sin B$$

$$\tan(A \pm B) = \frac{\tan A \pm \tan B}{1 \mp \tan A \tan B}$$

$$\sin 2A = 2 \sin A \cos A$$

$$\cos 2A = \cos^2 A - \sin^2 A = 2\cos^2 A - 1 = 1 - 2 \sin^2 A$$

$$\tan 2A = \frac{2 \tan A}{1 - \tan^2 A}$$

$$\sin \frac{1}{2} A = \sqrt{\frac{1 - \cos A}{2}}$$

$$\cos \frac{1}{2} A = \sqrt{\frac{1 + \cos A}{2}}$$

$$\tan \frac{1}{2} A = \sqrt{\frac{1 - \cos A}{1 + \cos A}} = \frac{1 - \cos A}{\sin A} = \frac{\sin A}{A + \cos A} = \csc A - \cot A$$

$$\sin A + \sin B = 2 \sin \frac{1}{2} S \cos \frac{1}{2} D,$$

where $S = A + B$ and $D = A - B$

$$\sin A - \sin B = 2 \cos \frac{1}{2} S \sin \frac{1}{2} D$$

$$\cos A + \cos B = 2 \cos \frac{1}{2} S \cos \frac{1}{2} D$$

$$\cos A - \cos B = -2 \sin \frac{1}{2} S \sin \frac{1}{2} D$$

$$\sin A \sin B = \frac{1}{2}(\cos D - \cos S)$$

$$\cos A \cos B = \frac{1}{2}(\cos S + \cos D)$$

$$\sin A \cos B = \frac{1}{2}(\sin S + \sin D)$$

$$\sin A = \frac{T}{\sqrt{1+T^2}} = \frac{2t}{1+t^2},$$

where $$T = \tan A \text{ and } t = \tan \frac{1}{2}A$$

$$\cos A = \frac{1}{\sqrt{1+T^2}} = \frac{1-t^2}{1+t^2}$$

$$\tan A = T = \frac{2t}{1-t^2}$$

$$s = \sin A,\ c = \cos A$$

$\cos A = c$ | $\sin A = s$

$\cos 2A = 2c^2 - 1$ | $\sin 2A = 2cs$

$\cos 3A = 4c^3 - 3c$ | $\sin 3A = 3s - 4s^3$

$\cos 4A = 8c^4 - 8c^2 + 1$ | $\sin 4A = 4c(s - 2s^2)$

$\cos 5A = 16c^5 - 20c^3 + 5c$ | $\sin 5A = 5s - 20s^3 + 16s^5$

$\cos 6A = 32c^6 - 48c^4 + 18c^2 - 1$ | $\sin 6A = 2c\,(3s - 16s^3 - 16s^5)$

$\cos 7A = 64c^7 - 112c^5 + 56c^3 - 7c$ | $\sin 7A = 7s - 56s^3 + 112s^5 - 64s^7$

$\cos 8A = 128c^8 - 256c^6 + 160c^4 - 32c^2 + 1$ | $\sin 8A = 8c(s - 10s^3 + 24s^5 - 16s^7)$

$$\cos^2 A = \frac{1}{2}(\cos 2A + 1)$$

$$\cos^3 A = \frac{1}{4}(\cos 3A + 3\cos A)$$

$$\cos^4 A = \frac{1}{8}(\cos 4A + 4\cos 2A + 3)$$

$$\cos^5 A = \frac{1}{16}(\cos 5A + 5\cos 3A + 10\cos A)$$

$$\cos^6 A = \frac{1}{32}(\cos 6A + 6\cos 4A + 15\cos 2A + 10)$$

$$\cos^7 A = \frac{1}{64}(\cos 7A + 7\cos 5A + 21\cos 3A + 35\cos A)$$

$$\cos^8 A = \frac{1}{128}(\cos 8A + 8\cos 6A + 28\cos 4A + 56\cos 2A + 35)$$

$$\sin^2 A = \frac{1}{2}(1 - \cos 2A)$$

$$\sin^3 A = \frac{1}{4}(3\sin A - \sin 3A)$$

$$\sin^4 A = \frac{1}{8}(\cos 4A - 4\cos 2A + 3)$$

$$\sin^5 A = \frac{1}{16}(\sin 5A - 5\sin 3A + 10\sin A)$$

$$\sin^6 A = \frac{1}{32}(10 - 15\cos 2A + 6\cos 4A - \cos 6A)$$

$$\sin^7 A = \frac{1}{64}(35\sin A - 21\sin 3A + 7\sin 5A - \sin 7A)$$

$$\sin^8 A = \frac{1}{128}(\cos 8A - 8\cos 6A + 28\cos 4A - 56\cos 2A + 35)$$

$$\sin A = A - \frac{A^3}{3!} + \frac{A^4}{5!} - \ldots$$

$$\cos A = 1 - \frac{A^2}{2!} + \frac{A^4}{4!} - \ldots$$

$\int_0^{\pi/2} \sin^m\theta \cos^n \theta d\theta = \frac{(m-1)!!(n-1)!!X}{(m+n)!!}$, where $X = \pi/2$ if m and n both even, and $X = 1$ otherwise.

$e^{ni\theta} = e^{in\theta}$ (de Moivre's theorem - the only one you need know. All others can be deduced from it.)

Plane Triangles:

$$\frac{a}{\sin A} = \frac{b}{\sin B} = \frac{c}{\sin C}$$

$$a^2 = b^2 + c^2 - 2bc \cos A$$

$$a \cos B + b \cos A = c$$

$$s = \frac{1}{2}(a + b + c)$$

$$\sin\frac{1}{2}A = \sqrt{\frac{(s-b)(s-c)}{s(s-a)}}$$

$$\cos\frac{1}{2}A = \sqrt{\frac{s(s-a)}{bc}}$$

$$\tan\frac{1}{2}A = \sqrt{\frac{(x-b)(s-c)}{s(s-a)}}$$

Spherical Triangles:

$$\frac{\sin a}{\sin A} = \frac{\sin b}{\sin B} = \frac{\sin c}{\sin C}$$

$$\cos a = \cos b \cos c + \sin b \sin c \cos A$$

$$\cos A = -\cos B \cos C + \sin B \sin C \cos a$$

$$\cos(IS)\cos(IA) = \sin(IS)\cot(OS) - \sin(IA)\cot(OA)$$

Chapter 2

Higher-Order Metrics

Given three points A,B,C, let dx^1 denote the distance between A and B, and let dx^2 denote the distance between B and C. Can we express the distance ds between A and C in terms of dx^1 and dx^2? Since dx^1, dx^2, and ds all represent distances with comensurate units, it's clear that any formula relating them must be homogeneous in these quantities, i.e., they must appear to the same power. One possibility is to assume that ds is a linear combination of dx^1 and dx^2 as follows

$$\delta\sigma = g_1 dx_1 + g_2 dx_2$$

where g_1 and g_2 are constants. In a simple one-dimensional manifold this would indeed be the correct formula for ds, with $|g_1| = |g_2| = 1$, except for the fact that it might give a negative sign for ds, contrary to the idea of an interval as a positive magnitude. To ensure the correct sign for ds, we might take the absolute value of the right hand side, which suggests that the fundamental equality actually involves the *squares* of the two sides of the above equation, i.e., the quantities ds, dx^1, dx^2 satisfy the relation

$$\begin{aligned}(ds)^2 &= (g_1 dx^1 + g_2 dx^2)^2 \\ &= g_{11}dx^1dx^1 + g_{12}dx^1dx^2 + g_{21}dx^2dx^1 + g_{22}dx^2dx^2\end{aligned}$$

where we have put $g_{ij} = g_i\, g_j$. Thus we have $g_{11}g_{22}\text{-}4(g_{12})^2 = 0$, which is the condition for factorability of the expanded form as the square of a linear expression. This will be the case in a one-dimensional manifold, but in more general circumstances we find that the values of the g_{ij} in the expanded form are such that the expression is not factorable into linear terms with real coefficients. In this way we arrive at the second-order metric form, which is the basis of Riemannian geometry.

Of course, by allowing the second-order coefficients g_{ij} to be arbitrary, we make it possible for $(ds)^2$ to be negative, analagous to the fact that ds in equation could be negative, which is what prompted us to square both sides of leading to equation. Now that $(ds)^2$ can be negative, we're naturally led to consider the possibility that the fundamental relation is actually the equality of the *squares*. This gives

$$(ds)^4 = \sum g_{\mu\nu\alpha\beta} dx^{\mu} dx^{n} dx^{\alpha} dx^{\beta}$$

where the sum is evaluated for μ,ν,α,β each ranging from 1 to n, where n is the dimension of the manifold. Once again, having arrived at this form, we immediately dispense with the assumption of factorability, and allow general fourth-order metrics. These are non-Riemannian metrics, although Riemann actually alluded to the possibility of fourth and higher order metrics in his famous inagural dissertation. He noted that

> The line element in this more general case would not be reducible to the square root of a quadratic sum of differential expressions, and therefore in the expression for the square of the line element the deviation from flatness would be an infinitely small quantity of degree two, whereas for the former manifolds [i.e., those whose squared line elements are sums of squares] it was an infinitely small quantity of degree four. This pecularity [i.e., this quantity of the second degree] in the latter manifolds therefore might well be called the planeness in the smallest parts...

It's clear even from his brief comments that he had given this possibility considerable thought, but he never published any extensive work on it. Finsler wrote a dissertation on this subject in 1918, so such metrics are now often called Finsler metrics.

To visualize the effect of higher order metrics, recall that for a second-order metric the locus of points at a fixed distance ds from the origin must be a conic, i.e., an ellipse, hyperbola, or parabola. In contrast, a fourth-order metric allows more complicated loci of equi-distant points. When applied in the context of Minkowskian metrics, these higher-order forms raise some intriguing possibilities. For example, instead of a spacetime structure with a single light-like characteristic c, we could imagine a structure with *two* null characteristics, c_1 and c_2. Letting x and t denote the spacelike and timelike coordinates respectively, this means $(ds/dt)^4$ vanishes for *two* values (up to sign) of dx/dt. Thus there are four roots given by $\pm c_1$ and $\pm c_2$, and we have

$$\left(\frac{ds}{dt}\right)^4 = \left(\left(\frac{dx}{dt}\right)^2 - (c_1)^2\right)\left(\left(\frac{dx}{dt}\right)^2 - (c_2)^2\right).$$

The resulting metric is

$$(ds)^4 = \left((dx)^2 - c_1^{\,2}(dt)^2\right)\left((dx)^2 - c_2^{\,2}(dt)^2\right)$$
$$= (dx)^4 - (c_1^{\,2} + c_2^{\,2})(dx)(dt) + (c_1 c_2)^2 (dt)^4.$$

The physical significance of this "metric" naturally depends on the physical meaning of the coordinates x and t. In Minkowski spacetime these represent what physical rulers and clocks measure, and we can translate these coordinates from one inertial system to another according to the Lorentz transformations while always preserving the form of the Minkowski metric with a fixed numerical value of c. The coordinates x and t are defined in such a way that c remains invariant, and this definition happily coincides with the physical measures of rulers and clocks. However, with *two* distinct light-like "eigenvalues", it's no longer possible for a single family of spacetime decompositions to preserve the values of both c_1 and c_2. Consequently, the metric will take the form of eeuation only with respect to one particular system of xt coordinates. In any other frame of reference at least one of c_1 and c_2 must be different.

Suppose that with respect to a particular inertial system of coordinates x,t the spacetime metric is given by equation with $c_1 = 1$ and $c_2 = 2$. We might also suppose that c_1 corresponds to the null surfaces of electromagnetic wave propagation, just as in Minkowski spacetime. Now, with respect to any other system of coordinates x',t' moving with speed v relative to the x,t coordinates, we can decompose the absolute intervals into space and time components such that $c_1 = 1$, but then the values of the other lightlines (corresponding to $\pm c_2'$) must be $(v + c_2)/(1 + v\,c_2)$ and $(v-c_2)/(1-v\,c_2)$. Consequently, for states of motion far from the one in which the metric takes the special form equation, the metric will become progressively more asymmetrical. This is illustrated in the figure below, which shows contours of constant magnitude of the squared interval.

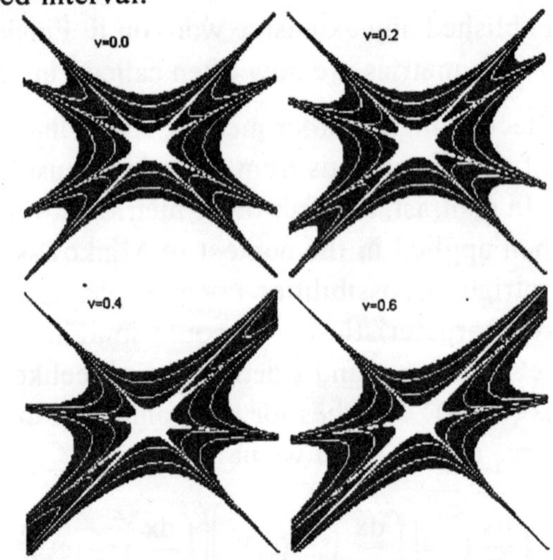

Clearly this metric does not correspond to the observed spacetime structure, even in the symmetrical case with v = 0, because it is not Lorentz-invariant. As an alternative to this structure containing "super-light" null surfaces we might consider metrics with some finite number of "sub-light" null surfaces, but the failure to exhibit even approximate Lorentz-invariance would remain.

However, it is possible to construct infinite-order metrics with infinitely many super-light and/or sub-light null surfaces, and in so doing recover a structure that in many respect is virtually identical to Minkowski spacetime, except for a set (of spacetime trajectories) of measure zero. This can be done by generalizing equation to include infinitely many discrete factors

$$\lim_{n\to\infty}\left[(ds)^2\right]^n = \prod_{i=1}^{n}\left[(dx)^2 - c_i^{\,2}(dt)^2\right].$$

where the values of c_i represent an infinite family of sub-light parameters given by

$$c_1 = \mu \quad c_j = \frac{\mu + c_{j-1}}{1+\mu c_{j-1}} \qquad j=2,3,...$$

A plot showing how this spacetime structure develops as n increases is shown below.

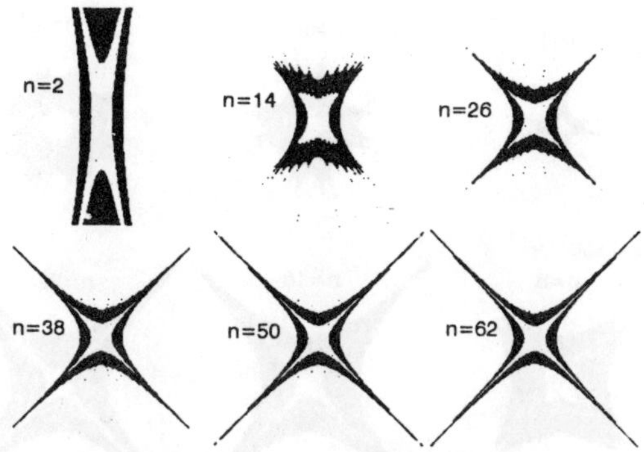

This illustrates how, as the number of sub-light cones goes to infinity, the structure of the manifold goes over to the usual Minkowski pseudometric, except for the discrete null sub-light surfaces which are distributed throughout the interior of the future and past light cones, and which accumulate on the light cones.

The sub-light null surfaces become so thin that they no linger show up on these contour plots for large n, but they remain present to all orders.

In the limit as n approaches infinity they become discrete null trajectories embedded in what amounts to ordinary Minkowski spacetime. To see this, notice that if none of the factors on the right hand side of equation is exactly zero we can take the natural log of both sides to give

$$\lim_{n\to\infty} \ln\left[(ds)^2\right] = \frac{1}{n}\sum_{i-1}^{n} \ln\left[(dx)^2 - c_i^{\,2}(dt)^2\right].$$

Thus the natural log of $(ds)^2$ is the asymptotic average of the natural logs of the quantities $(dx)^2 - c_i^{\,2}(dt)^2$. Since the values of c_i accumultate on 1, it's clear that this converges on the usual Minkowski metric (provided we are not precisely *on* any of the discrete sub-light null surfaces).

The preceding metric was based purely on sub-light null surfaces. We could also include n super-light null surfaces along with the n sub-light null surfaces, yielding an asymptotic family of metrics which, again, goes over to the usual Minkowski metric as n goes to infinity (except for the discrete null surface structure). This metric is given by the formula

$$\left[(ds)^2\right]^{2n} = \prod_{i=1}^{n}\left[(dx)^2 - c_i^{\,2}(dt)^2\right]\left[(dx)^2 - c_i^{\,-2}(dt)^2\right]$$

$$= \prod_{i=1}^{n}\left[(dx)^4 - \left(c_i^{\,2} + c_i^{\,-2}\right)(dx)^2(dt)^2 + (dx)^4\right]$$

where the values of c_i are generated as before. The results for various values of n are illustrated in the figure below.

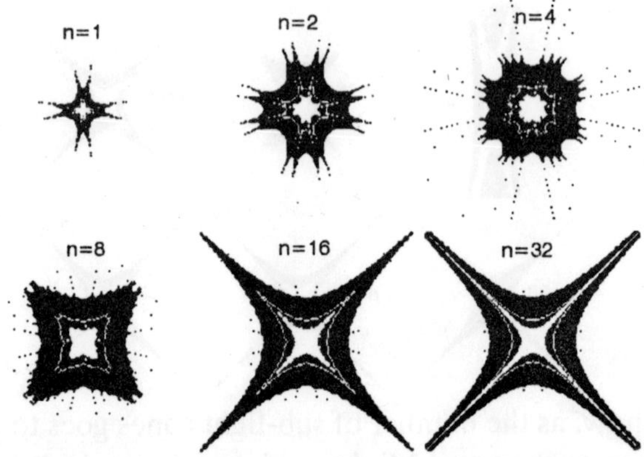

Notice that the quasi Lorentz-invariance of this metric has a subtle periodicity, because any one of the sublight null surfaces can be aligned with the time axis by a suitable choice of velocity, or the time axis can be placed "in between" two null surfaces. In a 1+1 dimensional spacetime the structure is perfectly symmetrical modulo this cycle from one null surface to the next. In other words, the set of exactly equivalent reference systems corresponds to a cycle with a period of μ, which is the increment between each c_i and c_{i+1}.

However, with more spatial dimensions the sub-light null structure is subtly less symmetrical, because each null surface represents a discrete cone, which associates two of the trajectories in the xt plane as the sides of a single cone. Thus there must be an absolutely innermost cone, in the topological sense, even though that cone may be far off center, i.e., far from the selected time axis. Similarly for the super-light cones (or spheres), there would be a single state of motion with respect to which all of those null surfaces would be spherically symmetrical. Only the accumulation shell, i.e., the actual light-cone itself, would be spherically symmetrical with respect to all states of motion.

Packing Universes In Spacetime: One of the interesting aspects of the Minkowski metric is that every lightcone (in principle) contains infinitely many nearly-complete lightcones. Consider just a single spatial dimension in which an infinite number of point particles are moving away from each other with mutual velocities as shown below:

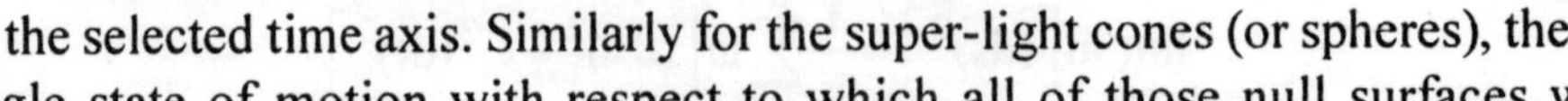

Each particle finds itself mid-way between its two nearest neighbors, which are receding at nearly the speed of light, so that each particle can be regarded as the origin of a nearly-complete lightcone. On the other hand, all of these particles emanate from a single point, and the entire infinite set of points (and nearly-complete lightcones) resides within the future lightcone of that single point.

More formally, a complete lightcone in a flat Lorentzian xt plane comprises the boundary of all points reachable from a given point P along world lines with speeds less than 1 relative to any and every inertial worldline through P. Also, relative to any specific inertial frame W we can define an "ε-complete lightcone" as the region reachable from P along world lines with speeds less than (1−ε) relative to W, for some arbitrarily small $\varepsilon > 0$. A complete lightcone contains infinitely many epsilon-complete lightcones, as illustrated above by the infinite linear sequence of particles in space, each receding with a speed of (1−ε) relative to its closest neighbors. Since we can never observe something infinitely red-shifted, it follows that our observable universe can fit inside an ε-complete lightcone just as well as in a truly complete lightcone. Thus a single lightcone in infinite flat Lorentzian spacetime encompasses infinitely many mutually exclusive ε-universes.

If we arbitrarily select one of the particles as the "rest" particle P_0, and number the other particles sequentially, we can evaluate the velocities of the other particles with respect to the inertial coordinates of P_0, whose velocity is $v_0 = 0$. If each particle has a mutual velocity u relative to each of its nearest neighbors, then obviously P_1 has a speed $v_1 = u$. The speed of P_2 is u relative to P_1, and its speed relative to P_0 is given by the relativistic speed composition formula $v_2 = (v_1 + u)/(uv_1 + 1)$. In general, the speed of P_k can be computed recursively based on the speed of P_{k-1} using the formula

$$V_k = \frac{V_{k-1} + u}{uV_{k-1} + 1}.$$

This is just a linear fractional function, so we can use the method described to derive the explicit formula

$$V_k = \frac{1 - \left(\frac{1-u}{1+u}\right)^k}{1 + \left(\frac{1-u}{1+u}\right)}.$$

Similarly, in full 3+1 dimensional spacetime we can consider packing ε-complete lightspheres inside a complete lightsphere. A flash of light at point P in flat Lorentzian spacetime emanates outward in a spherical shell as viewed from any inertial worldline through P. We arbitrarily select one such worldline W_0 as our frame of reference, and let the slices of simultaneity relative to this frame define a time parameter t. The points of the worldline W_0 can be regarded as the stationary center of a 3D expanding sphere at each instant t. On any given time-slice t we can set up orthogonal space coordinates x,y,z relative to W_0 and normalize the units so that the radius of the expanding lightsphere at time t equals 1. In these terms the boundary of the lightsphere is just the sphere

$$x^2 + y^2 + z^2 = 1$$

Now let W_1 denote another inertial worldline through the point P with a velocity $v = v_1$ relative to W_0, and consider the region R_1 surrounding W_1 consisting of the points reachable from P with speeds not exceeding $u = u_1$ relative to W_1. The region R_1 is spherical and centered on W_1 relative to the frame of W_1, but on any time-slice t (relative to W_0) the region R_1 has an ellipsoidal shape. If v is in the z direction then the cross-sectional boundary of R_1 on the xy plane is given parametrically by

$$x(\theta) = \frac{u\cos(\theta) + v}{1 + uv\cos(\theta)} \qquad y(\theta) = \frac{u\sin(\theta)\sqrt{1 - v^2}}{1 + uv\cos(\theta)}$$

as θ ranges from 0 to 2π. The entire boundary is just the surface of rotation of this ellipse about the z axis. If v_1 has a magnitude of (1-ε) for some arbitrarily small ε > 0, and if we set $u_1 = |v_1|$, then as ε goes to zero the boundary of the region R_1 approaches the limiting ellipsoid

$$2x^2 + 2y^2 + 4\left(z - \frac{1}{2}\right)^2 = 1.$$

Similarly if W_2 is an inertial worldline with speed $|v_2| = |v_1|$ in the negative z direction relative to W_0, then the boundary of the region R_2 consisting of the points reachable from P with speeds not exceeding $u_2 = |v_2|$ approaches the limiting ellipsoid

$$2x^2 + 2y^2 + 4\left(z + \frac{1}{2}\right)^2 = 1.$$

The regions R_1 and R_2 are mutually exclusive, meeting only at the point of contact [0,0,0]. Each of these regions can be called an "ε-complete" lightsphere.

Interestingly, beginning with R_1 and R_2 we can construct a perfect tetrahedral packing of eight epsilon-complete lightspheres by placing six more spheres in a hexagonal ring about the z axis with centers in the xy plane, such that each sphere just touches R_1 and R_2 and its two adjacent neighbors in the ring. Each of these six spheres represents a region reachable from P with speeds less than u_1 relative to one of six worldlines whose speeds are (1-4ε) relative to W_0. The normalized boundaries of these six ellipsoids on a time-slice t are given by

$$25\left(x\cos\left(\frac{k\pi}{3}\right) - y\sin\left(\frac{k\pi}{3}\right) - \frac{4}{5}\right)^2 + 5\left(x\cos\left(\frac{k\pi}{3}\right) - y\sin\left(\frac{k\pi}{3}\right)\right)^2 + 5z^2 = 1$$

for k = 0,1,..,5. In the limit as epsilon goes to zero the hexagonal cluster of e-spheres touching any given e-sphere becomes vanishingly small with respect to the given sphere's frame of reference, so we approach the condition that this hexagonal pattern tessellates the entire surface of each e-sphere in a perfectly symmetrical tetrahedral packing of identical epsilon-complete lightspheres. A cross-sectional side-view and top-view of this configuration are shown below.

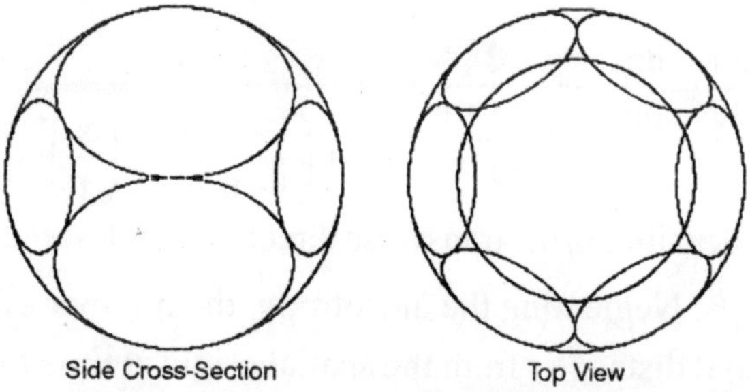

Side Cross-Section Top View

These considerations show that we can regard a single light cone as a cosmological model, taking advantage of the complete symmetry in Minkowski spacetime. Milne was the first to discuss this model in detail. He postulated a cloud of particles expanding in flat spacetime from a single event O, with a distribution of velocities such that the mutual velocities between neighboring particles was the same for every particle, just as in the one-dimensional case described at the beginning of this section. With respect to any particular system of inertial coordinates t,x,y,z whose origin is at the event O, the cloud of particles is spherically symmetrical with radially outward speed v = r/t. The density of the particles is also spherically symmetrical, but it is not isotropic. To determine the density with respect to the inertial coordinates t,x,y,z, we first consider the density in the radial direction at a point on the x axis at time t. If we let u denote the mutual speed between neighboring particles, then the speed v_n of the nth particle away from the center is

$$v_n = \frac{x_n}{t} = \frac{1-\left(\frac{1-u}{1+u}\right)^n}{1+\left(\frac{1-u}{1+u}\right)^n}$$

where x_n is the radial distance of the nth particle along the x axis. Solving for n gives

$$n = \frac{\ln\left(\frac{1-x/t}{1+x/t}\right)}{\ln\left(\frac{1-u}{1+u}\right)}.$$

Differentiating with respect to x gives the density of particles in the x directions

$$\frac{dn}{dx} = \frac{2}{t\ln\left(\frac{1+u}{1-u}\right)}\frac{1}{1-\left(\frac{x}{t}\right)^2}.$$

This confirms that the one-dimensional density at the spatial origin drops in proportion to 1/t. Also, by symmetry, the densities in the transverse directions y and z at any point are given by this same expression as a function of the proper time $\tau = t\sqrt{1-(x/t)^2}$ at that point

$$\frac{dn}{dy} = \frac{dn}{dz} = \frac{2}{\tau \ln\left(\frac{1+u}{1-u}\right)} = \frac{2}{t \ln\left(\frac{1+u}{1-u}\right)\sqrt{1-\left(\frac{x}{t}\right)^2}}.$$

This shows that the densities in the transverse directions are less than in the radial direction by a factor of $\sqrt{1-(r/t)^2}$. Neglecting the anisotropy, the number of particles in a volume element dxdydz at a radial distance r from the spatial origin at time t is proportional to

$$\frac{dn^3}{dxdydz} = \frac{2}{t^3 \ln\left(\frac{1+u}{1-u}\right)^3 \left[1-\left(\frac{r}{t}\right)^2\right]^2}.$$

This distribution applies to every inertial system or coordinates with origin at O, so this cosmology looks the same, and is spherically symmetrical, with respect to the rest frame of each individual particle.

The above analysis was based on a foliation of spacetime into slices of constant-t for some particular system of inertial coordinates, but this is not the only possible foliation, nor even the most natural. From a cosmological standpoint we might adopt as our time coordinate at each point the proper time of uniform worldline extending from O to that point. This would give hyperboloid spacelike surfaces consisting of the locus of all the points with a fixed proper age from the origin event O. One of these spacelike slices is illustrated by the "τ = k" line in the figure below.

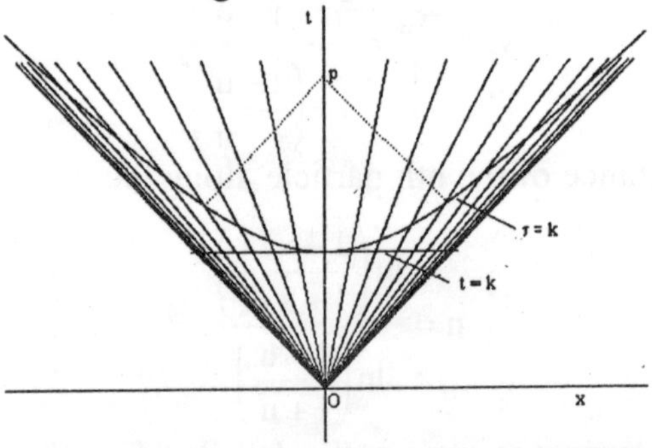

Rindler points out that if τ = k is the epoch at which the density of the expanding cloud drops low enough so that matter and thermal radiation decouple, we should expect at the present event "p" to be receiving an isotropic and highly red-shifted "background radiation" along the dotted lightlike line from that de-coupling surface as shown in the figure. As our present event p advances into the future we expect to see a progressively more red-shifted (i.e., lower temperature) background radiation. This simplistic model gives a surprisingly good representation of the 3°K microwave radiation that is actually observed.

It's also worth noting that if we adopt the hyperboloid foliation the universe of this expanding cloud is spatially infinite. The absolute radial distance along this surface from the spatial center to a point at r is

$$s = \frac{1}{2}\ln\left(\frac{\tau + r}{\tau - r}\right)$$

where $r^2 = x^2 + y^2 + z^2$ in terms of the inertial coordinates of the central spatial point. Furthermore, we can represent this hyperboloid spatial surface as existing over the flat Euclidean xy plane with the elevation $\sqrt{\tau^2 + (x^2 + y^2)}$. By making the elevation imaginary, we capture the indefinite character of the surface. In the limit near the origin we can expand h to give

$$h = i\tau\sqrt{1 + +x^2 + y^2} = i\tau\left(1 + \frac{1}{2}\left(\frac{x^2 + y^2}{\tau^2}\right) + \ldots\right) \approx i\tau + \frac{i}{2\tau}x^2 + \frac{i}{2\tau}y^2.$$

We have a surface tangent to the xy plane at the origin with elevation given by $h = ax^2 + bxy + cy^2$ where $a = c = i/2\tau$ and $b = 0$. Consequently the Gaussian curvature of this spatial surface is $K = 4ac\text{-}b^2 = -1/\tau^2$. By symmetry the same analysis is applicable at every point on the surface, so this surface has constant negative curvature. This applies to any two-dimensional spatial tangent plane in the three-dimensional space at each point for constant τ.

We can also evaluate the metric on this two-dimensional spacelike slice, by writing the total differential of h

$$dh = \frac{\partial h}{\partial x}dx + \frac{\partial h}{\partial y}dy = \frac{i}{\sqrt{1 + x^2 + y^2}}(xdx + ydy).$$

Squaring this and adding the result to $(dx)^2 + (dy)^2$ gives the line element for this surface in terms of the tangent xy plane coordinates projected onto the surface

$$(ds)^2 = \left(\frac{1 + x^2}{1 + x^2 + y^2}\right)(dx)^2 + \left(\frac{-2xy}{1 + x^2 + y^2}\right)(dx)(dy) + \left(\frac{1 + x^2}{1 + x^2 + y^2}\right)(dy)^2.$$

SPACETIME MEDIATION OF QUANTUM INTERACTIONS

According to general relativity the shape of spacetime determines the motions of objects while those objects determine (or at least influence) the shape of spacetime. Similarly in electrodynamics the fields determine the motions of charges in spacetime while the charges determine the fields in spacetime. This dualistic structure naturally arises when we replace action-at-a-distance with purely local influences in such a way that the interactions between "separate" objects are mediated by an entity extending between them. We must then determine the dynamical attributes of this mediating entity, e.g., the electromagnetic field in electrodynamics, or spacetime itself in general relativity.

However, many common conceptions regarding the nature and extension of these mediating entities are called into question by the apparently "non-local" correlations in quantum mechanics, as highlighted by EPR experiments.

The apparent non-locality of these phenomena arises from the fact that although we regard spacetime as metrically Minkowskian, we continue to regard it as topologically Euclidean. The observed phenomena are more consistent with a completely Minkowskian spacetime, in which physical locality is directly induced by the pseudo-metric of spacetime. According to this view, spacetime operates on matter via interactions, and matter defines for spacetime the set of allowable interactions, i.e., consistent with conservation laws. A quantum interaction is considered to originate on (or be "mediated" by) the locus of spacetime points that are null-separated from each of the interacting sites. In general this locus is a quadratic surface in spacetime, and its surface area is inversely proportional to the mass of the transferred particle.

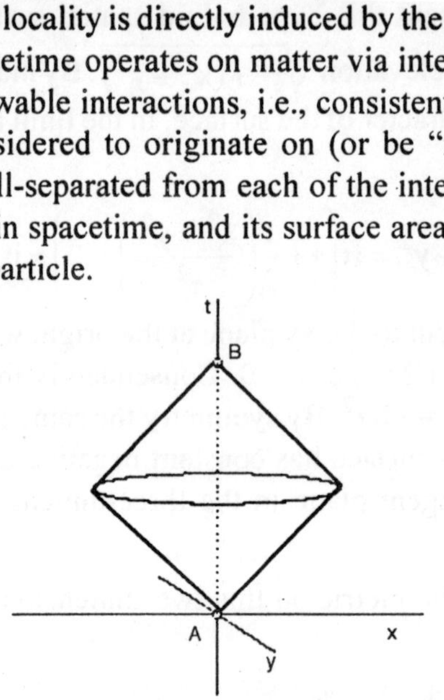

For two timelike-separated events A and B the mediating locus is a closed surface as illustrated below (with one of the spatial dimensions suppressed)

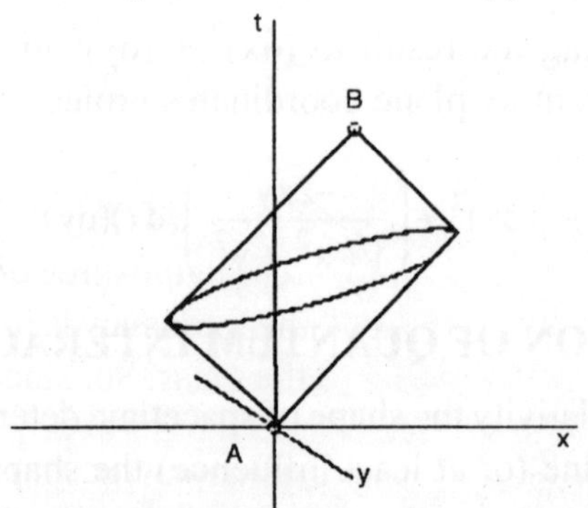

The mediating surface is shown here as a dotted circle, but in 4D spacetime it's actually a closed surface, spherical and purely spacelike relative to the frame of the interval AB. This type of interaction corresponds to the transit of massive real particles.

Of course, relative to a frame in which A and B are in different spatial locations, the locus of intersection has both timelike and spacelike extent, and is an ellipse (or rather an ellipsoidal surface in 4D) as illustrated below.

The surface is purely spacelike and isotropic only when evaluated relative to its rest frame (i.e., the frame of the interval AB), whereas this surface maps to a spatial ellipsoid, consisting of points that are no longer simultaneous, relative to any other co-moving frame.

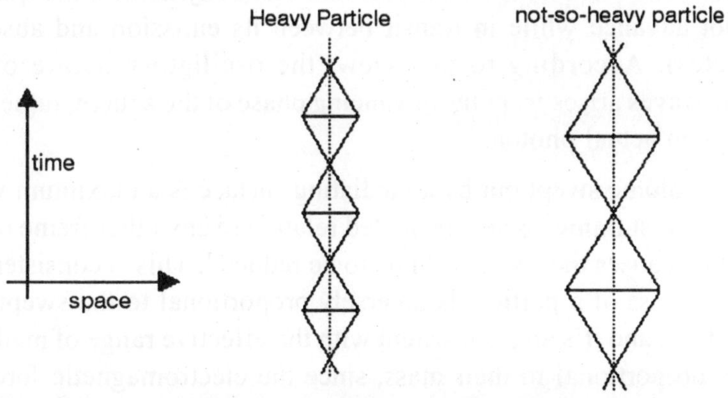

The directionally asymmetric aspects of the surface area correspond precisely to the "relativistic mass" components of the corresponding particles as a function of the relative velocity of the frames.

The propagation of a free massive particle along a timelike path through spacetime can be regarded as involving a series of surfaces, from which emanate inward-going "waves" along the nullcones in both the forward and backward direction, deducting the particle from the past focal point and adding it to the future focal point, as shown below for particles with different masses.

Recall that the frequency υ of the de Broglie matter wave of a particle of mass m is

$$\nu = \frac{c}{h}\sqrt{m^2c^2 + p_x^{\ 2} + p_y^{\ 2} + p_z^{\ 2}}$$

where p_x, p_y, p_z are the components of momentum in the three directions. For a (relatively) stationary particle the momentums vanish and the frequency is just $\upsilon = (mc^2)/h \text{ sec}^{-1}$. Hence the *time* per cycle is inversely proportional to the mass. So, since each cycle consists of an advanced and a retarded cone, the surface of intersection is a sphere (for a stationary mass particle) of radius r = h/mc, because this is how far along the null cones the wave propagates during one cycle. Of course, h/mc is just the Compton scattering wavelength of a particle of mass m, which characterizes the spatial expanse over which a particle tends to "scatter" incident photons in a characteristic way. This can be regarded as the effective size of a particle when "viewed" by means of gamma-rays. We may conceive of this effect being due to a high-energy photon getting close enough to the nominal worldline of the massive particle to interfere with the null surfaces of propagation, upsetting the phase coherence of the null waves and thereby diverting the particle from it's original path.

For a massless particle the quantum phase frequency is zero, and a completely free photon (if such a thing existed) would just be represented by an entire null-cone. On the other hand, real photons are necessarily emitted and absorbed, so they corresponds to bounded null intervals. Consistent with quantum electrodynamics, the quantum phase of photon does not advance while in transit between its emission and absorption (unlike massive particles). According to this view, the oscillatory nature of macroscopic electromagnetic waves arises from the advancing phase of the source, rather than from any phase activity of an actual photon.

The spatial volume swept out by a mediating surface is a maximum when evaluated with respect to it's rest frame. When evaluated relative to any other frame of reference, the spatial contraction causes the swept volume to be reduced. This is consistent with the idea that the effective mass of a particle is inversely proportional to the swept volume of the propagating surface, and it's also consistent with the effective range of mediating particles being inversely proportional to their mass, since the electromagnetic force mediated by massless photons has infinite range, whereas the strong nuclear force has a very limited range because it is mediated by massive particles. Schematics of a stationary and a moving particle are shown below.

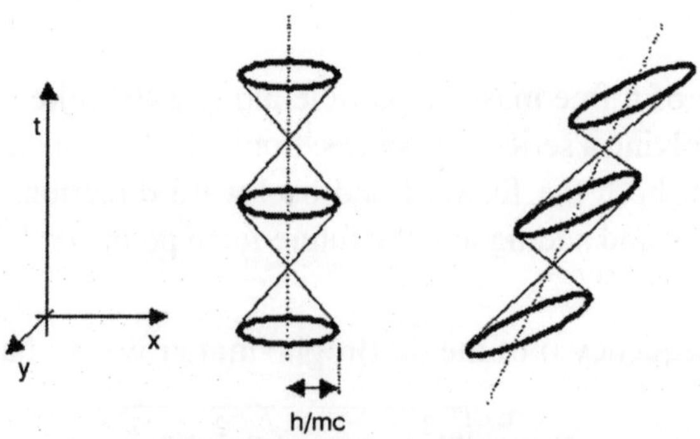

This is the same illustration that appeared in the discussion of Lorentz's "corresponding states", although in that context the shells were understood to be just electromagnetic waves, and Lorentz simply conjectured that all physical phenomena conform to this same structure and transform similarly. In a sense, the relativistic Schrodinger wave equation and Dirac's general argument for light-like propagation of all physical entities based on the combination of relativity and quantum mechanics provide the modern justification for Lorentz's conjecture.

Looking back even further, we see that by conceiving of a particle as a sequence of surfaces of finite extent, it is finally possible to answer Zeno's question about how a moving particle differs from a stationary particle in "a single instant". The difference is that the mediating surfaces of a moving particle are skewed in spacetime relative to those of a stationary particle, corresponding to their respective planes of simultaneity.

Some quantum interactions involve more than two particles. For example, if two coupled particles separate at point A and interact with particles at points B and C respectively, the interaction (viewed straight from the side) looks like this:

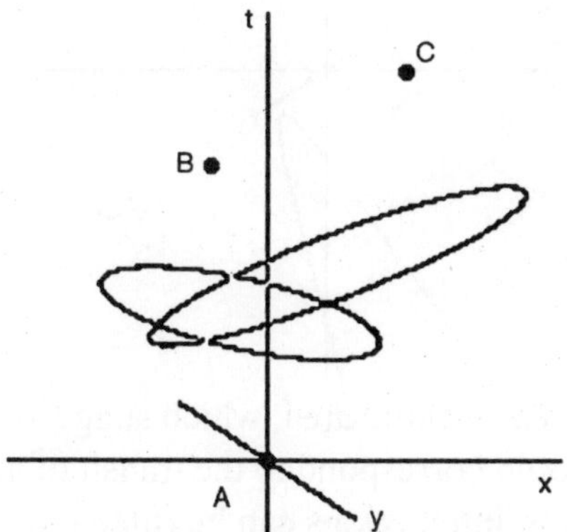

The mediating surface for the pair AB intersects with the mediating surface for AC at the two points of intersection of the dotted circles, but in full 4D spacetime the intersection of the two mediating spheres is a closed circle. (It's worth noting that these two surfaces intersect if and only if B and C are spacelike separated. This circle enforces a particular kind of consistency on any coherent waves that are generated on the two mediating surfaces, and are responsible for "EPR" type correlation effects.)

The locus of null-separated points for two lightlike-separated events is a degenerate quadratic surface, namely, a straight line as represented by the segment AB below:

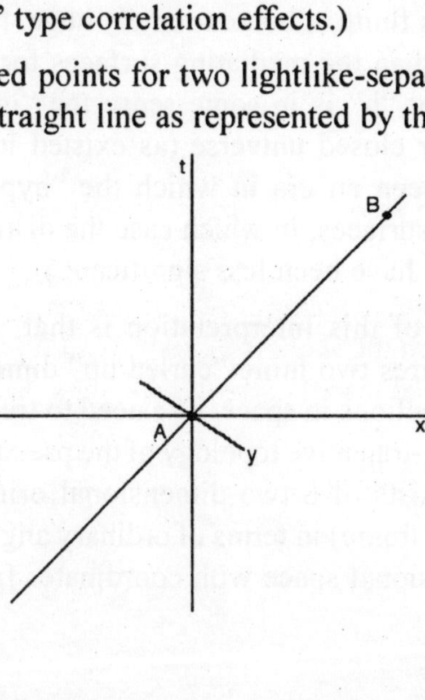

The "surface area" of this locus (the intersection of the two cones) is necessarily zero, so these interactions represent the transits of massless particles. For two spacelike-separated events the mediating locus is a two-part hyperboloid surface, represented by the hyperbola shown at the intersection of two null cones below.

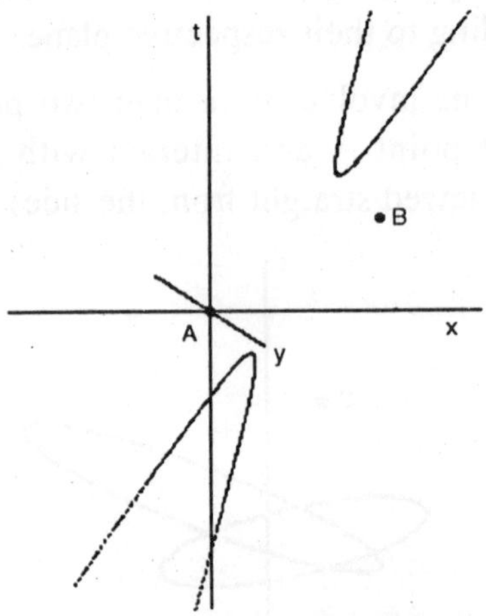

This hyperboloid surface has infinite area, which suggests that any interaction between spacelike separated events would correspond to the transit of an infinitely massive particle. On this basis it seems that these interactions can be ruled out. There is, however, a limited sense in which such interactions might be considered. Recall that a pseudosphere can be represented as a sphere with purely imaginary radius. It's conceivable that observed interactions involving virtual (conjugate) pairs of particles over spacelike intervals (within the limits imposed by the uncertainty relations) may correspond to hyperboloid mediating surfaces.

(It's also been suggested that in a closed universe the "open" hyperboloid surfaces might need to be regarded as finite, albeit extremely huge. For example, they might be 35 orders of magnitude larger than the mediating surfaces for timelike interactions. This is related to vague notions that "h" is in some sense the "inverse" of the size of a finite universe. In a much smaller closed universe (as existed immediately following the big bang) there may be have been an era in which the "hyperboloid" surfaces had areas comparable to the ellipsoid surfaces, in which case the distinction between spacelike and time-like interactions would have been less significant.)

An interesting feature of this interpretation is that, in addition to the usual 3+1 dimensions, spacetime requires two more "curled up" dimensions of angular orientation to represent the possible directions in space. The need to treat these as dimensions in their own right arises from the non-transitive topology of the pseudo-Riemannian manifold. Each point [t,x,y,z] actually consists of a two-dimensional orientation space, which can be parameterized (for any fixed frame) in terms of ordinary angular coordinates θ and ϕ. Then each point in the six-dimensional space with coördinates $[x,y,z,t,\theta,\phi]$ is a terminus for a unique pair of spacetime rays, one forward and one backward in time. A simple mechanistic visualization of this situation is to imagine a tiny computer at each of these points, reading its input from the two rays and sending (matched conservative) outputs on the two rays. This is illustrated below in the xyt space:

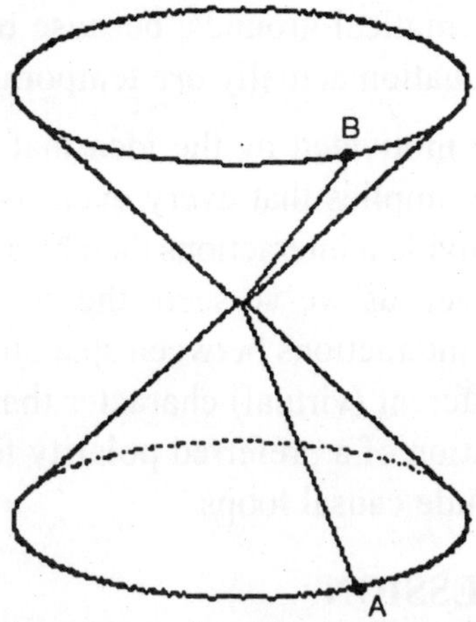

The point at the origin of these two views is on the mediating surface of events A and B. Each point in this space acts purely locally on the basis of purely local information. Specifying a preferred polarity for the two null rays terminating at each point in the 6D space, we automatically preclude causal loops and restrict information flow to the future null cone, while still preserving the symmetry of wave propagation.

(Note that an essential feature of spacetime mediation is that *both* components of a wave-pair are "advanced", in the sense that they originate on a spherical surface, one emanating forward and one backward in time, but both converge *inward* on the particles involved in the interaction.

According to this view, the "unoccupied points" of spacetime are elements of the 6D space, whereas an event or particle is an element of the 4D space (t,x,y,z). If effect an event is the union of all the pairs of rays terminating at each point (x,y,z). The transformations of θ and ϕ under Lorentzian boosts are beautifully handled by linear fractional functions applied to their stereometric mappings on the complex plane.

One common objection to the idea that quantum interactions occur locally between null-separated points is based on the observation that, although every point on the mediating surface is null-separated from each of the interacting events, they are spacelike-separated from each other, and hence unable to communicate or coordinate the generation of two equal and opposite outgoing quantum waves (one forward in time and one backward in time). The answer to this objection is that no communication is required, because the "coordination" arises naturally from the context.

The points on the mediating locus are not communicating with each other, but each of them is in receipt of identical bits of information from the two interaction events A and B.

Each point responds independently based on its local input, but the combined effect of the entire locus responding to the same information is a coherent pair of waves.

Another objection to the "spacetime mediation" view of quantum mechanics is that it relies on temporally symmetric propagation of quantum waves. Of course, this objection can't be made on strictly mathematical grounds, because both Maxwell's equations and the (relativistic) Schrodinger equation actually *are* temporally symmetric.

The objection seems to be motivated by the idea that the admittance of temporally symmetric waves automatically implies that every event is causally implicated in every other event, if not directly by individual interactions then by a chain of interactions, resulting in a non-sensical mess. However, as we've seen, the spacetime mediation view leads naturally to the conclusion that interactions between spacelike-separated events are either impossible or else of a very different (virtual) character than interactions along time-like intervals. Moreover, the stipulation of a preferred polarity for the ray pairs terminating at each point is sufficient to preclude causal loops.

ON RELATIVISTIC PRECESSION

The note on Anomalous Precession described how, from the Schwarzschild solution for a spherically symmetrical gravitational field, we can derive the equation of motion for a small orbiting test particle

$$\left(\frac{dr}{d\tau}\right)^2 = 2h^2m(1/r)^3 - (h/r)^2 + 2m(1/r) - (1-k^2)$$

where m is the mass of the central gravitating body in geometrical units, and h and k are constants of motion. Making use of the relation $d\theta/d\tau = h/r^2$, the left hand side of this equation can be written as

$$\left(\frac{dr}{d\tau}\right)^2 = \left(\frac{d\theta}{d\tau}\right)^2\left(\frac{dr}{d\theta}\right)^2 = \frac{h^2}{r^4}\left(\frac{dr}{d\theta}\right)^2 = h^2\left(\frac{d(1/r)}{d\theta}\right)^2.$$

Thus we can write equation in the form

$$\left(\frac{d(1/r)}{d\theta}\right)^2 = 2m(1/r)^2 - (1/r)^2 + \frac{2m}{h^2}(1/r) - \left(\frac{1-k}{h^2}\right).$$

Differentiating with respect to θ and dividing through by $2d(1/r)/d\theta$, we get

$$\frac{d(1/r)}{d\theta^2} = 3m(1/r)^2 - (1/r) + \frac{m}{h^2}.$$

If r is constant, the left side vanishes and the right side gives the analog of Kepler's third law. On the other hand, given a small periodic radial disturbance, this equation allows us to determine the period of that disturbance. Let $1/r = (1/r_0)(1 + \varepsilon\cos(\Omega\theta))$ for some small value of ε, where $1/r_0$ is the mean reciprocal radius of the orbit, and Ω is the frequency at which a small radial disturbance propagates around the orbit. Substituting this expression for $1/r$ into the above equation, evaluating the second derivative, and collecting terms by powers of ε, we get

$$\left[\frac{m}{h^2}-\frac{1}{r_0}\left(1-\frac{3m}{r_0}\right)\right]+\left[\frac{\cos(\Omega\theta)}{r_0}\left(\Omega^2-\left(1-\frac{6m}{r_0}\right)\right)\right]\varepsilon+\left[\frac{3m}{r_0{}^2}\cos(\Omega\theta)^2\right]\varepsilon^2=0.$$

If the amplitude ε of the radial disturbance is zero, the second two terms vanish, and the first term corresponds to Kepler's third law. On the other hand, for non-zero radial excursions we must choose Ω so that the second and third terms vanish. For sufficiently small ε the third term can be made arbitrarily small, so the vanishing of the second term represents the condition for infinitesimal radial disturbances. Thus an infinitesimal disturbance propagates around the orbit with the angular "frequency" given by

$$\Omega=\left(1-\frac{6m}{r_0}\right)^{1/2}$$

The orbital equation is

$$r(\theta)=\frac{r_0}{1+\varepsilon\cos(\Omega\theta)}$$

which would be a Keplerian ellipse if Ω were equal to unity, meaning that the radial disturbance has the same period as the angular position. One radian of the radial cycle takes place over $1/\Omega$ radians of the angular cycle, so the precession of the radial cycle relative to the angular cycle is

$$\frac{1}{\Omega}-1=\left(1-\frac{6m}{r_0}\right)^{-1/2}-1=3\left(\frac{m}{r_0}\right)+\frac{27}{2}\left(\frac{m}{r_0}\right)^2+\ldots$$

Hence to the first order the precession is simply $3m/r_0$. For example, the orbit of the planet Mercury has a harmonic mean radius of $r_0 = (55.44)10^6$ km, and the mass of the Sun in geometrical units is m = 1.475 km, so a small radial cycle will precess relative to the angular cycle by $(2.66)10^{-8}$ radians per radian, or equivalently revolutions per revolution. Multiplying this by (360)(60)(60) gives 0.1034 arc seconds per revolution. Since Mercury completes 414.9378 revolutions per century, we get the familiar result 42.92 arc seconds per century.

It might be objected that the eccentricity of Mercury's orbit ($\varepsilon = 0.206$) is rather large, whereas our derivation stipulated arbitrarily small values of ε to ensure that the third term in the previous equation to be negligible. However, the third term is also negligibly small, for *any* value of ε, if the ratio of m/r_0 is sufficiently small, because this is the ratio of the coefficients of the second and third terms. For astronomical orbits in our solar system this ratio is always quite small, so the above solution is valid for any eccentricity in these conditions. Nevertheless, it's worth bearing in mind that in general (i.e., for arbitrary values of m/r_0) the solution is valid only if the eccentricity is small. This requirement is somewhat obscured in many presentations, especially since they typically express the result in the form

$$\frac{24\pi^3 a^2}{\left(1-\varepsilon^2\right)c^2T^2}\ \mathrm{rad/rev}$$

where *a* is the semi-major axis of the orbit. Of course, this is equivalent to $3m/r_0$ with units of radians per radian, where r_0 is the harmonic mean radius of the orbit. The simpler form emphasizes the fact that the eccentricity doesn't really enter into the result in a meaningful way. It appears in the traditional form only because of the choice to express the result in terms of the semi-major axis, and in order to relate this to the harmonic mean radius it is necessary to specify the eccentricity. But all that matters is the mean radius, and for any given mean radius there can be infinitely many different eccentricities (with their corresponding semi-major axes). As explained above, the solution is generally valid only for orbits with sufficiently small eccentricities, for which the value of the eccentricity drops out of the result.

The is analogous to the speed of propagation of a pressure wave in air. We find that for sufficiently small amplitudes the speed of sound in a given medium is essentially independent of the amplitude of the wave. Likewise for sufficiently small radial waves, the speed of propagation around an orbit is independent of the amplitude, i.e., the eccentricity.

The traditional formula is misleading because it suggests that the physical effect of eccentricity has been taken into account in the solution, when in fact it has not. The value of ε appearing in the traditional formula merely serves to convert the semi-major axis (which is not a suitable parameter) to the harmonic mean radius.

The physical solution is strictly valid only for essentially circular orbits with just infinitesimal radial perturbations. The only reason the formula is accurate for typical astronomical orbits with high eccentricity (e.g., the asteroid Icarus) is that m/r_0 is extremely small for these orbits. If we needed to determine the precession for an orbit with large eccentricity and a large value of m/r_0 (i.e., in an extremely strong gravitational field) we would need a more complicated solution, taking into account the third term in equation.

People sometimes wonder about the meaning of precession for a perfectly circular orbit, i.e., when the eccentricity ε is precisely zero.

We should first point out that this question is not unique to relativistic precession, because the precession resulting from third bodies or non-spherical bodies in classical orbital mechanics also leads to values of precession that are independent of the eccentricity for nearly circular orbits, and hence that seemingly apply to perfectly circular orbits. In general there are several possible meanings of actual precession in a circle, such as precession of a cyclic variation in speed, or even in the position of the orbiting body, just as the faster of two runners on a circular track precesses in comparison with the slower runner.

In the latter case, the precession simply implies that the body is moving slightly faster than it would have been moving in the absence of the precession, so we would expect to find the orbiting body progressively further and further ahead of its expected position based on the Keplerian two-body relation between speed and orbital radius.

However, in neither the classical nor the relativistic cases can we make a direct comparison with the simple Keplerian two-body relation, because that relation assumes a single, perfectly spherical, gravitating body (which is not applicable to the conditions for classical precession), and it makes no distinction between coordinate and proper spacetime intervals (which makes it inapplicable to relativistic precession). In both cases the meanings of the parameters appearing in Kepler's law become ambiguous, so it isn't possible to single out any particular aspect of the orbit as *the* effect of precession.

The most useful way of interpreting precession for circular orbits is that, just as the speed of sound has a definite value even when there are no sound waves present, so too the speed of propagation of small radial disturbances around an orbit has a definite value even if there are no radial disturbances present. Of course, if ε is precisely zero, the second and third terms vanish identically, so no constraint is placed on Ω, but for any small non-zero ε, the value of Ω is determined.

In other common derivations of the precession (such as Einstein's original derivation, reproduced closely in Weinberg's 1972 text) the difference between the maximum and minimum radial distance appears in the denominator of several expressions, so it is strictly undefined when the radial distance is precisely constant. Nevertheless, the precession approaches a constant value in the limit as the radial perturbation goes to zero, just as the speed of sound approaches a constant value in the limit as the amplitude of the pressure disturbance goes to zero. Needless to say, there is nothing paradoxical about this.

This explanation suggests that it might be possible to express the angular speed Ω of infinitesimal radial disturbances directly in terms of the differences between coordinate and proper intervals (analogous to expressions for the speed of sound in terms of the ratio of the differentials of pressure and density). We recall that the relativistic form of Kepler's third law for circular orbits is

$$m = r^3 \omega_t^2$$

where $\omega_t = d\theta/dt$, and this can be expressed in terms of the proper angular speed $\omega_\tau = d\theta/d\tau$ by multiplying by the square of $dt/d\tau$ along the orbital path. We can infer this ratio directly from the Schwarzschild metric (with $dr = 0$) in the plane of the orbit

$$(d\tau)^2 = \left(1 - \frac{2m}{r}\right)(dt)^2 - r^2(d\theta)^2.$$

Dividing through by $(dt)^2$ and noting that $d\theta/dt = \omega_t$, we get

$$\left(\frac{d\tau}{dt}\right)^2 = 1 - \frac{2m}{r} - r^2\omega t^2 = 1 - \frac{3m}{r}.$$

Hence the coordinate angular speed is related to the proper angular speed by

$$\omega_t^{\ 2} = \omega_t^{\ 2}\left(1 - \frac{3m}{r}\right).$$

Of course, this doesn't signify a change in the speed of the orbiting body as seen from a distant observer, it merely expresses the fact that, in terms of a clock co-moving with the orbiting body, the angular speed has a slightly greater value than it does when evaluated in terms of a distant clock. Thus equation can be written equivalently as

$$m = r^3\omega_\tau^{\ 2}\left(1 - \frac{3m}{r}\right).$$

Now, we must also recognize that the radial Schwarzschild parameter r appearing in these equations cannot be strictly identified with the classical radial distance parameter. The Schwarzschild r is defined such that the area of the spatial surface at constant r is $4\pi r^2$, but the spatial manifold is not flat, so this r is not the integrated proper spatial distance to the origin. In order to express the relativistic analog of Kepler's third law entirely in terms of proper space and time variables for a circular orbit, we need to somehow convert the coordinate distance r to a "proper distance".

We might define the proper radial parameter as the integral of the proper spatial distance to the origin on a surface of constant time, but this is complicated by the fact that the Schwarzschild time foliation becomes singular at r = 2m. On the other hand, it can be argued, since we are interested only in infinitesimal radial disturbances, the relevant proper radial space parameter for a circular orbit ought to be in the same proportion to r as the differentials. At any given radial position, the differentials of the proper and the coordinate distances are related by the gravitational "boost" factor

$$d\rho = dr\left(1 - \frac{2m}{r}\right)^{-1/2}.$$

For any given circular orbit with coordinate radius r_0, we can simply define a "proper" radial coordinate ρ related to the Schwarzschild coordinate r by

$$\rho = r\left(1 - \frac{2m}{r_0}\right)^{-1/2}.$$

Thus when $r = r_0$ the differentials agree with the proper spatial measures. Now, beginning from the relativistic analog of Kepler's law in terms of Schwarzschild coordinates

$$\omega_t = \frac{m^{1/2}}{r^{3/2}}$$

we substitute the proper parameters τ and ρ for the coordinate parameters t and r to give the equivalent expression

$$\omega_\tau = \frac{1}{\left(1-\frac{3m}{r}\right)^{1/2}}\frac{1}{\left(1-\frac{2m}{r}\right)^{3/4}}\frac{m^{1/2}}{\rho^{3/2}}.$$

The quantity on the left side is simply the angular speed of a circular orbit in terms of a co-moving clock, whereas the rightmost factor on the right side *looks like* an angular speed of a classical circular orbit of radius ρ. What might this angular speed represent? Notice that, to the first order, each of the leading factors exceeds unity by (3/2)m/r, so the temporal and the spatial factors each contribute half of the relativistic precession, giving a total of 3m/r. Hence we are led to identify the rightmost factor with Ω, the angular speed of the radial perturbation in terms of the proper time. Thus we re-write the above equation as

$$\omega_\tau = \left(1-\frac{3m}{r}\right)^{1/2}\left(1-\frac{2m}{r}\right)^{3/4}\omega_\tau \approx \left(1-\frac{3m}{r}\right)\omega_\tau.$$

Just as before, we can now express the precession of the radial disturbance as

$$\frac{1}{\Omega_\tau}-1 = \left(1-\frac{3m}{r}\right)^{-1/2}-1 \approx \frac{3m}{r}.$$

The pseudo-derivation hinges on the identification of the frequency Ω_τ of the radial disturbance with the quantity $m^{1/2}/\rho^{3/2}$, which admittedly is not terribly well motivated, but it does at least suggest an alternate way of conceiving of the frequency of the propagating radial cycle in relativistic orbits. In other words, it suggests splitting the relativistic "Kepler's law" into two parts, one for the angular cycle, and one for the radial cycle. The latter can be expressed as

$$m = \Omega\rho^2 \rho^3$$

In the Newtonian two-body problem, a single frequency applies to both the angular cycle and the radial cycle, whereas in general relativity these are two different frequencies, so we need this second version of "Kepler's third law", to go along with the better-known angular version

$$m = \omega_t^{\,2} r^3$$

If $t = \tau$ and $r = \rho$ (as in Newtonian mechanics), these two are identical, but in general relativity they differ slightly, resulting in the relativistic orbital precession.

value. The mouse can still traverse the entire spiral in a finite span of time, because the path distance to the center is proportional to the radial distance, so we have

$$\frac{dt}{dr} = -\frac{k}{\sqrt{r}}$$

and hence the time to reach the center from a radial distance r is . But is this task really achievable? It still entails the mouse circling the center infinitely many times.

Maxwell points out that many such motions can essentially be represented by the motion of a particle subjected to two forces, one directed toward some central point with magnitude proportional to the distance from that point, and another retarding force proportional to the speed of the particle.

He shows that the path of a particle subjected to such forces is a logarithmic spiral. Letting ω denote the constant angular speed, and α the characteristic angle of the spiral (i.e., the angle between a radius and a tangent), he determined that the acceleration of the particle with speed v at a distance r from the center can be decomposed into a centrally directed acceleration of magnitude μr and a retarding acceleration (opposite the direction of travel) of magnitude 2kv where

$$\mu = \frac{\omega^2}{\sin(\alpha)^2} \qquad k = \frac{\omega}{\sin(\alpha)}.$$

These relations can also be written in the form

$$\frac{\omega}{\sqrt{\mu}} = \sin(\alpha) \qquad \frac{k}{\sqrt{\mu}} = \cos(\alpha).$$

Since Maxwell take w to be constant, the oscillations persist forever, with decreasing magnitude.

It would be interesting to determine what purely central force law (if any) applied to a point particle would yield a logarithmic spiraling orbit.

Chapter 3

Rectangles on Conic Section

A conic is a two-dimensional figure created by the intersection of a plane and a right circular cone. All conics can be written in terms of the following equation: $Ax^2 + Bxy + Cy^2 + Dx + Ey + F = 0$. The four conics we'll explore in this text are parabolas, ellipses, circles, and hyperbolas.

The equations for each of these conics can be written in a standard form, from which a lot about the given conic can be told without having to graph it. We'll study the standard forms and graphs of these four conics, as well as the polar equations for the conics, which are useful for certain applications.

As discussed earlier, there are no four squares in arithmetic progression. This fact can be expressed in terms of lattice points on the diagonal line x = y as shown in the figure below.

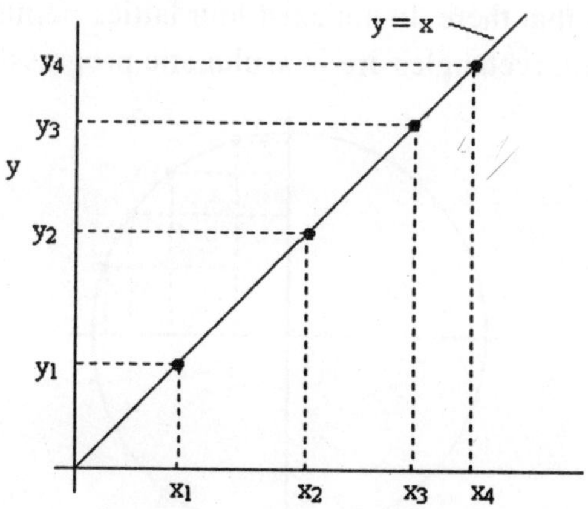

There do not exist four points with integer coordinates (x_j,y_j), j = 1,2,3,4, on the line y = x such that the four products x_jy_j are in arithmetic progression. This fact can be immediately generalized to lattice points on an arbitrary line y = ax + b for rational coefficients a,b with $a \neq 0$. To prove this, simply note that

$$x_jy_j = x_j\left(ax_j + b\right) = \frac{1}{a}\left(ax_j + \frac{b}{2}\right) - \frac{b^2}{4a}$$

which shows that if these four quantities are in arithmetic progression then we can add $b^2/(4a)$ to each number and multiply each number by "a", giving four rational squares in arithmetic progression, which we've already seen is impossible.

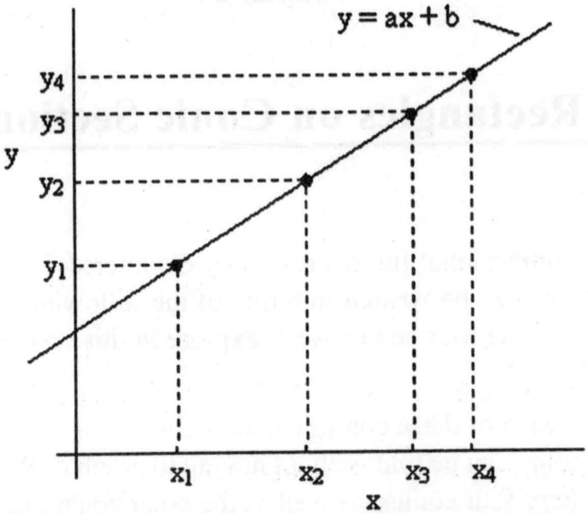

Can this proposition be generalized any further? The note entitled Discordance Impedes Square Magic included a table of right triangles with rational edge lengths on a common hypotenuse, and gives numerical evidence suggesting that no four such triangles exist with areas in arithmetic progression.

If true, this signifies that there do not exist four lattice points on a circle such that the areas of the corresponding rectangles are in arithmetic progression, as shown below.

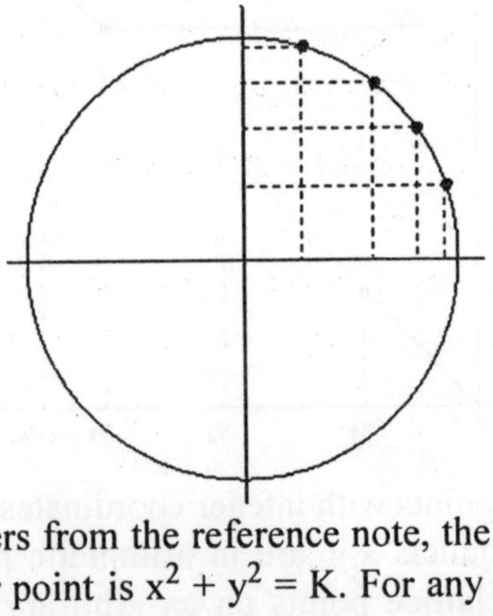

In terms of the parameters from the reference note, the radius of the circle is K, and the locus containing the four point is $x^2 + y^2 = K$. For any four such points with integer coefficients, the areas xy cannot be in arithmetic progression. Generalizing even further, we may conjecture that for any rational numbers a, b there do not exist positive integers x_j, y_j, j = 1,2,3,4 such that

$$x_j^2 + a y_j^2 = b$$

for j = 1,2,3,4 and such that the four products $(x_j\, y_j)$ are in arithmetic progression. In other words, for any four points (x_j, y_j) with integer coordinates on the conic $x^2 + ay^2 = b$ the four rectangles with those edge lengths cannot have areas in arithmetic progression. To illustrate, consider four rectangles inscribed in a quadrant of an ellipse as shown below.

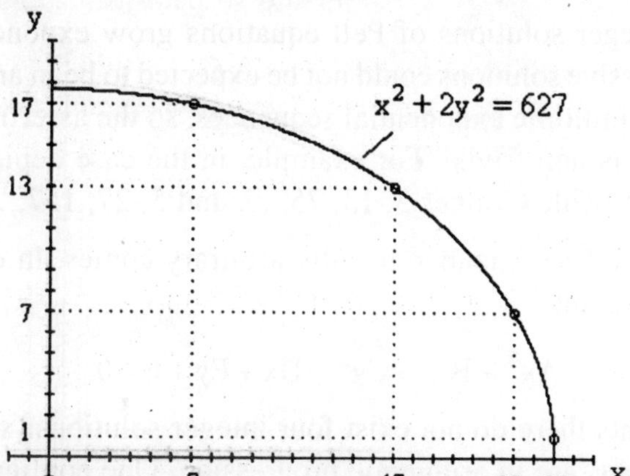

In this figure the four rectangles inscribed in the ellipse $x^2 + 2y^2 = 627$ have the edge lengths (25,1), (7,17), (23,7) and (17,13), so the four rectangular areas are 25, 119, 161, and 221. Obviously these are *not* in arithmetic progression. The conjecture is that no four rectangles with integer edge lengths inscribed in a quadrant of a central conic can have areas in arithmetic progression.

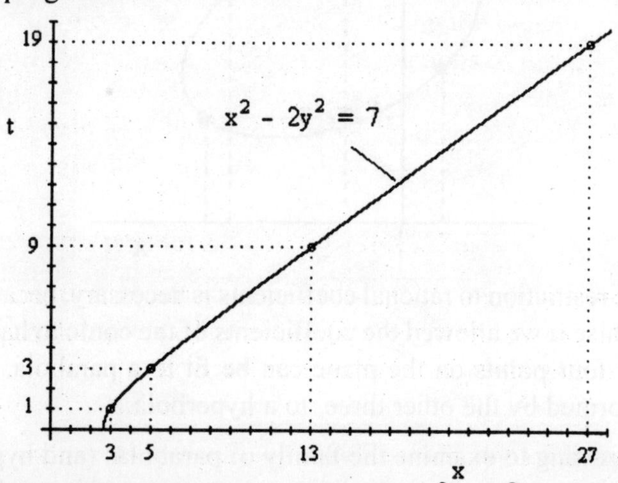

With b = 0 the conic degenerates to a straight line $x^2 = ay^2$. In this case the coefficient "a" must be a square, so it can be absorbed into y, and we have simply $x^2 = y^2$ and thus (in the first quadrant) x = y. The four rectangles are therefore squares, and the conjecture is equivalent to the well-known theorem on four squares. On the other hand, with a = 1 and b equal to an arbitrary positive integer constant, this gives the (seeming) fact that no four rectangles (with integer edges) on a common diagonal can have areas in arithmetic progression.

Another family of cases is given by setting a = –N and b = D, so the x,y pairs are solutions of a single Pell equation. $x^2 - Ny^2 = D$. In this case the conic is a hyperbola, as shown in the figure below with a = –2 and b = 7.

In general the integer solutions of Pell equations grow exponentially, so the areas corresponding to successive solutions could not be expected to be in arithmetic progression. However, there can be multiple exponential sequences, so the assertion that no four are in arithmetic progression is non-trivial. For example, in the case depicted above, there are two solution sequences, with x values 3, 13, 75, ... and 5, 27, 157, ...

The conjecture could be extended to fully arbitrary conics. In other words, we may conjecture that for any conic

$$Ax^2 + Bxy + Cy^2 + Dx + Ey + F = 0$$

with rational coefficients there do not exist four integer solutions (x_j, y_j), j = 1,2,3,4 such that the four products $x_j y_j$ are in arithmetic progression. One configuration is depicted in the figure below.

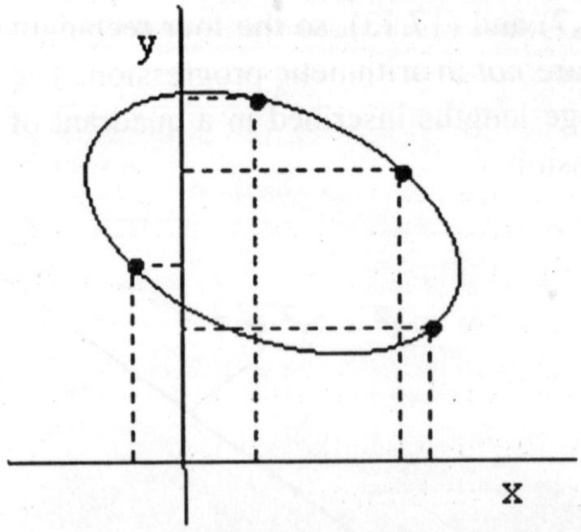

Incidentally, the restriction to rational coefficients is necessary, because the proposition would certainly be false if we allowed the coefficients of the conic to have any real values. This is because any four points on the plane can be fit to a parabola, or, if one point is inside the triangle formed by the other three, to a hyperbola.

It might be interesting to examine the family of parabolas (and hyperbolas) that pass through sets of four points in arithmetic progression, to see if it can be proven that the coefficients of such conics are necessarily irrational.

EQUATION OF A TRIANGLE

An ellipse can be expressed as the locus of points with orthogonal coordinates (x,y) such that $(x/a)^2 + (y/b)^2 = 1$ for some constants a,b. However, it's more difficult to give a succinct description of a piece-wise linear curve, e.g., a triangle. We seek a function f() such that f(x,y)=0 if and only if the point (x,y) is on the perimeter of a given triangle. It's not hard to see that no such function exists if we restrict ourselves to functions definable in terms of a finite number of basic arithmetical operations, i.e., additions, subtractions, multiplications, and divisions.

This raises the question of what other operation could be adjoined to the four basic operations that would enable us to define a general polygon. One possibility is the familiar "absolute value" function, defined as

$$|x| = \begin{cases} x \text{ if } x >= 0 \\ -x \text{ if } x <= 0 \end{cases}.$$

This function permits us to express a square as $|x| + |y| = 1$. More general quadralaterals can be expressed in the form

$$A|x| + Bx + C|y| + D = 1$$

Two of the four vertices of this quadralateral are on the x axis at the points x=-1/(A-B) and x=1/(A+B), while the other two are on the y axis at the points y=-1/(C-D) and y=1/(C+D). This NEARLY enables us to define a triangle by setting one of these vertices to zero; we can come arbitrarily close to zero using finite coefficients, but we can't quite define a true triangle.

A brute force approach would be to define three rays through the origin of the coordinate system:

$$y = ax \qquad y = bx \qquad y = cx$$

These divide the plane into six regions. We can then construct a function that reduces to one of three lines, depending on the region. The point (x,y) is above or below the line y = ax depending on whether the sign of y-ax is positive or negative. Therefore, a Boolean variable indicating whether (x,y) is above the line y=ax can be expressed as

$$Q[a] = \frac{1 + \dfrac{y - ax}{|y - ax|}}{2}.$$

Obviously the complement of this function

$$q[a] = \frac{1 - \dfrac{y - ax}{|y - ax|}}{2}$$

is a Boolean variable indicating whether (x,y) is BELOW the line y=ax. However, these functions are undefined for points on the line, because then |y-ax| = 0. One way of removing this ambiguity is to express Q and q in terms of a function that is arguably even simpler than the absolute value, namely, the sign function

$$\operatorname{sign}(x) = \begin{cases} +1 & \text{if } x >= 0 \\ -1 & \text{if } x <= 0 \end{cases}.$$

With this function we can define Q and q as

$$Q[a] = \frac{1+\operatorname{sign}(y-ax)}{2} \qquad q[a] = \frac{1-\operatorname{sign}(y-ax)}{2}.$$

where now Q signifies that (x,y) is above OR ON the line y=ax.

Using these boolean operators we can construct a function that reduces at any given point to the equation of one of the three edgelines of a triangle, depending on the relation of that that point to the other two edgelines. For example, consider a triangle whose vertices fall on the three rays shown below

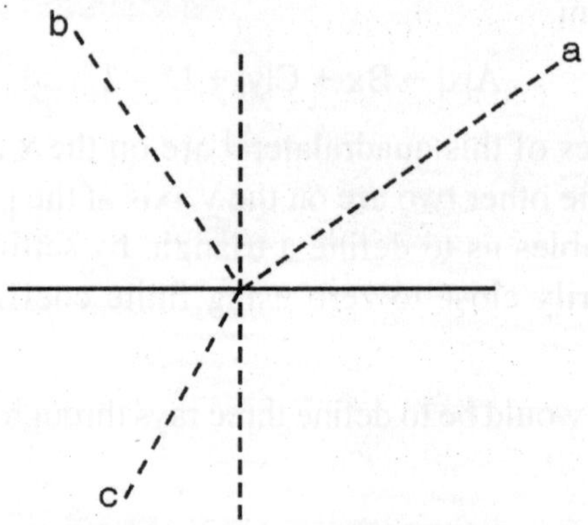

The edgeline connecting the vertices on rays a and b should be invoked whenever (x,y) is on or above the raylines a and b. The edgeline connecting a and c should be invoked when (x,y) is BELOW both of the raylines a and c.

The edgeline connecting b and c should be invoked when (x,y) is ON or ABOVE rayline c and BELOW rayline b. Thus, if the equations for the individual edgelines are A1x+B1y = 1, A2x+B2y=1, and A3x+B3y=1, then the overall equation for the triangle can be expressed in the form

$$\begin{aligned} &Q[a]\, Q[b]\, (A1x + B1y) \\ &\quad + q[a]\, q[c]\, (A2x + B2y) \\ &\qquad + q[a]\, Q[c]\, (A3x + B3y) = 1 \end{aligned}$$

To illustrate, if we take the triangle given by the rays

$y = (1/5)x \qquad y = -(8/5)x \qquad y = 2x$

with the edge lines

$$x/15 + y/6 = 1$$
$$x/6 - y/3 = 1$$
$$-x/3 - y/12 = 1$$

then equation is satisfied by the points on the perimeter of the triangle shown in Figure:

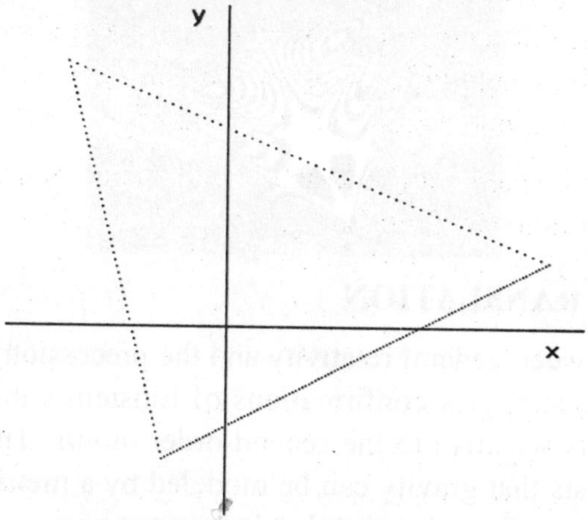

This figure assumes we have defined Q and q in terms of the "sign" function. Another way of dealing with the singularity of the absolute value versions of Q and q would be to multiply through equation by the factor 4|y-ax||y-bx||y-cx|. Letting R,S,T and r,s,t denote respectively the values and the absolute values of y-ax, y-bx, and y-cx, this gives the non-singular function

$$(R + r)(S + s)T\ (A1x + B1y)$$
$$(R - r)(T - t)S\ (A2x + B2y)$$
$$(S - s)(T + t)R\ (A3x + B3y) = 4$$

This is equivalent to (1) except that it is also satisfied by the points on any of the three rays. Thus the solution of equation is as illustrated in Figure.

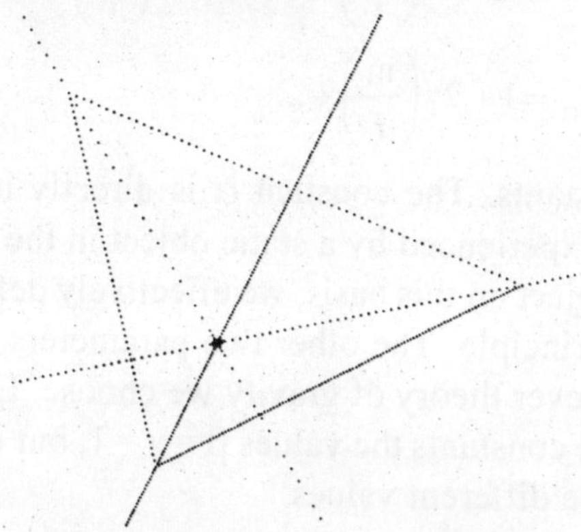

One interesting aspect of this "triangle function" is that we don't need to restrict our attention to just the points (x,y) that satisfy the equation. Instead, we can assign a color to every point on the plane depending on how close it comes to satisfying the equation. This results in some striking images, such as the "stealth fighter" shown in Figure.

AN ANOMALY IN TRANSLATION

The agreement between general relativity and the precession of Mercury's orbit was, and remains, one of the strongest confirmations of Einstein's theory because, of all the classical tests, it alone is sensitive to the second-order in m/r. The equivalence principle by itself strongly suggests that gravity can be modeled by a metrical theory of spacetime (in which particles follow extremal paths), but it does not necessarily single out Einstein's field equations as the laws governing the metric.

In general we can only say that the diagonalized coefficients of the metric in the vicinity of a spherically symmetrical gravitating body should be expressible (at least in the weak field) as power series in m/r. In terms of the usual Schwarzschild coordinates r and t the conventional way of writing this generalized metric is

$$(d\tau)^2 = g_{tt}(d\tau)^2 - g_{rr}(dr)^2 - r(d\theta)^2 - r^2 \sin(\theta)^2 (d\phi)^2$$

where

$$g_{tt} = 1 - 2\alpha\left(\frac{m}{r}\right) + 2(\beta - \alpha\gamma)\left(\frac{m}{r}\right)^2 + \ldots$$

$$g_{rr} = 1 + 2\gamma\left(\frac{m}{r}\right) + \ldots$$

where α, β and γ are constants. The constant α is directly measurable in terms of the gravitational acceleration experienced by a static object in the weak field limit, and since we define the mass of an object on this basis, we effectively define α = 1 in any theory that satisfied the equivalence principle. The other two parameters, β and γ, are dependent on the field equations of whatever theory of gravity we choose. Einstein's field equations of general relativity give these constants the values β = γ = 1, but in other metrical theories of gravity these constants have different values.

Of the three classical tests of general relativity, the gravitational redshift depends only on g_{tt} and can only be evaluated up to the first order, so it really verifies only the fact that $\alpha = 1$, which is to say, it verifies only the equivalence principle. Needless to say, this is an important verification, but it doesn't single out Einstein's field equations as opposed to the field equations of other possible metric theories. The second classical test was the deflection of starlight grazing the Sun during a solar eclipse, and it can be shown that this test verifies not only $\alpha = 1$, it also depends on g_{rr} up to the first order, so it verifies $\gamma = 1$ as well. Likewise the "Shapiro test" based on the time delay of radar echoes from the inner planets allows us to evaluate a and g, but none of these tests enable us to evaluate the g_{tt} coefficient to the second order, i.e., they do not constrain the value of β.

However, according to the general metrical theory, the precession per revolution of an orbit with semi-latus rectum L is $6\pi m/L$ times the factor $(2 - \beta + 2\gamma)/3$. Since the precession of Mercury's orbit is in close agreement with the value $6\pi m/L$, and since we can determine $\gamma = 1$ by other means, we can conclude that $\beta = 1$, consistent with Einstein's field equations. Thus the perihelion observations are by far the strongest confirmation we have of the correctness of general relativity.

Because of the importance of this observational verification, it is of interest to know whether, or how much, Einstein was influenced or guided by it in the formulation of general relativity. Throughout the years from 1908 to 1914 while he was working on the theory, he often assessed the redshift and the starlight deflection predictions of his current theory, but he never mentioned (in print) the precession of Mercury. But when he presented his completed theory late in 1915 the agreement with Mercury's precession was cited prominently as the crowning glory.

Actually we know Einstein was quite conscious of the Mercury precession anomaly throughout the years when he was developing general relativity. For one thing, we know he was an avid reader of the works of Henri Poincare, who wrote about the anomaly in his 1908 book Science and Method. In the section entitled "The New Mechanics and Astronomy" he notes that for a theory of gravity with a velocity dependent potential (along the lines suggested by Weber for electromagnetism)

There would result, in the perihelion of Mercury, a secular variation of 14" [seconds of arc per century], *in the same direction as that which has been observed and not explained*, but smaller, since the latter is 38". He goes on to say that Lorentz's theory of relativity (and therefore also Einstein's special relativity, although Poincare never mentioned Einstein in connection with relativity) predicts an advance of 7" seconds of arc per century for Mercury's perihelion. He concludes that

This cannot be regarded as an argument in favor of the new dynamics, since we still have to seek another explanation of the greater part of the anomaly connected with Mercury, but still less can it be regarded as an argument against it. This is interesting because it clearly shows that, at least in Poincare's mind, there was an anomaly connected with Mercury, and moreover that this anomaly was roughly 38". According to some accounts (e.g., Roseveare) there was no pressing anomaly perceived at this time, because Seeliger's hypothesis (1906) of a solar corona was thought to be adequate to account for the extra precession of Mercury's orbit. Roseveare says

I think that the reason for Nordstrom's attitude and the general neglect of the perihelion of Mercury as an anomaly to be explained by any new gravitational theory [in 1913] was that Seeliger's hypothesis was being taken very seriously... Since it was felt by both Einstein and Nordstrom that no empirical argument existed beyond the light deflection predictions, one can only assume that the perihelion motion of Mercury was not considered to be anomalous and that the prevailing hypothesis explaining it, Seelinger's hypothesis, was valid.

According to this account, the paper written by Einstein's friend Freundlich in February 1915 arguing against Seeliger's hypothesis was motivated by a desire to restore the precession of Mercury's orbit to the status of an anomaly so that it could be used as a test of gravitational theories. This seems like rather odd reasoning, and Roseveare himself admits that his argument is undermined by Einstein's later comment (in 1916) that Freundlich's attack on Seeliger's hypothesis was "forcing an open door", which clearly implies that Einstein (like Poincare) did not take Seeliger's hypothesis seriously.

Admittedly in 1906 (the year Seeliger published his hypothesis), Poincare stated that although the extra precession in Mercury's orbit was at that time the most grave discordance known for Newton's laws, he recognized that it could be explained by a ring of matter around the Sun. However, he may just have been acknowledging the latest hypothesis. In his review of astronomy and the new mechanics just two years later (quoted above) he made no mention of Seeliger's hypothesis or a circumsolar ring as a possible explanation for the anomaly, and he clearly treated the anomaly as a fact that could be used to discriminate between theories of gravity.

Similarly, in a review of gravitation theories, Walter Ritz wrote in 1909. Astronomical observations carried out over many centuries have revealed some deviations between observation and calculation, which cannot be explained by Newton's law up to now, and which a new theory will have to explain. Of these anomalies by far the largest is of the planet Mercury, whose ellipse precesses slowly, under the effect of the remaining planets; but the observed precession is larger by approximately 42 arc-seconds *per* century than the computed. The difference is small, but nevertheless unquestionable and unexplained. Again this clearly indicates not only that the precession of Mercury's orbit was considered anomalous, but that it was widely suspected that its resolution would come from a new theory of gravity. Of course, Einstein was very familiar with Ritz's work, having engaged him in a public debate in 1909 on the subject of the advanced solutions of Maxwell's equations.

In any case, we do have one definite piece of evidence for the fact that Einstein (like Poincare) regarded Mercury's precession as anomalous, and as something to be explained by a new theory of gravity – even before Poincare's 1908 book. In December 1907, just as Einstein was beginning to work seriously on his ideas about a relativistic theory of gravity (and just after having "the happiest thought of my life", i.e., the equivalence principle), he wrote in a letter to his friend Conrad Habicht. At the moment I am working on a relativistic analysis of the law of gravitation by means of which I hope to explain the still unexplained secular changes in the perihelion of Mercury.

We also know that around 1913 Einstein and his friend Michele Besso worked out the perihelion advance implied by the so-called *Entwurf* theory of gravity that Einstein had developed with the help of Marcel Grossmann. They were disappointed to find that the theory actually predicted a negative value for the precession, making the anomaly even worse. They decided not to publish the derivation. It wasn't until 1915 that Einstein developed a theory that gave the correct precession value of 43" of arc per century. Of course, his reasons for believing that he had finally arrived at the correct theory were mainly related to the logical coherence of it. "The sense of the thing is too evident." Nevertheless, the quantitative agreement with Mercury's precession seems to have been psychologically very powerful, and certainly contributed to his impression that the theory was correct. Pais says

This discovery was, I believe, by far the strongest emotional experience of Einstein's scientific life, perhaps in all his life. Nature had spoken to him. He had to be right. 'For a few days I was beside myself with joyous excitement'. Later he told Fokker that his discovery had given him palpitations of the heart. What he told de Haas is even more profoundly significant: when he saw that his calculations agreed with the unexplained astronomical observations, he had the feeling that something actually snapped in him…

Oddly enough Pais' translation of the 1907 letter to Habicht differs from the translations of Anna Beck (Einstein's Collected Papers) and Roseveare. Pais translated the passage as:

At this time I am [again] busy with considerations on relativity theory in connection with the law of gravitatation… I hope to clear up the so-far unexplained secular changes of the perihelion length of Mercury… [but] so far it does not seem to work.

This is strange for several reasons. First, why did Pais split the quote with ellipses? The other translations give no indication that anything is missing. (In Anna Beck's translation of this letter the reference to Mercury is just the single sentence quoted previously.) Second, did Einstein really refer to the *length* of the perihelion? A perihelion doesn't have a length. (Perhaps he meant the length of the period between perihelia?) Third, and most puzzling, where did Pais get the final phrase "so far it does not seem to work"? This phrase doesn't appear in any of the other translations. It is such a meaningful phrase that it's hard to imagine a translator leaving it out, but it's equally hard to imagine it being casually inserted without justification. What does the original German text say?

Incidentally, the popular book "The Evolution of Physics" (1938) by Einstein and Infeld says:

The deviation of the motion of the planet Mercury from the ellipse was known before the general relativity theory was formulated, and no explanation could be found. On the other hand, *general relativity developed without any attention to this special problem* [my emphasis]. Only later was the conclusion about the rotation of the ellipse in the motion of a planet around the sun drawn from the new gravitational equations.

Considering that Einstein began his search for a new gravitational theory in 1907 with the expressed purpose (as stated in his letter to Habicht) of explaining the anomalous precession of Mercury, and kept this objective in view throughout the intermediate development (including the Entwurf of 1913), and considering that Einstein listed the failure of the Entwurf theory to give the correct perihelion of Mercury as one of the three reasons that led him to lose faith in that theory, which then led him to the fully covariant theory of general relativity, it seems hard to justify the claim that general relativity was developed without *any attention* to this problem.

This is somewhat similar to Einstein's assertions that special relativity was developed without any attention to the Michelson and Morley experiment–despite the fact that at other times (notably his 1922 talk in Japan on how he developed the theory of relativity) he acknowledged that this experiment had been an important factor in his thinking. Of course, in both cases it's perfectly correct to say that the theories followed logically and unavoidably from very broad and fundamental principles, so they were in no way ad hoc explanations of the respective experimental facts, but it's still factually quite inaccurate to say that Einstein paid no attention to Mercury's anomalous precession during the years when he was developing general relativity.

In the case of "Evolution of Physics", the explanation for this clearly erroneous statement may be that Einstein had very little to do with the actual writing of the book. He apparently agreed to lend his name to the project in order to help his friend Infeld raise money so that Infeld might be allowed to remain in the United States after his grant expired. The prospect of being exported back to Poland in 1938 was not very appealing. Jefferson Weaver says Einstein's actual contribution to the book was "negligible". Pais says that Einstein was "not enthusiastic" about the book, and then concludes his brief discussion by quoting Einstein's cryptic comment in reference to the project:

One should not undertake anything which endangers the tenuous bridge of confidence between people.

ROUND TRIPS AND ONE-WAY SPEEDS

Given two mutually resting and unaccelerated clocks, denoted as A and B, separated by a distance L, let Clock A emit a pulse of light at time t_1, and let this pulse be reflected off clock B at time t_2 and arrive back at Clock A at time t_3. The round-trip speed of light is equal to $2L/(t_3 - t_1)$, but this does not tell us the one-way speed in each direction. To determine the one-way speed we need to synchronize clocks A and B, but therein lies an ambiguity, because there is more than one possible way of synchronizing two spatially separate clocks.

One method is to stipulate that the pulse reflects off Clock B at the time $t_2 = t_1 + (1/2)(t_3 - t_1)$, which is equivalent to the stipulation of isotropic light speed. In other words, with this synchronization, the speed of the pulse is the same in both directions (in terms of this coordinate system). However, nothing prevents us from defining synchronization by the more general relation

$$t_2 = t_1 + \varepsilon\,(t_3 - t_1)$$

for any value of ε in the range from 0 to 1. The case ε = 1/2 corresponds to inertial coordinates, in terms of which inertia is homogeneous and isotropic, but we aren't required to regard inertial coordinates as "true". To this extent the choice of synchronization methods, and therefore the one-way speed of light, is conventional, in a way that the round-trip speed of light is not.

Given any inertial coordinate system x,y,z,t, we are free to apply a coordinate transformation of the form

$$x' = x \quad y' = y \quad z' = z \quad t' = t + \frac{kx}{c}$$

where $k = 2\varepsilon - 1$. In terms of the primed coordinates the speed of light is then dependent on the angle of the light ray with respect to the x axis. For an arbitrary pulse of light emanating from the origin and reaching the point x,y,z at the time t, we have the squared speed

$$c^2 = \frac{x^2 + y^2 + z^2}{t^2}$$

and in terms of the primed coordinates we have

$$(c)^2 = \frac{(x')^2 + (y')^2 + (z')^2}{(t')^2} = \frac{x^2 + y^2 + z^2}{\left(t - \frac{kx}{c}\right)^2}.$$

Thus we have the ratio

$$\frac{c'}{c} = \frac{t}{t - \frac{t}{kx}} = \frac{1}{1 - k\left(\frac{x}{ct}\right)}$$

Also, multiplying through equation by t^2 and dividing by x^2 gives

$$\left(\frac{ct}{x}\right)^2 = 1 + \frac{y^2 + z^2}{x^2} = 1 + \tan(\theta)^2 = \frac{1}{\cos(\theta)^2}$$

where θ is the angle between the light ray and the positive x axis. Hence $x/(ct) = \cos(\theta)$, and we can substitute into the previous equation to give

$$\frac{C'}{c} = \frac{1}{1 - k\cos(\theta)}$$

which is the equation of an ellipse. This is the directional dependence that the anisotropic speed of light must have in order to be consistent with the primed coordinates. It corresponds to a simple tilt of the plane of simultaneity.

Furthermore, this dependence is sufficient to ensure that the average speed of light over any closed loop path is precisely c (for any constant value of k in the range from –1 to +1). To prove this, we can partition any continuous path into a large number n of essentially linear segments of equal incremental length ds, so the total path length is nds.

Each segment is characterized by coordinate increments dx, dy, and dz, and the cosine of the angle θ between this segment and the positive x axis is given by $\cos(\theta) = dx/ds$. Substituting into the above expression and re-arranging terms gives

$$c' = \frac{c}{1 - k\dfrac{dx}{ds}}$$

so the time to traverse the segment is

$$\frac{ds}{c'} = \frac{ds}{c}\left(1 - k\frac{dx}{ds}\right).$$

The average speed over the entire path is therefore the total distance traveled, nds, divided by the sum of all the times, which gives

$$c_{path} = \frac{n}{\sum \frac{1}{c'}} = \frac{nc}{\sum\left(1 - k\dfrac{dx}{ds}\right)} = \frac{nc}{\sum\left(1 - k\dfrac{dx}{ds}\right)} = \frac{c}{1 - \dfrac{k}{n}\sum\left(\dfrac{dx}{ds}\right)}.$$

In the limit the summation becomes n/L times the integral of $(dx/ds)ds = dx$ over the path, where L is the total path length, so we have

$$c_{path} = \frac{c}{1 - \dfrac{k}{L}\displaystyle\int_{path} dx}.$$

The integral vanishes for any path that begins and ends at the same x coordinate, so the average speed of light over any such path is simply c. It follows that the average speed of light over any closed loop is c. The preceding discussion shows that an ellipsoidal dependence of light-speed on direction is *sufficient* to ensure the average speed c for all closed paths, but it doesn't show necessity. We arrived at that function by considering the effect of simply shifting the simultaneity convention by tilting the locus of simultaneity. This leaves us with the question of whether any other directional dependence give invariant round-trip light speed. To prove that the ellipsoidal dependence is unique, consider the fact that, for an out and back trip along a segment of length L oriented at an angle θ (relative to the direction of maximum light speed), the round trip speed is

$$c = \frac{2L}{\frac{L}{c(\theta)} + \frac{L}{c(\theta+\pi)}} = \frac{2}{\frac{1}{c(\theta)} + \frac{1}{c(\theta+\pi)}}$$

so it is necessary for c to be the harmonic mean of the speeds in every pair of opposite directions. This, in itself, is easy to accomplish, since we can choose the speeds arbitrarily over one hemisphere of directions, and then compute the required speeds on the remaining hemisphere accordingly. Strictly speaking, this is all we need to ensure a null result for the experiment of Michelson and Morley, because they compared the speed of light for two "back and forth" paths.

However, this is not enough to ensure the same average speed for all paths. For example, to give an average speed of c for a regular triangular path, we must have

$$c = \frac{1}{\frac{1}{c(\theta)} + \frac{1}{c(\theta + 2\pi/3)} + \frac{1}{c(\theta = 4\pi/3)}}.$$

Indeed, to ensure an average speed of c for *all* regular polygonal paths, we must have

$$\frac{n}{c} = \sum_{j=0}^{n-1} \frac{1}{c(\theta + 2\pi_j / n)}$$

for all integers n greater than 1, and for all angles θ. Since c is always positive, without loss of generality we can define a continuous function f(θ) such that

$$c(\theta) = \frac{c}{1 - f(\theta)}.$$

The preceding equation is satisfied if and only if

$$\sum_{j=0}^{n-1} f\left(\theta + \frac{2\pi}{n} j\right) = 0.$$

An arbitrary function $f(\theta)$ can be expressed as a cosine Fourier series as follows

$$f(\theta) = a_0 + a_1 \cos(\omega\theta) + a_2 \cos(2\omega\theta) + a_3 \cos(3\omega\theta) + \ldots$$

so the previous summation can be written as

$$\sum_{j=0}^{n-1} f\left(\theta + \frac{2\pi}{n} j\right) = a_0 + a_1 \cos(\phi) + a_2 \cos(2\phi) + a_3 \cos(3\phi) + \ldots$$

$$+ a_0 + a_1 \cos\left(\phi + \frac{2\pi}{n}\right) + a_2 \cos\left(2\phi + 2\frac{2\pi}{n}\right) + a_3 \cos\left(3\phi + \frac{2\pi}{n}\right) + \ldots$$

$$+ a_0 + a_1 \cos\left(\phi + \frac{2\pi}{n} 2\right) + a_2 \cos\left(2\phi + 2\frac{2\pi}{n} 2\right) + a_3 \cos\left(3\phi + 3\frac{2\pi}{n} 2\right) + \ldots + \ldots$$

$$+ a_0 + a_1 \cos\left(\phi + \frac{2\pi}{n}(n-1)\right) + a_2 \cos\left(2\phi + 2\frac{2\pi}{n}(n-1)\right) + a_3 \cos\left(3\phi + 3\frac{2\pi}{n}(n-1)\right) + \ldots$$

where $\phi = \pi\theta/L$ is arbitrary. Collecting terms by coefficients, and expanding the cosines of sums, we get

$$\sum_{j=0}^{n-1} f\left(\theta + \frac{2\pi}{n} j\right) = n a_0$$

$$+ a_1 \left[\cos(\phi) \sum_{j=0}^{n-1} \cos\left(\frac{2\pi}{n} j\right) - \sin(\phi) \sum_{j=0}^{n-1} \sin\left(\frac{2\pi}{n} j\right)\right]$$

$$+ a_2 \left[\cos(2\phi) \sum_{j=0}^{n-1} \cos\left(2\frac{2\pi}{n} j\right) - \sin(2\phi) \sum_{j=0}^{n-1} \sin\left(2\frac{2\pi}{n} j\right)\right]$$

$$+ a_3 \left[\cos(3\phi) \sum_{j=0}^{n-1} \cos\left(3\frac{2\pi}{n} j\right) - \sin(3\phi) \sum_{j=0}^{n-1} \sin\left(3\frac{2\pi}{n} j\right)\right] + \ldots$$

Each of the sine summations vanishes identically, so we are left with

$$\sum_{j=0}^{n-1} f\left(\theta + \frac{2\pi}{n} j\right) = \sum_{k=0}^{\infty} \left(a_k \cos(k\phi) \sum_{j=0}^{n-1} \cos\left(k \frac{2\pi}{n} j\right)\right).$$

This quantity must vanish for each integer n greater than 1, and for arbitrary values of ϕ. It follows that each coefficient a_k times the corresponding inner summation must individually vanish. For $k = 0$ the inner summation equals n, so we must have $a_0 = 0$. With $k = 1$ the inner summation vanishes for all n greater than 1, so a_1 can be any constant. However, for $k = 2$ the inner summation equals 2 for $n = 2$. In other words, the requirement for the two-way (back and forth) average speed to be c forces $a_2 = 0$. Similarly for $k = 3$ the inner summation equals 3 for $n = 3$, meaning that the regular triangular path forces $a_3 = 0$. Likewise the requirement for the average speed of light to be c on square paths forces $a_4 = 0$, and so on. Thus, in order to have an average speed of c for all regular polygonal paths, it is necessary for $f(\theta)$ to be of the form $a_1 \cos(\theta)$, and we have already seen that this is sufficient to give the average speed c for all closed paths.

CONVENTIONAL WISDOM

It is a matter of empirical fact that the speed of light is invariant in terms of inertial coordinates, and yet the invariance of the speed of light is often said to be a matter of convention-as indeed it is. The empirical fact refers to the speed of light *in terms of inertial coordinates,* but the decision to define speeds in terms of inertial coordinates is conventional. It's trivial to define systems of space and time coordinates in terms of which the speed of light is not invariant, but we ordinarily choose to describe events in terms of inertial coordinates, partly because of the invariance of light speed based on those coordinates. Of course, this invariance would be tautological if inertial coordinate systems were simply *defined* as the systems in terms of which the speed of light is invariant. The class of inertial coordinate systems is actually defined in purely mechanical terms, without reference to the propagation of light.

They are the coordinate systems in terms of which mechanical inertia is homogeneous and isotropic (which are the necessary and sufficient conditions for Newton's three laws of motion to be valid, at least quasi-statically). The empirical invariance of light speed with respect to this class of coordinate systems is a non-trivial empirical fact, but nothing *requires* us to define "velocity" in terms of inertial coordinate systems. Such systems cannot claim to have any *a priori* status as the "true" class of coordinates. Despite the undeniable success of the principle of inertia as a basis for organizing our understanding of the processes of nature, it is nevertheless a convention.

The conventionalist view can be traced back to Poincare, who wrote in "The Measure of Time" in 1898... we have no direct intuition about the equality of two time intervals. The simultaneity of two events or the order of their succession, as well as the equality of two time intervals, must be defined in such a way that the statements of the natural laws be as simple as possible. In the same paper, Poincare described the use of light rays, together with the convention that the speed of light is invariant and the same in all directions, to give an operational meaning to the concept of simultaneity. In his book "Science and Hypothesis" (1902) he summarized his view of time by saying. There is no absolute time. When we say that two periods are equal, the statement has no meaning, and can only acquire a meaning by a convention.

Poincare's views had a strong influence on the young Einstein, who avidly read "Science and Hypothesis" with his friends in the self-styled "Olympia Academy". Solovine remembered that this book "profoundly impressed us, and left us breathless for weeks on end". Indeed we find in Einstein's 1905 paper on special relativity the statement. We have not defined a common time for A and B, for the latter cannot be defined at all unless we establish by definition that the time required by light to travel from A to B equals the time it requires to travel from B to A. In a later popular exposition, Einstein tried to make the meaning of this definition more clear by saying. That light requires the same time to traverse the path A to M (the midpoint of AB) as for the path B to M is in reality neither a supposition nor a hypothesis about the physical nature of light, but a stipulation which I can make of my own freewill in order to arrive at a definition of simultaneity.

Of course, this concept of simultaneity is also embodied in Einstein's second "principle", which asserts the invariance of light speed. Throughout the writings of Poincare, Einstein, and others, we see the invariance of the speed of light referred to as a convention, a definition, a stipulation, a free choice, an assumption, a postulate, and a principle... as well as an empirical fact. There is no conflict between these characterizations, because the convention (definition, stipulation, free choice, principle) that Poincare and Einstein were referring to is nothing other than the *decision* to use inertial coordinate systems, and once this decision has been made, the invariance of light speed is an empirical fact. As Poincare said in 1898, we naturally choose our coordinate systems "in such a way that the statements of the natural laws are as simple as possible", and this almost invariably means inertial coordinates. It was the great achievement of Galileo, Descartes, Huygens, and Newton to identify the principle of inertia as the basis for resolving and coordinating physical phenomena.

Unfortunately this insight is often disguised by the manner in which it is traditionally presented. The beginning physics student is typically expected to accept uncritically an intuitive notion of "uniformly moving" time and space coordinate systems, and is then told that Newton's laws of motion happen to be true with respect to those "inertial" systems. It is more meaningful to say that we define inertial coordinate systems as those systems in terms of which Newton's laws of motion are valid. We naturally coordinate events and organize our perceptions in such a way as to maximize symmetry, and for the motion of material objects the most important symmetries are the isotropy of inertia, the conservation of momentum, the law of equal action and re-action, and so on. Newtonian physics is organized entirely upon the principle of inertia, and the basic underlying hypothesis is that for any object in any state of motion there exists a system of coordinates in terms of which the object is instantaneously at rest and inertia is homogeneous and isotropic (implying that Newton's laws of motion are at least quasi-statically valid).

The empirical validity of this remarkable hypothesis accounts for all the tremendous success of Newtonian physics. The specification of a particular state of motion, combined with the requirement for inertia to be homogeneous and isotropic, completely determines a system of coordinates (up to insignificant scale factors, rotations, etc), and such a system is called an inertial system of coordinates. Such coordinate systems can be established unambiguously by purely mechanical means (neglecting the equivalence principle and associated complications in the presence of gravity). The assumption of inertial isotropy with respect to a given state of motion suffices to establishes the loci of inertial simultaneity for that state of motion. Poincare and Einstein rightly noted the conventionality of this simultaneity definition because they were not pre-supposing the choice of inertial simultaneity. In other words, we are not required to use inertial coordinates. We simply choose, of our own free will, to use inertial coordinates-with the corresponding inertial definition of simultaneity-because this renders the statement of physical laws and the descriptions of physical phenomena as simple and perspicuous as possible, by taking advantage of the maximum possible symmetry.

In this regard it's important to remember that inertial coordinates are *not* entirely characterized by the quality of being unaccelerated, i.e., by the requirement that isolated objects move uniformly in a straight line. It's also necessary to require the unique simultaneity convention that renders mechanical inertial isotropic (the same in all spatial directions), which amounts to the stipulation of equal one-way speeds for the propagation of physically identical actions. These comments are fully applicable to the Newtonian concept of space, time, and inertial reference frames. Given two objects in relative motion we can define two systems of inertial coordinates in which the respective objects are at rest, and we can orient these coordinates so the relative motion is purely in the x direction. Let t,x and T,X denote these two systems of inertial coordinates. That such coordinates exist is the main physical hypothesis underlying Galilean physics. An auxiliary hypothesis – one that was not always clearly recognized – concerns the relationship between two such systems of inertial coordinates, given that they exist.

Galileo assumed that if the coordinates x,t of an event are known, and if the two inertial coordinate systems are the rest frames of objects moving with a relative speed of v, then the coordinates of that event in terms of the other system (with suitable choice of origins) are $T = t$, $X = x-vt$. Viewed in the abstract, this is a rather peculiar and asymmetrical assumption, although it is admittedly borne out by experience-at least to the precision of measurement available to Galileo. However, we now know, *empirically*, that the relation between relatively moving systems of inertial coordinates has the symmetrical form $T = (t-vx)/\gamma$ and $X = (x-vt)/\gamma$ where $\gamma = (1-v^2)^{1/2}$ when the time and space variables are expressed in the same units such that the constant $(3)10^8$ meters/second equals unity. It follows that the one-way (not just the two-way) speed of light is invariant and isotropic with respect to any and every system of inertial coordinates.

The empirical content of this statement is simply that the propagation of light is isotropic with respect to the same class of coordinate systems in terms of which mechanical inertia is isotropic. This is consistent with the fact that light itself is an inertial phenomena, e.g., it conveys momentum. In fact, the inertia of light can be seen as a common thread running through three of the famous papers published by Einstein in 1905. In the paper entitled "On a Heuristic Point of View Concerning the Production and Transformation of Light" Einstein advocated a conception of light as tiny quanta of energy and momentum, somewhat reminiscent of Newton's inertial corpuscles of light. It's clear that Einstein already understood that the conception of light as a classical wave is incomplete. In the paper entitled "Does the Inertia of a Body Depend on its Energy Content?" he explicitly advanced the idea of light as an inertial phenomenon, and of course this was suggested by the fundamental ideas of the special theory of relativity presented in the paper "On the Electrodynamics of Moving Bodies".

The Galilean conception of inertial frames assumed that all such frames share a unique foliation of spacetime into "instants". Thus the relation "in the present of" constituted an equivalence relation across all frames of reference. If A is in the present of B, and B is in the present of C, then A is in the present of C. However, special relativity makes it clear that there are infinitely many distinct loci of inertial simultaneity through any given event, because inertial simultaneity depends on the velocity of the worldline through the event. The inertial coordinate systems *do* induce a temporal ordering on events, but only a partial one. With respect to any given event we can still partition all the other events of spacetime into distinct causal regions, including "past", "present" and "future", but in addition we have the categories "future null" and "past null", and none of these constitute equivalence classes. For example, it is possible for A to be in the present of B, and B to be in the present of C, and yet A is not in the present of C. Being "in the present of" is not a transitive relation.

It could be argued that a total unique temporal ordering of events is a more useful organizing principle than the isotropy of inertia, and so we should adopt a class of coordinate systems that provides a total ordering. We can certainly do this, as Einstein himself described in his 1905 paper.

To be sure, we could content ourselves with evaluating the time of events by stationing an observer with a clock at the origin of the coordinates who assigns to an event to be evaluated the corresponding position of the hands of the clock when a light signal from that event reaches him through empty space. However, we know from experience that such a coordination has the drawback of not being independent of the position of the observer with the clock.

The point of this "drawback" is that there is no physically distinguished "origin" on which to base the time coordination of all systems of reference, so from the standpoint of assessing possible causal relations we must still consider the full range of possible "absolute" temporal orderings. This yields the same partial ordering of events as does the set of inertial coordinates, so the "total ordering" that we can achieve by imposing a single temporal foliation on all frames of reference is only formal, and not physically meaningful. Nevertheless, we could make this choice, especially if we regard the total temporal ordering of events as a requirement of intelligibility. This seems to have been the view of Lorentz, who wrote in 1913 about the comparative merits of the traditional Galilean and the new Einsteinian conceptions of time.

> It depends to a large extent on the way one is accustomed to think whether one is attracted to one or another interpretation. As far as this lecturer is concerned, he finds a certain satisfaction in the older interpretations, according to which... space and time can be sharply separated, and simultaneity without further specification can be spoken of... one may perhaps appeal to our ability of imagining arbitrarily large velocities. In that way one comes very close to the concept of absolute simultaneity.

Of course, the idea of "arbitrarily large velocities" already pre-supposes a concept of absolute simultaneity, so Lorentz's rationale is not especially persuasive, but it expresses the point of view of someone who places great importance on a total temporal ordering, even at the expense of inertial isotropy. Indeed one of Poincare's criticisms of Lorentz's early theory was that it sacrificed Newton's third law of equal action and re-action. (This can be formally salvaged by assigning the unbalanced forces and momentum to an undetectable ether, but the physical significance of a conservation law that references undetectable elements is questionable.) Oddly enough, even Poincare sometimes expressed the opinion that a total temporal ordering would always be useful enough to out-weigh other considerations, and that it would always remain a safe convention. The approach taken by Lorentz and most others may be summarized by saying that they sacrificed the physical principles of inertial relativity, isotropy, and homogeneity in order to maintain the assumed Galilean composition law. This approach, although technically serviceable, suffers from a certain inherent lack of conviction, because while asserting the ontological reality of anisotropy in all but one (unknown) frame of reference, it unavoidably requires us to disregard that assertion and arbitrarily assume one particular frame as being "the" rest frame.

Poincare and Einstein recognized that in our descriptions of events in spacetime in terms of separate space and time coordinates we're free to select our "basis" of decomposition. This is precisely what one does when converting the description of events from one frame to another using Galilean relativity, but, as noted above, the Galilean composition law yields anisotropic results when applied to actual observations. So it appeared (to most people) that we could no longer maintain isotropy and homogeneity in all inertial frames together with the ability to transform descriptions from one frame to another by simply applying the appropriate basis transformation. But Einstein realized this was too pessimistic, and that the new observations were fully consistent with both isotropy in all inertial frames *and* with simple basis transformations between frames, *provided* we adjust our assumption about the effective metrical structure of spacetime. In other words, he brilliantly discerned that Lorentz's anisotropic results totally vanish in the context of a different metrical structure.

Even a metrical structure is conventional in a sense, because it relies on our ontological premises. For example, the magnitude of the interval between two events may *seem* to be one thing but actually be another, due (perhaps) to variations in our means of observation and measurement. However, once we have agreed on the physical significance of inertial coordinate systems, the invariance of the quantity $(dt)^2 - (dx)^2 - (dy)^2 - (dz)^2$ also becomes physically significant.

Let us take a system of co-ordinates in which the equations of Newtonian mechanics hold good.

Suitably qualified, this immediately establishes not only the convention of simultaneity, but also the means of operationally establishing it, and its physical significance. Any observer in any state of inertial motion can throw two identical particles in opposite directions with equal force (i.e., so there is no net disturbance of the observer's state of motion), and the convention that those two particles have the same speed suffices to fully specify an entire system of space and time coordinates, which we call inertial coordinates. It is then an empirical fact-not a definition, convention, assumption, stipulation, or postulate-that the speed of light is isotropic in terms of inertial coordinates. This obviously doesn't imply that inertial coordinates are "true" in any absolute sense, but the principle of inertia has proven to be immensely powerful for organizing our knowledge of physical events, and for discerning and expressing the apparent chains of causation.

If a flash of light emanates from the geometrical midpoint between two spatially separate particles at rest in an inertial frame, the arrival times of the light wave at those two particles are simultaneous in terms of that rest frame's inertial coordinates. Furthermore, we find empirically that all other physical processes are isotropic with respect to those inertial coordinates, e.g., if a sound wave emanates from the midpoint of a uniform steel beam at rest in an inertial frame, the sound reaches the two ends simultaneously in accord with this definition. If we adopt any other convention we introduce anisotropies in our descriptions of physical processes, such as sound in a uniform stationary steel beam propagating more rapidly in one direction than in the other. The isotropy of physical phenomena-including the propagation of light-*is* strictly a convention, but it was not introduced by special relativity, it is one of the fundamental principles which we use to organize our knowledge, and it leads us to choose inertial coordinates for the description of events. On the other hand, the isotropy of multiple distinct physical phenomena in terms of inertial coordinates is *not* purely conventional, because those coordinates can be defined in terms of just one of those phenomena. The value of this definition is due to the fact that a wide variety of phenomena are (empirically) isotropic with respect to the same class of coordinate systems.

Of course, it could be argued that all these phenomena are, in some sense, "the same". For example, the energy conveyed by electromagnetic waves has momentum, so it is an inertial phenomenon, and therefore it is not surprising that the propagation of such energy is isotropic in terms of inertial coordinates. From this point of view, the value of the definition of inertial coordinates is that it reveals the underlying unity of superficially dissimilar phenomena, e.g., the inertia of energy. This illustrates that our conventions and definitions are not empty, because they represent ways of organizing our knowledge, and the efficiency and clarity of this organization depends on choosing conventions that reflect the unity and symmetries of the phenomena. We could, if we wish, organize our knowledge based on the assumption of a total temporal ordering of events, but then it would be necessary to introduce a whole array of unobservable anisotropic "corrections" to the descriptions of physical phenomena.

As we've seen, the principle of relativity constrains, but does not uniquely determine, the form of the mapping from one system of inertial coordinates to another. In order to fix the observable elements of a spacetime theory with respect to every member of the equivalence class of inertial frames we require one further postulate, such as the invariance of light speed. However, we should distinguish between the strong and weak forms of the light-speed invariance postulate. The strong form asserts that the *one-way* speed of light is invariant with respect to the natural space-time basis associated with any inertial state of motion, whereas the weak form asserts only that the *round-trip* speed of light is invariant. To illustrate the different implications of these two different assumptions, consider an experiment of the type conducted by Michelson and Morley in their efforts to detect a directional variation in the speed of light, due to the motion of the Earth through the aether, with respect to which the absolute speed is light was presumed to be referred.

To measure the speed of light along a particular axis they effectively measured the elapsed time at the point of origin for a beam of light to complete a round trip out to a mirror and back. At first we might think that it would be just as easy to measure the one-way speed of light, by simply comparing the time of transmission of a pulse of light from one location to the time of reception at another location, but of course this requires us to have clocks synchronized at two spatially separate locations, whereas it is precisely this synchronization that is at issue. Depending on how we choose to synchronize our separate clocks we can measure a wide range of light speeds. To avoid this ambiguity, we must evaluate the time interval for a transit of light at a single spatial location (in the coordinate system of interest), which requires us to measure a round trip, just as Michelson and Morley did.

Incidentally, it might seem that Roemer's method of estimating the speed of light from the variations in the period between eclipses of Jupiter's constituted a one-way measurement. Similarly people sometimes imagine that the one-way speed of light could be discerned by (for example) observing, from the center of a circle, pulses of light emitted uniformly by a light source moving at constant speed around the perimeter of the circle. Such methods are indeed capable of detecting certain kinds of anisotropy, but they cannot detect the anisotropy entailed by Lorentz's ether theory, nor any of the other theories that are observationally indistinguishable from Lorentz's theory (which itself is indistinguishable from special relativity). In any theory of this class, there is an ambiguity in the definition of a "circle" in motion, because circles contract to ellipses in the direction of motion. Likewise there is ambiguity in the definition of "uniformly-timed" pulses from a light source moving around the perimeter of a moving circle (ellipse). The combined effect of length contraction and time dilation in a Lorentzian theory is to render the anisotropies unobservable.

The empirical indistinguishability between the theories in this class implies that there is no unambiguous definition of "the one-way speed of light". We can measure without ambiguity only the lapses of time for closed-loop paths, and such measurements cannot establish the "open-loop" speed. The ambiguity in the one-way speed remains, because over any closed loop, by definition, the net change in each and every direction is zero. Hence it is possible to consistently interpret all observations based on the assumption of non-isotropic light speed.

Admittedly the resulting laws take on a somewhat convoluted appearance, and contain unobservable parameters, but they can't be ruled out empirically. To illustrate, consider a measurement of the round-trip speed of light, assuming light travels at a constant speed c relative to some absolute medium with respect to which our laboratory is moving with a speed v. Under these assumptions, we would expect a pulse of light to travel with a speed c+v (relative to the lab) in one direction, and c−v in the opposite direction. So, if we send a beam of light over a distance L out to a mirror in the "c+v" direction, and it bounces back over the same distance in the "c−v" direction, the total elapsed time to complete the round trip of length 2L is

$$\Delta t = \frac{L}{c+v} + \frac{L}{c-v} = L\left(\frac{2c}{c^2 - v^2}\right).$$

Therefore, the average round-trip speed relative to the laboratory would be

$$\frac{2L}{\Delta t} = \frac{c^2 - v^2}{c} = c\left(1 - \left(\frac{v}{c}\right)^2\right).$$

This shows why a round-trip measurement of the speed of light would not be expected to reveal any dependency on the velocity of the laboratory unless the measurement was precise enough to resolve second-order effects in v/c. The ability to detect such small effects was first achieved in the late 19th century with the development of precision interferometry (exploiting the wave-like properties of light.) The experiments of Michelson and Morley showed that, despite the movement of the Earth in its orbit around the Sun (to say nothing of the movement of the solar system, and even of the galaxy), there was no $(v/c)^2$ term in the round-trip speed of light. In other words, they found that $2L/\Delta t$ is always equal to c, at least to the accuracy they could measure, which was more than adequate to rule out a second-order deviation. Thus we have a firm empirical basis for asserting that the round-trip speed of light is independent of the motion of the source.

This is the weak form of the invariant light speed postulate, but in his 1905 paper Einstein asserted something stronger, namely, that we should adopt the convention of regarding the *one-way* speed of light as invariant. This stronger postulate doesn't follow from the results of Michelson and Morley, nor from any other conceivable experiment or observation-but there is also no conceivable observation that could conflict with it. The invariant round-trip speed of light fixes the observable elements of the theory, but it does not uniquely determine the presumed ontological structure, because multiple different interpretations can be made to fit the same set of appearances. The one-way speed of light is necessarily an interpretative element of our experience.

To illustrate the ambiguity, notice that we can ensure a null result for the Michelson and Morley experiment while maintaining non-constant light speed, merely by requiring that the speed of light v_1 and v_2 in the two opposite directions of travel (out and back) satisfy the relation

$$c = \frac{2L}{\frac{L}{v_1} + \frac{L}{v_2}} = \frac{2}{\frac{1}{v_1} + \frac{1}{v_2}}.$$

In other words, a linear round-trip measurement of light speed will yield the constant c in every direction provided only that the *harmonic mean* of the one-way speeds in opposite directions always equals c.

This is easily accomplished by defining the one-way velocity v_1 as a function of direction arbitrarily for all directions in one hemisphere, and then setting the velocities in the opposite directions the velocities v_2 in the opposite directions as $v_2 = cv_1 / (2v_1\text{-}c)$. However, we also wish to cover more complicated round-trips, rather than just back and forth on a single line. To ensure that a circuit of light around an equilateral triangle with edges of length L yields a round-trip speed of c, the speeds v_{-1}, v_2, v_3 in the three equally spaced directions must satisfy

$$c = \frac{3L}{\frac{L}{v_1} + \frac{L}{v_2} + \frac{L}{v_3}} = \frac{3}{\frac{1}{v_1} + \frac{1}{v_2} + \frac{1}{v_3}}$$

so again we see that the light speeds must have a harmonic mean of c. In general, to ensure that every closed loop of light, regardless of the path, yields the average speed c, it's necessary (and also sufficient) to have light speed $v = C(\theta)$ as a function of angle θ in a principal plane such that, for any positive integer n,

$$\frac{1}{C(0)} + \frac{1}{C(2\pi/n)} + ... + \frac{1}{C(2(n-1)\pi/n)} = \frac{n}{c}.$$

In units with c = 1, we need the n terms on the left side to sum to n, so the velocity function must be such that $1/C(\theta) = 1 + f(\theta)$ where the function $f(\theta)$ satisfies

$$f\left(\theta+0\frac{2\pi}{n}\right)+f\left(\theta+1\frac{2\pi}{n}\right)+f\left(\theta+2\frac{2\pi}{n}\right)+....+f\left(\theta+(n-1)\frac{2\pi}{n}\right)=0$$

for all θ. The canonical example of such a function is simply $f(\theta) = k\cos(\theta)$ for any constant k. Thus if we postulate that the speed of light varies as a function of the angle of travel θ relative to some primary axis according to the equation

$$C(\theta)=\frac{c}{1+k\cos(\theta)}$$

then we are assured that all closed-loop measurements of the speed of light will yield the constant c, despite the fact that the one-way speed of light is distinctly non-isotropic (for non-zero k). This equation describes an ellipse, and no measurement can disprove the hypothesis that the one-way speed of light.

It is, strictly speaking, a matter of convention. If we choose to believe that light has the same speed in all directions, then we assume k = 0, and in order to send a synchronizing signal to two points we would locate ourselves midway between them (i.e., at the location where round trips between ourselves and those two points take the same amount of time.) On the other hand, if we choose to believe light travels twice as fast in one direction as in the other, then we would assume k = 1/3, and we would locate ourselves 2/3 of the way between them (i.e., twice as far from one as the other, so round trip times are two to one). The latter case is illustrated in the figure below.

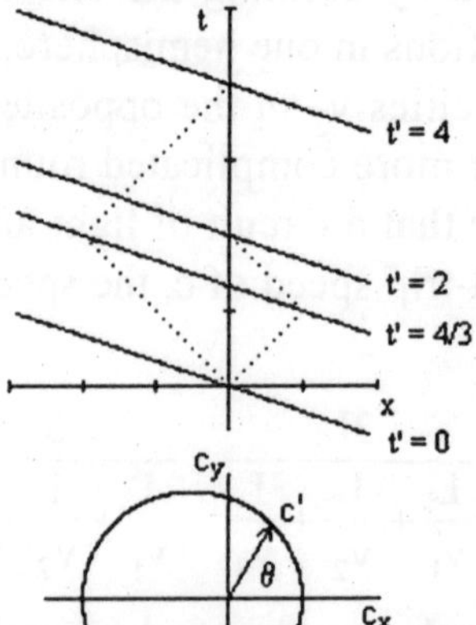

Regardless of what value we assume for k (in the range from -1 to +1), we can synchronize all clocks according to our belief, and everything will be perfectly consistent and coherent. Of course, in any case it's necessary to account consistently for the lapse of time for information to get from one clock to another, but the lapse of time between any two clocks separated by a distance L can be anything we choose in the range from virtually

0 to 2L/c. The only real constraint is that that the speed be an elliptical function of the direction angle.

The velocity profile is simply the polar equation of an ellipse (or ellipsoid is revolved about the major axis), with the pole at one focus, the semi-latus rectum equal to c, and eccentricity equal to k. This just projects the ellipse given by cutting the light cone with an oblique plane. Interestingly, there are really *two* light cones that intersect on this plane, and they are the light cones of the two events whose projections are the two foci of the ellipse-for timelike separated events. Recall that all rays emanating from one focus of an ordinary ellipse and reflecting off the ellipse will re-converge on the other focus, and that this kind of ray optics is time-symmetrical. In this context our projective ellipse is the intersection of two null-cones, i.e., it is the locus of all points in spacetime that are null-separated from *both* of the "foci events". This was to be expected in view of the time-symmetry of Maxwell's equations.

Our main reason for assuming k = 0 is our preference for symmetry, simplicity, and consistency with inertial isotropy. Within our empirical constraints, k can be interpreted as having any value between -1 and +1, but the principle of sufficient reason suggests that it should not be assigned a non-zero value in the absence of any rational justification. Nevertheless, it remains a convention (albeit a compelling one), but we should be clear about what precisely is – and what is not – conventional. The invariance of lightspeed is a convention, but the invariance of lightspeed in terms of inertial coordinates is an empirical fact, and this empirical fact is *not* a formal tautology, because inertial coordinates are determined by the mechanical inertia of material objects, independent of the propagation of light. Recall that Einstein's 1905 paper states that if a pulse of light is emitted from an unaccelerated clock at time t_1, and is reflected off some distant object at time t_2, and is received back at the original clock at time t_3, then the inertial coordinate synchronization is given by stipulating that

$$t_2 = t_1 + \frac{1}{2}(t_3 - t_1).$$

Reichenbach noted that the formally viable simultaneity conventions correspond to the assumption

$$t_2 = t_1 + \varepsilon(t_3 - t_1)$$

where ε is any constant in the range from 0 to 1. This describes the same class of "elliptical speed" conventions as discussed above, with ε = (k+1)/2 where k ranges from –1 to +1. The corresponding coordinate transformation is a simple time skew, i.e., x' = x, y' = y, z' = z, t' = t + kx/c. This describes the essence of the Lorentzian "absolutist" interpretation of special relativity. Beginning with the putative absolute rest frame inertial coordinates x,y, Lorentz associates with each state of motion v a system of coordinates x',t' related to x,y by a Galilean transformation with parameter v. In other words, x' = x – vt and t' = t. He then re-scales the x',t' coordinates to account for what he regards as the physical contraction of the lengths of stable object and the slowing of the durations of stable physical processes, to arrive at the coordinates x" = x/ γ and t" = t γ where $\gamma = (1-v^2/c^2)^{1/2}$. These he regards as the proper rest frame coordinates for objects moving with speed v in terms of the absolute frame. There is nothing logically unacceptable about these coordinate systems, but we must realize that they do not constitute inertial coordinate systems in the full sense. Mechanical inertia and the speed of light are not isotropic in terms of such coordinates, precisely because the time foliation (i.e., the simultaneity convention) is skewed relative to the $\varepsilon = 1/2$ convention.

If we begin with the *inertial* rest frame coordinates for the state of motion v (which Lorentz and Einstein agree are related to the putative absolute rest frame coordinate by a Lorentz transformation), and then apply the time skew transformation with parameter k = -v/c, we arrive at these Lorentzian rest frame coordinates. Needless to say, our choice of coordinate systems does not affect the outcome of any physical measurement, except that the outcome will be expressed in different terms. For example, by the Einsteinian convention the speed of light is isotropic in terms of the rest frame coordinates of any material object, whereas by the Lorentzian convention it is not.

This difference is simply due to different definitions of "rest frame coordinates". If we specify inertial coordinate systems (i.,e., coordinates in terms of which inertia is isotropic and Newton's laws are quasi-statically valid) then there is no ambiguity, and both Lorentz and Einstein agree that the speed of light is isotropic in terms of all inertial coordinate systems.

In subsequent sections we'll see that the standard formalism of general relativity provides a convenient means of expressing the relations between spacetime events with respect to a larger class of coordinate systems, so it may appear that inertial references are less significant in the general theory.

In fact, Einstein once hoped that the general theory would not rely on the principle of inertia as a primitive element. However, this hope was not fulfilled, and the underlying physical basis of the spacetime manifold in general relativity remains the set of primitive inertial paths (geodesics) through spacetime. Not only do these inertial paths determine the equivalence class of allowable coordinate systems (up to diffeomorphism), it even remains true that at each event we can construct a (local) system of inertial coordinates with respect to which the speed of light is c in all directions.

Thus the empirical fact of lightspeed invariance and isotropy with respect to inertial coordinates remains as a primitive component of the theory. The difference is that in the general theory the convention of using inertial coordinates is less prevalent, because in general there is no single global inertial coordinate system, and non-inertial coordinate systems are often more convenient on a curved manifold.

PASCAL'S MYSTIC HEXAGRAM

One of the first applications of modern mathematics to emerge during the Renaissance was the study of perspective for the purposes of painting and architecture. This led naturally to a consideration of projection, i.e., the mapping of images from one plane surface S to another plane surface S' by projection from a point O.

The point p on S maps to a point p' on S' such that O, p and p' are colinear. Obviously the distances between points and the angles between lines are not preserved under a projective mapping, but certain properties of figures are preserved. For example, straight lines map to straight lines, and conics map to conics. In addition to these qualitative invariants, there are also quantitative invariants. Consider four points A,B,C,D on a line in S, projected from the point O to the points A', B', C', D' on a line in S' as shown below.

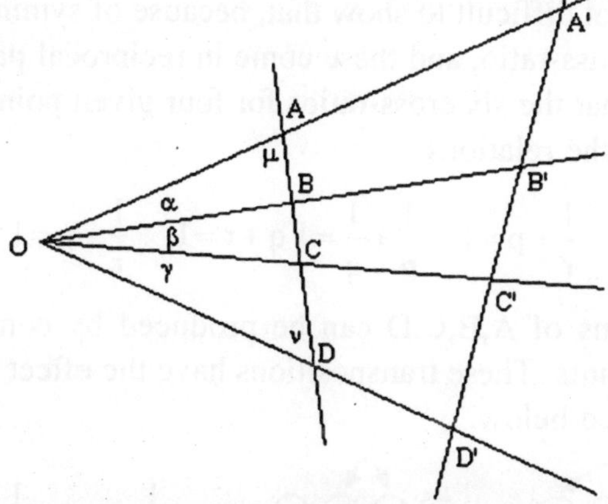

Applying the law of sines to the triangles OBA and OBD, we have

$$\frac{AB}{OB}=\frac{\sin(\alpha)}{\sin(\mu)} \quad \frac{DB}{OB}=\frac{\sin(\beta+\gamma)}{\sin(\nu)}.$$

The ratio of these two is

$$\frac{AB}{BD}=\frac{\sin(\alpha)}{\sin(\beta+\gamma)}\frac{\sin(\nu)}{\sin(\mu)}.$$

Likewise if we apply the law of sines to the triangles OCD and OCA we get

$$\frac{DC}{CA}=\frac{\sin(\gamma)}{\sin(\alpha+\beta)}\frac{\sin(\mu)}{\sin(\nu)}.$$

Thus if we multiply AB/BD by DC/CA the sines of m and n cancel out, and are left with

$$\frac{(AB)(CD)}{(AC)(BD)}=\frac{\sin(\alpha)\sin(\gamma)}{\sin(\alpha+\beta)\sin(\beta+\gamma)}.$$

This quantity is called a cross-ratio, and we see that it depends only on the angles between the projecting rays emanating from the point O, so it has the same value if we project the four points A,B,C,D onto any other line. In other words, we have

$$\frac{(AB)(CD)}{(AC)(BD)} = \frac{(A'B')(C'D')}{(A'C')(B'+D')}.$$

Furthermore, since the distances between the points don't change if we change the position of the projecting point O, it follows that the cross-ratio is the same for any projection point. Also, given the positions of any three of the co-linear points A,B,C,D and the value of their cross-ratio, the position of the fourth point is fully determined.

It's worth mentioning that although we speak of "the" cross-ratio of four points, the value depends on the order in which we take the points. There are 4! = 24 possible permutations, but it's not difficult to show that, because of symmetries, there are only six distinct values of the cross-ratio, and these come in reciprocal pairs. Thus there are three real values p,q r such that the six cross-ratios for four given points are p, 1/p, q, 1/q, r, 1/r. In addition, we have the relations

$$q+r=1 \qquad \frac{1}{r}+p=1 \qquad \frac{1}{p}+\frac{1}{q}=1 \; q+r=1 \qquad \frac{1}{r}+p=1 \qquad \frac{1}{p}+\frac{1}{q}=1$$

All 24 permutations of A,B,C,D can be produced by combining the six possible transpositions of the points. These transpositions have the effect of cycling between pairs of values, as summarized below.

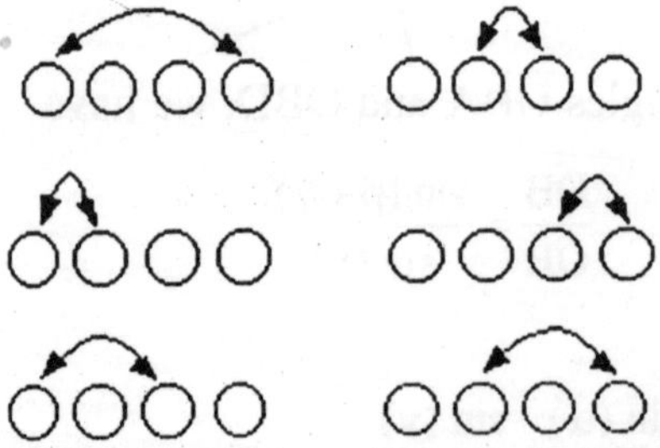

$$p \leftrightarrow \frac{1}{p} \qquad q \leftrightarrow \frac{1}{q} \qquad r \leftrightarrow \frac{1}{r}$$

$$p \leftrightarrow q \qquad \frac{1}{q} \leftrightarrow \frac{1}{r} \qquad r \leftrightarrow \frac{1}{p}.$$

$$p \leftrightarrow r \qquad \frac{1}{q} \leftrightarrow \frac{1}{p} \qquad \frac{1}{r} \leftrightarrow p$$

So, to define a single cross-ratio we need to specify four co-linear points (or, equivalently, four lines that intersect at a given point) in a particular order. We will use the notation [ABCD] to indicate the cross-ratio of the co-linear points A,B,C,D in that order. When referring to four lines emanating from the point O and passing through the points A,B,C,D we will designate the origin point with a subscript, as in $[ABCD]_O$.

Incidentally, the cross-ratio is also an important quantity in complex analysis. Given any four complex numbers z_1, z_2, z_3, z_4, the cross-ratio defined as

$$[z_1, z_2, z_3, z_4] = \frac{(z_1 - z_2)(z_3 - z_4)}{(z_3 - z_2)(z_1 - z_4)}$$

is invariant under arbitrary Mobius (i.e., linear fractional) transformations. This quantity is a purely real number if and only if z_1 through z_4 lie on a straight line or a circle.

Geometrically we can define a cross-ratio either in terms of four points on a line or in terms of an origin and four arbitrary points in a plane with the origin, since these four points specify a set of four lines through the origin. For example, we could specify a cross-ratio in terms of a given origin point O and four points A,B,C D on the perimeter of a circle as shown below.

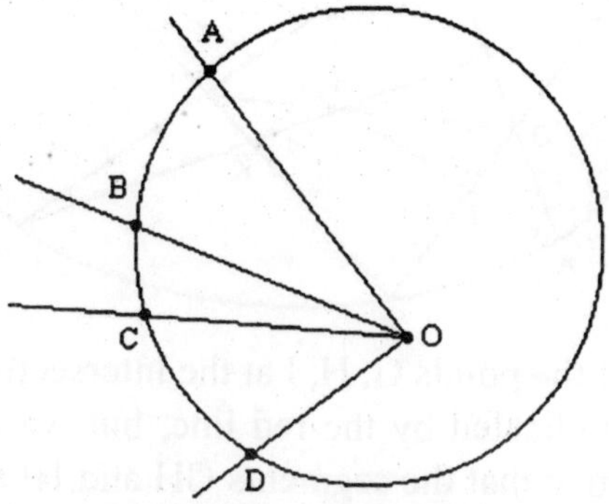

Since the points A,B,C,D are not co-linear, the cross-ratio depends on the position of the origin point O. Thus if we take the same four points but consider the lines through them emanating from a different origin located, say, on the circle itself, the cross-ratio is different, because the angles between the lines are different. However, it's easy to see that the cross-ratio for four given points on a circle is the same for an origin located anywhere on the same circle. This is illustrated in the figure below with the origin point O located at an arbitrary point on the circle.

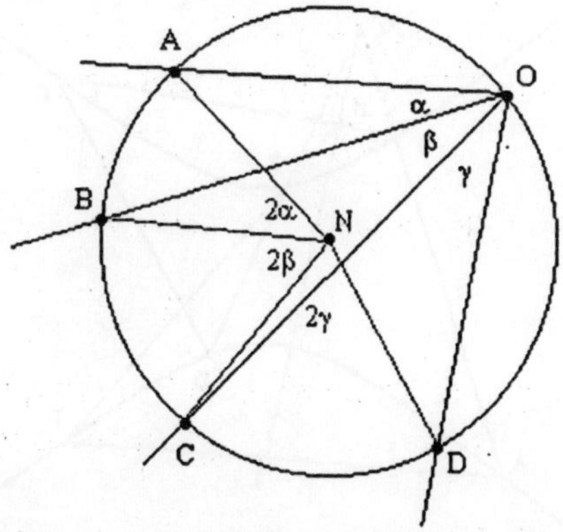

We know from elementary geometry that if N is the center of the circle then the angle ANB is twice the angle AOB for any point O on the circle. Therefore, if the points A through D are fixed, the cross-ratio for any origin point on the circle is invariant. Furthermore, since the cross-ratio is invariant under arbitrary projections, this proposition is valid for arbitrary conics as well. In other words, given any four points on a conic (such as an ellipse, hyperbola, or parabola), the cross-ratio for the lines through those points is invariant for any origin point located anywhere on the same conic.

To illustrate the usefulness of these ideas, consider six points A through F placed arbitrarily on a conic, such as on an ellipse as shown below, draw the lines AB, BC,..., EF, FA, making a hexagon.

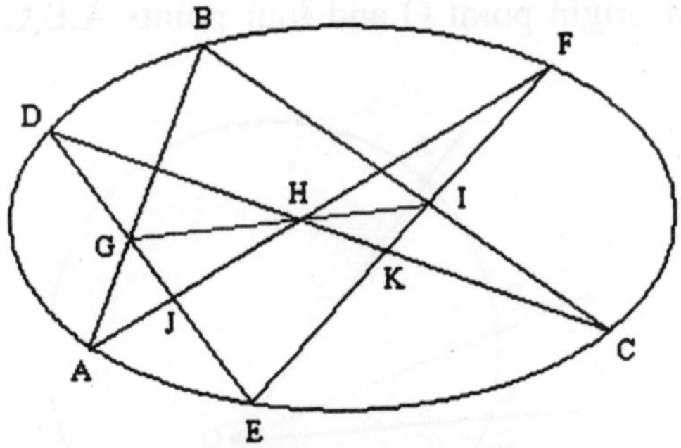

Visually it appears that the points G, H, I at the intersections of opposite edges of this hexagon are co-linear, as indicated by the red line, but we would like to prove this. In other words, we wish to prove that the segments GH and HI really lie along a single line.

A typical proof making use of cross-ratios begins with the fact that $[EKIF]_H$ is equal to $[EKIF]_C$. Now, the lines emanating from C and passing through K and I also pass through D and B respectively, so $[EKIF]_C$ is equal to $[EDBF]_C$, and we can switch origin points from C to A, because they are both on the ellipse, so these cross-ratios equal $[EDBF]_A$. We notice that the lines emanating from A and passing through the points B and F also pass through G and J respectively, so the cross-ratio equals $[EDGJ]_A$.

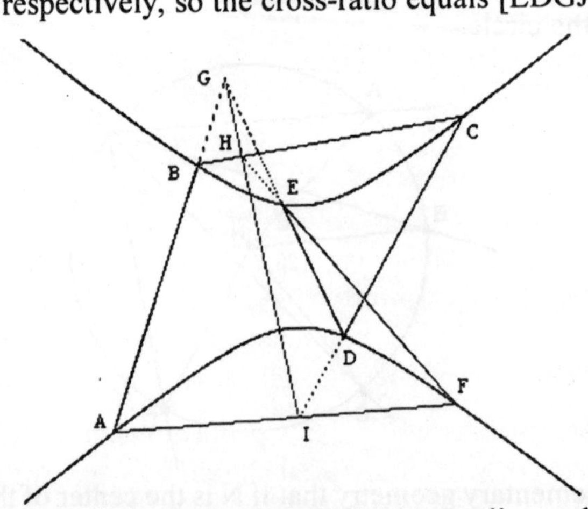

Since these points are co-linear, the cross-ratio also applies to the origin H, so it equals $[EDGJ]_H$. Thus we've shown that $[EKIF]_H$ equals $[EDGJ]_H$. The points K and D represent the same line through H, and the points F and J represent the same line through H, so it follows that the points I and G fall on a single line through H, which was to be shown.

As noted previously, a conic curve is projected to a conic curve, because a projection maps curves of second degree to curves of second degree. Therefore, the above theorem applies to any hexagon inscribed in any conic. For example, if we place the vertices of a hexagon on a hyperbola we arrive at a figure as shown below.

Notice that we extend the edges of the hexagon as necessary to find the intersection points of opposite edges. We can also arrange the vertices so that the points of intersection lie entirely outside the figure, as shown for the ellipse below.

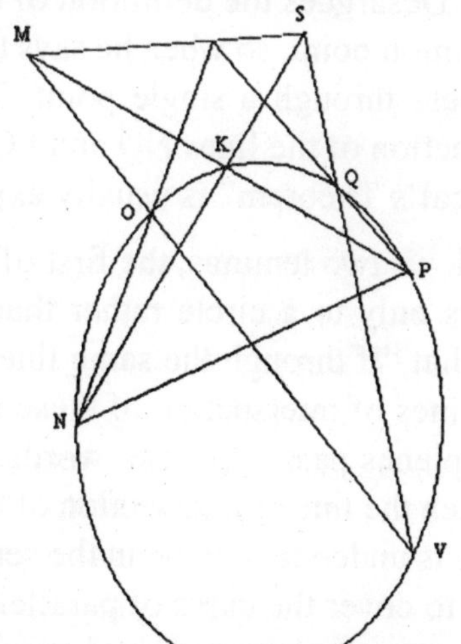

This theorem was first stated in 1640 by Blaise Pascal when he was just 16 years of age. His father, Etienne, had retired from his civil service job in 1631 to devote himself to the education of Blaise and his two older sisters. At the age of 14 Pascal joined his father as a member of the group of mathematicians and scientists associated with Marin Mersenne. In addition to the Pascals, this remarkable group included (either in the weekly meetings or by correspondence) Descartes, Fermat, Gassendi, Desargues, Roberval, Beeckman, Peiresc, and Hobbes, along with extensive communications with Huygens, Torricelli, Galileo, and many others. Girard Desargues (1593-1662) originated the idea of projective geometry, and Pascal wrote that he (Pascal) owed everything he had found on the subject to the writings of Desargues. Unfortunately Desargues' brilliant innovation occurred simultaneously with the introduction of analytic geometry by Descartes, so the ideas of projective geometry were over-shadowed for nearly two hundred years.

The younger Pascal was one of the few people to appreciate the power and beauty of Desargues' approach to geometry, but Pascal himself soon gave up mathematics and devoted most of the rest of his short life to theology.

Oddly enough, Pascal didn't actually present "his theorem" as a theorem, nor did he ever publish a proof of it. The only thing he ever published on the subject was a brief summary, announcing several results without proof. This was to be followed by a comprehensive treatise on conics, which Pascal apparently worked on for many years, but then abandoned. No copies have survived, so we have only the brief statements from the 1640 "Essay on Conics" (along with notes that Leibniz took when he saw a copy of Pascal's unpublished treatise). In terms of the lettering in the figure above, Pascal stated that:

If through points K,V any conic section whatever passes cutting the lines MK, MV, SK, SV in points P,O,N,Q, then the lines MS, NO, PQ will be of the same order.

Pascal carried over from Desargues the definition of an *ordonnance* of lines as a set of lines passing through a common point, so when he says the lines MS, NO, PQ are of the same order, he means they pass through a single point. This of course is equivalent to saying that the point of intersection of the lines NO and PQ is co-linear with the points M and S, which is the way "Pascal's Theorem" is usually expressed today.

The above statement follows two lemmas, the first of which is identical to the above statement except that it refers only to a circle rather than a general conic section. The second lemma states simply that "If through the same line several planes are passed, and are cut by another plane, all lines of intersection of these planes are of the same order as the line through which these planes pass". In other words, if a plane containing a certain line is cut by another plane, then the line of intersection of these two planes passes through the given line. (Of course, this is understood to be in the sense of projective geometry with its lines and points at infinity to cover the cases of parallel lines and planes.) Then Pascal says "On the basis of these two lemma's and several easy deductions from them, we can demonstrate" the general proposition about conics quoted above. After listing several other results, the brief essay concludes

There are many other problems and theorems, and many deductions which can be made from what has been stated above, but the lack of confidence which I have, owing to my little experience and capacity, does not allow me to go further into the subject until it has passed the examination of able men who may be willing to take this trouble. After that if someone thinks the subject worth continuing, we shall endeavor to extend it as far as God gives us the strength.

Alas these were the last words Pascal ever published on the subject.

Of course, Pascal's celebrated theorem is a generalization of Pappus' Theorem. Ironically, it can be argued that a true understanding of it comes only when viewed from the algebraic standpoint of Descartes' analytic geometry, as pointed out by Julius Plücker (1801-1868) using the fact that two algebraic plane curves of degree m and n with no common factor have mn points of intersection (counting multiplicities and points at infinity). This is called Bezout's Theorem for obscure reasons, since Bezout apparently never proved it, and the proposition had been stated and used by earlier mathematicians such as Newton. In any case, the equation of a line L(x,y) = 0 has degree 1, and the union of m lines is just the product of the equations of the individual lines, so it has degree m. Thus for both the theorem of Pappus and the theorem of Pascal the inscribing figure is a locus Q(x,y) = 0 of degree 2, and we can split the six lines into two sets of three to give the two functions $L_{AB}L_{CD}L_{EF}$ and $L_{BC}L_{DE}L_{FA}$, each of degree 3. Therefore, these loci intersect with each other in nine points, six of which are on the inscribing locus and three of which are not.

Now, any linear combination of the two cubics also intersects with Q at the six given points, but in addition we can choose the coefficients (of the linear combination) so that the combined function intersects Q at a seventh point, which implies that Q has a common factor with the combination of cubics. If Q is an irreducible conic, the common factor can only be Q itself. If Q is the product of two linear factors (as in Pappus' Theorem), then we need simply note that the combination of cubics can be forced to pass through the intersection of the two factors, so the combination of degree 3 intersects with each line in four points, and therefore it must have a common factor with each of them, which again implies that the entire inscribing function Q divides some linear combination of the two cubics. Thus there must be constants α,β and a linear function L(x,y) such that

$$\alpha L_{AB}L_{CD}L_{EF} + \beta L_{BC}L_{DE}L_{FA} = Q(x,y)L(x,y).$$

Therefore, the three zeros of the two cubics that do not lie on Q(x,y) = 0 must lie on the straight line L(x,y) = 0. This proves Pappus' and Pascal's theorems together, and immediately points to higher-order generalizations.

For a simple example, consider the instance of Pappus' theorem in the figure below.

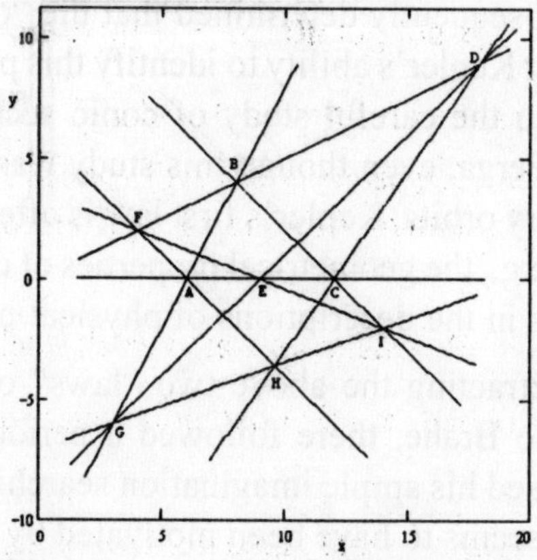

Algebraically, the inscribing locus consists of the two lines y = 0 and x-2y = 0, so it is represented by the quadratic

$$Q(x,y) = (y)(x-2y).$$

The edges of the hexagon ABCDEF are the lines

$$L_{AB}(x,y) = 2x - y - 12 \quad L_{BC}(x,y) = x + y - 12$$

$$L_{CD}(x,y) = 3x - 2y - 36 \quad L_{DE}(x,y) = x - y - 9.$$

$$L_{EF}(x,y) = 2x - 5y - 18 \quad L_{FA}(x,y) = x + y - 6$$

So, by Bezout's Theorem, we know there exist real numbers α,β and a linear function L(x,y) such that equation is satisfied. We find that $\alpha = 1$ and $\beta = -12$ gives

$$(2x-y-12)(3x-2y-36)(2x+5y-18)-12(x+y-12)(x-y-9)(x+y-6)$$

$$=(y)(x-2y)(4x-11y-78)$$

Therefore, the three points G, H, I must lie on the straight line $4x-11y-78 = 0$.

KEPLER, NAPIER, AND THE THIRD LAW

By the year 1605 Johannes Kepler, working with the relativistic/inertial view of the solar system suggested by Copernicus, had already discerned two important mathematical regularities in the orbital motions of the planets:

I. Planets move in ellipses with the Sun at one focus.

II. The radius vector describes equal areas in equal times.

This shows the crucial role that interpretations and models sometimes play in the progress of science, because it's obvious that these profoundly important observations could never even have been formulated in terms of the Ptolemaic earth-centered model.

Oddly enough, Kepler arrived at these conclusions in reverse order, i.e., he first determined that the radius vector of a planet's "oval shaped" path sweeps out equal areas in equal times, and only subsequently determined that the "ovals" were actually ellipses. It's often been remarked that Kepler's ability to identify this precise shape from its analytic properties was partly due to the careful study of conic sections by the ancient Greeks, particularly Apollonius of Perga, even though this study was conducted before there was even any concept of planetary orbits. Kepler's first law is often cited as an example of how purely mathematical ideas (e.g., the geometrical properties of conic sections) can sometimes find significant applications in the descriptions of physical phenomena.

After painstakingly extracting the above two "laws" of planetary motion from the observational data of Tycho Brahe, there followed a period of more than twelve years during which Kepler exercised his ample imagination searching for any further patterns or regularities in the data. He seems to have been motivated by the idea that the orbits of the planets must satisfy a common set of simple mathematical relations, analogous to the mathematical relations which the Pythagoreans had discovered between harmonious musical tones. However, despite all his ingenious efforts during these years, he was unable to discern any significant new pattern beyond the two empirical laws which he had found in 1605. Then, as Kepler later recalled, on the 8th of March in the year 1618, something marvelous "appeared in my head". He suddenly realized that

III. The proportion between the periodic times of any two planets is precisely one and a half times the proportion of the mean distances.

In the form of a diagram, his insight looks like this:

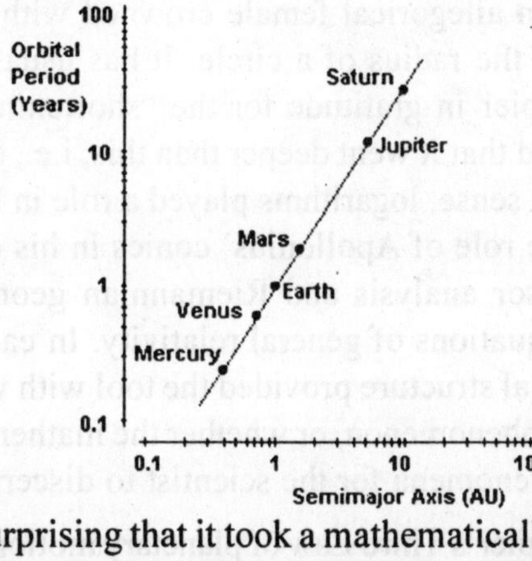

At first it may seem surprising that it took a mathematically insightful man like Kepler over twelve years of intensive study to notice this simple linear relationship between the logarithms of the orbital periods and radii. In modern data analysis the log-log plot is a standard format for analyzing physical data. However, we should remember that logarithmic scales had not yet been invented in 1605. A more interesting question is why, after twelve years of struggle, this way of viewing the data suddenly "appeared in his head" early in 1618.

Is it just coincidental that John Napier's "Mirifici Logarithmorum Canonis Descripto" was first seen by Kepler towards the end of the year 1616? We know that Kepler was immediately enthusiastic about logarithms, which is not surprising, considering the masses of computation involved in preparing the Rudolphine Tables. Indeed, he even wrote a book of his own on the subject in 1621. It's also interesting that Kepler initially described his "Third Law" in terms of a 1.5 ratio of proportions, exactly as it would appear in a log-log plot, rather than in the more familiar terms of squared periods and cubed distances. It seems as if a purely mathematical invention, namely logarithms, whose intent was simply to ease the burden of manual arithmetical computations, may have led directly to the discovery/formulation of an important physical law, i.e., Kepler's third law of planetary motion. (Ironically, Kepler's academic mentor, Michael Maestlin, chided him-perhaps in jest?-for even taking an interest in logarithms, remarking that "it is not seemly for a professor of mathematics to be childishly pleased about any shortening of the calculations".) By the 18th of May, 1618, Kepler had fully grasped the logarithmic pattern in the planetary orbits:

Now, because 18 months ago the first dawn, three months ago the broad daylight, but a very few days ago the full Sun of a most highly remarkable spectacle has risen, nothing holds me back. It's interesting to compare this with Einstein's famous comment about "...years of anxious searching in the dark, with their intense longing, the final emergence into the light—only those who have experienced it can understand it".

Kepler announced his Third Law in Harmonices Mundi, published in 1619, and also included it in his "Ephemerides" of 1620. The latter was actually *dedicated* to Napier, who had died in 1617. The cover illustration showed one of Galileo's telescopes, the figure of an elliptical orbit, and an allegorical female crowned with a wreath consisting of the Naperian logarithm of half the radius of a circle. It has usually been supposed that this work was dedicated to Napier in gratitude for the "shortening of the calculations", but Kepler obviously recognized that it went deeper than this, i.e., that the Third Law is purely a logarithmic harmony. In a sense, logarithms played a role in Kepler's formulation of the Third Law analogous to the role of Apollonius' conics in his discovery of the First Law, and with the role that tensor analysis and Riemannian geometry played in Einstein's development of the field equations of general relativity. In each of these cases we could ask whether the mathematical structure provided the tool with which the scientist was able to describe some particular phenomenon, or whether the mathematical structure effectively selected an aspect of the phenomena for the scientist to discern.

Just as we can trace Kepler's Third Law of planetary motion back to Napier's invention of logarithms, we can also trace Napier's invention back to even earlier insights. It's no accident that logarithms have applications in the description of Nature. Indeed in his introduction to the tables, Napier wrote:

> A logarithmic table is a small table by the use of which we can obtain a knowledge of all geometrical dimensions and motions in space...

The reference to motions in space is very appropriate, because Napier originally conceived of his "artificial numbers" (later renamed logarithms, meaning number of the ratio) in purely kinematical terms. In fact, his idea can be expressed in a form that Zeno of Elea would have immediately recognized. Suppose two runners leave the starting gate, travelling at the same speed, and one of them maintains that speed, whereas the speed of the other drops in proportion to his distance from the finish line. The closer the second runner gets to the finish line, the slower he runs. Thus, although he is always moving forward, the second runner never reaches the finish line. This is exactly the kind of scenario that Zeno exploited to illustrate paradoxes of motion.

Here, 2000 years later, we find Napier making very different use of it, creating a continuous mapping from the real numbers to his "artificial numbers". With an appropriate choice of units we can express the position x of the first runner as a function of time by $x(t) = t$, and the position X of the second runner is defined by the differential equation $dX/dt = 1-X$ where the position "1" represents the finish line. The solution of this equation is $X(t) = 1-e^{-t}$, where e^x is the function that equals its own derivative. Then Napier defined $x(t)$ as the "logarithm" of $1-X(t)$, which is to say, he defined t as the "logarithm" of □e^{-t}. Of course, the definition of logarithm was subsequently revised so that we now define t as the logarithm of e^t, the latter being the function that equals its own derivative.

The logarithm was one of many examples throughout history of ideas that were "in the air" at a certain time. It had been known since antiquity that the exponents of numbers in a geometric sequence are additive when terms are multiplied together, i.e., we have $a^n a^m = a^{(m+n)}$. In fact, there are ancient Babylonian tablets containing sequences of powers and problems involving the determination of the exponents of given numbers. In the 1540's Stifel's "Arithmetica integra" included tables of the successive powers of numbers, which was very suggestive for Napier and others searching for ways to reduce the labor involved in precise manual computations.

In the 1580's Viete derived several trigonometric formulas such as

$$\cos(x)\cos(y) = \frac{\cos(x+y)+\cos(x-y)}{2}$$

If we have a table of cosine values this formula enables us to perform multiplication simply by means of addition. For example, to find the product of 0.7831 and 0.9348 we can set cos(x) = 0.7831 and cos(y) = 0.9348 and then look up the angles x,y with these cosines in the table. We find x = 0.67116 and y = 0.36310, from which we have the sum x+y = 1.03426 and the difference x–y = 0.30806. The cosines of the sum and difference can then be looked up in the table, giving cos(x+y) = 0.51116 and cos(x-y) = 0.95292. Half the sum of these two numbers equals the product 0.73204 of the original two numbers. This technique was called prosthaphaeresis (the Greek word for addition and subtraction), and was quickly adopted by scientists such as the Dane Tycho Brahe for performing astronomical calculations. Of course, today we recognize that the above formula is just a disguised version of the simple exponent addition rule, noting that $\cos(x) = (e^{ix} + e^{-ix})/2$.

At about this same time (1594), John Napier was inventing his logarithms, whose purpose was also to reduce multiplication and division to simple addition and subtraction by means of a suitable transformation. However, Napier might never have set aside his anti-Catholic polemics to work on producing his table of logarithms had it not been for an off-hand comment made by Dr. John Craig, who was the physician to James VI of Scotland (later James I of England and Ireland). In 1590 Craig accompanied James and his entourage bound for Norway to meet his prospective bride Anne, who was supposed to have journeyed from Denmark to Scotland the previous year, but had been diverted by a terrible storm and ended up in Norway. (The storm was so severe that several supposed witches were held responsible and were burned.) James' party, too, encountered severe weather, but eventually he met Anne in Oslo and the two were married.

On the journey home the royal party visited Tycho Brahe's observatory on the island of Hven, and were entertained by the famous astronomer, well known as the discoverer of the "new star" in the constellation Cassiopeia. During this stay at Brahe's lavish Uraneinborg ("castle in the sky") Dr. Craig observed the technique of prosthaphaeresis that Brahe and his assistants used to ease the burden of calculation. When he returned to Scotland, Craig mentioned this to his friend the Baron of Murchiston (aka John Napier), and this seems to have motivated Napier to devote himself to the development of his logarithms and the generation of his tables, on which he spent the remaining 25 years of his life. During this time Napier occasionally sent preliminary results of Brahe for comment.

Several other people had similar ideas about exploiting the exponential mapping for purposes of computation. Indeed, Kepler's friend and assistant Jost Burgi evidently devised a set of "progress tables" (basically anti-logarithm tables) around 1600, based on the indices of geometric progressions, and made some use of these in his calculations. However, he didn't fully perceive the potential of this correspondence, and didn't develop it very far.

Incidentally, if the story of a group of storm-tossed nobles finding themselves on a mysterious island ruled over by a magician sounds familiar, it may be because of Shakespeare's "The Tempest", written in 1610. This was Shakespeare's last complete play and, along with Love's Labor's Lost, his only original plot, i.e., these are the only two of his plays whose plots are not known to have been based on pre-existing works.

It is commonly believed that the plot of "The Tempest" was inspired by reports of a group of colonists bound for Virginia who were shipwrecked in Bermuda in 1609. However, it's also possible that Shakespeare had in mind the story of James VI and his marriage expedition, arriving after a series of violent storms on the island of the Danish astronomer and astrologer Tycho Brahe and his castle in the sky (which, we may recall, included a menagerie of exotic animals).

We know "The Tempest" was produced at the royal court in 1611 and again in 1612 as part of the festivities preceding the marriage of the King's daughter, and it certainly seems likely that James and Anne would associate any story involving a tempest with their memories of the great storms of 1589 and 1590 that delayed Anne's voyage to Scotland and prompted James' journey to meet her. The providential aspects of Shakespeare's "The Tempest" and its parallels with their own experiences could hardly have been lost on them.

Shakespeare's choice of the peculiar names Rosencrantz and Guildenstern for two minor characters in "Hamlet, Prince of Denmark" gives further support to the idea that he was familiar with Tycho, since those were the names of two of Tycho's ancestors appearing on his coat of arms.

There is also evidence that Shakespeare was personally close to the Digges family (e.g., Leonard Digges contributed a sonnet to the first Folio), and Thomas Digges was an English astronomer and mathematician who, along with John Dee, was well acquainted with Tycho. Digges was an early supporter and interpreter of Copernicus' relativistic ideas, and was apparently the first to suggest that our Sun was just an ordinary star in an infinite universe of stars.

Considering all this, it is surely not too farfetched to suggest that Tycho may have been the model for Prospero, whose name, being composed of Providence and sparrow, is an example of Shakespeare's remarkable ability to weave a variety of ideas, influences, and connotations into the fabric of his plays, just as we can see in Kepler's three laws the synthesis of the heliocentric model of Copernicus, Apollonius' conics, and the logarithms of Napier.

ANOMALOUS PRECESSIONS

The Earth's equatorial plane maintains a nearly constant absolute orientation in space throughout the year due to the gyroscopic effect of spinning about its axis. Similarly the plane of the Earth's orbit around the Sun remains essentially constant. These two planes are tilted by 23.5 degrees with respect to each other, so they intersect along a single line whose direction remains constant, assuming the planes themselves maintain fixed attitudes. At the Spring and Autumn equinoxes the Sun is located precisely on this fixed line in opposite directions from the Earth. Since this line is a highly stable directional reference, it has been used by astronomers since ancient times to specify the locations of celestial objects. (Of course, when we refer to "the location of the Sun" we are speaking somewhat loosely. With the increased precision of observations made possible by the invention of the telescope, it is strictly necessary to account for the Sun's motion about the center of mass of the solar system.

It is this center of mass of the Sun and planets, rather than just of the Sun, that is taken as the central inertial reference point for the most precise astronomical measurements and calculations.) By convention, the longitude of celestial objects is referenced from the direction of this line pointing to the Spring equinox, and this is called the "right ascension" of the object. In addition, the "declination" specifies the latitude, i.e., the angular position North or South of the Earth's equatorial plane.

This system of specifying positions is quite stable, but not perfect. Around 150 BC the Greek astronomer Hipparchus carefully compared his own observations of certain stars with observations of the same stars recorded by Timocharis 169 years earlier (and with some even earlier measurements from the Babylonians), and noted a slight but systematic difference in the longitudes.

Of course, these were all referenced to the supposedly fixed direction of the line of intersection between the Earth's rotational and orbital planes, but Hipparchus was led to the conclusion that this direction is not perfectly stationary, i.e., that the direction of the Sun at the equinoxes is not constant with respect to the fixed stars, but precesses by about 0.0127 degrees each year. This is a remarkably good estimate, considering the limited quality of the observations that were available to Hipparchus.

The accepted modern value for the precession of the equinoxes is 0.01396 degrees per year, which implies that the line of the equinoxes actually rotates completely around 360 degrees over a period of about 26,000 years. Interpreting this as a gradual change in the orientation of the Earth's axis of rotation, the precession of the equinoxes is the third of what Copernicus called the "threefold movement of the Earth", the first two being a rotation about its axis once per day, and a revolution about the Sun once per year. Awareness of this third motion is arguably a distinguishing feature of human culture, since it can only be discerned on the basis of information spanning multiple generations.

The reason for mentioning this, aside from expressing admiration for human ingenuity, is that when we observe the axis of the elliptical orbit of a planet such as Mercury (for example) over a long period of time, referenced to our equinox line, we must expect to find an apparent precession of about 0.01396 degrees per year, which equals 5025 arc seconds per century, assuming Mercury's orbital axis is actually stationary. However, astronomers have actually observed a precession rate of 5600 arc seconds per century for the axis of Mercury's orbit, so evidently the axis is not truly stationary. This might seem like a problem for Newtonian gravity, until we remember that Newton predicted stable elliptical orbits only for the idealized two-body case. When analyzing the actual orbit of Mercury we must also take into account the gravitational pull of the other planets, especially Venus and Earth (because of their proximity) and Jupiter (because of its size).

It isn't simple to work out these effects, and unfortunately there is no simple analytical solution to the n-body problem in Newtonian mechanics, but using the calculational techniques developed by Lagrange, Laplace, and others, it is possible to determine that the effects of all the other planets should contribute an additional 532 arc seconds per century to the precession of Mercury's orbit. Combined with the precession of our equinox reference line, this accounts for 5557 arc seconds per century, which is close to the observed value of 5600, but still short by 43 arc seconds per century. The astronomers assure us that their observations can't be off by more than a fraction of an arc second, so there seems to be a definite problem here.

A similar problem had appeared in the 1840's when the newly discovered planet Uranus began to deviate noticeably from the precise course that Newtonian theory prescribed. On that occasion, the astronomer Le Verrier and the mathematician Adams had (independently) inferred the existence of a previously unknown planet beyond the orbit of Uranus, and even gave instructions where it could be found. Sure enough, when that indicated region of the sky was searched by Johann Galle at the Berlin Observatory, the planet that came to be called Neptune was discovered in 1846, astonishingly close to the predicted location. This was a tremendous triumph for Le Verrier, and surely gave him confidence that all apparent anomalies in the planetary orbits could be explained on the basis of Newtonian theory, and could be used as an aid to the discovery of new celestial objects. He soon turned his attention to the anomalous precession of Mercury's orbit (which he estimated at 38 arc seconds per century, somewhat less than the modern value), and suggested that it must be due to some previously unknown mass near the Sun, possibly a large number of small objects, or perhaps even another planet, inside the orbit of Mercury.

At one point there were reports that a small planet orbiting very near the Sun had actually been sighted, and it was named Vulcan, after the Roman God of fire. Le Verrier became convinced that the new planet existed, but subsequent attempts to observe the hypothetical planet failed to find any sign of it. Even the original sightings were cast into doubt, since they had been made by an amateur, and other astronomers reported that they had been observing the Sun at the very same time and had seen nothing. Another popular theory to explain Mercury's anomalous precession, championed by the astronomer Simon Newcomb, was that the small particles of matter that cause the "zodiacal light" might account for Mercury's anomalous precession, but Newcomb soon realized that if there were enough matter to affect Mercury's perihelion so significantly there would also be enough to cause other effects on the orbits of the inner planets-effects which are not observed. Similar inconsistencies undermined the "Vulcan" hypothesis.

As a result of the failures to arrive at a realistic Newtonian explanation for the anomalous precession, some researchers, notably Asaph Hall and Newcomb, began to think that perhaps Newtonian theory was at fault, and that perhaps gravity isn't exactly an inverse square law. Hall noted that he could account for Mercury's precession if the law of gravity, instead of falling off as $1/r^2$, actually falls of as $1/r^n$ where the exponent n is 2.00000016. However, most people didn't (and still don't) find that idea to be very appealing, since it conflicts with basic conservation laws, e.g., Gauss's Law, unless we also postulate a correspondingly modified metric for space (ironically enough).

More recently, efforts have been made to explain some or all of Mercury's precession by oblateness in the shape of the sun. In 1966 Dicke and Goldenberg reported that the sun's polar axis is shorter than its equatorial axes by about 50 parts per million. If true that would account for 3.4" per century, so the unexplained part would be only 39.6", significantly different from GR's prediction of 43".

The Brans-Dicke theory of gravity can account for 39.6" precisely by adjusting a free parameter of the theory. However, Dicke's and Goldenberg's solar oblateness data was contradicted by a number of other heliometric measurements, all of which showed that the solar axes differ by no more than about 4 parts per million. In addition, the sun doesn't appear to rotate nearly fast enough to be as oblate as Dicke and Goldenberg thought, so their results could only be right if the interior of the sun is spinning about 25 times faster than the visible exterior, which is highly implausible. The current consensus is that the Sun is not nearly oblate enough to upset the agreement between Mercury's observed precession and the predictions of GR. This is all the more impressive considering that, in contrast to the Brans-Dicke and other alternative theories, GR has almost no "freedom" to adjust its predictions. It is highly constrained by its own logic, so it's remarkable that it continues to survive experimental challenges.

It should be noted that Mercury isn't the only object in the solar system that exhibits anomalous precession. The effect is most noticeable for objects near the Sun with highly elliptical orbits, but it can be seen even in the nearly circular orbits of Venus and Earth, although the discrepancy isn't nearly so large as for Mercury. In addition, the asteroid Icarus is ideal for studying this effect, because it has an extremely elliptical orbit and periodically passes very close to the Sun. Here's a table showing the anomalous precession of four objects in the inner solar system, based on direct observations:

	Arc Seconds Per Century	
	GR	***osbserved***
Mercury	43.0	43.1 ± 0.5
Venus	8.6	8.4 ± 4.8
Earth.	3.8	5.0 ± 1.2
Icarus (asteroid)	10.3	9.8 ± 0.8

The large tolerances for Venus and Earth are mainly due to the fact that their orbits are so nearly circular, making it difficult to precisely determine the axes of their elliptical orbits. Incidentally, Icarus periodically crosses the Earth's path, and has actually passed within a million kilometers of us-less than 3 times the distance to the Moon. It's about 1 mile in diameter, and may eventually collide with the Earth-reason enough to keep an eye on its precession.

One hope that Einstein had throughout the time he was working on the general theory was that it would explain the anomalous precession of Mercury. Of course, as we've seen, "explanations" of this phenomenon were never in short supply, but none of them were very compelling, all seeming to be ad hoc. In contrast, Einstein found that the extra precession arises unavoidably from the fundamental principles of general relativity. To determine the relativistic prediction for the advance of an elliptical orbit, let's work in the single plane $\theta = \pi/2$, so of course $d\theta/dt$ and all higher derivatives also vanish, and we have $\sin(\theta) = 1$. Thus the term involving θ in the Schwarzschild metric drops out, leaving just

$$(d\tau)^2 = \left(\frac{r-2m}{r}\right)(dt)^2 - \left(\frac{r}{r-2m}\right)(dt)^2 - r^2(d\phi)^2.$$

Taking the parameter λ equal to the proper time τ, those equations are

$$\frac{d^2t}{d\tau^2} = \frac{2m}{r(r-2m)}\left(\frac{dr}{d\tau}\right)\left(\frac{dr}{d\tau}\right)$$

$$\frac{d^2t}{d\tau^2} = -\frac{m(r-2m)}{r^3}\left(\frac{dr}{d\tau}\right)^2 + \frac{m}{r(r-2m)}\left(\frac{dr}{d\tau}\right)^2 + (r-2m)\left(\frac{d\phi}{d\tau}\right)^2$$

$$\frac{d^2\phi}{d\tau^2} = -\frac{2}{r}\left(\frac{dr}{d\tau}\right)\left(\frac{d\phi}{d\tau}\right).$$

We can immediately integrate equations to give

$$\frac{dt}{d\tau} = \frac{kr}{r-2m} \qquad \frac{d\phi}{d\tau} = \frac{h}{r^2}$$

where k and h are constants of integration, determined by the initial conditions of the orbit. We can now substitute for these derivatives into the basic Schwarzschild metric divided by $(d\tau)^2$ to give

$$1=\left(\frac{k^2r}{r-2m}\right)-\left(\frac{r}{r-2m}\right)\left(\frac{dr}{d\tau}\right)^2-\frac{h}{r^2}.$$

Solving for $(dr/d\tau)^2$, we have

$$\left(\frac{dr}{d\tau}\right)^2=\left(2h^2m\right)\frac{1}{r^3}-\left(h^2\right)\frac{1}{r^2}+(2m)\frac{1}{r}-(1-k^2).$$

Differentiating this with respect to τ and dividing by $2(dr/d\tau)$ gives

$$\frac{d^2r}{d\tau^2}=\frac{m}{r^2}+\frac{h^2}{r^3}-\frac{3h^2m}{r^4}.$$

(We arrive at this same equation if we insert the squared derivatives of the coordinates into equation, because one of the geodesic equations is always redundant to the line element.) Letting $\omega = d\phi/d\tau$ denote the proper angular speed, we have $h = \omega r^2$, and the above equation can be written as

$$\frac{d^2r}{d\tau^2}=\frac{m}{r^2}+\omega^2(r-3m).$$

Obviously if $\omega = 0$ this gives the "proper" analog of Newton's inverse-square law for radial gravitational acceleration. With non-zero ω the term $\omega^2 r$ corresponds to the Newtonian centripetal acceleration which, if we defined the tangential velocity $v = \omega r$, would equal the classical v^2/r. This term serves to offset the inward pull of gravity, but in the relativistic version we find not $\omega^2 r$ but $\omega^2(r-3m)$. (To avoid confusion, it's worth nothing that the quantity $\omega^2(1-3m/r)$ would be simply ω^2 if ω were defined as the derivative of ϕ with respect to the Schwarzschild coordinate time t instead of the proper time τ. The relativistic version of Kepler's third law for circular orbits is *formally* identical to the Newtonian version – but only if we identify the Newtonian coordinates with the Schwarzschild coordinates.) For values of r much greater than 3m this difference can be neglected, but clearly if r approaches 3m we can expect to see non-classical effects, and of course if r ever becomes less than 3m we would expect completely un-classical behavior. In fact, this corresponds to the cases when an orbiting particle spirals into the center, which never happens in classical theory.

Since the above equations involve powers of $(1/r)$ it's convenient to work with the parameter $u = 1/r$. Differentiating u with respect to ϕ gives $du/d\phi = -(1/r^2)\, dr/d\phi$. Also, since $r^2 = h/(d\phi/d\tau)$, we have $dr/d\tau = -h\,(du/d\phi)$. Substituting for $dr/d\tau$ and $1/r$ into equation gives the following differential equation relating u to ϕ

$$h^2\left(\frac{du}{d\phi}\right)^2 = (2h^2m)u^3 - (h^2)u^2 + (2m)u - (1-k^2).$$

Differentiating again with respect to ϕ and dividing by $2h^2$ $(du/d\phi)$, we arrive at

$$3mu^2 - u + \left(\frac{m}{h^2} - \ddot{u}\right) = 0$$

where denotes $d^2u/d\phi^2$. Solving this quadratic for u gives

$$u = \frac{1}{6m}\left(1 - \sqrt{1 - 12\left(\frac{m^2}{h^2} - m\ddot{u}\right)}\right).$$

The quantity in the parentheses under the square root is typically quite small compared with 1, so we can approximate the square root by the first few terms of its expansion

$$u \approx \frac{1}{6m}\left[\frac{1}{2}\left(12m\left(\frac{m}{h^2} - \ddot{u}\right)\right) + \frac{1}{8}\left(12m\left(\frac{m}{h^2} - \ddot{u}\right)\right)^2\right].$$

Expanding the right hand side and re-arranging terms gives

$$\left(1 + 6\left(\frac{m}{h}\right)^2\right)\ddot{u} + u = \left(\frac{m}{h^2} + 3\frac{m^3}{h^4}\right) + 3m\ddot{u}^2.$$

The value of in typical astronomical problems is numerically quite small (many orders of magnitude less than 1), so the quantity on the right hand side will be negligible for planetary motions.

Therefore, we're left with a simple harmonic oscillator of the form where M and F are constants. For some choice of initial ϕ the general solution of this equation can be expressed as $u(\phi) = F\left[1 + k\cos\left(\phi\sqrt{M}\right)\right]$ where k is a constant of integration. Therefore, reverting back to the parameter $r = 1/u$, the relation between r and ϕ is

$$r(\phi) = \left(\frac{h^2/m}{1 + 3(m/h^2)}\right)\frac{1}{1 + k\cos(\Omega\phi)} \quad \text{where } \Omega = \frac{1}{\sqrt{1 + 6\left(\frac{m}{h}\right)}}.$$

If the "frequency" Ω was equal to unity, this would be the polar equation of an ellipse with the pole at one focus, and the constant k would signify the eccentricity. Also, the leading factor would be the radial distance from the focus to the ellipse at an angle of $\pi/2$ from the major axis, i.e., it would represent the semilatus rectum.

However, the value of Ω is actually slightly less than 1, which implies that ϕ must go slightly beyond 2π in order to complete one cycle of the radial distance. Consequently, for small values of m/h the path is approximately a Keplerian ellipse, but the axis of the ellipse precesses slightly, as illustrated below.

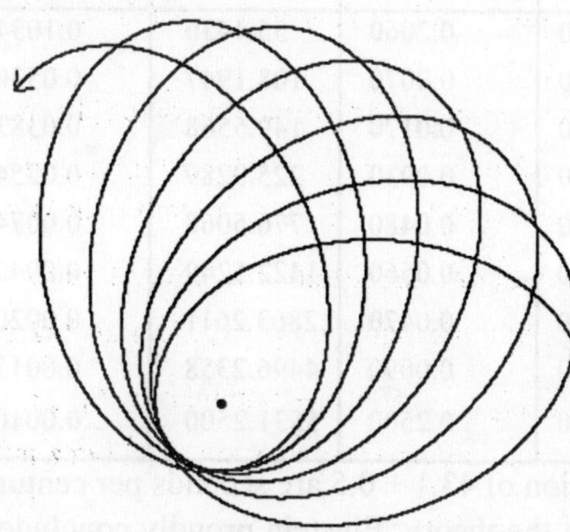

This illustration depicts a much more severe case than could exist for any planet in our solar system, because the perihelion of the orbit is only 200m where m is the gravitational radius (in geometrical units) of the central object, which means it is only 100 times the corresponding "black hole radius". Our Sun's mass is not nearly concentrated enough to permit this kind of orbit, since the Sun's gravitational radius is only m = 1.475 kilometers, whereas it's matter fills a sphere of radius 696,000 kilometers.

To determine the relativistic prediction for the orbital precession of the planetary orbits, we can expand the expression for Ω as follows

$$\Omega = \frac{1}{\sqrt{1+6\left(\frac{m}{h}\right)}} \approx 1-3\left(\frac{m}{h}\right)^2 + \frac{27}{2}\left(\frac{m}{h}\right)^4.$$

Since m/h is so small, we can take just the first-order term, and noting that one cycle of the radial function will be completed when $\Omega\phi = 2\pi$, we see that ϕ must increase by $2\pi/\Omega$ for each radial cycle, so the precession per revolution is

$$\frac{2\pi}{\Omega} - 2\pi = 2\pi\left(\frac{1}{1-3(m/h)^2} - 1\right) \approx 6\pi\left(\frac{m}{h}\right)^2.$$

We saw above that the semilatus rectum L is approximately h^2/m, so the amount of precession per revolution (for slow moving objects in weak gravitational fields, such as the planets in our solar system) can be written as simply $6\pi m/L$, where m is the gravitational radius of the central body. As noted above, the gravitational radius of our Sun is 1.475 kilometers, so based on the elements of the planetary orbits we can construct the following table of relativistic precession.

Planet	Semimajor Axis (10^6 km)	Orbital Eccentricity k	Semilatus Rectum, L (10^6 km)	Precession per rev (arc sec)	Revolutions per century	Precession per century (arc sec)
Mercury	57.9000	0.2060	55.4430	0.1034	414.9378	42.9195
Venus	108.2000	0.0070	108.1947	0.0530	162.6016	8.6186
Earth	149.6000	0.0170	149.5568	0.0383	100.0000	3.8645
Mars	227.9000	0.0930	225.9289	0.0254	53.1915	1.3502
Jupiter	778.3000	0.0480	776.5068	0.0074	8.4317	0.0623
Satum	1427.0000	0.0560	1422.5249	0.0040	3.3944	0.0137
Uranus	2869.6000	0.0470	2863.2611	0.0020	1.1903	0.0024
Neptune	4496.6000	0.0090	4496.2358	0.0013	0.6068	0.0008
Pluto	5900.0000	0.2500	5531.2500	0.0010	0.4032	0.0004

The observed precession of 43.1 ± 0.5 arc seconds per century for the planet Mercury is in close agreement with the theory. Einstein proudly concluded his presentation of the vacuum field equations in his 1916 paper on general relativity by pointing out that they explained the anomalous precession of Mercury. He returned to this subject at the end of the paper, giving the precession formula and closing his masterpiece with the words

Calculation gives for the planet Mercury a rotation of the orbit of 43" per century, corresponding exactly to the astronomical observation (Leverrier); for the astronomers have discovered in the motion of the perihelion of this planet, after allowing for disturbances by the other planets, an inexplicable remainder of this magnitude.

We mentioned previously that the small eccentricities of Venus and Earth make it difficult to determine their lines of apsides with precision, but modern measurement techniques (including the use of interplanetary space probes and radar ranging) and computerized analysis of the data have enabled the fitting of the entire solar system to a parameterized post-Newtonian (PPN) model that encompasses a fairly wide range of theories (including general relativity).

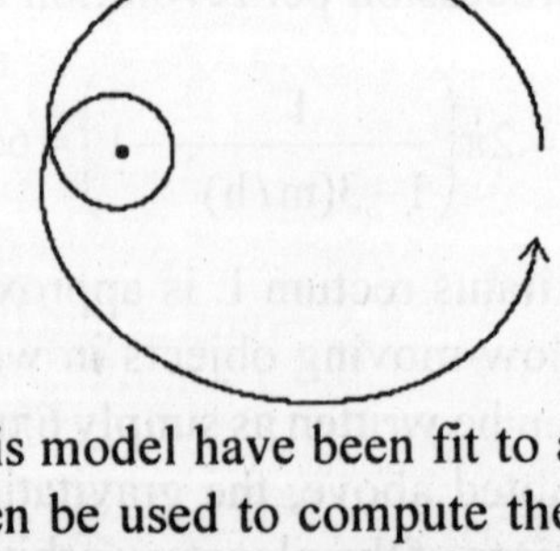

Once the parameters of this model have been fit to all the available data for the Sun and planets, the model can then be used to compute the "best observational fit" for the precessions of the individual planets based on the PPN formalism. This gives precessions (in excess of the Newtonian predictions) of 43.1, 8.65, 3.85, and 1.36 arcseconds per century for the four inner planets respectively, in remarkable agreement with the predictions of general relativity.

If we imagine an extremely dense central object, whose mass is concentrated inside it's gravitational radius, we can achieve much greater deviations from conventional Newtonian orbits. For example, if the precession rate is roughly equal to the orbital rate, we have an orbit as shown below:

For an orbit with slightly less energy the path looks like this:

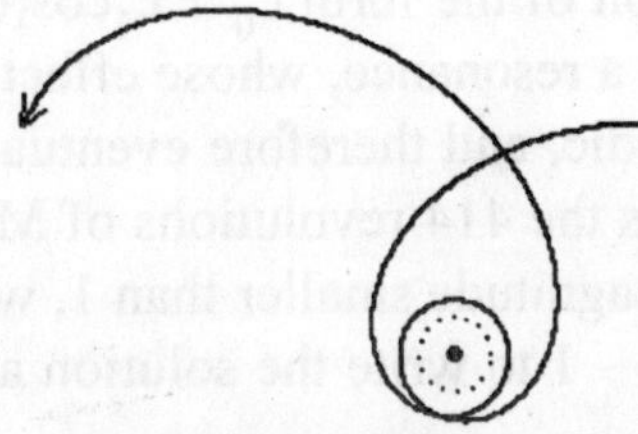

where the dotted circle signifies the "light orbit" radius r = 3m. With sufficient angular momentum it's possible to arrange for persistent timelike orbits periodically descending down to any radius greater than 3m, which is the smallest possible radius of a circular orbit (but note that a circular orbit with radius less than 6m is unstable). If a timelike geodesic ever passes inside that radius it must then spiral in to the central mass, as illustrated below.

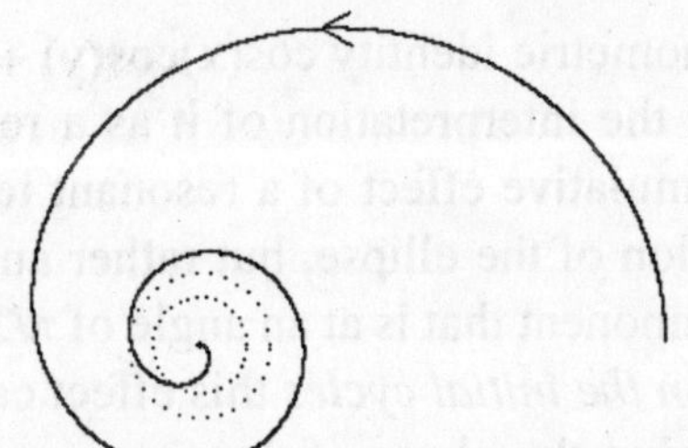

Here the outer dotted circle is at 3m, and the inner circle is at the event horizon, 2m. Once a worldline has fallen within 2m, whether geodesic or not, it's radial coordinate must (according to the Schwarzschild solution) thereafter decrease monotonically to zero.

Regarding these spiral solutions there is an ironic historical precedent. A few years before writing the Principia Newton once described in a letter to Robert Hooke the descent of an object along a spiral path to the center of a gravitating body. Several years later, after the Principia had established Newton's reputation, the two men became engaged in a bitter priority dispute over the discovery of universal gravitation, and Hooke used this letter as evidence that Newton hadn't understood gravity at that time, because the classical inverse-square law of gravity permits no such spiral solutions. Newton replied that it had simply been a "negligent stroke with his pen". Interestingly, although people sometimes credit Newton with originating the idea of photons based on his erroneous corpuscular theory of light, it's never been suggested that his "negligent spiral" was a premonition of the Schwarzschild solution of Einstein's field equations.

Incidentally, the relativistic contribution to a planet's orbital precession rate is often derived as a "resonance" effect. Recall that the general solution of an ordinary *linear* differential equation contains a term proportional to $e^{\lambda x}$ for each root λ of the characteristic polynomial, and a resonance occurs when the characteristic polynomial has a repeated root, in which case the solution has a term proportional to $xe^{\lambda x}$.

If there is another repetition of the root it is represented by a term proportional to $x^2e^{\lambda x}$, and so on. As a means of approximating the solution of the *non*-linear equation, many authors introduce a trial solution of the form $c_0 + c_1\cos(\phi) + c_2\phi\sin(\phi)$, suggesting that the last term is to be regarded as a resonance, whose effect grows cumulatively over time because the factor ϕ is not periodic, and therefore eventually has observable effects over a large number of orbits, such as the 414 revolutions of Mercury per century. Now, provided $(c_2/c_1)\phi$ is many orders of magnitude smaller than 1, we can use the small-angle approximations $\sin(x) \sim x$ and $\cos(x) \sim 1$ to write the solution as

$$c_0 + c_1\left(\cos(\phi) + \left(\frac{c_2}{c_1}\phi\right)\sin(\phi)\right) \approx c_0 + c_1\left(\cos\left(\frac{c_2}{c_1}\phi\right)\cos(\phi) + \sin\left(\frac{c_2}{c_1}\phi\right)\sin(\phi)\right)$$

$$= c_0 + c_1\left(\phi - \frac{c_2}{c_1}\phi\right).$$

where we've used the trigonometric identity $\cos(x)\cos(y) + \sin(x)\sin(y) = \cos(x-y)$. This yields the correct result, but the interpretation of it as a resonance effect is misleading, because the predominant cumulative effect of a resonant term proportional to $\phi\sin(\phi)$ in the long run is not a precession of the ellipse, but rather an increase in the magnitude of the radial excursions of a component that is at an angle of $\pi/2$ relative to the original major axis. It just so happens that *on the initial cycles* this effect causes the overall perihelion to precess slightly, simply because the phase of the sine component is beginning to assert itself over the phase of the cosine component.

In other words, the apparent "precession" resulting from the $\phi\sin(\phi)$ term on the initial cycles is really just a one-time phase shift corresponding to a secular increase in the radial amplitude, and does not actually represent a change in the frequency of the solution. It can be shown that a term involving $\phi\sin(\phi)$ appears in the second-order power series expansion of the solution to equation, which explains why it is a useful curve-fitting function for small ϕ, but it does not represent a true resonance effect, as shown by the fact that the ultimate cumulative effect of this term is discarded when we apply the small-angle approximation to estimate the frequency shift.

ABERRATION OF FORCES AND WAVES

Consider an illuminated charged sphere resting at the origin of a system x,y,z of inertial coordinates, and a small test particle moving with speed v in the positive x direction at a fixed y coordinate as shown below.

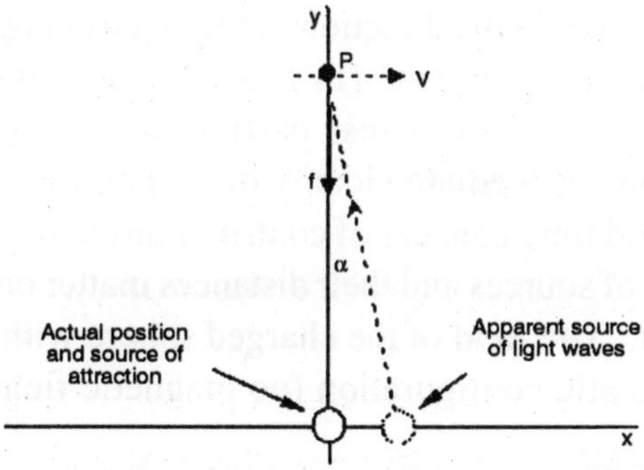

If the distance between the two objects is sufficiently great, the light (electromagnetic waves) emanating from the sphere will consist of essentially planar horizontal waves when it reaches the test particle. Since the particle is moving tangentially with speed v, the angle of the incoming light will be affected by aberration, such that the apparent source of the light (from the point of view of the test particle as it crosses the y axis) is at an angle α = arcsin(v/c) ahead of the actual position of the sphere.

However, the direction of the electrical force exerted by the sphere on the test particle points directly toward the actual position of the sphere. Thus, the incoming electromagnetic waves from the sphere experience aberration, but the electromagnetic force of attraction to the sphere does not. This sometimes misleads people into thinking that the force somehow propagates instantaneously (to account for the absence of aberration).

Of course we can just as well consider the test particle to be at rest, and the charged sphere to be moving in the negative x direction with speed v. From this point of view, if D denotes the distance from the sphere to the particle, then at any time t the particle "sees" the sphere at the location it occupied at a time t-D/c, because D/c is how long it take for light to travel the distance D. This is illustrated in the figure below.

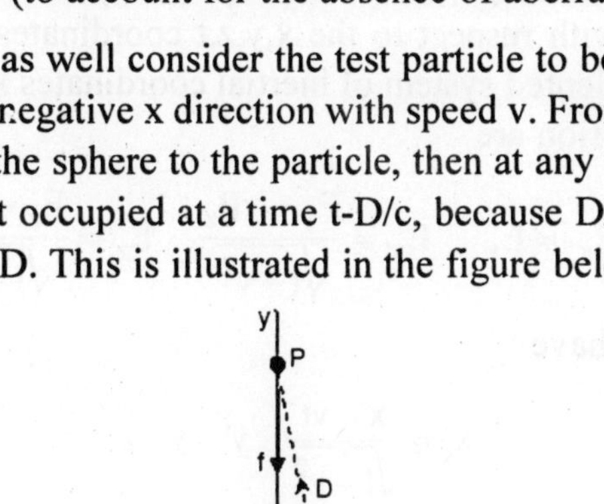

Just as before, the light arrives at the (stationary) particle P from a direction differing from the true current direction of the source at time t by the angle α. Also, since we have simply changed coordinate systems, which can have no effect on any physical attributes, we know the electromagnetic force on the particle at the time t points directly toward the sphere's actual (not apparent) position at that time.

The absence of aberration in the direction of the electromagnetic force does *not* indicate that the force propagates infinitely fast. (In fact, the concept of a "moving force" is not even well defined.) The force on a test particle at any given instant is due to the electromagnetic field in the immediate vicinity of the particle at that instant. In general the field at any given place and time consists of contributions from multiple sources at a variety of distances. The number of sources and their distances matter only insofar as they determine the electromagnetic field. The field of the charged sphere with respect to the rest frame of the sphere is an electro-static configuration (no magnetic field) with spherical symmetry centered on the source.

A uniformly moving charged test particle in this field is subjected to a force proportional to (and therefore pointing in the same direction as) the electric field vector at its present location, so the force obviously points directly towards the source at all times. On the other hand, in terms of the rest frame of the test particle the charged sphere is in uniform motion and the electromagnetic field has both electric and magnetic components. However, since the test particle is at rest with respect to these coordinates, it does not experience any magnetic force, so again the force on the particle is proportional to the electric field vector.

To determine the direction of this force we need to know how the components of the electric field transform from one system of inertial coordinates to another. As explained in Force Laws and Maxwell's Equations, if E_x, E_y, and E_z are the components of the electric field at a given point with respect to the x,y,z,t coordinates, then the components with respect to a similarly oriented system of inertial coordinates x',y',z',t' moving with speed v in the positive x direction are

$$E_{x'} = E_x \quad E_{y'} = \frac{E_y - vB_z}{\sqrt{1-v^2}} \cdot E_{z'} = \frac{E_z - vB_y}{\sqrt{1-v^2}}.$$

Of course we also have

$$x' = \frac{x - vt}{\sqrt{1-v^2}} \quad y' = y \quad z' = z.$$

In the unprimed coordinates (the rest frame of the charged sphere) we know the electric field components at the location of the test particle point directly toward the origin, which means E_x, E_y, and E_z are proportional to the coordinates x, y, and z of the test particle. Also, since the magnetic field is zero with respect to the unprimed coordinates, and since the origins of the two coordinate systems coincide at t = 0, we have

$$\frac{E_{x'}}{E_{y'}} = \frac{E_x}{E_y}\sqrt{1-v^2} = \frac{x}{y}\sqrt{1-v^2} = \frac{x'}{y'}.$$

Similarly it follows that $E_{x'}/E_{z'} = x'/z'$ and $E_{y'}/E_{z'} = y'/z'$, confirming that the electric field vector at every location points directly toward the instantaneous source with respect to the rest frame of the test particle (relative to which the source is moving with a speed v). Thus the absence of "force aberration" for objects in *fully developed* inertial motion is an immediate consequence of Lorentz covariance.

The qualifier "fully developed" is necessary, because every object is instantaneously at rest with respect to some inertial frame, but the object's field in its current rest frame is spherical and satisfies the steady-state relations only out to a distance $D = c\Delta t$ where Δt is the length of time the object has been unaccelerated. This highlights the fact that although the field of an object exists and acts at a distance from the object, changes in the field propagate at the finite speed c. When an object changes its state of motion, the field must change accordingly, and these changes propagate outward from the source at the speed c. In the far field these changes propagate in the form of waves.

One thing that sometimes puzzles people about the lack of force aberration is that they tend to regard the electric field as the gradient of a potential, and they know the equi-potential surfaces for a uniformly moving charged particle are contracted in the direction of motion so they form ellipsoids instead of spheres, and clearly the spatial gradient of this potential does not point towards the center (except for lines parallel or perpendicular to the axis of motion).

The explanation is that the electric field vector equals the spatial gradient of the potential field only if the field is stationary, i.e., unchanging with time. If the field is changing with time, the full expression for the electric field must include an additional term to account for this, i.e., we have

$$E = -\Delta\phi - \frac{1}{c}\frac{\partial A}{\partial t}$$

where A signifies the vector potential of the electromagnetic field. The second term on the right hand side represents the effect of the changing potential with time. Using the Lorentz gage

$$\Delta . A = -\frac{1}{c}\frac{\partial \phi}{\partial t}$$

the field equations for the electromagnetic potentials are

$$\nabla^2\phi - \frac{1}{c^2}\frac{\partial \phi}{\partial t^2} = -4\pi\rho \qquad \nabla^2 A - \frac{1}{c^2}\frac{\partial^2 A}{\partial t^2} = -4\pi\rho\frac{v}{c}.$$

It follows that, if v is constant (and has been for a sufficiently long time), and we are given a solution $\phi(x,y,z,t)$ for the scalar potential, we can multiply this solution by v/c to give a solution of the vector potential

$$A = \frac{\phi}{c}v.$$

Therefore, under these conditions, the time-dependent term in the last equation for E can be written as

$$\frac{1}{c}\frac{\partial A}{\partial t}=\frac{v}{c^2}\frac{\partial \phi}{\partial t}.$$

Now, by definition, the total derivative of ϕ along any incremental path dx,dy,dz,dt is

$$d\phi=\frac{\partial \phi}{\partial x}dx+\frac{\partial \phi}{\partial y}dy+\frac{\partial \phi}{\partial z}dz+\frac{\partial \phi}{\partial t}dt.$$

Dividing by dt and solving for the partial of ϕ with respect to t gives

$$\frac{\partial \phi}{\partial t}=\frac{\partial \phi}{\partial t}-\left(\frac{\partial \phi}{\partial x}\frac{dx}{dt}+\frac{\partial \phi}{\partial y}\frac{dy}{dt}+\frac{\partial \phi}{\partial z}\frac{dz}{dt}\right).$$

Taking dx/dt etc. as the components of the sphere's velocity v, the total derivative $d\phi/dt$ represents the change in ϕ with time along a co-moving worldline, and since the (fully developed) field is stationary with respect to the rest frame of the sphere, we have $d\phi/dt = 0$. Therefore the partial of f with respect to t equals the negative of the dot product of the spatial gradient of ϕ with the velocity v, so the previous expression for the time-dependent electric field is

$$E=-\nabla\phi+\frac{v}{c^2}(v.\nabla\phi).$$

We are considering the case when the sphere's motion is in the positive x direction, so we have v = (v,0,0) and the above expression becomes

$$E=-\left[\left(1-\frac{v^2}{c^2}\right)\frac{\partial \phi}{\partial x},\frac{\partial \phi}{\partial y},\frac{\partial \phi}{\partial z}\right].$$

A surface of constant ϕ is a stationary sphere in the rest frame of the source, so it transforms to an ellipsoid due to contraction in the direction of motion as shown below.

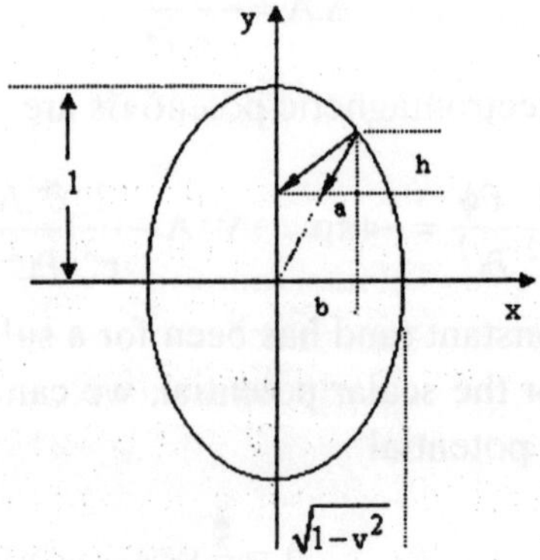

The equation of this cross-section is

$$\left(\frac{x}{\sqrt{1-v^2}}\right)^2 + y^2 = 1$$

where we have chosen units so that c = 1. Taking the differential of both sides gives

$$\frac{2x}{1-v^2}dx + 2ydy = 0.$$

The slope of the normal to the ellipse at the point (x,y) is the negative reciprocal of dy/dx,, which is

$$-\frac{dx}{dy} = \frac{h}{b} = \left(1-v^2\right)\frac{y}{x}.$$

According to our expression for E(r,t) we begin with the gradient of ϕ and then reduce the x component by the factor $(1\text{-}v^2)$, where we still have c = 1. Thus we have

$$\frac{h}{a} = \left(1-v^2\right)\frac{y}{x\left(1-v^2\right)} = \frac{y}{x}.$$

This confirms that the electromagnetic force exerted by the field of the moving charged sphere on the test particle at time t is directed toward the position of the sphere at the same time t. This is a natural consequence of Lorentz covariance, and does not imply any instantaneous transfer of energy or information.

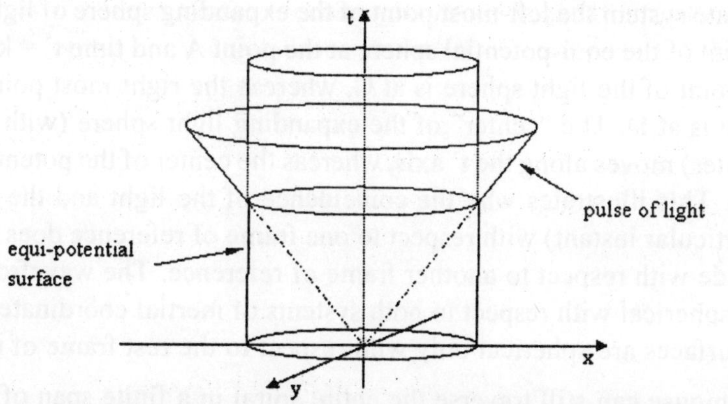

It's true that, in quantum theory, the electromagnetic force can be considered to be mediated by photons, but these are *virtual* photons, which are actually just analytical components of the field. In effect these virtual photons form a cloud around the source particle, and they "exist" only within the uncertainty envelope. An electromagnetic interaction between two electrons, for example, is modeled as an exchange of photons between the overlapping fields of two particles. It is not represented by a photon traversing from one particle to the other. Virtual photons don't even possess definite trajectories through space and time. They are conceptual entities arising in the quantization of the electromagnetic field.

Another point that sometimes puzzles people is why an equi-potential sphere transforms to an ellipsoid under a Lorentz transformation, whereas a spherical wave of light transforms to a spherical wave under the same transformation. The reason is that an equi-potential sphere is stationary, whereas a wave of light is expanding, as illustrated below.

A side view showing the intersections of these two surfaces with two difference planes of simultaneity is shown below.

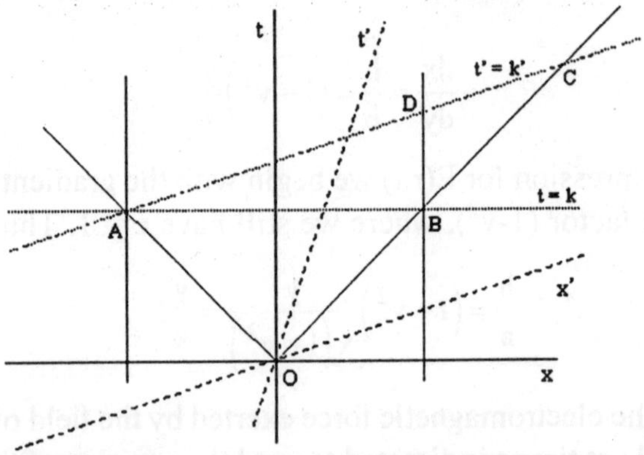

The source of the light pulse and the potential field moves along the t axis. With respect to the x,t coordinate system the expanding spherical shell of light coincides with an equi-potential sphere at the time t = k with the diameter AB. However, with respect to the x',t' coordinate system the left-most point of the expanding sphere of light just touches the left-most point of the equi-potential sphere at the point A and time t' = k'. At this time the right-most point of the light sphere is at C, whereas the right-most point of the equi-potential surface is at D. The "center" of the expanding light sphere (with respect to the primed coordinates) moves along the t' axis, whereas the center of the potential still moves along the t axis. This illustrates why the coincidence of the light and the equi-potential spheres (at a particular instant) with respect to one frame of reference does not imply that they ever coincide with respect to another frame of reference. The wavefront of the light pulse is always spherical with respect to both systems of inertial coordinates, whereas the equi-potential surfaces are spherical only with respect to the rest frame of the source.

value. The mouse can still traverse the entire spiral in a finite span of time, because the path distance to the center is proportional to the radial distance, so we have

$$\frac{dt}{dr} = -\frac{k}{\sqrt{r}}$$

and hence the time to reach the center from a radial distance r is. But is this task really achievable? It still entails the mouse circling the center infinitely many times.

Maxwell points out that many such motions can essentially be represented by the motion of a particle subjected to two forces, one directed toward some central point with magnitude proportional to the distance from that point, and another retarding force proportional to the speed of the particle.

He shows that the path of a particle subjected to such forces is a logarithmic spiral. Letting ω denote the constant angular speed, and α the characteristic angle of the spiral (i.e., the angle between a radius and a tangent), he determined that the acceleration of the particle with speed v at a distance r from the center can be decomposed into a centrally directed acceleration of magnitude μr and a retarding acceleration (opposite the direction of travel) of magnitude 2kv where

$$\mu = \frac{\omega^2}{\sin(\alpha)^2} \qquad k = \frac{\omega}{\sin(\alpha)}.$$

These relations can also be written in the form

$$\frac{\omega}{\sqrt{\mu}} = \sin(\alpha) \qquad \frac{k}{\sqrt{\mu}} = \cos(\alpha).$$

Since Maxwell take w to be constant, the oscillations persist forever, with decreasing magnitude.

It would be interesting to determine what purely central force law (if any) applied to a point particle would yield a logarithmic spiraling orbit.

Chapter 4

Prisms

A prism is defined as a geometric solid with two congruent bases that lie in parallel planes. You can create a prism by dragging any two-dimensional triangle or polygon through space without rotating or tilting it. The three-dimensional space defined by the moving triangle or polygon is the body of the prism. The planes at which the movement of the two-dimensional shape begins and ends are the two bases of the prism. The perpendicular distance between the parallel planes in which the bases lie is the height of the prism.

The figures below are all prisms. The bases of these prisms are shaded, and the height (altitude) of each prism marked by a dashed line:

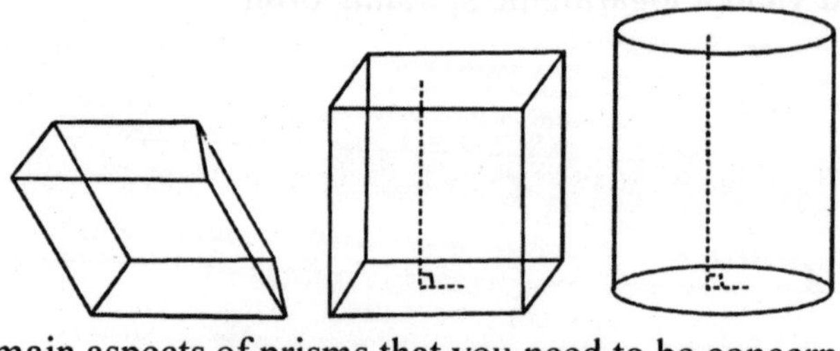

There are two main aspects of prisms that you need to be concerned with volume and surface area. Some prisms have additional characteristics you should know about, but we'll cover them when we discuss specific kinds of solids.

Volume of a Prism

The volume of a prism measures the amount of space taken up by that prism. The general formula for calculating the volume of a prism is very simple:

$$\text{Volume} = Bh$$

where B is the area of the base and h is the height.

Surface Area of a Prism

The surface area of a prism is the sum of the areas of all the sides of the prism. The formula for the surface area of a prism therefore depends on the type of prism you are dealing with. As with volume, we cover the specifics of calculating surface area as we cover each type of geometric solid.

RECTANGULAR SOLIDS

A rectangular solid is a prism with a rectangular base and lateral edges that are perpendicular to its base. In short, a rectangular solid is shaped like a box.

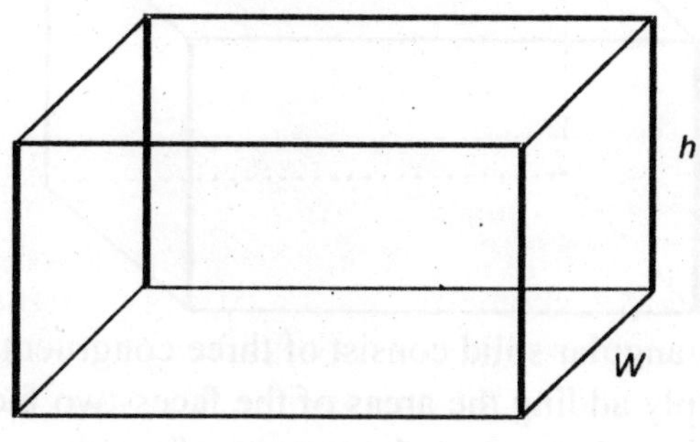

A rectangular solid has three important dimensions: length (l), width (w), and height (h). If you know these measurements, you can find the solid's surface area, volume, and diagonal length.

Volume of a Rectangular Solid

The volume of a rectangular solid is given by the following formula:

$$\text{Volume} = lwh$$

where l is the length, w is the width, and h is the height. Notice how this formula corresponds with the general formula for the volume of a prism: the product lw is the area of the base. Now, try to find the volume of the prism in the following example:

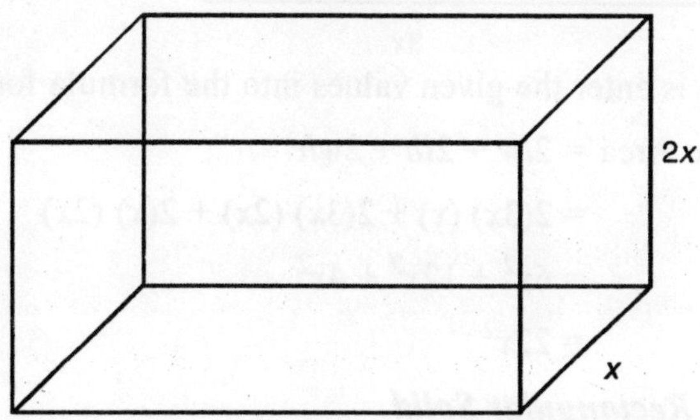

In this solid, $l = 3x$, $w = x$, and $h = 2x$. Simply plug the values into the formula given for volume, and you would find Volume = $(3x)(2x)(x) = 6x3$.

Surface Area of a Rectangular Solid

The surface area of a rectangular solid is given by the following formula:

$$\text{Surface Area} = 2lw + 2lh + 2wh$$

where l is the length, w is the width, and h is the height.

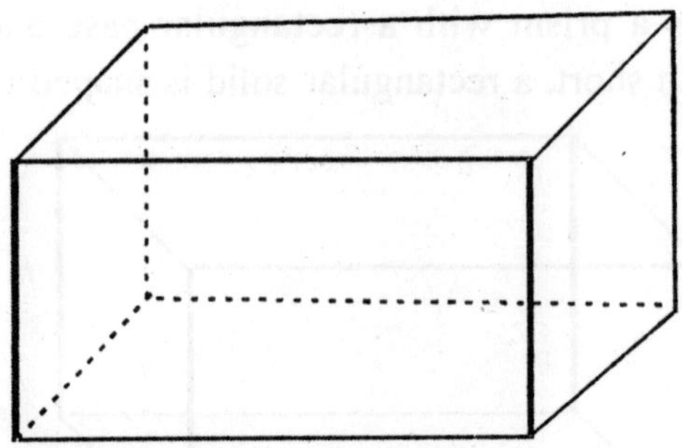

The six faces of a rectangular solid consist of three congruent pairs. The surface area formula is derived by simply adding the areas of the faces-two faces have areas of $l \times w$, two faces have areas of $l\ h$, and two faces have areas of $w \times$ h.

To practice, try to find the surface area of the rectangular solid we used as an example for volume. Here's the figure again:

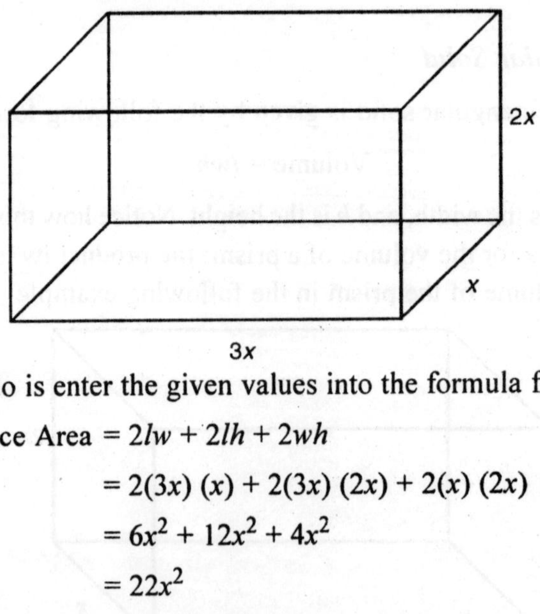

All you have to do is enter the given values into the formula for surface area:

$$\begin{aligned}\text{Surface Area} &= 2lw + 2lh + 2wh \\ &= 2(3x)\ (x) + 2(3x)\ (2x) + 2(x)\ (2x) \\ &= 6x^2 + 12x^2 + 4x^2 \\ &= 22x^2\end{aligned}$$

Diagonal Length of a Rectangular Solid

The diagonal of a rectangular solid, *d*, is the line segment whose endpoints are opposite corners of the solid. Every rectangular solid has four diagonals, each with the same length, that connect each pair of opposite vertices. Here's one diagonal drawn in:

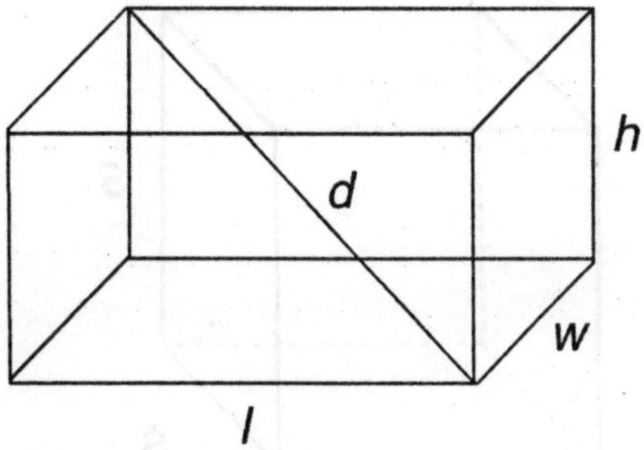

The formula for the length of a diagonal is:

$$d = \sqrt{l^2 + w^2 + h^2}$$

where *l* is the length, *w* is the width, and *h* is the height.

You can look at this formula as the Pythagorean theorem in three dimensions. In fact, you can derive this formula using the Pythagorean theorem. First, find the length of the diagonal along the base. This is $\sqrt{l^2 + w^2}$. Then use the Pythagorean theorem again, incorporating height to find the length of the diagonal from one corner to the other: $d^2 = (\sqrt{l^2 + w^2}) + h^2$. Thus, $d^2 = l^2 + w^2 + h^2$ and $d = \sqrt{l^2 + w^2 + h^2}$. A Math IIC question might ask you:

What is the length of diagonal *AH* in the rectangular solid below if $AC = 5$, $GH = 6$, and $CG = 3$?

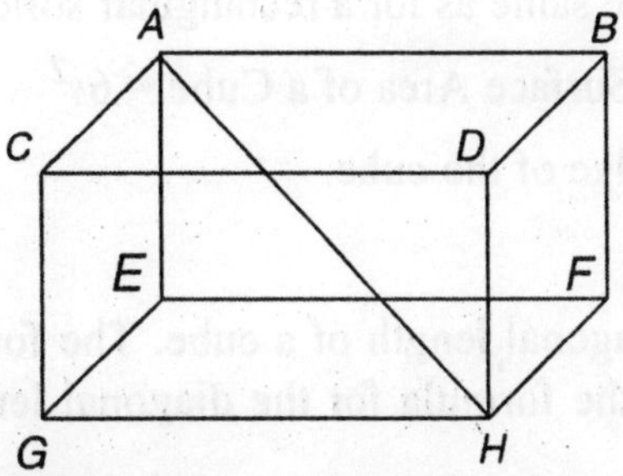

The question gives the length, width, and height of the rectangular solid, so you can just plug those numbers into the formula:

$$AH = \sqrt{5^2 + 6^2 + 3^2} = \sqrt{25 + 36 + 9} = \sqrt{70}$$

The length of the diagonal *AH* (as well as *BG*, *CF*, and *DE*) is $\sqrt{70}$.

CUBES

Just as a square is a special kind of rectangle, a cube is a special kind of rectangular solid. A cube is a rectangular solid that has edges with all the same length. In other words, the length, width, and height are equal, and each of its six faces is a square.

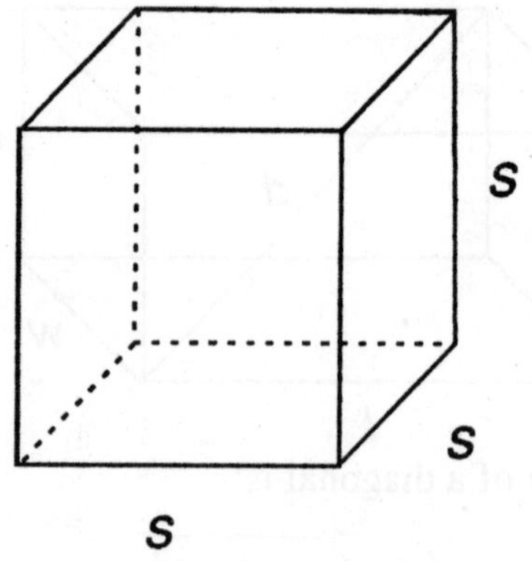

Volume of a Cube

The formula for finding the volume of a cube is essentially the same as the formula for the volume of a rectangular solid.

However, in a cube the length, width, and height are all equal, so that the cube volume formula is:

$$\text{Volume of Cube} = s^3$$

where s is the length of one edge of the cube.

Surface Area of a Cube

Since a cube is just a rectangular solid whose sides are all equal, the formula for finding the surface area of a cube is the same as for a rectangular solid, except with $s = l = w = h$:

$$\text{Surface Area of a Cube} = 6s^2$$

where s is the length of one edge of the cube.

Diagonal Length of a Cube

The same goes for the diagonal length of a cube. The formula for the diagonal of a cube is simply adapted from the formula for the diagonal length of a rectangular solid, with $s = l = w = h$:

$$d = \sqrt{3s^2} = s\sqrt{3}$$

where s is the length of one edge of the cube.

CYLINDERS

A cylinder is a prism with a circular base.

The height of a cylinder, h, is the length of the line segment whose endpoints are the centers of the bases. The radius of a cylinder, r, is the radius of its base. If you know the height and radius of a cylinder, you can calculate its volume and surface area quite easily.

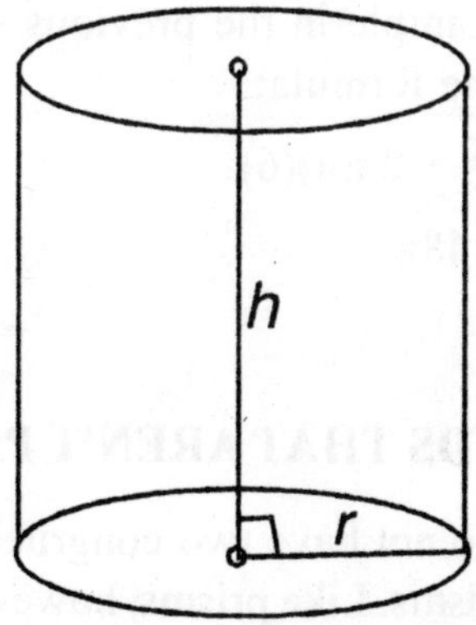

Volume of a Cylinder

The volume of a cylinder is the product of the area of its base with its height. Because a cylinder has a circular base, the volume of a cylinder is:

Volume of a Cylinder = $\pi r^2 h$

where r is the radius of the circular base and h is the height. Try to find the volume of the cylinder below.

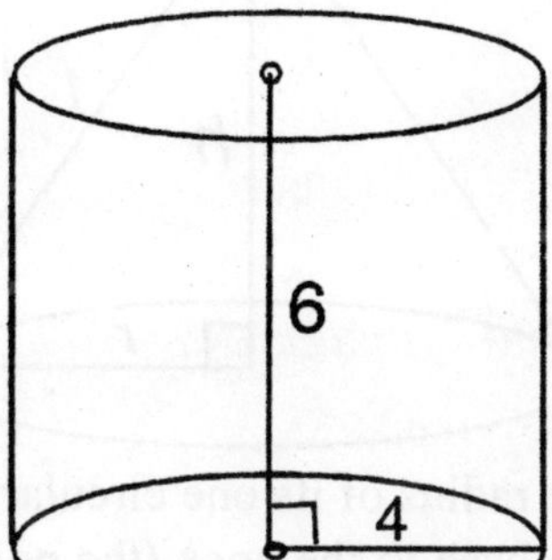

This cylinder has a radius of 4 and a height equal to 6. Using the formulas from the previous page:

$$\text{Volume} = \pi (4)^2 (6) = 96\pi$$

Surface Area of a Cylinder

The surface area of a cylinder is the sum of the areas of the two bases and the lateral face of the cylinder. The bases are congruent circles, so their areas can be found easily. The lateral face is the tubing that connects the two bases. When "unrolled," the lateral base is simply a rectangle whose length is the circumference of the base and whose width is the height of the cylinder. Therefore, the surface area of a cylinder is given by this formula:

$$\text{Surface Area} = 2\pi r^2 + 2\pi rh$$

where r is the radius and h is the height.

Go back to the cylinder example in the previous section, and now find the surface area. Plugging the values into the formula:

$$\begin{aligned}\text{Surface Area} &= 2\pi\,(4)^2 + 2\pi(4)(6)\\ &= 32\pi + 48\pi\\ &= 80\pi\end{aligned}$$

SOLIDS THAT AREN'T PRISMS

The solids in this section do not have two congruent bases that lie in parallel planes and thus cannot be considered prisms. Like prisms, however, the two most important aspects you'll need to know for the test are volume and surface area.

CONES

A cone is not a prism, but it is similar to a cylinder. A cone is essentially a cylinder in which one of the bases is collapsed into a single point at the center of the base.

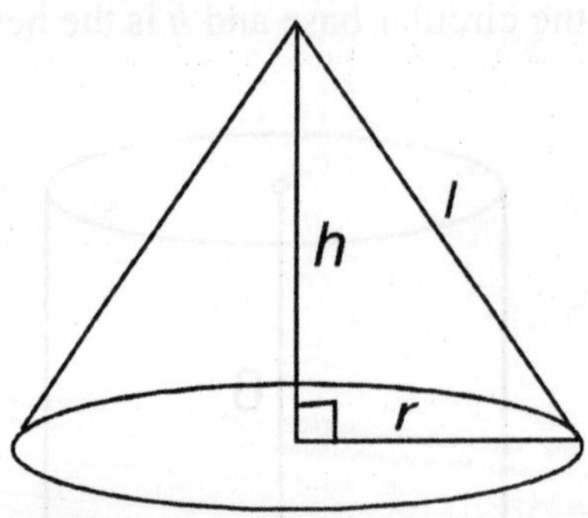

The radius of a cone is the radius of its one circular base. The height of a cone is the distance from the center of the base to the apex (the point on top). The lateral height, or slant height, of a cone is the distance from a point on the edge of the base to the apex. In the previous figure, these three measurements are denoted by r, h, and l, respectively.

Notice also that the height, radius, and lateral height of a cone make up a right triangle. This means that if you know the value for any two of these measurements, you will always be able to find the third by using the Pythagorean theorem.

Volume of a Cone

Since a cone is similar to a cylinder, except collapsed to a single point at one end, the formula for the volume of a cone is a fraction of the formula for the volume of a cylinder:

$$\text{Volume of Coue} = \frac{1}{3}\text{ Volume of a Cylinder} = \frac{1}{3}\pi r^2 h$$

where r is the radius and h is the height.

For practice, find the volume of the cone pictured below:

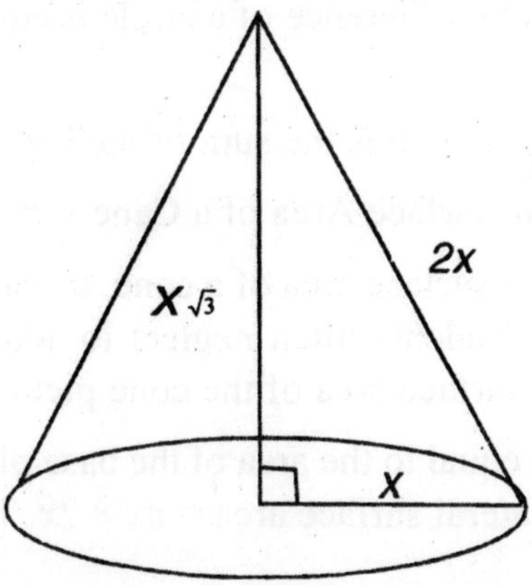

To answer this question, just use the formula for the volume of a cone with the following values plugged in: $r = x$, $l = 2x$, and $h = x\sqrt{3}$. The volume is 1/3 $\pi(x^2)(x\sqrt{3}) = x\ \pi/3$.

Surface Area of a Cone

The surface area of a cone consists of the lateral surface area and the area of the base. The base is a circle and therefore has an area of πr^2. The lateral surface is the cone "unrolled." Depending on the shape of the cone, it can be the shape of a triangle with a curved base, a half circle, or a "Pacman" shape. The area of the lateral surface is a length that is related to the circumference of the circle times the lateral height, l. This is the formula:

$$\text{Lateral Surface Area of a Coue} = \pi rl$$

where r is the radius and l is the lateral height. An alternate formula for lateral surface area is:

$$\text{Lateral Surface Area of a Coue} = \pi\frac{1}{3}cl$$

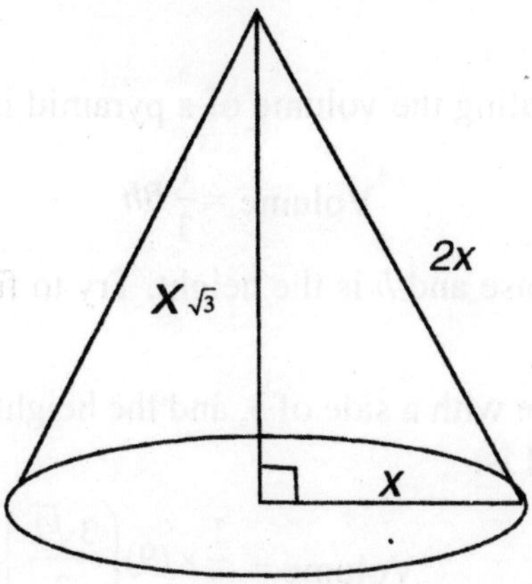

where c is the circumference and l is still the lateral height. The two formulas for surface area are the same, since the circumference of a circle is equal to $2\pi r$. But knowing one or the other can save time.

The *total* surface area of a cone is the sum of the base area and lateral surface area:

$$\text{Total Surface Area of a Cane} = \pi r^2 + \pi rl$$

When you are finding the surface area of a cone, be careful not to just find the lateral surface area and then stop. Students often neglect to add the area of the circular base. Practice by finding the total surface area of the cone pictured below:

The total surface area is equal to the area of the base plus the lateral surface area. The area of the base $= \pi x^2$. The lateral surface area $= \pi x \times 2x$. The total surface area therefore equals $\pi x^2 + \pi 2x^2 = 3\pi x^2$.

PYRAMIDS

A pyramid is like a cone, except that it has a polygon for a base. Though pyramids are not tested very often on the Math IIC test, you should be able to recognize them and calculate their volume.

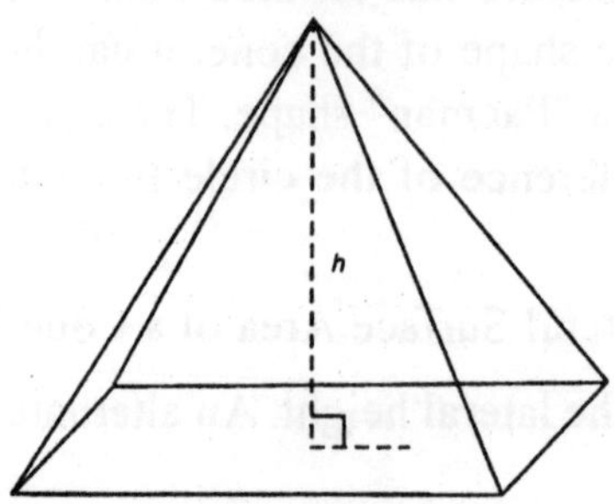

The shaded area in the figure above is the base, and the height is the perpendicular distance from the apex of the pyramid to its base.

Volume of a Pyramid

The formula for calculating the volume of a pyramid is:

$$\text{Volume} = \frac{1}{3}Bh$$

where B is the area of the base and h is the height. Try to find the volume of the pyramid below:

The base is just a square with a side of 3, and the height is $3\sqrt{2}/2$. $B = 3^2 = 9$, and the total volume of the pyramid is:

$$\text{Volume} = \frac{1}{3} \times (9)\left(\frac{3\sqrt{2}}{2}\right)$$

$$= \frac{1}{3} \times \frac{27}{\sqrt{2}}$$

$$= \frac{9\sqrt{2}}{2}$$

Surface Area of a Pyramid

The surface area of a pyramid is rarely tested on the Math IIC test. If you do come across a question that covers the topic, you can calculate the area of each face individually using techniques from plane geometry, since the base of the pyramid is a square and the sides are triangles.

SPHERES

A sphere is the collection of points in three-dimensional space that are equidistant from a fixed point, the center of the sphere. Essentially, a sphere is a 3-D circle. The main measurement of a sphere is its radius, r, the distance from the center to any point on the sphere.

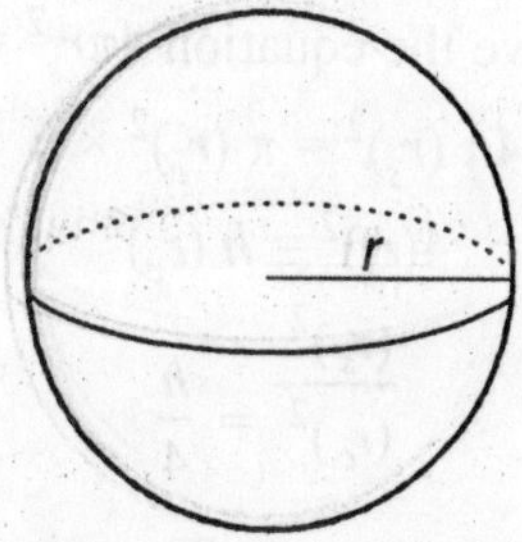

If you know the radius of a sphere, you can figure out both its volume and surface area. The equation for the volume of a sphere is:

$$\text{Volume of a Sphere} = \frac{4\pi r^2}{3}$$

The equation for the surface area of a sphere is:

$$\text{Surface Area of Sphere} = 4\pi r^2$$

RELATING LENGTH, SURFACE AREA, AND VOLUME

The Math IIC tests not only whether you've memorized the formulas for the different geometric solids, but also if you understand those formulas.

The test gauges your understanding by asking you to calculate the lengths, surface areas, and volumes of various solids. The Math IIC will ask you about the relationship between these three properties. The exam includes two kinds of questions that cover these relations.

Comparing Dimensions

The first way the Math IIC will test your understanding of the relationship between the basic measurements of geometric solids is by giving you the length, surface area, or volume of various figures and asking you to compare their dimensions. The math needed to answer comparing dimensions questions isn't all that hard. But in order to get to the math, you need to have a good grasp of the formulas for each type of solid and be able to relate those formulas to each other algebraically. For example:

The surface area of a sphere is the same as the volume of a cylinder. What is the ratio of the radius of the sphere to the radius of the cylinder?

This question tells you that the surface area of a sphere and the volume, a cylinder are equal. A sphere's surface area is $4\pi(r_s)^2$, where r_s is the radius of the sphere.

A cylinder's volume is $\pi(r_c)^2 \times h$, where r_c is the radius of the cylinder and h is its height. Therefore:

$$4\pi (r_s)^2 = \pi (r_c)^2 \times h$$

The question asks for the ratio between the radii of the sphere and cylinder. This ratio is given by r_s/r_c. Now you can solve the equation $4\pi r_s^{\,2} = \pi r_c^{\,2} \times h$ for the ratio r_s/r_c.

$$4\pi (r_s)^2 = \pi (r_c)^2 \times h$$

$$4(r_s)^2 = h (r_c)^2$$

$$\frac{(r_s)^2}{(r_c)^2} = \frac{h}{4}$$

$$\frac{r_s}{r_c} = \sqrt{\frac{h}{4}}$$

Changing Measurements

The second way the Math IIC will test your understanding of the relationships between length, surface area, and volume is by changing one of these measurements by a given factor and then asking how this change will influence the other measurements.

When the solids in the question are increased by a single constant factor, a simple rule can help you come to an answer.

1. If a solid's length is multiplied by a given factor, then its surface area is multiplied by the square of that factor, and its volume is multiplied by the cube of that factor.

Remember that this rule holds true if *all* of a solid's dimensions increase in length by a given factor. So for a cube or a sphere, just a side or the radius need change for the rule to hold, but for a rectangular solid, cylinder, or other solid, all of the dimensions must change. If the dimensions of the object do not increase by a constant factor, for instance, if the height of a cylinder doubles but the radius of the base triples, you will have to go back to the equation for the dimension you are trying to determine and calculate by hand.

Example: If you double the length of the side of a square, by how much do you increase the area of that square?

If you understand the formula for the area of a square, this question is not difficult. The formula for the area of a square is $A = s^2$, where s is the length of a side. Replace s with $2s$, and you see that the area of a square quadruples when the length of its sides double: $(2s)^2 = 4s^2$.

Example: If a sphere's radius is halved, by what factor does its volume decrease?

The radius of the sphere is multiplied by a factor of 1/2 (or divided by a factor of 2), and so its volume multiplies by the cube of that factor: $(1/2)^3 = 1/8$. Therefore, the volume of the sphere is multiplied by a factor of 1/8 (divided by 8), which is the same as decreasing by a factor of 8.

Example: A rectangular solid has dimensions $x \times y \times z$ (these are its length, width, and height) and has a volume of 64. What is the volume of a rectangular solid of dimensions $x/2 \times y/2 \times z$?

If this rectangular solid had dimensions that were all one-half as large as the dimensions of the solid whose volume is 64, then its volume would be $(1/2)^3 \times 64 = 1/8 \times 64 = 8$. But dimension z is not multiplied by 1/ 2 like x and y. Hence, to answer a question like this one, you should turn to the volume formula for rectangular solids: volume = $l \times w \times h$. It is given in the question that $xyz = 64$. So, $x/2 \times y/2 \times z = 1/4\ xyz = 1/4 \times 64 = 16$.

INSCRIBED SOLIDS

An inscribed solid is a solid fit inside another solid, with the edges of the two solids touching. The figures below are, from left to right, a cylinder inscribed in a sphere, a sphere inscribed in a cube, and a rectangular solid inscribed in a sphere.

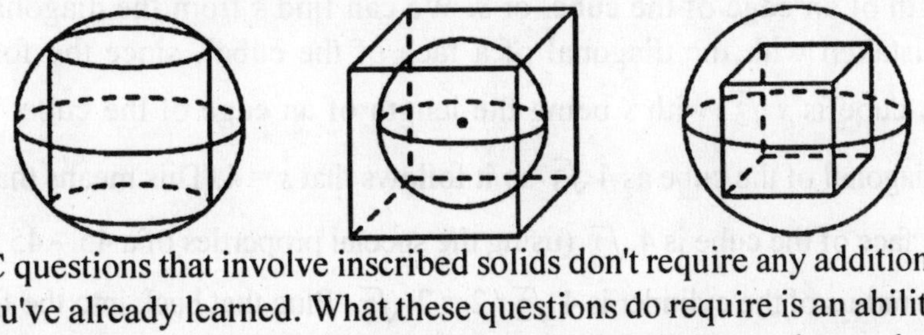

Math IIC questions that involve inscribed solids don't require any additional techniques than those you've already learned. What these questions do require is an ability to visualize inscribed solids and the awareness of how certain line segments relate to both solids in a given figure.

Most often, an inscribed solid question will present a figure of an inscribed solid and give you information about one of the solids. For example, you may be given the radius of a cylinder and then be asked to find the volume of the other solid using the figure as your guide. Use what you know about the radius of the cylinder to find the dimensions of the other solid. Take a look at this example: In the figure below, a cube is inscribed in a cylinder. If the length of the diagonal of the cube is 4 and the height of the cylinder is 5, what is the volume of the cylinder?

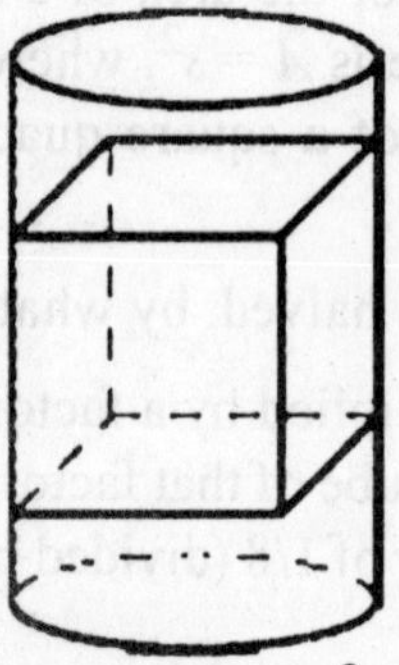

The formula for the volume of a cylinder is $\pi r^2(h)$. It is given in the question that $h = 5$, but there is no value given for r. So in order to solve for the volume of the cylinder, we need to first find the value of r. The key step in this problem is to recognize that the diagonal of a face of the cube is also the diameter, or twice the radius, of the cylinder. To see this, draw a diagonal, d, in either the top or bottom face of the cube.

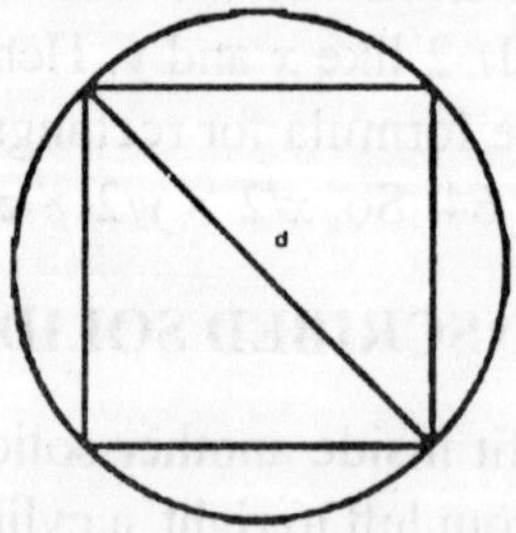

In order to find this diagonal, which is the hypotenuse in a 45 – 45 – 90 triangle, we need the length of an edge of the cube, or s. We can find s from the diagonal of the cube (not to be mistaken with the diagonal of a face of the cube), since the formula for the diagonal of a cube is $s\sqrt{3}$ with s being the length of an edge of the cube. The question gives us the diagonal of the cube as $4\sqrt{3}$ so it follows that $s = 4$. This means that the diagonal along a single face of the cube is $4\sqrt{2}$ (using the special properties of a 45 – 45 – 90 triangle). Therefore, the radius of the cylinder is $4\sqrt{2}/2 = 2\sqrt{2}$. Plug that back into the formula for the volume of the cylinder, and you get $\pi \times (2\sqrt{2})^2 \times 5 = 40\pi$.

THE RULES OF INSCRIBED SOLIDS

Math IIC questions involving inscribed solids are much easier to solve when you know how the lines of different solids relate to one another. For instance, the previous example showed that when a cube is inscribed in a cylinder, the diagonal of a face of the cube is equal to the diameter of the cylinder.

The better you know the rules of inscribed solids, the better you'll do on these questions.

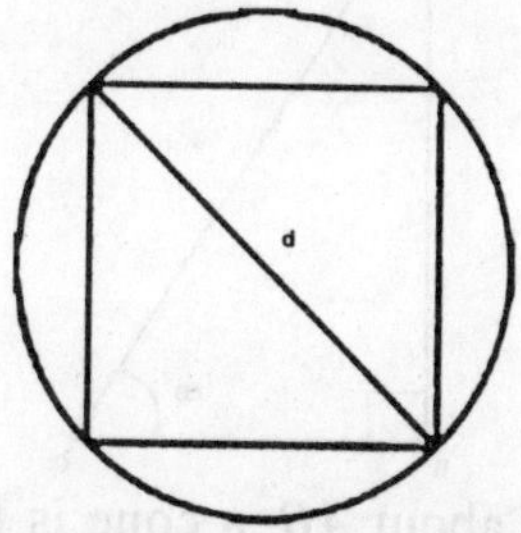

Fig. Cylinder Inscribed in a Sphere

The diameter of the sphere is equal to the diagonal of the cylinder's height and diameter.

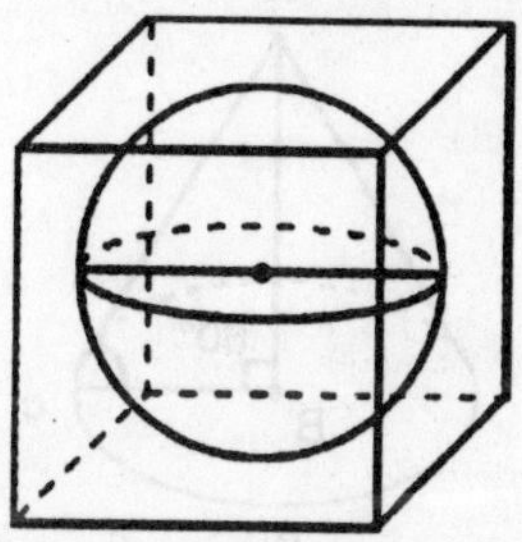

Fig. Sphere Inscribed in a Cube

The diameter of the sphere is equal to the length of cube's edge.

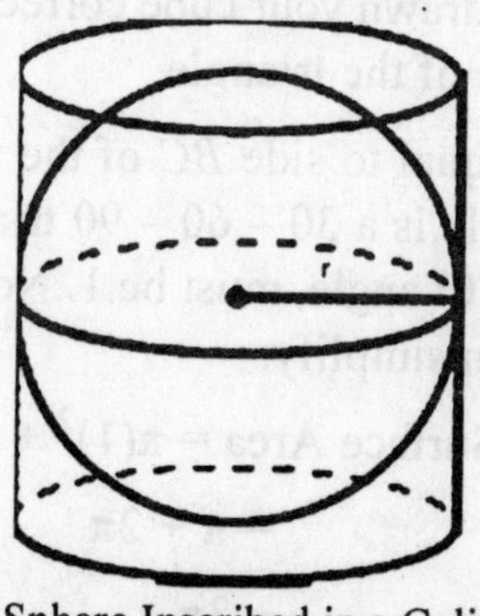

Fig. Sphere Inscribed in a Cylinder

The cylinder and the sphere have the same radius.

SOLIDS PRODUCED BY ROTATING POLYGONS

Another type of Math IIC question that you may come across involves a solid produced by the rotation of a polygon. The best way to explain this type of problem is with an example:

What is the surface area of the geometric solid produced by the triangle below when it is rotated 360° about the axis *AB*?

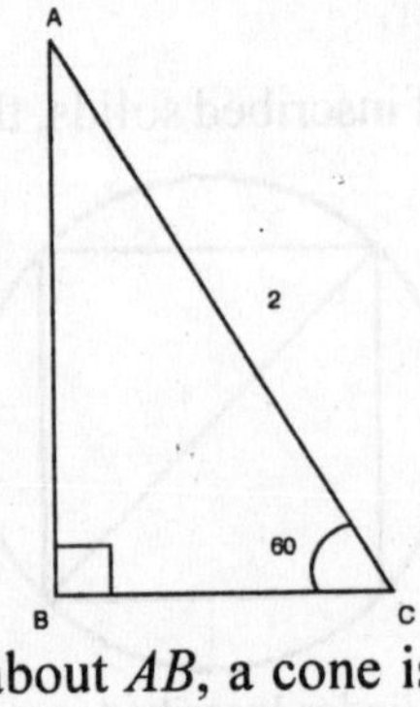

When this triangle is rotated about *AB*, a cone is formed. To solve the problem, the first thing you should do is sketch the cone that the triangle will form.

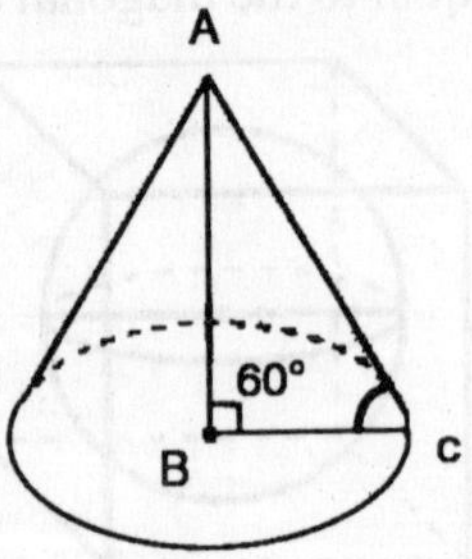

The question asks you to figure out the surface area of the cone. The formula for surface area is $\pi r^2 + \pi rl$, which means you need to know the lateral height of the cone and the radius of the circle. If you've drawn your cube correctly, you should see that the lateral height is equal to the hypotenuse of the triangle.

The radius of the circle is equal to side *BC* of the triangle. You can easily calculate the length of *BC* since the triangle is a 30 – 60 – 90 triangle. If the hypotenuse is 2, then *BC*, being the side opposite the 30° angle, must be 1. Now plug both values of *l* and *r* into the surface area formula and then simplify:

$$\text{Total Surface Area} = \pi(1)^2 + \pi(1)(2)$$
$$= \pi + 2\pi$$
$$= 3\pi$$

COMMON ROTATIONS

You don't need to learn any new techniques or formulas for problems dealing with rotating figures. You just have to be able visualize the described rotation and be aware of which parts of the polygons become which parts of the geometric solid. Below is a summary of which polygons, when rotated a specific way, produce which solids.

A rectangle rotated about its edge produces a cylinder.

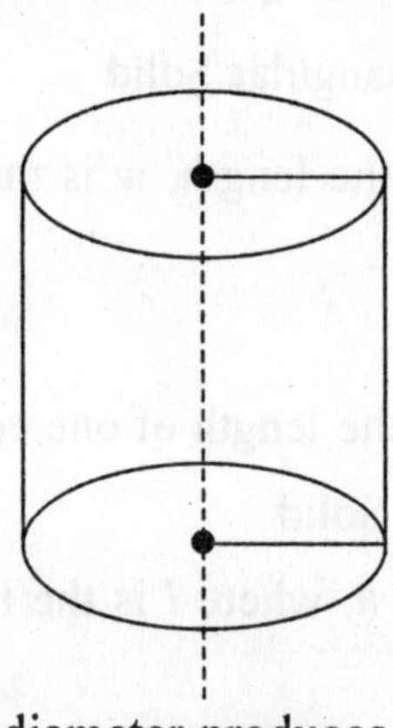

A semicircle rotated about its diameter produces a sphere.

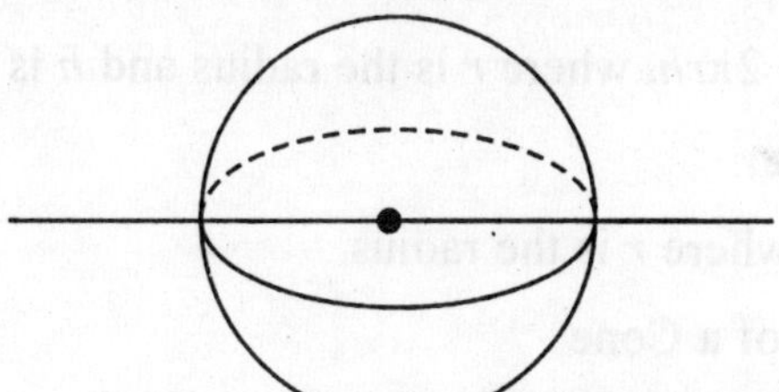

A right triangle rotated about one of its legs produces a cone.

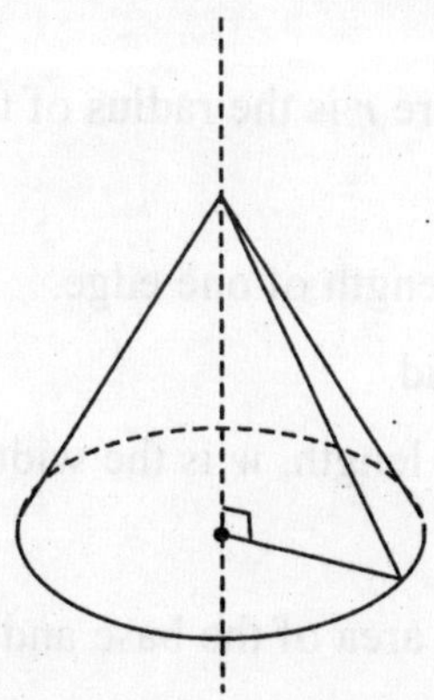

A rectangle rotated about a central axis (which must contain the midpoint of both of the sides that it intersects) produces a cylinder. A circle rotated about its diameter produces a sphere. An isosceles triangle rotated about its axis of symmetry (the altitude from the vertex of the non-congruent angle) produces a cone.

Key Formulas

Length of a Diagonal of a Cube

$d = \sqrt{3s^2} = s\sqrt{3}$. where s is the length of one edge of the cube.

Length of a Diagonal of a Rectangular Solid

$d = \sqrt{l^2 + w^2 + h^2}$, where l is the length, w is the width, and h is the height of the rectangular solid.

Surface Area of a Cube

Surface area = $6s^2$, where s is the length of one edge.

Surface Area of a Rectangular Solid

Surface area = $2lw + 2lh + 2wh$, where l is the length, w is the width, and h is the height.

Surface Area of a Cylinder

Surface area = $2\pi r^2 + 2\pi rh$, where r is the radius and h is the height.

Surface Area of a Sphere

Surface area = $4\pi r^2$, where r is the radius.

Lateral Surface Area of a Cone

Lateral surface area = πrl, where r is the radius and l is the lateral height of the cone.

Surface Area of a Cone

Surface area = $\pi r^2 + \pi rl$, where r is the radius of the base and l is the lateral height.

Volume of a Cube

Volume = s^3, where s is the length of one edge.

Volume of a Rectangular Solid

Volume = lwh, where l is the length, w is the width, and h is the height.

Volume of a Prism

Volume = Bh, where B is the area of the base and h is the height.

Volume of a Cylinder

Volume = $\pi r^2 h$, where r is the radius of the circular base and h is the height.

Volume of a Cone

Volume $= \frac{1}{3} \pi r^2 h$, where r is the radius and h is the height.

Volume of a Pyramid

Volume $= \frac{1}{3} Bh$, where B is the area of the base and h is the height.

Volume of a Sphere

Volume $=\dfrac{4\pi r^3}{3}$, where r is the radius.

EXPLANATIONS

1. C

The formula for the surface area of a rectangular solid is:

$$\text{Surface Area} = 2lw + 2lh + 2wh$$

where l, w, and h are the dimensions of the solid. To answer the question, substitute $2l$, $2w$, and $2h$ for l, w, and h:

$$\begin{aligned}\text{Surface Area} &= 2(2l)(2w) + 2\,(2w)(2h) + 2\,(2l)(2h)\\ &= 2(4lw) + 2(4wh) + 2(4lh)\\ &= 4(2lw + 2wh + 2lh)\end{aligned}$$

When the measurements of the rectangular solid are doubled, the surface area of the solid is multiplied by 4.

2. D

The formula for the volume of a cylinder is $V = \pi r^2 h$, where r is the radius of its circular base and h is its height. The height of the cylinder is given, so we just need to find out its radius. The best way to do this is to sketch a diagram. If you look at the figure below, you can see that the diameter of the sphere, d, is the diagonal of the cylinder. The diagonal of the cylinder and its height, together with the diameter of the cylinder's circular base as the third side, make up a right triangle:

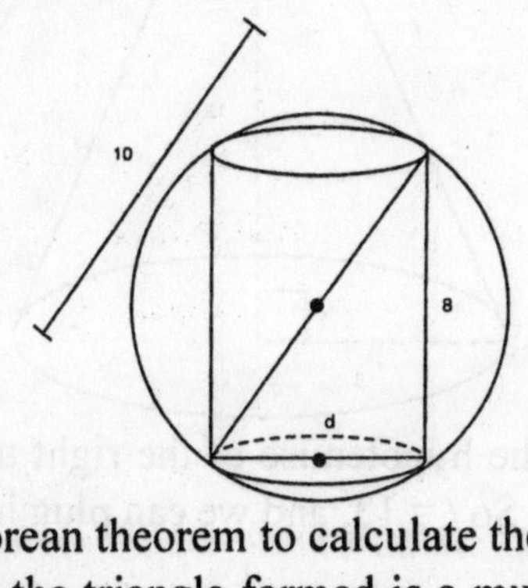

We could use the Pythagorean theorem to calculate the diameter of the cylinder base, but you should recognize that the triangle formed is a multiple of a 3 – 4 – 5 triangle, a Pythagorean triple. Therefore, since $h = 2 \times 4 = 8$ and $d = 2r = 2 \times 5 = 10$, the diameter of the cylinder base is $2 \times 3 = 6$. This means that the radius of the base is 3, and the volume of the cylinder can now be calculated:

$$\pi r^2 h = \pi(3^2)(8) = 72\pi = 226.19$$

3. C

The formula for the volume of a cube is $V_c = s^3$, where s is the length of a side of the cube. The formula for the volume of a sphere is $V_s = 4\pi r^3/3$, where r is the radius of the sphere. Therefore, since we know that $V_c = 3V_s$, we can substitute in the formulas for the two solids and solve for s/r:

$$s^3 = 3\left(\frac{4\pi r^3}{3}\right)$$

$$s^3 = 4\pi r^3$$

$$\frac{s^3}{r^3} = 4\pi$$

$$\frac{s}{r} = \sqrt[3]{4\pi}$$

4. C

When the right triangle is rotated around the leg of length 12, a cone is formed with radius 5 and height 12. The formula for the surface area of a cone is $\pi r^2 + \pi rl$, where r is the radius and l is the lateral length of the cone. So, in order to calculate the surface area, we need to first find l:

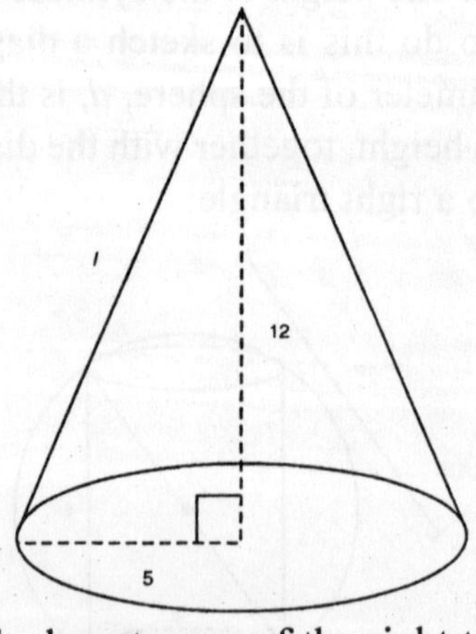

As you can see, l is just the hypotenuse of the right triangle, which is a 5 – 12 – 13 triangle (a Pythagorean triple). So $l = 13$, and we can plug in values to find the surface area of the cone:

$$\text{Surface Area} = \pi(5)^2 + \pi(5)(13) = 25\pi + 65\pi = 90\pi$$

Chpater 5

Multi-Layered Polyhedra

We usually regard an ordinary polyhedron as a 3-dimensional solid with a single-layered 2-dimensional surface. On this basis we can easily count the number of vertices, edges, and faces, and compute the Euler characteristic of the surface

$$e(S) = V - E + F$$

It can be shown that any simply-connected closed 2-dimensional surface (i.e., a surface that has the same topology as a sphere) has an Euler characteristic of 2. This applies to all the most familiar polyhedra, such as the Platonic and Archimedian solids, and many others as well. (Apparently Descartes was the first person to ever notice the simple numerical fact that

$$V - E + F = 2$$

for all these solid shapes, which is remarkable, considering how numero-logically inclined the ancient Pythagoreans were.)

To be a bit more rigorous about the meaning of "simply-connected", note that a surface S is defined to be "connected" if any pair of its points can be joined by a continuous curve lying entirely within the surface. Further, a surface is said to be "simply-connected" if any CLOSED curve C on the surface divides the surface into TWO distinct regions, each of which is internally "connected" in the sense just described, and such that any continuous curve which joins a point in one of those regions to a point in the other must cross the closed curve C.

Another, somewhat less formal, way of characterizing a simply-connected surface is to say that any closed curve on such a surface can be "shrunk" continuously to a point, without ever departing the surface. We mentioned previously that the surface of a sphere is simply-connected, and obviously any circle drawn on a sphere can be shrunk to a point. On the other hand, the surface of a torus (a doughnut) is NOT simply connected, because we can draw a closed curve around a torus shape in such a way that the curve cannot be shrunk any further. Also, from our previous definition, we can see that some closed curves C on the surface of a torus do NOT bisect the surface into disconntected parts, as shown below.

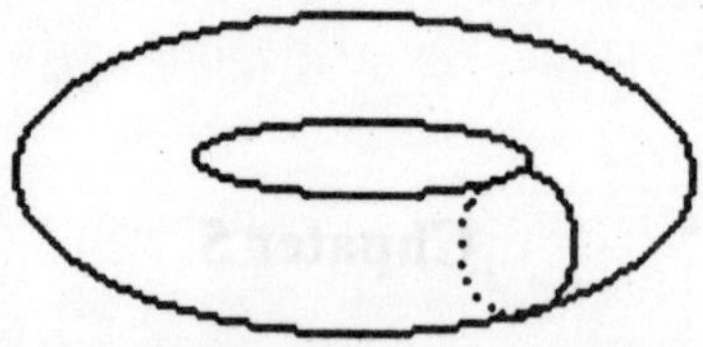

In this sense, the torus is topologically distinct from the surface of a sphere, and consequently it has a different Euler characteristic. If we triangulate the surface of a torus (in any way we like) and count the number of edges, faces, and vertices, we find that the Euler characteristic is 0.

Incidentally, the Euler characteristic of a surface is directly related to the total Gaussian curvature of that surface. Specifically, the Gauss-Bonnet theorem states that for any compact oriented 2-dimensional Riemannian manifold S, if we let K denote the Gaussian curvature as a function of position on the manifold, then we have

$$\sum_{S} K\ dA = 2pi\ e(s).$$

In other words, the integral of the Gaussian curvature over the entire area of the manifold equals 2pi times the Euler characteristic. For example, the Gaussian curvature of a sphere of radius r is a constant $1/r^2$, so the integral of this over the entire surface of the sphere is simple $1/r^2$ times the total surface area A of the sphere. Now, since e(S)=2 for a sphere, we must have

$$\frac{1}{r^2}A = 2pi\ (2) = 4pi$$

and therefore the surface area of a sphere must be 4 pi r^2, which of course it is. Compare this with the reasoning that Archimedes originally used to deduce the surface area of a sphere. The Gauss-Bonnet theorem is a truly beautiful fact, relating a metrical property (curvature) to a purely topological property (the Euler characteristic). Notice that this theorem can be applied to polyhedra as well, if we agree to assign the appropriate amount of "curvature" to the vertices (since there is no curvature on any face or edge).

It follows from this theorem that the "net" Gaussian curvature of a torus is zero, so it's possible to define a perfectly flat metric globally over the entire surface of a torus. On the other hand, it's obviously not possible to define a flat metric globally on a sphere, or on any surface homeomorphic to a sphere. This is true even though the Gaussian curvature on the surface of a polyhedron is zero at all but a finite number of points (the vertices).

But this leads us to an interesting question, namely, whether it is possible to "wrap" a completely flat closed surface (i.e., a torus) onto the surface of a convex polyhedron if we allow multi-layered surfaces. For example, if we create a surface out of squares, we must have a certainly number of squares meeting at each vertex, and each square represents a right angle. Now, for a single-layered flat surface there must be 4 right angles meeting at each vertex, whereas for a single-layered cube there are only 3 right angles meeting at each vertex.

However, our surface need not be connected after just a single loop around a given vertex. Indeed we can imagine a "spiral staircase" consisting of a surface that circles a vertex many times before re-connecting with the original sheet. Of course, if this 2D surface is embedded in 3D space we won't be able to re-connect without self-intersecting, but this is just a restriction related to the embedding in 3D space, similar to the fact that a "Klein bottle can't be realized in 3D space without self-intersecting, whereas it CAN be realized in 4D space. By the same token, we can imagine a sheet circling a vertex arbitrarily many times before finally re-connecting to the original sheet.

The underlying "shape" of this multi-layered manifold will still depend on the angular travel around the vertex that brings us back to the same location in the manifold, but now we can allow multiples of the basic angular requirements. If we must travel through some multiple of 4 right angles to get back where we started, then we can imagine that these sheets are molded in a flat shape, whereas if we must travel through a multiple of 3 right angles to get back where we started, we can imagine that that this surface is molded in the shape of a corner of a cube. But what if the total angular travel around a vertex is a multiple of BOTH 3 right angles AND 4 right angles, i.e., what if it is a multiple of 12 right angles? In that case we have a multi-sheeted surface that can be interpreted as either flat or as the vertex of a convex solid, i.e., as a point with non-zero curvature.

Actually it isn't difficult to construct this surface out of paper. Just cut three identical squares, and draw two perpendicular lines though the center of each square so that each is divided into four faces. Then cut along one of those lines from one edge to the center, and tape the squares together in a spiral pattern, as indicated below.

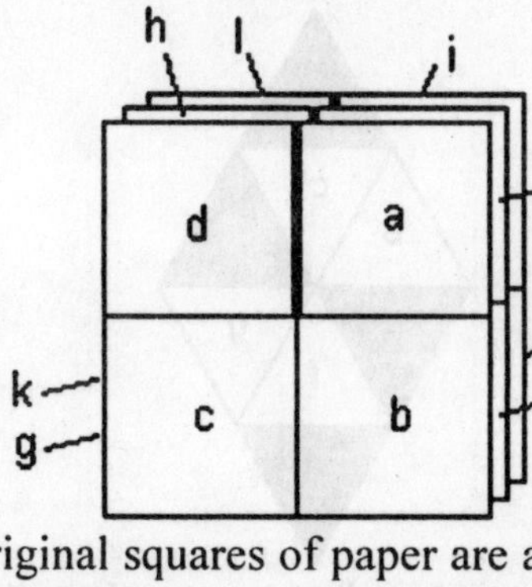

The cuts in the three original squares of paper are along the thick vertical lines, and the right-hand edge of "d" is connected to the left-hand edge of "e" (which is underneath "a"). Similarly, the edge of "h" is connected to the edge of "i". Lastly, after making three complete circuits around the central vertex, the edge of "l" on the bottom sheet is connected to the left-hand edge of "a" on the top sheet. This last connection obviously can't be achieved in 3D without the surface intersecting itself, but just as with the Klein bottle we are free to imagine an embedding in a higher dimensional space that allows us to adjoin all the edges as described. As an alternative, we can neglect the embedding entirely, and simply conceive of this as a 2D surface with its own intrinsic properties.

The interesting thing about this surface is that it "works" as a 3-layered flat plane, bit it also works as a 4-layered corner of a cube. To see this, it's helpful to first crease the sheets along the orthogonal lines, and then simply "wind up" the faces around the vertex, so that three (rather than four) square faces meet at the vertex. Notice that the intrinsic connections between all the faces of this "wound up" version are identical to the connections of the "flat" version, including the alignment of the adjoining edges of the "l" and "a" faces. In effect, we began with a 3-layered flat surface, but this can also be interpreted as a 4-layered surface that follows the shape of a cubical vertex. If we consider the surface "modulo the cube", we see that it has intrinsic curvature. If we consider the surface modulo the "flat" arrangement we group the faces into the four equivalence classes

{a,e,i}, {b,f,j}, {c,g,k}, {d,h,l}

whereas if we consider the surface modulo the curved arrangement we group the faces into the three equivalence classes

{a,d,g,j}, {b,e,h,k}, {c,f,i,l}

Of course, neither of these interpretations yields a closed surface. This raises the interesting question of whether it's possible to devise a closed surface that can be interpreted equally well as flat or curved, depending on which "shape modulus" we select. It turns out that it IS possible. The simplest example is based on the tetrahedron. To construct a model of this, cut out a piece of paper in the diamond shape shown below, comprised of 8 equilateral triangles

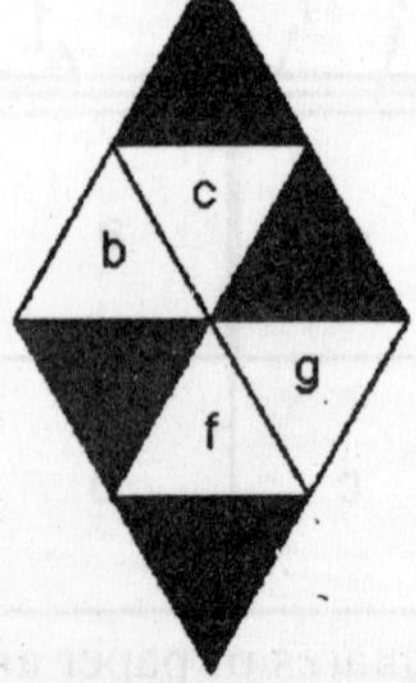

Now make a cut along the boundary between regions "f" and "g", and wind up the surface around the central point, so that three triangles meet at that point. This results in a

2-layered tetrahedron, and the four faces correspond to the four colors on the above figure. Thus the eight original triangles are grouped into the four equivalence classes

{a,h}, {b,g}, {c,f}, {d,e}

The important point to notice is that this arrangement brings all the edges into alignment in a way that is perfectly consistent with a FLAT arrangement of the faces. Note that if we take many copies of the above diamond figure and set them side-by-side, we have the doubly-periodic tiling illustrated below

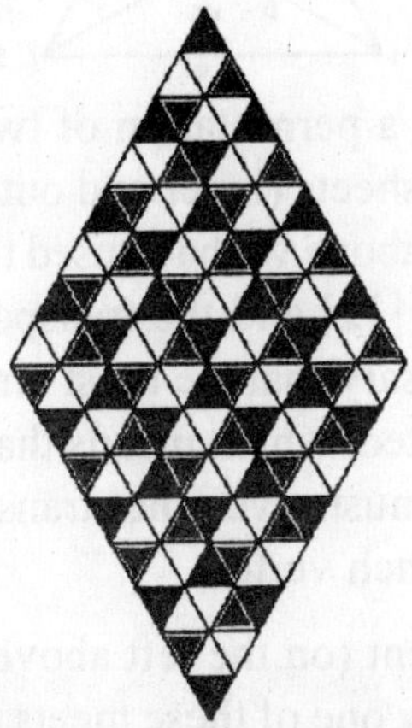

This shows how all the 8 regions of the original diamond must be interconnected to form a torus, which then is given by identifying the upper left boundary of the original diamond with the lower right, and identifying the upper right with the lower left. In this flat interpretation we have only a single layer, so we have the 8 equivalence classes

{a}, {b}, {c}, {d}, {e}, {f}, {g}, {i}

The connections between these eight triangles are identical, whether we view them as comprising a single-layered torus or a 2-layered tetrahedron. Of course, the curvature of the latter arises only when we reduce the surface modulo the tetrahedron. By the way, notice that we have the extrinsic arrangement of layers (i.e., "inner" and "outer") indicated below

{b,e,f,h} {a,c,d,g}

Of course, we can make either of these sets the "inner" surfaces and the other the "outer" surfaces, depending on how we choose to "wind up" the faces around the original central vertex.

Another way of constructing the "tetratorus" is to begin with the same 8-part diamond shaped piece of paper as above, but this time make a cut along the boundary between regions "d" and "g". This results in the same instrinsic arrangement as before, but the extrinsic knotting of the surfaces is different. With this construction we get the "outer\inner" partition {e,f,g,h} {a,b,c,d}

The actual extrinsic distinction between these two arrangements is clarified by considering the left-most and middle graphs of the tetrahedron shown below.

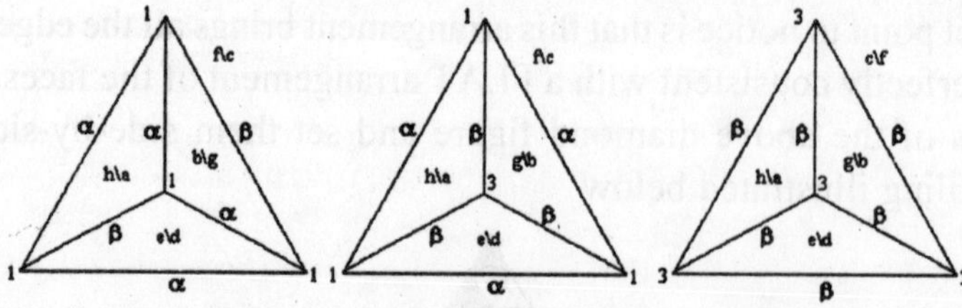

Each edge can be regarded as a permutation of two elements, since we must decide for each edge how to map the two sheets (inner and outer) of one face to the two sheets of the adjoining face. In the diagrams above we have used the Greek letters "alpha" and "beta" to denote the identity permutation [12] and the transposition [21] respectively. Now, for each vertex to be "flat", it is necessary that we must circle around it through 6 equilateral triangles to get back where we started, which means that a single circuit around the vertex (as seen modulo the tetrahedron) must give a net transposition, so we must have an odd number of [21] edges meeting at each vertex.

Indeed, for the first arrangement (on the left above) there are two edges carrying the [21] (beta) permutation, and exactly one of these meets each vertex. On the other hand, for the second arrangement (in the center above) we have swapped the outer\inner positions of the congruent faces "b" and "g", which implies that every "alpha" on the edge of this face becomes a "beta", and vice versa. This still works, as we now have 3 transpositions meeting at one vertex, and 1 transposition at each of the remaining vertices, i.e., we still have an odd number of "beta's" meeting at each vertex. From an intrinsic standpoint these two surfaces are equivalent, but they are knotted differently in the embedding space. To see this, observe that it's possible in the first arrangement to make a complete loop around all the outer (or all the inner) faces without ever crossing the same edge twice and always remaining on the same level. In contrast, this is not possible with the second arrangement.

An exploded view of the tetratorus is shown below

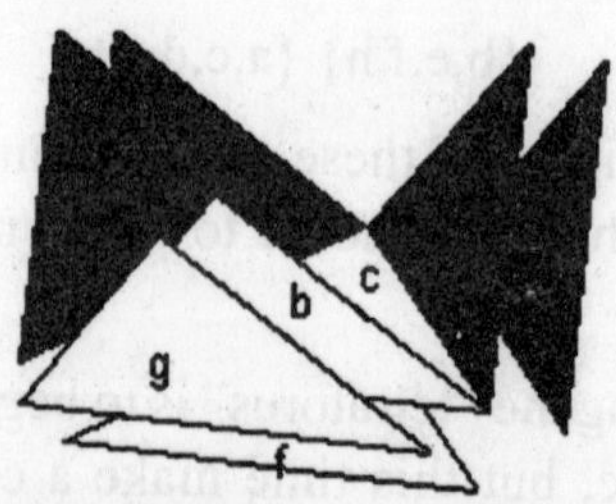

Depending on which of the embeddings we choose, we can make the appropriate identifications of edges so that the surface of the two-layered tetrahedron has the same connectivity as the flat torroidal surface. For example, to give the embedding corresponding to the middle graph in the preceding schematics we would make the natural connections of the outer layers to the outer base (yellow) triangle, and then make transposing identifications along the remaining three edges. In other words, connect g,e, and h with f, and then make the connections gd,be and ga,bh and ae,dh.

One other possible embedding arrangement now immediately presents itself. If we swap the outer\inner positions of the congruent faces "c" and "f" in the middle graph above, we produce the right-hand graph, which must have transpositions for ALL 6 of its edges. In this case we have 3 "beta" edges meeting at each vertex, and this embedding is obviously distinct from each of the other two, because we can't travel from any face to any other face without switching layers. Nevertheless, all three of these multi-layered tetrahedral surfaces are intrinsically equivalent, not only to each other, but also to the single-layered torus.

Recall that in order to make a surface that corresponds to both a multi-layered plane and a multi-layered CUBE we needed the flat arrangement to have 3 layers while the cubical arrangement had 4 layers. This was necessary in order for a circuit around a vertex to be a multiple both of 2pi and of (3/2)pi. In contrast, to make a tetratorus we found that the flat surface was single-layered and the tetrahedron was 2-layered. In general it's not difficult to see what the requirements are for the various regular polyhedra. We have the basic attributes

	Vertices	***Edges***	***Faces***
tetrahedron	4	6	4
octahedron	6	12	8
cube	8	12	6
icosahedron	12	30	20
dodecahedron	20	30	12

and the Euler characteristic is V-(E-F). Also, the number of vertices doesn't change when we transform the surface from flat to curved, so the total number of vertices will be Vj if we have j layers in the flat interpretation. On the other hand, the numbers of edges and faces is proportional to the number of layers we have in the polyhedral interpretation. Letting k denote this number, we see that the intrinsic Euler characteristic is Vj -(E-F)k, and we need to make this vanish. The minimal solutions for each type of solid are

	j	***k***
tetrahedron	1	2
octahedron	2	3
cube	3	4
icosahedron	5	6
dodecahedron	9	10

For example, here's an image of "flat" vertex where pentagons meet.

It's easy to see that these are the numbers of layers necessary, in each case, to arrange around a single vertex so that it can be interpreted as a multiple of a flat surface AND as a multiple of the vertex of the respective polyhedron.

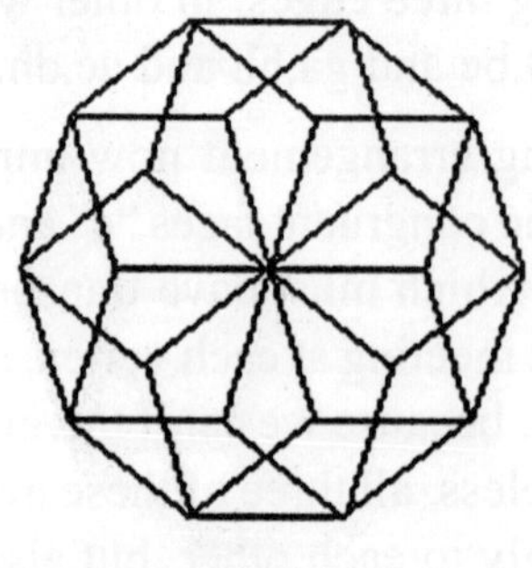

It would be interesting to know if complete closed surfaces equivalent to multi-layered tori can be constructed in cases beyond the tetrahedron. In each case it would be a matter of identifying the arrangement(s) of permutations of k elements that would yield the appropriate layer changes for each edge, in such a way that the entire closed surface "works" as both a flat surface and (modulo the polyhedron) a curved surface. This involves considerations that are closely related to those discussed in the note on Permutation Loops.

DIFFERENTIAL OPERATORS AND THE DIVERGENCE THEOREM

One of the most important and useful mathematical constructs is the "del operator", usually denoted by the symbol ∇ (which is called the "nabla"). This can be regarded as a vector whose components in the three principle directions of a Cartesian coordinate system are partial differentiations with respect to those three directions. Of course, the partial differentiations by themselves have no definite magnitude until we apply them to some function of the coordinates. Letting i, j, k denote the basis vectors in the x,y,z directions, the del operator can be expressed as

$$\nabla = i\frac{\partial}{\partial x} + j\frac{\partial}{\partial y} + k\frac{\partial}{\partial z}.$$

All the main operations of vector calculus, namely, the divergence, the gradient, the curl, and the Laplacian can be constructed from this single operator. The entities on which we operate may be either scalar fields or vector fields. A scalar field is just a single-valued function of the coordinates x,y,z. For example, the static pressure of air in a certain region could be expressed as a scalar field p(x,y,z), because there is just a single value of static pressure p at each point. On the other hand, a vector field assigns a vector v to each point in space. An example of this would be the velocity v(x,y,z) of the air throughout a certain region.

If we simply multiply a scalar field such as p(x,y,z) by the del operator, the result is a vector field, and the components of the vector at each point are just the partial derivatives of the scalar field at that point, i.e.,

$$\nabla_p = i\frac{\partial p}{\partial x} + j\frac{\partial p}{\partial y} + k\frac{\partial p}{\partial z}.$$

This is called the gradient of p. On the other hand, if we multiply a vector field v(x,y,z) by the del operator we first need to decide what kind of "multiplication" we want to use, because there are two different kinds of vector multiplication, commonly called the dot product and the cross product. For two arbitrary vectors $a = a_x i + a_y j + a_z k$ and $b = b_z i + b_y j + b_z k$ the dot product a·b and the cross product a×b are defined as

$$z.b = a_x b_x + a_y b_y + a_z b_z$$

$$z \times b = \left(a_y b_z - a_z b_y\right)i + \left(a_z b_x - a_x b_z\right)j + \left(a_x b_y - a_y b_x\right)k.$$

Intuitively, the dot product is a scalar equal to the product of the magnitudes of a and b times the cosine of the angle between them, and the cross product is a vector perpendicular to both a and b (with direction determined conventionally by the "right hand rule") and whose magnitude is equal to the product of the magnitudes of a and b times the sine of the angle between them. Since the dot product yields a scalar, it is often called the "scalar product". Likewise the cross product is often called the "vector product".

The dot product of ∇ and a vector field $v(x,y,z) = v_x(x,y,z)i + v_y(x,y,z)j + v_z(x,y,z)k$ gives a scalar, known as the divergence of v, for each point in space:

$$\nabla.v = \frac{\partial v_x}{\partial x} + \frac{\partial v_y}{\partial y} + \frac{\partial v_z}{\partial z}.$$

The cross product of ∇ and a vector field v(x,y,z) gives a vector, known as the curl of v, for each point in space:

$$\nabla \times v = \left(\frac{\partial v_z}{\partial y} - \frac{\partial v_y}{\partial z}\right)i + \left(\frac{\partial v_x}{\partial z} - \frac{\partial v_z}{\partial x}\right)j + \left(\frac{\partial v_y}{\partial x} - \frac{\partial v_x}{\partial y}\right)k.$$

Notice that the gradient of a scalar field is a vector field, the divergence of a vector field is a scalar field, and the curl of a vector field is a vector field.

Can we construct from the del operator a natural differential operator that creates a scalar field from a scalar field? Actually we already have the ingredients for such an operator, because if we apply the gradient operator to a scalar field to give a vector field, and then apply the divergence operator to this result, we get a scalar field. This is sometimes called the "div grad" of a scalar field, and is given by

$$\nabla.\nabla = \frac{\partial^2}{\partial y^2} + \frac{\partial^2}{\partial y^2} + \frac{\partial^2}{\partial z^2}.$$

For convenience we usually denote this operator by the symbol ∇^2, and it is usually called the Laplacian operator, because Laplace studied physical applications of scalar fields $\varphi(x,y,z)$ (such as the potential of an inverse-square force law) that satisfy the equation $\nabla^2\varphi = 0$, i.e.,

$$\nabla^2\varphi = \frac{\partial^2\varphi}{\partial x^2} + \frac{\partial^2\varphi}{\partial y^2} + \frac{\partial^2\varphi}{\partial z^2} = 0.$$

This formula arises not only in the context of potential fields (such as the electrostatic potential in an electric field, and the velocity potential in a frictionless ideal fluid), it also appears in Poisson's equation $\nabla^2\varphi = -4\pi\rho$ and in the fundamental wave equation

$$\nabla^2\varphi = \frac{1}{c^2}\frac{\partial^2\varphi}{\partial t^2}$$

where c is the speed of propagation of the wave. Incidentally, this suggests another useful differential operator, formed by bringing the right hand term over to the left side to give

$$= \frac{\partial^2}{\partial x^2} + \frac{\partial^2}{\partial y^2} + \frac{\partial^2}{\partial z^2} - \frac{1}{c^2}\frac{\partial^2}{\partial t^2}.$$

This operator is known as the d'Alembertian (because the wave equation was first studied by Jean d'Alembert), and it could be regarded as the dot product of two "dal" operators

$$i\frac{\partial}{\partial x} + j\frac{\partial}{\partial y} + k\frac{\partial}{\partial z} + l\frac{1}{ic}\frac{\partial}{\partial t}$$

where i,j,k,l are basis vectors of a four-dimensional Cartesian coordinate system.

One of the most important theorems in vector analysis is known as the Divergence Theorem, which is also sometimes called Gauss' Theorem. This is essentially just an application of the fundamental theorem of calculus

$$\int_a^b \frac{df(x)}{dx}dx = f(b) - f(a).$$

This enables us to express the integral of the quantity df/dx along an interval in terms of the values of f itself at the endpoints of that interval. Now suppose we are given a scalar function f(x,y,z) throughout a region V enclosed by a surface S, and we want to evaluate the integral of the quantity ∂f/∂x over this entire region. This can be written as

$$\int_a^b \frac{\partial f}{dx}dV = \int_z\int_y\int_z \frac{\partial f}{\partial x}dxdydz$$

where the three integrals are evaluated over suitable ranges to cover the entire region V. For any fixed values of $y = y_0$ and $z = z_0$ the function $f(x,y_0,z_0)$ is totally dependent on x, so we can evaluate the integral along any line parallel to the x axis through the region V for any particular y_0,z_0 using the fundamental theorem of calculus

$$\int_a^b \frac{df(x,y_0,z_0)}{dx}dx = f(b,y_0,z_0) - f(a,y_0,z_0)$$

where a,b are the x values at which the line intersects the surface S. (For ease of description we are assuming every line parallel to the x axis intersects the surface in only two points, but this restriction turns out to be unnecessary.) We now just need to integrate the above quantity over suitable ranges of y_0 and z_0 to give the integral throughout the entire region V. Notice that the two points (a,y_0,x_0) and (b,y_0,x_0) are both, by definition, on the surface S, at opposite ends of a line interval parallel to the x axis. Hence, we can cover all the contributions by integrating the value of the function f(x,y,z) over the entire surface S.

Of course, we recognize that $f(a,y_0,z_0)$ is subtracted from the total, whereas $f(b,y_0,z_0)$ is added, so it's clear that we need to apply a weighting factor to f(x,y,z) at each point when we integrate.

The value for points on the "low" end of the intervals must be subtracted, and the value for points on the "high" end of the intervals must be added. Furthermore, if we want to evaluate the integration over the surface area of S by integrating over dS, we need to scale the magnitude of the contribution of each point to give the appropriate weight to each incremental region dS of the surface, because a portion of the surface that is very oblique to the yz plane doesn't "contribute as many line segments" parallel to the x axis as does an equal increment of the surface that is parallel to the yz plane.

Clearly the necessary scale factor at each point is the cosine of the angle between the positive x axis and the normal to the surface at that point. Notice that this automatically gives us the appropriate sign for each contribution as well, because (for a convex surface) the normal to the surface will have a positive x component on one end of the interval and a negative x component on the other, which means the cosines of the respective angles will have opposite signs. (Again, for ease of description we are assuming a convex surface, but it's not hard to show that this restriction is unnecessary, since a given line parallel to the x axis can have multiple segments which can be treated separately.)

Consequently, we can write our original integral as

$$\int_V \frac{\partial f}{\partial x}dV = \int_S f\cos(n,i)dS$$

where cos(n,i) denotes the cosine of the angle between a unit vector normal to the surface and a basis vector in the positive x direction. Naturally, since the coordinate axes are symmetrical, we could arrive at analogous results with x replaced by either y or z, and i replaced by j or k respectively.

Now suppose we are given a vector field F(x,y,z) with the components $F_x(x,y,z)$, $F_y(x,y,z)$, and $F_z(x,y,z)$. Using the above results and simple additivity we have

$$\int_V \left(\frac{\partial F_x}{\partial x} + \frac{\partial F_y}{\partial y} + \frac{\partial F_z}{\partial z} \right) dV = \int_S \left[F_x \cos(n,i) + F_y \cos(n,j) + F_z \cos(n,k) \right] dS.$$

The integrand on the left side is ∇·F, i.e. the divergence of F. Also, notice that cos(n,i), cos(n,j), and cos(n,k) are the components of the normal unit vector n, so the integrand on the right side is simply F·n, i.e., the dot product of F and the unit normal to the surface. Hence we can express the Divergence Theorem in its familiar form

$$\int_V \nabla.F = \int_S F.n \, dS.$$

Several interesting facts can be deduced from this theorem. For example, if we define F as the gradient of the scalar field φ(x,y,z) we can substitute ∇φ for F in the above formula to give

$$\int_V \nabla^2 \varphi \, dV = \int_S \nabla \varphi . n \, dS.$$

The integrand of the volume integral on the left is the Laplacian of φ, so if φ is harmonic (i.e., a solution of Laplace's equation) the left side vanishes. The integrand of the right hand integral is the normal "flux" through the surface, so we see that the integral of the normal flux over any closed surface (in a region that everywhere satisfies Laplace's equation) is zero. It follows that there can be no local maximum or minimum inside a region where Laplace's equation is satisfied, because such a point would, by definition, be completely enclosed by a surface of everywhere positive (or everywhere negative) normal flux, making it impossible for the integral of the flux over the surface to vanish.

This, in turn, implies that if φ is constant over an entire closed surface (where Laplace's equations is satisfied) then the value of φ is constant throughout the enclosed volume, because otherwise the volume would have to contain a local maximum or minimum. Combining these facts with the additivity of harmonic functions, we can conclude that there is a unique harmonic function within an enclosed region that has specified values on the enclosing surface, because if φ_1 and φ_2 are two functions satisfying those boundary conditions, the harmonic field $\varphi_1 - \varphi_2$ is zero over the entire boundary, and therefore it vanishes throughout the interior as well.

If we allow non-zero charge or mass density (or sources or sinks) in the enclosed region, such that Poisson's equation $\nabla^2\varphi = -4\pi\rho$ is satisfied, then the left hand integral is the net charge contained within the volume, and this equals the integral of the normal flux over the enclosing surface. This is often called Gauss' law of electrostatics, and it constitutes one of Maxwell's equations.

Another important consequence of the Divergence Theorem can be seen by noting that the scalar quantity $\nabla\varphi\cdot n$ at any given point on the surface equals the partial derivative $\partial\varphi/\partial n$ where n is the displacement parameter in the direction normal to the surface. Furthermore, if the surface in question is a sphere -which we can assume without loss of generality is centered at the origin -then the normal displacement parameter n is equal to the radial parameter r. So for such a spherical surface we have

$$\int_S \frac{\partial\varphi}{\partial n}\, dS = \int_S \frac{\partial\varphi}{\partial r}\, dS = 0.$$

Recall that, in terms of spherical coordinates r, θ, and ϕ, where θ is latitude (zero at the North Pole) and ϕ is longitude (zero on the positive x axis) the basic line element in space is

$$(ds)^2 = (dr)^2 + r^2(d\theta)^2 + r^2 \sin(\theta)^2 (d\phi)^2$$

so an incremental change dθ at constant r and ϕ corresponds to a change ds = r dθ, whereas an incremental change dϕ at constant r and θ gives ds = r sin(θ) dϕ. Hence the surface element in terms of θ and ϕ is $dS = r^2 \sin(\theta)\, d\theta\, d\phi$, so the preceding integral can be written as

$$r^2 \int_0^{2\pi}\int_0^{\pi} \frac{\partial[\varphi(r,\theta,\phi)\sin(\theta)]}{\partial r}\, d\theta d\phi = 0.$$

Applying Leibniz's Rule for the derivative of an integral, and multiplying by 4π/4π, this becomes

$$4\pi r^2 \frac{d}{dr}\left[\frac{1}{4\pi}\int_0^{2\pi}\int_0^{\pi} \varphi(r,\theta,\phi)\sin(\theta)\, d\theta d\phi\right] = 0.$$

The quantity in the square brackets is just the mean value of φ(r,θ,ϕ) on the surface of the sphere, and this equation shows that the derivative of the mean value with respect to the radius r vanishes, so the mean value of φ on the surface of a sphere centered at any fixed point is independent of the radius. Considering the limit as r approaches zero, it's clear that the mean value of a harmonic function φ on the surface of a sphere (of any radius) is equal to the value of φ at the center of the sphere.

EIGEN VALUES

Many problems in mathematics and physics reduce "eigenvalue problems", and people often expend lots of effort trying to determine the eigenvalues of a general NxN system, and this is sometimes reduced to finding the roots of the Nth degree characteristic polynomial. The task of efficiently finding all the complex roots of a high-degree polynomial is non-trivial, and is complicated by the need to consider various special cases such as repeated roots. However, depending on what we're really trying to accomplish, we may not need to find the eigenvalues at all. Of course, the description of an action is usually simplest when its components are resolved along the directions of the eigenvectors, e.g., the eigenvectors of a rotation matrix define its axis of rotation. As a result, it's easy to write equations for the positions of points on a rotating sphere if we work in coordinates that are aligned with the eigenvectors of the rotation, whereas if we work in some other arbitrary coordinates the equations describing the motion will be more complicated.

However, there are many situations in which the characteristic axes of the phenomenon can be exploited without ever explicitly determining the eigenvalues. For example, suppose we have the system of three continuous variables x(t), y(t), and z(t) that satisfy the equations

$$ax + by + cz = x'$$

$$dx + ey + fz = y'$$

$$gx + hy + iz = z'$$

and we want to determine the exact root-matched recurrence formulas for x(nT), y(nT), and z(nT) for a given discrete time interval T. The characteristic polynomial is

$$r^3 - Ar^2 + Br - C = 0$$

where

$A = a+e+i$

$B = ae-bd+ai-cg+ei-fh$

$C = aei+bfg+cdh-ceg-bdi-afh$

The root-matched recurrence coefficients are proportional to the elementary symmetric functions (with alternating signs) of the quantities $\exp(r_k T)$, where r_k, k=1,2,3 are the eigenvalues, so a common approach is to solve the characteristic equation for the eigenvalues and then compute the recurrence coefficients. Even for a simple cubic the precise determination of the (complex) roots in embedded code, covering all possible cases, can be tricky and time-consuming. For higher order systems it's even worse.

However, we don't really need to solve for the eigenvalues. Let w(k) denote the sum of the kth powers of the roots r_1, r_2, r_3. Clearly w(0)=3, and the values of w(k) for k=1,2,... are given by

$$w(1) + A = 0$$

$$w(2) + w(1)\,A + 2\,B = 0$$

and

$$w(k) + A\ w(k\text{-}1) + B\ w(k\text{-}2) + C\ w(k\text{-}3) = 0$$

for all k > N-1. Now let E(k) denote the sum of the kth powers of the values $\exp(r_k\ T)$, k=1,2,3. Expanding each of the exponential functions as a Taylor series and combining terms by powers of T gives

$$E(k) = \underset{j=0}{\overset{\text{inf}}{\text{SUM}}} \frac{(kT)^3}{j!} w(j) \qquad k=1,2,3.$$

Letting s_k denote $(-1)^k$ times the kth elementary symmetric function of the quantities $\exp(r_j\ T)$, we can now compute s_0, s_1, s_2, and s_3 using the identities

$$s_0 = 1$$

$$E(1) + 1s_1 = 0$$

$$E(2) + E(1)s_1 + 2s_2 = 0$$

$$E(3) + E(2)s_1 + E(1)s_2 + 3s_3 = 0$$

Then the exact homogeneous recurrence formula for x(t) is

$$s_0\ x(nT) + s_1\ x((n-1)T) + s_2\ x((n-2)T) + s_3\ x((n-3)T) = 0$$

and the recurrences for y(t) and z(t) are identical. The summation for E(k) is quite well-behaved and can be implemented quite easily in embedded code. This method, which is directly applicable to systems of any order, requires no root-finding algorithm and no special consideration of repeated roots, etc.

RECONSTRUCTING BROUWER

Any continuous mapping of a compact convex set into itself contains at least one fixed point. In other words, given any continuous function f(x) of an n-cell into itself, there exists at least one point x_0 such that $f(x_0) = x_0$. This is known as Brouwer's fixed-point theorem, and it has been used to establish fundamental existence theorems in many different branches of mathematics (e.g., the theory of differential equations).

L. E. J. Brouwer (Luitzen Egbertus Jan, 1881-1966) was a Dutch mathematician and the founder of intuitionism, a philosophy which admits only things that can be constructed by finitary means. In particular, intuitionism does not accept the validity of the Law of the Excluded Middle, nor the Axiom of Choice. Ironically, Brouwer is best known (outside the philosophy of mathematics) for his fixed-point theorem, which he proved by means of conventional (non-constructive) analysis, i.e., techniques that he philosophically considered to be questionable if not downright invalid.

For a simple one-dimensional interval, the fixed-point theorem is fairly obvious, because any continuous function x' = f(x) of an interval, say from 0 to 1, into that same range must somewhere meet the line representing x' = x, as illustrated below.

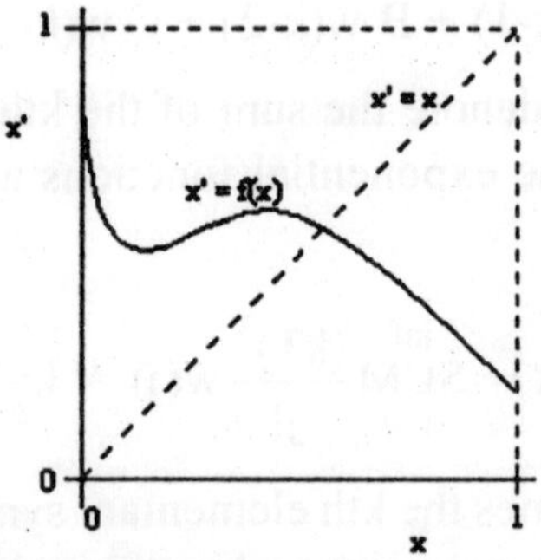

Notice that we can assert the function x' = f(x) crosses the line x' = x for an arbitrary continuous function f, even without explicitly constructing the point of intersection. In fact, if f(x) is a higher algebraic or transcendental function, it may be impossible to give a strict finite construction of the fixed point, but we can nevertheless assert its existence - according to conventional mathematical principles.

A two-dimensional n-cell can be represented as the Cartesian product of two unit intervals of this same type, which we can denote by x_1 and x_2, and the function can be split into two components $x_1' = f_1(x_1,x_2)$ and $x_2' = f_2(x_1,x_2)$. From the one-dimensional case we know that for any given value of x_2 there exists at least one value of x_1 such that $x_1 = f_1(x_1,x_2)$. If there was always just one such value of x_1 for any given x_2, the existence of an overall fixed point would follow immediately. To show how, let $F(x_2)$ denote this (putatively unique) fixed value of x_1 for any given x_2. Clearly F must be a continuous function of x_2, so the composite function $x_2' = f_2(F(x_2),x_2)$ is a continuous function of x_2, and hence we can invoke the one-dimensional case again to assert the existence of a value of x_2 such that $x_2 = f_2(F(x_2),x_2)$, which is a completely fixed point. Of course, we could also have proceeded in the opposite direction, defining a function $G(x_1)$ that gives the (putatively unique) fixed value of x_2 as a function of x_1. On a simple two-dimensional plot, the function $x_2 = G(x_1)$ must pass continuously from the left edge to the right end of the cell, and the function $x_1 = F(x_2)$ must pass continuously from the top edge to the bottom edge, and the argument is that they must intersect at some point, as illustrated below.

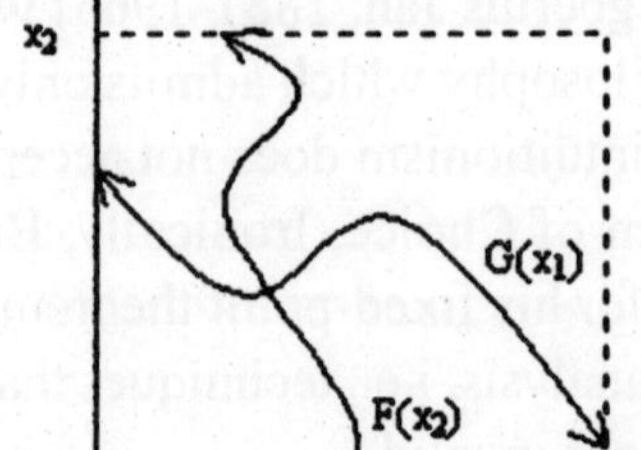

However, this argument is not adequate, because for any given value of x_2 there may be more than one value of x_1 such that $x_1 = f_1(x_1,x_2)$, and the number of such fixed values may change as we vary x_2, so we cannot automatically claim single continuous functions F and G passing between their respective edge pairs. Therefore, in order to prove that there is a point of intersection we must consider the multi-valued functions more carefully. In general, for any given value of x_2, the function $x_1' = f_1(x_1,x_2)$ can have multiple fixed points, as illustrated below.

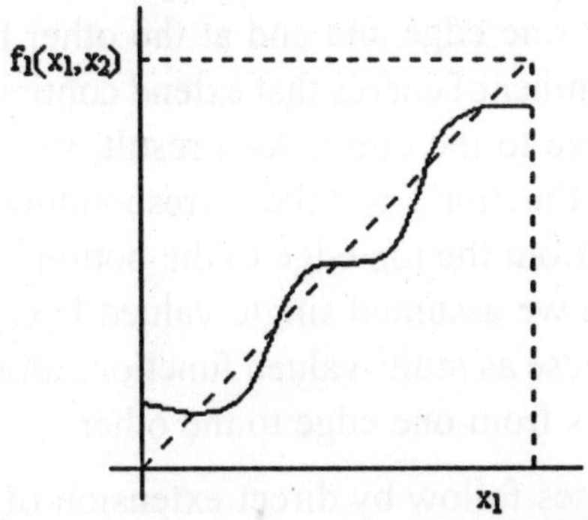

Furthermore, as x_2 is varied, this curve can shift and change shape in such a way that some of the fixed points "disappear" and others "appear" elsewhere. The four discrete corner-points can easily be treated separately, so we need consider only single-valued functions f_1 and f_2 that map to the interiors of the intervals. For such functions the number of times the function f crosses the diagonal is obviously odd, and the fixed points "appear" and "disappear" in pairs when two of them coincide.

Of course, each simple crossing represents one distinct fixed point, whereas any occurrence (including contiguous spans) of the function being tangent to the diagonal is counted as either zero or one crossing accordingly as the function departs from the diagonal on the same side or on opposite sides. These observations suffice to establish the topology of the locus of fixed points. A plot of the multi-valued function $F(x_2)$ of the x_1 fixed points as a function of x_2 for an illustrative case is shown below.

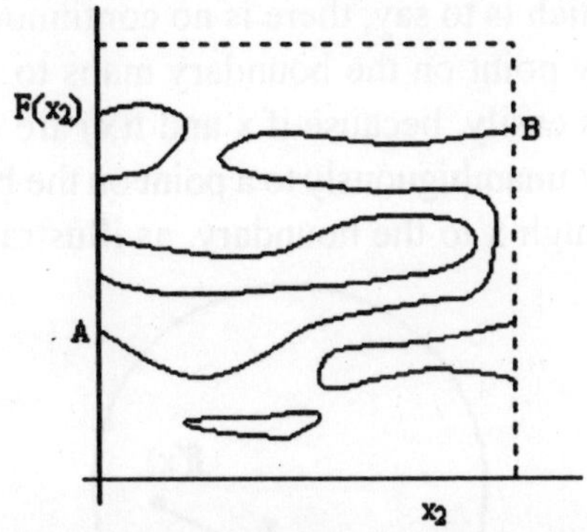

Notice that we cannot simply select the lowest fixed point, or the highest fixed point, or any other locally identifiable one of the fixed points at any given x_2, because those can change discontinuously, and therefore don't yield a continuous curve. However, notice also that at each value of x_2 there is an odd number of x_1-fixed points, and this implies that there *is* a continuous path extending all the way from the left edge to the right edge. (In the figure above, we have the path from "A" to "B".) This is because any closed loop, and any path that starts and ends on the same edge, has even parity, i.e., contributes an even number of crossings at each x_2.

Only paths that start at one edge and end at the other have odd parity, so it follows that there must be an odd number of curves that extend continuously (though not necessarily single-valued) from one edge to the other. As a result, we are assured of an intersection between this (multi-valued) function F and the corresponding function G (since G similarly includes a contiguous path from the top edge to the bottom edge). In effect, the argument is the same as before, when we assumed single-valued $F(x_2)$ and $G(x_1)$ functions, except that now we parameterize these as multi-valued functions along the continuous curve (such as from A to B) that extends from one edge to the other.

Higher dimensional cases follow by direct extension of this argument. For example, in three dimensions we have three continuous functions $x_1' = f_1(x_1,x_2,x_3)$, $x_2' = f_2(x_1,x_2,x_3)$, and $x_3' = f_3(x_1,x_2,x_3)$, and for any given x_2 and x_3 we can define $F(x_2,x_3)$ as the parameterized x1-fixed point curve, i.e., a continuous curve of x_1 values such that $x_1 = f_1(x_1,x_2,x_3)$. This allows us to "eliminate" x_1 and reduce the problem to the two-dimensional case given by $x_2' = f_2(F(x_2,x_3),x_2,x_3)$, and $x_3' = f_3(F(x_2,x_3),x_2,x_3)$. (Visually we can imagine the continuous multi-valued sheet of x_1 fixed points as a function of x_2 and x_3, and the sheet of x_2 fixed points as a function of x_1 and x_3, and the sheet of x_3 fixed points as a function of x_1 and x_2. Since there is at least one complete sheet in each direction that bifurcates the n-cell along that axis, these three sheets unavoidably have at least one point of intersection.) Continuing on in this way, the general result for all n follows.

The usual proof of Brouwer's fixed-point theorem makes use of some machinery from simplical homology theory. First we establish that there does not exist a "retraction" of an n-cell onto its boundary, which is to say, there is no continuous mapping from an n-cell to its boundary such that every point on the boundary maps to itself. Given this, Brouwer's fixed-point theorem follows easily, because if x and f(x) are everywhere distinct in the n-cell, we can map each point x unambiguously to a point on the boundary by simply projecting along the ray from f(x) through x to the boundary, as illustrated below for a disk.

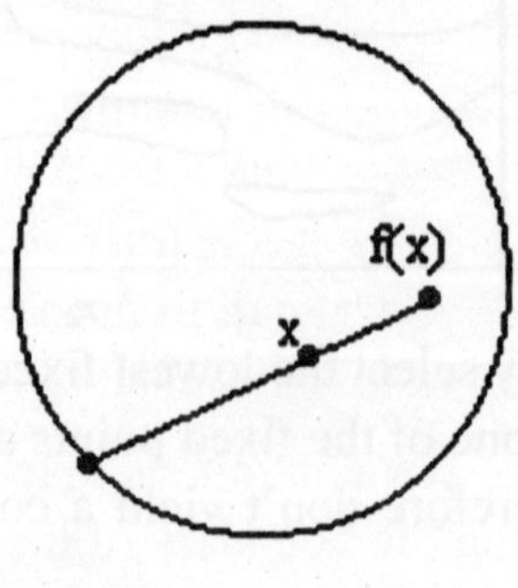

This mapping is continuous, and every point on the boundary maps to itself, so the mapping is a "retraction", which contradicts the fact that no retraction of an n-cell exists. Of course, we haven't actually shown this yet. The proof of the no-retraction theorem is not trivial. To prove it, first recall that a *homotopy* between two mappings f(x) and g(x) can be represented by a function H(x,t) that is continuous in both x and t, and such that H(x,0) = f(x) and H(x,1) = g(x). In other words, the parameter "t" serves to smoothly interpolate from f(x) to g(x). Now suppose there exists a retraction R(x) from the interior of a unit sphere (for example) to its boundary, and consider the parametric family of mappings given by H(p,t) = R((1-t)p) where "p" is a unit vector corresponding to an arbitrary point on the boundary of the unit sphere.

Thus H is a family of mappings from the surface of a unit sphere into itself, parameterized by t. With t = 1, we have H(p,1) = R(0), meaning that every point on the surface maps to a single point. On the other hand, with t = 0 we have H(p,0) = R(p), meaning that every point on the surface maps to itself. These facts enable us to rule out the existence of the retraction R(x), because a singular mapping f(p) = p_0 cannot be homotopic (i.e., cannot be related by a homotopy H) to the identity mapping g(p) = p. In other words, we cannot smoothly interpolate between a singular mapping and the identity mapping.

The last assertion may be fairly intuitive, but to establish it rigorously requires some effort. In general, Brouwer characterized mappings from one n-sphere to another according to the number of times the image of former wraps around the latter. He called this integer the *degree* of the mapping. For example, the identity mapping f(p) = p of points on the surface of a sphere has degree 1. On the other hand, we could define a mapping such that two points on the source sphere map to each single point on the destination sphere, and the degree of such a mapping would be 2. The degree of a singular mapping f(p) = p_0 is zero, because the image of the source sphere does not wrap around the destination sphere at all. Brouwer then proved that if two mappings from one n-sphere to another are homotopic, they must be of the same degree, which is the result needed to complete the classical topological proof of Brouwer's fixed-point theorem. Incidentally, H. Hopf subsequently proved the converse, i.e., if two mappings from one n-sphere to another have the same degree, they are homotopic.

Notice that Brouwer's topologic proof of the fixed-point theorem is non-constructive, just as was our earlier algebraic proof. It posits a mapping with no fixed point, and then deduces a contradiction, proving that no such mapping is possible, but it does not give a constructive procedure for actually determining a fixed point for any given mapping... or does it? Ron Maimon proposed the following constructive analog of Brouwer's fixed-point theorem: Given any uniformly continuous map from the disk to the disk, and any $\varepsilon > 0$, there exists a point which is not moved by more than ε from it's original position.

As a proof of this assertion for any continuous map f(x), he asks us to consider the value of the vector field x–f(x) along the boundary of the disk. He asserts that the vector must point inward at every point. (We should perhaps note that it's possible for a point on the boundary to map to a point on the boundary, so the vector need not be strictly inward.) Therefore the image of boundary is another closed curve in the interior of the disk, as illustrated below.

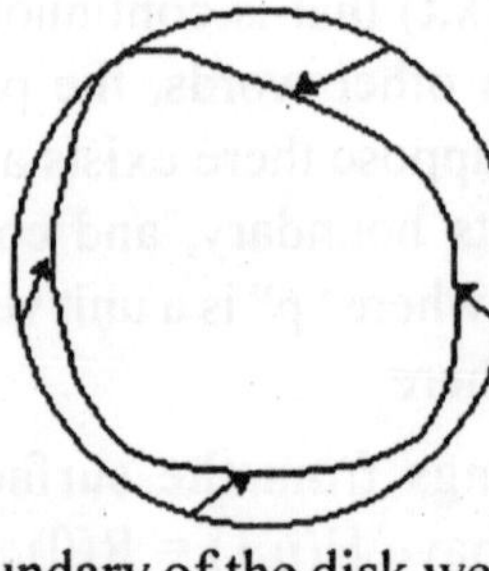

As we proceed around the boundary of the disk we can keep track of how the direction of the vector field changes, like the hand of a clock, and count how many times this hand rotates as we complete one trip around the boundary. Call this integer the *winding number*. To make this procedure finitistic and constructive, Maimon proposes to split up the boundary into segments of size δ, small enough so that the vector field is constant to a precision of ε over each segment.

(This actually requires a limitation on the scale of variations, since otherwise we could never be sure of having identified a sufficiently small scale for the evaluations.) The winding number ought to be -1 (i.e., the vectors rotate once around in the clockwise direction.) He argues that the only way we can fail to determine the winding number unambiguously is if there is a point where the magnitude of the vector field is so close to zero (i.e., within ε of zero) that we cannot establish its direction, but in that case we have found a point that is moved less than ε under the mapping, so we are done.

Assuming we have found an unambiguous non-zero winding number around the loop, we then cut the enclosed region in half, and evaluate the winding numbers around the loops enclosing these two halves.

If we assume the mapping is single-valued (which precludes non-simple loops in the image), the winding number around each region must be either 0 or -1, and since the sum of the winding numbers for the partitions equals the winding number around the total region, we must have one half with a winding number of 0 and the other with a winding number of -1. Taking the latter, we can split it in half, and repeat the process. Eventually we must *either* reach a stage where we are unable to evaluate the winding number because of the presence of an ε-fixed-point, or else we must converge on a region smaller than ε with winding number -1, which Maimon argues must contain an ε-fixed-point.

This procedure is clarified by considering the 1-dimensional case of the Brouwer theorem, i.e., the intermediate value theorem. We bisect the interval and check to see which half still satisfies the original conditions of the problem. Eventually we reach a stage where we cannot decide up to our limit of resolution, and we declare that the midpoint of that interval is a fixed point. So we are done. But does this constitute a valid constructive argument? According to the constructivist point of view, arguments of the "either/or" are not automatically accepted, because one of the basic tenets of constructivism is the rejection of the free use of the "law of the excluded middle". A well-known example is the "proof" that there exist irrational numbers x and y such that x^y is rational. We start with the algebraic identity

$$\left[\sqrt{2}^{\sqrt{2}}\right]^{\sqrt{2}} = 2.$$

We know that the square root of 2 is irrational. Also, either the quantity in square brackets is rational or it is not rational. If it's not rational, then the above identity proves the result. On the other hand, if it *is* rational, then the fact that it is rational proves the result. Thus, one way or the other, there exist two irrational numbers x,y such that x^y is rational. However, we have not provided a finite procedure for determining which of the options actually leads to the x,y in question, so we haven't really constructed an example. We have only "proved" that such an example must exist, so this is not regarded as a valid constructive proof.

What does this imply for Maimon's version of Brower's fixed-point proof? His algorithm proves that any of the possible outcomes of the "algorithm" must lead to an ε-fixed point, but it doesn't tell us which of these outcomes must occur. On the other hand, Maimon's algorithm is explicitly finite, so it can be said to conform to the constructivist requirements. This is possible because Maimon is not actually proving Brouwer's fixed point theorem, he is proving the ε-fixed-point theorem. He argues that this is the only meaningful form of the theorem, because "there is no constructive definition of irrational numbers anyway, except for a small number of special cases". This is important, because the actual fixed point (in the classical sense) may have irrational coordinates, which could obviously never be determined in a finite number of steps.

Clearly, the most this algorithm can do (in a finite number of steps) is to narrow the range of possibilities down to what remains an infinite set. It seems that constructivism must assume the universe of concepts has an absolute scale, relative to which the concept of "sufficiently small" can be defined. Lacking that, all finite regions are in a sense equivalent, so knowing a fixed point is within a certain region is no more constructive than knowing that either

$$\left[\sqrt{2},\sqrt{2}\right] \quad \text{or} \quad \left[\sqrt{2}^{\sqrt{2}},\sqrt{2}\right]$$

is a pair of irrational numbers [x,y] such that x^y is rational. On the xy plane this "construction" has narrowed down the location of [x,y] to a set of measure 0 (actually just two points). Why isn't this as good as saying we have "constructed" a fixed point to accuracy epsilon? It can only be the absolute distance between these two points that causes us to be dis-satisfied with the reduction.

Also, it's important to recognize that Maimon's guarantee of an ε-fixed point in a finite number of steps was proven classically, not intuitionistically, since it relies on the truth of statements of the form A or B without proving either A or B. In addition, the procedure gives us no assurance that we have defined a sufficiently small scale δ to resolve the variations in the mapping. In other words, we are presumably required to adopt some finitistic definition of continuity, with an absolute scale. This is indeed what Maimon was proposing, based on ideas about quantum effects in physics, but whether this is a valid foundation for a mathematical argument is debatable.

Perhaps most importantly, we should ask if it is legitimate to simply dismiss any consideration of the true fixed point by declaring that infinite precision is meaningless. Suppose we perform the algorithm to an accuracy of ε, and we verify a fixed point at that accuracy. What does that tell us about the result of the algorithm if carried out with accuracy of $\varepsilon/10^6$? Are the two results necessarily close, or might we get a completely different point if our ε was different?

It's not difficult to give examples, even in one dimension, of functions with very low slopes, such that many point are *almost* fixed points (i.e., they are ε-fixed points) without actually being fixed points. I've heard that Brouwer himself, when he became a constructivist, stopped believing his theorem; is this why?

Incidentally, there exist well-known "constructive" proofs of Brouwer's theorem, based on Sperner's coloring lemma. In one dimension, this lemma says that if we have a line segment subdivided into smaller segments, and if we color the left-most vertex red, the right-most vertex green, and use red or green for each interior vertex, then there is an odd number of segments with one red and one green vertex.

In two dimensions, we begin with a large triangle, and we sub-divide it into smaller triangles such that no vertex is in the middle of an edge of some other triangle except for the edges of the original large triangle. Now color the three outer corner vertices red, blue, and green respectively.

Then color the vertices along the edges of the original large triangle, using on each edge only the colors of the two end vertices of that edge. Then color the interior vertices arbitrarily using the three colors. Sperner's lemma tells us that there is an odd number of triangles with red-blue-green vertices. To prove this, construct the "dual" graph, with a dot inside each sub-triangle, one outside each outer edge, and connect two adjacent dots if the edge between them has two distinct colors. Degree counting will show there is an odd number of degree-3 vertices on the inside. This can be extended by induction to higher dimensional cases.

Chapter 6

Rolling Spheres and Cones

Let us consider a sphere of unit radius rolling (at a constant rate, without slipping or spinning) clockwise around a circle of radius R on a flat plane. For what values of R will the sphere periodically be returned to its initial configuration (position and orientation)? We can show that the sphere's configuration will be periodic if and only if R = sqrt(q^2–1) where q is a rational number greater than 1. Also, q is the ratio of the plane circle's circumference to the circumference of the circular locus of contact points on the sphere. Thus, if q=m/n the sphere will resume its initial configuration every n times around the circle.

As an example, setting q = 13/7, we find that a unit sphere rolling around a plane circle of radius 2sqrt(30)/7 = 1.56492... will return to its initial configuration every 7 times around the circle. Obviously the smallest circle that has a period of 1 (i.e., for which the sphere's configuration repeats with each trip around the circle) has a radius of sqrt(3) = 1.73205..., whereas the smallest circle with a period of 2 has a radius of sqrt(5)/2 = 1.11803... In general, the smallest circle with period n has radius sqrt(2n+1)/n.

The equation R = sqrt(q^2–1) can be derived very simply if we recognize that the pure rolling motion of a sphere in a circular path on the plane is identical to the motion of a sphere inscribed in a cone rolling on the plane, as illustrated below.

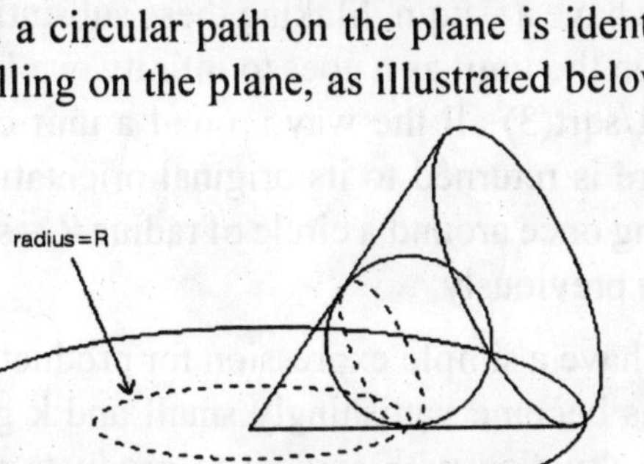

The circular locus of contact points is just the points of tangency between the sphere and the cone, and the relations between R and the circumference of that contact circle can be inferred directly from the diagram. Incidentally, this shows that for a "right" cone-which means the sphere and the plane circle have the same radius-the ratio of contact circumferences is q = sqrt($1+1^2$) = sqrt(2), so the sphere will never return precisely to its initial configuration.

Now consider an arbitrary curve drawn on a flat surface, and a sphere rolling (without slipping or spinning) along this curve. We can parameterize the curve by giving the tangent angle "a" (relative to an arbitrarily chosen axis) as a function of the distance along the curve. If the sphere rolls along a straight line with angle "a" for a finite distance L, the change in the orientation of the sphere is given by multiplying the initial orientation by the rotation matrix

$$M = \begin{bmatrix} ccC + ss & cs(1-C) & -cS \\ cs(1-C) & ssC + cc & sS \\ cS & -sS & C \end{bmatrix}$$

where c=cos(a), s=sin(a), C=cos(L/r), and S=sin(L/r) and r is the radius of the sphere.

The net change in orientation after the sphere has rolled along k linear segments, each of length L with the path angles a1, a2,...ak, is given by the matrix product

$$M(net) = M(a1)\, M(a2) \, ... \, M(ak)$$

As L approaches the infintesimal dL we see that S goes to dL/r and C differs from 1 by only a second-order term in (dL/r), so the infintesimal rotation matrix is

$$m(a) = \begin{bmatrix} 1 & 0 & -\cos(a)(dL/r) \\ 0 & 1 & \sin(a)(dL/r) \\ \cos(a)(dL/r) & -\sin(a)(dL/r) & 1 \end{bmatrix}.$$

For a unit circle path (i.e., R=1) we have L = a, and therefore dL=da. If we take an arc of angular size q on this path, and divide it into n equal segments, we have da = q/n, and for the kth segment we have a = kq/n. Making these substutions and evaluating the product m(kq/n) for k = 1 to n in the limit as n goes to infinity, we find that the net effect of rolling a sphere of radius r=1/sqrt(3) all the way around a unit circle is the identity matrix. In other words, the sphere is returned to its original orientation, in agreement with the fact that a unit sphere rolling once around a circle of radius R = sqrt(3) is returned to its original orientation, as we saw previously.

We would like to have a simple expression for products of matrices in the limit as the individual path lengths become vanishingly small and k goes to infinity. To do this we might recall a similar situation with respect to products of real numbers. For example, consider how we would determine the geometric mean of the real numbers between 0 and 1. The geometric mean G of a finite set of values v1, v2,..,vk is simply the quantity G = $(v1*v2*...*vk)^{(1/k)}$, so the natural logarithm of G is

$$\ln(G) = \frac{\ln(v1) + \ln(v2) + \ldots + \ln(vk)}{k}.$$

This makes it clear that the logarithm of the geometric mean of a continuous function v(x) from x = a to b is

$$\ln(G) = \frac{1}{b-a} \int_{x=a}^{b} \ln(v(x))\, dx.$$

For the simple case v(x) = x with x between 0 and 1 this gives ln(G) = -1, and so the geometric mean of the real numbers from 0 to 1 is G = 1/e.

Now, by analogy, it might be tempting to suppose that the logarithm of M(net) in is just the sum of the logarithms of the matricies M(aj), j=1,2,..,k. If this were true, we could proceed to the continuous case, letting m(a) denote the infintessimal rotation matrix for an incremental movement along the path, and we would have

$$\ln(M(net)) = \int_{L_0}^{L_1} \ln[m(a(L))]\, dL$$

which would imply that the net re-orientation of the sphere after rolling along the curve from L0 to L1 is given by the exponential of this ln(M(net)). However, this assumes the logarithmic property log(AB) = log(A)+log(B) applies to matrices, which is not generally true, because the logarithmic property does not apply to non-commuting entities-such as general 3D rotation matricies. To see this, notice that exp(A)exp(B) does not generally equal exp(A+B), because this would imply exp(A)exp(B) = exp(B)exp(A).

In view of this, it seems that some sort of non-commuting "addition" would be needed in order to recapture the familiar logarithmic property for arbitrary matrices. It would then be necessary to define the analagous "ordered" integration, so the integral along a path between any two points would depend on the direction in which the path was traversed. Furthermore, the net effect of rolling once around a given closed path will generally depend on the starting point, because if we consider two points P1 and P2 on a closed loop, and let A denote the rotation matrix given by rolling the sphere (clockwise) along the curve from P1 to P2, and B the rotation matrix given by rolling on from P2 around the rest of the loop to P1, then the net re-orientation going clockwise around the loop from P1 to P1 is AB, whereas the net re-orientation going clockwise around the loop from P2 to P2 is BA, which need not equal AB for general paths.

Nevertheless, the above formulas are applicable to rolling along a restricted class of self-similar curves (such as straight lines), so it's interesting to develop the formulas for rolling along an arbitrary path based on the premise (which admittedly has only limited validity) that the logarithmic property applies to the rotation matrices. We could also consider the cases where rolling along a given path length yields the same re-orientation of the sphere, and the cases where (for closed loops) the net re-orientation around the loop is independent of the starting point. For example, this condition is true for perfectly circular paths on the plane.

Notice that the infintesimal rotation matrix (3) can be split into two parts

$$m = \begin{bmatrix} 1 & 0 & 0 \\ 0 & 1 & 0 \\ 0 & 0 & 1 \end{bmatrix} + \frac{1}{r}\begin{bmatrix} 0 & 0 & -\cos(a) \\ 0 & 0 & \sin(a) \\ \cos(a) & -\sin(a) & 0 \end{bmatrix} dL.$$

This fits nicely with the power series expansion for the logarithm

$$x^2 \; x^3 \ln(a+x) = x - \frac{x^2}{2} + \frac{x^2}{3} - \text{etc.}$$

Letting "I" denote the identity matrix, and u(a) denote the right hand matrix, we can write

$$\ln(m) = \ln - \left\langle I + \frac{u(a)}{r} dL \right\rangle - \frac{u(a)}{r} dL - \frac{u(a)^2}{2r^2}(dL) + ...$$

so all terms after the first are second or higher order in dL/r. Thus, in the limit as dL -> 0 we have simply ln(m) = u(a)/r dL. As a result, if L is the path-length parameter of a plane curve that makes an angle a(L) relative to some arbitrary axis, then (under the assumption of the logarithmic property) the net rotation of a sphere rolling along that curve from L0 to L1 is given by the matrix

$$\exp\left[\int_{L_0}^{L_1} \frac{u(a)(L))}{r} dL\right].$$

If a(L) is constant we recover equation for finite rolling along a straight line. Note that these rotations are indeed commutative, so the formula is applicable. On the other hand, if we take a(L)=L (yielding a circular path of radius R=1) and evaluate for this case (with L0 = 0 and L1 = q), we arrive at the net rotation

$$\exp\begin{bmatrix} 0 & 0 & \frac{-\sin(q)}{r} \\ 0 & 0 & \frac{1-\cos(q)}{r} \\ \frac{\sin(q)}{r} & \frac{1+\cos(q)}{r} & 0 \end{bmatrix}$$

Unfortunately, due to the inapplicability of the logarithmic property for matrices, this does not correspond to the actual rotation of a sphere rolling around a circular path, as can be seen from the fact that it gives the identity rotation whenever q = 2pi, regardless of the value of r, which (as we saw previously) is not correct.

RANDOM TILING OF A SPHERE

Suppose we place an equilateral triangle with edge length 1 on the surface of a sphere of radius R >> 1 (say, 100). Now randomly select one of the three edges of this triangle and construct an identical triangle on that edge (also on the surface of the sphere). Now we have four exposed edges, each of which is eligible for the construction of another equilateral triangle. We choose one of these edges at random and construct another triangle. This leaves us with 5 exposed edges, all eligible, from which we randomly select one and construct another equilateral triangle, and so on.

Before long there will be some "exposed edges" on which it's not possible to construct an equilateral triangle (without overlapping an existing triangle), so we say those edges are not eligible, and we exclude them from our random selection list. At the beginning of this process our list of eligible edges will obviously increase by 1 at each step, but then it will occassionally not increase on some steps, and eventually it will start to decrease as we continue to adjoin more triangles. Ultimately we will have no more eligible edges.

What fraction of the sphere's surface area would we expect to cover in this way? What is the asymptotic behavior of this fraction as R increases? (Since this process would completely tile the plane, it might seem that the coverage ratio limit as R approaches infinity must be 1, but perhaps there's a tendancy for each local region to "waste" a certain amount of area due to certain patterns of inter-ference that are repeated throughout the surface for ANY finite R, even though all the interference vanishes when the surface is perfectly flat.) How many ineligible exposed edges would be present at the end? If, instead of randomly deciding where to add the next triangle, we are allowed to plan our moves, is there a simple algorithmic strategy that maximizes the covered area and/ or minimizes the final number of ineligible exposed edges?

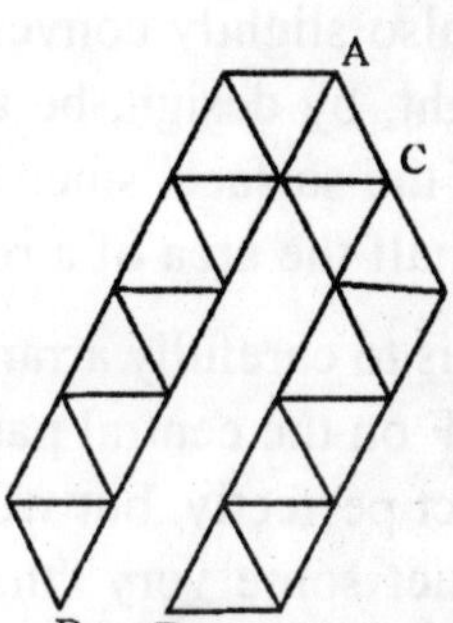

Some people have suggested that, since it's possible to fit 5 but not 6 equilateral triangles to a common vertex, the coverage must be around 5/6, regardless of whether the pattern was random or not. However, I think the coverage may actually be less than that, and it can depend on the sequence in which the tiles are placed. We could ask the opposite question, namely, what is the LEAST number of tiles that leave no eligible edges remaining. Suppose we have a configuration like this on the surface of a very large sphere:

The lines AB and CD are nearly geodesics and essentially parallel as they cross the segment AC. However, they don't remain parallel (as they would on a perfectly flat plane). Instead they are slowly converging, I think, based on the fact that the centerlines of those two rows of tiles must be geodesics, and by symmetry the cross-wise row of tiles along AC must be the point of maximum separation between those geodesics.

As a result, not only is it impossible to fill in the 6th edge of the upper "hexagon", it's also impossible to place a triangle anywhere between the lines AB and CD anywhere below C. If we continue this pattern, the lines AB and CD will eventually intersect, but it will take a very long time if the curvature is very slight.

Also, we could repeat this pattern, creating a series of "stripes" alternating between tiled and untilable regions, like this:

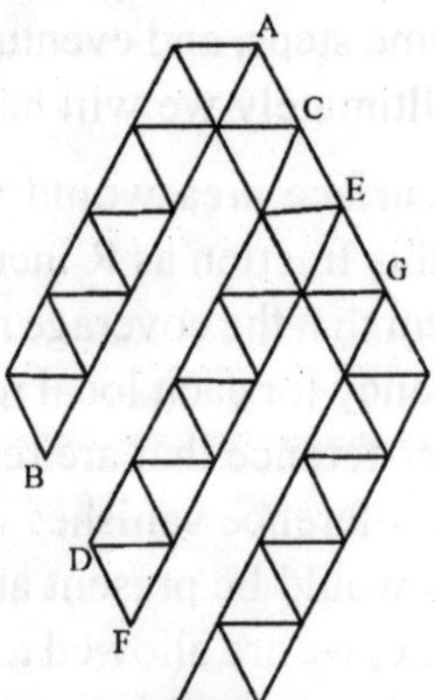

Here the lines EF and GH are also slightly converging, making the region between them unfillable. In this way we might, by design, be able to exhaust our eligible edges while having tiled only about 2/3 of the surface, since the unfillable slice between every two rows of tiles would have about half the area of a row of tiles.

Of course, this pattern requires us to carefully arrange our construction. For example, if we didn't tile all the way out to DF on the central path, we could build a tile path from H to B, which wouldn't quite connect perfectly, but would block the central row of tiles. In general I think we could construct some very "maze-like" patterns, and if we just randomly selected our sites for each successive triangle, we would (or at least COULD) get some fairly jagged arrangements with extensive unfillable "slices". As a result, I'd expect to end up with less than 5/6 coverage... and for it to be possible to get significantly different coverage by design versus random selection.

In fact, I think it's possible to show that we can actually get arbitrarily low coverage simply by constructing one geodesic "belt" around a great cirlce of the sphere, and then constructing a triangle on each of the exposed faces of that belt. We'll end up with an arrangement that looks like this

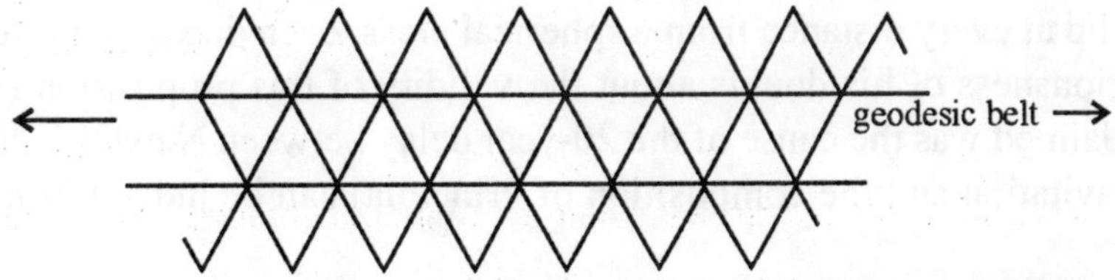

Notice that it's not possible to construct any more triangles on this configuration, since the only exposed edges are subtending and angle just slightly less than what is required to accommodate another triangle. Thus, for an arbitrarily large sphere we have a saturated configuration of tiles that covers just an arbitrarily thin slice around a great circle. Naturally we wouldn't expect to reach this configuration by chance, but I think it illustrates how this type of tiling can easily become saturated even while larges regions of the sphere's surface are totally uncovered. Of course, this construction assumes the "belt" fits perfectly, but there are infinitely many values of R for which this is true.

Along these same lines, we could tile along any of the geodesic projections of the edges of a regular solid, and then "saturate" those geodesic segments by placing triangles on each exposed face, just as we did with the geodesic belt. In this way, we can get just about any amount of coverage, from the max possible all the way down to virtually zero.

Speaking of the max possible, I'm wondering if it might be possible to achieve coverages close to 1 in many cases (aside from the obvious Pythagorean cases), because it might be possible to construct a spiral belt, with "dog legs" as needed in order to circle the sphere and end up just skimming the previous loop, without leaving much empty space. Anyway, it's clear that we can have saturated tilings of the sphere with a wide range of coverages, so this again raises the original question, i.e., what is the expected coverage for a random tiling?

NEWTON'S PROPOSITION LXXI

Newton's earliest speculations (in the 1660's) about universal gravitation involved a rough comparison between the acceleration of a falling apple near the Earth's surface and the acceleration required to keep the Moon in its orbit. He found the accelerations to be "pretty near" inversely proportional to the squares of the distances from the Earth's center. This is an interesting observation, but it isn't clear whether this is consistent with the idea of gravity as an attractive force exerted by every particle of matter on every other particle of matter. At great distances it's quite plausible that the attraction between two spherical objects should be closely expressible in terms of the distances between their centers, but it's not intuitively obvious that the net force near the surface of a massive sphere should be inversely proportional to the square of the distance from the center of the sphere, as if all the mass of the sphere were located at its center. In fact, Newton told Halley in 1686 (while still at work on the Principia) that he had only succeeded the previous year in convincing himself that the "duplicate proportion" (i.e., the inverse-square force of gravity) was actually valid at every distance from a spherical mass, even down to the surface of the sphere. The seriousness of his doubts about the validity of this proposition (which some scholars have claimed was the cause of the 20-year delay between Newton's first thoughts on universal gravitation and the composition of Principia) can be judged from his remark to Halley that

"There is so strong an objection against the accurateness of this proposition that without my demonstrations... it cannot be believed by a judicious philosopher to be any where near accurate."

The problem can obviously be reduced to one involving just a thin spherical shell of matter, since we can build up a solid sphere from such shells. Interestingly it's fairly easy to see that the net force exerted by an inverse-square attraction to the individual particles of a shell on a point inside the shell is identically zero, and this is suggestive of the unique correspondence between spherical distributions of matter and inverse-square force laws. When treated as a simple problem in integration the exterior points can be handled just as easily as the interior points. The usual approach is to consider a cross-section of a spherical shell as shown below:

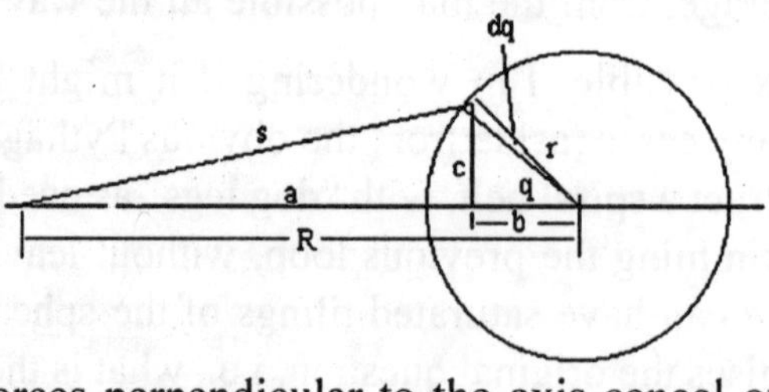

By symmetry the forces perpendicular to the axis cancel out, so we need consider only the forces in the direction of the axis. The incremental force exerted by the ring of matter in the region dq, when rotated about the horizontal axis, is

$$dF = spi\, r^2 \frac{\sin(q)\,[R - r\cos(q)]}{S^3}\, dq$$

We also have

$$\cos(q) \frac{R^2 + r^2 - s^2}{2Rr}$$

and we can take the differentials of both sides to give

$$-\sin(q)\, dq = \frac{s}{Rr}\, ds.$$

Solving this for dq and substituting back into (1), along with the expression for cos(q) from, gives

$$dF = \frac{pi\ r}{R^2}\left(1 + \frac{R^2 - r^2}{s^2}\right)ds.$$

Integrating this from s = R-r to R+r gives

$$F = \frac{pi\ r}{R^2}\int_{R-r}^{R+r}\left(1 + \frac{R^2 - r^2}{s^2}\right)ds = \frac{4pi\ r^2}{R^2}.$$

For an interior point the same expression applies, except that the limits of integration are r-R to r+R, in which case the integral vanishes and we have F = 0.

Since Newton was in possession of his own version of calculus at this time, we might expect to find something like this integration in the Principia, but in fact Newton composed the Principia entirely without the (exlicit) use of his method of "fluents and fluxions". This is usually attributed to his preference for the synthetic methods of the ancients, such as Archimedes, Euclid, and so on, but it's sometimes suggested that although the Principia contains no explicit calculus (setting aside the material the Newton added in later editions when trying to solidify his priority of invention), it nevertheless is based on the thought processes of calculus, merely disguised. It's certainly true that Newton made use of reasoning about "ultimate ratios", etc., but such reasoning has been employed since ancient times, notably by Archimedes. What distinguishes the modern calculus is precisely the formalized algorithms and systematic procedures for dealing with continuous functions (with or without a rigorous foundation), and these are almost entirely absent from the Principia.

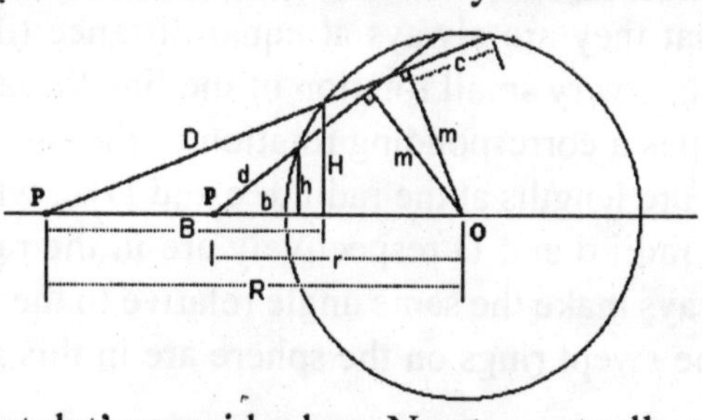

To illustrate this point, let's consider how Newton actually proved his Proposition 71 concerning the net attraction of a spherical distribution of matter on an external particle. First we note that in Proposition 70 he treated the case of interior points simply by matching the opposite regions on the sphere and showing that they cancel, but this doesn't apply directly to the case of an exterior point. So, to finally settle his doubts about the attraction of spherical bodies, in 1685 he developed a synthetic demonstration by comparing the forces applied by a spherical shell on two external points at different distances, and showed that the forces are inversely proportional to the squares of their distances from the center of the sphere.

Consider a cross-section of a spherical shell of matter, and two external particles at P and p, as shown below.

Newton draws lines from P and p through the sphere such that they cut off equal arcs, which implies that both lines are an equal distance, denoted by m, from the center of the sphere. In terms of the distances indicated on this drawing, we know that the forces along the horizontal axis exerted on these two particles by the ring-shaped slices of the spheres (on the near side) swept out by rotating about the axis are

$$F = \frac{W\,2pi{\cdot}H}{D^2}\,\frac{B}{D} \qquad f = \frac{W\,2pi\,h}{d^2}\,\frac{b}{d}$$

where W and w are the incremental widths of the rings. Hence the ratio of these forces is

$$\frac{F}{f} = \frac{W}{w}\,\frac{H/D}{m/d}\,\frac{B/D}{h/d}\,\frac{B/D}{b/d}\,\frac{d}{D}.$$

Also we know the triangle with edges H,B,D is similar to the triangle with edges m,D+c,R; and likewise the triangle with edges h,b,d is similar to the triangle with edges m,d+c,r. Therefore B/D = (D+c)/R and b/d = (d+c)/r. In addition, H/D = m/R and h/d = m/ r. Making these substitutions, we have

$$\frac{F}{f} = \frac{W}{w}\,\frac{m/R}{m/r}\,\frac{(D+c)/R}{(d+c)/r}\,\frac{d}{D} = \frac{r^2}{R^2}\,\frac{W}{w}\,\frac{(D+c)d}{(d+c)D}.$$

Now imagine the line through p rotating in the plane and sweeping over the entire circle to the point of tangency, and likewise the line through P can rotate and sweep over the entire circle to its point of tangency. The key idea is that we correlate the rotation rates of these two lines such that they are always at equal distance (denoted by m) from the center of the sphere. Hence, a very small rotation of the line through p results in a certain change in m, and this implies a corresponding rotation of the line through P. Hence these lines sweep through equal arc lengths at the radii d+c and D+c, which implies that the arc lengths they sweep out at radii d and D respectively are in the ratio d/(d+c) to D/(D+c). Also, since these lines always make the same angle relative to the surface of the circle, we know that the widths of the swept rings on the sphere are in this same proportion, i.e.,

$$\frac{W}{w} = \frac{(d+c)D}{(D+c)d}.$$

Substituting this into the preceding expression gives the ratio of the forces

$$\frac{F}{f} = \frac{r^2}{R^2}.$$

It's easy to show that essentially the same analysis applies to the rings on the opposite side of the tangency point. Hence Newton was finally able to prove to himself (twenty years after the question first occurred to him) that, in fact, the force exerted by a spherical mass on any external particle varies inversely as the square of the distance of that particle from the center of the sphere, and this remains exactly true all the way down to the surface of the sphere. This created a logically coherent foundation for the theory of universal gravitation.

It's hard not to admire the ingenuity that Newton displayed here and throughout the Principia, producing time after time a clever synthetic demonstration of a fact that would (today) be treated by calculus. On the other hand, it's interesting to note that although Newton was in possession of his fluents and fluxions since the mid 1660's, he never took the occasion to settle his question about the force of attraction of a sphere with an inverse-square law-at least not until 20 years later, and then he resolved the question by means of a synthetic demonstration, rather than by explicit use of fluents and fluxions.

It has sometimes been suggested that Newton wrote two versions of the Principia, first using calculus, and then re-writing it in the synthetic style. It seems to me that Newton's published method is not very suggestive of the way this problem would be approached using calculus.

The whole strategy of taking the ratios of the forces on two distinct points, and of splitting up the demonstration into two parts for internal and external points, seems (to me) to imply that this synthetic demonstration was the one that Newton referred to in his comment to Halley, i.e., this is the thought process by which he (after 20 years) arrived at this conclusion. Despite being one of the inventors of modern analysis, the classical synthetic mode of thought seems to have been more to his liking.

POTENTIAL FLOW AND D'ALEMBERT'S PARADOX

For a vector function V(x,y,z) in space, let v_x, v_y, and v_z denote the components of V. The *circulation* of the vector field V around any simple closed path S is defined as the integral of the tangential component of V around that path (in the "right-handed" direction). If the path is defined parametrically as a function of the path length parameter s by equations for $x_p(s)$, $y_p(s)$, $z_p(s)$, then the unit tangent vector as a function of s is

$$p_x(s) = \frac{dx_p}{ds} \quad p_y(s) = \frac{dx_p}{ds} \quad p_z(s) = \frac{dx_p}{ds}.$$

This is automatically a unit vector because $(ds)^2 = (dx_p)^2 + (dy_p)^2 + (dz_p)^2$ by definition. Since we know the components of V as a function of x,y,z, and we know the values of x_p, y_p, z_p as a function of s around the closed path, we can express the components of V as functions of s on the path.

Thus we have the vector functions V(s) and p(s) for points on the path. Recall that the dot product of two vectors is a scalar whose magnitude equals the product of the magnitudes of the arguments times the cosine of the angle between them. Thus the dot product of V and p is the projection of V along the tangent direction, and so the circulation of V around the path S is simply the integral

$$\text{Circ}(V,S) = \int_s^0 [V(s).p(s)]\, ds.$$

If the path S is planar, we can also determine the enclosed area by means of a simple integration. For example, if the path lies entirely in the xy plane, the area is given by

$$A(S) = \frac{1}{2}\int_s^0 \left(y\frac{dx}{ds} - x\frac{dy}{ds} \right) ds.$$

At every point in space we can define "curl V" as a vector whose components in each of the three principle directions equal the circulation around a closed path, in the plane normal to that direction, divided by the area enclosed by the path in the limit as the area goes to zero. Without loss of generality we can assume square paths, so the z component of the curl of V can be expressed as the limit of the circulation around a square path in a plane parallel to the xy plane.

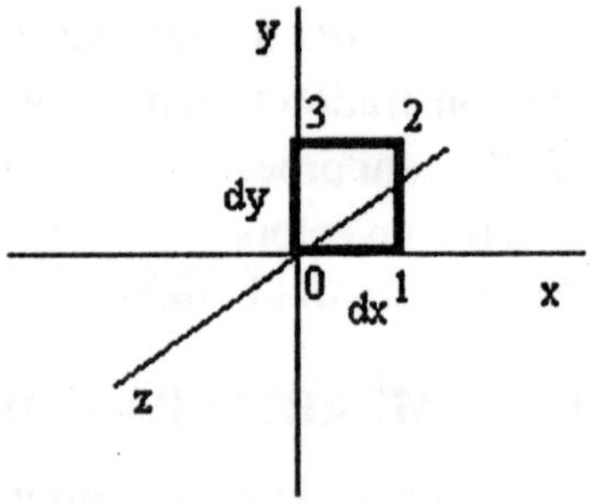

Since we are going to take the limit as the area goes to zero, and all the partial derivatives f the vector field are finite (by assumption), we can choose a small enough square so that the field is planar over this region. In other words, the change in the components of V from 0 to 1 is the same as the change from 2 to 3, and the change from 1 to 2 is the same as from 0 to 3. Also, the components vary linearly along each edge. In addition, notice that the projection of V along each edge of the square is just the respective component of V. Therefore, the circulation around this square path is

$$\text{circ} = \left(\frac{v_{x0} + v_{x1}}{2}\right) dx + \left(\frac{v_{y1} + v_{y2}}{2}\right) dy - \left(\frac{v_{x1} + v_{x3}}{2}\right) dx - \left(\frac{v_{y3} + v_{y0}}{2}\right) dy$$

$$= \frac{1}{2}\left(\frac{v_{y1} + v_{y0}}{dx} dx + \frac{v_{x2} + v_{x3}}{dx} dx\right) dy - \frac{1}{2}\left(\frac{v_{x3} + v_{x0}}{dy} dy + \frac{v_{x2} - v_{x1}}{dy} dy\right) dx.$$

Since the changes in the components along opposite edges are equal, this reduces to

$$\mathrm{circV} = \left(\frac{\partial v_y}{\partial x} - \frac{\partial v_x}{\partial y} \right) dxdy.$$

Obviously the area enclosed by this path is simply dxdy, so the limit of the circulation in the plane normal to the z direction as the area of the path goes to zero is

$$\lim_{A \to 0} \frac{\mathrm{circV}}{A} = \left(\frac{\partial v_y}{\partial x} - \frac{\partial v_x}{\partial y} \right).$$

(It can be shown that the limiting circulations are independent of the shape of the path.) Hence this is the z component of the curl of V. In the same way we can evaluate the limiting circulation for closed paths around this same point in planes normal to the y and x directions to give the y and x components of the curl. Letting i,j,k denote basis vectors in the x,y,z directions respectively, the curl can be written as

$$\mathrm{circV} = \left(\frac{\partial v_z}{\partial y} - \frac{\partial v_y}{\partial z} \right) i + \left(\frac{\partial v_x}{\partial z} - \frac{\partial v_z}{\partial x} \right) j + \left(\frac{\partial v_y}{\partial x} - \frac{\partial v_x}{\partial y} \right) k.$$

In fluid mechanics if the vector field V represents the velocity of the fluid at each point, then curl V is called the *vorticity* of the fluid at that point. The vorticity is just twice the angular velocity vector.

If the vorticity of a flowing fluid is everywhere zero, then the flow is said to be irrotational. Kelvin showed that vorticity is conserved in an ideal non-viscous incompressible fluid, so if the fluid is irrotational at any instant, it remains irrotational thereafter.

This is important because we often imagine a large reservoir of essentially static or uniformly flowing (and therefore irrotational) fluid as the upstream boundary conditions of a problem, and we can therefore assume the flow is irrotational throughout the process. Of course, the assumption of zero viscosity, and the absence of boundary layer effects, is never perfectly correct, and it's obviously possible to introduce or remove vorticity due to real-world boundary layer effects. Nevertheless, in a large class of realistic situations the assumption of irrotational flow is valid.

Under these conditions there exists a scalar field $\phi(x,y,z)$ such that the fluid velocity vector is the negative gradient of ϕ. In other words, letting v_x, v_y, v_z denote the components of the fluid velocity in the principle directions, we have

$$v_x = -\frac{\partial \phi}{\partial x} \qquad v_y = -\frac{\partial \phi}{\partial y} \qquad v_z = -\frac{\partial \phi}{\partial z}.$$

The scalar ϕ is called the velocity potential. To prove that such a scalar function ϕ exists for irrotational flow, recall that partial differentiation is commutative on a flat manifold, which means that for *any* scalar field ϕ we have (for example)

$$\frac{\partial^2\phi}{\partial x\partial y} - \frac{\partial^2\phi}{\partial y\partial x} = 0.$$

Substituting $-v_x$ for $\partial\phi/\partial x$ and $-v_y$ for $\partial\phi/\partial x$, we have

$$\frac{\partial v_y}{\partial x} - \frac{\partial v_x}{\partial y} = 0$$

which represents the vanishing of the z component of the vorticity. Likewise the vanishing of the other vorticity components is equivalent to the commutation of partial differentiation with respect to the other two pairs of coordinates.

For a continuous fluid medium we have the continuity condition

$$\nabla.(\rho V) = -\frac{\partial\rho}{\partial t}$$

where ρ is the fluid density. For an incompressible fluid the density is a constant, so the continuity condition reduces to simply the vanishing of the divergence, i.e., $\nabla \cdot V = 0$. Inserting the expressions for the components of V in terms of the potential, this gives Laplace's equation

$$\frac{\partial^2\phi}{\partial x^2} + \frac{\partial^2\phi}{\partial y^2} + \frac{\partial^2\phi}{\partial z^2} = \nabla^2\phi = 0.$$

Using this equation, together with appropriate boundary conditions, we can solve for the potential function throughout the flow field, and this yields the streamlines as well, since they are everywhere normal to the equi-potential surfaces. Also, given the velocity of the fluid at each point, we can immediately compute the static pressure as well, and we can integrate this pressure over any surface to find the net force on that surface. Of course, this is all based on the neglect of boundary layers, which can have some surprising consequences as we'll see below.

For a relatively simple example, consider the flow of an ideal incompressible irrotational fluid past a stationary sphere. Letting U denote the free-stream velocity of the fluid, it's clear that the potential field is axially symmetrical about the axis through the sphere parallel to the direction of flow, so we need only specify the potential ϕ as a function of the polar coordinates r (the distance from the sphere's center) and θ (the angle from the direction of flow. Laplace's equation reduces to

$$\frac{\partial}{\partial r}\left(r^2\frac{\partial\phi}{\partial r}\right) + \frac{1}{\sin(\theta)}\frac{\partial}{\partial\theta}\left(\sin(\theta)\frac{\partial\phi}{\partial\theta}\right) = 0.$$

For flow past a sphere we require that the radial component of the flow velocity is zero at the surface of the sphere. (Notice that we do *not* require the tangential velocity to vanish at the surface, because we are assuming perfectly inviscid, i.e., frictionless, flow.) If the radius of the sphere is R, the general solution for the potential field is

$$\phi = U\cos(\theta)\left(r + \frac{R}{2r^2}\right)$$

so the components of the flow velocity are

$$v_r = U\cos(\theta)\left(1 - \left(\frac{R}{r}\right)^3\right) \qquad v_\theta = -U\cos(\theta)\left(1 + \frac{1}{2}\left(\frac{R}{r}\right)^3\right).$$

Since $v_r = \partial\phi/\partial r$ and $v_\theta = (1/r)\partial\phi/\partial\theta$, the streamlines must satisfy the condition

$$\frac{d\theta}{dr} = \frac{v_r}{rv_\theta} = \frac{\tan(\theta)}{2}\frac{2r^3 + R^3}{r(r^3 - R^3)}.$$

This leads to the integral equation

$$\int_{\theta_1}^{\theta_2} \frac{2}{\tan(\theta)}d\theta = \int_{r_1}^{r_2} \frac{2r^3 + R^3}{r(r^3 - R^3)}dr.$$

Carrying out the integration gives

$$2\ln\left(\frac{\sin(\theta_2)}{\sin(\theta_1)}\right) = \ln\left(\frac{r_2(r_1 - R)(r_1^2 + r_1R + R^2)}{r_1(r_2 - R)(r_2^2 + r_2R + R^2)}\right).$$

Taking the exponential of both sides and setting $\theta_1 = \pi/2$ and r_1 equal to the closest approach of the streamline to the sphere, we have

$$\sin(\theta_2)^2 = \frac{r_2(r_1 - R)(r_1^2 + r_1R + R^2)}{r_1(r_2 - R)(r_2^2 + r_2R + R^2)}.$$

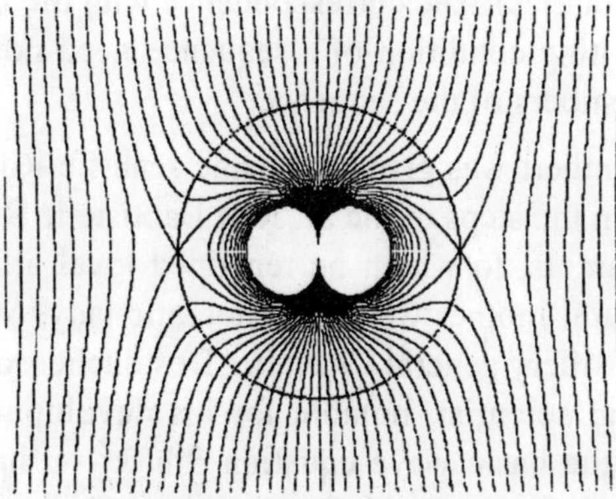

So, for any given r_1 we can compute $\sin(\theta_2)$ as a function of r_2, and from this we also have $\cos(\theta_2)$, so we can compute the streamlines as the locus of points with $x = r_2\cos(\theta_2)$

and $y = r_2 \sin(\theta_1)$. A plot of the equi-potential lines and streamlines for the flow past a sphere is shown below.

Notice that there is no boundary layer in this model, so the flow velocity does not go to zero at the surface of the sphere. Consequently the velocity field is symmetrical on the upstream and downstream sides of the sphere. This means the total momentum of the flow is unchanged, so the flow cannot have imparted any momentum to the sphere. Thus the net force exerted by the fluid on the sphere is zero.

This can also be seen by noting that since the velocities are symmetrical on the upstream and downstream sides, so are the pressures (using Bernoulli's formula $(1/2)\rho v^2 + p_0 =$ constant to give the static pressure as a function of the flow velocity). Jean Le Rond d'Alembert (1717-1783) performed a series of experiments to measure the drag on a sphere in a flowing fluid, and on the basis of the potential flow analysis he expected that the force would approach zero as the viscosity of the fluid approached zero. However, this was not the case. The net force seemed to converge on a non-zero value as the viscosity approached zero. Hence the vanishing of the net force in the potential flow analysis is known as d'Alembert's Paradox.

Interestingly, this is really just another form of the well-known "limit paradox" in elementary calculus, a paradox that is also historically associated with d'Alembert. The resolution becomes clear when we realize that any non-zero viscosity, no matter how small, will result in a boundary layer that forces the tangential flow velocity to vanish at the surface of the sphere.

As we lower the viscosity, the thickness of the boundary layer is reduced, but the flow velocity still drops to zero across that layer (the "no-slip" condition), and the results of this boundary layer, with its vorticity, mixing, and possible separation, lead to losses in the momentum of the flowing fluid and the transference of momentum to the sphere, i.e., to a net unbalanced force, roughly proportional to the velocity of the freestream flow (for relatively slow flow velocities). The correspondence with the limit paradox from calculus is obvious, i.e., the limiting condition of a sequence need not possess all the properties possessed by all the members of the sequence.

Incidentally, d'Alembert was given the name Jean Le Rond because as an infant he was found abandoned on the steps of the St. Jean Baptiste le Rond church. He was placed in the care of foster parents, to whom he remained loyal all his life, even after it was discovered that his natural mother was an aristocratic lady, Madame de Tencin, and his natural father was the artillery general Chevalier Destouches. d'Alembert became famous as a mathematician and scientist, notable for his development of partial differential equations, particularly the basic wave equation $\partial^2 h/\partial t^2 - \partial^2 h/\partial x^2 = 0$. During the years from 1751 to 1772 he collaborated with Denis Diderot on the famous "Encyclopedia", to which he contributed the preface and most of the articles on science and mathematics.

NON-EUCLIDEAN SPACES IN EUCLIDEAN SPACES

Is it possible to isometrically embed a non-Euclidean manifold in a Euclidean manifold of higher dimension? If we limit ourselves to just ONE new dimension the answer is no. This was proved around 1901 by Hilbert, who showed that the original non-Euclidean space (the 2D hyperbolic plane of Lobachevski, Bolyai, et al) cannot be isometrically embedded in its entirety in 3D Euclidean space. However, it CAN be embedded in 6D Euclidean space, and I think even in 5D Euclidean space. Apparently the question of whether there exists a complete isometric embedding in 4D Euclidean space remains open. In any case, we can always embed a smooth metrical non-Euclidean space in a higher-dimensional Euclidean space, but it usually takes more than just one extra dimension.

Of course, if we consider semi-Riemannian manifolds it complicates the question a bit. If we're willing to allow imaginary "distances" (i.e., distances whose squares are negative), then it IS possible to embed the hyperbolic plane in 3D Euclidean space -as the surface of a sphere with imaginary radius. In fact, as early as 1766 Lambert observed that if you assumed there are at least two distinct lines through a given point that don't intersect a given line, then the area of a triangle with angles a,b,c would be $-R^2$ (a+b+c-pi) for some constant R. He knew that the area of a triangle on a real sphere of radius R was R^2 (a+b+c-pi), so he wrote "one could almost conclude that the new geometry would be true on a sphere of imaginary radius". It turns out that if you just plug in iR as the radius in any formula for spherical geometry you get the corresponding formula for hyperbolic geometry.

On the other hand, it isn't clear that a formally Euclidean space with imaginary distances is any more intuitive than a curved space with strictly real distances. In fact, I would argue that the most un-intuitive aspect of the relativistic "metric" of spacetime is not it's curvature, but the fact that it isn't really a metric at all.

A metric space, in the strict sense of the term, is a manifold that satisfies the triangle inequality, which is the property that leads to our intuitive impressions of "locality". In particular, locality is transitive in a metric space, meaning that if A is close to B, and B is close to C, then A can't be too far from C. Spacetime doesn't satisfy this condition. There exist points A,B,C such that the absolute distances AB and BC are both less than the Planck length (10^{-35} meters), and yet the distance AC is the radius of the observeable universe. This has nothing to do with curvature; it's strictly a consequence of the fact that spacetime is not a metrical space and does not satisfy the triangle inequality. In view of this, a little bit of curvature should be the least of our worries.

CONSTANT HEADINGS AND RHUMB LINES

The metric on the surface of a sphere of radius R in terms of the latitude and longitude angles θ and ϕ is

$$(ds)^2 = R^2(d\theta)^2 + R^2 \sin(\theta)^2(d\phi)^2.$$

Note that $\theta = 0$ at the North Pole, and $\theta = \pi/2$ at the equator. At any given location if we define x,y distance coordinates with x pointing east-west and y pointing north-south, then $dy = R d\theta$ and $dx = R \sin(\theta)\, d\phi$. Therefore, if we fix $dy/dx = K$ for some constant non-zero K, we have

$$\frac{d\phi}{d\theta} = \frac{1}{K \sin(\theta)}.$$

Dividing the metric by $(ds)^2$ and inserting $d\phi/ds = (d\phi/d\theta)(d\theta/ds)$ gives

$$1 = R^2 \left(\frac{d\theta}{ds}\right)^2 \left(1 + \frac{1}{K^2}\right)$$

and so we have the equations of motion

$$\frac{d\theta}{ds} = \frac{K}{R\sqrt{1+K^2}} \qquad \frac{d\phi}{ds} = \frac{K}{R \sin(\theta)\sqrt{1+K^2}}$$

which can then be integrated to give explicit expressions for θ and ϕ as a function of the path length s. The left-hand equation gives

$$\theta(s) = \theta_0 + \frac{K}{R\sqrt{1+K^2}} s.$$

Inserting this into the right hand equation and integrating gives

$$\phi(s) = \phi_0 + \frac{1}{K} \ln\left(\frac{[1-\cos(ks+\theta_0)]\sin(\theta_0)}{[1-\cos(\theta_0)]\sin(ks+\theta_0)}\right)$$

where we have put

$$k = \frac{K}{R\sqrt{1+K^2}}.$$

This curve on the surface of a sphere is sometimes called a loxodrome, rhumb line, or spherical helix. A path beginning on the equator and maintaining a constant heading of K=0.2 is shown in the figure below.

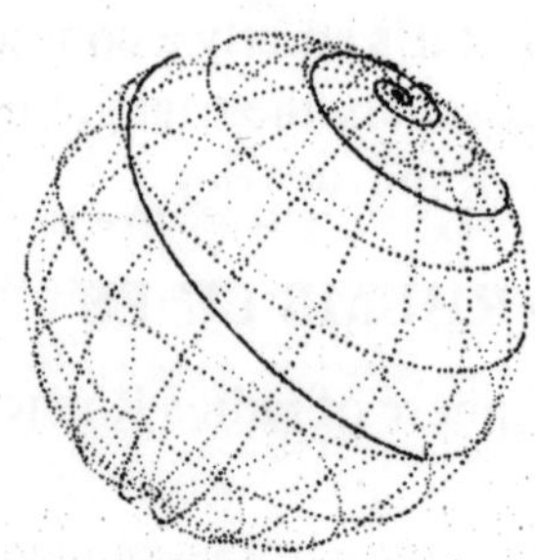

Curves like this were first discussed by Pedro Nunes in 1550. He is said to have believed that this curve represents the shortest distance between two points on the surface of a sphere, but considering the spiral nature of the curve near the poles, it's hard to imagine how anyone could have thought it was the shortest distance. Interestingly, a constant northwest heading (for example) will reach the pole in a finite distance, but it will circle the pole infinitely many time as it approaches. To see this, we can integrate the above equation for dq/ds for a rhumb line going from the North Pole ($\theta = 0$) to the South Pole ($\theta = \pi$). This gives the length of the complete rhumb line as

$$s = \pi R \frac{\sqrt{1+K^2}}{K}.$$

If K=0 the length becomes infinite, because that represents a purely "westward" heading. On the other hand, as K approaches infinity, the heading becomes purely "northward", and the total distance between the poles is simply half the circumference, i.e., $s = \pi R$. To find the total longitude travelled by a rhumb line as it moves from the latitude θ_1 to the latitude θ_2, we can integrate the equation for $d\phi/d\theta$ to give

$$\phi = \frac{1}{K}\ln\left(\frac{\tan(\theta_2/2)}{\tan(\theta_1/2)}\right).$$

Obviously if either θ_1 or θ_2 equals 0 or π, the argument of the natural logarithm is either 0 or infinite, and in either case the natural log is infinite. Therefore, a rhumb line extending to either of the poles must pass through infinite longitude, i.e., it must circle the pole infinitely many times.

DISTANCES IN BOUNDED REGIONSS

One of the standard problems in introductory calculus courses is to find the average distance between two randomly selected points inside a unit sphere.

The popularity of this particular problem is probably due to the fact that it happens to lead to an integral that can be evaluated in "closed form" to give a nice explicit answer, but it's interesting to consider the average distances (and powers of distances) within various other bounded regions in various dimensions. For regions other than a sphere, the solution is not always so simple, but we'll find that a parametric formulation of the distance density can be used to simplify the analysis.

We'll begin by considering circles (spheres) and squares (cubes) of different dimensions, starting with the classic problem of the solid unit sphere in 3D space. Let R and r be the radial distances from the origin to two randomly chosen points in a unit sphere, and let θ be the angle between these vectors. The distance between the two points is

$$s = \sqrt{R^2 - 2Rr\cos(\theta) + r^2}.$$

Covering just the case $R > r$, we need to triple-integrate this quantity over r, R, and θ, and we need to weight the quantity in proportion to the fraction of the two-point state-space. corresponding to each set of parameters. For given values of r and R, the angle θ defines a circle whose circumference is proportional to sin(θ), so this is the weight for the θ integration. Similarly each value of r and R defines a sphere with surface area proportional to r^2 and R^2 respectively, so these are the weights for the r and R integrations.

We can restrict our analysis to just the case R > r because the other case (R < r) is symmetrical and has the same distribution of distances. Hence we need only integrate the distance function with the appropriate weights for the ranges r = [0,1], R = [r,1], and θ = [0,π], and then divide the result by the triple integral of just the weights $(rR)^2 \sin(\theta)$, which is 1/9. This gives the mean distance

$$\overline{s} = 9\int_0^1 r^2 \int_r^1 R^2 \int_0^{\pi} \sin(\theta) = \sqrt{R^2 + r^2 - 2Rr\cos(\theta)} = \frac{36}{35}.$$

We can also evaluate this integral with the distance function raised to any integral power, to give the average of the nth powers of the distances

$$\overline{s^n} = \frac{(72)2^n}{(n+2)(n+4)(n+6)}.$$

Notice that the weight for θ is very fortuitous, because the factor of sin(θ) enables us to evaluate the integral in closed-form, using the easy integral for

$$\sin(x)\sqrt{A - B\cos(x)}.$$

In contrast, the seemingly simpler case of a unit *disk* is actually more difficult because it lacks this convenient weight factor. It's also worth noting that if we try to cover both the cases R > r and R < r at once by integrating over r = [0,1] and R = [0,1] we are led to a result like 21/20 instead of 36/35. The subtle problem here is that the integral of

$$\sin(\theta)\sqrt{R^2 + r^2 - 2Rr\cos(\theta)}$$

over the range θ = [0,π], divided by the integral of the weight, is

$$\frac{\left[(R+r)^2\right]^{3/2} - \left[(R-r)^2\right]^{3/2}}{6Rr}$$

which is formally symmetrical in R and r. Now, since R+r is always non-negative there's not much ambiguity in evaluating the left hand term in the numerator; we just take the positive value $(R+r)^3$, glossing over the fact that there's really a square root there in the 3/2 power (which means we *could* have taken the negative root). However, the right hand term requires a bit of thought: Do we evaluate it as $(R-r)^3$ or $(r-R)^3$? The answer depends on whether R > r or R < r. In these two cases the above expression reduces to

$$R+\frac{r^2}{3R} \text{ if } R>r \qquad r+\frac{R^2}{3r} \text{ if } R<r$$

so it's necessary (not just convenient) to treat the cases separately. Fortunately, due to symmetry, the cases give the same distribution of distances, so we need only evaluate one of them.

Now let's consider the distances on a unit disk. This case can be formulated in essentially the same way as with the unit sphere, except that the weights are different. Each angle θ now represents only a single point, so it's weight is just 1. Each radius r and R now represents a circle with length proportional to r and R respectively, so these are the weights. The triple integral of the product of these weights is π/8, so the average distance on a unit disk can be expressed as

$$\overline{s}=\frac{8}{\pi}\int_0^1 r\int_0^1 R\int_0^{\pi}\sqrt{R^2+r^2-2Rr\cos(\theta)}\,d\theta\,dR\,dr.$$

Unfortunately this integral isn't as easy to evaluate as the one for the sphere, because it lacks the sin(θ) factor. We will describe a much simpler approach later, so rather than presenting the elaborate integration here, suffice it to say that the result is 128/(45π). We can also evaluate the above integral for any integer power of the distance function. For odd powers of the form 2k–1 the general result is whereas for even powers of the form 2k the general result is

$$\overline{\left(s^{2k}\right)}=\frac{(2k+2)!}{[(k+1)!]^2\,(k+1)(k+2)}.$$

Thus the average squared distance between two point in a unit disk is 1, and the average cubed distance is 2048/(525π). Incidentally, this last formula shows that if C_n denotes the nth Catalan number, then the average (2k)th power of distances on a unit disk is just $C_{k+1}/(k+1)$.

As an alternative to the preceding triple integral, we could just integrate the unit disk by scanning two points (x,y) and (X,Y) orthogonally across the disk. The weight factors are all 1 in this case, and their product integrates to the squared area of the disk, π^2, so we have the quadruple integral

$$\overline{s}=\frac{1}{\pi^2}\int_{-1}^{1}\int_{-a}^{a}\int_{-1}^{1}\int_{-b}^{b}\sqrt{(x-X)^2+(y-Y)^2}\,dY\,dX\,dy\,dx$$

where $a=(1-x^2)^{1/2}$ and $b=(1-X^2)^{1/2}$. This gives the same results as the previous triple integral, but it's even more laborious to integrate. We will show later a single integral that gives the same results.

As a way of introducing the simpler method for evaluating the average distances in arbitrary bounded regions, let's first consider the distribution of distances on a 1×1 unit square.

We could use the rectangular approach of the preceding equation and just integrate the distances between two points (x,y) and (X,Y) as each parameter ranges from 0 to 1, but the resulting quadruple integral is not very easy to evaluate.

A much more efficient approach is to notice that the distances and densities (weight factors) on a unit square are distributed according to the parametric formulas

$$s(u,v) = \sqrt{u^2 + v^2} \qquad \delta(u,v) = 4(1-u)(1-v).$$

Therefore, we just need to integrate the product δ(u,v) s(u,v) as u and v range from 0 to 1. The orthogonal way of formulating this double-integral is

$$\overline{s} = 4\int_0^1 (1-u)\int_0^1 (u,v)\sqrt{u^2+v^2}\ dv\,du.$$

This is a big improvement over the quadruple integral, but it still is not easily evaluated in closed form.

Let's try polar coordinates in this parametric space by setting u = r cos(θ) and v = r sin(θ) and integrating over the region u < v by letting r range from 0 to 1/cos(θ) at each θ from 0 to π/4.

For any incremental slice of θ the weight at r is proportional to r, and of course the integral of r over this region u < v is just 1/2, so we have the integral

$$\overline{s} = 8\int_0^{\pi/8}\int_0^{1/\cos(\theta)} (1 - r\sin(\theta))(1 - r\cos(\theta))r^2 dr d\theta.$$

This value of this double-integral is

$$\frac{2+\sqrt{2}+5\ln\left(1+\sqrt{2}\right)}{15} = 0.521405433...$$

By incrementing the exponent of r in the integral we can evaluate the average of the nth powers of distances on the unit square. In general the results are of the form

$$\overline{s^n} = \frac{A_n + B_n\sqrt{2} + C_n \ln\left(1+\sqrt{2}\right)}{D_n}$$

where the values of A, B, C, D are as shown below etc.

n	*A*	*B*	*C*	*D*
1	2	1	5	15
2	1	0	0	3
3	8	17	21	210
4	17	0	0	90
5	16	73	45	1008
6	29	0	0	210
7	384	3239	1155	47520
8	187	0	0	1575

Now, what about the unit cube, or the unit 4D hyper-cube? The nice thing about the parametric distance density equations is that they immediately generalize to higher dimensions. The parametric equations for the distance density of a d-dimensional unit cube are

$$s(x_1, x_2, \ldots, x_d) = \sqrt{x_1^2 + x_2^2 + \ldots + x_4^2}$$

$$\delta(x_1, x_2, \ldots, x_d) = 2^d(1 - x_1)(1 - x_2)\ldots(1 - x_d).$$

For even powers of the distance, we can immediately evaluate the d-dimensional analogs of equation. We find that the average squared distance in a d-dimensional unit cube is simply d/6. (This gives a nice trivia question: In what dimensional space is the average squared distance in a unit cube equal to unity?) The average nth powers of distances in a d-dimensional unit cube are given by the following formulas for the first few even values of n:

n	average nth power of distance in unit d – cube
2	$(d)/[(1)(6)]$
4	$(5d^2 + 7d)/[(2)(90)]$
6	$(35d^3 + 147d^2 + 88d)/[(6)(1260)]$
8	$(175d^4 + 1470d^3 + 2789d^2 + 606d)/[(24)(9450)]$

Returning to the parametric density formulas for the distances inside a d-dimensional cube, notice that in each case the density of the vector $v = [x_1, x_2, \ldots x_d]$ within a bounded region B is proportional to the intersection of B with a copy of B shifted by the vector v. More generally, given *any* finite bounded region B in k-dimensional space, and any k-vector v, let B[→v] denote the image of B translated by v. Then the density of distances with the magnitude and direction of v is proportional to the content (volume) of the intersection of B with B[→v]. This is because B is the set of possible starting points for the vector v, and B[→v] is the corresponding set of end points. A particular instance of v (with a particular starting point in B) is a distance between two points of B if and only if its end point is also in B.

Thus, the intersection of B[→v] with B (divided by the content of B itself) is the proportion of points in B for which the vector v is a distance to another point in B. This is illustrated for a few simple shapes below.

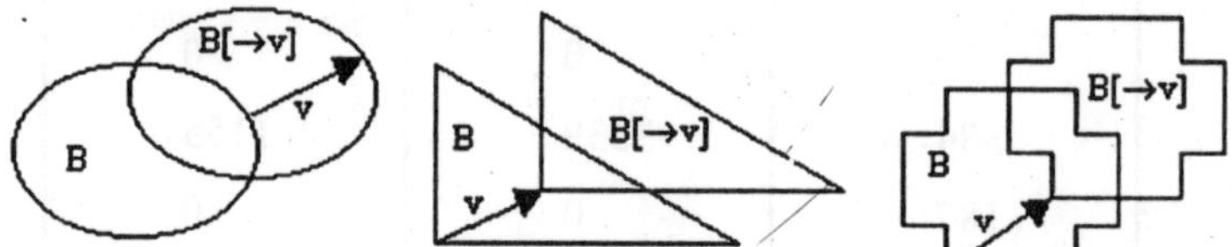

Let's apply this principle to determine the distribution of distances (between two randomly chosen points from a uniform distribution) on a unit disk (radius 1). For any 2-vector v = [r sin(θ), r cos(θ)] the density is proportional to the overlap of two unit circles with centers a distance r apart. This is given by

$$4\int_{r/2}^{1}\sqrt{1-x^2}\,dx = 2\,\mathrm{inv}\tan\left(\frac{\sqrt{4-r^2}}{r}\right)-\frac{r}{2}\sqrt{4-r^2}.$$

Now, in terms of the parametric lattice of distance vectors (which can be seen as the first quadrant of a circle of radius 2), the number of vectors with magnitude r is proportional to r, so the overall density of distances of magnitude r is proportional to r times the above quantity. The integral of this resulting function from r = 0 to 2 is simply π/2, so the distribution density of distances on a unit disk is

$$\delta(r)=\frac{r}{\pi}\left[4\,\mathrm{inv}\tan\left(\frac{\sqrt{4-r^2}}{r}\right)-\sqrt{4-r^2}\right].$$

The mean value of any function f of the point-to-point distances is then given by

$$\overline{f}=\int_{0}^{2}f(r)\delta(r)\,dr.$$

If we make the change of variables r = 2cos(θ/2), the density function can be simplified and we have

$$\overline{f}=\frac{2}{\pi}\int_{0}^{\pi}f\left(2\cos\left(\frac{\theta}{2}\right)\right)\sin(\theta)[\theta-\sin(\theta)]\,d\theta.$$

In particular, if f is just the nth power of the distances, we replace the f function with $[2\cos(\theta/2)]^n$. Thus for n = 1 we just need the elementary integrals

$$\int \cos\left(\frac{\theta}{2}\right)\sin(\theta)\theta\, d\theta = \frac{2}{9}\sin\left(\frac{3\theta}{2}\right) - \frac{1}{3}\theta\cos\left(\frac{3\theta}{2}\right) + 2\sin\left(\frac{\theta}{2}\right) - \theta\cos\left(\frac{\theta}{2}\right)$$

$$\int \cos\left(\frac{\theta}{2}\right)\sin(\theta)^2\, d\theta = \sin\left(\frac{\theta}{2}\right) - \frac{1}{6}\sin\left(\frac{3\theta}{2}\right) - \frac{1}{10}\sin\left(\frac{5\theta}{2}\right)$$

which gives

$$\overline{s} = \frac{2}{\pi}\int_0^{\pi} 2\cos\left(\frac{\theta}{2}\right)\sin(\theta)[\theta - \sin(\theta)]\, d\theta = \frac{128}{45\pi}$$

in agreement with the mean distance computed previously by triple and quadruple integrals. An important advantage of this approach is that it not only gives a simpler integration, it gives the entire distribution explicitly, rather than just the mean. Using this same density distribution we can compute the mean value of any function of the distances. For example, the mean of the inverse distances 1/s in a unit disk is given by

$$\overline{(s^{-1})} = \frac{2}{\pi}\int_0^{\pi}\left(2\cos\left(\frac{\theta}{2}\right)\right)^{-1}\sin(\theta)[\theta - \sin(\theta)]\, d\theta$$

$$= \frac{2}{\pi}\int_0^{\pi}\sin\left(\frac{\theta}{2}\right)[\theta - \sin(\theta)]\, d\theta = \frac{16}{3\pi}.$$

(The integral is even simpler in this case.) For a disk of radius r this is simply 16/(3πr). Thus, given n points uniformly distributed on a disk of radius r, there are n(n–1)/2 point-to-point distances, which implies that the expected sum of the inverse distances is 8n(n–1)/(3πr). It's interesting that while the mean reciprocal distance is finite, the mean squared reciprocal distance is not. This suggests a geometrical version of the Petersburg paradox: the game is to choose two points randomly from a uniform distribution inside a unit circle, and pay an amount proportional to the inverse square of the distance between them. Intuitively it might seem as if the payout would not be very high, but the expected value is actually infinite.

The proposition that the density of v is proportional to the intersection of B with B[→v] is very general, and applies to non-convex (and even non-connected regions), not just simple regions like spheres and cubes. For example, it can be used to determine the distance distribution for points of a torus.

N POINTS ON SPHERE ALL IN ONE HEMISPHERE

What is the probability that n randomly chosen points on a sphere will all lie in a single hemisphere? More generally, we can consider the same question for points chosen randomly from a uniform distribution on a d-dimensional spherical surface, denoted as S^d.

The trivial case is S^0, which consists of the points in a one-dimensional space (i.e., a line) equi-distant from a given point. This 0-sphere surface consists of just two points, so the probability that all n randomly placed points would all be in the same "hemisphere" is simply $2/2^n = 1/2^{n-1}$. (The "2" in the numerator is due to the fact that all the points can be in either of the two hemispheres.)

This "binary" approach can be extended to higher dimensions, if we realize that the selection of n points randomly on the surface of a d-sphere can be performed in two separate steps. First, we can select n axes through the center of the d-sphere. The directions of these axes are taken from a uniform distribution of directions. Each axis strikes the surface at two opposite points, so we need to make a binary choice for each axis to determine the position of the point. Thus, regardless of the directions of the axes, there are 2^n possible configurations, each of which is equally probable. A certain number of these configurations have all the points in a single hemisphere. The key insight is that this number is the same, regardless of the angles of the axes. Once we have established this, it follows that the probability of all n randomly placed points being in a single hemisphere is this number divided by 2^n.

Each "polar" axis corresponds to a partitioning of the d-sphere into two hemispheres. By choosing n axes, we partition the surface into a certain number of subsets. (We can exclude special cases such as axes the coincide or points where more than two slices coincide, because those have zero probability.) Each of these subsets lies entirely inside or entirely outside the hemisphere centered on each of the n points. So, each subset can be uniquely assigned an n-bit binary number signifying whether it is inside or outside the hemisphere centered on each of the n selected points.

By changing the pole selections as needed, we make cause any one of these regions to be inside the hemispheres centered on *all* n points, which implies that all n points are within a single hemisphere. Thus, each subset corresponds uniquely to a set of pole choices (for a given set of axes) that place all n points in a single hemisphere. We only need to determine the number of subsets into which the surface is partitioned by n general slices into hemispheres.

For a 0-sphere, which consists of just two points, the answer is trivially 2. For a 1-sphere (i.e., a circle), it's easy to see that cutting it into n semi-circles results in 2n segments. In general, if we let $f_d(n)$ denote the number of subsets of a d-sphere produced by n bisections, we can say that $f_d(n)$ is a polynomial in n of degree d. As noted above, the probability of n points all lying within a single hemisphere is $f_d(n)/2^n$. For any given d we know that this probability equals unity for n = 1, 2, ..., d+1. For example, with d = 0, a single point obviously has probability 1 of being in a single hemisphere. Likewise with d = 1 it's clear that any one or two points lie in a single hemisphere (since the special case of opposite polar points has probability zero). With d = 2 we can see that any one, two, or three points lie in a single hemisphere, because any two points define a "great circle" and the third point is in one of the two bounded hemispheres. In the same way, for arbitrary d, any d points define a d-dimensional slice through the origin, and the $(d+1)^{th}$ point must then lie in one of the bounding hemispheres.

This information is sufficient to determine all the functions $f_d(n)$. For example, with d = 2, we know $f_2(n)$ is a polynomial in n of degree 2, so we have

$$f_2(n) = c_0 + c_1 n + c_2 n^2$$

We also have the three conditions $f_2(1) = 2^1$, $f_2(2) = 2^2$, and $f_2(3) = 2^3$, so we can solve for the three coefficients with the system of equations

$$\begin{bmatrix} 1^0 & 1^1 & 1^2 \\ 2^0 & 2^1 & 2^2 \\ 3^0 & 3^1 & 3^2 \end{bmatrix} \begin{bmatrix} c_0 \\ c_1 \\ c_2 \end{bmatrix} = \begin{bmatrix} 2^1 \\ 2^2 \\ 2^3 \end{bmatrix}.$$

This gives $c_0 = 2$, $c_1 = -1$, and $c_2 = 1$, so we have $f_2(n) = n^2 - n + 2$, and therefore the probability of n points chosen randomly from a uniform distribution on the surface of a 2-sphere all lying on a single hemisphere is

$$\Pr(n,2) = \frac{n^2 - n + 2}{2^n} \frac{1}{2^{n-1}} \left[\binom{n}{0} + \binom{n}{2} \right].$$

Likewise for a 3-sphere (embedded in a 4-dimensional space) we can determine the coefficients of the cubic polynomial $f_3(n)$ by solving the system of equations

$$\begin{bmatrix} 1^0 & 1^1 & 1^2 & 1^3 \\ 2^0 & 2^1 & 2^2 & 2^3 \\ 3^0 & 3^1 & 3^2 & 3^3 \\ 4^0 & 4^1 & 4^2 & 4^3 \end{bmatrix} \begin{bmatrix} c_0 \\ c_1 \\ c_2 \\ c_3 \end{bmatrix} = \begin{bmatrix} 2^2 \\ 2^2 \\ 2^3 \\ 2^4 \end{bmatrix}.$$

This gives $c_0 = 0$, $c_1 = 8/3$, $c_2 = -1$, and $c_3 = 1/3$, so the probability of n points chosen randomly from a uniform distribution on the surface of a 3-sphere all lying on a single hemisphere is

$$\Pr(n,3) = \frac{\frac{1}{3}n^3 - n^2 + \frac{8}{3^2}}{2^n} = \frac{1}{2^{n-1}} \left[\binom{n}{1} + \binom{n}{3} \right].$$

In the same way we can determine the probabilities

$$\Pr(n, 4) = \frac{1}{2^{n-1}} \left[\binom{n}{0} + \binom{n}{2} + \binom{n}{4} \right]$$

$$\Pr(n, 5) = \frac{1}{2^{n-1}}\left[\binom{n}{1}+\binom{n}{3}+\binom{n}{5}\right]$$

$$\Pr(n, 6) = \frac{1}{2^{n-1}}\left[\binom{n}{0}+\binom{n}{2}+\binom{n}{4}+\binom{n}{6}\right]$$

and so on. Using the recurrence relation for binomial coefficients we can replace each C(n,d) with C(n–1,d) + C(n–1,d–1) and express all these results as

$$\Pr(n,d) = \frac{1}{2^{n-1}}\sum_{j=0}^{d}\binom{n-1}{j}.$$

Notice that if d = (n–2)/2 where n is even, the terms cover half of the row of binomial coefficients, and the sum of alternate coefficients over this half-row is $(2^n)/4$, so we have Pr{2d+2,d} = 1/2. For example, the probability that four point placed randomly on a circle will all lie in a single semi-circle is 1/2, as is the probability that six points placed randomly on the surface of a 2-sphere will all lie in a single hemisphere. Also, from the recurrence property for binomial coefficients, we have

$$\Pr\{n,d\} = \frac{\Pr\{n-1,d-1\}+\Pr\{n-1,d\}}{2}.$$

The same recurrence also applies to the complementary function 1 – Pr{n,d}.

Incidentally, the above formula for S^2 disagrees with the formula given in the rec.puzzles FAQ, which is $1 - [\,1 - (1/2)^{n-2}\,]^n$. To double-check the validity of the formula derived above, I wrote a "Monte Carlo" simulation program to randomly pick n points on a sphere and check to see if they are all in one hemisphere. The results are summarized below:

n	*Simulation*	*Pr(n, 2)*	*FAQ(n, 2)*	*FAQ(n-1, 2)*
1	1.0000	1.0000	2.0000	0.0000
2	1.0000	1.0000	1.0000	2.0000
3	1.0000	1.0000	0.8750	1.0000
4	0.8750	0.8750	0.6835	0.8750
5	0.6862	0.6875	0.6835	0.6835
6	0.4954	0.5000	0.4871	0.4871
7	0.3365	0.3438	0.3211	0.3211
8	0.2295	0.2266	0.1993	0.1993
9	0.1500	0.1445	0.1183	0.1183
10	0.0889	0.0898	0.0681	0.0681

These results (10000 samples per n) seem to agree well with the formula for Pr{n,2} derived above, but disagree with the FAQ formula. If we adjust the FAQ formula by replacing n with n-1 the discrepancy is reduced (and actually agrees exactly with the correct formula for n = 3 and 4), but it still does not give the correct results in general.

The general overall solution given above rests on the assertion that $f_d(n)$ has a definite value independent of the chosen axes (assuming no special coincidences), and that it is a polynomial in n of degree d. This was only presented as plausible, but without a rigorous proof. To prove these assertions for the case d = 2, notice that any two great circles intersect each other at two points, so there are four "edges", and they partition the spherical surface into four regions. (Again, we can neglect repeated circles, and points where more than two circles intersect, because these have zero probability.) If we then add another great circle to n given circles, the number of vertices will increase by 2n. Each vertex is the meeting point of four edges, but each edge meets at two vertices, so there are twice as many edges as vertices. Hence, the number of vertices increases by 4n.

The Euler characteristic of a spherical polyhedron is 2 = V – E + F, where V, E, and F are the numbers of vertices, edges, and faces, respectively. It follows that the number of faces must increase by 2n when adding the (n+1)th great circle. Thus we have the two conditions F(2) = 4 and F(n+1) = F(n) + 2n. It follows that F(n) is a quadratic polynomial, so $F(n) = An^2 + Bn + C$, and the condition implies 2n(A–1) = -(A+B). This must hold for all n, so we must have A = 1, and so B = -1.

Combining these with the first condition gives C = 2, so we get $F(n) = n^2 - n + 2$, and therefore the probability of all n points falling in a single hemisphere is $(n^2 - n + 2) / 2^n$, in agreement with the previous derivation. Similar arguments can be made for higher dimensions.

The preceding derivations are quite satisfactory, but it's interesting to explore other possible methods. One approach can be based on a result contained in William Feller's book "An Introduction to Probability Theory and Its Applications", which gives the probability that n arcs of length A chosen independently and randomly on the perimeter of a circle of circumference C cover the whole circle as

$$\Pr(\text{cover}) = \sum_{k=0}^{m} (-1)^k \binom{n}{k} \left(1 - k\frac{A}{C}\right)^{n-1}$$

where m is the greatest integer less than C/A. The negation of this covering event is that the arcs leave at least one gap, meaning that the largest separation between the center points of two neighboring arcs is greater than A. This is equivalent to the condition that all n center points fall on an arc of length less than C–A. Thus, the probability that all points fall on an arc of length less than C–A is just 1 –Pr{cover}. Letting x denote the fraction (C–A)/C we have

$$\Pr(x,n) = \sum_{k=1}^{m} (-1)^k \binom{n}{k} [1 - k(1-x)]^{n-1}.$$

In particular, with x = 1/2 (meaning all n points fall in some *half* of the circle) this gives again the result Pr{1/2; n} = n / 2^{n-1}. It isn't obvious how this approach can be generalized to higher dimensions.

Another approach is to proceed by "brute force", splitting the surface of S^d in half along various axes, and using inclusion-exclusion to compute the probability of all the points falling into one of those discrete halves. We can then carry this over to the limit as the number of halves approaches all possible halves. This is admittedly not very elegant, but some of the results for the finite numbers of discrete halves are interesting. To illustrate, consider the simple case of S^1, i.e., points distributed on a circle. Divide the perimeter into 2k equal arcs denoted sequentially by $a_1, a_2, \ldots, a_{2k}$, and let A_i denote the event of all n randomly placed points lying on the semi-circle composed of the arcs $a_i, a_{i+1}, \ldots a_{i+k-1}$. To determine the probability that at least one A_i is true, apply the inclusion-exclusion principle step-by-step as follows:

$$\Pr(A_1) = \left(\frac{1}{2}\right)^n.$$

$$\Pr(A_1 \cup A_2) = (1/2)^n + (1/2)^n - \left(\frac{k-1}{2k}\right)^n$$

$$\Pr\{A_1 \cup A_2 \cup A_3\} = \Pr\{A_1 \cup A_2\} + \Pr\{A_3\} - \Pr\{(A_1 \cup A_2) \cap A_3\}$$

and so on. As each additional A_i up to A_{k+1} is included in the union, the probability of the union is increased by

$$\left(\frac{1}{2}\right)^n - \left(\frac{k-1}{2k}\right)^n$$

so we have

$$\Pr\{A_1 \cup A_2 \cup \ldots \cup A_{k+1}\} = \left(\frac{1}{2}\right)^n + k\left[\left(\frac{1}{2}\right)^n - \left(\left(\frac{k-1}{2k}\right)^n\right)\right].$$

As each of the remaining terms A_{k+i}, i = 2 to k, is added to the union, the probability is increased by

$$\left(\frac{1}{2}\right)^n - \left(\frac{k-1}{2k}\right)^n - \left(\frac{i-1}{2k}\right)^n + \left(\frac{i-2}{2k}\right)^n.$$

The 3rd and 4th terms in these quantities cancel on consecutive values of i, so their net effect is just a single term $-[(k-1)/(2k)]^n$. Thus, the probability of all n points falling in k contiguous arcs is

$$\Pr\{A_1 \cup A_2 \cup \ldots \cup A_{k+1}\} = \left[\left(\frac{1}{2}\right)^n - \left(\frac{k-1}{2k}\right)^n\right] = \frac{k}{2^{n-1}}\left[\left(\frac{1}{2}\right)^n - \left(\frac{k-1}{2k}\right)^n\right].$$

Expanding the binomial, we have

$$\Pr\{A_1 \cup A_2 \cup \ldots \cup A_{2k}\} = \frac{n}{2^{n-1}}\left[1 - \frac{n-1}{2!}\left(\frac{1}{k}\right) + \frac{(n-1)(n-2)}{3!}\left(\frac{1}{k}\right)^2 - \ldots\right]$$

which gives the result $n/2^{n-1}$ in the limit as k increases to infinity. It's interesting that the formulas for Pr{n,d} seem to be closely related to the formula for the probability of n points all falling within some k contiguous arcs of a circle (S^1) that has been divided into 2k equal arcs. This was given previously as

$$\frac{1}{2^{n-1}}\left[n - \frac{n(n-1)}{2!}\left(\frac{1}{k}\right) + \frac{n(n-1)(n-2)}{3!}\left(\frac{1}{k}\right)^2 - \ldots\right].$$

It's almost as if, as k approaches 1, this probability on the discretized circle approaches

$$\Pr\{n, 2_{j+1}\} - \Pr\{n, 2_j\} + \left(\frac{1}{2}\right)^{n-1}$$

for sufficiently large j. It's also interesting that the formula for Pr{n,d} is fundamentally different -but complementary -for odd and even dimensions, somewhat like sine and cosine functions.

Since the discretized "brute force" approach to the original 1-sphere problem yielded such interesting results, we might also investigate the application of the discretized approach to the 2-sphere problem. In principle it's certainly possible to draw several great circles on a sphere and then use inclusion-exclusion to find the probability of all n points falling in one of those hemispheres. However, it isn't obvious how to add more great circles in a uniform way, except for certain special configurations, i.e., normal to the axes of Platonic solids.

With just *one* great circle the probability of all n points falling in one of the two hemispheres bounded by that circle is obviously $(1/2)^{n-1}$. We can write this as

$$P_1\{n\} = 2\left(\frac{1}{2}\right)^n.$$

We can also choose the *three* mutually orthogonal axes of a regular octahedron, and draw the corresponding great circles normal to those three axes. These circles divide the surface into 8 identical "triangular" regions that define six distinct hemispheres. By inclusion-exclusion we find that the probability of all n points falling in one of these six hemispheres is exactly

$$P_3\{n\} = 6\left(\frac{1}{2}\right)^n - 12\left(\frac{1}{2}\right)^n + 8\left(\frac{1}{8}\right)^n.$$

In the case of *four* great circles, drawn normal to the four axes of a cube, the derivation is a bit more challenging. In this case the surface is divided into 6 "square" regions and 8 "triangular" regions. If we let s and t denote the areas of these two types of regions as a fraction of the total area, then each of the 8 hemispheres has area 4t + 3s = 1/2. The non-zero overlaps between two hemispheres come in two types: there are 12 of area 2t + 2s and 12 of area 2t + s. The non-zero overlaps between three hemispheres also come in two types: there are 24 of area t + s and 8 of area t. Finally, the non-zero overlaps between four hemispheres come in two types: there are 8 of area t and 6 of area s. Thus, the probability of all n points falling within one of these 8 hemispheres is

$$P_4\{n\} = 8(4t+3s)^n - 12(2t+2s)^n - 12(2t+s)^n + 24(t+s)^n - 6(s)^n$$

where t = 0.043869914... and s = 0.108173448... Incidentally, noting that the spherical angle at each corner of the triangular regions is $2\,\text{inv}\tan(1/\sqrt{2})$, this probability can also be expressed as

$$P_4\{n\} = 2\left(\frac{1}{2}\right)^n\left[4 - 6(1-q)^n - 6q^n + 12\left(\frac{1-q}{2}\right)^n - 3(1-2q)^n\right]$$

where

$$q = \frac{2}{\pi}\text{inv}\tan\left(\frac{1}{\sqrt{2}}\right).$$

The four great circles in this example are the projections of the edges of an inscribed "cuboctahedron", which is one of the 13 semi-regular Archimedean solids. The only other Archimedian solid whose projected edges are great circles is the "icosidodeca-hedron", which has *six* great circles, one normal to each of the six axes of the icosahedron. These circles partition the surface into triangular and pentagonal regions with fractional areas t and p respectively. Carrying out the same kind of analysis for this case, we find

$$P_6\{n\} = 12(6p+10t)^n - 30(4p+6t)^n + 20(3p+4t)^n$$
$$-30(2p+4t)^n + 60(2p+3t)^n - 30(2p+2t)^n$$

where t = 0.014312286762175... and p = 0.059479522063042...

The next case to consider is with *ten* great circles on a sphere, normal to the *ten* axes of a dodecahedron. This gives 20 distinct hemispheres. (The corresponding solid is not one of the Archimedean solids, because it has two distinct kinds of vertices.) These ten great circles divide the surface of the sphere into a total of 92 regions: 60 triangles, 20 hexagons, and 12 pentagons. Letting t, h, and p denote these individual areas as fractions of the total sphere's area, each hemisphere has area 6p + 10h + 30t = 1/2. The inclusion-exclusion formula gives the overall probability that all n points fall in one of these 20 hemispheres as

$$\begin{aligned} P_{10}\{n\} = {} & 20(6p+10h+30t)^n - 30(4p+8h+22t)^n - 60(4p+6h+18t)^n \\ & +60(3p+6h+16t)^n + 60(3p+5h+14t)^n - 60(2p+5h+12t)^n \\ & -60(2p+4h+12t)^n + 120(2p+4h+11t) - 60(2p+4h+10t)^n \\ & -30(2p+2h+8t)^n + 60(2p+2h+7t)^n - 30(2p+2h+6t)^n \\ & +12(p+5h+10t)^n \end{aligned}$$

where the normalized areas are p = 0.013028988120384..., h = 0.029670196644958..., and t = 0.004170754471287....

The coefficients for each of the discrete cases are listed in the table below. The solids listed indicate that the great circles are normal to the axes of that solid.

G	
1	2 = 2
3 octahedron	6-12+8 = 2
4 hexahedron/tetrahedron	8-12-12+24-6 = 2
6 icosahedron	12-30+20-30+60-30 = 2
10 dodecahedron	20-30-60+60+60-60-60+120-60-30+60-30+12 = 2

The hexahedron (cube) and tetrahedron give the same set of axes, because the four "axes" of the tetrahedron, taken to be the lines from the center to each vertex, coincide with the four axes of the cube, where each axis goes through two vertices. Notice that the sum of the coefficients is always 2. In contrast, the sum of the coefficients on S^1 is always 0. This sum is apparently equal to the Euler topological invariant of the surface, the same quantity appearing in the formula V – E + F = 2 discussed previously.

GRAVITY OF A TORUS

What is the gravitational field of a torus? In which direction would the net force of gravity point at various locations near a ring of matter? People sometimes think that, perhaps for reasons of symmetry, an object in the interior of a ring of matter would be drawn toward the center, but this is not the case – at least not for objects in the plane of the ring. To see why, consider a very thin ring of mass treated as a circle of radius R in the plane, and a particle inside this ring at a distance r from the center. Construct an arbitrary line passing through this particle, striking the ring in two opposite directions at distances L_1 and L_2. If we rotate this line about the particle through an incremental angle $d\theta$, it will sweep out sections of the ring proportional to $L_1\cos(\alpha)d\theta$ and $L_2\cos(\alpha)d\theta$, where α is the angle the chord makes with the normals to the circle at the points of intersection.

The net gravitational force exerted by these two opposing sections of the ring is proportional to the masses in these small sections divided by the squares of the distances, i.e., the force is proportional to $d\theta \cos(\alpha)\,(1/L_1 - 1/L_2)$ in the direction of the L_1 intersection point. Hence the net force is in the direction of the closest point on the ring, directly *away* from the center.

Of course if the particle was inside a spherical *shell* instead of a circular ring, the conical region surrounding each ray through the particle would sweep out a section of the shell proportional to the *square* of the distances L_1 and L_2, which would then cancel out with the inverse squares in the force expressions, leaving a net zero force on the particle. However, in the case of a ring of matter, the distances L_1 and L_2 appear only to the first power in the mass terms, so they cancel only one of the powers in the inverse squares. This is why the net force along every ray is positive in the direction of the closer point on the ring, which implies that a particle in the interior plane of the ring would move outward toward the ring. There would be an equilibrium point at the center of the ring, but this point would not be stable, i.e., it would be a local maximum of the potential energy.

To quantitatively determine the Newtonian gravitational force for a massive ring of radius R, we can note first that the field has cylindrical symmetry, so all that matters is the radial distance (from the ring's center) of a test particle and its elevation above the plane of the ring. It's convenient to use cylindrical coordinates, r,θ,z, where r is the radial distance from the center of the ring, θ is the angle in the plane of the ring, and z is the elevation above (or below) the plane of the ring. Without loss of generality we can focus on a test particle at a distance r from the origin and at an angle $\theta = 0$. Letting ρ denote the mass per unit length along the circumference of the ring, each increment of the ring contributes the incremental mass $dm = \rho R d\theta$. This mass is at a distance s from the test particle, given by

$$s = \sqrt{[R\cos(\theta) - r]^2 + [R\sin(\theta)]^2 + z^2} = \sqrt{R^2 - 2Rr\cos(\theta) + r^2 + z^2}.$$

The corresponding incremental potential is $-dm/s$, so the total gravitational potential is given by the integral

$$\Psi(r,z) = \int_0^{2\pi} \frac{-\rho R\, d\theta}{\sqrt{R^2 - 2Rr\cos(\theta) + r^2 + z^2}}.$$

The gravitational force in the r and z directions is given by the derivatives $d\psi/dr$ and $d\psi/dz$. There is no closed-form expression for the above integral, but we can bring the differentiations inside the integrals, giving the following expressions for the radial and axial forces

$$F_r = -\frac{\partial \Psi}{\partial r} = \rho R \int_0^{2\pi} \frac{R\cos(\theta) - r}{\sqrt{\left(R^2 - 2Rr\cos(\theta) + r^2 + z^2\right)^{3/2}}} d\theta$$

$$F_z = -\frac{\partial \Psi}{\partial z} = \rho R \int_0^{2\pi} \frac{-z}{\sqrt{\left(R^2 - 2Rr\cos(\theta) + r^2 + z^2\right)^{3/2}}} d\theta$$

Using these equations, with R = ρ = 1, the radial gravitational force near a ring of matter at various elevations above the plane of the ring is plotted in the figure below.

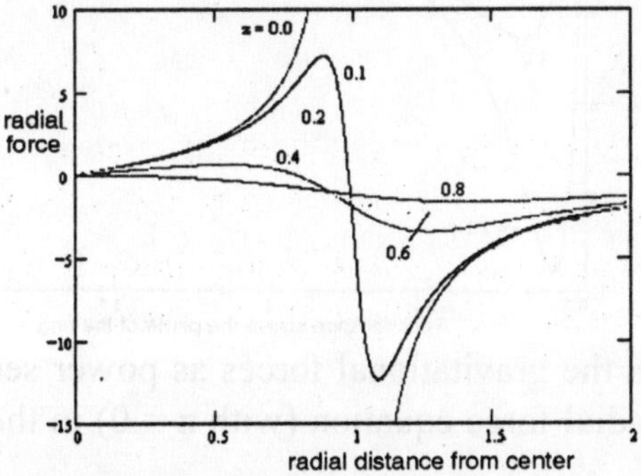

The curve labeled "z = 0.0" is for points in the plane of the ring. As discussed previously, for points away from the center the net gravitational force is positive, i.e., outward, directly away from the center.

However, at an elevation of 0.8R above the plane of the ring, the net radial force is negative as we move away from the central axis, meaning the radial force tends to push a test particle back toward the central axis. Even for elevations of just 0.1R, there is a point inside the ring's radius where the radial force passes through zero and then becomes negative. For each elevation (other than zero) there is a radial position of stable equilibrium. Of course, these locations are not equilibrium positions in the axial direction, as shown by the plot below, which gives the axial force as a function of axial distance (from the plane of the ring) for various radial positions.

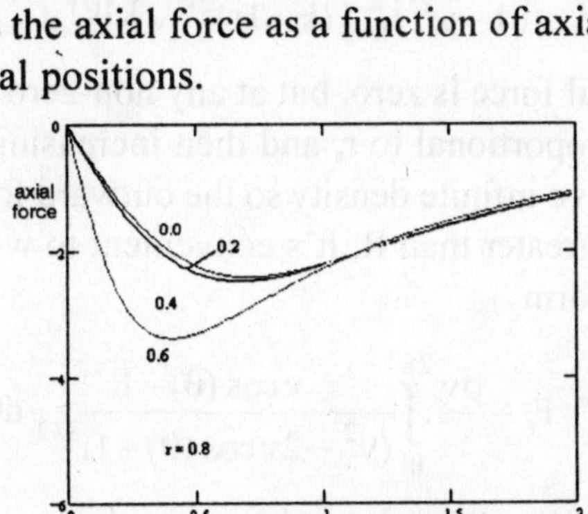

This plot shows that there is always a net axial force toward the plane of the ring. Hence a test particle is always driven toward the z = 0 plane, on which it is always driven toward the r = R radial position, i.e., it is propelled toward the ring. The r = 0 curve in this plot represents the axial force as a function of axial position for a test particle on the central axis. For each radial position there is a unique elevation of maximum force, pulling the test particle back toward the plane of the ring. A similar plot for radial positions outside the radius of the ring is shown below.

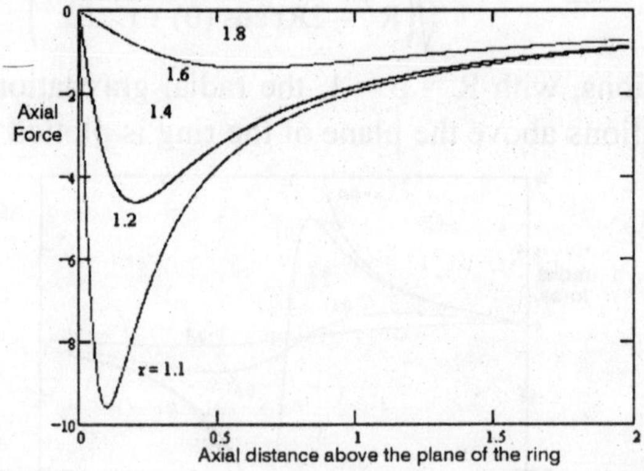

We can also express the gravitational forces as power series. If r is less than R it's convenient to write the radial force equation (with z = 0) in the form

$$F_r = \frac{\rho}{R}\int_0^{2\pi}\frac{\cos(\theta)-\mu}{\left(1-2\mu\cos(\theta)+\mu^2\right)^{3/2}}d\theta$$

where μ = r/R. Expanding this in powers of μ and dividing out the geometric series gives

$$F_r = \left(\frac{\pi r\rho}{R}\right)\left(\frac{1}{1-\mu^2}\right)\left[1+\frac{1}{8}\mu^2+\frac{3}{64}\mu^4+\frac{25}{1024}\mu^6+\frac{1323}{131072}\mu^8+\ldots\right]$$

$$= \left(\frac{\pi r\rho}{R}\right)\left(\frac{1}{1-\mu^2}\right)\sum_{k=0}^{\infty}\frac{1}{(k+1)4^{2k}}\left(\frac{(2k)!}{k!k!}\right)^2\mu^{2k}.$$

Thus at r = 0 the radial force is zero, but at any non-zero value of r there will be a net outward force, initially proportional to r, and then increasing sharply as r approaches R. (The ring is assumed to have infinite density so the outward force goes to infinity at r = R.) On the other hand, if r is greater than R, it's convenient to write the radial force equation (again with z = 0) in the form

$$F_r - \frac{\rho\nu}{r}\int_0^{2\pi}\frac{\nu\cos(\theta)-1}{(\nu^2-2\nu\cos(\theta)+1)^{3/2}}d\theta$$

where ν = R/r. Expanding this gives the series

$$\text{Fr} = -\left(\frac{2\pi R\rho}{r^2}\right)\left(\frac{1}{1-\nu^2}\right)\left[1-\frac{1}{4}\nu^2-\frac{3}{64}\nu^2-\frac{5}{256}\nu^8+\ldots\right]$$

$$= \left(\frac{2\pi R\rho}{r^2}\right)\left(\frac{1}{1-\nu^2}\right)\sum_{k=0}^{\infty}\frac{1}{(2k-1)4^{2k}}\left(\frac{(2k)!}{k!k!}\right)^2 \nu^{2k}.$$

So this starts out infinite on the outer surface (where $\nu = 1$), drops off steeply in the near field, and then approach an inverse-square relation in the far field. It's interesting to consider how closely the gravitational field of a general torus is to that of the same mass distributed on a thin ring. Obviously if we arrange a set of spheres in a circle, their field is the same as if their masses were located at points on the ring. We can also arrange a set of overlapping spheres with most of the overlap removed by internal spheres, all of which are centered on the ring, so in this way we can approximate very closely a general torus with just positive and negative spheres that can be treated as points on the central ring.

THE LAPLACE EQUATION AND HARMONIC FUNCTIONS

Given a scalar field φ, the Laplace equation in Cartesian coordinates is

$$\frac{\partial^2\varphi}{\partial x^2}+\frac{\partial^2\varphi}{\partial y^2}+\frac{\partial^2\varphi}{\partial z^2}=0.$$

This equation arises in many important physical applications, such as potential fields in gravitation and electro-statics, velocity potential fields in fluid dynamics, etc. Typically we are given a set of boundary conditions and we need to solve for the (unique) scalar field φ that is a solution of the Laplace equation and that satisfies those boundary conditions. A solution of Laplace's equation is called a "harmonic function" (for reasons explained below). Since the Laplace equation is linear, the sum of two or more individual solutions is also a solution.

The usual approach to solving the Laplace equation is to seek a "separable" solution given by the product of independent function of x, y, and z, as follows

$$\varphi(x,y,z) = A(x)\,B(y)\,C(z).$$

Every solution can be expressed as a sum of (possibly infinitely many) solutions of this form. Inserting this into the Laplace equation and evaluating the derivatives gives

$$\frac{d^2A(x)}{dx^2}B(y)\,C(z)+A(x)\frac{d^2A(x)}{dx^2}C(z)\;+A(x)B(y)\frac{d^2A(x)}{dx^2}=0.$$

Dividing through by the product A(x)B(y)C(z), this can be written in the form

$$\frac{1}{A(x)}\frac{d^2A(x)}{dx^2}+\frac{1}{B(x)}\frac{d^2B(y)}{dy^2}+\frac{1}{C(z)}\frac{d^2C(z)}{dy^2}=0.$$

Since x, y, and z can be varied independently, this equation can be identically satisfied only if each of the three terms is a constant, and these three constants sum to zero. It will be convenient to denote these constants by squared quantities, α^2, β^2, γ^2 respectively. Thus we have three separate ordinary differential equations

$$\frac{1}{A(x)}\frac{d^2A(x)}{dx^2} = \alpha^2 \quad \frac{1}{B(y)}\frac{d^2B(y)}{dy^2} = \beta^2 \quad \frac{1}{C(z)}\frac{d^2C(z)}{dy^2} = \gamma^2$$

with the condition $\alpha^2 + \beta^2 + \gamma^2 = 0$. Taking just the first of these, we can re-write it in the form

$$\frac{d^2A(x)}{dx^2} - \alpha^2 A(x) = 0$$

which the familiar equation of harmonic motion. This is the reason solutions of Laplace's equation are called "harmonic functions". The characteristic polynomial is

$$s^2 - \alpha^2 = 0$$

whose roots are $\pm\alpha$, so the general solution of the differential equation is

$$A(x) = a_1 e^{\alpha x} + a_2 e^{-\alpha x}$$

for constants a_1 and a_2. We arrive at similar solutions for B(y) and C(z), so the general solution of the Laplace equation of the form A(x)B(y)C(z) is

$$\varphi(x, y, z) = (a_1 e^{\alpha x} + a_2 e^{-\alpha x})(b_1 e^{\beta y} + b_2 e^{-\beta y})(c_1 e^{\gamma z} + c_2 e^{-\gamma z})$$

where the a_i and b_i are arbitrary constants, and α, β, γ are any complex constants such that $\alpha^2 + \beta^2 + \gamma^2 = 0$. By summing terms of this general form using the techniques of Fourier analysis we can determine the values of these constants necessary to match any given boundary conditions.

The above analysis is useful when the boundary conditions are easily expressible in terms of Cartesian coordinates, but we often encounter problems that are easier to express in terms of other, possibly curvilinear, coordinate systems. The general form of the Laplacian operator for an arbitrary (differentiable) coordinate system x^k, k = 1,2,3, with the metric tensor g_{ij} is

$$\nabla^2\varphi = \frac{1}{\sqrt{g}}\sum_{m=1}^{3}\sum_{n=1}^{3}\frac{\partial}{\partial x^m}\left(\sqrt{g}g^{mn}\frac{\partial\varphi}{\partial x^n}\right)$$

where g is the determinant of the covariant metric tensor. If the basis vectors of the coordinate system are mutually orthogonal at each point, the fundamental line element has the diagonal form

$$(ds)^2 = (h_1 dx^1)^2 + (h_2 dx)^2 + (h_3 dx^3)^2$$

where $h_i = \sqrt{g_{ii}}$. In this case we have $\sqrt{g} = h_1h_2h_3$ and $g_{ij} = 0$ for all $i \neq j$. Also, the non-zero components of the contravariant metric are $g^{ii} = 1/g_{ii}$, so the Laplacian expressed in terms of orthogonal coordinates has the form

$$\nabla^2\varphi = \frac{1}{h_1h_2h_3}\left[\frac{\partial}{\partial x^1}\left(\frac{h_2h_3}{h_1}\frac{\partial\varphi}{\varphi x^1}\right) + \frac{\partial}{\partial x^2}\left(\frac{h_1h_3}{h_2}\frac{\partial\varphi}{\partial x^2}\right) + \frac{\partial}{\partial x^3}\left(\frac{h_1h_2}{h_3}\frac{\partial\varphi}{\partial x^3}\right)\right].$$

For example, ordinary spherical coordinates r, θ, φ where

$$x = r\sin(\theta)\cos(\phi) \quad y = r\sin(\theta)\cos(\phi) \quad z = r\sin(\theta)\cos(\theta)$$

are orthogonal, i.e., the basis vectors in the r, θ and φ directions are mutually orthogonal at each point, so we can express the Laplacian in terms of spherical coordinates by means of the preceding expression. The line element for spherical coordinates is found by simply evaluating the derivatives ds/dr, etc., along the coordinate axes. This gives

$$(ds)^2 = (dr)^2 + r^2(d\theta)^2 + r^2\sin(\theta)^2(d\phi)^2$$

and we have $h_r = 1$, $h_\theta = r$, $h_\phi = r\sin(\theta)$. Inserting these into the general expression for the Laplacian in orthogonal coordinates gives

$$\nabla^2\varphi = \frac{1}{r^2}\frac{\partial}{\partial r}\left(r^2\frac{\partial\varphi}{\partial r}\right) + \frac{1}{r^2\sin(\theta)}\frac{\partial}{\partial\theta}\left(\sin(\theta)\frac{\partial\varphi}{\partial\theta}\right) + \frac{1}{r^2\sin(\theta)^2}\left(\frac{\partial^2\varphi}{\partial\phi^2}\right).$$

We are often interested in situations with axial symmetry, such as when determining the potential flow around a sphere. In such cases we can take the axis of symmetry over the angle φ, so we have $\partial\varphi/\partial\phi = 0$, and Laplace's equation reduces to

$$\frac{\partial}{\partial r}\left(r^2\frac{\partial\theta}{\partial r}\right) + \frac{1}{\sin(\theta)}\frac{\partial}{\partial\theta}\left(\sin(\theta)\frac{\partial\phi}{\partial\theta}\right) = 0.$$

As in the case of Cartesian rectangular coordinates, we can focus on separable solutions, i.e., solutions of the form $\varphi(r,\theta) = A(r)B(\theta)$. Evaluating the Laplacian of this, equating this to zero, and dividing through by $A(r)B(\theta)$ gives

$$r^2\frac{\ddot{A}(r)}{A(r)} + 2r\frac{\dot{A}(r)}{A(r)} + \frac{\ddot{B}(\theta)}{B(\theta)} + \frac{\cos(\theta)\dot{B}(\theta)}{\sin(\theta)B(\theta)} = 0$$

where dots signify derivatives of the function with respect to the argument. Since r and θ are independently variable, this equation can be identically satisfied only if the sum of the first two terms is constant, and the sum of the last two terms is constant. Denoting these constant by μ,ν, we have the two separate ordinary linear differential equations

$$r^2A''(r) + 2rA'(r) - \mu A(r) = 0$$
$$\sin(\theta)B''(\theta) + \cos(\theta)B'(\theta) - \nu\sin(\theta)B(\theta) = 0$$

with the condition $\mu + \nu = 0$. We could solve these explicitly, but notice that the first is satisfied by setting $A(r) = r^n$, provided we set $\mu = n(n+1)$. Also, if we set $B(\theta) = \cos(\theta)$ the second equation is satisfied with $\nu = 0$. Hence the function $\varphi(r,\theta) = r^n \cos(\theta)$ is a solution of Laplace's equation provided $n(n+1) = 0$, which implies either $n = 0$ or $n = -1$.

More generally, we can consider solutions that are linear combinations of functions of the form

$$\varphi(r,\theta) = r^n P_n(\cos(\theta))$$

where $P_n(x)$ is a polynomial. Inserting this into the axially symmetric form of Laplace's equation gives

$$n(n+1)r^n P_n(\cos(\theta)) + r^n(\sin(\theta)^2 \ddot{P}_n)(\cos(\theta)) - 2\cos(\theta)\dot{P}_n(\cos(\theta))) = 0$$

where dotted functions signify derivatives of $P_n(x)$ with respect to x. Dividing through by r^n, replacing $\cos(\theta)$ with "x", and noting that $\sin(\theta)^2 = 1-\cos(\theta)^2$, this becomes

$$(1-x^2)\ddot{P}_n(x) - 2x\dot{P}_n(x) + n(n+1)P_n(x) = 0.$$

There is a unique (up to a constant factor) polynomial P_n for each value of $n(n+1) = k$, which implies that each such polynomial is compatible with two distinct values of n, i.e., the two roots n and n' of $n^2 + n = k$. Of course, n need not be an integer, but if it is, we can easily determine the polynomial $P_n(x)$ that satisfies the above equation, simply by inserting a polynomial with undetermined coefficients into the equation and solving for the coefficients. The results for the first few values of n are as shown below:

n	n′	P(x)
0	−1	1
1	−2	x
2	−3	$3x^2 - 1$
3	−4	$5x^2 - 3x$
4	−5	$35x^4 - 30x^2 + 3$

Up to a scale factor, these are the Legendre polynomials. We can express an axially symmetrical solution of the Laplace equation as a linear combination of these individual solutions as follows

$$\varphi(r,\theta) = \left(a_0 + b_0\frac{1}{r}\right) + \left(a_0 r + b_1\frac{1}{r^2}\right)\cos(\theta) + \left(a_2 r^2 + b_2\frac{1}{r^3}\right)(3\cos(\theta)^2 - 1) + \ldots$$

where the a_i and b_i are arbitrary constants. The factors of the form $P_n(\cos(\theta))$ in this expansion are called zonal harmonics. As an example, the velocity potential of the flow of an ideal, irrotational, frictionless fluid past a stationary sphere of radius R is given by the second zonal harmonic with $a_1 = U$ and $b_1 = UR^3/2$, where U is the free-stream velocity of the fluid moving in the positive x direction (assuming the x axis is our axis of symmetry). Thus we have

$$\varphi(r,\theta) = Ur\cos(\theta) + \frac{UR^3}{2r^2}\cos(\theta).$$

The first term is obviously just Ux in Cartesian coordinates, which represents a uniform flow field of constant velocity U in the positive x direction. The second term is a dipole potential, proportional to x/r^3. This is the form of the "far field" potential associated with two nearby oppositely charged particles in electrostatics (aligned along the axis of symmetry). It's interesting that the superposition of these two simple fields gives the potential flow field around a sphere, as shown by the fact that the radial velocity vanishes at r = R.

One of the interesting properties of harmonic functions, i.e., continuous functions φ(x,y,z) that satisfy Laplace's equation, is that the mean value of φ on *any* sphere centered on the point p is equal to the value of φ at the point p. This is called the *mean value theorem*, and a formal proof of it is presented in Differential Operators and the Divergence Theorem. This theorem is often invoked to prove that a continuous differentiable harmonic function cannot contain a local maximum or minimum, because by definition such points have values of φ properly greater then (resp. less than) the values of φ for all the surrounding points, which would make it impossible for the average on the surface of a sphere surrounding the point to equal the value of φ at that point. In fact, by the same reasoning, if the value of a harmonic function at any given point is increasing in one direction then it must be decreasing in some other direction. (Of course, we don't actually need the mean value theorem to prove these facts; we can see at a glance that the Laplace equation automatically precludes local maxima and minima, because at any such point all three of the second partial derivatives with respect to x, y, and z would have the same sign, making it impossible for their sum to be zero.)

From this it follows that if the value of a harmonic function φ has the constant value c over an entire surface that completely encloses a region of space, then the value of φ must be c throughout that region. This is because if the value was anything other than c at any point in the interior of the region, then the region must contain a properly highest (or lowest) value, so there must be a point from which the function is decreasing (or increasing) in some directions but not increasing (or decreasing) in any directions, which is impossible.

Having established this, we can also see that the solution of Laplace's equation satisfying a complete set of boundary conditions is unique. To prove this, suppose two distinct harmonic functions φ_1 and φ_2 have the same values on a closed surface, but have different values in the interior of the enclosed region. We know that harmonic functions are additive, so $\varphi_1 - \varphi_2$ is also a harmonic function, and it's value is zero over the closed surface. But this implies the value of this difference is zero throughout the interior region enclosed by the surface (by the preceding proposition), so φ_1 and φ_2 are identical throughout the region.

It's also interesting to consider the mean value theorem in the context of the general power series of a harmonic function. If we expand $\varphi(x,y,z)$ into a power series about the point p (after shifting the coordinates to place p at the origin), we get an expression of the form

$$\begin{aligned}\varphi(r,\theta) &= c_{000} + c_{100}x + c_{010}y + c_{001}z + c_{110}xy \\ &\quad + c_{101}xz + c_{011}yz + c_{200}x^2 + c_{002}z^2 + \dots \\ &= \sum c_{\alpha\beta\gamma}\alpha_y\beta_z\gamma.\end{aligned}$$

Recall that if θ denotes the "latitude" (zero at the equator) and ϕ the "longitude", then the mean value of a function $f(\theta,\phi)$ over the surface of a sphere of radius R is given by the double integral

$$f_{mean} = \frac{1}{4}\int_{-\pi}^{\pi}\int_{-\pi/2}^{\pi/2} f(\theta,\phi)\cos(\theta)d\theta\, d\phi.$$

Making the substitutions $x = R\cos(\phi)\cos(\theta)$, $y = R\sin(\phi)\cos(\theta)$, $z = R\sin(\theta)$ into the expression for φ and evaluating the mean value over this sphere gives

$$\begin{aligned}\varphi_{mean} &= c_{000} + \frac{1}{3}(c_{200} + c_{200} + c_{002})R^2 \\ &\quad + \frac{1}{5}\left(c_{400} + c_{040} + c_{004} + \frac{1}{3}(c_{220} + c_{202} + c_{022})\right)R^4 \\ &= \frac{1}{7}\left(c_{600} + c_{060} + c_{006} + \frac{1}{5}((c_{420} + c_{402} + c_{042} + c_{024} + c_{204}) + \frac{1}{3}(c_{222})\right)R^6 + \dots\end{aligned}$$

In general we find that only even indices appear, and the mean value can be written as

$$\varphi_{mean} = \sum_{\alpha,\beta,\gamma\geq 0} \frac{(\alpha+\beta+\gamma)!}{\alpha!\beta!\gamma!}\frac{(2\alpha)!(2\beta)!(2\gamma)!}{(2(\alpha+\beta+\gamma)+1)!} c_{2\alpha,2\beta,2\gamma} R^{2(\alpha+\beta+\gamma)}.$$

The constant coefficient in this expression is c_{000}, which equals the value of φ at the central point p, corresponding to $R = 0$. For all other values of R, the mean value theorem asserts that the mean value remains constant, so the coefficients of all non-zero powers of R must vanish. To show that they do, we can substitute the original power series for φ into the Laplace equation and collect terms by powers of x,y,z. We find the coefficient of $x^q y^r z^s$ is

$$C_{qrs} = (q+2)(q+1)c_{q+2,r,s} + (r+2)(r+1)c_{q,r+2} + (s+2)(s+1)c_{q,r,s+2}.$$

Each of these coefficients must vanish. With q = r = s = 0 this gives

$$2(c_{200}+c_{020}+c_{002})=0$$

which proves that the coefficient of R^2 in the series for φ_{mean} is zero. Likewise for the (q,r,s) combinations (2,0,0), (0,2,0), and (0,0,2) we have the three conditions

$$12c_{400}+2c_{220}+2c_{202}=0$$
$$12c_{040}+2c_{220}+2c_{022}=0$$
$$12c_{004}+2c_{220}+2c_{202}=0.$$

Summing these three conditions and dividing by 12 gives

$$c_{004}+c_{040}+c_{004}+\frac{1}{3}(c_{220}+c_{202}+c_{022})=0$$

which proves that the coefficient of R^4 in the double integral for φ_{mean} is zero. Proceeding on to the next term, with the (q,r,s) combinations (4,0,0), (0,4,0), (0,0,4), (2,2,0), (2,0,2), and (0,2,2) we get the six conditions

$$30c_{600}+2c_{420}+2c_{402}=0$$
$$30c_{060}+2c_{240}+2c_{042}=0$$
$$30c_{006}+2c_{024}+2c_{204}=0$$
$$12c_{420}+12c_{240}+2c_{222}=0$$
$$12c_{402}+12c_{204}+2c_{222}=0$$
$$12c_{042}+12c_{024}+2c_{222}=0.$$

Multiplying the first three by 3 and then adding all these expressions together and dividing by 90 gives

$$\left(c_{600}+c_{060}+c_{006}+\frac{1}{5}\left((c_{420}+c_{402}+c_{042}+c_{240}+c_{024}+c_{204})+\frac{1}{2}(c_{222})\right)\right)=0$$

which proves that the coefficient of R^6 in the double integral for φ_{mean} is zero. This process can be continued indefinitely, corresponding to the fact that the coefficient of each positive power of R is zero, and hence $\varphi_{mean} = c_{000}$ for all R.

STOKES' MISTAKE

The original interpretation of stellar aberration when it was first discovered by Bradley was in terms of a corpuscular theory of light, but when the wave theory gained prominence in the 19th century the explanation for stellar aberration became more difficult. The basic problem was that no one knew if the putative luminiferous ether was immobile or participated in the motions of ordinary material objects. On the assumption that it was just slightly dragged along by ordinary objects, Fresnel had given a fairly successful account of aberration and other first order optical phenomena, but many physicists were dissatisfied with the idea that the ether was so little affected by ordinary matter.

In 1845 George Gabriel Stokes sought to interpret the effect of stellar aberration in terms of an ether that was totally dragged along with material bodies. He assumed that, for macroscopic frequencies, the ether acted as an incompressible and irrotational fluid with no viscosity (and hence no drag), and that the ether's velocity matches the velocity of any solid object at the boundary of the object.

On the basis of these assumptions he derived the amount by which incoming plane waves would be tilted due to the flow of the carrying medium, and found that the result matched the observed aberration of starlight.

However, Hendrik Lorentz later pointed out that Stokes' premises are mutually exclusive, because the velocity field of a potential flow past a spherical object does not match the sphere's speed at the boundary. Nevertheless, it's interesting to review the derivation, to understand precisely how the a flow field with a velocity potential leads to deflection of rays. Consider a portion of a plane wave, centered at the origin of xyz coordinates, with the normal pointing in the positive x direction as shown below.

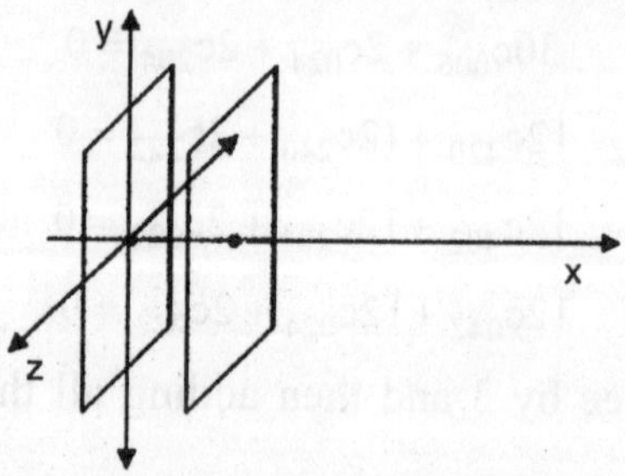

Since the wave is assumed to be propagating through the ether, and the ether itself is moving (relative to the frame of the quiescent background ether far from any bodies), the wavefront acquires additional components of velocity from the ether. If the motion of the ether varies from place to place, it can have the effect of tilting the wavefronts. To the first order the only significant effects are due to the component of the ether's velocity in the direction of the wave propagation. Thus, letting $u = [u_x, u_y, u_z]$ denote the velocity vector of the ether, we will focus on the component $u_x(x,y,z)$. In the xy plane the motion of a wave front from the time t to the time t + dt is illustrated below.

This shows that if the x-component of the ether speed is increasing with y, then the wavefront will be tilted downward, i.e., toward the direction of decreasing y. This tilt can be represented by an increment dy applied to the normal vector N at time t to give the new normal vector N at time t + dt. Likewise we can determine the tilt in the z direction based on the partial derivative of u_x with respect to z. Quantitatively we have

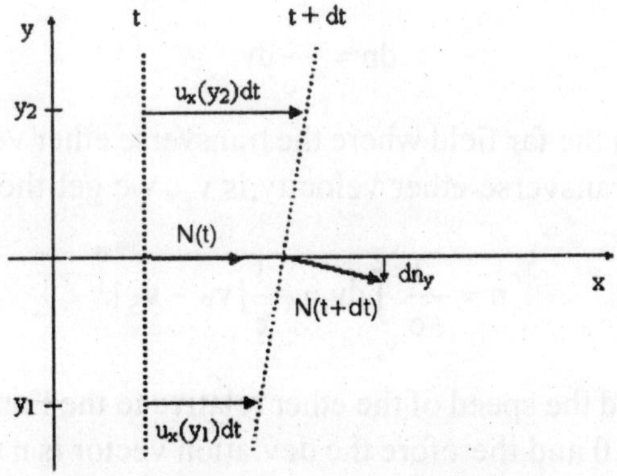

$$dn_y = -\frac{\partial u_x}{\partial y}dt \qquad dn_z = -\frac{\partial u_x}{\partial z}dt.$$

Changes in u_x with respect to x don't have any first-order tilting effect, so these are the only two components that contribute to tilting the wavefront. Thus the incremental deviation of the normal vector is a vector parallel to the yz plane

$$dn = (dn_y, dn_z) = \left(-\frac{\partial u_x}{\partial y}, -\frac{\partial u_x}{\partial z}\right)dt.$$

At this point we face the seemingly challenging task of integrating these incremental deviations over the entire path of the wave from the emitting star to the Earth.

In addition, it may seem unlikely to yield an answer that agrees with observation, because these equations involve just the derivatives of the speed component u_x of the ether in the direction of the wave, whereas the aberration angle ought to be determined by the transverse components u_y and u_z of the ether.

However, at this point Stokes assumed the ether is incompressible (so that the speed of light is homogeneous), inviscid (to avoid drag), and irrotational. The latter assumption ensures that the ether supports a velocity potential. By definition the vorticity is zero in an irrotational fluid, so we have

$$\frac{\partial u_x}{\partial y} = \frac{\partial u_y}{\partial z} \qquad \frac{\partial u_x}{\partial z} = \frac{\partial u_z}{\partial x}.$$

Thus the incremental tilting vector can be written as

$$dn = \left(-\frac{\partial u_y}{\partial x}, -\frac{\partial u_z}{\partial x}\right)dt = -\frac{dv}{dx}dt$$

where the transverse ether velocity v is the projection of the ether velocity vector u onto the yz plane (i.e., transverse to the direction of the wave). Since dx/dt = c to the first order, we have

$$dn = -\frac{1}{c}dv.$$

Integrating this from the far field where the transverse ether velocity is v_S to the surface of the Earth where the transverse ether velocity is v_E, we get the total angular deflection

$$n = -\frac{1}{c}\int_{v_s}^{v_E} dv = -\frac{1}{c}[v_E - v_S].$$

Now Stokes assumed the speed of the ether relative to the Earth was zero at the Earth's surface, so we have $v_E = 0$ and therefore the deviation vector is $n = v_S/c$. The magnitude of v_S is simply $\sin(\theta)$ times the magnitude of u_S, where θ is the elevation angle of the star, so the aberration angle is given by

$$\sin(\alpha) = \sin(\theta)\frac{|u_S|}{c}$$

in agreement with observation.

Unfortunately, as Lorentz pointed out, there is a serious flaw in this reasoning. Potential flow past a sphere does not result in zero relative flow velocity at the surface of the sphere. The flow velocity at the surface is parallel to the surface, but it is non-zero, and varies significantly from one point to another on the surface. Therefore, if the luminiferous ether was a fluid of the type assumed by Stokes, the aberration of starlight would differ significantly from what is observed. Any variation of Stokes' assumptions in order to achieve zero surface velocity, such as Planck's suggestion of allowing the ether to be compressible, also introduces other effects that invalidate the above analysis and lead to aberration angles that differ from what is observed. Therefore, Stokes' theory of a completely dragged ether was unsuccessful. Oddly enough, many popular attempts to explain aberration in terms of a totally dragged fluidic ether are even more seriously flawed than Stokes' theory. They typically imagine simple transverse sliding between layers of the ether, with the incoming rays deflected accordingly. There are two problems with this simplistic explanation. First, with only transverse motion there is no tilting of the wavefronts, so this is really just a disguised ballistic theory rather than a true wave theory. Second, although it's possible to arrive at the correct magnitude of aberration with this approach, the proponents of this explanation tend to overlook the fact that it predicts aberration in the opposite direction from what is actually observed.

Incidentally, Lorentz's derivation of the aberration angle in Stokes' model, although it arrives at the correct conclusion, contains a little slight of hand. The derivation in Lorentz's book "Theory of Electrons" begins by saying that a particle of ether at the point (x_0,y_0,z_0) at time t will have moved after an increment of time dt to the point with coordinates

$$x_1 = x_0 + u_x dt \qquad y_1 = y_0 + u_y dt \qquad z_1 = z_0 + u_x dt.$$

Lorentz then notes that the velocity components (in any small region) can be expressed at linear functions of the coordinates, so we have (for example)

$$u_x(x,y,z) = \alpha x + \beta x + \gamma y + \delta z$$

for a small region around a given particle. In this expression the origins of the x,y,z coordinates are subsumed in the constant α. Of course, the constants β, γ, and δ are just the partial derivatives of u_x with respect to x, y, and z respectively. Substituting this expression for u_x into the equation for x_1, he gets

$$x_1 = x_0 + (\alpha + \beta x + \gamma y + \delta z)dt.$$

He then considers the set of points beginning on the plane x = a with various values of y and z, to determine the new plane formed by these points after a time dt. Thus we have $x_0 = a$, and Lorentz says we might as well substitute x_1,y_1,z_1 in place of x,y,z because the partial derivatives don't change much over this incremental distance. After making these substitutions, he re-write the equation in the form

$$(1-\beta dt)x_1 - (\gamma dt)y_1 - (\delta dt)z_1 = (x_0 + \alpha)$$

and from this he concludes the direction cosines of the normal to the plane of particles at the time t + dt are proportional to the coefficients (1−βdt), −γdt, and −δdt. This agrees with the y and z components we derived previously, but what are we to make of the x component? This was reached by replacing x with x_1 in the previous equation, but we could just as well have set $x = x_0$, in which case the coefficient of x_1 would be simply 1. Essentially the normal to the plane (to the first order in u/c) is independent of how the ether speed in the x direction changes as a function of x, because this simply advances the entire wavefront at the same rate. Thus the partial derivative of u_x with respect to x has no effect on the tilting vector. It's convenient to include this component in the formulation, as Lorentz did, because it enables us to deal with the entire velocity vector u, without needing to introduce the projection of this vector onto the yz plane, but it works only because the effect of the x component is negligible.

SOMETHING LIKE QUANTUM MECHANICS

Suppose one of the observable attributes X of a physical system has the form of a binary variable x which can be either true or false, and suppose we have a physical law that tells us how the value of x changes with time in specified circumstances. Letting a numerical value of x = +1 signify "true" and a numerical value of x = −1 signify "false", the equation(s) expressing our physical law may be of the form x = f(t) where the function f takes on only the values -1 or +1. In this case we say the physical law ranges exactly over the observable states of the system.

However, it's conceivable that a physical law might range over a larger set than the set of observable states. For example, the function f(t) representing the physical law governing the behavior of x might range over the entire set of real numbers on the interval from −1 to +1, even though the only observable states correspond to the two specific values −1 and +1, as depicted below.

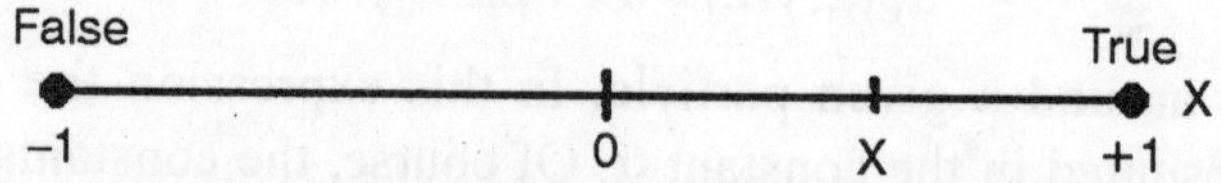

In order for the physical law to be meaningful we need some interpretation of the intermediate values of x, values which do not correspond to observable states. One possible interpretation is that the numerical value of x indicates the propensity of the system to be observed in State +1 if a measurement of the X attribute is made. If the numerical value of x is −1, 0, or +1, the probability that X will be observed in State +1 should be 0, 1/2, and 1 respectively. Naturally the probability of being observed in State −1 is the complement. This leads us to define

$$P\{x=+1\}=\frac{1+x}{2} \qquad P\{x=-1\}=\frac{1-x}{2}.$$

If we find that repeating the same measurement twice in a row invariably yields the same value each time, we may conclude that the state of the system "jumps" from the intermediate value to the observed value when a measurement is made. (Needless to say, this raises a number of questions about what constitutes a measurement.)

We define the *uncertainty* in the attribute X as zero if x has the value -1 or +1, and as infinite if x has the value 0. The simplest function giving these values is

$$U(X)=\frac{1-x^2}{x^2}.$$

It's convenient to represent these ideas by allowing the state of the system to be not just the points between +1 and -1 on the X axis, but at any of the points on the unit circle in the x,y plane as shown below.

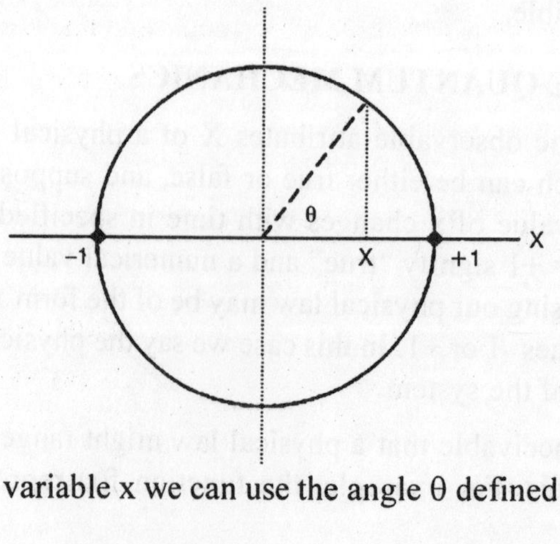

In place of the state variable x we can use the angle θ defined by x = cos(θ). Then the probability of observing the system in State +1 is $[1+\cos(\theta)]/2$ and the uncertainty in X is $U_X(\theta) = \tan(\theta)^2$.

In this representation each variable's real axis is accompanied by a perpendicular axis, which might be regarded as the axis of another observable variable, which we will call Y, as shown below.

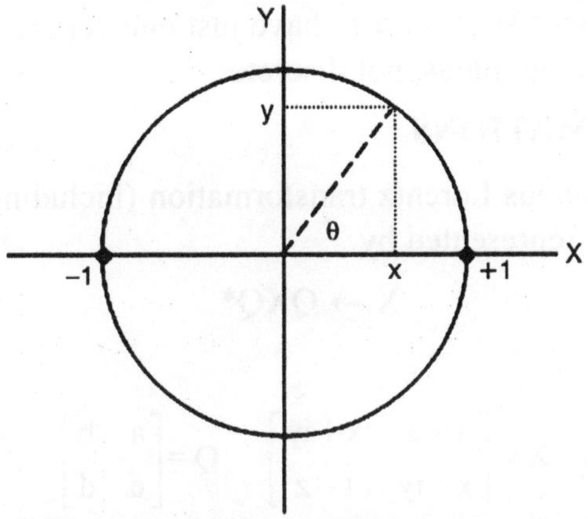

Just as with the X variable, we can also make a measurement of the Y variable, with the result that the system jumps to +1 on the Y axis (if Y is observed to be true) or −1 on the Y axis (if Y is observed to be false). In terms of the parameter θ the probability of State +1 is $[1+\sin(\theta)]/2$ and the uncertainty in Y is $U_Y(\theta) = \tan(\theta + \pi/2)^2$. The product of the uncertainties in the X and Y variables is therefore

$$U_x(\theta)U_Y(\theta) = \tan(\theta)^2 \tan(\theta + \pi/2)^2 = 1.$$

This is the minimum amount of uncertainty that we can have in the joint state of these two variables, because by measuring one of them, reducing its individual uncertainty to zero, the uncertainty in the other value becomes infinite. We can imagine a measurement process that is intermediate between X and Y, essentially selecting another axis on which to project the state of the system, but after performing this intermediate measurement the product of uncertainties will still be 1.

This representation can immediately be generalized to three variables by allowing the state of the system to reside at any point on a unit sphere in three dimensions. Then depending on which measurement we perform, the system will be projected onto one of the three axes. The equations of motion (i.e., the physical laws) governing the system range over the entire surface of the sphere, but the only observable states are the six points [±1,0,0], [0, ±1,0], and [0,0, ±1]. These are the eigenvalues for rotations about the x, y, and z axes respectively.

The two non-classical features of the model described above are (1) the laws of motion range over a larger set of states than the set of observable states, and (2) some sets of observable variables are inherently inter-dependent, in the sense that they can be represented by a smaller number of state variables. Thus the system variables are over-determined in one sense and under-determined in another. They are over-determined because there are more states than observable states, but they are under-determined because there are more observable variables than state variables. For example, our original X variable has only two observable values (-1 or +1) but infinitely many state values (the real numbers from -1 to +1), and the same is true for the Y variable, but these two variables can actually be represented by a single variable, θ, so they have just one degree of freedom, not two, but this degree of freedom is continuous, not discrete.

MOBIUS TRANSFORMATIONS

Any proper orthochronous Lorentz transformation (including ordinary rotations and relativistic boosts) can be represented by

$$X \rightarrow QXQ^*$$

where

$$X = \begin{bmatrix} t+z & x+iy \\ x-iy & t-z \end{bmatrix} \qquad Q = \begin{bmatrix} a & b \\ c & d \end{bmatrix}$$

and Q* is the transposed conjugate of Q. The coefficients a,b,c,d of Q are allowed to be complex numbers, normalized so that ad –bc = 1. Just to be explicit, this implies that if we define

$$a = a_1 + a_2 \qquad b = b_1 + ib_2 \qquad c = c_1 + ic_2 \qquad d = d_1 + id_2$$

then the Lorentz transformation (1) is

$$\begin{aligned} t' &= [a_1b_1 + a_2b_2 + c_1d_1 + c_2d_2]x + [a_1b_2 - a_2b_1 + c_1d_2 - c_2d_1]y \\ &\quad + (1/2)\left[|a|^2 - |b|^2 + |c|^2 - |d|^2\right]z + (1/2)\left[|a|^2 + |b|^2 + |c|^2 + |d|^2\right]t \\ x' &= [a_1d_1 + a_2d_2 + b_1c_1 + b_2c_2]x + [a_1d_2 - a_2d_1 + b_1c_2 + b_2c_1]y \\ &\quad + [a_1c_1 + a_2c_2 - b_1d_1 - b_2d_2]z + [a_1c_1 + a_2c_2 - b_1d_1 - b_2d_2]t \\ y' &= [-a_1d_1 + a_2d_1 + b_1c_2 + b_2c_1]x + [a_1d_1 - a_2d_2 + b_1c_1 + b_2c_2]y \\ &\quad + [-a_1c_2 + a_2c_1 - b_1d_2 - b_2d_1]z + [-a_1c_2 + a_2c_1 - b_1d_2 - b_2d_1]t \\ z' &= [a_1b_1 + a_2b_2 - c_1d_1 - c_2d_2]x + [a_1b_2 - a_2b_1 + c_1d_2 + c_2d_1]y \\ &\quad + (1/2)\left[|a|^2 - |b|^2 - |c|^2 + |d|^2\right]z + (1/2)\left[|a|^2 + |b|^2 - |c|^2 - |d|^2\right]t \end{aligned}$$

Two observers at the same point in spacetime but with different orientations and velocities will "see" incoming light rays arriving from different relative directions with respect to their own frames of reference, due partly to ordinary rotation, and partly to the aberration effect described in the previous section. This leads to the remarkable fact that the combined effect of any proper orthochronous (and homogeneous) Lorentz transformation on the incidence angles of light rays at a point corresponds precisely to the effect of a particular linear fractional transformation on the Riemann sphere via ordinary stereographic projection from the extended complex plane. The latter is illustrated below:

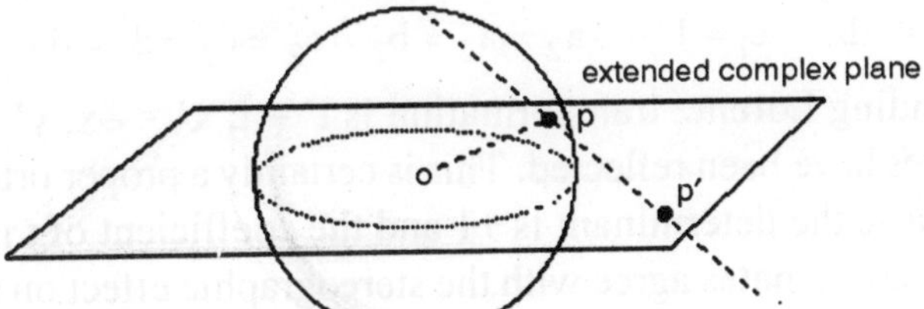

The complex number p in the extended complex plane is identified with the point p' on the unit sphere that is struck by a line from the "North Pole" through p. In this way we can identify each complex number uniquely with a point on the sphere, and vice versa. (The North Pole is identified with the "point at infinity" of the extended complex plane, for completeness.)

Relative to an observer located at the center of the Riemann sphere, each point of the sphere lies in a certain direction, and these directions can be identified with the directions of incoming light rays at a point in spacetime.

If we apply a Lorentz transformation of the form (1) to this observer, specified by the four complex coefficients a,b,c,d, the resulting change in the directions of the incoming rays of light is given exactly by applying the linear fractional transformation (also known as a Mobius transformation)

$$w \rightarrow \frac{aw+b}{cw+d} \qquad a,b,c,d \in C, \quad ab - bc = 1$$

to the points of the extended complex plane. Of course, our normalization ad −bc = 1 implies the two conditions

$$a_1 d_1 - a_2 d_2 - b_1 c_1 + b_2 c_2 = 1$$
$$a_1 d_2 - a_2 d_1 - b_1 c_2 + b_2 c_1 = 0$$

so of the eight coefficients needed to specify the four complex numbers a,b,c,d, these two constraints reduce the degrees of freedom to six, which is precisely the number of degrees of freedom of Lorentz transformations (namely, three velocity components v_x, v_y, v_z, and three angular specifications for the longitude and latitude of our line of sight and orientation about that line).

To illustrate this correspondence, first consider the "identity" Mobius transformation $w \rightarrow w$. In this case we have

$$a_1 = d_1 = 1 \qquad a_2 = b_1 = b_2 = c_1 = c_2 = d_2 = 0$$

so our Lorentz transformation reduces to t' = t, x' = x, y' = y, z' = z as expected. None of the points move on the complex plane, so none move on the Riemann sphere under stereographic projection, and nothing changes in the sky's appearance. Now let's consider the Mobius transformation w → −1/w. In this case we have

$$b_1 = -1 \qquad c_1 = 1 \qquad a_2 = a_2 = b_2 = c_2 = c_2 = d_1 = d_2 = 0$$

and so the corresponding Lorentz transformation is t' = t, x' = −x, y' = y, z' = −z . Thus the x and z coordinates have been reflected. This is certainly a proper orthochronous Lorentz transformation, because the determinant is +1 and the coefficient of t is positive. But does reflecting the x and z coordinates agree with the stereographic effect on the Riemann sphere of the transformation w → −1/w? Note that the point w = r + 0i maps to −1/r + 0i.

There's a nice little geometric demonstration that the stereographic projections of these points have coordinates (x,0,z) and (−x,0,−z) respectively, noting that the two projection lines have negative inverse slopes and so are perpendicular in the xz plane, which implies that they must strike the sphere on a common diameter (by Pythagoras' theorem).

A similar analysis shows that points off the real axis with projected coordinates (x,y,z) in general map to points with projections (−x,y,−z) points.

The two examples just covered were both trivial in the sense that they left t unchanged. For a more interesting example, consider the Mobius transformation w → w + p, which corresponds to the Lorentz transformation

$$t' = (1 + p^2/2)t + px - (p^2/2)z$$
$$x' = pt + x - pz$$
$$y' = y$$
$$z' = (p^2/2)t + px + (1 - p^2/2)z$$

If we denote our spacetime coordinates by the column vector X with components x_0 = t, x_1 = x, x_2 = y, x_3 = z, then the transformation can be written as

$$X' = LX$$

where

$$L = \begin{bmatrix} 1+p^2/2 & p & 0 & -p^2/2 \\ p & 1 & 0 & -p \\ 0 & 0 & 1 & 0 \\ p^2/2 & p & 0 & 1-p^{2/2} \end{bmatrix}$$

To analyze this transformation it's worthwhile to note that we can decompose any Lorentz transformation into the product of a simple boost and a simple rotation. For a given relative velocity with magnitude |v| and components v_1, v_2, v_3, let γ denote the "boost factor"

$$\gamma = \frac{1}{\sqrt{1-|v|^2}}.$$

It's clear that

$$L_{00} = \gamma \qquad L_{10} = \gamma v_1 \qquad L_{20} = \gamma v_2 \qquad L_{30} = \gamma v_3.$$

Thus, these four components of L are fixed purely by the boost. The remaining components depend on the rotational part of the transformation. If we define a "pure boost" as a Lorentz transformation such that the two frames see each other moving with velocities (v_1,v_2,v_3) and $(-v_1,-v_2,-v_3)$ respectively, then there is a unique pure boost for any given relative velocity vector v_1,v_2,v_3. This boost has the components

$$\begin{bmatrix} \gamma & \gamma v_1 & \gamma v_2 & \gamma v_3 \\ Qv_1v_0 & 1+Qv_1^2 & Qv_1v_2 & Qv_1v_3 \\ Qv_2v_0 & Qv_2v_0 & 1+Qv_2^2 & Qv_2v_3 \\ Qv_3v_0 & Qv_3v_1 & Qv_3v_2 & 1+Qv_2^2 \end{bmatrix}$$

where $Q = (\gamma-1)/|v|^2$. From our expression for L we can identify the components to give the boost velocity in terms of the Mobius parameter p

$$v_1 = \frac{2p}{2+p^2} \qquad v_2 = 0 \qquad v_3 = \frac{p^2}{2+p^2}$$

and

$$|v|^2 = \left(\frac{p^2}{2+p^2}\right)(4+p^2) \quad \gamma = 1+\frac{p^2}{2} \qquad Q = \frac{(2+p^2)^2}{2(4+p^2)}.$$

From these we write the pure boost part of L as follows

$$B = \begin{bmatrix} 1+\frac{p^3}{2} & p & 0 & \frac{p^3}{2} \\ p & 1+\frac{2p^2}{4+p} & 0 & \frac{p^3}{4+p^2} \\ 0 & 0 & 1 & 0 \\ p^2/2 & \frac{p^3}{4+p} & 0 & 1+\frac{p^4}{2(4+p^2)} \end{bmatrix}.$$

We know that our Lorentz transformation L can be written as the product of this pure boost B times a pure rotation R, i.e., L = BR, so we can determine the rotation

$$R = B^{-1}L$$

which in this case gives

$$B=\begin{bmatrix}1 & 0 & 0 & 0\\ 0 & \dfrac{4-p^2}{4+p^2} & 0 & \dfrac{-4p}{4+p^2}\\ 0 & 0 & 1 & 0\\ 0 & \dfrac{4p}{4+p^3} & 0 & 1+\dfrac{4-p^2}{4+p^2}\end{bmatrix}.$$

In terms of Euler angles, this represents a rotation about the y axis through an angle of

$$\theta=\operatorname{inv}\cos\left(\frac{4-p^2}{4+p^2}\right).$$

The correspondence between the coefficients of the Mobius transformation and the Lorentz transformation described above assumes stereographic projection from the North pole to the equatorial plane. More generally, if we're projecting from the North Pole of the Riemann sphere to a complex plane parallel to (but not necessarily on) the equator, and if the North Pole is at a height h above the plane, then every point in the plane is a factor of h further away from the origin than in the case of equatorial projection (h=1), so the Mobius transformation corresponding to the above Lorentz transformation is w → (Aw+B)/(Cw+D) where

$$A=a\sqrt{h}\qquad B=\frac{b}{\sqrt{h}}\qquad C=c\sqrt{h}\qquad D=\frac{d}{\sqrt{h}}.$$

It's also worth noting that the instantaneous aberration observed by an *accelerating* observer does not differ from that observed by a momentarily co-moving inertial observer. We're referring here to the null (light-like) rays incident on a *point* of zero extent, so this is not like a finite spinning body whose outer edges have significant velocities relative to their centers. We're just referring to different coordinate systems whose origins coincide at a given point in spacetime, and describing how the light rays pass through that point in terms of the different coordinate systems at that instant. In this context the acceleration (or spinning) of the systems make no difference to the answer. In other words, as long as our inertial coordinate system has the same velocity and orientation as the (ideal point-like) observer at the moment of the observation, it doesn't matter if the observer is in the process of changing his orientation or velocity. (This is a corollary of the "clock hypothesis" of special relativity, which asserts that a traveler's time dilation at a given instant depends only on his velocity and not his acceleration at that instant.)

In general we can classify Mobius transformations (and the corresponding Lorentz transformations) according to their "squared trace", i.e., the quantity

$$\sigma=\frac{(a+d)^2}{(ad-bc)}.$$

This is also the "conjugacy parameter", i.e., two linear fractional transformations are conjugate if and only if they have the same value of σ. The different kinds of transformations are listed below:

$$0 \le \sigma < 4 \quad \text{elliptic}$$
$$\sigma = 4 \quad \text{parabolic}$$
$$\sigma > 4 \quad \text{hyperbolic}$$
$$\sigma < 0 \text{ or not real} \quad \text{loxodromic}$$

For example, the class of pure rotations are a special case of elliptic transformations, having the form

$$w \to \frac{aw + b}{-\bar{b}w + \bar{a}} \qquad \text{with} \qquad a\bar{a} + b\bar{b} = 1$$

where an overbar denotes complex conjugation. Also, it's not hard to show that the compositions of an arbitrary linear fractional transformation f(z) are cyclical with a period m if and only if $\sigma = 4\cos(2k\pi/m)^2$.

We've seen that the general finite transformation of the incoming null rays can be expressed naturally in the form of a finite Mobius transformation of the complex plane (under sterographic projection). This is a very simple algebraic operation, given by the function

$$f(z) = \frac{az + b}{cz + d}$$

for complex constants a,b,c,d. This generates the discrete sequence $f_1(z) = f(z)$, $f_2(z) = f(f(z))$, $f_3(z) = f(f(f(z)))$, and so on for all $f_n(z)$ where n is a positive integer. It's also possible to parameterize a Mobius transformation to give the corresponding infinitesimal generator, which can be applied to give "fractional iterations" such as $f_{1/2}(z)$, or more generally the continuously parameterized transformation $f_p(z)$ for any real (or even complex) value of p. To accomplish this we must (in general) first map the discrete generator f(z) to a domain in which it has some convenient exponential form, then apply the pth-order transformation, and then map back to the original domain. There are several cases to consider, depending on the character of the discrete generator.

In the degenerate case when ad = bc with c ≠ 0, the pth iterate of f(z) is simply the constant $f_p(z) = a/c$. On the other hand, if c = 0 and a = d ≠ 0, then $f_p(z) = z + (b/d)p$. The third case is with c = 0 and a ≠ d. The pth iterate of f(z) in this case is

$$f_p(z) = \left(\frac{a}{d}\right)^p \left(z - \frac{b}{d-a}\right) + \left(\frac{b}{d-a}\right)$$

Notice that the second and third cases are really linear transformations, since c = 0.

The fourth case is with $c \neq 0$ and $(a+d)^2/(ad-bc) = 4$, which leads to the following closed form expression for the pth iterate

$$f_p(z) = \frac{1}{\left(\frac{2b}{d-a+2cz}\right)+\left(\frac{2c}{a+d}\right)} + \left(\frac{a-d}{2c}\right).$$

This corresponds to the case when the two fixed points of the Mobius transformation are co-incident. In this "parabolic" case, if $a+d = 0$ then the Mobius transformation reduces to the first case with $ad-bc = 0$.

Finally, in the most general case we have $c \neq 0$ and $(a+d)^2/(ad-bc) \neq 4$, and the pth iterate of f(z) is given by

$$f_p(z) = \frac{A}{1-K^p\left(1-\frac{A}{z-B}\right)} + B$$

where

$$A = \pm\sqrt{\left(\frac{a-d}{c}\right)^2 + \frac{4b}{c}} \qquad B = \left(\frac{a-d}{2c}\right) \mp \sqrt{\left(\frac{a-d}{c}\right)^2 + \frac{b}{c}}$$

$$K = \frac{1 \pm \sqrt{1-(4/\sigma)}}{1 \mp (4/\sigma)} \qquad \sigma = \frac{(a+d)^2}{ad-bc}.$$

This is the general case with two distinct fixed points. (If $a+d = 0$ then $\sigma = 0$ and $K = -1$.)

The parameters A and B are the coefficients of the linear transformation that maps real line to the locus of points with real part equal to 1/2. Notice that the pth composition of f satisfies the relation

$$\left(1-\frac{A}{f_p(z)-B}\right) = K^p\left(1-\frac{A}{z-B}\right)$$

so we have

$$h(f_p(z)) = K^p h(z)$$

where

$$h(z) = \pm\frac{z-(A+B)}{z-B}.$$

Thus $h(f(z)) = Kh(z)$, which shows that f(z) is conjugate to the simple function Kz. Since A+B is the complex conjugate of B, we see that h(z) can be expressed as

$$h(z) = \pm \frac{2cz - \rho}{2cz - \bar{\rho}}$$

where

$$\rho = (a-d) \pm \sqrt{(ad-bc)(\sigma-4)} \quad \bar{\rho} = (a-d) \mp \sqrt{(a-d)+4bc}.$$

This enables us to express the pth composition of any linear fractional transformation with two fixed points, and therefore any corresponding Lorentz transformation, in the form

$$f_p(z) = h^{-1}(Kph(z)).$$

This shows that there is a particular oriented frame of reference, represented by h(z), with respect to which the relation between the oriented frames z and f(z) is purely exponential. (We must refer to *oriented* frames rather than merely frames because the Mobius transformation represented the effects of general orientation as well as velocity boost.)

To show explicitly how the action of fp(z) on the complex plane varies with p, consider the relatively simple linear fractional transformation f(z) with fixed points at 0 and 1 on the real axis, which implies A = 1 and B = 0. In parameterized form the pth composition of this transformation is of the form

$$f_p(z) = \frac{z}{K^p + (1-K^p)z} = \frac{1}{\left(\frac{1-z}{z}\right)K^p + 1}$$

for some complex constant K, and the similarity parameter for this transformation is $\sigma = (1+K)^2/K$. For any given K and complex initial value z = x + iy, let

$$K = Re^{ie} \quad u_r = \frac{x}{x^2+y^2} - 1 \quad u_i = \frac{y}{x^2+y^2}.$$

Then the real and imaginary components of $f_p(z)$ are given by

$$f_p(z)_{real} = \frac{1 + R^p(u_r \cos(p\theta) - u_i \sin(p\theta))}{1 + R^{2p}(u_r^2 + u_i^2) + 2R^p(u_r \cos(p\theta) - u_i \sin(p\theta))}$$

$$f_p(z)_{imag} = \frac{R^p(u_r \sin(p\theta) + u_i \cos(p\theta))}{1 + R^{2p}(u_r^2 + u_i^2) + 2R^p(u_r \cos(p\theta) - u_i \sin(p\theta))}.$$

This makes explicit how the action of $f_p(z)$ on the complex plane is entirely determined by the magnitude and phase angle of the constant K, which, as we saw previously, is given by

$$K = \frac{1 \pm \sqrt{1-(4/\sigma)}}{1 \mp \sqrt{1-(4/\sigma)}}.$$

If a,b,c,d are all real, then σ is real, in which case either K is real (σ>4 or σ<0) or K is complex (0<σ<4) with a magnitude (norm) of 1. However, if a,b,c,d are allowed to be complex, then K can be complex with a magnitude other than 1. Of course, if K is real, we can set R = K and θ = 0, so pθ = 0 for all p, and the above equations reduce to

$$x_n = \frac{1 + K^n u_r}{1 + K^{2n}\left(u_r^{\,2} + u_i^{\,2}\right) + 2K^n u_r}$$

$$y_n = \frac{K^n u_i}{1 + K^{2n}\left(u_r^{\,2} + u_i^{\,2}\right) + 2K^n u_r}$$

Clearly the computational complexity of the continuous parameterized transformation exceeds that of the discrete transformation. This raises an interesting question, at least from a neo-Platonic perspective. Is it possible that nature prefers the simplicity of the discrete form over the continuous? In other words, are all physically realizable Lorentz transformations actually discrete? If so, what determines the "size" of the minimum transformation? What is the "size" of a Mobius transformation?

We know that every Mobius is conjugate to a pure exponential, and the effect of which is a rotation and a re-scaling. In addition, the conjugation itself may impose some kind of *size*. It's interesting that the elements now are not frames but differences between frames, including rotations. Thus, rather than the ontological objects of consideration being events of the form x,y,z,t, or even coordinate systems, the objects are the transformations with complex coefficients a,b,c,d. This again introduces the octonion space, though restricted by the fact that there are only three (complex) degrees of freedom.

Chapter 7

Solid Models

Most geometric objects we see every day are solids. That is, they are geometric objects with interior. Solids can be very simple like a cube or very complex like a piston engine. To be processed by computers, solids must have some representations that can describe the geometry and characteristics completely. In fact, a good representation should address the following issues:

1. *Domain:* While no representation can describe all possible solids, a representation should be able to represent a useful set of geometric objects.
2. *Unambiguity:* When you see a representation of a solid, you will know what is being represented without any doubt. An unambiguous representation is usually referred to as a complete one.
3. *Uniqueness:* That is, there is only one way to represent a particular solid. If a representation is unique, then it is easy to determine if two solids are identical since one can just compare their representations.
4. *Accuracy:* A representation is said accurate if no approximation is required.
5. *Validness:* This means a representation should not create any invalid or impossible solids. More precisely, a representation will not represent an object that does not correspond to a solid.
6. *Closure:* Solids will be transformed and used with other operations such as union and intersection. "Closure" means that transforming a valid solid always yields a valid solid.
7. *Compactness and Efficiency:* A good representation should be compact enough for saving space and allow for efficient algorithms to determine desired physical characteristics.

These issues may be contradictory with each other. For efficiency purpose, a curvilinear solid may be approximated by a polyhedron. There are many efficient and robust algorithms for handling polyhedra; however, accuracy may not be maintained in the process of approximation. For example, given two curvilinear solids that are tangent to each other, this tangency may disappear after converting to a polyhedron.

Problems occur even for the polyhedra world. Many graphics APIs such as PHIGS PLUS and OpenGL have built-in data structures for representing polyhedra; but, these representations could generate invalid solids. There are representations that can always represent valid solids; but, these representations are in general more complex than those available in graphics APIs.

In summary, designing representations for solids is a difficult job and compromises are often necessary. This course will only discuss the following representations: wireframes, boundary representations and constructive solid geometry.

WIREFRAME MODELS

The wireframe model is perhaps the oldest way of representing solids. A wireframe model consists of two tables, the vertex table and the edge table. Each entry of the vertex table records a vertex and its coordinate values, while each entry of the edge table has two components giving the two incident vertices of that edge. A wireframe model does not have face information. For example, to represent a cube defined by eight vertices and 12 edges, one needs the following tables

Vertex Table

Vertex #	*x*	*y*	*z*
1	1	1	1
2	1	-1	1
3	-1	-1	1
4	-1	1	1
5	1	1	-1
6	1	-1	-1
7	-1	-1	-1
8	-1	1	-1

Edge Table

Edge #	*Start Vertex*	*End Vertex*
1	1	2
2	2	3
3	3	4
4	4	1
5	5	6
6	6	7
7	7	8
8	8	5
9	1	5
10	2	6
11	3	7
12	4	8

The following figure shows all eight vertices and twelve edges and their numbers in white and yellow, respectively.

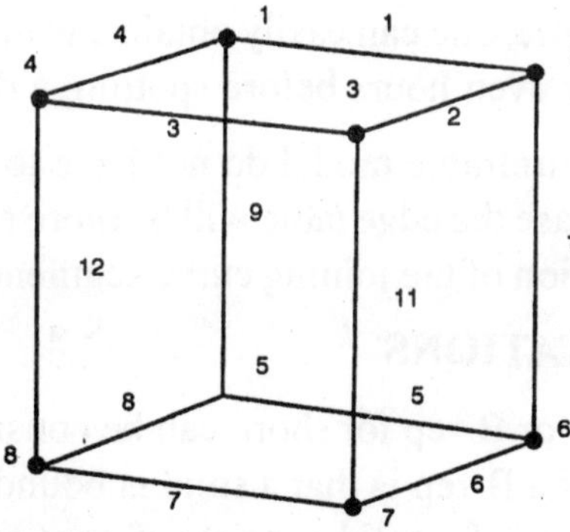

Wireframe Models Are Ambiguous

While wireframe uses the simplest data structures, it is ambiguous. The following is a well-known example that consists of 16 vertices and 32 edges. We know it represents a solid and each of the quadrilaterals (some of them are squares) defines a face of the solid.

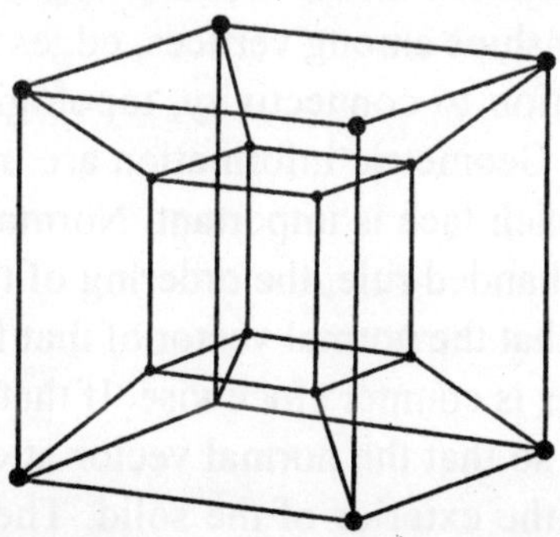

The inner cube represents a hole; but, we cannot tell the direction of the opening of the cube. As shown in the following figure, there are three possibilities for this opening. While the other two can be obtained by rotating the remaining one, in general we will not be so lucky because the outer boundary may be a box of different side lengths and because this model is part of a big one which does not allow free interpretation.

Because of wireframe models are ambiguous, their uses are limited. However, wireframe models are popular, because they are efficient (i.e., only vertices and edges are displayed and processed) when they work. For example, wireframe models can be used for preview purpose. Rendering a complex model or an animation sequence could be very time consuming if all objects are to be rendered. If wireframe models (usually including its face information) are available, one can easily obtain a general feeling of the final result without waiting for minutes or even hours before spotting a design flaw.

Note that the edges in a wireframe model do not have to be line segments. They can be curve segments and in this case the edge table will be more complicated since in addition to the two endpoints a description of the joining curve segment (e.g., equation) is required.

BOUNDARY REPRESENTATIONS

Boundary Representation, or B-rep for short, can be considered as an extension to the wireframe model. The merit of a B-rep is that a solid is bounded by its surface and has its interior and exterior. The surface of a solid consists of a set of well-organized faces, each of which is a piece of some surface (.e.g., a surface patch). Faces may share vertices and edges that are curve segments. Therefore, a B-rep is an extension to the wireframe model by adding face information to the latter.

There are two types of information in a B-rep: topological and geometric. Topological information provide the relationships among vertices, edges and faces similar to that used in a wireframe model. In addition to connectivity, topological information also include orientation of edges and faces. Geometric information are usually equations of the edges and faces. The orientation of each face is important. Normally, a face is surrounded by a set of vertices. Using the right-handed rule, the ordering of these vertices for describing a particular face must guarantee that the normal vector of that face is pointing to the exterior of the solid. Normally, the order is counter clockwise. If that face is given by an equation, the equation must be rewritten so that the normal vector at every point on the part that is being used as a face points to the exterior of the solid. Therefore, by inspecting normal vectors one can immediately tell the inside and outside of a solid under B-rep. This orientation must be done for all faces. The following shows three faces and their outward pointing normal vectors. To describe the top surface, the vertices should be 6, 7, 2, 1 or 7, 2, 1, 6 or 2, 1, 6, 7 or 1, 6, 7, 2. To describe the left face, the order should be 1, 2, 3, 4 or 2, 3, 4, 1 or 3, 4, 1, 2 or 4, 1, 2, 3.

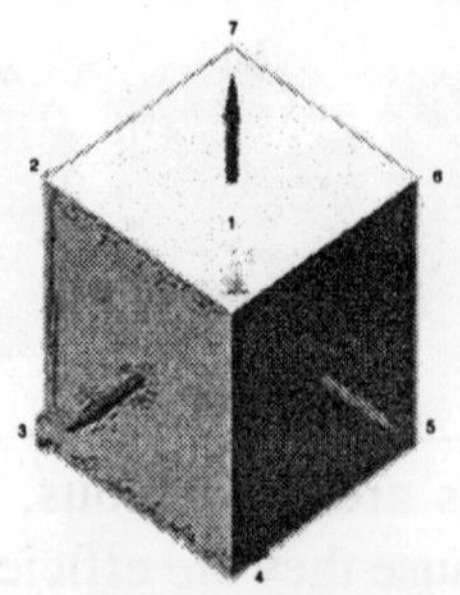

Unfortunately, not all surfaces can be oriented this way. If the surface of a solid can be oriented this way, it is called *orientable*; otherwise, it is non-*orientable*. The following shows the well-known *Mobius* band which is one-sided and non-*orientable*.

MANIFOLDS

The surface of a solid must satisfy some conditions so that the resulting solid is well-behaved. This is the so-called manifold condition. While recent research has shown that the surface of a solid does not have to be a manifold, we shall restrict to manifold surfaces to simply our discussion.

A surface is a 2-manifold if and only if for each point x on the surface there exists an open ball with center x and sufficiently small radius so that the intersection of this ball and the surface can be continuously deformed to an open disk.

An open ball with center at the coordinate origin and radius r is defined by equation $x^2 + y^2 + z^2 < r^2$. It contains all points inside of the sphere $x^2 + y^2 + z^2 = r^2$. An open disk is defined similarly $x^2 + y^2 < r^2$. By continuously deformed, it means one can twist or bend the shape without cutting (i.e. adjacency relations must be maintained) and gluing (i.e. a one-to-one relation is required).

Let us take a look at the following figure:

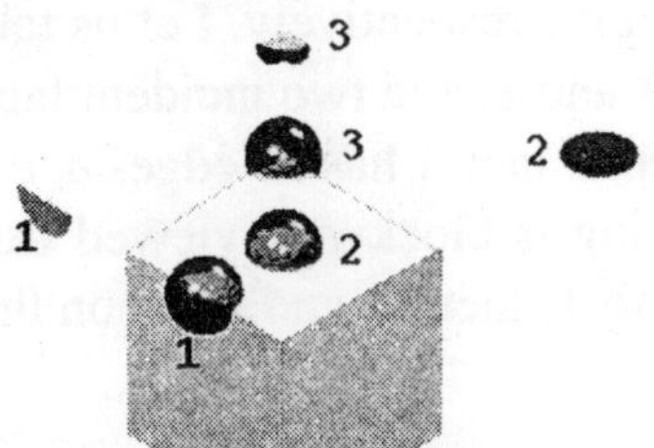

There is a cube and three open balls. Ball 2 has its center on the top face. The intersection of this ball and the surface of the cube is an open disk (shown in red). Ball 1 has its center on an edge. Its intersection with the surface of the cube is a "bent" open disk, which of course can be "unbent" to make it an open disk. Ball 3 has its center at a corner. Its intersection with the cube's surface is a three-way bent open disk.

The following shows a solid whose bounding surface is not a manifold. The intersection of the open ball and the surface of the solid is the union of two intersecting open disks. This intersection cannot be deformed to an open disk without "gluing." Consequently, the surface is not a manifold.

THE WINGED-EDGE DATA STRUCTURE

Perhaps the oldest data structure for a B-rep is Baumgart's winged-edge data structure. It is quite different from that of a wireframe model, because the winged-edge data structure uses edges to keep track almost everything. *In what follows, we shall assume there is no holes in each face* and later extend it to cope with holes. Moreover, we shall assume edges and faces are line segments and polygons. Topologically, one can always stretch curvilinear edges and faces so that they become flat without changing the relationships among them.

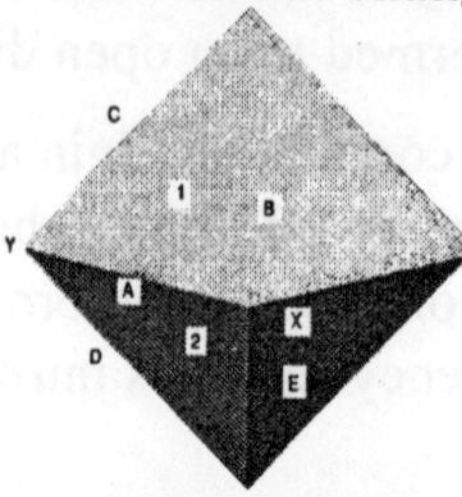

The above figure shows a polyhedron with vertices, edges and faces indicated with upper cases, lower cases and digits, respectively. Let us take a look at edge $a = XY$. This edge has two incident vertices X and Y, and two incident faces 1 and 2. A face is a polygon surrounded by edges. For example, face 1 has its edges a, c and b, and face 2 has its edges a, e and d. Note that the ordering is clockwise viewed from outside of the solid. If the direction of the edge is from X to Y, faces 1 and 2 are on the right and left side of edge a, respectively.

To capture the ordering of edges correctly, we need four more pieces of information. Since edge a is traversed once when traversing face 1 and traversed a second time when traversing face 2, it is used twice in different directions. For example, when traversing the edges of face 1, the predecessor and successor of edge a are edge b and edge c, and when traversing the edges of face 2, the predecessor and successor of edge a are edge d and edge e. Note that although there are four edges incident to vertex X, only three of them are used when finding faces incident to edge a. Therefore, for each edge, the following information are important:

1. Vertices of this edge,
2. Its left and right faces,
3. The predecessor and successor of this edge when traversing its left face, and
4. The predecessor and successor of this edge when traversing its right face.

The Edge Table

Each entry in the edge table contains those information mentioned earlier: edge name, start vertex and end vertex, left face and right face, the predecessor and successor edges when traversing its left face, and the predecessor and successor edges when traversing its right face. Note that clockwise ordering (viewing from outside of the polyhedron) is used for traverse. Note also that the direction of edge a is from *X* to *Y*. If the direction is changed to from Y to X, all entries but the first one in the following table must be changed accordingly.

Edge	*Vertices*		*Faces*		*Left Traverse*		*Right Traverse*	
Name	*Start*	*End*	*Left*	*Right*	*Pred*	*Succ*	*Pred*	*Succ*
a	X	Y	1	2	*b*	*d*	*e*	*c*

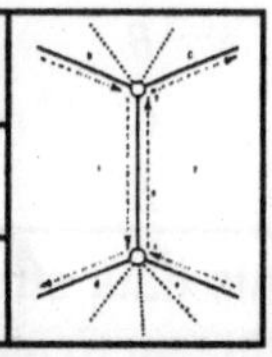

The above shows the information for the entry of edge a. The four edges b, c, *d* and *e* are the wings of edge a and hence edge a is "winged."

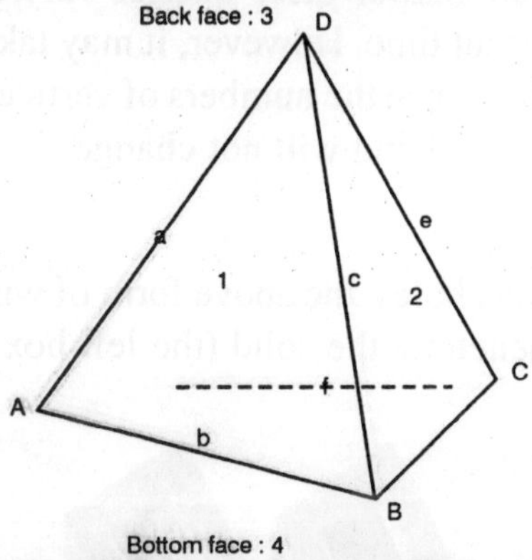

The above is a tetrahedron with four vertices A, B, C and D, six edges a, b, c, d, *e* and f, and four faces 1, 2, 3 (back) and 4 (bottom). Its edge table is the following. Please use the above diagram to verify this table.

Edge	*Vertices*		*Faces*		*Left Traverse*		*Right Traverse*	
Name	*Start*	*End*	*Left*	*Right*	*Pred*	*Succ*	*Pred*	*Succ*
a	*A*	*D*	3	1	*e*	*f*	*b*	*c*
b	*A*	*B*	1	4	*c*	*a*	*f*	*d*
c	*B*	*D*	1	2	*a*	*b*	*d*	*e*
d	*B*	*C*	2	4	*e*	*c*	*b*	*f*
e	*C*	*D*	2	3	*c*	*d*	*f*	*a*
f	*A*	*C*	4	3	*d*	*b*	*a*	*e*

OTHER TABLES

The winged-edge data structure requires two more tables, the vertex table and the face table. These two are very simple. The vertex table has one entry for each vertex which contains an edge that is incident to this vertex. The face table has one entry for each face which contains an edge that is one of this face's boundary edges. Therefore, we have the following table. Note that since there are multiple choices of edges, you may come up with different tables:

Vertex Name	*Incident Edge*
A	*a*
B	*b*
C	*d*
D	*e*

Face Name	*Incident Edge*
1	*a*
2	*c*
3	*a*
4	*b*

With this data structure, one can easily answer the question: which vertices, edges, faces are adjacent to each face, edge, or vertex. There are nine of these adjacency relations. For example, is vertex X adjacent to face 5? Are faces 3 and 5 adjacent to each other? The winged-edge data structure can answer these queries very efficiently and some of them may even be answered in constant time. However, it may take longer time to answer other adjacency queries. Note also that once the numbers of vertices, edges and faces are known, the size of all three tables are fixed and will not change.

FACES WITH HOLES

If some faces of a solid have holes, the above form of winged-edge data structure does not work. These holes may penetrate the solid (the left box below) or just like a pothole (the right box below).

There are two ways for resolving this problem:

1. For a face with inner loops, the outer boundary is ordered clockwise, while its inner loops, if any, are ordered counter clockwise.

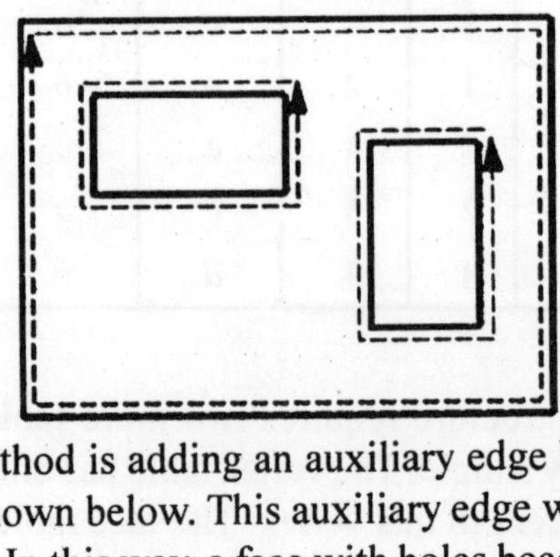

2. Another simple method is adding an auxiliary edge between each inner loop and the outer loop as shown below. This auxiliary edge will have the same face for its left and right faces. In this way, a face with holes becomes a single loop which can

be represented with the winged-edge data structure. When traversing a loop, auxiliary edges can be identified easily since its left and right faces are the same.

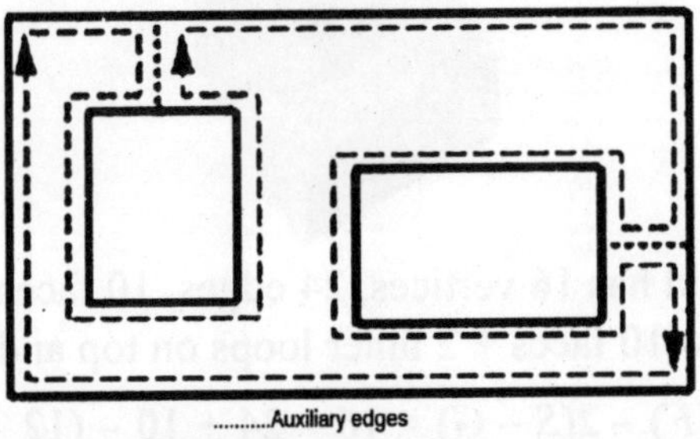

THE EULER-POINCARÉ FORMULA

The Euler-Poincaré formula describes the relationship of the number of vertices, the number of edges and the number of faces of a manifold.

It has been generalized to include potholes and holes that penetrate the solid. To state the Euler-Poincaré formula, we need the following definitions:

1. *V:* the number of vertices
2. *E:* the number of edges
3. *F:* the number of faces
4. *G:* the number of holes that penetrate the solid, usually referred to as genus in topology
5. *S:* the number of shells. A shell is an internal void of a solid. A shell is bounded by a 2-manifold surface, which can have its own genus value. Note that the solid itself is counted as a shell. Therefore, the value for S is at least 1.

- *L:* the number of loops, all outer and inner loops of faces are counted.

Then, the Euler-Poincaré formula is the following:

$$V - E + F - (L - F) - 2(S - G) = 0$$

Examples:

1. A cube has eight vertices ($V = 8$), 12 edges ($E = 12$) and six faces ($F = 6$), no holes and one shell ($S = 1$); but $L = F$ since each face has only one outer loop. Therefore, we have

   $$V - E + F - (L - F) - 2(S - G) = 8 - 12 + 6 - (6 - 6) - 2(1 - 0) = 0$$

2. The following solid has 16 vertices, 24 edges, 11 faces, no holes, 1 shell and 12 loops (11 faces + one inner loop on the top face). Therefore,

   $$V - E + F - (L - F) - 2(S - G) = 16 - 24 + 11 - (12 - 11) - 2(1 - 0) = 0$$

3. The following solid has 16 vertices, 24 edges, 10 faces, 1 hole (i.e., genus is 1), 1 shell and 12 loops (10 faces + 2 inner loops on top and bottom faces). Therefore,
$V - E + F - (L - F) - 2(S - G) = 16 - 24 + 10 - (12 - 10) - 2(1 - 1) = 0$

4. The following solid has a penetrating hole and an internal cubic chamber as shown by the right cut-away figure. It has 24 vertices, 12*3 (cubes) = 36 edges, 6*3 (cubes)- 2 (top and bottom openings) = 16 faces, 1 hole (i.e., genus is 1), 2 shells and 18 loops (16 faces + 2 inner loops on top and bottom faces). Therefore,
$V - E + F - (L - F) - 2(S - G) = 24 - 36 + 16 - (18 - 16) - 2(2 - 1) = 0$

5. The following solid has two penetrating holes and no internal chamber as shown by the right cut-away figure. It has 24 vertices, 36 edges, 14 faces, 2 hole (i.e., genus is 2), 1 shells and 18 loops (14 faces + 4). Therefore,
$V - E + F - (L - F) - 2(S - G) = 24 - 36 + 14 - (18 - 14) - 2(1 - 2) = 0$

Part of the information recorded in a B-rep is topological (i.e., adjacency relations). Invalid solids may be generated if the representation is not carefully constructed. One way of checking this topological invalidity is to use the Euler-Poincaré formula. If its value is not zero, we are sure something must be wrong in the representation. However, this is only a one-side test. More precisely, a zero value of the Euler-Poincaré formula does not mean the solid is valid.

The figure above has a box and an additional sheet which is simply a rectangle. This object has 10 vertices, 15 edges, 7 faces, 1 shell and no hole. Its loop number is equal to the number of faces. The value of the Euler-Poincaré formula is zero as shown below,

$$V - E + F - (L - F) - 2(S - G) \qquad = 10 - 15 + 7 - (7 - 7) - 2(1 - 0) = 0$$

but this is not a valid solid! Therefore, if the value of Euler-Poincaré formula is non-zero, the representation is definitely not a valid solid. However, the value of the Euler-Poincaré formula being zero does not guarantee the representation would yield a valid solid.

HOW TO COUNT GENUS CORRECTLY?

It is not always easy to count genus G correctly. For example, suppose we have a sphere punched by three tunnels as shown below. The left object shows the outside look. The right one shows the inside by cutting half of the sphere off. What is the genus value G? We know that the genus value counts the number of penetrating holes. But, in this example, it is somewhat ambiguous. In fact, by combining any two center-going tunnels we will have three penetrating holes. However, this is incorrect!

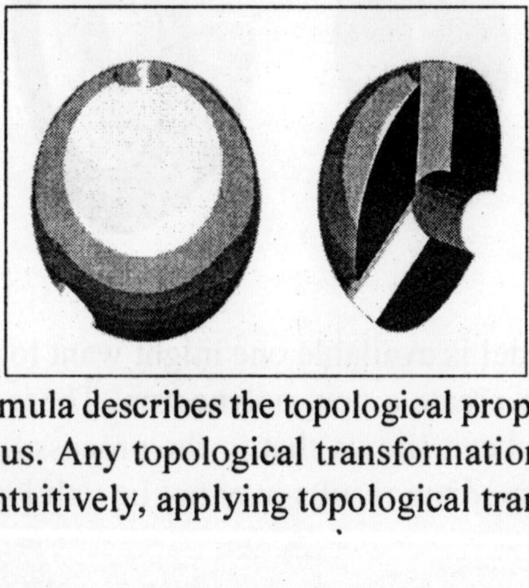

The Euler-Poincaré Formula describes the topological property amount vertices, edges, faces, loops, shells and genus. Any topological transformation applied to the model will not alter this relationship. Intuitively, applying topological transformations means we can twist, stretch and squash the model but we cannot cut some parts off nor glue some parts together. Let us apply some intuitive topological transformations to this model for computing the genus. We can push the "walls" surrounding the three tunnels so that the interior of the model becomes a thin shell. This is shown in the image below.

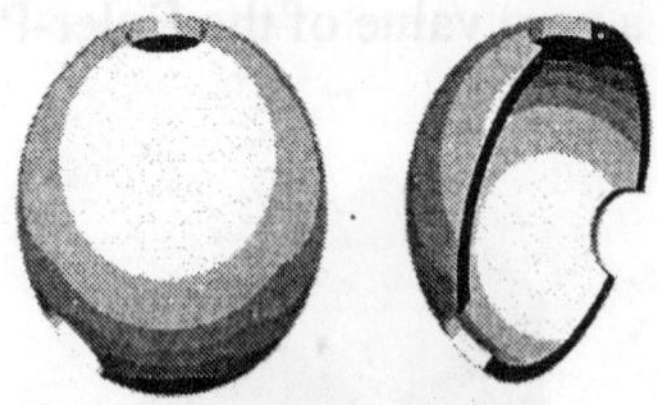

Stretch the top hole so that it is large enough (below left). Then, collapse the top portion to flatten the model. This is shown in the right model below. How many penetrating holes are there? Two! Therefore, G is 2.

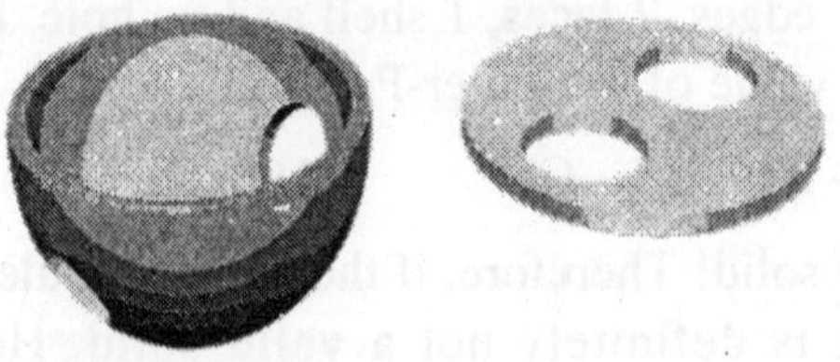

Sometimes, penetrating holes may appear in an unlikely situation. Consider the following model which is obtained by taking out a torus and tube from the interior of a sphere. What is the genus of this model?

It does not look like there is a penetrating hole. So, is genus equal to? In fact, $G = 1$! Figure it out yourself. Do some twisting, stretching and squashing.

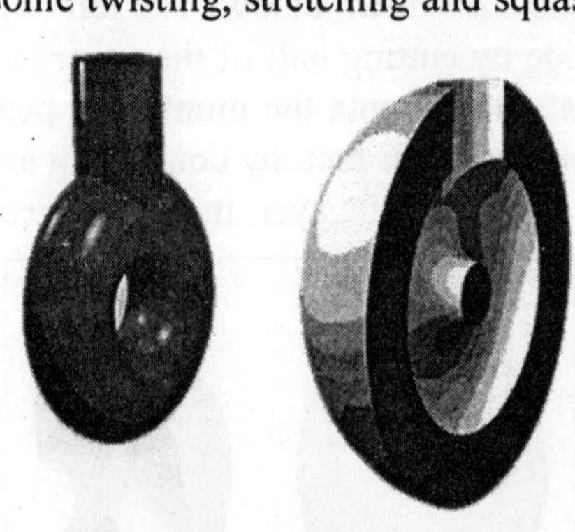

EULER OPERATORS

Once a polyhedron model is available one might want to edit it by adding or deleting vertices, edges and faces to create a new polyhedron. These operations are called Euler Operators. However, it has been shown that in the process of editing a polyhedron with Euler operators, some intermediate results may not be valid solids at all.

Recall from the discussion of the Euler-Poincaré formula that the following holds for all polyhedra:

$$V - E + F - (L - F) - 2(S - G) = 0$$

where V, E, F, L, S and G are the numbers of vertices, edges, faces, loops, shells and genus, respectively. Based on this relation, some Euler operators have been selected for editing a polyhedron so that the Euler-Poincaré formula is always satisfied. There are two groups of such operators: the Make group and the Kill group. Operators start with M and K are operators of the Make and Kill groups, respectively.

Euler operators are written as Mxyz and Kxyz for operations in the Make and Kill groups, respectively, where x, y and z are elements of the model (e.g., a vertex, edge, face, loop, shell and genus). For example, MEV means adding an edge and a vertex while KEV means deleting an edge and a vertex.

It has been proved by Mantyla in 1984 that Euler operators form a complete set of modeling primitives for manifold solids. More precisely, every topologically valid polyhedron can be constructed from an initial polyhedron by a finite sequence of Euler operators. Therefore, Euler operators are powerful operations.

The Make Group of Euler Operators

The Make group consists of four operators for adding some elements into the existing model creating a new one, and a Make-Kill operator for adding and deleting some elements at the same time. Here are the operators:

Operator Name	*Meaning*	*V*	*E*	*F*	*L*	*S*	*G*
MEV	Make an edge and a vertex	+1	+1				
MFE	Make a face and an edge		+1	+1	+1		
MSFV	Make a shell, a face and a vertex	+1		+1	+1	+1	
MSG	Make a shell and a hole					+1	+1
MEKL	Make an edge and kill a loop		+1		-1		

The above table shows the change of values of V, E, F, L, S and G. Note that adding a face produces a loop, the outer loop of that face. Therefore, when F is increased, L should also be increased. This new loop and the new face will cancel each other in the subexpression $L - F$. Please verify that none of these operators would cause the Euler-Poincaré formula to fail.

The following table illustrates the way of using Euler operators to construct a tetrahedron. Vertices, edges, and faces are in red, blue and green; the only shell is in transparent gray.

The first step uses *MSFV* to obtain a shell with a face and a vertex. The next three steps use *MEVs*, each of which adds an edge and a vertex. The last three steps use *MFE*, each of which adds a face and an edge. Thus, in seven steps or seven Euler operators a tetrahedron is constructed.

Operator Name	*Meaning*	*V*	*E*	*F*	*L*	*S*	*G*	*Result*
MSFV	Make a shell, a face and a vertex	+1		+1	+1	+1		
MEV	Make an edge and a vertex	+1	+1					
MEV	Make an edge and a vertex	+1	+1					
MEV	Make an edge and a vertex	+1	+1					
MFE	Make a face and an edge		+1	+1		+1		
MFE	Make a face and an edge		+1	+1		+1		
MFE	Make a face and an edge		+1	+1		+1		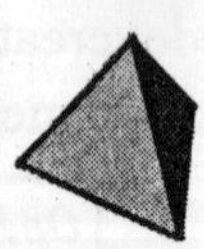

MSG simply makes a shell with a hole. After this, one can add vertices, edges, faces, loops. There must be loops, because the new hole penetrates at least one face.

MEKL makes an edge and at the same time kills a loop. A commonly used *MEKL* is adding an edge connecting the outer loop and the inner loop of a face. In this case, the number of edges E is increased by 1 and the number of loops L is decreased by 1 since that loop is killed. The left figure below shows two loops of the top face while the right one shows the new edge added after performing a *MEKL*. Thus, the inner and outer are "joint" together with the new edge becoming a single face.

The Kill Group of Euler Operators

The Kill group just performs the opposite of what the Make group does. In fact, replacing the *M* and *K* in all Make operators with *K* and *M*, respectively, would get the operators of the Kill group. Therefore, the Kill group consists of the following five operators:

Operator Name	*Meaning*	*V*	*E*	*F*	*L*	*S*	*G*
KEV	Kill an edge and a vertex	–1	–1				
KFE	Kill a face and an edge		–1	–1	–1		
KSFV	Kill a shell, a face and a vertex	–1		–1	–1	–1	
KSG	Kill a shell and a hole					–1	–1
KEML	Kill an edge and make a loop		–1		+1		

With these operators, one can start with a tetrahedron and reduce it to nothing. Since these operators are the opposites of the Make operators, we shall not go further here.

CONSTRUCTIVE SOLID GEOMETRY

Constructive Solid Geometry, or CSG for short, is yet another way of representing solids. A CSG solid is constructed from a few primitives with Boolean operators (i.e., set union, intersection and difference). Thus, a CSG solid can be written as a set equations and can also be considered a design methodology.

CSG Primitives

The standard CSG primitives consist of the block (i.e., cube), triangular prism, sphere, cylinder, cone and torus. These six primitives are in some normal or generic form and must be instantiated by the user to be used in his/her design. Moreover, the instantiated primitive may require transformations such as scaling, translation and rotation to be positioned at the desired place.

Suppose the block primitive is defined by its "lower left" corner < –1, –1, –1 > and "upper right" corner < 1, 1, 1 >. To produce a rectangular box with center at < 3, 2, 3 > and height and width 3 and length 5, a user may first scale the block primitive 1.5 times in the y- and z-direction and 2.5 times in the x-direction, and then translate the result to < 3, 2, 3 >. If the block primitive is called Block in a CSG system, the result may be obtained as follows:

translate(scale(Block, < 2.5, 1.5, 1.5 >), < 3, 2, 3 >)

In the above, the object to be transformed and the transformation data are the first and second arguments, respectively.

Boolean Operators

We can combined two instantiated and perhaps transformed primitives into one with set union, set intersection and set difference operators. However, simple set operators may create problems as will be discussed in regularized Boolean operators, modifications are required. Let us just use set operations on this page.

Given two sets, *A* and *B*, its union consists of all points from either *A* or *B*; its intersection consists of all points in both sets; and its difference, written as *A* – *B* (resp., *B* – *A*), consists of all points in *A* but not in *B* (resp., in *B* but not in *A*). In the following, A is the vertical cylinder and *B* is the horizontal cylinder. From left to right, the four solids are the union and intersection of *A* and *B*, *A* – *B* and *B* – *A*.

Therefore, a solid can be considered as the result of applying Boolean operators to a set of instantiated and transformed CSG primitives.

Let us take a look at a simple example. We want to design a bracket-like shape with a hole shown on the right-most figure below. We start with two instantiations of blocks and one instantiation of a cylinder (the left-most figure). Then, the two blocks are scaled and one of them is rotated to a vertical position.

The cylinder is also scaled so that its radius matches that of the hole. These three instantiations are than transformed to their desired positions. The final product is obtained by computing the union of the two blocks and then subtracting from it the cylinder.

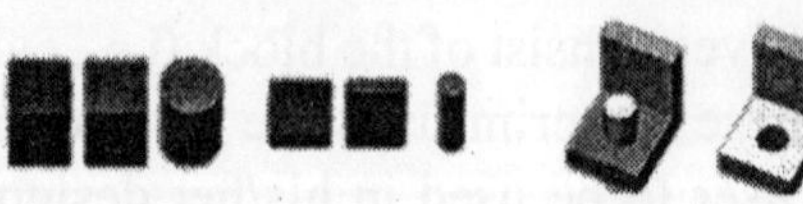

Please note that the design of the above solid is not unique. For example, the *L* shape can be constructed from subtracting a cube from another one.

CSG Expressions

The design procedure of the above bracket can be written as an expression:

diff(union(trans(Block1), trans(Block2)), trans(Cylinder))

where union(*A*, *B*) and diff(*A*, *B*) are the union and difference of *A* and *B*, and trans indicates appropriate transformations. Or, if we use + and – for set union, intersection and difference, the above function calls can be rewritten as a set expression as follows:

(trans(Block1) + trans(Block2)) – trans(Cylinder)

This expression can be converted to an expression tree, the *CSG* Expression, of the design:

In fact, every solid constructed using the *CSG* technique has a corresponding *CSG* expression which in turn has an associated *CSG* tree.

The expression of the *CSG* tree is a representation of the final design. Recall that the same solid may have different *CSG* expressions/trees. For example, one might punch a hole from Block1 first and then compute the union of this result with Block2. As a result, *CSG* representations are not unique.

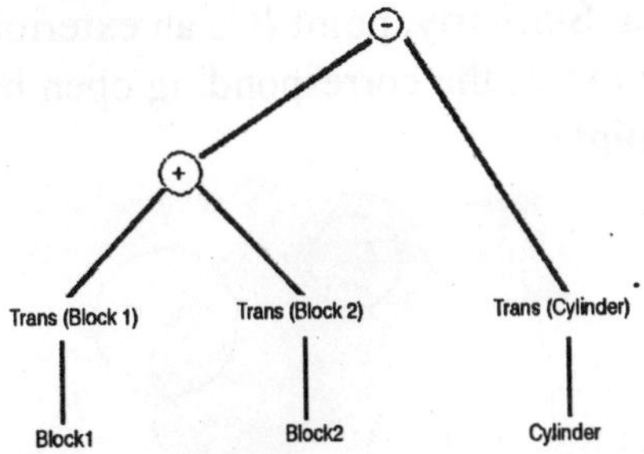

INTERIOR, EXTERIOR AND CLOSURE

We need the concept of interior, exterior and closure to fully appreciate the discussion of regularized Boolean operators. Intuitively, the interior of a solid consists of all points lying inside of the solid; the closure consists of all interior points and all points on the solid's surface; and the exterior of a solid is the set of all points that do not belong to the closure.

Consider a sphere, $x^2 + y^2 + z^2 = 1$. Its interior is the set of all points that satisfy $x^2 + y^2 + z^2 < 1$, while its closure is $x^2 + y^2 + z^2 = 1$. Therefore, the closure is the union of the interior and the boundary (its surface $x^2 + y^2 + z^2 = 1$). Obviously, its exterior is $x^2 + y^2 + z^2 > 1$. A solid is a three-dimensional object and so does its interior and exterior. However, its boundary is a two-dimensional surface.

Formal Definitions

Recall that the open ball with center (a, b, c) and radius r consists of all points that satisfy the following relation:

$$(x - a)^2 + (y - b)^2 + (z - c)^2 < r^2$$

A point *P* is an interior point of a solid *S* if there exists a radius r such that the open ball with center *P* and radius r is contained in the solid *S*. The set of all interior points of solid *S* is the interior of *S*, written as int(*S*). Based on this definition, the interior of an open ball is the open ball itself.

On the other hand, a point *Q* is an exterior point of a solid *S* if there exists a radius r such that the open ball with center *Q* and radius r does not intersect *S*. The set of all exterior point of solid S is the exterior of solid *S*, written as ext(*S*).

Those points that are not in the interior nor in the exterior of a solid *S* constitutes the boundary of solid *S*, written as *b*(*S*). Therefore, the union of interior, exterior and boundary of a solid is the whole space. The closure of a solid *S* is defined to be the union of *S*'s interior and boundary, written as closure(*S*). Or, equivalently, the closure of solid *S* contains all points that are not in the exterior of *S*.

Examples: Here is an example in the plane. You should change all open balls to open disks. Point A is an interior point of the shaded area since one can find an open disk that is contained in the shaded area. Similarly, point B is an exterior point. Point C is a boundary point because whatever the radius the corresponding open ball will contain some interior points and some exterior points.

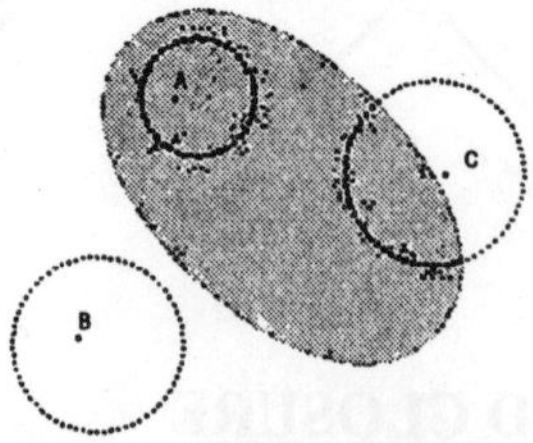

Note that a surface (a two-dimensional object) is never a solid (a three-dimensional object). In fact, a surface does not have any interior point. Take any point of the surface, the open ball with arbitrary radius and center at that point always intersects the sphere in an open disk (in pale green in the lower right corner). Therefore, no open ball can be contained in the sphere, and, as a result, that point is not an interior point of the sphere. Thus, we conclude that a surface does not have any interior point.

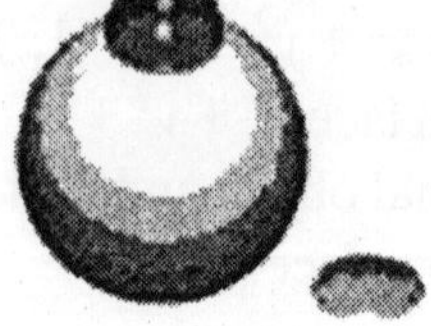

REGULARIZED BOOLEAN OPERATORS

We certainly expect that the union, intersection and difference of two solids is a solid. Unfortunately, in many cases this is not always true. In the following figure, two cubes touch each other and their intersection is a rectangle shown on the right. A rectangle is not a three-dimensional object and hence not a solid!

To eliminate these lower dimensional branches, the three set operations are regularized as follows. The idea of regularing is very simple.

- Compute the result as usual and lower dimensional components many be generated.
- Compute the interior of the result. This step removes all lower dimensional components. An example has been shown in previous page. The result is a solid without its boundary.

- Compute the closure of the result obtained in the above step. This adds the boundary back.

Let +, and – be the regularized set union, intersection and difference operators. Let A and B be two solids. Then, $A + B$, $A \wedge B$ and $A - B$ can be defined mathematically based on the above description:

$$A + B = \text{closure(int(the set union of } A \text{ and } B)$$

$$A \wedge B = \text{closure(int(the set intersection of } A \text{ and } B)$$

$$A - B = \text{closure(int(the set difference of } A \text{ and } B)$$

where closure() and int() are the closure and interior discussed in previous page. Based on this definition, the intersection of the two cubes shown at the beginning of this page is empty. As mentioned earlier, the set intersection of these two cubes is a rectangle, which is a two dimensional object and has no interior. Hence, after taking interior (i.e., int()), we get an empty set, whose closure is also empty. Consequently, the intersection is empty.

A CSG DESIGN EXAMPLE

Let us deign three perpendicular tubes as shown below left.

An immediate reaction is that let us subtract a smaller cylinder from a larger one, yielding a tube which is shown in below right. Then, take two more instances of this tube, each of which is rotated an appropriate angle; compute the union of these three tubes; and the result is shown in the figure below.

But, the result is not quite right, because the inner junctions of the tubes are blocked as seen in the right figure above. A correct solution is to design two instances of three perpendicular cylinders, one larger and the other smaller. Then, subtracting the smaller from the larger yields the designed result as shown below.

Chapter 8

Bezier Curves

Our first question is: why do we need new forms of parametric curves? An immediate answer is that those parametric curves discussed in the previous unit are not very geometric. More precisely, given such a parametric form it is difficult to know the underlying geometry it represents without some furth er analysis. The coefficients of the equations do not have any geometric meaning, and it is almost impossible to predict the change of shape if one or more coefficients are modified. As a result, designing a curve that follows certain outline is very difficult.

In practice, designers or users usually do not care about the underlying mathematics (and equations, of course). They are more or less focusing on getting their jobs done. To do so, a system that supports users to design curves must be

1. *Intuitive:* We expect that every step and every algorithm will have an intuitive and geometric interpretation.
2. *Flexible:* The system should provide the users with more control for designing and editing the shape of a curve. The way of creating and editing a curve should be easy, intuitive and geometric rather than by manipulating equations.
3. *Unified Approach:* The way of representing, creating and editing different types of curves (e.g., lines, conic sections and cubic curves) must be the same. That is, it does not require different techniques for manipulating different curves (i.e., conics and cubics).
4. *Invariant:* The represented curve will not change its geometry under geometric transformations such as translation, rotation and affine transformations.
5. *Efficiency and Numerically Stability:* A user of a curve design system may not care about the beauty of the underlying geometry; but, he/she expects the system to deliver the curve he/she wants fast and accurately. Moreover, a large amount of computations will not "distort" the shape of the curve (i.e., numerical stability).

This unit focuses on some techniques for curve design that can fulfill the above criteria. We shall discuss Bézier curves here, and B-spline and NURBS curves in the next two units. The unified theme of these techniques consists of the following advantages:

1. A user layouts a set of control points for the system to come up with a curve that more or less follows the trend of the set of control points.
2. A user can change the positions of some control points and some other characteristics for modifying the shape of the curve. No equation is required, because the equation of a curve is usually not stored.
3. If necessary, a user can add control points and other vital information without changing the shape of the curve. In this way, a user has more freedom of editing a curve because adding control points and other information increases the degree of freedom of the curve.
4. A user can even break a curve into two pieces for "micro" editing and then join them back into one piece.
5. There are very geometric, intuitive and numerically stable algorithms for finding points on the curve without knowing the equation of the curve.
6. Once you know curves, the surface counterpart is a few steps away. More precisely, the transition from curve to surface will not cause much difficulty, since what you learn for curves applies directly to surfaces.

We shall start with the most fundamental one in this unit: the Bézier curves. Bézier curves were discovered simultaneously by Paul de Casteljau at Citroen and Pierre E. Bézier at Renault around late 50s and early 60s. Basis splines, or B-splines for short, were known and studied by N. Lobachevsky whose major contribution to mathematics is perhaps the so-called hyperbolic (non-Euclidean) geometry in late eighteenth century. However, we shall adopt a modern version developed by C. de Boor, M. Cox and L. Mansfield in late 70s. Note that Bézier curves are special cases of B-splines.

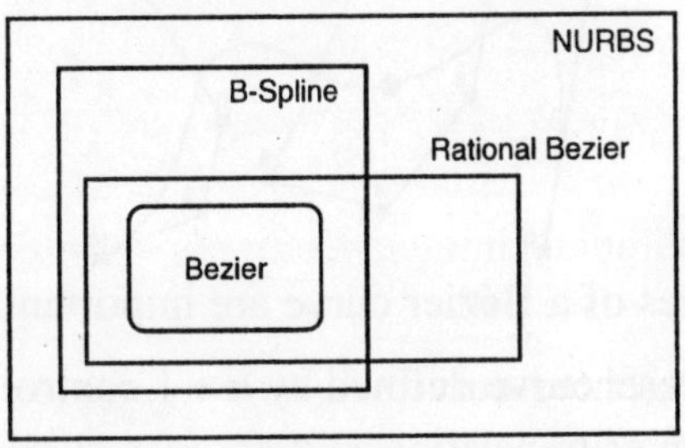

Both Bézier curves and B-splines are polynomial parametric curves. As discussed in the previous unit, polynomial parametric forms cannot represent some simple curves such as circles. As a result, Bézier curves and B-splines can only represent what polynomial parametric forms can. By introducing homogeneous coordinates making them rational, Bézier curves and B-splines are generalized to rational Bézier curves and Non-Uniform

Rational B-splines, or NURBS for short. Obviously, rational Bézier curves are more powerful than Bézier curves since the former now can represent circles and ellipses. Similarly, NURBS are more powerful than B-splines. The relationship among these four types of curve representations is shown below.

CONSTRUCTION OF BÉZIER CURVES

Given $n + 1$ points $P_0, P_1, P_2,...$ and P_n in space, the control points, the Bézier curve defined by these control points is

$$C(u) = \sum_{i=0}^{n} B_{n,i}(u)P_i$$

where the coefficients are defined as follows:

$$Bn, i(u) = \frac{n!}{i!(n-i)} u^i (1-u)^{n-i}$$

Therefore, the point that corresponds to u on the Bézier curve is the "weighted" average of all control points, where the weights are the coefficients $B_{n,i}(u)$. The line segments P_0P_1, $P_1P_2,..$, $P_{n-1}P_n$, called legs, joining in this order form a control polyline. Many authors prefer to call this control polyline as control polygon. Functions $B_{n,i}(u)$, $0 = i = n$, are referred to as the Bézier basis functions or Bernstein polynomials.

Note that the domain of u is [0, 1]. As a result, all basis functions are non-negative. In the above, since u and i can both be zero and so do $1 - u$ and $n - i$, we adopt the convention that 0^0 is 1. The following shows a Bézier curve defined by 11 control points, where the blue dot is a point on the curve that corresponds to $u = 0.4$. As shown in the figure, the curve more or less follows the polyline.

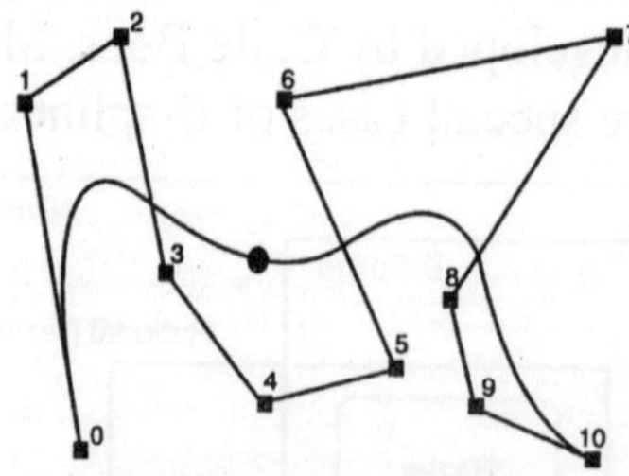

The following properties of a Bézier curve are important:

1. The degree of a Bézier curve defined by $n + 1$ control points is n:

   In each basis function, the exponent of u is $i + (n - i) = n$. Therefore, the degree of the curve is n.

2. *$C(u)$ passes through P_0 and P_n:* This is shown in the above figure. The curve passes though the first and the last control point. Please verify this yourself with some simple algebraic manipulation.

3. *Non-negativity:* All basis functions are non-negative. We have mentioned this earlier.

4. *Partition of Unity:* The sum of the basis functions at a fixed u is 1. It is not difficult to verify that the basis functions are the coefficients in the binomial expansion of the expression $1 = (u + (1 - u))^n$. Hence, their sum is one. Moreover, since they are nonnegative, we conclude that the value of any basis function is in the range of 0 and 1.

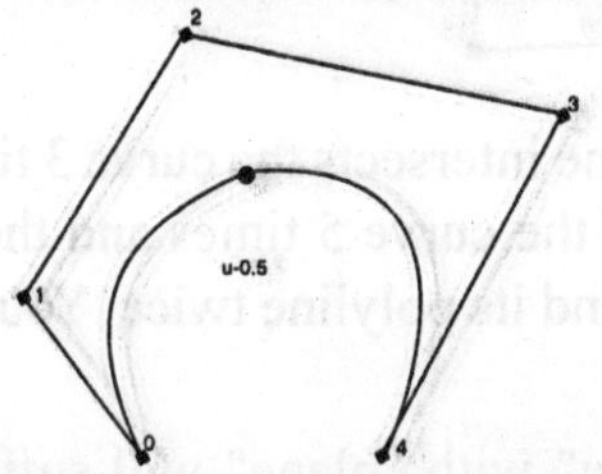

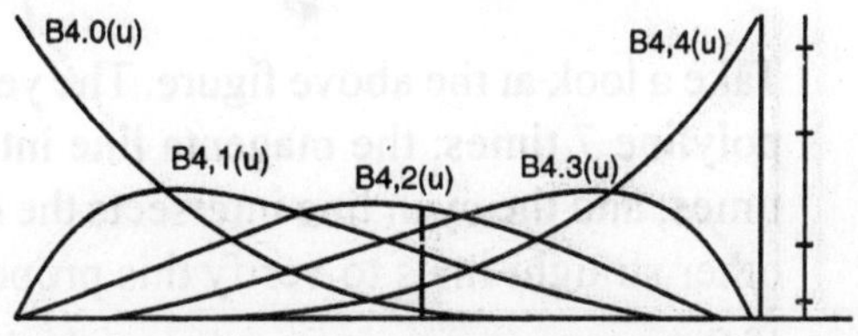

In the left figure above, we have a Bézier curve defined by five control points. Its five basis functions are functions in u as shown in the right figure above. The figures show u=0.5 and all five basis functions. The right-most vertical bar shows the way of partitioning 1 into five intervals and hence the name of partition of unity. Note that the color of a partition is identical to the color used to draw its corresponding basis functions.

Since all basis functions are in the range of 0 an 1 and sum to one, they can be considered as weights in the computation of a weighted average. More precisely, we could say "to compute C(u), one takes the weight Bn,i(u)for control point Pi and sum them together."

5. *Convex Hull Property:* This means the Bézier curve defined by the given $n + 1$ control points lies completely in the convex hull of the given control points. The convex hull of a set of points is the smallest convex set that contains all points. In the following figure, the convex hull of the 11 control points is shown in color gray. Note that not all control points are on the boundary of the convex hull. For example, control points 3, 4, 5, 6, 8 and 9 are in the interior. The curve, except for the first two endpoints, lies completely in the convex hull.

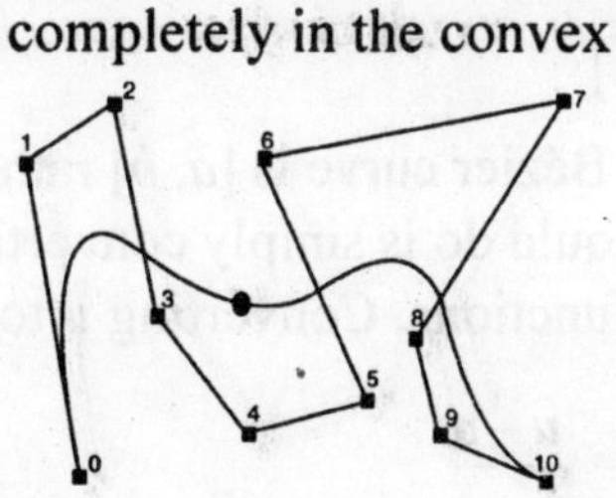

This property is important because we are guaranteed that the generated curve will be in an understood and computable region and will not go outside of it.

6. *Variation Diminishing Property:* If the curve is in a plane, this means no straight line intersects a Bézier curve more times than it intersects the curve's control polyline.

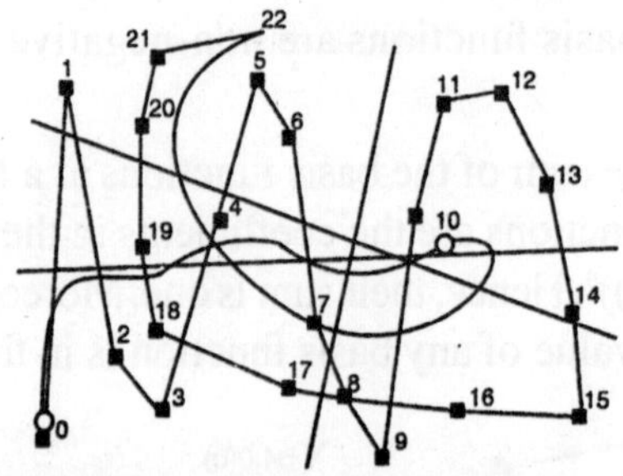

Take a look at the above figure. The yellow line intersects the curve 3 times and the polyline 7 times; the magenta line intersects the curve 5 times and the polyline 7 times; and the cyan line intersects the curve and its polyline twice. You could draw other straight lines to verify this property.

If the curve is a space curve, replacing "line" with "plane" will suffice. Special cases may occur; however, it will not be a difficult task to come up with a proper counting scheme that takes these special cases into consideration.

So, what is the meaning of this property? It tells us that the complexity (i.e., turning and twisting) of the curve is no more complex than the control polyline. In other words, the control polyline twists and turns more frequently than the Bézier curve does, because an arbitrary line hits the control polyline more often than it hits the curve. Take a look at the above figure,the control polyline is more complex than the curve it defines.

7. *Affine Invariance:* If an affine transformation is applied to a Bézier curve, the result can be constructed from the affine images of its control points. This is a nice property. When we want to apply a geometric or even affine transformation to a Bézier curve, this property states that we can apply the transformation to control points, which is quite easy, and once the transformed control points are obtained the transformed Bézier curve is the one defined by these new points. Therefore, we do not have to transform the curve.

If the Domain of *u* Is Not [0.1]

Sometimes the domain of a Bézier curve is [*a*, *b*] rather than [0, 1]. Thus, a change of variable is required. What we should do is simply converting a *u* in [*a*, *b*] to a new *u* in [0, 1] and using this *u* in the basis functions. Converting *u* to [0, 1] can be done as follows:

$$u = \frac{u-a}{b-a}$$

Plugging this *u* into the basis function $B_{n,i}(u)$gives the following:

$$B_{n,i}(u) = \frac{n!}{i!(n-i)!}\left(\frac{u-a}{b-a}\right)^{i}\left(1-\frac{u-a}{b-a}\right)^{n-i}$$

These new basis functions define a Bézier curve on the domain of $[a, b]$. We shall use this fact in later sections.

MOVING CONTROL POINTS

Changing the position of a control point will change the shape of the defined Bézier curve. Our question is:

How does the shape of the curve change if a control point is moved to a new position?

Suppose a control point Pk is moved to a new position $P_k + v$, where vector v gives both the direction and length of this move. This is shown as follows:

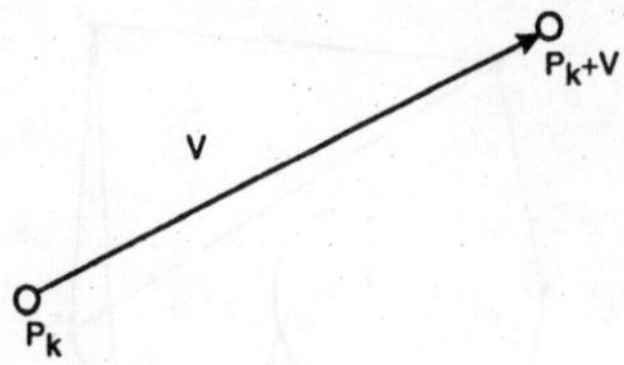

Let the original Bézier curve be as follows:

$$C(u) = \sum_{i=0}^{n} B_{n,i}(u)P_i$$

Since the new Bézier curve is defined by $P_0, P_1 \ldots, P_k + v, \ldots, P_n$, its equation $D(u)$ is

$$D(u) = \sum_{i=0}^{n} B_{n,i}(u)P_i + B_{n,k}(u)(P_k + v) + \sum_{i=k+i}^{n} B_{n,i}(u)P_i$$

$$= \sum_{i=0}^{n} B_{n,i}(u)P_i + B_{n,k}(u)v$$

$$= C(u) + B_{n,k}(u)v$$

In the above, since only the k-th term uses a different control point $Pk + v$, after regrouping we know that the new curve is the sum of the original curve and an extra term $B_{n,k}(u)v$. This means:

The corresponding point of u on the new curve is obtained by translating the corresponding point of u on the original curve in the direction of v with a distance of $|B_{n,k}(u)v|$.

More precisely, given a u, we have point $C(u)$ on the original curve and $D(u)$ on the new curve and $D(u) = C(u) + B_{n,k}(u)v$. Since v gives the direction of movement, $D(u)$ is the result of moving $C(u)$ in the same direction.

The length of this translation is, of course, the length of vector $B_{n,k}(u)v$. Therefore, when $B_{n,k}(u)$ reaches its maximum, the change from $C(u)$ to $D(u)$ is the largest.

The following figure illustrates this effect. Both the black and red curves are Bézier curves of degree 8 defined by 9 control points. The black one is the original curve. If its control point 3 is moved to a new position as indicated by the blue vector, the black curve changes to the red one. On each of these two curves there is the point corresponding to u=0.5. It is clear that $C(0.5)$ moves in the same direction to D(0.5).

The distance between C(0.5) and D(0.5) is the length of vector $B_{8,3}(0.5)v = \frac{8!}{(3!(8-3)!) \times 0.5^3(1-0.5)^{8-3}} v = 0.22v$. Hence, the distance is about 22% of the distance between the original control point 3 and the new control point 3 as shown in the figure below.

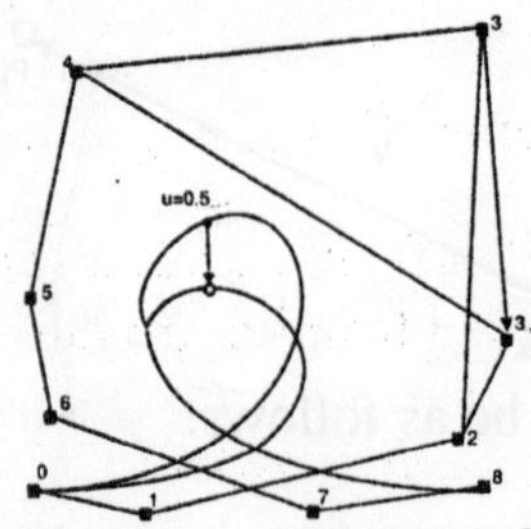

We can obtain one more important conclusion from the above discussion. Since $B_{n,k}(u)$is non-zero in the open interval (0, 1), $B_{n,k}(u)$v is not a zero vector in (0,1). This means that except for the two endpoints $C(0)$ and $C(1)$ all points on the original curve are moved to new locations. Therefore, we have

Changing the position of a control point causes the shape of a Bézier curve to change globally.

FINDING A POINT ON A BÉZIER CURVE: DE CASTELJAU'S ALGORITHM

Following the construction of a Bézier curve, the next important task is to find the point $C(u)$ on the curve for a particular u. A simple way is to plug u into every basis function, compute the product of each basis function and its corresponding control point, and finally add them together.

While this works fine, it is not numerically stable (*i.e.*, could introduce numerical errors during the course of evaluating the Bernstein polynomials).

In what follows, we shall only write down the control point numbers. That is, the control points are 00 for P_0, 01 for P_1,.. ., 0*i* for P_i,.. ., 0*n* for P_n. The 0*s* in these numbers indicate the initial or the 0-*th* iteration. Later on, it will be replaced with 1, 2, 3 and so on.

The fundamental concept of de Casteljau's algorithm is to choose a point C in line segment *AB* such that *C* divides the line segment *AB* in a ratio of $u : 1 - u$ (i.e., the ratio of the distance between A and C and the distance between *A* and *B* is *u*). Let us find a way to determine the point *C*.

The vector from A to B is $B - A$. Since u is a ratio in the range of 0 and 1, point C is located at $u(B - A)$. Taking the position of A into consideration, point C is $A + u(B - A) = (1 - u)A + uB$. Therefore, given a u, $(1 - u)A + uB$ is the point C between A and B that divides AB in a ratio of $u : 1 - u$.

The idea of de Castelja u's algorithm goes as follows. Suppose we want to find $C(u)$, where u is in [0, 1]. Starting with the first polyline, 00-01-02-03...-0n, use the above formula to find a point 1i on the leg (i.e. line segment) from 0i to $0(i + 1)$ that divides the line segment 0i and $0(i + 1)$ in a ratio of u:1-u. In this way, we will obtain n points 10, 11, 12,.. .., $1(n - 1)$. They define a new polyline of $n - 1$ legs.

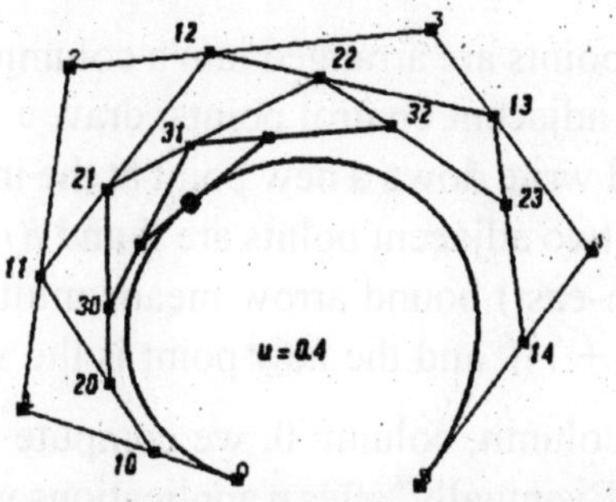

In the figure above, u is 0.4. 10 is in the leg of 00 and 01, 11 is in the leg of 01 and 02,.. ., and 14 is in the leg of 04 and 05. All of these new points are in blue.

The new points are numbered as 1*i*'s. Apply the procedure to this new polyline and we shall get a second polyline of $n - 1$ points 20, 21,.. ., $2(n - 2)$ and n - 2 legs. Starting with this polyline, we can construct a third one of $n - 2$ points 30, 31,.. ., $3(n - 3)$ and $n - 3$ legs. Repeating this process n times yields a single point n0. De Casteljau proved that this is the point $C(u)$on the curve that corresponds to u.

Let us continue with the above figure. Let 20 be the point in the leg of 10 and 11 that divides the line segment 10 and 11 in a ratio of u:1-u. Similarly, choose 21 on the leg of 11 and 12, 22 on the leg of 12 and 13, and 23 on the leg of 13 and 14. This gives a third polyline defined by 20, 21, 22 and 23. This third polyline has 4 points and 3 legs. Keep doing this and we shall obtain a new polyline of three points 30, 31 and 32. From this fourth polyline, we have the fifth one of two points 40 and 41. Do it once more, and we have 50, the point $C(0.4)$ on the curve.

This is the geometric interpretation of de Casteljau's algorithm, one of the most elegant result in curve design.

Actual Computation

Given the above geometric interpretation of de Casteljau's algorithm, we shall present a computation method, which is shown in the following figure.

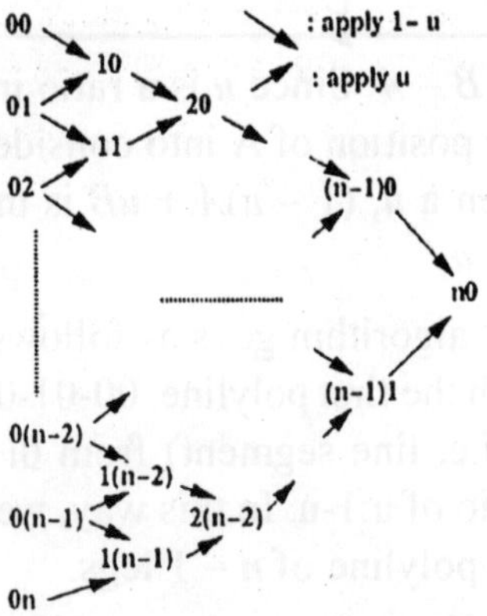

First, all given control points are arranged into a column, which is the left-most one in the figure. For each pair of adjacent control points, draw a south-east bound arrow and a north-east bound arrow, and write down a new point at the intersection of the two adjacent arrows. For example, if the two adjacent points are ij and $i(j + 1)$, the new point is $(i + 1)j$. The south-east (*resp.*, north-east) bound arrow means multiplying $1 - u$ (resp., u) to the point at its tail, ij (resp., $i(j + 1)$), and the new point is the sum.

Thus, from the initial column, column 0, we compute column 1; from column 1 we obtain column 2 and so on. Eventually, after n applications we shall arrive at a single point $n0$ and this is the point on the curve.

The following algorithm summarizes what we have discussed. It takes an array P of $n + 1$ points and a u in the range of 0 and 1, and returns a point on the Bézier curve $C(u)$.

Input: array $P[0 : n]$ of $n + 1$ points and real number u in [0, 1]

Output: point on curve, $C(u)$

Working: point array $Q[0 : n]$

```
for        i = 0 to n do
                Q[i] = P[i]; // save input
for        k = 1 to n do
for        i = 0 to n – k do
                Q[i] = (1 – u) Q[i] + u Q[i + 1];
return Q[0];
```

A Recurrence Relation

The above computation can be expressed recursively. Initially, let $P_{0,j}$ be P_j for $j = 0, 1, \ldots, n$. That is, $P_{0,j}$ is the j-th entry on column 0. The computation of entry j on column i is the following:

$$Pi,j = (1 - u)\ Pi\!-\!1,j + uwP_{i-1,j+1} \begin{cases} i = 1, 2, \ldots, n \\ j = 0, 1, \ldots, n - i \end{cases}$$

More precisely, entry $P_{i,j}$ is the sum of $(1 - u)P_{i-1,}j$ (upper-left corner) and $uP_{i-1}, _{j+1}$ (lower-left corner). The final result (i.e., the point on the curve) is Pn,0. Based on this idea, one may immediately come up with the following recursive procedure:

function deCasteljau (i, j)

begin if $i = 0$ then

return $P_{0,j}$

else return $(1 - u)$* deCasteljau $(i - 1, j) + u$* deCasteljau $(i - 1, j + 1)$

end

This procedure looks simple and short; however, it is extremely inefficient. Here is why. We start with a call to deCastelja $u(n, 0)$ for computing $P_{n,0}$. The else part splits this call into two more calls, deCastelja $u(n - 1, 0)$ for computing $P_{n-1,0}$ and deCastelja $u(n - 1, 1)$ for computing $P_{n-1,1}$.

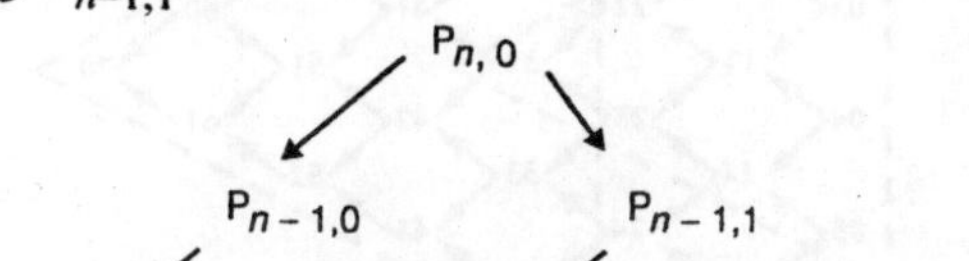

Consider the call to deCastelja $u(n - 1, 0)$. It splits into two more calls, deCastelja $u(n - 2, 0)$ for computing $P_{n-2,0}$ and deCastelja $u(n - 2, 1)$ for computing $P_{n-2,1}$. The call to deCastelja $u(n - 1, 1)$ splits into two calls, deCastelja $u(n - 2, 1)$ for computing $P_{n-2,1}$ and deCastelja $u(n - 2, 2)$ for computing $P_{n-2,2}$.

Thus, deCastelja $u(n - 2, 1)$ is called twice. If we keep expanding these function calls, we should discover that almost all function calls for computing $P_{i,j}$ are repeated, not once but many times.

How bad is this? In fact, the above computation scheme is identical to the following way of computing the n-th Fibonacci number:

function Fibonacci(n)

begin if $n = 0$ or $n = 1$ then

return 1

else

return Fibonacci $(n - 1)$ + Fibonacci $(n - 2)$

end

This program takes an exponential number of function calls (an exercise) to compute Fibonacci (n). Therefore, the above recursive version of de Casteljau's algorithm is not suitable for direct implementation, although it looks simple and elegant!

An Interesting Observation

The triangular computation scheme of de Casteljau's algorithm offers an interesting observation. Take a look at the following computation on a Bézier curve of degree 7 defined by 8 control points 00, 01,..., 07. Let us consider a set of consecutive points on the same column as the control points of a Bézier curve. Then, given a u in [0, 1], how do we compute the corresponding point on this Bézier curve? If de Casteljau's algorithm is applied to these control points, the point on the curve is the opposite vertex of the equilateral's base formed by the selected points!

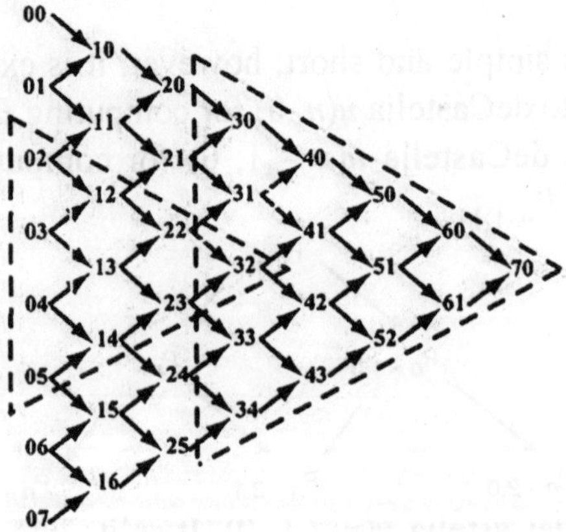

For example, if the selected points are 02, 03, 04 and 05, the point on the curve defined by these four control points that corresponds to u is 32. If the selected points are 11, 12 and 13, the point on the curve is 31. If the selected points are 30, 31, 32, 33 and 34, the point on the curve is 70.

By the same reason, 70 is the point on the Bézier curve defined by control points 60 and 61. It is also the point on the curve defined by 50, 51 and 52, and on the curve defined by 40, 41, 42 and 43. In general, if we select a point and draw an equilateral as shown above, the base of this equilateral consists of the control points from which the selected point is computed.

DE CASTELJAU'S ALGORITHM

De Casteljau's algorithm looks so different from the naive method that computes each Bn,i(u). A natural question is: does de Casteljau's algorithm correctly compute C(u)? The answer is a "yes" and we will justify this claim.

Let us look at the computation of a Bézier curve defined by 7 control points 00, 01, 02, 03, 04, 05 and 06 as shown below. The point on the curve is 60. Since 60 is computed from the seven control points 00,.. ., 06, each of the input control point 0i contributes to the computation of 60. What is this contribution?

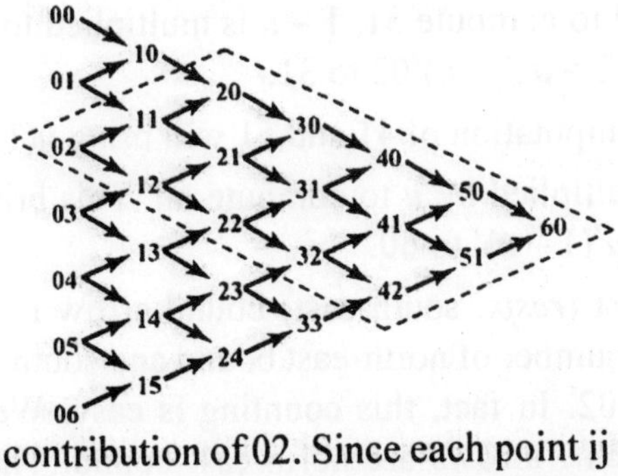

Let us investigate the contribution of 02. Since each point ij is multiplied by 1-u (resp., u) to compute its south-east (resp., north-east) neighbor, 02 is used by the computation of 11 and 12. Since 11 (resp., 12) has a component of 02, when 11 (resp., 12) is used to compute 20 and 21 (resp., 21 and 22), the contribution of 02 propagates to 20, 21 and 22. Therefore, the contribution of 02 can only reach the intermediate points in the diamond shape area as shown above.

If we follow the north-east bound arrow starting with 0i, each step causes the first (resp., second) index to be increased (resp., reduced) by one. After i steps, we will reach point i0. Similarly, if we follow the south-east bound arrow starting with $0i$, each step causes the first index to be increased by 1; however, the second index never changes. Therefore, after $n - i$ steps, we will reach point $(n - i)i$. Now, we have three corners of the diamond shape. The fourth one is, of course, $n0$. This is shown in the figure below. Verify this claim with the above figure.

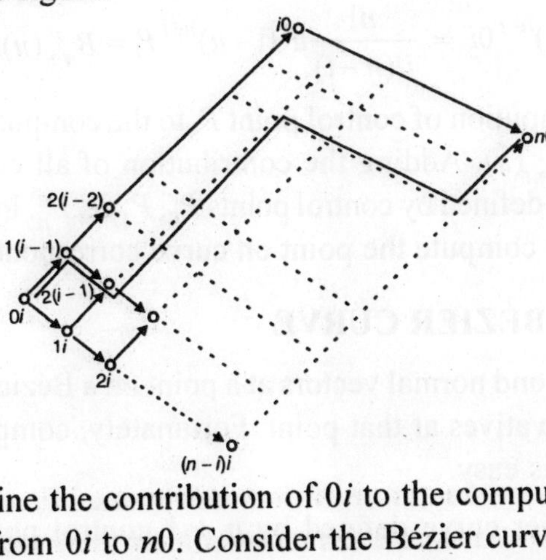

How do we determine the contribution of $0i$ to the computation of $n0$? This can be obtained by "walking" from $0i$ to $n0$. Consider the Bézier curve of degree 6 above as an example. If we follow the directional path 02, 12, 21, 31, 41, 51 and 60, then we have the following:

1. 12 has the contribution of $(1 - u)02$.

2. When 12 is used to compute 21, 12 contributes u12 to 21. Since 12 contains $(1 - u)02$, 21 will include $u(1 - u)02$.
3. When 21 is used to compute 31, $1 - u$ is multiplied to 21, and, hence, this brings $u(1 - u)(1 - u)02 = u(1 - u)^2 02$ to 31.
4. Similarly, the computation of 41 and 51 will bring $u(1 - u)^4 02$ to 51.
5. Finally, 51 is multiplied by u to compute 60. This brings the contribution of 02, along this path, $u^2(1 - u)^4$ to 60.

Since each north-east (*resp.*, south-east) bound arrow is associated with a u (resp., $(1 - u)$), by counting the number of north-east bound and south-east bound arrows, we will know the coefficient of 02. In fact, this counting is easy. We claim that each path has exactly n arrows of which i arrows are north-east bound. This is because the diamond shape is a $i \times (n - i)$ grid with n-i south-east bound arrows per "column" and i north-east bound arrows per "row." Therefore, from $0i$ to $n0$, on a single path, there are always $n - i$ $(1 - u)$'s and i u's, and, hence, the contribution of $0i$ to $n0$ is $ui(1 - u)^{n-i}$.

With this idea in mind, proving the correctness of de Casteljau's algorithm is very simple. Control point 0i contributes to the computation of $n0$ through all possible paths from $0i$ to $n0$. How many paths are there? Think this problem in the following way. Each path always has n arrows. Of these n arrows, i arrows are north-east bound. Hence, the number of ways of inserting these i arrows into n positions is equal to the number of different paths from 0i to n0. This is exactly the combinatorial coefficient $C(n, i) = (n!)/(i!(n - i)!)$. Since the contribution of each path is always $ui(1 - u)^{n-i}$ and since we have $(n!)/(i!(n - i)!)$ paths, the total contribution of $0i$ to $n0$ is the following:

$$\frac{n!}{i!(n-i)}u^i(1-u)^{n-i}0i = \frac{n!}{i!(n-i)}u^i(1-u)^{n-i}P_i = B_{n,i}(u)P_i$$

Therefore, the contribution of control point P_i to the computation of $C(u)$is exactly the Bernstein polynomial $B_{n,i}(u)$. Adding the contribution of all control points together, we have the the Bézier curve defined by control points $P_0, P_1, \ldots, P_n$. In conclusion, de Casteljau's algorithm does correctly compute the point on curve corresponding to a given u.

DERIVATIVES OF A BÉZIER CURVE

To compute tangent and normal vectors at a point on a Bézier curve, we must compute the first and second derivatives at that point. Fortunately, computing the derivatives at a point on a Bézier curve is easy.

Recall that the Bézier curve defined by $n + 1$ control points $P_0, P_1, \ldots, P_n$ has the following equation:

$$C(u) = \sum_{i=0}^{n} B_{n,i}(u)P_i$$

where $B_{n,i}(u)$ is defined as follows:

$$B_{n,\,i}(u) = \frac{n!}{i!(n-i)} u^i (1-u)^{n-i}$$

Since the control points are constants and independent of the variable u, computing the derivative curve $C'(u)$reduces to the computation of the derivatives of $B_{n,i}(u)$'s. With some simple algebraic manipulations, we have the following result for $B'_{n,i}(u)$:

$$\frac{d}{du} B_{n,i}(u) = B'_{n,i}(u) = n(B_{n-1,\,i-1}(u) - B_{n-1,\,i}(u))$$

Then, computing the derivative of the curve $C(u)$yields:

$$\frac{d}{du}(C) = C'(u) = \sum_{i=0}^{n-1} B_{n-1,i}(u)\{n(p_{i+1} - p_i)\}$$

Let $Q0 = n(P_1 - P_0)$, $Q_1 = n(P_2 - P_1)$, $Q_2 = n(P_3 - P_2),\ldots, Q_{n-1} = n(P_n - P_{n-1})$. The above equation reduces to the following:

$$C\text{¢}(u) = \sum_{i=0}^{n-1} B_{n-1,i}(u) Q_i$$

Therefore, the derivative of $C(u)$is a Bézier curve of degree $n - 1$ defined by n control points $n(P_1 - P_0)$, $n(P_2 - P_1)$, $n(P_3 - P_2),\ldots, n(P_n - P_{n-1})$. This derivative curve is usually referred to as the hodograph of the original Bézier curve. Note that $P_{i+1} - P_i$ is the direction vector from P_i to P_{i+1} and $n\ (P_{i+1} - P_i)$ is n times longer than the direction vector. Once the control points are known, the control points of its derivative curve can be obtained immediately. The left figure below shows a Bézier curve of degree 7 and the right figure shows its derivative which is a degree 6 Bézier curve.

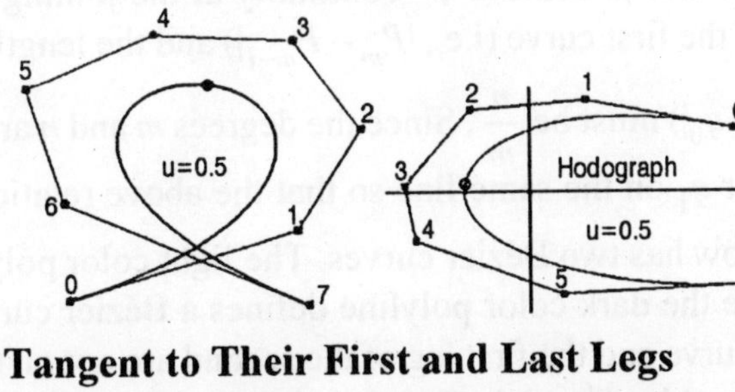

Bézier Curves Are Tangent to Their First and Last Legs

Letting $u = 0$ and $u = 1$ gives $C'(0) = n\ (P_1 - P_0)$ and $C'(1) = n\ (P_n - P_{n-1})$ The first means that the tangent vector at $u = 0$ is in the direction of $P_1 - P_0$ multiplied by n. Therefore, the first leg in the indicated direction is tangent to the Bézier curve. The second means that the tangent vector at $u = 1$ is in the direction of $P_n - P_{n-1}$ multiplied by n. Therefore, the last leg in the indicated direction is tangent to the Bézier curve. The following figures show this property well.

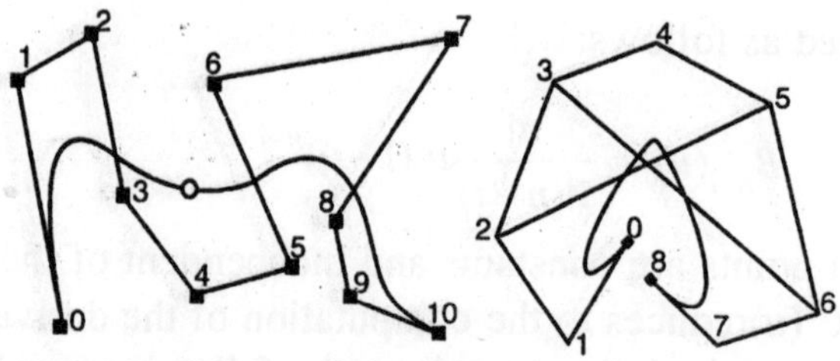

Joining Two Bézier Curves with C1-Continuity

That a Bézier curve being tangent to its first and last legs provides us with a technique for joining two or more Bézier curves together for designing a desired shape. Let the first curve $C(u)$be defined by $m + 1$ control points $P_0, P_1, P_2,.. ., P_m$. Let the second curve $D(u)$be defined by $n + 1$ control points $Q_0, Q_1, Q_2,..., Q_n$. If we want to join these two Bézier curves together, then P_m must be equal to Q_0. This guarantees a C0 continuous join. Recall that the first curve is tangent to its last leg and the second curve is tangent to its first leg. Consequently, to achieve a smooth transition, P_{m-1}, $P_m = Q_0$, and Q_1 must be on the same line such that the directions from P_{m-1} to Pm and the direction from Q_0 to Q_1 are the same. This is shown below.

While joining two Bézier curves in this way looks smooth, it is still a C^0 join and is not yet C^1. However, it is G^1, because they have the same tangent vector directions. To achieve C^1 continuity, we have to make sure that the tangent vector at $u = 1$ of the first curve, $C'(1)$, and the tangent vector at $u = 0$ of the second curve, $D'(0)$, are identical. That is, the following must hold:

$$C\text{¢}(1) = m(P_m - P_{m-1}) = D\text{¢}(0) = n(Q_1 - Q_0)$$

This relation states that to achieve C^1 continuity at the joining point the ratio of the length of the last leg of the first curve (i.e., $|P_m - P_{m-1}|$) and the length of the first leg of the second curve (i.e., $|q_1 - q_0|$) must be $\frac{n}{m}$. Since the degrees m and n are fixed, we can adjust the positions of P_{m-1} or q_1 on the same line so that the above relation is satisfied.

The left figure below has two Bézier curves. The light color polyline defines a Bézier curve of degree 4 while the dark color polyline defines a Bézier curve of degree 5. Since the last leg of the first curve and the first leg of the second are not on the same line, the two curves are not joint smoothly. The right figure shows two Bézier curves that are tangent to a line at the joining point. However, they are not C^1 continuous. The left curve is of degree 4, while the right curve is of degree 7. But, the ratio of the last leg of the left curve and the first leg of the second curve seems near 1 rather than $\frac{7}{4} = 1.75$. To achieve C^1 continuity, we should increase (resp., decrease) the length of the last (resp. first) leg of the left (resp., right). However, they are G^1 continuous.

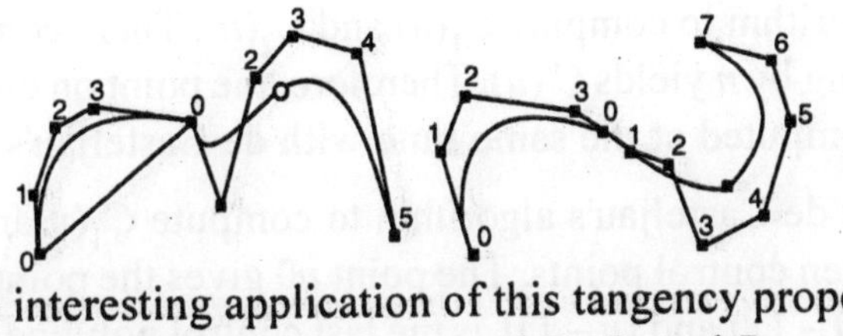

There is one more interesting application of this tangency property. If we let the first and last control points be identical (i.e., $P_0 = P_n$) and P_1, P_0 and P_{n-1} collinear, the generated Bézier curve will be a closed one and is G^1 continuous at the joining point as shown below. To achieve C^1 continuous at P0, we must have $P_1 - P_0 = P_n - P_{n-1}$ (i.e., the first and last legs have the same length and P_1, $P_0 = P_n$, P_{n-1} are collinear)

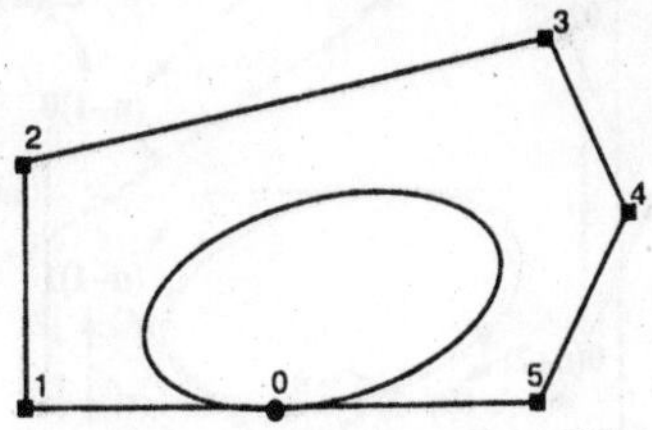

Note that although the above curve looks like an ellipse, it is not because this curve is of degree 6 and Bézier curves are polynomials which cannot represent circles and ellipses.

A Relation Between the Derivative and de Casteljau's Algorithm

Let us rewrite the derivative of a Bézier curve as follows:

$$C¢(u) = \sum_{i=0}^{n-1} B_{n-1,i}(u)\{n(P_{i+1} - P_i)\}$$

$$= n\left[\left(\sum_{i=0}^{n-1} B_{n,i}(u)P_{i+1}\right) - \left(\sum_{i=0}^{n-1} B_{n,1}(u)P_i\right)\right]$$

Thus, the derivative of a Bézier curve is the difference of two Bézier curves of degree n-1. For simplicity, let these two curves be $C_1(u)$and $C_2(u)$:

$$C_1(u) = \sum_{i=0}^{n-1} B_{n,i}(u)P_{i+1}$$

$$C_2(u) = \sum_{i=0}^{n-1} B_{n,i}(u)P_i$$

From these definitions, we know that the first curve $C_1(u)$ is defined by control points $P_1, P_2, \ldots, P_n$, that the second curve $C_2(u)$ is defined by control points $P_0, P_1, \ldots, P_{n-1}$, and that the derivative is:

$$C'(u) = n\{C_1(u) - C_2(u)\}$$

Therefore, theoretically, to compute the derivative of $C(u)$ at a particular u, we can use de Casteljau's algorithm to compute $C_1(u)$ and $C_2(u)$. Then, computing their difference followed by multiplying by n yields $C'(u)$. Therefore, the point on curve $C(u)$ and its tangent vector $C'(u)$can be computed at the same time with de Casteljau's algorithm.

Now let us apply de Casteljau's algorithm to compute $C_1(u)$and $C_2(u)$. The left-most column has all the given control points. The point $n0$ gives the point $C(u)$ on the curve, and the segment joining $(n-1)0$ and $(n-1)1$ is the last control polyline of the de Casteljau net. In fact, n0 lies in the segment of $(n-1)0$ and $(n-1)1$ and divides it in a ratio of $u : 1-u$.

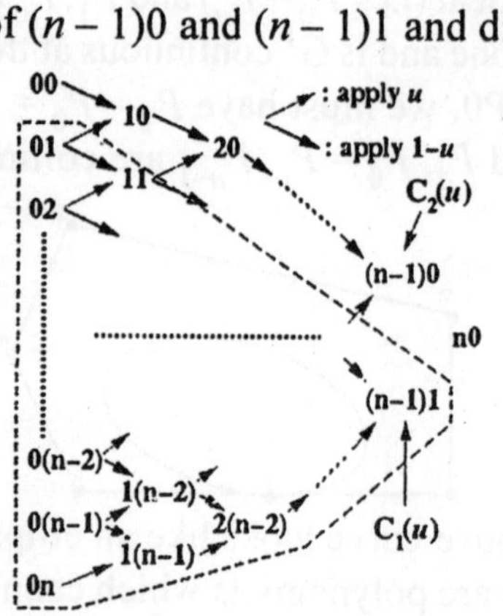

Since curve $C_1(u)$ is defined by control points 01, 02,.. ., 0n, using the above diagram we know that $(n-1)1$ is the point on curve $C_1(u)$. Similarly, since $C_2(u)$ is defined by control points 00, 01,.. ., $0(n-1)$, using the above diagram we know that $(n-1)0$ is the point on curve $C_2(u)$.

Therefore, vector $C_1(u) - C_2(u)$ is the vector from point $(n-1)0$ to point $(n-1)1$, and the derivative of $C(u)$ is n times $C_1(u) - C_2(u)$.

What does this mean? Since the line segment from $(n-1)0$ to $(n-1)1$ of the de Casteljau net has the same direction as the tangent vector at C(u), it is tangent to the curve at $C(u)$! In summary, we have

The last polyline, actually a line segment, of a de Casteljau net is tangent to the Bézier curve at C(u).

The following figure shows an example. The point shown corresponds to $u = 0.5$. It is clear that the last line segment of a de Casteljau net is tangent to $C(0.5)$.

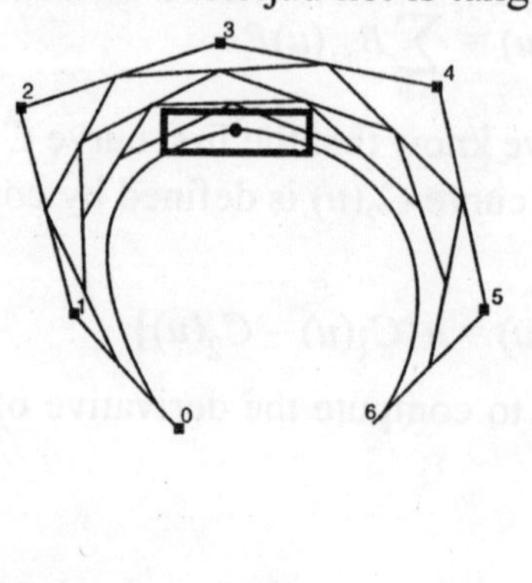

Higher Derivatives

Computing higher order derivatives of a Bézier curve is a simple matter. Recall that the derivative of $C(u)$is the following:

$$C\text{¢}(u) = \sum_{i=0}^{n-1} B_{n-1,i}(u) Q_i$$

Applying the derivative formula to the above Bézier curve yields the following, which gives the second derivative of the original Bézier curve:

$$C^2(u) = \sum_{i=0}^{n-2} B_{n-2,i}(u)\{(n-1)(Q_{i+1} - Q_i)\}$$

$$= \sum_{i=0}^{n-2} B_{n-2,i}(u)\{n(n-1)(Q_{i+2} - 2Q_{i+1} + Q_i)\}$$

After obtaining $C'(u)$and $C''(u)$, the moving triad and curvature at $C(u)$can be computed easily. Higher derivatives can be found by recursively applying the formula of derivative.

To express higher derivative concisely, we shall use finite difference. Let us define D_i^0 to be control point P_i for $0 \le i \le n$. Then, we define the first level difference D_i^1 to be the difference of the previous level:

$$D_0^1 = D_1^0 - D_0^0 = P_1 - P_0$$
$$D_1^1 = D_2^0 - D_1^0 = P_2 - P_1$$
$$\vdots$$
$$D_{n-2}^1 = D_{n-1}^0 - D_{n-2}^0 = P_{n-1} - P_{n-2}$$
$$D_{n-1}^1 = D_n^0 - D_{n-1}^0 = P_n - P_{n-1}$$

Note that the first level difference has only n points, one less than the number of given control points. More precisely, the first level of difference is the following:

$$D_i^1 = D_{i+1}^0 - D_i^0 \qquad 0 \text{ £ } i \text{ £ } n-1$$

The second level difference is defined to be the difference of the first level points:

$$D_i^2 = D_{i+1}^1 - D_i^1 \qquad 0 \text{ £ } i \text{ £ } n-2$$

This second level difference has n-1 points. Repeating this procedure, we can define the k–th level difference as follows. The k-th level difference has $n - k + 1$ point.

$$D_i^k = D_{i+1}^{k-1} - D_i^{k-1} \qquad 0 \text{ £ } i \text{ £ } n-k$$

With these notation in hand, higher derivatives can be expressed concisely. First of all, let us rewrite $C'(u)$as follows, using the D_i^1's:

$$C¢(u) = n\sum_{i=0}^{n-1} B_{n-1,i}(u)(P_{i+1} - P_i)$$

$$= n\sum_{i=0}^{n-1} B_{n-1,i}(u)(D_{i+1}^0 - D_i^0)$$

$$= n\sum_{i=0}^{n-1} B_{n-1,i}(u)D_i^1$$

Applying the derivative to $C'(u)$to compute $C''(u)$gives us a formula of $C''(u)$in D_i^2's:

$$C^2(u) = n(n-1)\sum_{i=0}^{n-2} B_{n-2,i}(u)(D_{i+1}^2 - D_i^1)$$

$$= n(n-1)\sum_{i=0}^{n-2} B_{n-2,i}(u)D_i^2$$

Do the same to $C''(u)$and we have a formula of $C'''(u)$in D_i^3's:

$$C^2¢ = n(n-1)(n-2)\sum_{i=0}^{n-3} B_{n-3,i}(u)(D_{i+1}^2 - D_i^2)$$

$$= n(n-1)(n-2)\sum_{i=0}^{n-3} B_{n-3,i}(u)D_i^3)$$

Continue this process and we can compute the *k*-th derivative $C_k(u)$in D_i^k's:

$$Ck(u) = n(n-1)(n-2)...(n-k+1)\sum_{i=0}^{n-k} B_{n-k,i}(u)(D_{i+1}^{k-1} - D_i^{k-1})$$

$$= n(n-1)(n-2)...(n-k+1)\sum_{i=0}^{n-k} B_{n-k,i}(u)D_i^k$$

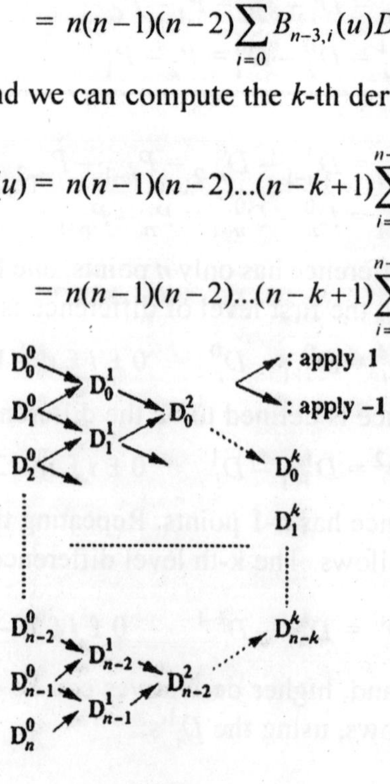

Therefore, to compute the *k*-th derivative of $C(u)$at a particular u, we first compute the D_i^{k}'s and apply de Casteljau's algorithm to compute the point corresponding to u on the Bézier curve defined by D_i^{k}'s. More precisely, we can arrange all D_i^{0}'s on column 0, assign the south-east and north-east arrows –1 and 1, respectively, to compute column 1 of D_i^{1}'s. Repeat this until we reach column k as shown below:

Finally, applying de Casteljau's algorithm to the points on column k with u and multiply the result by $n(n-1)(n-2)...(n-k+1)$ gives us the vector of $C^k(u)$!

SUBDIVIDING A BÉZIER CURVE

The meaning of subdividing a curve is to cut a given Bézier curve at $C(u)$for some u into two curve segments, each of which is still a Bézier curve. Because the resulting Bézier curves must have their own new control points, the original set of control points is discarded. Moreover, since the original Bézier curve of degree n is cut into two pieces, each of which is a subset of the original degree n Bézier curve, the resulting Bézier curves must be of degree n.

Our problem is: Given a set of $n + 1$ control points $P_0, P_1, P_2,..., P_n$ and a parameter value u in the range of 0 and 1, we want to find two sets of $n + 1$ control points $Q_0, Q_1, Q_2,..., Q_n$ and $R_0, R_1, R_2,..., R_n$ such that the Bézier curve defined by Q_i's (resp., R_i's) is the piece of the original Bézier curve on $[0, u]$ (resp., $[u, 1]$).

While a correctness proof is a little tedious, the algorithm is actually very simple. In fact, de Casteljau's algorithm for evaluating the point $C(u)$on the curve has provided all necessary information.

In the left figure below, all intermediate steps of applying de Casteljau's algorithm for computing $C(u)$are shown and the right one shows the subdivisions of the curve at $C(u)$and their corresponding control polylines.

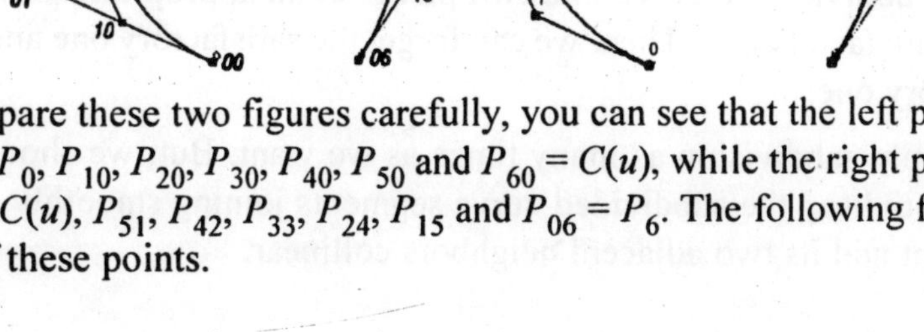

If you compare these two figures carefully, you can see that the left polyline consists of points $P_{00} = P_0, P_{10}, P_{20}, P_{30}, P_{40}, P_{50}$ and $P_{60} = C(u)$, while the right polyline consists of points $P_{60} = C(u), P_{51}, P_{42}, P_{33}, P_{24}, P_{15}$ and $P_{06} = P_6$. The following figures illustrate the selection of these points.

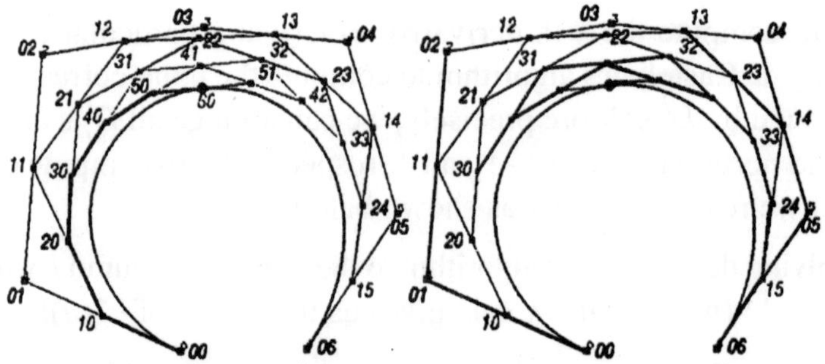

Recall from the triangular computation scheme of de Casteljau's algorithm. For a given u, it takes n iterations to compute $C(u)$. In the course of computation, one can collect the first and the last points on each column and, at the end, the collection of the first (resp., last) points gives the subdivision corresponding to the piece of the original curve defined on $[0, u]$ (resp., $[u, 1]$). Thus, in the following triangular scheme, the top edge in the direction of the arrows and the lower edge in the reversed direction of the arrows provide the control points of the first and the second curve segments, respectively.

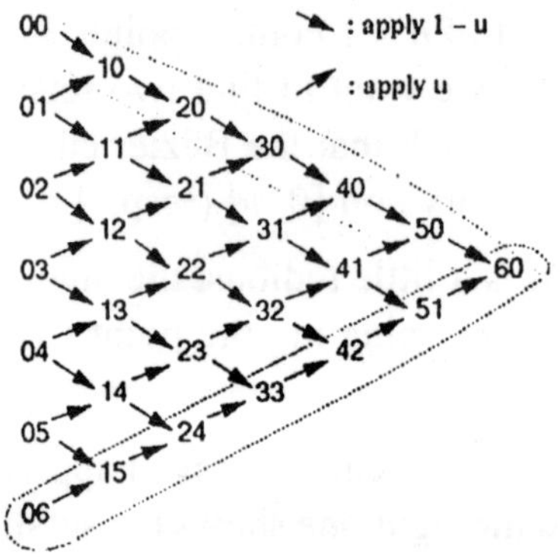

Note that since the line segment defined by 50 and 51 is tangent to the curve at point 60, the last leg of the left curve (i.e., point 50 to point 60) is tangent to the left curve, and the first leg of the right curve (i.e., point 60 to point 51) is tangent to the right curve.

Need of Curve Subdivision

Curve subdivision has many applications. For example, it can be used for computing the intersection of two Bézier curves, rendering Bézier curves, and making curve design easier. Suppose we have designed a curve, but not up to our expectation. In this case, we might want to subdivide the curve into two pieces at an appropriate point, a satisfactory one and an unsatisfactory one. Then, we can forget the satisfactory one and concentrate on the unsatisfactory one.

We can apply subdivision as many times as we want. But, we should keep in mind that if we want to keep the subdivided curve segments joining smoothly, we should keep the joining point and its two adjacent neighbors collinear.

SUBDIVISION ALGORITHM

We shall establish the correctness of the subdivision algorithm. It requires some tedious, but simple, calculations. Suppose the given Bézier curve of degree n is defined by n+1 control points $P_0, P_1, \ldots, P_n$. Suppose further the curve is subdivided at a fixed u in [0,1]. The subdivision algorithm states that the curve segment on [0, u] is defined by points $P_{00} = P_0, P_{10}, P_{20}, \ldots, P_{n0}$. To establish the correctness of the subdivision algorithm, our strategy is

1. Find the Bézier curve of degree n defined by control points $P_{00} = P_0, P_{10}, P_{20}, \ldots, P_{n0}$
2. Verify that this curve is the given curve restricted on domain [0, u]. The same technique applies to the second curve segment on [u, 1].

Let the original curve be

$$C(u) = \sum_{i=0}^{n} B_{n,i}(u) P_i$$

According to the observations discussed on the de Casteljau's algorithm page, we know that point P_{k0} is computed from control points $P_0, P_1, \ldots, P_k$ at the fixed value u:

$$P_{k0} = \sum_{i=0}^{k} B_{k,i}(u) P_i$$

The Bézier curve of degree n defined by $P_{00} = P_0, P_{10}, P_{20}, \ldots, P_{n0}$ is

$$D(t) = \sum_{k=0}^{n} B_{n,k}(t) P_{k0}$$

where t is in [0,1]. Note that we use a different parameter value t to avoid confusion. Note also that the domain of this curve is [0, u] rather than [0, 1]. This can be done easily with a simple change of parameter. We will do this at the end of this proof. Plugging P_{k0} into the curve $D(t)$, we have

$$D(t) = \sum_{k=0}^{n} B_{n,k}(t) \left[\sum_{i=0}^{k} B_{k,i}(u) P_i \right]$$

Therefore, the curve $D(t)$ is still represented by the original control points $P_0, P_1, \ldots, P_n$. If we can figure out the coefficient of each P_i, we will have a proof! Consider point P_h. When k is greater than or equal to h, the term $B_{k,h}(u)$is the coefficient of P_h. Therefore, the coefficient of P_h in the curve $D(u)$is:

$$P_h\text{'s coefficient} = B_{n,h}(t)\, B_{h,h}(u) + B_{n,h+1}(t)\, B_{h+1,}\, h(u) + \ldots + B_{n,n}(t)\, B_{n,h}(u)$$

$$= \sum_{j=0}^{n-h} B_{n,h+j}(t) B_{h+j,h}(u)$$

$$= \sum_{j=0}^{n-h}\left[\frac{n!}{(h+j)!(n-(h+j)!}t^{h+j}(1-t)^{n-(h+j)}\right]\left[\frac{(h+j)!}{h!j!}u^{h}(1-u)^{8}\right]$$

Terms $(tu)_h$, $n!$ and $j!$ can be moved to the outside of the summation:

$$Ph\text{'s coefficient} = \frac{n!}{h!}(tu)^h\sum_{j=0}^{n-h}\frac{1}{((n-h)-j)!\,j!}[t(1-u)]^j[(1-t)^{(n-h)-j}]$$

We are almost there because we have $(tu)^h$, $n!$ and j, the missing part is a $(n - h)!$. So, let us multiply $(n - h)!$ to each term in the summation and divide the outside term by $(n - h)!$. This yields:

$$P_h\text{'s coefficient} = \frac{n!}{h!(n-h)!}(tu)^n\sum_{j=0}^{n-h}\frac{(n-h)!}{((n-h)-j)!\,j!}[t(1-u)]^j[(1-t)^{(n-h)-j}]$$

Recall the following binomial expansion of $(a + b)^m$:

$$(a+b)^m = \sum_{s=0}^{m}\frac{m!}{s!(m-s)!}a^s b^{m-s}$$

The summation can be simplified with the above binomial expansion formula to the following:

$$\sum_{j=0}^{n-h}\frac{(n-h)!}{((n-h)-j)!\,j!}[(1-u)]^j[(1-t)^{n-h)-j}]$$

$$= [t(1-u)+(1-t)]^{n-h} = [1-(tu)]^{n-h}$$

Hence, the coefficient of Ph is:

$$P_h\text{'s coefficient} = \frac{n!}{h!(n-h)!}(tu)^h(1-(tu))^{n-h}$$

The relationship between the original curve $C(u)$and the subdivision curve on [0,u] is shown below:

$$D(t) = \sum_{h=0}^{n} B_{n,h}(tu)P_h$$

This means that when t traces the curve $D(u)$on [0, 1], tu traces the original curve $C(u)$ in the interval [0, u]. Moreover, these two components are identical, and, as a result, curve $D(t)$ is the curve segment of $C(u)$on [0, u]. This establishes the correctness of the subdivision algorithm.

DEGREE ELEVATION OF A BEZIER CURVE

Many applications that involve two or more Bézier curves require all involved curves to have the same degree. Moreover, although higher degree Bézier curves require longer time to process, they do have higher flexibility for designing shapes. Therefore, it would be very helpful to increase the degree of a Bézier curve without changing its shape.

Note that "without changing the curve's shape" is the key point; otherwise, just increasing the degree of a Bézier curve does not make any practical sense. Increasing the degree of a Bézier curve without changing its shape is referred to as degree elevation. In what follows, only an algorithm will be discussed.

Suppose we have a Bézier curve of degree n defined by $n + 1$ control points P_0, P_1, P_2,.. ., P_n and we want to increase the degree of this curve to $n + 1$ without changing its shape. Since a degree $n + 1$ Bézier curve is defined by $n + 2$ control points, we need to find such a new set of control points. Obviously, P0 and Pn must be in the new set because the new curve also passes through them. Therefore, what we need is only n new control points. Let the new set of control points be Q_0, Q_1, Q_2,..., Q_{n+1}. As mentioned above, $Q_0 = P_0$ and $Q_{n+1} = P_n$. The other control points are computed as follows:

$$Qi = \frac{i}{n+1}P_{i-1} + \left(1 - \frac{i}{n+1}\right)P_i \quad 1 \le i \le n$$

If you are not comfortable with the above general formula, the following are formulae for each control points from Q_1 to Q_n.

$$Q_1 = \frac{1}{n+1}P_0 + \left(1 - \frac{1}{n+1}\right)P_1$$

$$Q_2 = \frac{2}{n+1}P_1 + \left(1 - \frac{2}{n+1}\right)P_2$$

$$Q_3 = \frac{3}{n+1}P_2 + \left(1 - \frac{3}{n+1}\right)P_3$$

$$\vdots$$

$$Q_{n-1} = \frac{n-1}{n+1}P_{n-2} + \left(1 - \frac{n-1}{n+1}\right)P_{n-1}$$

$$Q_n = \frac{n}{n+1}P_{n-1} + \left(1 - \frac{n}{n+1}\right)P_n$$

Each leg of the original polyline contains exactly one new control point. More precisely, leg P_{i-1} P_i contains new control point Q_i. Recall from the discussion of de Casteljau's algorithm that a point C on a line segment AB that divides AB in a ratio of $u : 1 - u$ can be written as $C = (1 - u)A + uB$. From the formulae for the new control points, we see that Q_i divides the segment P_{i-1} P_i in a ratio of $\frac{1-i}{(n+1)} : \frac{i}{(n+1)}$. However, unlike de Casteljau's algorithm, this ratio is not a constant but varying with the value of i. This computation is very similar to that of de Casteljau's algorithm, though.

Once the new set of control points is obtained, the original set can be discarded. Since each leg of the original control polyline contains a new control point, the process of replacing the old set with the new one can be viewed as cutting off the corners at the original control points. The following figure illustrates this corner-cutting effect.

The figure shows a Bézier curve of degree 4 whose control points are shown in red rectangles and control polyline in blue dashed line segments. After increasing its degree to 5, the new control polyline is shown in solid line segments. It is clear that all corners are cut. The left table gives the ratio on each leg of the original control polyline.

i	$1 - i/(n+1)$
1	0.8
2	0.6
3	0.4
4	0.2

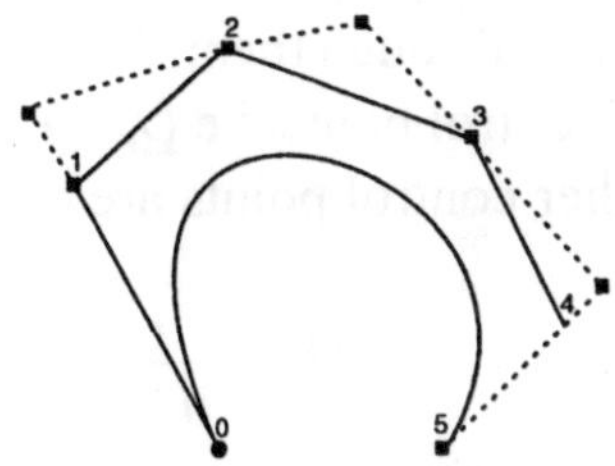

Note that degree elevation can be used repeatedly as long as your system permits. Note also that as the degree increases the number of control points increases. Moreover, the new control polyline moves toward the curve. In the following figures, we start with a Bézier curve of degree 6 with 7 control points. Then, its degree is increased to 7, 8, 10, 15 and 29. The shape of the curve is not changed as its degree increases and the control polyline moves closer and closer to the curve. Eventually, as the degree keeps increasing to infinity, the control polyline approaches to the curve and has it as a limiting position.

DEGREE ELEVATION ALGORITHM

We shall establish the correctness of the degree elevation algorithm. Again, it requires simple, but tedious, calculations.

Suppose the given Bézier curve $C(u)$of degree n is defined by control points $P_0, P_1,...,$ P_n, and the degree elevated Bézier curve $D(u)$of degree $n + 1$ is defined by control points $Q_0 = P_0, Q_1, Q_2,.. ., Q_n, Q_{n+1} = P_n$. We shall find the n unknown control points Q_i's.

Our strategy is very simple. Since $C(u)$ and $D(u)$ are the same curve, their derivatives at any u should be equal.

Hence, by comparing the first, second, third,..., and n-th derivatives at 0, we should be able to determine what Q_i's are. To simplify our calculation, we shall use mathematical indiction. More precisely, we find Q_1 by comparing the first derivatives at 0. Then, we assume that $Q_1, Q_2,.. ., Q_{k-1}$ are computed and use these known results to compute Q_k. Iterating this process n times, we will compute all new control points.

The Base Case

The first derivative of $C(u)$ is

$$\frac{d}{du}C(u) = C'(u) = \sum_{i=0}^{n-1} B_{n-1,i}(u)\{n(P_{i+1} - P_i)\}$$

Thus, the derivative at $u = 0$ is $n(P_1 - P_0)$. Similarly, the derivative of $D(u)$ at $u = 0$ is $(n + 1)(Q_1 - Q_0) = (n + 1)(Q_1 - P_0)$. Because these two derivatives are equal, we have

$$n(P_1 - P_0) = (n + 1)(Q_1 - P_0)$$

Solving for Q_1, we have our first result:

$$Q_1 = \frac{1}{n+1}P_0 + \left(1 + \frac{1}{n+1}\right)P_1$$

The Induction Step

For the induction step, we shall assume that the computation of $Q_0 = P_0, Q_1, \ldots, Q_{k-1}$ are correct and show that Q_k is also correct. This is done by comparing the k-th derivatives of the original curve $C(u)$and the degree elevated curve $D(u)$.

Just like in the base case, we actually do not need the complete curves. Let the k-th derivative of $C(u)$and $D(u)$ be $C^k(u)$and $D^k(u)$, respectively. The k-th derivative vectors at $u = 0$ are $C^k(0)$ and $D^k(0)$. Because $C(u)$and $D(u)$are the same curve, their k-th derivatives at $u = 0$ should be identical. More precisely, we should have $C^k(0) = D^k(0)$. From this relation and the known $Q_0 = P_0, Q_1, \ldots, Q_{k-1}$, we shall compute Q_k.

In the discussion of higher derivatives, we know that the k-th derivative function of curve $C(u)$ is

$$\begin{aligned} C^k(u) &= n(n-1)(n-2)\ldots(n-k+1)\sum_{i=0}^{n-k} B_{n-k,i}(u)\left(D_{i+1}^{k-1} - D_i^{k-1}\right) \\ &= n(n-1)(n-2)\ldots(n-k+1)\sum_{i=0}^{n-k} B_{n-k,i}(u)\,D_i^k \end{aligned}$$

where D_k^i is

$$D_i^k = \sum_{j=0}^{k}(-1)^{k-j}C(k,j)P_{i+j}$$

and the combinatorial coefficient $C(k, j)$ is

$$C(k,j) = \frac{k!}{j!(k-j)!}$$

The computation of the D_i^k's is an exercise at the end of this unit. Therefore, $C^k(0)$ is computed as follows:

$$Ck(0) = n(n-1)...(n-k+1)D_0{}^k$$

$$= n(n-1)...(n-k+1)\sum_{j=0}^{k}(-1)^{k-j}C(k,j)P_j$$

Similarly, we can compute $D^k(0)$. Note that curve $D(u)$ is of degree $n + 1$. This is the reason that the multiplier is $(n + 1)\,n(n - 1)...(n - k + 2)$.

$$D^k(0) = (n+1)n(n-1)...(n-k+k)\sum_{j=0}^{k}(-1)^{k-j}C(k,j)Q_j$$

Setting $C^k(0)$ equal to $D^k(0)$ and canceling some common factors yields the following:

$$(n+1)\sum_{j=0}^{k}(-1)^{k-j}C(k,j)Q_j = (n-k+1)\sum_{j=0}^{k}(-1)^{k-j}C(k,j)P_j$$

In fact, we can do more. Since we know Q_k is unknown and $Q_0 = P_0$, we can take these two terms out of the summation of $D^k(0)$.

Now, the summation runs from 1 to $k-1$. To make the summation in $C^k(0)$ compatible, we also take the terms of P_k and P_0 out of the summation. This gives the following:

$$(n+1)Q_k + (-1)^k(n+1)P_0 + \sum_{j=1}^{k-1}(-1)^{k-j}(n+1)C(k,j)Q_j$$

$$= (n-k+1)P_k + (-1)^k(n-k+1)P_0 + \sum_{j=1}^{k-1}(-1)^{k-j}(n-k+1)C(k,j)P_j$$

Rearranging the terms yields the following:

$$(n+1)Qk = (n-k+1)P_k - (-1)^k kP_0 + \sum_{j=1}^{k-1}(-1)^{k-j}C(k,j)\left[(n-k+1)P_j - (n+1)Q_j\right]$$

By the induction hypothesis, for j in the range of 0 and $k - 1$, Q_j is known and is computed as follows:

$$Q_j = \frac{j}{n+1}P_{j-1} + \left(1 - \frac{j}{n+1}\right)P_j$$

Plug Qj into the term in the summation and we have:

$$(n - k + 1)P_j - (n + 1)Q_j = -(k - j)P_j - jP_{j-1}$$

Plug this result back into the expression of $(n + 1)Q_k$ and split the summation into two:

$$(n+1)Qk = (n-k+1)P_k - (-1)^k kP_0 + \sum_{j=1}^{k-1}(-1)^{k-j}C(k,j)\left[-(j-j)P_j - jP_{j-1}\right]$$

$$= (n-k+1)P_k - (-1)^k kP_0 + \sum_{j=1}^{k-1}(-1)^{k-j+1}C(k,j)\left[(k-j)P_j + jP_{j-1}\right]$$

$$= (n-k+1)P_k - (-1)^k kP_0 + \sum_{j=1}^{k-1}(-1)^{k-j+1}C(k,j)\,(k-j)P_j$$

$$+\sum_{j=1}^{k-1}(-1)^{k-j+1}C(k,j)\,(k,j)jP_{j-1}$$

After taking P_{k-1} out of the first summation, the first summation becomes:

$$\sum_{j=1}^{k-1}(-1)^{k-j+1}C(k,j)\,(k-j)jP_j = kP_{k-1} + \sum_{j=1}^{k-2}(-1)^{k-j+1}\frac{k!}{j!(k-j-1)!}P_j$$

Then, we take the P_0 term out of the second summation:

$$\sum_{j=1}^{k-1}(-1)^{k-j+1}C(k,j)jP_{j-1} = (-1)^k kP_0 + \sum_{j=2}^{k-1}(-1)^{k-j+1}\frac{k!}{(j-1)!(k-j)!}P_{j-1}$$

However, in this expression, the index j runs from 2 to $k-1$. To make it compatible with that of the first summation, we can use a new index $h = j-1$. After this substitution, the second summation becomes the following:

$$\sum_{j=0}^{k-1}(-1)^{k-j+1}C(k,j)jP_{j-1} = (-1)^k kP_0 + \sum_{h=1}^{k-2}(-1)^{k-h}\frac{k!}{h!(k-h-1)!}P_{j-1}$$

Note that the corresponding terms in these two summations are equal with different signs (i.e., $(-1)^{k-j+1} = (-1) \times (-1)^{k-j}$ in the first summation and $(-1)^{k-h}$ in the second). Hence, adding these two summations together will cancel out all terms in the summation and yield the following:

$$\sum_{j=1}^{k-1}(-1)^{k-j}C(k,j)\left[(n-k+1)P_j - (n+1)Q_j\right] = kP_{k-1} + (-1)^k kP_0$$

Putting this back into the expression of (n+1)Qk yields

$$(n+1)Q_k = kP_{k-1} + (n-k+1)P_k$$

Dividing the expression by $n+1$ gives us the desired result:

$$Q_k = \frac{k}{n+1}P_{k-1} + \left(1 - \frac{k}{n+1}\right)P_k$$

Therefore, the degree elevation computation for k also holds, and the degree elevation algorithm is correct.

Chapter 9

B-Spline Curves

In the mathematical subfield of numerical analysis, a B-spline is a spline function that has minimal support with respect to a given degree, smoothness, and domain partition. A fundamental theorem states that every spline function of a given degree, smoothness, and domain partition, can be represented as a linear combination of B-splines of that same degree and smoothness, and over that same partition. The term *B-spline* was coined by Isaac Jacob Schoenberg and is short for basis spline. B-splines can be evaluated in a numerically stable way by the de Boor algorithm.

In the computer science subfields of computer-aided design and computer graphics the term B-spline frequently refers to a spline curve parametrized by spline functions that are expressed as linear combinations of B-splines. A B-spline is simply a generalisation of a Bézier curve, and it can avoid the Runge phenomenon without increasing the degree of the B-spline.

MOTIVATION

Consider designing the profile of a vase. The left figure below is a Bézier curve of degree 11; but, it is difficult to bend the "neck" toward the line segment P4P5. Of course, we can add more control points near this segment to increase the weight to that region. However, this will increase the degree of the curve. In many cases, it is not worth to use such a high degree polynomial.

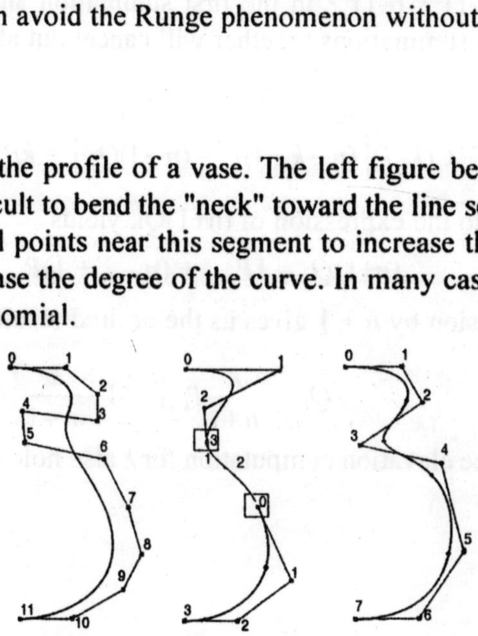

As discussed in a previous page about the derivatives of a Bézier curve, we can join two Bézier curves together. As long as the last leg of the first curve and the first leg of the second have the same direction, we can at least achieve G^1 continuity because the tangent vectors have the same direction but may have different length (*i.e.*, if the lengths are the same, it becomes C^1 continuous).

The middle figure above uses this idea. It has three Bézier curve segments of degree 3 with joining points marked with yellow rectangles. This shows that with multiple low degree Bézier curve segments satisfying the G^1 continuous condition, we still can design complex shapes. But, maintaining this G^1 continuous condition may be tedious and undesirable.

Is it possible that we still can use lower degree curve segments without worrying about the G^1 continuous condition? B-spline curves are generalizations of Bézier curves and are developed to answer this question.

The right figure above is a B-spline curve of degree 3 defined by 8 control points. In fact, there are five Bézier curve segments of degree 3 joining together to form the B-spline curve defined by the control points. In the above, those little dots subdivide the B-spline curve into Bézier curve segments. One can move control points for modifying the shape of the curve just like what we do to Bézier curves. We can also modify the subdivision of the curve. Therefore, B-spline curves have higher degree of freedom for curve design.

Subdividing the curve directly is difficult to do. Instead, we subdivide the domain of the curve. Thus, if the domain of a curve is [0, 1], this closed interval is subdivided by points called knots. Let these knots be $0 \le u_0 \le u_1 \le ... \le u_m \le 1$. Then, points $C(u_i)$'s subdivide the curve as shown in the figure below and, consequently, modifying the subdivision of [0, 1] changes the shape of the curve.

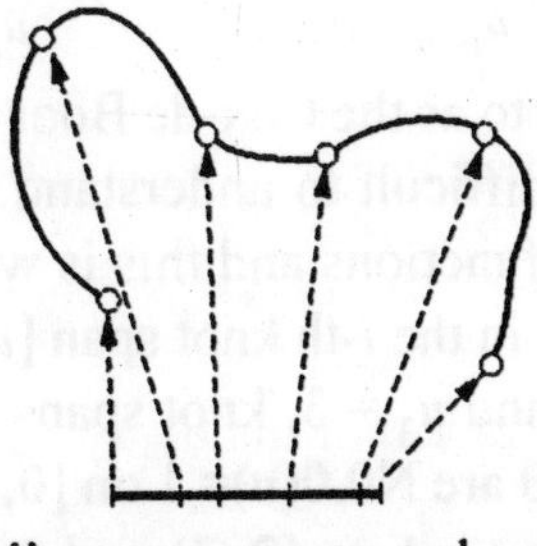

In summary, to design a B-spline curve, we need a set of control points, a set of knots and a set of coefficients, one for each control point, so that all curve segments are joined together satisfying certain continuity condition.

The computation of the coefficients is perhaps the most complex step because they must ensure certain continuity conditions. Fortunately, this computation is usually not needed in this course. We only need to know their characteristics for reasoning about B-spline curves.

B-SPLINE BASIS FUNCTIONS

Bézier basis functions are used as weights. B-spline basis functions will be used the same way; however, they are much more complex. There are two interesting properties that are not part of the Bézier basis functions, namely: (1) the domain is subdivided by knots, and (2) basis functions are not non-zero on the entire interval. In fact, each B-spline basis function is non-zero on a few adjacent subintervals and, as a result, B-spline basis functions are quite "local".

Let U be a set of $m + 1$ non-decreasing numbers, u_0 £ u_2 £ u_3 £...£ u_m. The u_i's are called knots, the set U the knot vector, and the half-open interval $\{u_i, u_i + 1)$ the i-th knot span. Note that since some ui's may be equal, some knot spans may not exist. If a knot ui appears k times (*i.e.*, $u_i = u_i + 1 = u_i + k - 1$), where $k > 1$, ui is a multiple knot of multiplicity k, written as ui(k). Otherwise, if ui appears only once, it is a simple knot. If the knots are equally spaced (*i.e.*, $u_i + 1 - u_i$ is a constant for 0 £ i £ $m - 1$), the knot vector or the knot sequence is said uniform; otherwise, it is non-uniform.

The knots can be considered as division points that subdivide the interval [u0, um] into knot spans. All B-spline basis functions are supposed to have their domain on $[u_0, u_m]$. In this note, we use $u_0 = 0$ and $u_m = 1$ frequently so that the domain is the closed interval [0, 1].

To define B-spline basis functions, we need one more parameter, the degree of these basis functions, p. The i-th B-spline basis function of degree p, written as $N_{i,p}(u)$, is defined recursively as follows:

$$N_{i,0}(u) = \begin{cases} 1 & \text{if } u_i \le u < u_{i+1} \\ 0 & \text{otherwise} \end{cases}$$

$$N_{i,p}(u) = \frac{u - u_i}{u_{i+p} - u_i} N_{i,p-1}(u) + \frac{u_{i+p+1} - u}{u_{i+p+1} - u_{i+1}} N_{i+1,p-1}(u)$$

The above is usually referred to as the Cox-de Boor recursion formula. This definition looks complicated; but, it is not difficult to understand. If the degree is zero (*i.e.*, $p = 0$), these basis functions are all step functions and this is what the first expression says. That is, basis function $N_{i,0}(u)$is 1 if u is in the i-th knot span $[u_i, u_i + 1)$. For example, if we have four knots $u_0 = 0$, $u_1 = 1$, $u_2 = 2$ and $u_3 = 3$, knot spans 0, 1 and 2 are (0, 1), (1, 2), (2, 3) and the basis functions of degree 0 are N0,0(u)= 1 on [0,1) and 0 elsewhere, N1,0(u)= 1 on [1,2) and 0 elsewhere, and N_2, 0(u) = 1 on [2, 3) and 0 elsewhere. This is shown below:

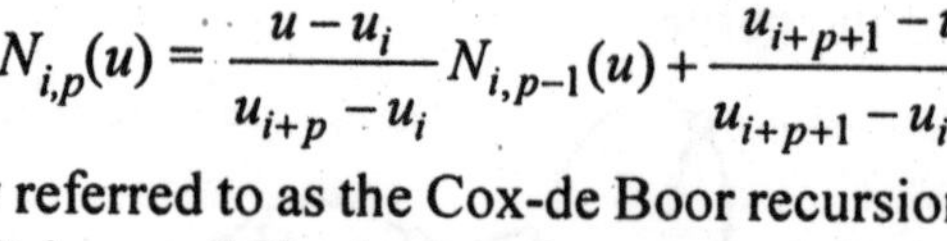

To understand the way of computing $N_{i,p}(u)$ for p greater than 0, we use the triangular computation scheme. All knot spans are listed on the left (first) column and all degree zero basis functions on the second. This is shown in the following diagram.

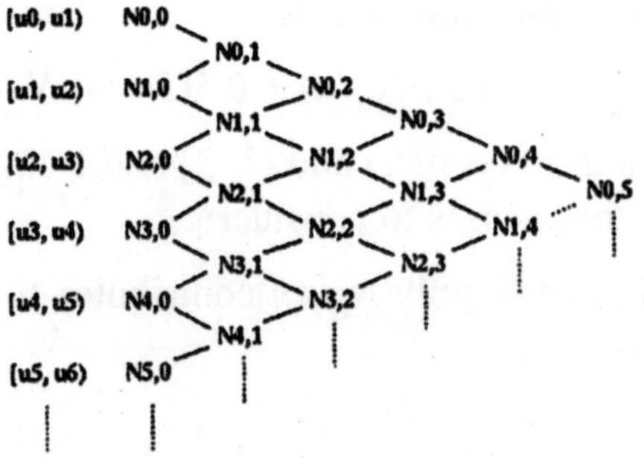

To compute $N_{i,1}(u)$, $N_{i,0}(u)$and $N_{i+1,0}(u)$are required. Therefore, we can compute $N_{0,1}(u)$, $N_{1,1}(u)$, $N_{2,1}(u)$, $N_{3,1}(u)$and so on. All of these $N_{i,1}(u)$'s are written on the third column. Once all $N_{i,1}(u)$'s have been computed, we can compute $N_{i,2}(u)$'s and put them on the fourth column. This process continues until all required Ni,$p(u)$'s are computed. In the above, we have obtained $N_{0,0}(u)$, $N_{1,0}(u)$ and $N_{2,0}(u)$ for the knot vector $U = \{0, 1, 2, 3\}$. Let us compute $N_{0,1}(u)$ and $N_{1,1}(u)$. To compute $N_{0,1}(u)$, since $i = 0$ and $p = 1$, from the definition we have

$$N0,1(u) = \frac{u-u_0}{u_1-u_0}N_{0,0}(u)+\frac{u_2-u}{u_2-u_1}N_{1,0}(u)$$

Since $u_0 = 0$, $u_1 = 1$ and $u_2 = 2$, the above becomes

$$N_{0,1}(u) = uN_{0,0}(u) + (2 - u)\,N_{1,0}(u)$$

Since N0,0(u)is non-zero on [0, 1) and $N_{1,0}(u)$is non-zero on (1,2), if u is in (0,1) (*resp.*, 1,2), only $N_{0,0}(u)$ (*resp.*, $N_{1,0}(u)$ contributes to $N_{0,1}(u)$.

Therefore, if u is in (0, 1), $N_{0,1}(u)$ is $uN_{0,0}(u) = u$, and if u is in (1, 2), $N_{0,1}(u)$ is $(2 - u)$ $N_{1,0}(u) = (2 - u)$. Similar computation gives $N_{1,1}(u) = u - 1$ if u is in (1, 2), and $N_{1,1}(u) = 3 - u$ if u is in (2, 3). In the following figure, the black and red lines are $N_{0,1}(u)$ and $N_{1,1}(u)$, respectively. Note that $N_{0,1}(u)$ (resp., $N_{1,1}(u)$ is non-zero on (0, 1) and (1, 2) (resp., 1, 2) and (2,3).

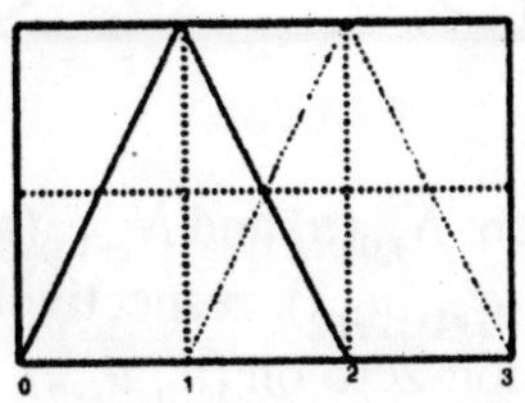

Once $N_{0,1}(u)$ and $N_{1,1}(u)$ are available, we can compute $N_{0,2}(u)$. The definition gives us the following:

$$N0,2(u) = \frac{u-u_0}{u_2-u_0}N_{0,1}(u)+\frac{u_3-u}{u_3-u_1}N_{1,1}(u)$$

Plugging in the values of the knots yields

$$N_{0,2}(u) = 0.5uN_{0,1}(u) + 0.5(3 - u)\, N_{1,1}(u)$$

Note that $N_{0,1}(u)$is non-zero on [0,1) and (1, 2) and $N_{1,1}(u)$ is non-zero on (1, 2) and (2, 3). Therefore, we have three cases to consider:

1. u is in (0,1): In this case, only $N_{0,1}(u)$contributes to the value of $N_{0,2}(u)$. Since $N_{0,1}(u)$is u, we have
   $N_{0,2}(u) = 0.5u^2$
2. u is in (1, 2): In this case, both $N_{0,1}(u)$ and $N_{1,1}(u)$ contribute to $N_{0,2}(u)$. Since
   $N_{0,1}(u)= 2 - u$
   and $N_{1,1}(u) = u - 1$
   on (1, 2), we have $N_{0,2}(u) = (0.5u)\,(2 - u) + 0.5(3 - u)\,(3 - u) = 0.5(-3 + 6u - 2u^2)$
3. u is in (2, 3): In this case, only $N_{1,1}(u)$contributes to $N_{0,2}(u)$. Since
   $N_{1,1}(u) = 3 - u$
   on (2, 3), we have $N_{0,2}(u) = 0.5(3 - u)\,(3 - u) = 0.5(3 - u)^2$

If we draw the curve segment of each of the above three cases, we shall see that two adjacent curve segments are joined together to form a curve at the knots. More precisely, the curve segments of the first and second cases join together at $u = 1$, while the curve segments of the second and third cases join at $u = 2$. Note that the composite curve shown here is smooth. But in general it is not always the case if a knot vector contains multiple knots.

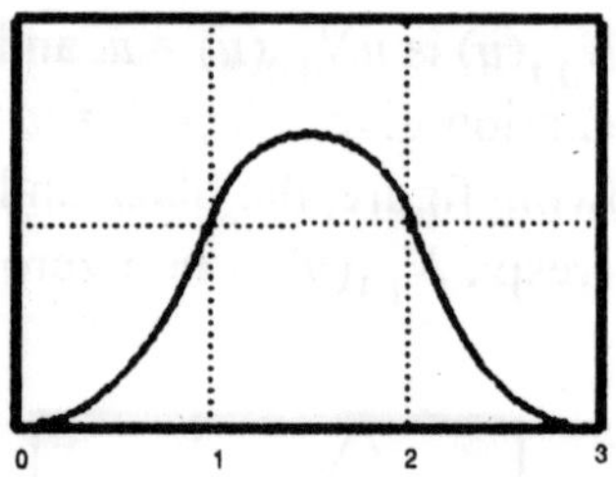

Two Important Observations

Since $N_{i,1}(u)$ is computed from $N_{i,0}(u)$ and $N_{i+1,0}(u)$ and since $N_{i,0}(u)$ and $N_{i+1,0}(u)$ are non-zero on span $[u_i, u_{i+1})$ and $[u_{i+1}, u_{i+2})$, respectively, $N_{i,1}(u)$ is non-zero on these two spans. In other words, $N_{i,1}(u)$ is non-zero on (u_i, u_{i+2}). Similarly, since $N_{i,2}(u)$ depends on $N_{i,1}(u)$ and $N_{i+1,1}(u)$ and since these two basis functions are non-zero on (u_i, u_{i+2}) and (u_{i+1}, u_{i+3}), respectively, $N_{i,2}(u)$ is non-zero on $[u_i, u_{i+3})$. In general, to determine the non-zero domain of a basis function $N_{i,p}(u)$, one can trace back using the triangular computation scheme until it reaches the first column. The covered spans are the non-zero domain of this basis function. For example, suppose we want to find out the non-zero domain of

$N_{1,3}(u)$. Based on the above discussion, we can trace back in the north-west and south-west directions until the first column is reached as shown with the blue dotted line in the following diagram. Thus, $N_{1,3}(u)$is non-zero on $[u_1, u_2)$, $[u_2, u_3)$, $[u_3, u_4)$ and $[u_4, u_5)$. Or, equivalently, it is non-zero on (u_1, u_5).

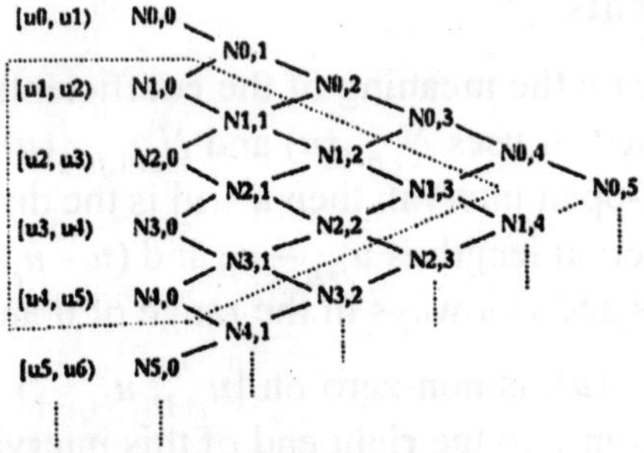

In summary, we have the following observation:

Basis function $N_{i,p}(u)$ is non-zero on (u_i, u_{i+p+1}). Or, equivalently, $N_{i,p}(u)$ is non-zero on $p + 1$ knot spans (u_i, u_{i+1}), (u_{i+1}, u_{i+2}),..., (u_{i+p}, u_{i+p+1}).

Next, we shall look at the opposite direction. Given a knot span (u_i, u_{i+1}), we want to know which basis functions will use this span in its computation. We can start with this knot span and draw a north-east bound arrow and a south-east bound arrow.

All basis functions enclosed in this wedge shape use $N_{i,0}(u)$(why?) and hence are non-zero on this span. Therefore, all degree p basis functions that are non-zero on [ui, ui+1) are the intersection of this wedge and the column that contains all $N_{i,p}(u)$'s. In fact, this column and the two arrows form an equilateral triangle with this column being the vertical side. Counting from $N_{i,0}(u)$ to $N_{i,p}(u)$ there are p +1 columns. Therefore, the vertical side of the equilateral triangle must have at most $p + 1$ entries, namely $N_{i,p}(u)$, $N_{i-1,p}(u)$, $N_{i-2,p}(u)$,..., $N_{i-p+2,p}(u)$, N_{i-p+1},$p(u)$ and $N_{i-p,p}(u)$.

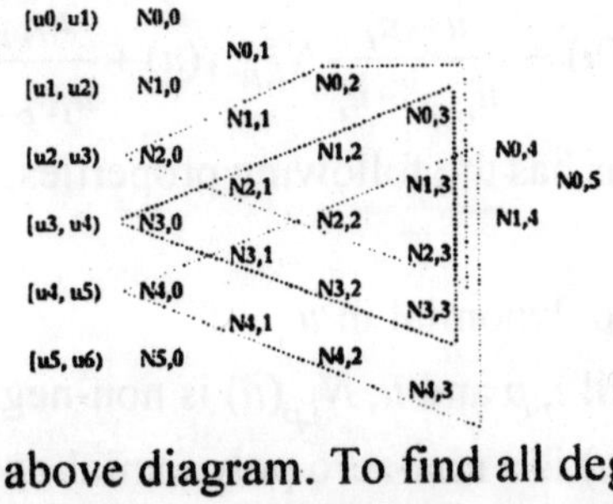

Let us take a look at the above diagram. To find all degree 3 basis functions that are non-zero on (u_4, u_5), draw two arrows and all functions on the vertical edges are what we want. In this case, they are $N_{1,3}(u)$, $N_{2,3}(u)$, $N_{3,3}(u)$, and $N_{4,3}(u)$.

This is shown with the orange triangle. The blue (resp., red) triangle shows the degree 3 basis functions that are non-zero on $[u_3, u_4)$ (resp., u_2, u_3). Note that there are only three degree three basis polynomials that are non-zero on (u_2, u_3).

In summary, we have observed the following property.

On any knot span $[u_i, u_{i+1})$, at most $p+1$ degree p basis functions are non-zero, namely: $N_{i-p,p}(u)$, $N_{i-p+1,p}(u)$, $N_{i-p+2,p}(u)$,..., $N_{i-1,p}(u)$ and $N_{i,p}(u)$,

Meaning of the Coefficients

Finally, let us investigate the meaning of the coefficients in the definition of $N_{i,p}(u)$. As $N_{i,p}(u)$ is being computed, it uses $N_{i,p-1}(u)$ and $N_{i+1,p-1}(u)$. The former is non-zero on $[u_i, u_{i+p})$. If u is in this half-open interval, then u – ui is the distance between u and the left end of this interval, the interval length is $u_{i+p} - u_i$, and $(u - u_i)/(u_{i+p} - u_i)$ is the ratio of the above mentioned distances and is always in the range of 0 and 1.

The second term, $N_{i,p-1}(u)$, is non-zero on $[u_{i+1}, u_{i+p+1})$. If u is in this interval, then $u_{i+p+1} - u$ is the distance from u to the right end of this interval, $u_{i+p+1} - u_{i+1}$ is the length of the interval, and $(u_{i+p+1} - u)/(u_{i+p+1} - u_{i+1})$ is the ratio of these two distances and its value is in the range of 0 and 1. Therefore, $N_{i,p}(u)$ is a linear combination of $N_{i,p-1}(u)$and $N_{i+1,p-1}(u)$ with two coefficients, both linear in u, in the range of 0 and 1.

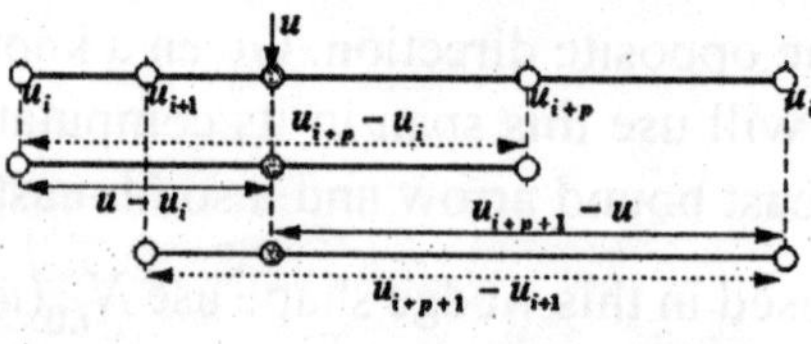

PROPERTIES OF B-SPLINE BASIS FUNCTIONS

Let us recall the definition of the B-spline basis functions as follows:

$$N_{i,0}(u) = \begin{cases} 1 & \text{if } u_i \le u < u_{i+1} \\ 0 & \text{otherwise} \end{cases}$$

$$Ni,p(u) = \frac{u - u_i}{u_{i+p} - u_i} N_{i,p-1}(u) + \frac{u_{i+p+1} - u}{u_{i+p+1} - u_{i+1}} N_{i+1,p-1}(u)$$

This set of basis functions has the following properties, many of which resemble those of Bézier basis functions.

1. $N_{i,p}(u)$ is a degree p polynomial in u
2. *Nonnegativity:* For all i, p and u, $N_{i,p}(u)$ is non-negative
3. *Local Support:* $N_{i,p}(u)$ is a non-zero polynomial on $[u_i, u_{i+p+1})$

   This has been discussed on previous page.
4. On any span $[u_i, u_{i+1})$, at most $p+1$ degree p basis functions are non-zero, namely: $N_{i-p,p}(u)$, $N_{i-p+1,p}(u)$, $N_{i-p+2,p}(u)$,..., and $N_{i,p}(u)$
5. *Partition of Unity:* The sum of all non-zero degree p basis functions on span (u_i, u_{i+1}) is 1:

The previous property shows that $N_{i-p,p}(u)$, $N_{i-p+1,p}(u)$, $N_{i-p+2,p}(u)$,..., and $N_{i,p}(u)$ are non-zero on (u_i, u_{i+1}). This one states that the sum of these $p + 1$ basis functions is 1.

6. If the number of knots is m+1, the degree of the basis functions is p, and the number of degree p basis functions is $n + 1$, then $m = n + p + 1$:

   This is not difficult to see. Let $N_{n,p}(u)$ be the last degree p basis function. It is non-zero on $[u_n, u_{n+p+1})$. Since it is the last basis function, u_{n+p+1} must be the last knot um. Therefore, we have u_{n+p+1} = um and $n + p + 1 = m$. In summary, given m and p, let $n = m - p - 1$ and the degree p basis functions are $N_{0,p}(u)$, $N_{1,p}(u)$, $N_{2,p}(u)$,..., and $N_{n,p}(u)$.

7. Basis function $N_{i,p}(u)$ is a composite curve of degree p polynomials with joining points at knots in $[u_i, u_{i+p+1})$

   The example shown on the previous page illustrates this property well. For example, $N_{0,2}(u)$, which is non-zero on (0, 3), is constructed from three parabolas defined on (0, 1), (1, 2) and (2, 3). They are connected together at the knots 2 and 3.

8. At a knot of multiplicity k, basis function $N_{i,p}(u)$ is C_{p-k} continuous.

   Therefore, increasing multiplicity decreases the level of continuity, and increasing degree increases continuity. The above mentioned degree two basis function $N_{0,2}(u)$is C_1 continuous at knots 2 and 3, since they are simple knots ($k = 1$).

The Impact of Multiple Knots

Multiple knots do have significant impact on the computation of basis functions and some "counting" properties. We shall look at two of them and a calculation example will be provided on the next page.

1. Each knot of multiplicity k reduces at most k-1 basis functions' non-zero domain.

   Consider $N_{i,p}(u)$and $N_{i+1,p}(u)$. The former is non-zero on $[u_i, u_{i+p+1})$ while the latter is non-zero on $[u_{i+1}, u_{i+p+2})$. If we move u_{i+p+2} to u_{i+p+1} so that they become a double knot. Then, $N_{i,p}(u)$ still has $p + 1$ knot spans on which it is non-zero; but, the number of knot spans on which $N_{i+1,p}(u)$is non-zero is reduced by one because the span (u_{i+p+1}, u_{i+p+2}) disappears.

   This observation can be generalized easily. In fact, ignoring the change of knot span endpoints, to create a knot of multiplicity k, $k - 1$ basis functions will be affected. One of them loses one knot span, a second of them loses two, a third of them loses three and so on.

   The figures below show the basis functions of degree 5 where the left end and right end knots have multiplicity 6, while all knots in between are simple [Figure (a)]. Figure (b) is the result of moving u_5 to u_6. Those basis functions ended at u_6 have fewer knot spans on which they are non-zero. Then, u_4 and then u_3 are moved to u_6, making u_6 a knot of multiplicity 4 (Figure (c) and (d). Figure (e) shows the result after moving u_2 to u_6, creating a knot of multiplicity 5.

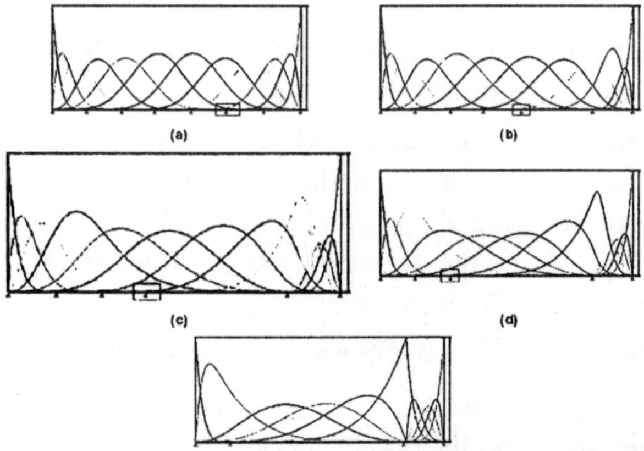

2. At each internal knot of multiplicity k, the number of non-zero basis functions is at most $p - k + 1$, where p is the degree of the basis functions.

   Since moving u_{i-1} to ui will pull a basis function whose non-zero ends at u_{i-1} to end at ui, this reduces the number of non-zero basis function at ui by one. More precisely, increasing u_i's multiplicity by one will reduce the number of non-zero basis functions by one. Since there are at most $p + 1$ basis functions can be non-zero at u_i, the number of non-zero basis functions at a knot of multiplicity k is at most $(p + 1) - k = p - k + 1$.

   In the above figures, since the multiplicity of knot u_6 are 1 (simple), 2, 3, 4 and 5, the numbers of non-zero basis function at u6 are 5, 4, 3, 2 and 1.

COMPUTATION EXAMPLES OF B-SPLINE BASIS FUNCTIONS

Two examples, one with all simple knots while the other with multiple knots, will be discussed in some detail on this page.

Simple Knots

Suppose the knot vector is $U = \{0, 0.25, 0.5, 0.75, 1\}$. Hence, $m = 4$ and $u_0 = 0$, $u_1 = 0.25$, $u_2 = 0.5$, $u_3 = 0.75$ and $u_4 = 1$. The basis functions of degree 0 are easy. They are $N_{0,0}(u)$, $N_{1,0}(u)$, $N_{2,0}(u)$ and $N_{3,0}(u)$ defined on knot span (0, 0.25,), (0.25, 0.5), (0.5, 0.75) and (0.75,1), respectively, as shown below.

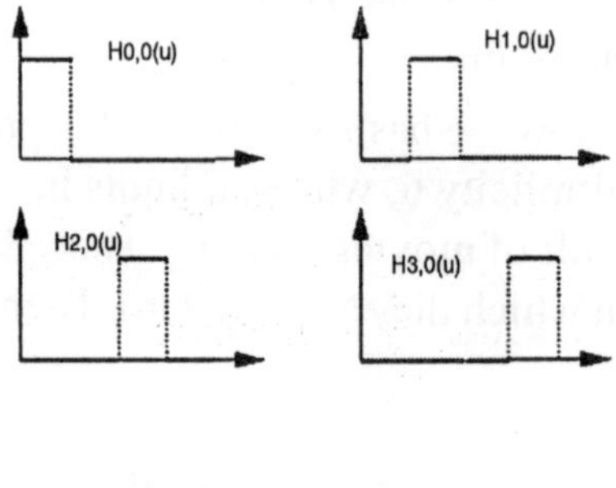

The following table gives the result of all $N_{i,1}(u)$'s:

Basis Function	Range	Equation
$N_{0,1}(u)$	[0, 0.25) [0.25, 0.5)	$4u$ $2(1-2u)$
$N_{1,1}(u)$	[0.25, 0.5) [0.5, 0.75)	$4u-1$ $3-u$
$N_{2,1}(u)$	[0.5, 0.75) [0.75, 1)	$2(2u-1)$ $4(1-u)$

The following shows the graphs of these basis functions. Since the internal knots 0.25, 0.5 and 0.75 are all simple (i.e., $k = 1$) and $p = 1$, there are $p - k + 1 = 1$ non-zero basis function and three knots. Moreover, $N_{0,1}(u)$, $N_{1,1}(u)$ and $N_{2,1}(u)$ are C_0 continuous at knots 0.25, 0.5 and 0.75, respectively.

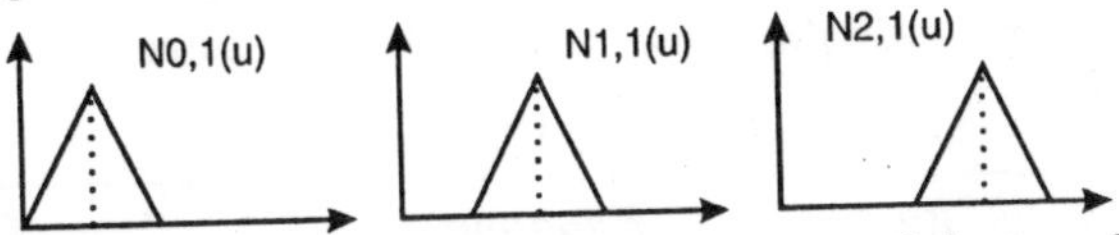

From $N_{i,1}(u)$'s, one can compute the basis functions of degree 2. Since $m = 4$, $p = 2$, and $m = n + p + 1$, we have $n = 1$ and there are only two basis functions of degree 2: $N_{0,2}(u)$and $N_{1,2}(u)$. The following table is the result:

Basis Function	Range	Equation
$N_{0,2}(u)$	[0, 0.25) [0.25, 0.5) [0.5, 0.75)	$8u^2$ $-1.5 + 12u - 16u^2$ $4.5 - 12u + 8u^2$
$N_{1,2}(u)$	[0.25, 0.5) [0.5, 0.75) [0.75, 1)	$0.5 - 4u + 8u^2$ $-1.5 + 8u - 8u^2$ $8(1-u)^2$

The following figure shows the two basis functions. The three vertical blue lines indicate the positions of knots. Note that each basis function is a composite curve of three degree 2 curve segments.

For example, $N_{0,2}(u)$ is the green curve, which is the union of three parabolas defined on (0, 0.25), [0.25, 0.5) and (0.5, 0.75). These three curve segments join together forming a smooth bell shape. Please verify that $N_{0,2}(u)$ (resp., $N_{1,2}(u)$ is C^1 continuous at its knots 0.25 and 0.5 (resp., 0.5 and 0.75). As mentioned on the previous page, at the knots, this composite curve is of C^1 continuity.

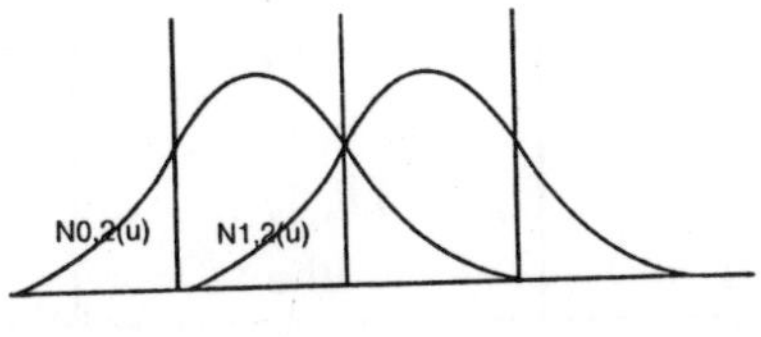

Knots with Positive Multiplicity

If a knot vector contains knots with positive multiplicity, we will encounter the case of 0/0. Therefore, we shall define 0/0 to be 0. Fortunately, this is only for hand calculation. For computer implementation, there is an efficient algorithm free of this problem. Furtherore, if u_i is a knot of multiplicity k (i.e., $u_i = u_{i+1} = ... = u_{i+k-1}$), then knot spans (u_i, u_{i+1}), (u_{i+1}, u_{i+2}),..., (u_{i+k-2}, u_{i+k-1}) do not exist, and, as a result, $N_{i,0}(u)$, $N_{i+1,0}(u)$,..., $N_{i+k-1,0}(u)$are all zero functions. Consider a knot vector $U = \{0, 0, 0, 0.3, 0.5, 0.5, 0.6, 1, 1, 1\}$. Thus, 0 and 1 are of multiplicity 3 [i.e., 0(3) and 1(3)] and 0.5 is of multiplicity 2 (i.e., 0.5(2)). As a result, $m = 9$ and the knot assignments are

u_0	u_1	u_2	u_3	u_4	u_5	u_6	u_7	u_8	u_9
0	0	0	0.3	0.5	0.5	0.6	1	1	1

Let us compute $N_{i,0}(u)$'s. Note that since $m = 9$ and $p = 0$ (degree 0 basis functions), we have $n = m - p - 1 = 8$. As the table below shows, there are only four non-zero basis functions of degree 0: $N_{2,0}(u)$, $N_{3,0}(u)$, $N_{5,0}(u)$ and $N_{6,0}(u)$.

Basis Function	*Range*	*Equation*	*Comments*
$N_{0,0}(u)$	all u	0	since $[u_0, u_1) = [0,0)$ does not exist
$N_{1,0}(u)$	all u	0	since $[u_1, u_2) = [0,0)$ does not exist
$N_{2,0}(u)$	(0, 0.3)	1	
$N_{3,0}(u)$	(0.3, 0.5)	1	
$N_{4,0}(u)$	all u	0	since $[u_4, u_5) = [0.5,0.5)$ does not exist
$N_{5,0}(u)$	(0.5, 0.6)	1	
$N_{6,0}(u)$	(0.6, 1)	1	
$N_{7,0}(u)$	all u	0	since $[u_7, u_8) = [1,1)$ does not exist
$N_{8,0}(u)$	all u	0	since $[u_8, u_9) = [1,1)$ does not exist

Then, we proceed to basis functions of degree 1. Since p is 1, $n = m - p - 1 = 7$. The following table shows the result:

Basis Function	*Range*	*Equation*
$N_{0,1}(u)$	all u	0
$N_{1,1}(u)$	(0, 0.3)	$1 - (10/3)u$
$N_{2,1}(u)$	(0, 0.3)	$(10/3)u$
	(0.3, 0.5)	$2.5(1 - 2u)$
$N_{3,1}(u)$	(0.3, 0.5)	$5u - 1.5$
$N_{4,1}(u)$	(0.5, 0.6)	$6 - 10u$
$N_{5,1}(u)$	(0.5, 0.6)	$10u - 5$
	(0.6, 1)	$2.5(1 - u)$
$N_{6,1}(u)$	(0.6, 1)	$2.5u - 1.5$
$N_{7,1}(u)$	all u	0

The following figure shows the graphs of these basis functions.

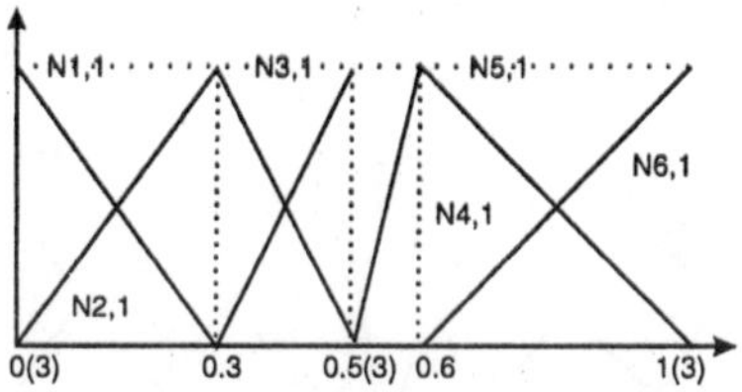

Let us take a look at a particular computation, say N1,1(u). It is computed with the following expression:

$$N_{1,1}(u) = \frac{u-u_1}{u_2-u_1}N_{1,0(u)} + \frac{u_3-u}{u_3-u_2}N_{2,0}(u)$$

Plugging $u_1 = u_2 = 0$ and $u_3 = 0.3$ into this equation yields the following:

$$N_{1,1}(u) = \frac{u}{0}N_{1,0}(u) + \left(1-\frac{10}{3}u\right)N_{2,0}(u)$$

Since $N_{1,0}(u)$is zero everywhere, the first term becomes 0/0 and is defined to be zero. Therefore, only the second term has an impact on the result. Since $N_{2,0}(u)$is 1 on (0,0.3), $N_{1,1}(u)$is $1 - (10/3)u$ on (0,0.3).

Next, let us compute all $N_{i,2}(u)$'s. Since $p = 2$, we have $n = m - p - 1 = 6$. The following table contains all $N_{i,2}(u)$'s:

Basis Function	***Range***	***Equation***
$N_{0,2}(u)$	(0, 0.3)	$(1 - (10/3)u)^2$
$N_{1,2}(u)$	(0, 0.3) (0.3, 0.5)	$(20/3)(u - (8/3)u^2)$ $2.5(1 - 2u)^2$
$N_{2,2}(u)$	(0, 0.3) (0.3, 0.5)	$(20/3)u^2$ $-3.75 + 25u - 35u^2$
$N_{3,2}(u)$	(0.3, 0.5) (0.5, 0.6)	$(5u - 1.5)^2$ $(6 - 10u)^2$
$N_{4,2}(u)$	(0.5, 0.6) (0.6, 1)	$20(-2 + 7u - 6u^2)$ $5(1 - u)^2$
$N_{5,2}(u)$	(0.5, 0.6) (0.6, 1)	$12.5(2u - 1)^2$ $2.5(-4 + 11.5u - 7.5u^2)$
$N_{6,2}(u)$	(0.6, 1)	$2.5(9 - 30u + 25u^2)$

The following figure shows all basis functions of degree 2. Let us pick a typical computation as an example, say $N_{3,2}(u)$. The expression for computing it is

$$N3,2(u) = \frac{u-u_3}{u_5-u_3}N_{3,1}(u) + \frac{u_6-u}{u_6-u_4}N_{4,1}(u)$$

Plugging in $u_3 = 0.3$, $u_4 = u_5 = 0.5$ and $u_6 = 0.6$ yields

$$N_{3,2}(u) = (5u - 1.5)\, N_{3,1}(u) + (6 - 10u)N_{4,1}(u)$$

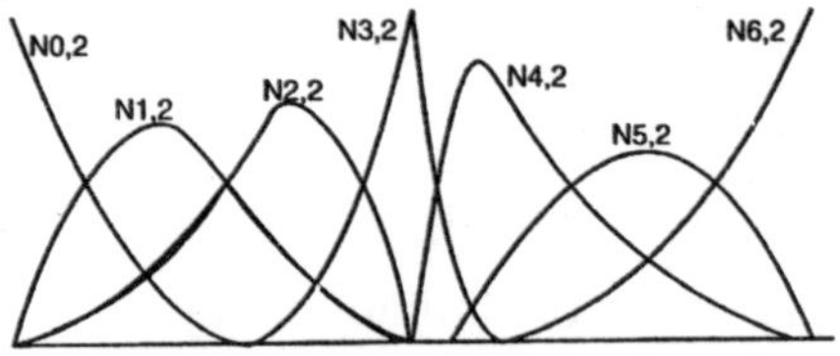

Since $N_{3,1}(u)$is non-zero on (0.3, 0.5) and is equal to $5u - 1.5$, $(5u - 1.5)^2$ is the non-zero part of $N_{3,2}(u)$ on (0.3, 0.5). Since $N_{4,1}(u)$ is non-zero on (0.5, 0.6) and is equal to $6- 10u$, $(6 - 10u)^2$ is the non-zero part of $N_{3,2}(u)$ on (0.5, 0.6).

Let us investigate the continuity issues at knot 0.5(2). Since its multiplicity is 2 and the degree of these basis functions is 2, basis function $N_{3,2}(u)$ is C^0 continuous at 0.5(2). This is why $N_{3,2}(u)$ has a sharp angle at 0.5(2). For knots not at the two ends, say 0.3, C^1 continuity is maintained since all of them are simple knots.

B-SPLINE CURVES

Given $n + 1$ control points $P_0, P_1,\ldots, P_n$ and a knot vector $U = \{u_0, u_1,\ldots, u_m\}$, the B-spline curve of degree p defined by these control points and knot vector U is

$$C(u) = \sum_{i=0}^{n} N_{i,p}(u)P_i$$

where $N_{i,p}(u)$'s are B-spline basis functions of degree p. The form of a B-spline curve is very similar to that of a Bézier curve. Unlike a Bézier curve, a B-spline curve involves more information, namely: a set of n+1 control points, a knot vector of m+1 knots, and a degree p. Note that n, m and p must satisfy $m = n + p + 1$.

More precisely, if we want to define a B-spline curve of degree p with $n + 1$ control points, we have to supply $n + p + 2$ knots $u_0, u_1,\ldots, u_{n+p+1}$. On the other hand, if a knot vector of $m + 1$ knots and $n + 1$ control points are given, the degree of the B-spline curve is $p = m - n - 1$. The point on the curve that corresponds to a knot u_i, $C(u_i)$, is referred to as a knot point. Hence, the knot points divide a B-spline curve into curve segments, each of which is defined on a knot span. We shall show that these curve segments are all Bézier curve of degree p on the curve subdivision page.

Although $N_{i,p}(u)$ looks like $B_{n,i}(u)$, the degree of a B-spline basis function is an input, while the degree of a Bézier basis function depends on the number of control points. To change the shape of a B-spline curve, one can modify one or more of these control parameters: the positions of control points, the positions of knots, and the degree of the curve.

If the knot vector does not have any particular structure, the generated curve will not touch the first and last legs of the control polyline as shown in the left figure below. This type of B-spline curves is called open B-spline curves. We may want to clamp the curve so that it is tangent to the first and the last legs at the first and last control points, respectively, as a Bézier curve does. To do so, the first knot and the last knot must be of multiplicity $p + 1$. This will generate the so-called clamped B-spline curves. By repeating some knots and control points, the generated curve can be a closed one. In this case, the start and the end of the generated curve join together forming a closed loop as shown in the right figure below. In this note, we shall use clamped curve.

The above figures have $n + 1$ control points ($n = 9$) and $p = 3$. Then, m must be 13 so that the knot vector has 14 knots. To have the clamped effect, the first $p + 1 = 4$ and the last 4 knots must be identical. The remaining $14 - (4 + 4) = 6$ knots can be anywhere in the domain. In fact, the curve is generated with knot vector $U = \{0, 0, 0, 0, 0.14, 0.28, 0.42, 0.57, 0.71, 0.85, 1, 1, 1, 1\}$. Note that except for the first four and last four knots, the middle ones are almost uniformly spaced. The figures also show the corresponding curve segment on each knot span. In fact, the little triangles are the knot points.

OPEN CURVES OF B-SPLINE CURVES

As mentioned earlier, if the first and last knots do not have multiplicity $p + 1$, where p is the degree of a B-spline curve, the curve will not be tangent to the first and last legs at the first and last control points, respectively. The curve is an open B-spline curve. In this case, we should be careful about one additional restriction. On a previous page we showed an example of basis function computation using knot vector $U = \{0, 0.25, 0.5, 0.75, 1\}$, where $m = 4$. If the basis functions are of degree 1 (i.e., $p = 1$), there are three basis functions $N_{0,1}(u)$, $N_{1,1}(u)$and $N_{2,1}(u)$as shown below.

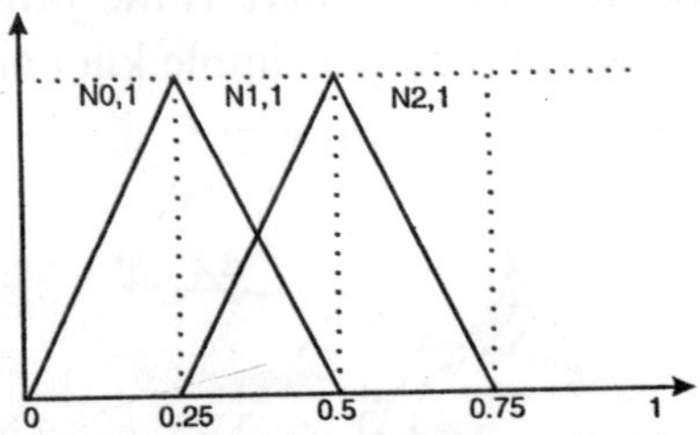

Since this knot vector is not clamped, the first and the last knot spans (*i.e.*, 0, 0.25) and (0.75, 1)) have only one non-zero basis functions while the second and third knot spans (*i.e.*, 0.25, 0.5) and (0.5, 0.75) have two non-zero basis functions. Recall from the B-spline basis function important properties that on a knot span (u_i, u_{i+1}), there are at most p+1 degree p non-zero basis functions of degree p. Therefore, in this example, knot spans (0,0.25) and (0.75,1) do not have "full support" of basis functions. In general, for degree p, intervals [u0, up) and [um-p, um] will not have "full support" of basis functions and are ignored when a B-spline curve is open. Consequently, we have the following important note:

Open B-spline Curves, the Domain is $[u_p, u_{m-p}]$

Consider a B-spline curve of degree 6 (*i.e.*, $p = 6$) defined by 14 control points (i.e., $n = 13$). The number of knots is 21 (*i.e.*, $m = n + p + 1 = 20$). If the knot vector is uniform, the knots are 0, 0.05, 0.10, 0.15,..., 0.90, 0.95 and 1.0. The open curve is defined on $(u_p, u_{n-p}) = (u_6, u_{14}) = (0.3, 0.7)$ and is not tangent to the first and last legs. The left figure below shows the curve and the right figure gives the B-spline basis functions.

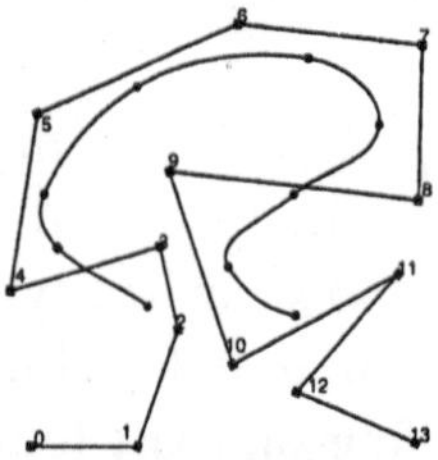

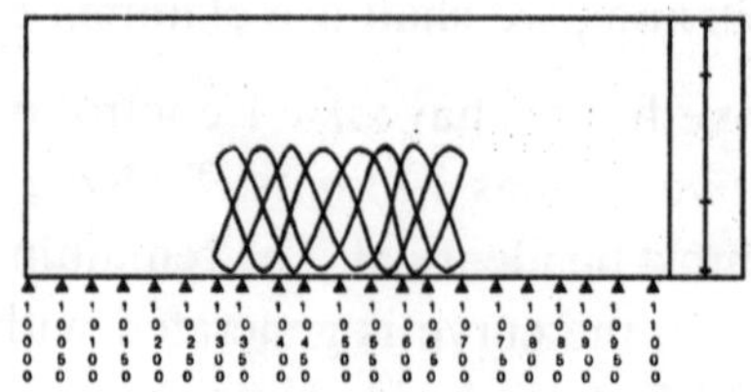

Even though some knot spans at both end are not used, the B-spline curve still is defined by all control points. As discussed on the B-spline basis functions page that there are at most $p + 1$ basis functions $N_{i-p,p}(u)$, $N_{i-p,p+1}(u)$,..., $N_{i,p}(u)$ non-zero on knot span (u_i, u_{i+1}). Therefore, there are $p + 1$ non-zero functions on (u_p, u_{i+1}): $N_{0,p}(u)$, $N_{1,p}(u)$,..., $N_{p,p}(u)$. This is a "full" count. Note that $N_{0,p}(u)$ has its tail in (u_p, u_{p+1}) and, hence, the contribution of control point P_0 to the construction of the open B-spline curve is smaller than those of most other control points. A similar argument can be used to show that P_n is also involved in the definition of the curve.

Let us use an example to illustrate the change between an open curve and a clamped one. We start with an open B-spline curve of degree 4 defined by 9 control points (*i.e.*, $n = 8$) and a uniform knot vector { 0, 1/13, 2/13, 3/13,..., 12/13, 1 }. If we change the second knot 1/13 to 0 making 0 a double knot, the curve is the yellow one in the figure. In fact, this curve and the original one with 0 being a simple knot are almost identical.

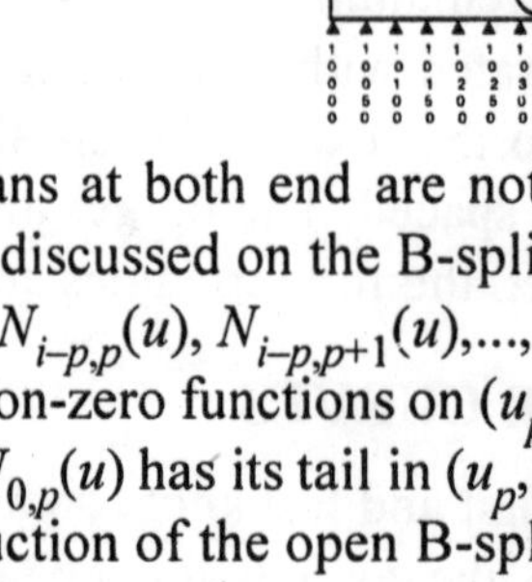

Now, if we change the third knot 2/13 to 0 making 0 a knot of multiplicity 3, the result is the red curve. If the fourth knot 3/13 is change to 0 (multiplicity 4), the resulting curve is the blue one. Now, let us make the fifth knot 4/13 to 0. Since 0 is now a knot of multiplicity 5 (*i.e.*, $p + 1$), the curve not only passes through the first control point but also is tangent to the first leg of the control polyline (*i.e.*, clamped). The shape of the curve changes drastically by pulling one end of it to the first control point. The same holds true if we make the last 5 knots to 1.

CLOSED CURVES OF B-SPLINE CURVES

There are many ways to generate closed curves. The simple ones are either wrapping control points or wrapping knot vectors.

Wrapping Control Points

Suppose we want to construct a closed B-spline curve $C(u)$ of degree p defined by n+1 control points $P_0, P_1, \ldots, P_n$. The number of knots is $m + 1$, where $m = n + p + 1$. Here is the construction procedure:

1. Design an uniform knot sequence of $m + 1$ knots: $u_0 = 0$, $u_1 = 1/m$, $u_1 = 2/m, \ldots$, $u_m = 1$. Note that the domain of the curve is (u_p, u_{n-p}).
2. Wrap the first p and last p control points. More precisely, let P0 = Pn-p+1, P1 = $P_{n-p+2}, \ldots, P_{p-2} = P_{n-1}$ and P_{p-1} = Pn. This is shown in the figure below.

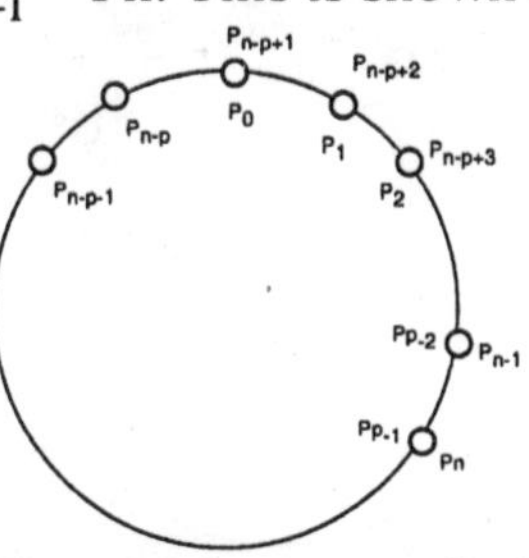

The constructed curve is C_{p-1} continuous at the joining point $C(u_p) = C(u_{n-p})$. The following is an example. Figure (a) shows an open B-spline curve of degree 3 defined by 10 ($n = 9$) control points and a uniform knot vector. In the figure, control point pairs 0 and 7, 1 and 8, and 2 and 9 are placed close to each other to illustrate the construction. Figure (b) shows the result of making control points 0 and 7 identical. The shape of the curve does not change very much. Then, control points 1 and 8 are made identical as shown in Figure (c). It is clear that the gap between the first and last points of the curve is closer. Finally, the curve becomes a closed on when control points 2 and 9 are made identical as shown in Figure (d).

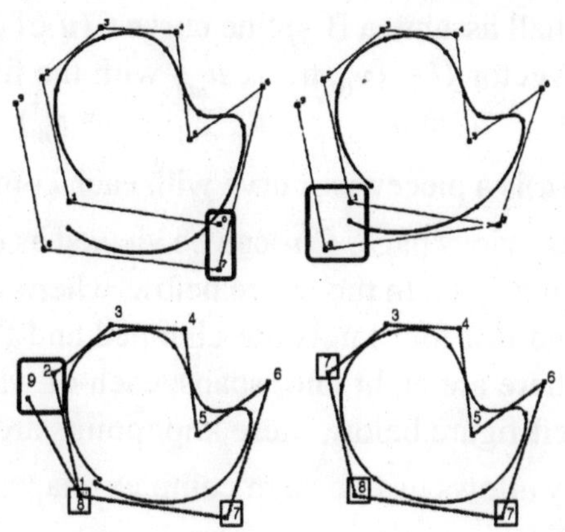

Wrapping Knots

Another way of constructing closed B-spline curves is by wrapping knots. Suppose we want to construct a closed B-spline curve $C(u)$of degree p defined by $n + 1$ control points $P_0, P_1,..., P_n$. The following is the construction procedure:

1. Add a new control point $P_{n+1} = P_0$. Therefore, the number of control points is $n + 2$.
2. Find an appropriate knot sequence of $n + 1$ knots $u_0, u_1,..., u_n$. These knots are not necessarily uniform, an advantage over the method discussed above.
3. Add $p + 2$ knots and wrap around the first $p + 2$ knots: $u_{n+1} = u_0, u_{n+2} = u_1,..., u_{n+p} = u_{p-1}, u_{n+p+1} = u_p, u_{n+p+2} = u_{p+1}$ as shown in the following diagram. In this way, we have $n + p + 2 = (n + 1) + p + 1$ knots.

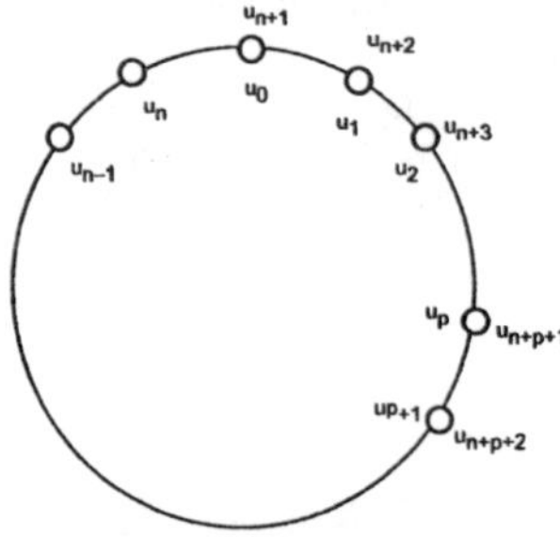

4. The open B-spline curve $C(u)$of degree p defined on the above constructed $n + 1$ control points and n+p+2 knots is a closed curve with C_{p-1} continuity at the joining point $C(u_0) = C(u_{n+1})$. Note that the domain of this closed curve is $[u_0, u_{n+1}]$.

IMPORTANT PROPERTIES OF B-SPLINE CURVES

B-spline curves share many important properties with Bézier curves, because the former is a generalization of the later. Moreover, B-spline curves have more desired properties than Bézier curves. The list below shows some of the most important properties of B-spline curves.

In the following we shall assume a B-spline curve $C(u)$of degree p is defined by $n + 1$ control points and a knot vector $U = \{u_0, u_1,..., u_m\}$ with the first $p + 1$ and last p+1 knots "clamped" (*i.e.*, $u_0 = u_1 =... = u_p$ and $u_{m-p} = u_{m-p+1} =... = u_m$).

1. B-spline curve C(u)is a piecewise curve with each component a curve of degree p.

   As mentioned in previous page, $C(u)$can be viewed as the union of curve segments defined on each knot span. In the figure below, where $n = 10$, $m = 14$ and $p = 3$, the first four knots and last four knots are clamped and the 7 internal knots are uniformly spaced. There are eight knot spans, each of which corresponds to a curve segment. In the left figure below, these knot points are shown as triangles.

   This nice property allows us to design complex shapes with lower degree polynomials. For example, the right figure below shows a Bézier curve with the same set of control points. It still cannot follow the control polyline nicely even though its degree is 10!

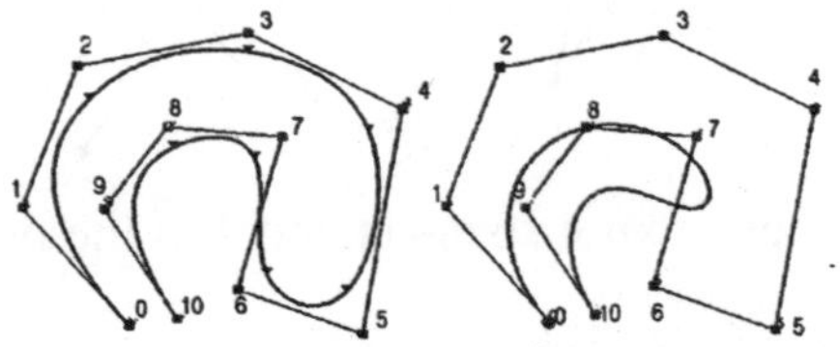

In general, the lower the degree, the closer a B-spline curve follows its control polyline. The following figures all use the same control polyline and knots are clamped and uniformly spaced. The first figure has degree 7, the middle one has degree 5 and the right figure has degree 3. Therefore, as the degree decreases, the generated B-spline curve moves closer to its control polyline.

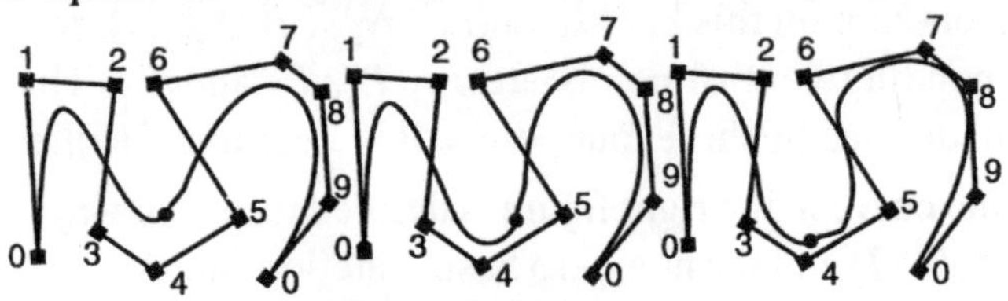

2. Equality $m = n + p + 1$ must be satisfied.

   Since each control point needs a basis function and the number of basis functions satisfies $m = n + p + 1$.

3. Clamped B-spline curve $C(u)$ passes through the two end control points P_0 and P_n.

   Note that basis function $N_{0,p}(u)$ is the coefficient of control point P_0 and is non-zero on $[u_0, u_{p+1})$. Since $u_0 = u_1 = ... = u_p = 0$ for a clamped B-spline curve, $N_{0,0}(u)$, $N_{1,0}(u)$,..., $N_{p-1},0(u)$ are zero and only $N_{p,0}(u)$ is non-zero (recall from the triangular computation scheme). Consequently, if $u = 0$, then $N_{0,p}(0)$ is 1 and $C(0) = P_0$. A similar discussion can show $C(1) = P_n$

4. Strong Convex Hull Property: A B-spline curve is contained in the convex hull of its control polyline. More specifically, if u is in knot span (u_i, u_{i+1}), then $C(u)$ is in the convex hull of control points $P_{i-p}, P_{i-p+1},..., P_i$.

   If u is in knot span (u_i, u_{i+1}), there are only p+1 basis functions [*i.e.*, $N_{i,p}(u),..., N_{i-p+1,p}(u)$, $N_{i-p,p}(u)$] non-zero on this knot span. Since $N_{k,p}(u)$ is the coefficient of control point P_k, only $p + 1$ control points $P_i, P_{i-1}, P_{i-2},..., P_{i-p}$ have non-zero coefficients. Since on this knot span the basis functions are non-zero and sum to 1, their "weighted" average, $C(u)$, must lie in the convex hull defined by control points $P_i, P_{i-1}, P_{i-2},.., P_{i-p}$. The meaning of "strong" is that while $C(u)$still lies in the convex hull defined by all control points, it lies in a much smaller one.

   The above two B-spline curves have 11 control points (*i.e.*, $n = 10$), degree 3 (*i.e.*, $p = 3$) and 15 knots ($m = 14$) with first four and last four knots clamped.

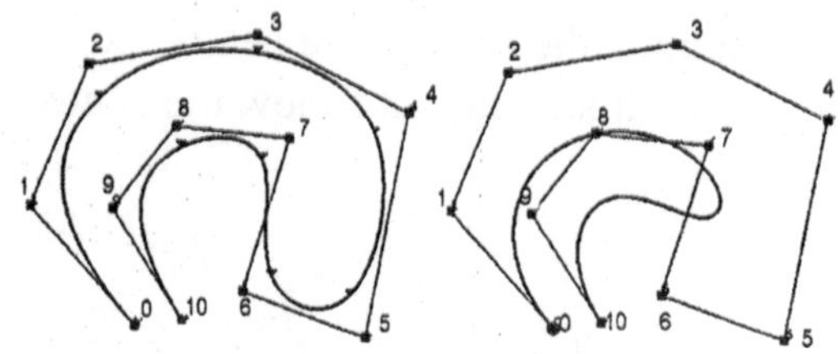

Therefore, the number of knot spans is equal to the number curve segments. The knot vector is

u_0	u_1	u_2	u_3	u_4	u_5	u_6	u_7	u_8	u_9	u_{10}	u_{11}	u_{12}	u_{13}	u_{14}
0	0	0	0	0.12	0.25	0.37	0.5	0.62	0.75	0.87	1	1	1	1

The left figure has u in knot span $(u_4, u_5) = (0.12, 0.25)$ and the corresponding point [*i.e.* $C(u)$] in the second curve segment. Therefore, there are p+1 = 4 basis functions non-zero on this knot span (*i.e.*, $N_{4,3}(u)$, $N_{3,3}(u)$, $N_{2,3}(u)$and $N_{1,3}(u)$) and the corresponding control points are P_4, P_3, P_2 and P_1. The shaded area is the convex hull defined by these four points. It is clear that $C(u)$lies in this convex hull.

The B-spline curve in the right figure is defined the same way. However, u is in $(u_9, u_{10}) = (0.75, 0.87)$ and the non-zero basis functions are $N_{9,3}(u)$, $N_{8,3}(u)$, $N_{7,3}(u)$ and $N_{6,3}(u)$. The corresponding control points are P_9, P_8, P_7 and P_6.

Consequently, as u moves from 0 to 1 and crosses a knot, a basis functions becomes zero and a new non-zero basis function becomes effective. As a result, one control point whose coefficient becomes zero will leave the the definition of the current convex hull and is replaced with a new control point whose coefficient becomes non-zero.

5. Local Modification Scheme: changing the position of control point P_i only affects the curve $C(u)$on interval $[u_i, u_{i+p+1})$.

This follows from another important property of B-spline basis functions. Recall that $N_{i,p}(u)$ is non-zero on interval (u_i, u_{i+p+1}). If u is not in this interval, $N_{i,p}(u)\ P_i$ has no effect in computing $C(u)$ since $N_{i,p}(u)$ is zero. On the other hand, if u is in the indicated interval, $N_{i,p}(u)$ is non-zero. If P_i changes its position, $N_{i,p}(u)\ P_i$ is changed and consequently $C(u)$is changed.

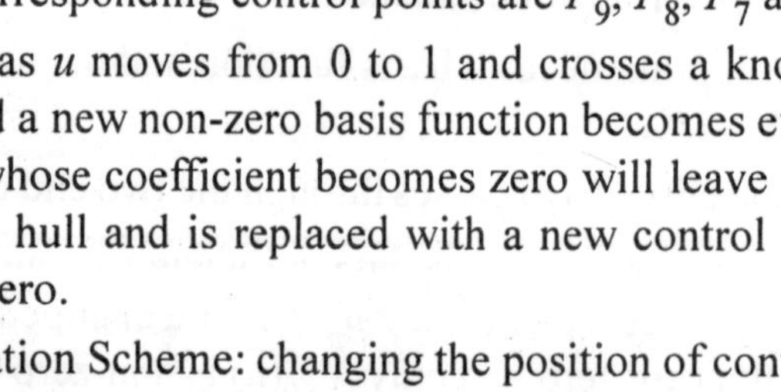

The above B-spline curves are defined with the same parameters as in the previous convex hull example. We intent to move control point P_2. The coefficient of this control point is $N_{2,3}(u)$and the interval on which this coefficient is non-zero is $(u_2, u_{2+3+1}) = (u_2, u_6) = (0,0.37)$. Since $u_2 = u_3 = 0$, only three segments that correspond to (u_3, u_4) (the domain of the first curve segment), (u_4, u_5) (the domain of the second curve segment) and (u_5, u_6) (the domain of the third curve segment) will be affected. The right figure shows the result of moving P_2 to the lower right corner. The first, second and third curve segments change their shapes and all remaining curve segments stay in their original place without any change.

This local modification scheme is very important to curve design, because we can modify a curve locally without changing the shape in a global way. This will be elaborated on the moeing control point page. Moreover, if fine-tuning curve shape is required, one can insert more knots (and therefore more control points) so that the affected area could be restricted to a very narrow region. We shall talk about knot insertion later.

6. *$C(u)$ is C_{p-k} continuous at a knot of multiplicity k:* If u is not a knot, $C(u)$is in the middle of a curve segment of degree p and is therefore infinitely differentiable. If u is a knot in the non-zero domain of $N_{i,p}(u)$, since the latter is only C_{p-k} continuous, so does $C(u)$.

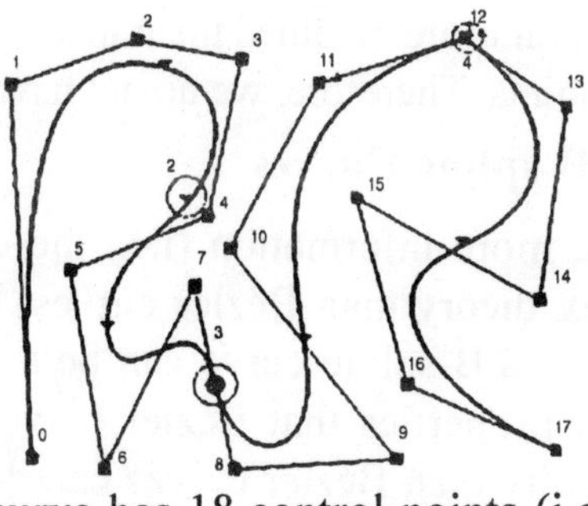

The above B-spline curve has 18 control points (i.e., $n = 17$), degree 4, and the following clamped knot vector

u_0 to u_4	u_5	u_6 and u_7	u_8	u_9 to u_{11}	u_{12}	u_{13} to u_{16}	u_{17}	u_{18} to u_{22}
0	0.125	0.25	0.375	0.5	0.625	0.75	0.875	1

Thus, u_6 is a double knot, u_9 is a triple knot and u_{13} is a quadruple knot. Consequently, $C(u)$is of C^4 continuous at any point that is not a knot, C^3 continuous at all simple knots, C^2 continuous at u6, C^1 continuous at u_9, C^0 continuous at u_{13}.

All points on the curve that correspond to knots are marked with little triangles. Those corresponding to multiple knots are further marked with circles and their multiplicities. It is very difficult to visualize the difference between C^4, C^3 and even C^2 continuity. For the C^1 case, the corresponding point lies on a leg, while the C^0 case forces the curve to pass through a control point. We shall return to this issue later when discussing modifying knots.

7. *Variation Diminishing Property:* The variation diminishing property also holds for B-spline curves. If the curve is in a plane (resp., space), this means no straight line (resp., plane) intersects a B-spline curve more times than it intersects the curve's control polyline.

   In the above figure, the blue line intersects both the control polyline and the B-spline curve 6 times, while the yellow line also intersects the control polyline and the B-spline curve 5 times. However, the orange line intersects the control polyline 6 times and the curve 4 times.
8. *Bézier Curves Are Special Cases of B-spline Curves:* If $n = p$ (i.e., the degree of a B-spline curve is equal to n, the number of control points minus 1), and there are $2(p + 1) = 2(n + 1)$ knots with $p + 1$ of them clamped at each end, this B-spline curve reduces to a Bézier curve.
9. *Affine Invariance:* The affine invariance property also holds for B-spline curves. If an affine transformation is applied to a B-spline curve, the result can be constructed from the affine images of its control points. This is a nice property. When we want to apply a geometric or even affine transformation to a B-spline curve, this property states that we can apply the transformation to control points, which is quite easy, and once the transformed control points are obtained the transformed B-spline curve is the one defined by these new points. Therefore, we do not have to transform the curve.

The Advantage of Using B-spline Curves

B-spline curves require more information (i.e., the degree of the curve and a knot vector) and a more complex theory than Bézier curves. But, it has more advantages to offset this shortcoming. First, a B-spline curve can be a Bézier curve. Second, B-spline curves satisfy all important properties that Bézier curves have. Third, B-spline curves provide more control flexibility than Bézier curves can do. For example, the degree of a B-spline curve is separated from the number of control points.

More precisely, we can use lower degree curves and still maintain a large number of control points. We can change the position of a control point without globally changing the shape of the whole curve (local modification property). Since B-spline curves satisfy the strong convex hull property, they have a finer shape control. Moreover, there are other techniques for designing and editing the shape of a curve such as changing knots. However, keep in mind that B-spline curves are still polynomial curves and polynomial curves cannot represent many useful simple curves such as circles and ellipses. Thus, a generalization of B-spline, NURBS, is required. We will discuss NURBS later.

COMPUTING THE COEFFICIENTS OF B-SPLINE CURVES

Although de Boor's algorithm is a standard way for computing the point on a B-spline curve that corresponds to a given u, we really need these coefficients in many cases (e.g., curve interpolation and approximation). We shall illustrate a simple way to do this. Given a clamped

B-spline curve of degree p defined by n+1 control points P_0, P_1,..., P_n, and m+1 knots $u_0 = u_1$ =... = u_p = 0, u_{p+1}, u_{p+2},..., u_{m-p-1}, u_{m-p} = um $-p$ +1 =... = u_m = 1, we want to compute the coefficients $N_{0,p}(u)$, $N_{1,p}(u)$,..., $N_{n,p}(u)$ for any given u in (0, 1). A naive way is the use of the following recurrence relation:

$$Ni,0(u) = \begin{cases} 1 & \text{if } u_i \le u < u_{i+1} \\ 0 & \text{otherwise} \end{cases}$$

$$Ni,p(u) = \frac{u - u_i}{u_{i+p} - u_i} N_{i,p-1}(u) + \frac{u_{i+p+1} - u}{u_{i+p+1} - u_{i+1}} N_{i+1,p-1}(u)$$

However, this is a very time consuming process. To compute $N_{i,p}(u)$, we need to compute $N_{i,p-1}(u)$ and $N_{i+1,p-1}(u)$. To compute $N_{i,p-1}(u)$, we need to compute $N_{i,p-2}(u)$ and $N_{i+1,p-2}(u)$. To compute $N_{i+1,p-1}(u)$, we need $N_{i+1,p-2}(u)$ and $N_{i+2,p-2}(u)$.

As you can see, $N_{i+1,p-2}(u)$ appears twice, and, as a result, its recursive computations will also be repeated. As the recursion goes deeper, more duplicated computations will occur. This is very similar to the recursive version of de Casteljau's algorithm discussed on a previous page. Consequently, the computation speed is very slow.

There is an easy way. Suppose u is in knot span (u_k, u_{k+1}). An important property of B-spline basis functions states that at most p + 1 basis functions of degree p are non-zero on (u_k, u_{k+1}), namely: $N_{k-p,p}(u)$, $N_{k-p+1,p}(u)$, $N_{k-p+2,p}(u)$,..., $N_{k-1,p}(u)$, $N_{k,p}(u)$. By definition, the only non-zero basis function of degree 0 on (u_k, u_{k+1}) is $N_{k,0}(u)$. As a result, the coefficients can be computed from Nk,0(u)in a "fan-out" triangular form as shown below:

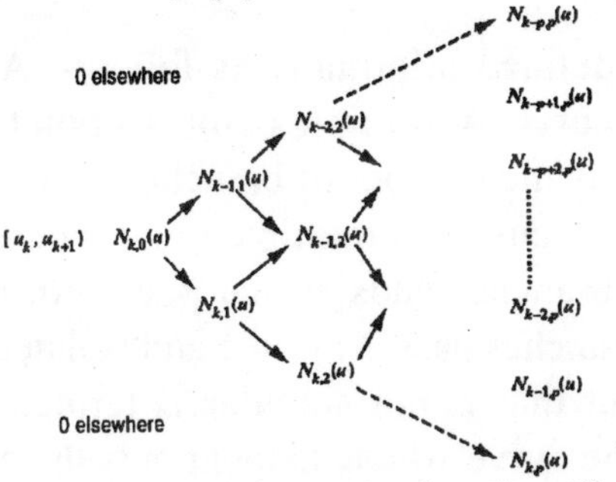

Since $N_{k,0}(u) = 1$ on knot span (u_k, u_{k+1}) and other B-spline basis functions of degree 0 are zero on (u_k, u_{k+1}), we can start from $N_{k,0}(u)$and compute the basis functions of degree 1 $N_{k-1,1}(u)$ and $N_{k,1}(u)$. From these two values, we can compute the basis functions of degree 2 $N_{k-2,2}(u)$, $N_{k-1,2}(u)$ and $N_{k,2}(u)$. This process repeats until all p + 1 non-zero coefficients are computed. In this computation, "internal" values such as $N_{k-1,2}(u)$has a north-west predecessor [*i.e.*, $N_{k-1,1}(u)$] and a south-west predecessor [*i.e.*, $N_{k,1}(u)$]; values on the north-east direction boundary of the above triangle such as $N_{k-1,1}(u)$ has only a south-west predecessor [*i.e.*, $N_{k,0}(u)$]; and values on the south-east direction boundary of this triangle such as $N_{k,2}(u)$ has only a north-west predecessor [i.e., $N_{k,1}(u)$]. Thus, values that are on the north-east (resp., south-east) direction boundary use the second (resp., first) term of the recurrence relation in the definition.

Chapter 10

Special Plane Curves

The locus of points in the euclidean plane that satisfy some geometric or algebraic definition. Not all sets of points deserve to be called a curve, but the distinction is somewhat arbitrary. For most of this article, a curve is considered to be the locus of a set of points that satisfy an algebraic or transcendental equation in two variables. The most interesting geometric properties are those preserved by linear transformations, especially translations, rotations, reflections, and magnifications. Useful geometric properties include the number of branches into which the curve is divided; the number and degree of nodes, cusps, isolated points, and flex points; the number of loops; symmetries; branches that go to infinity; and asymptotes.

These terms can be defined informally as follows. A branch is a maximal smooth continuous portion of the curve. A multiple point is a point in the plane that lies on two or more branches; its degree is the number of branches involved. A node is a multiple point where the branches cross. A cusp is a multiple point where the branches meet but do not pass; that is, each of the branches ends at that point. An isolated point is a point of the curve through which no branches pass. Multiple and isolated points are collectively termed singular points. Any point that is not singular is termed ordinary. A flex (or point of inflection) is a point on the curve whose tangent cuts the curve. A smooth closed branch forms a loop. A curve is symmetric about a line L if every line perpendicular to L intersects the curve at equal distances from L on opposite sides of L; that is, portions of the curve form mirror images about L.

The curve is symmetric about a point P if every line through P intersects the curve at equal distances from P in opposite directions. An asymptote is a line toward which a branch approaches as it moves to infinity from the origin; the curve and line are said to intersect at infinity. In addition to these geometric properties, the form of the defining equation is of interest. This can be an algebraic (polynomial in x and y) or transcendental equation. In the former case, quadratic, cubic, and quartic equations are of special interest. Once a coordinate system has been chosen, and the defining equation is known (in any of the forms, though the parametric form is usually the most useful), various properties of the curve can be defined in terms of the equation. These include the locations of x- and y-intercepts, local maxima and minima, flexes, nodes, and cusps.

ALGEBRAIC CURVES

Curves that can be given in implicit form as $f(x,y) = 0$, where f is a polynomial, are called algebraic. The degree of f is called the degree or order of the curve.

Thus conics are algebraic curves of degree two. Curves of degree three already have a great variety of shapes, and only a few common ones will be given here. The simplest case is when the curve is the graph of a polynomial of degree three: $y = ax^2 + bx^2 + cx + d$, with $a \neq 0$. This curve is a (general) cubic parabola. It is symmetric with respect to the point B where $x = - b/3a$.

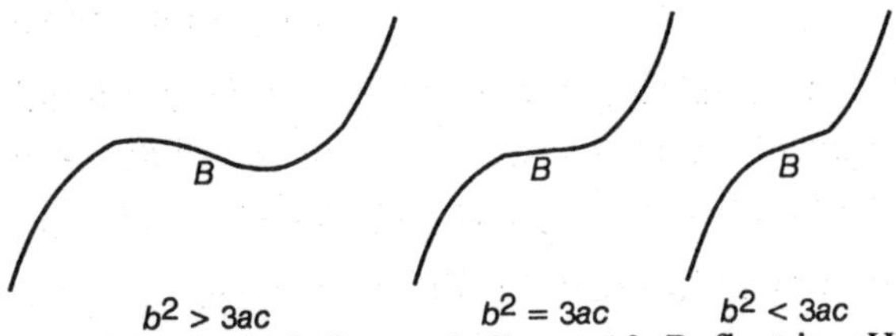

Fig: The General Cubic Parabola for $a > 0$. For $a < 0$, Reflect in a Horizontal Line.

The semicubic parabola has equation $y^2 = kx^3$; by proportional scaling one can take $k = 1$.

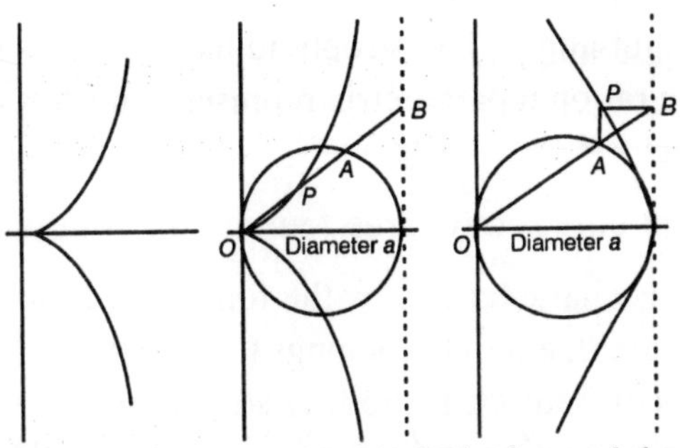

Fig: The Semicubic Parabola, the Cissoid of Diocles, and the witch of Agnesi

This curve should not be confused with the cissoid of Diocles, which has equation $(a - x)y^2 = x^3$ with $a \neq 0$. The latter is asymptotic to the line $x = a$, while the semicubic parabola has no asymptotes. The cissoid's points are characterized by the equality $OP = AB$. One can take $a = 1$ by proportional scaling.

More generally, any curve of degree three with equation $(x - x^0)y^2 = f(x)$, where f is a polynomial, is symmetric with respect to the x-axis and asymptotic to the line $x = x^2$. In addition to the cissoid, the following particular cases are important:

1. The witch of Agnesi has equation $xy^2 = a^2 (a - x)$, with $a \neq 0$, and is characterized

by the geometric property, right. The same property provides the parametric representation $x = a(1+\sin 2\theta)$, $y = a \tan\theta$. Once more, proportional scaling reduces to the case $a = 1$.

2. The folium of Descartes has equation $(x-a)y^2 = -x^2 (1/3\, x + a)$, with $a \neq 0$ (reducible to $a = 1$ by proportional scaling). By rotating 135°(right) we get the alternative and more familiar equation $x^3 + y^3 = cxy$, where $c = 1/3\sqrt{2}\, a$. The folium of Descartes is a rational curve, this is, it has a parametric representation by rational functions. In the tilted position such a representation is $x = ct/(1+t^3)$, $y = ct^2/(1+t^3)$ (so that $t = y/x$).

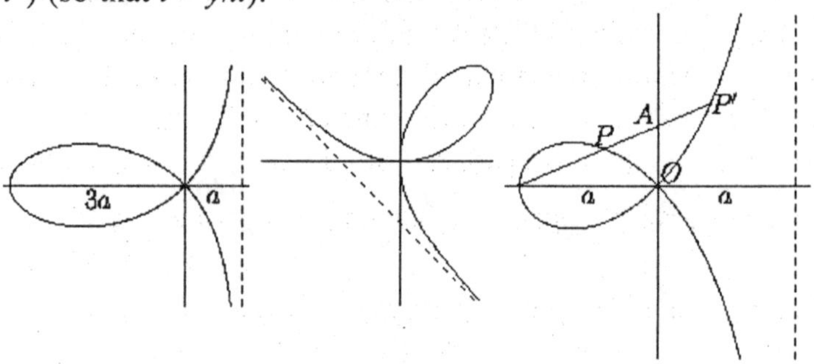

Fig. The Folium of Descartes in Two Positions, and the Strophoid

3. The strophoid has equation $(x - a)y^2 = - x^2 (x + a)$, with $a \neq 0$ (reducible to $a = 1$ by proportional scaling). It satisfies the property $AP=AP'= OA$, right; this means that POP' is a right angle. The strophoid has the polar representation $r = - a \cos 2\theta \sec\theta$, and the rational parametric representation $x = a(t^2 -1)/(t^2 +1)$, $y = at(t^2 - 1)/(t^2 +1)$ (so that $t = y/x$).

Among the important curves of degree four are the following:

1. A Cassini's oval is characterized by the following condition: given two foci F and F', a distance $2a$ apart, a point P belongs to the curve if the product of the distances PF and PF' is a constant k^2. If the foci are on the x-axis and equidistant from the origin, the curve's equation is $(x^2 + y^2 + a^2)$ $-4a^2 x^2 = k^4$.

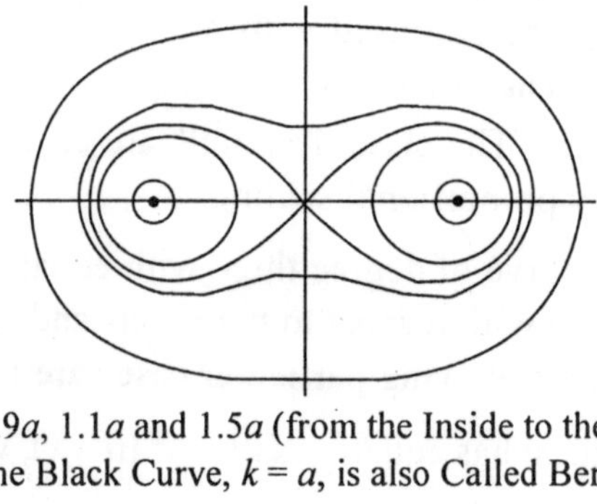

Fig. Cassini's Ovals for $k = 5a, .9a, 1.1a$ and $1.5a$ (from the Inside to the Outside). The Foci (Dots) are *at* x=*a* and x=- *a*. The Black Curve, $k = a$, is also Called Bernoulli's Lemniscate

Changes in a correspond to rescaling, while the value of k/a controls the shape: the curve has one smooth piece, one piece with a self-intersection, or two pieces depending on whether k is greater than, equal to, or smaller than Figure. The case $k = a$ is also known as the lemniscate (of Jakob Bernoulli); the equation reduces to $(x^2 + y^2)^2 = a^2 (x^2 - y^2)$, and upon a 45° rotation to $(x^2 + y^2)^2 = 2a\,xy$. Each Cassini's oval is the section of a torus of revolution by a plane parallel to the axis of revolution.

2. A conchoid of Nichomedes is the set of points such that the signed distance AP in Figure, left, equals a fixed real number k (the line L and the origin O being fixed). If L is the line $x = a$, the conchoid has polar equation

$$r = a \sec \theta + k.$$

Once more, a is a scaling parameter, and the value of k/a controls the shape: when $k > -a$ the curve is smooth, when $k = -a$ there is a cusp, and when $k < -a$ there is a self-intersection. The curves for k and $-k$ can also be considered two leaves of the same conchoid, with cartesian equation $(x - a)^2 (x^2 + y^2) = k^2 x^2$.

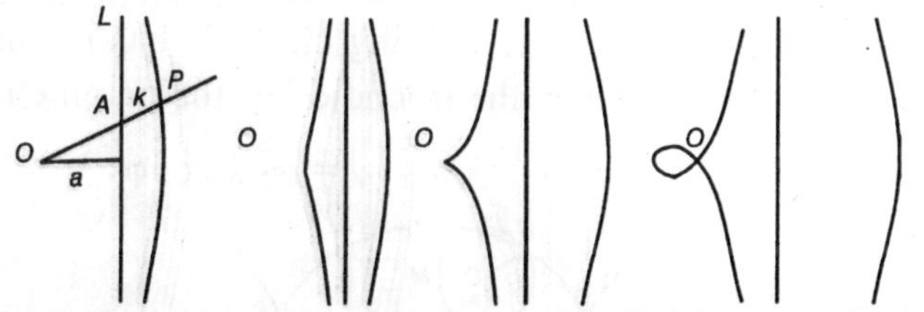

Fig: Defining Property of the Conchoid of Nichomedes (Left), and Curves for $k = \pm .5a$, $k = \pm a$, and $k = \pm 1.5a$ (Right)

A limaçon of Pascal is the set of points such that the distance AP in Figure, left, equals a fixed positive number k measured on either side (the circle C and the origin O being fixed). If C has diameter a and center at $(0, \frac{1}{2}a)$, the limaçon has polar equation

$$r = a \cos \theta + k,$$

and cartesian equation $(x^2 + y^2 - ax)^2 = k^2 (x^2 + y^2)$. The value of k/a controls the shape, and there are two particularly interesting cases. For $k = a$ we get a cardioid. For $a = \frac{1}{2}k$ we get a curve that can be used to trisect an arbitrary angle α : if we draw a line L through the center of the circle C and making an angle α with the positive x-axis, and if we call P the intersection of L with the limaçon $a = 2k$, the line from O to P makes an angle $1/3\,\alpha$ with L.

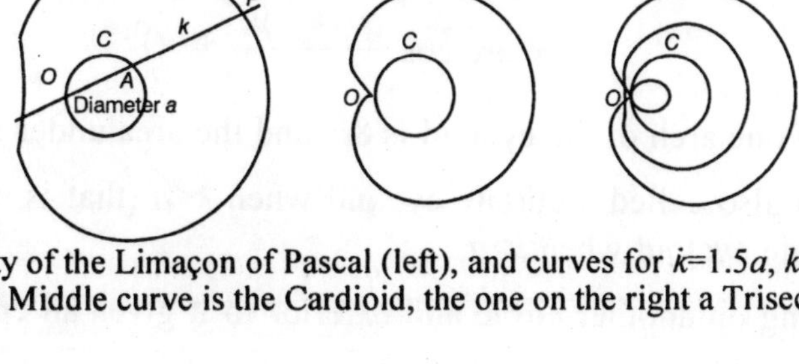

Fig. Defining Property of the Limaçon of Pascal (left), and curves for k=1.5a, k=a, and k=.5a (right). The Middle curve is the Cardioid, the one on the right a Trisectrix

Hypocycloids and epicycloids with rational ratio are also algebraic curves, generally of higher degree.

ROULETTES (SPIROGRAPH CURVES)

Suppose given a fixed curve C and a moving curve M, which rolls on C without slipping. The curve drawn by a point P kept fixed with respect to M is called a roulette, of which P is the pole.

The most important examples of roulettes arise when M is a circle and C is a straight line or a circle, but an interesting additional example is provided by the catenary $y = a \cosh(x/a)$, which arises by rolling the parabola $y = x^2/(4a)$ on the x-axis with pole the focus of the parabola (that is, $P = (0,a)$ in the initial position). The catenary is the shape taken under the action of gravity by a chain or string of uniform density whose ends are held in the air.

A circle rolling on a straight line gives a trochoid, with the cycloid as a special case when the pole P lies on the circle. If the moving circle M has radius a and the distance from the pole P to the center of M is k, the trochoid has the parametric equation

$$x = a\varphi - k\sin\varphi,\ y = a - k\cos\varphi.$$

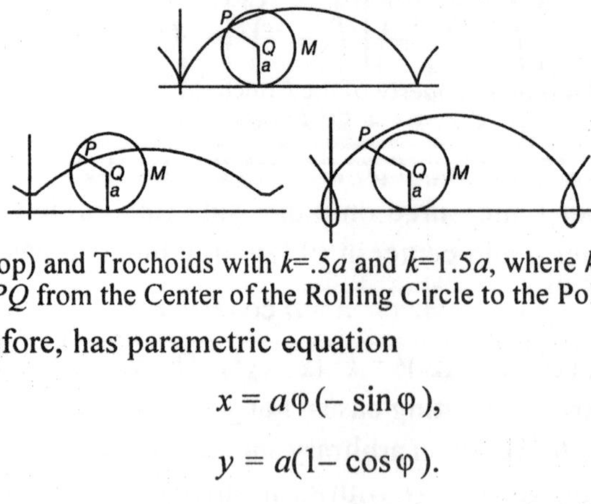

Fig. Cycloid (top) and Trochoids with k=.5a and k=1.5a, where k is the Distance PQ from the Center of the Rolling Circle to the Pole

The cycloid, therefore, has parametric equation

$$x = a\varphi(-\sin\varphi),$$

$$y = a(1 - \cos\varphi).$$

One can eliminate φ to get x as a (multivalued) function of y, which takes the following form for the cycloid:

$$x = \text{arc}\cos\frac{a - y}{a} \pm \hat{G}$$

The length of one arch of the cycloid is $8a$, and the area under the arch is $3\pi a^2$.

A trochoid is also called a curtate cycloid when $k<a$ (that is, when P is inside the circle) and a prolate cycloid when $k>a$.

A circle rolling on another circle and exterior to it gives an epitrochoid. If a is the radius of the fixed circle, b that of the rolling circle and k is the distance from P to the center of the rolling circle, the parametric equation of the epitrochoid is

$$x = (a+b)\cos\theta - k\cos((1+a/b)\,\theta),$$
$$y = (a+b)\sin\theta - k\sin((1+a/b)\,\theta).$$

These equations assume that at the start everything is aligned along the positive x-axis, left.

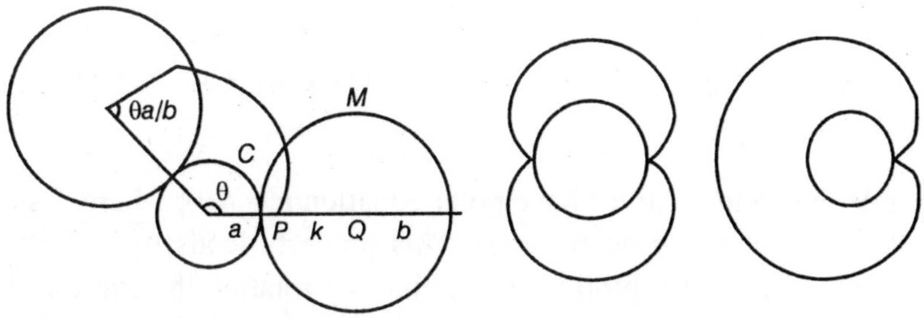

Fig. *Left:* Initial Configuration for Epicycloid (black) and Configuration at Parameter value (red). Middle: Epicycloid with $b = \frac{1}{2}a$ (nephroid). Right: Epicycloid with $b = a$ (Cardioid).

Usually one considers the case when a/b is a rational number, say $a/b = p/q$ where p and q are relatively prime. Then the rolling circle returns to its original position after rotating q times around the fixed circle, and the epitrochoid is a closed curve---in fact, an algebraic curve. One also usually takes $k = b$, so that P lies on the rolling circle; the curve in this case is called an epicycloid. The middle diagram shows the case $b = k = \frac{1}{2}a$, called the nephroid; this curve is the cross section of the caustic of a spherical mirror. The diagram on the right shows the case $b = k = a$, which gives the cardioid.

Hypotrochoids and hypocycloids are defined in the same way as epitrochoids and epicycloids, but the rolling circle is inside the fixed one. The parametric equation of the hypotrochoid is

$$x = (a - b)\cos\theta + k\cos((a/b - 1)\,\theta),$$
$$y = (a - b)\sin\theta - k\sin((a/b - 1)\,\theta),$$

where the letters have the same meaning as for the epitrochoid. Usually one takes a/b rational and $k = b$. There are several interesting particular cases:

1. $b = k = a$ gives a point.
2. $b = k = \frac{1}{2}a$ gives a diameter of the circle C.
3. $b = k = a$ gives the deltoid which has the algebraic equation

   $$(x^2 + y^2)^2 - 8ax + 24axy^2 + 18a(x^2 + y^2) - 27a^4 = 0.$$

4. $b = k = \frac{1}{4}a$ gives the astroid, an algebraic curve of degree six whose equation can be reduced to $x^{2/3} + y^{2/3} = a^{2/3}$. The figure illustrates another property of the astroid: its tangent intersects the coordinate axes at points that are always the same distance

a apart. Otherwise said, the astroid is the envelope of a moving segment of fixed length whose endpoints are constrained to lie on the two coordinate axes.

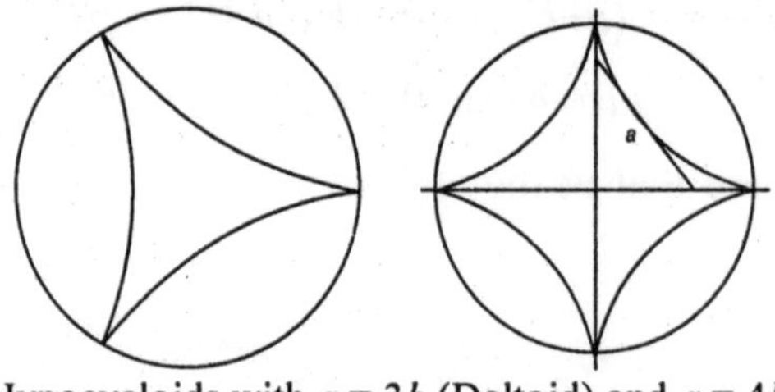

Fig. The Hypocycloids with $a = 3b$ (Deltoid) and $a = 4b$ (Astroid)

SPIRALS

A number of interesting curves have polar equation $r = f(\theta)$, where f is a monotonic function (always increasing or decreasing). This property leads to a spiral shape. The logarithmic spiral or Bernoulli spiral is self-similar: by rotation the curve can be made to match any scaled copy of itself. Its equation is $r = k^{2\theta}$; the angle between the radius from the origin and the tangent to the curve is constant, and equal to φ = arccot a. A curve parametrized by arclength and such that the radius curvature is proportional to the parameter at each point is a Bernoulli spiral.

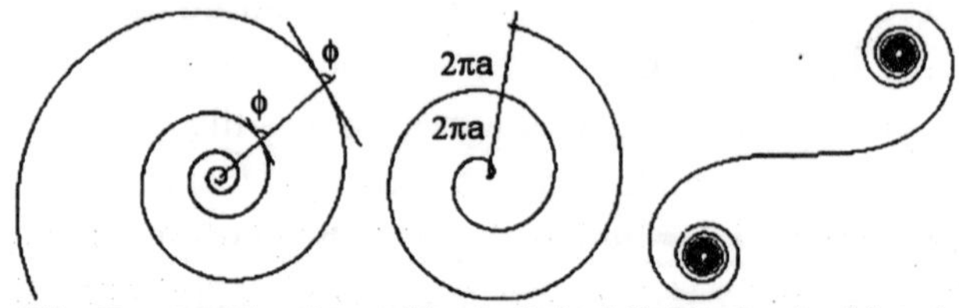

Fig. The Bernoulli cr Logarithmic Spiral (left), the Archimedes or Linear S Piral (Middle) and the Cornu Spiral (Right)

In the Archimedean spiral or linear spiral, it is the spacing between intersections along a ray from the origin that is constant. The equation of this spiral is $r = a\theta$; by scaling one can take $a = 1$. It has an inner endpoint, in contrast with the logarithmic spiral, which spirals down to the origin without reaching it. The Cornu spiral or clothoid important in optics and engineering, has the following parametric representation in Cartesian coordinates:

$$x = aC\,(t) = a\int_0^t \cos\left(\frac{1}{2}\pi s^2\right)ds,\quad y = aS\,(t) = a\int_0^t \sin\left(\frac{1}{2}\pi s^2\right)ds,$$

(C and S are the so-called Fresnel integrals) A curve parametrized by arclength and such that the radius curvature is inversely proportional to the parameter at each point is a Cornu spiral (compare the Bernoulli spiral).

THE PEANO CURVE AND FRACTAL CURVES

There are examples of curves (in the sense of continuous maps from the real line to the plane) that completely cover a two-dimensional region of the plane. We give a construction of such a Peano curve, adapted from David Hilbert's example. The construction is inductive, and is based on replacement rules. We consider building blocks of six shapes,

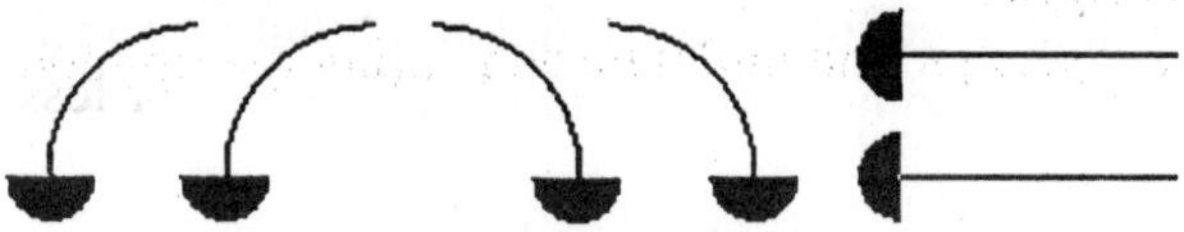

Fig. The Plano Curve

the length of the straight pieces being twice the radius of the curved ones. A sequence of these patterns end-to-end represents a curve, if we disregard the red and green half-disks. The replacement rules are the following:

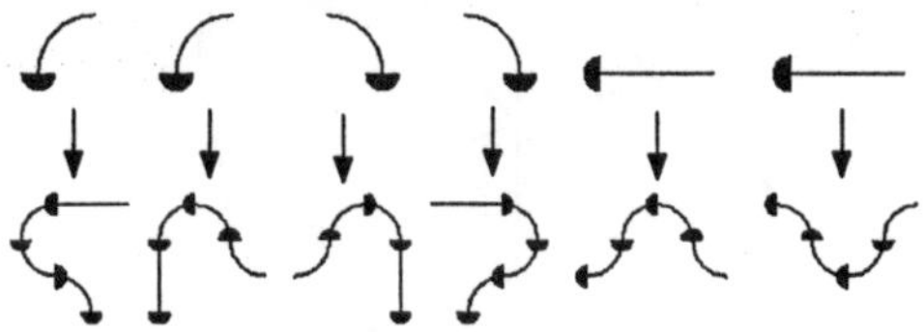

Fig. Fractal Curves

The rules are applied taking into account the way each piece is turned. Here we apply the replacement rules to a particular initial pattern:

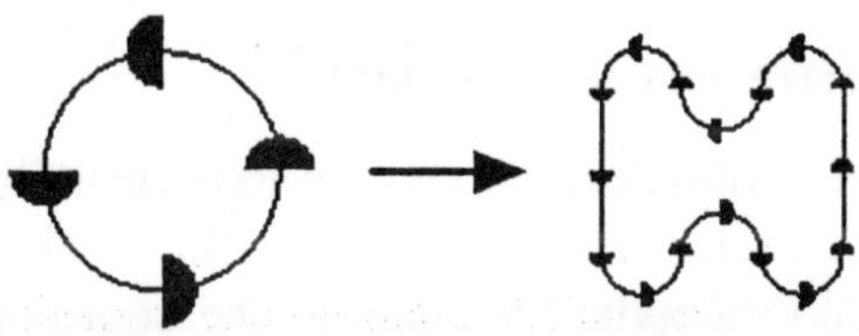

(We scale the result so it has the same size as the original.) Applying the process repeatedly gives, in the limit, the Peano curve. Here are the first five steps:

The same idea of replacement rules leads to many interesting fractal, and often self-similar, curves. For example, the substitution leads to the Koch snowflake when applied to an initial equilateral triangle, like this (the first three stages and the sixth are shown):

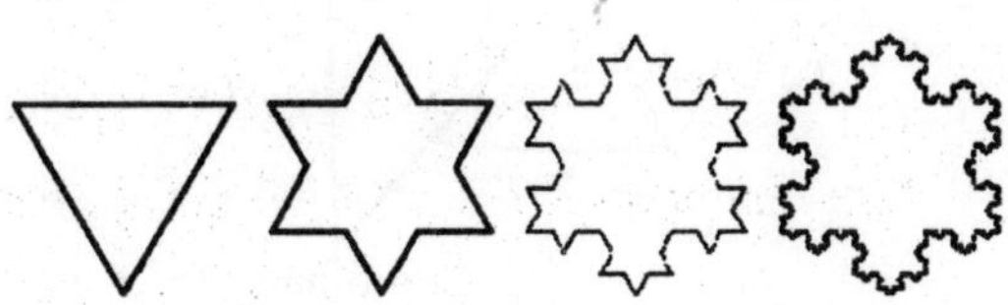

CARTESIAN COORDINATES IN SPACE

In cartesian coordinates (or rectangular coordinates), a point *P* is referred to by three real numbers, indicating the positions of the perpendicular projections from the point to three fixed, perpendicular, graduated lines, called the axes.

If the coordinates are denoted *x*, *y*, *z*, in that order, the axes are called the *x*-axis, etc., and we write $P = (x,y,z)$.

Often the *x*-axis is imagined to be horizontal and pointing roughly toward the viewer (out of the page), the *y*-axis also horizontal and pointing more or less to the right, and the *z*-axis vertical, pointing up.

The system is called right-handed if it can be rotated so the three axes are in this position. The point $x = 0$, $y = 0$, $z = 0$ is the origin, where the three axes intersect.

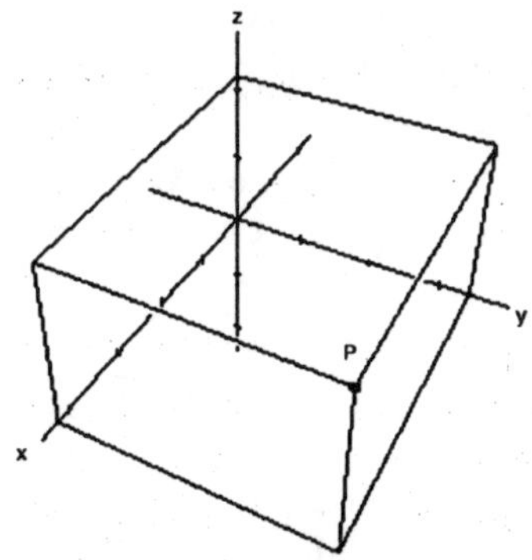

Fig. In Cartesian Coordinates, $P = (4.2,3.4,2.2)$

CYLINDRICAL COORDINATES IN SPACE

To define cylindrical coordinates, we take an axis (usually called the *z*-axis) and a perpendicular plane, on which we choose a ray (the initial ray) originating at the intersection of the plane and the axis (the origin). The coordinates of a point *P* are: the polar coordinates (r,θ) of the projection of *P* on the plane, and the coordinate *z* of the projection of *P* on the axis.

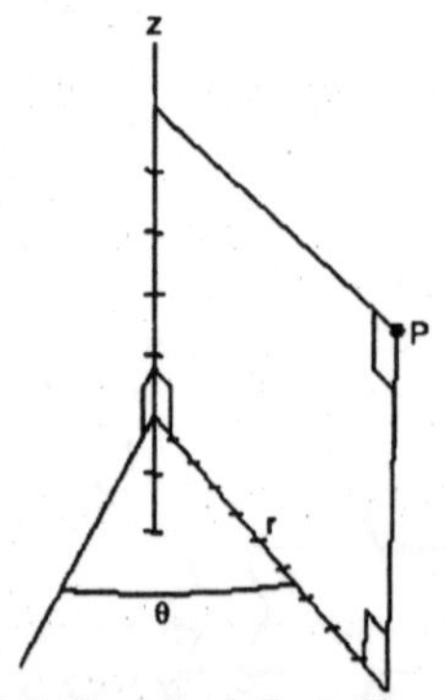

Fig. Among the Possible Sets (r,θ,z) of Cylindrical Coordinates for *P* are $(10,30°,5)$ and $(10,390°,5)$

SPHERICAL COORDINATES IN SPACE

To define spherical coordinates, we take an axis (the polar axis) and a perpendicular plane (the equatorial plane), on which we choose a ray (the initial ray) originating at the intersection of the plane and the axis (the origin *O*).

The coordinates of a point *P* are: the distance θ from *P* to the origin; the angle φ (zenith) between the line *OP* and the positive polar axis; and the angle θ (azimuth) between the initial ray and the projection of *OP* to the equatorial plane. As in the case of polar and cylindrical coordinates, θ is only defined up to multiples of 360°, and likewise φ.

Usually φ is assigned a value between 0 and 180°, but values of φ between 180° and 360° can also be used; the triples (*P*, φ ,θ) and (*P*, 360°- φ, 180°+θ) represent the same point. Similarly, one can extend *P* to negative values; the triples (*P*, φ ,θ) and (-*P*, 180°- φ, 180°+ θ) represent the same point.

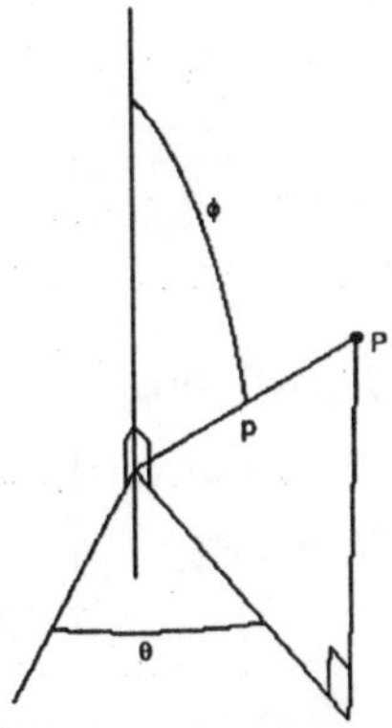

Fig. A set of Spherical Coordinates for *P* is (***P***,θ, φ) = (10,60°,30°)

Relations between Cartesian, Cylindrical, and Spherical Coordinates

Consider a cartesian, a cylindrical, and a spherical coordinate system, related as shown in Figure.

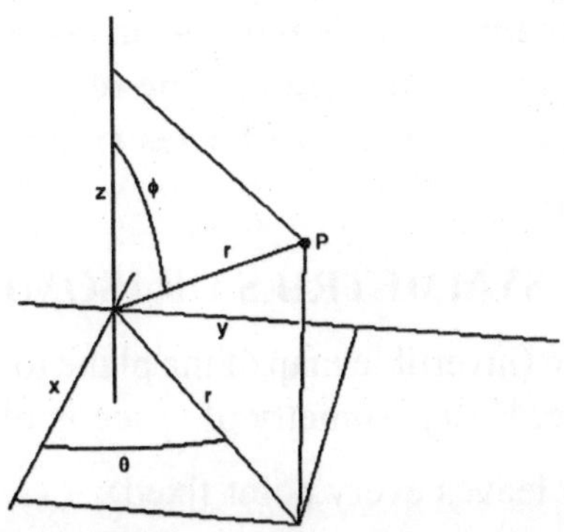

Fig. Standard relations between cartesian, cylindrical, and spherical coordinate systems. The origin is the same for all three. The positive *z*-axes of the cartesian and cylindrical systems coincide with the positive polar axis of the spherical system. The initial rays of the cylindrical and spherical systems coincide with the positive x-axis of the cartesian system, and the rays θ = 90° coincide with the positive y-axis

Then the cartesian coordinates (*x*,*y*,*z*), the cylindrical coordinates (*r*,θ,*z*), and the spherical coordinates (*P*, φ ,θ) of a point are related as follows:

$$\text{cart} \leftrightarrow \text{cyl}\begin{cases} x = r\cos\theta, \\ y = r\sin\theta, \end{cases} \begin{cases} r = \sqrt{x^2+y^2}, \\ \theta = \arctan\dfrac{y}{x}, \end{cases} \begin{cases} \sin\phi = \dfrac{x}{\sqrt{r^2+z^2}}, \\ \cos\phi = \dfrac{y}{\sqrt{r^2+z^2}} \end{cases}$$

$$\text{cyl} \leftrightarrow \text{sph}\begin{cases} r = P\sin\phi, \\ z = P\cos\phi, \end{cases} \begin{cases} P = \sqrt{r^2+z^2}, \\ \phi = \arctan\dfrac{r}{z}, \end{cases} \begin{cases} \sin\phi = \dfrac{r}{\sqrt{r^2+z^2}}, \\ \cos\phi = \dfrac{z}{\sqrt{r^2+z^2}} \end{cases}$$

$$\text{cart} \leftrightarrow \text{sph}\begin{cases} x = P\cos\theta\sin\phi, \\ y = P\sin\theta\sin\phi, \\ z = P\cos\phi, \end{cases} \begin{cases} P = \sqrt{x^2+y^2+z^2} \\ \theta = \arctan\dfrac{y}{x}, \\ \phi = \arctan\dfrac{\sqrt{x^2+y^2}}{z} \\ = \arccos\dfrac{z}{\sqrt{x^2+y^2+z^2}}. \end{cases}$$

HOMOGENEOUS COORDINATES IN SPACE

A quadruple of real numbers $(x{:}y{:}z{:}t)$, with $t \neq 0$, is a set of homogeneous coordinates for the point P with cartesian coordinates $(x/t,\ y/t,\ z/t)$.

Thus the same point has many sets of homogeneous coordinates: $(x{:}y{:}z{:}t)$ and $(x'{:}y'{:}z'{:}t')$ represent the same point if and only if there is some real number α such that $x'= \alpha x$, $y'= \alpha y$, $z'= \alpha\, z$, $t'= \alpha\, t$. If P has cartesian coordinates (x^0,y^0,z^0), one set of homogeneous coordinates for P is $(x^0,y^0,z^0,1)$.

SPACE SYMMETRIES OR ISOMETRIES

A transformation of space (invertible map of the plane to itself) that preserves distances is called an isometry of space. Every isometry of space is of one of the following types:

- The identity (which leaves every point fixed);
- A translation by a vector v;
- A rotation through an angle α around a line L;
- A screw motion through an angle α around a line L, with displacement d;
- A reflection in a line P;
- A glide-reflection in a line P with displacement vector v.

The identity is a particular case of a translation and of a rotation; rotations are particular cases of screw motions; and reflections are particular cases of glide-reflections. However, as in the plane case, is more intuitive to consider each case separately.

FORMULAS FOR SYMMETRIES IN CARTESIAN COORDINATES

In the formulas below, a multiplication between a matrix and a triple of coordinates should be carried out regarding the triple as a column vector (or a matrix with three rows and one column).

Translation by (x^0,y^0,z^0):

$$(x,y,z) \mapsto (x+x^0,\ y+y^0,\ z+z^0)$$

Rotation through α (counterclockwise) around the line through the origin with direction cosines a, b, c:

$$(x,y,z) \mapsto M(x,y,z),$$

where M is the matrix

$$\begin{bmatrix} a^2(1-\cos\alpha)+\cos\alpha & ab(1-\cos\alpha)-c\sin\alpha & ac(1-\cos\alpha)+b\sin\alpha, \\ ab(1-\cos\alpha)+c\sin\alpha & b^2(1-\cos\alpha)+\cos\alpha & bc(1-\cos\alpha)-a\sin\alpha \\ ac(1-\cos\alpha)-b\sin\alpha & bc(1-\cos\alpha)+a\sin\alpha & c^2(1-\cos\alpha)+\cos\alpha \end{bmatrix}$$

Rotation through α (counterclockwise) around the line with direction cosines a,b,c through an arbitrary point (x^0,y^0,z^0):

$$(x,y,z)\ \_\ (x^0,y^0,z^0)+M(x-x^0,\ y-y^0,\ z-z^0),$$

where M is given by (1).

Arbitrary rotations and Euler angles. Any rotation of space fixing the origin can be decomposed as a rotation by φ about the z-axis, followed by a rotation by θ about the y-axis, followed by rotation by about the z-axis. The numbers φ, θ and ψ are called the Euler angles of the composite rotation, which acts as follows:

$$(x,y,z) \mapsto M(x,y,z),$$

where M is the matrix given by

$$\begin{bmatrix} \cos\phi\cos\theta\cos\psi-\sin\phi\sin\psi & -\sin\phi\cos\theta\cos\psi-\cos\phi\sin\psi & \sin\theta\cos\psi \\ \sin\phi\cos\psi+\cos\phi\cos\theta\sin\psi & -\sin\phi\cos\theta\sin\psi+\cos\phi\cos\psi & \sin\theta\sin\psi \\ -\cos\phi\sin\theta & \sin\theta\sin\phi & \cos\theta \end{bmatrix}.$$

(An alternative decomposition, more natural if we think of the coordinate system as a rigid trihedron that rotates in space, is the following: a rotation by ψ about the z-axis, followed by a rotation by θ about the rotated y-axis, followed by a rotation by φ about the rotated z-axis. Note that the order is reversed.)

Provided that θ is not a multiple of 180°, the decomposition of a rotation in this form is unique (apart from the ambiguity arising from the possibility of adding a multiple of 360°to any angle). Figure shows how the Euler angles can be read off geometrically.

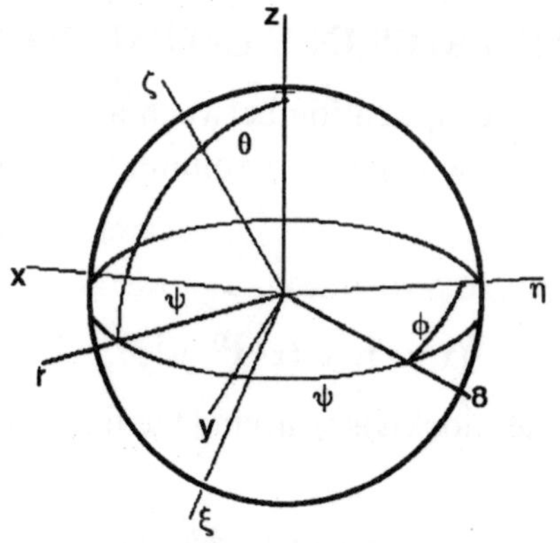

Fig. The coordinate rays Ox, Oy, Oz, Together with their Images $O\xi$, $O\eta$, $O\zeta$ under a rotation, fix the Euler angles associated with that Rotation, as follows: $\theta = zO\zeta$, $\psi = xOr = y\ Os$, and $= sO\eta$. (Here the ray Or is the projection of O to the xy-plane. The ray Os is Determined by the Intersection of the xy- and -planes.)

Warning: Some references define Euler angles differently; the most common variation is that the second rotation is taken about the x-axis instead of about the y-axis.

Screw motion: With angle α and displacement d around the line with direction cosines a, b, c through an arbitrary point (x^0,y^0,z^0):

$$(x,y,z) \mapsto (x^0 + ad,\ y^0 + bd,\ z^0 + cd) + M(x-x^0,\ y-y^0,\ z-z^0),$$

where M is given by (1).

REFLECTION

Reflection in a plane with equation $ax + by + cz + d = 0$:

$$(x,\ y) \mapsto \frac{1}{a^2 + b^2 + c^2}\ (M(x^0,\ y^0,\ z^0) - (2ad,\ 2bd,\ 2cd)),$$

where R is the matrix

$$M = \begin{pmatrix} -a^2 + b^2 + c^2 & -2ab & -2ac \\ -2ab & a^2 - b^2 + c^2 & -2bc \\ -2ac & -2bc & a^2 + b^2 - c^2 \end{pmatrix}.$$

Reflection: In a plane going through (x^0,y^0,z^0) and whose normal has direction cosines a, b, c:

$$(x,y)\_\ (x^0 + y^0 + z^0) + M(x-x^0,\ y-y^0,\ z-z^0),$$

where M is as in (3).

Glide-reflection in a plane P with displacement vector v: Apply first a reflection in P, then a translation by the vector v.

FORMULAS FOR SYMMETRIES IN HOMOGENEOUS COORDINATES

All isometries of space can be expressed in homogeneous coordinates in terms of multiplication by a matrix. As in the case of plane isometries, this means that the successive application of transformations reduces to matrix multiplication. (In the formulas below, $\begin{bmatrix} M & 0 \\ 0 & 1 \end{bmatrix}$ is the 4 × 4 projective matrix obtained from the 3 × 3 matrix M by adding a row and a column as stated.)

Translation: by (x^0, y^0, z^0):

$$\begin{bmatrix} 1 & 0 & 0 & x_0 \\ 0 & 1 & 0 & y_0 \\ 0 & 0 & 1 & z_0 \\ 0 & 0 & 0 & 1 \end{bmatrix}$$

Rotation through the origin: $\begin{bmatrix} M & 0 \\ 0 & 1 \end{bmatrix}$ where M is given in (10.1.1) or (10.1.2), as the case may be.

Reflection in a plane through the origin: $\begin{bmatrix} M & 0 \\ 0 & 1 \end{bmatrix}$, where M is given in (10.1.3).

From this one can deduce all other transformations, as in the case of plane transformations.

OTHER TRANSFORMATIONS OF SPACE

SIMILARITIES

A transformation of space that preserves shapes is called a similarity. Every similarity of the plane is obtained by composing a proportional scaling transformation (also known as a homothety) with an isometry. A proportional scaling transformation centered at the origin has the form

$$(x,y,z) \mapsto (ax,ay,az),$$

where $a \neq 0$ is the scaling factor (a real number). The corresponding matrix in homogeneous coordinates is

$$H_2 = \begin{bmatrix} a & 0 & 0 & 0 \\ 0 & 1 & 0 & 0 \\ 0 & 0 & a & 0 \\ 0 & 0 & 0 & 1 \end{bmatrix}.$$

In cylindrical coordinates, the transformation is

$$(r,\theta,z) \mapsto (ar,\varphi,\theta az).$$

In spherical coordinates, it is

$$(r,\varphi,\theta) \mapsto (ar,\varphi,\theta).$$

AFFINE TRANSFORMATIONS

A transformation that preserves lines and parallelism (maps parallel lines to parallel lines) is an affine transformation. There are two important particular cases of such transformations: A nonproportional scaling transformation centered at the origin has the form

$$(x,y,z) \mapsto (ax,by,cz),$$

where $a,b,c \neq 0$ are the scaling factors (real numbers). The corresponding matrix in homogeneous coordinates is

$$H\,a,b,c = \begin{bmatrix} a & 0 & 0 & 0 \\ 0 & b & 0 & 0 \\ 0 & 0 & c & 0 \\ 0 & 0 & 0 & 1 \end{bmatrix}.$$

A shear in the x-direction and preserving horizontal planes has the form

$$(x,y,z) \mapsto (x+rz,y,z),$$

where r is the shearing factor. The corresponding matrix in homogeneous coordinates is

$$S_r = \begin{bmatrix} 1 & 0 & r & 0 \\ 0 & 1 & 0 & 0 \\ 0 & 0 & 1 & 0 \\ 0 & 0 & 0 & 1 \end{bmatrix}.$$

Every affine transformation is obtained by composing a scaling transformation with an isometry, or one or two shears with a homothety and an isometry.

PROJECTIVE TRANSFORMATIONS

A transformation that maps lines to lines (but does not necessarily preserve parallelism) is a projective transformation. Any spatial projective transformation can be expressed by an invertible 4×4 matrix in homogeneous coordinates; conversely, any invertible 4×4 matrix defines a projective transformation of the plane. Projective transformations (if not affine) are not defined on all of space, but only on the complement of a plane (the missing plane is ``mapped to infinity").

The following particular case is often useful, especially in computer graphics, in projecting a scene from space to the plane. Suppose an observer is at the point $E = (x^0, y^0, z^0)$ of space, looking toward the origin $O=(0,0,0)$. Let P, the screen, be the plane through O and perpendicular to the ray EO. Place a rectangular coordinate system $\xi\eta$ on P with origin at O and such that the positive η -axis lies in the half-plane determined by E and the positive z-axis of space (that is, the z-axis is pointing ``up" as seen from E). Then consider the transformation that associates to a point $X=(x,y,z)$ the triple (ξ, η, ξ), where (ξ, η) are the coordinates of the point where the line EX intersects P (the screen coordinates of X as seen from E), and ξ is the inverse of the signed distance from X to E along the line EO (the depth of X as seen from E). This is a projective transformation, given by the matrix

$$\begin{bmatrix} -r^2 y_0 & r^2 x_0 & 0 & 0 \\ -r x_0 z_0 & -r y_0 z_0 & rP^2 & 0 \\ 0 & 0 & 0 & rp \\ -p x_0 & -p y_0 & -p x_0 & r^2 p \end{bmatrix}$$

with $p = \sqrt{x_0^2 + y_0^2}$ and $r = \sqrt{x_0^2 + y_0^2 + z_0^2}$.

DIRECTIONS, PLANES AND LINES

DIRECTION ANGLES AND DIRECTION COSINES

Given a vector (a,b,c) in three-space, the direction cosines of this vector are

$$\cos\alpha = \frac{a}{\sqrt{a^2 + b^2 + c^2}},$$

$$\cos\beta = \frac{b}{\sqrt{a^2 + b^2 + c^2}},$$

$$\cos\chi = \frac{c}{\sqrt{a^2 + b^2 + c^2}},$$

Here the direction angles α, β, χ are the angles that the vector makes with the positive x-, y- and z-axes, respectively. In formulas, it is usually the direction cosines that occur, rather than the direction angles. We have

$$\cos^2\alpha + \cos^2\beta + \cos^2\chi = 1$$

PLANES

The (cartesian) equation of a plane is linear in the coordinates x and y, that is, of the form $ax + by + cz + d = 0$. The normal direction to this plane is (a,b,c). The intersection of this plane with the x-axis, or x-intercept, is $x = -d/a$; the y-intercept is $y = -d/b$, and the z-intercept is $z = -d/c$.

The plane is vertical (perpendicular to the xy-plane) if $c = 0$; it is perpendicular to the x-axis if $b = c = 0$; and likewise for the other coordinates.

When $a^2 + b^2 + c^2 = 1$ and $d \leq 0$ in the equation $ax + by + cz + d = 0$, the equation is said to be in normal form. In this case d is the distance of the plane to the origin, and (a,b,c) are the direction cosines of the normal.

To reduce an arbitrary equation $ax + by + cz + d = 0$ to normal form, divide by $\pm\sqrt{a^2 + b^2 + c^2}$, where the sign of the radical is chosen opposite the sign of d when $d \neq 0$, the same as the sign of c when $d = 0$ and $c \neq 0$, and the same as the sign of b otherwise.

PLANES WITH PRESCRIBED PROPERTIES

Plane through (x^0,y^0,z^0) and perpendicular to the direction (a,b,c):

$$a(x-x^0)+b(y-y^0)+c(z-z^0) = 0$$

Plane through (x^0,y^0,z^0) and parallel to the directions (a^1,b^1,c^1) and (a^2,b^2,c^2):

$$\begin{vmatrix} x-x_0 & y-y_0 & z-z_0 \\ a_1 & b_1 & c_1 \\ a_2 & b_2 & c_2 \end{vmatrix} = 0$$

Plane through (x^0,y^0,z^0) and (x^1,y^1,z^1) and parallel to the direction (a,b,c):

$$\begin{vmatrix} x-x_0 & y-y_0 & z-z_0 \\ x_1-x_0 & y_1-y_0 & z_1-z_0 \\ a & b & c \end{vmatrix} = 0$$

Plane going through (x^0,y^0,z^0), (x^1,y^1,z^1) and (x^2,y^2,z^2):

$$\begin{vmatrix} x & y & z & 1 \\ x_0 & y_0 & z_0 & 1 \\ x_2 & y_2 & z_2 & 1 \end{vmatrix} = 0 \text{ or } \begin{vmatrix} x-x_2 & y-y_0 & z-z_0 \\ x_1-x_0 & y_1-y_0 & z_1-z_0 \\ x_2-x_0 & y_2-y_0 & z_2-z_0 \end{vmatrix}$$

(The last three formulas remain true in oblique coordinates.)

The distance from the point (x^0, y^0, z^0) to the plane $ax + by + cz + d = 0$ is

$$\left| \frac{ax_0 + by_0 + cz_0 + d}{\sqrt{a^2 + b^2 + c^2}} \right|$$

The angle between two planes $a^0x + b^0y + c^0z + d^0 = 0$ and $a^1 x + b^1 y + c^1z + d^1 = 0$ is

$$\text{arcoos} \quad \frac{a_0x_1 + b_0b_1 + c_0c_1}{\sqrt{a_0^1 + b_0^2 + c_0^2}\sqrt{a_1^2 + b_1^2 + c_1^2}}$$

In particular, the two planes are parallel when $a^0 : b^0 : c^0 = a^1 : b^1 : c^1$, and perpendicular when $a^1a^1 + b^1b^1 + c^1c^1 = 0$.

CONCURRENCE AND COPLANARITY

Four planes $a^0x + b^0y + c^0z + \mathrm{d}^0 = 0$, $a^1x + b^1 y + c^1 z + d^1 = 0$, $a^2x + b^2y + c^2z + d^2 = 0$ and $a^3x + b^3y + c^3z + d = 0$ are concurrent if and only if

$$\begin{vmatrix} a_0 & b_0 & c_0 & d_0 \\ a_1 & b_1 & c_1 & d_1 \\ a_2 & b_2 & c_2 & d_2 \\ a_3 & b_3 & c_3 & d_3 \end{vmatrix} = 0.$$

Four points (x^0, y^0, z^0), (x^1, y^1, z^1), (x^2, y^2, z^2) and (x^{3s}, y^3, z^3) are coplanar if and only if

$$\begin{vmatrix} x_0 & y_0 & z_0 & 1 \\ x_1 & y_1 & z_1 & 1 \\ x_2 & y_2 & z_2 & 1 \\ x_3 & y_3 & z_3 & 1 \end{vmatrix} = 0$$

(Both of these assertions remain true in oblique coordinates.)

LINES

Two planes that are not parallel or coincident intersect in a straight line, so one can express a line by a pair of linear equations

$$ax + by + cz + d = 0 \text{ and } a'x + b'y + c'z + d' = 0$$

such that $bc' - cb'$, $ca' - ac'$, and $ab' - ba'$ are not all zero. The line thus defined is parallel to the vector

$$(bc' - cb', ca' - ac', ab' - ba').$$

The direction cosines of the line are those of this vector. (The direction cosines of a line are only defined up to a simultaneous change in sign, since the opposite vector still gives the same line.)

The following particular cases are important:

Line through (x^0, y^0, z^0) parallel to the vector (a, b, c):

$$\frac{x-x_0}{a}=\frac{y-y_0}{b}=\frac{z-z_0}{c}$$

Line through $(x^0, y^0, z^0))$ and (x^1, y^1, z^1):

$$\frac{x-x_0}{x_1-x_0}=\frac{y-y_0}{y_1-y_0}=\frac{z-z_0}{z_1-z_0}$$

This line is parallel to the vector $(x^1 - x^0, y^1 - y^0, z^1 - z^0)$.

DISTANCES

The distance between two points in space is the length of the line segment joining them. The distance between the points (x^0, y^0, z^0) and (x^1, y^1, z^1) is is

$$\sqrt{(x_1-x_0)^2+(y_1-y_0)^2+(z_1-z_0)^2}$$

The point k% of the way from $P^0 = x^0, y^0, z^0)$ to $P^1 = (x^1, y^1, z^1)$ is

$$\left(\frac{kx_1+(100-k)x_2}{100}, \frac{ky_1+(100-k)y_2}{100}, \frac{kz_1+(100-k)z_2}{100}\right)$$

(The same formula works also in oblique coordinates.) This point divides the segment P^1P^0 in the ratio k: $(100 - k)$. As a particular case, the midpoint of P^0P^1 is given by

$$(½(x^1+x^2), ½(y^1+y^2)+½(z^1+z^2)).$$

The Distance between the point (x^0, y^0, z^0) and the line through (x^1, y^1, z^1) in direction (a, b, c):

$$\sqrt{\frac{\begin{vmatrix} y_0-y_1 & z_0-z_1 \\ b & c \end{vmatrix}^2+\begin{vmatrix} z_0-z_1 & x_0-x_1 \\ c & a \end{vmatrix}^2+\begin{vmatrix} x_0-x_1 & y_0-y_1 \\ a & b \end{vmatrix}^2}{a^2+b^2+c^2}}$$

Distance between the line through (x^0, y^0, z^0) in direction (a^0, b^0, c^0) and the line through (x^1, y^1, z^1)in direction (a^1, b^1, c^1):

$$\left|\frac{\begin{vmatrix} x_1-x_0 & y_1-y_0 & z_1-z_0 \\ a_0 & b_0 & c_0 \\ a_1 & b_1 & c_1 \end{vmatrix}}{\sqrt{\begin{vmatrix} b_0 & c_0 \\ b_1 & c_1 \end{vmatrix}^2+\begin{vmatrix} c_0 & a_0 \\ a_1 & b_1 \end{vmatrix}^2+\begin{vmatrix} a_0 & b_0 \\ a_1 & b_1 \end{vmatrix}^2}}\right|$$

ANGLES

Angle between lines with directions (x^0,y^0,z^0) and (x^1,y^1,z^1):

$$\arccos \frac{a_0a_1+b_0b_1+c_0c_1}{\sqrt{a_0^2+b_0^2+c_0^2}\sqrt{a_1^2+b_1^2+c_1^2}}$$

In particular, the two lines are parallel when $a^0:b^0:c^0 = a^1:b^1:c1^2$, and perpendicular when $a^0\, a^1 + b^0\, b^1 + c^0\, c^1 = 0$.

Angle between lines with direction angles α^0, β_0, γ_0 and α^1, β_1, γ_1:

$$\arccos(\cos\alpha^0 \cos\alpha^1 + \cos\beta_0 \cos\beta_1 + \cos\gamma_0 \cos\gamma_1)$$

CONCURRENCE, COPLANARITY, PARALLELISM

Two lines specified by point and direction are coplanar if and only if the determinant in the numerator of (1) is zero. In this case they are concurrent (if the denominator is nonzero) or parallel (if the denominator is zero).

Three lines with directions (a^0,b^0,c^0), (a^1,b^1,c^1) and (a^2,b^2,c^2) are parallel to a common plane if and only if

$$\begin{vmatrix} a_0 & b_0 & c_0 \\ a_1 & b_1 & c_1 \\ a_2 & b_2 & c_2 \end{vmatrix} = 0.$$

POLYHEDRA

For any polyhedron topologically equivalent to a sphere---in particular, for any convex polyhedron---the Euler formula holds: $v - e + f = 2$, where v is the number of vertices, e is the number of edges, and f is the number of faces.

Many common polyhedra are particular cases of cylinders or cones. A cylinder with a polygonal base (directrix) is called a prism. A cone with a polygonal base is called a pyramid. A frustum of a cone with a polygonal base is called a trucated pyramid. A prism whose base is a parallelogram is a parallelepiped. The volume of a parallelepiped with one vertex at the origin and adjacent vertices at (x^1,y^1,z^1), (x^2,y^2,z^2) and (x^3,y^3,z^3) is given by

$$\frac{1}{6}\begin{vmatrix} x_1 & y_1 & z_1 \\ x_2 & y_2 & z_2 \\ x_3 & y_3 & z_3 \end{vmatrix}$$

The rectangular parallelepiped is a particular case: all its faces are rectangles. If the side lengths are a, b, c, the volume is abc, the total area is $2(ab + ac + bc)$, and each diagonal has length $\sqrt{a^2+b^2+c^2}$. When $a = b = c$ we get the cube:

A pyramid whose base is a triangle is a tetrahedron. The volume of a tetrahedon with one vertex at the origin and the other vertices at (x^1,y^1,z^1), (x^2,y^2,z^2) and (x^3,y^3,z^3) is given by

$$\frac{1}{6}\begin{vmatrix} x_1 & y_1 & z_1 \\ x_2 & y_2 & z_2 \\ x_3 & y_3 & z_3 \end{vmatrix}$$

In a tetrahedron with vertices P^0, P^1, P^2, P^3, let d_{ij} be the distance (edge length) from P^i to P_j. Form the determinants

$$\Delta = \begin{vmatrix} 0 & 1 & 1 & 1 & 1 \\ 1 & 0 & d_{01}^2 & d_{02}^2 & d_{03}^2 \\ 1 & d_{01}^2 & 0 & d_{12}^2 & d_{13}^2 \\ 1 & d_{02}^2 & d_{12}^2 & 0 & d_{23}^2 \\ 1 & d_{03}^2 & d_{13}^2 & d_{23}^2 & 0 \end{vmatrix} \text{ and } \Gamma = \begin{vmatrix} 0 & d_{01}^2 & d_{02}^2 & d_{03}^2 \\ d_{01}^2 & 0 & d_{12}^2 & d_{13}^2 \\ d_{02}^2 & d_{12}^2 & 0 & d_{23}^2 \\ d_{03}^2 & d_{13}^2 & d_{23}^2 & 0 \end{vmatrix}.$$

Then tne volume of the tetrahedron is $\sqrt{|\Delta|/288}$, and the radius of the circumscribed sphere is $\sqrt{\frac{1}{2}|\Gamma/\Delta|}$.

REGULAR POLYHEDRA

Figure shows the five regular polyhedra, or Platonic solids.

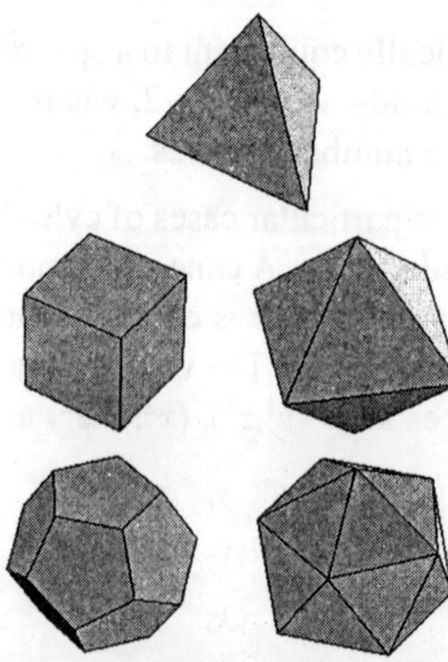

Fig. *The Platonic solids. Top:* The Tetrahedron (self-dual). Middle: the cube and the Octahedron (dual to one another). Bottom: the Dodecahedron and the Icosahedron (Dual to one another).

In the following tables and formulas, a is the length of an edge, θ the dihedral angle at each edge, R the radius of the circumscribed sphere, r the radius of the inscribed sphere, V the volume, S the total area, v the total number of vertices, e the total number of edges, f the total number of faces, p the number of edges in a face, and q the number of edges meeting at a vertex.

$$\theta = 2 \arcsin \frac{\cos(180°/q)}{\sin(180°/p)}$$

$$\frac{R}{a} = \frac{\frac{1}{2}\sin(180°/q)}{\sin(180°/p)\cos\frac{1}{2}\theta}$$

$$\frac{R}{r} = \tan\frac{180°}{p}\tan\frac{180°}{q}$$

$$\frac{S}{a^2} = \frac{fp}{4}\cot\frac{180°}{p}$$

$$V = \frac{1}{3}rS$$

Name	v	e	f	p	q	**sin θ**	**θ**
Regular tetrahedron	4	6	4	3	3	$2\sqrt{2}/3$	70°31′44″
Cube	8	12	6	4	3	1	90°
Regular octahedron	6	12	8	3	4	$2\sqrt{2}/3$	109°28′16″
Regular dodecahedron	20	30	12	5	3	$2\sqrt{5}$	116°33′54″
Regular icosahedron	12	30	20	3	5	2/3	138°11′23″

Name	*R/a*		*r/a*	
Tetrahedron	$\sqrt{6}/4$	0.612372	$\sqrt{6}/12$	0.204124
Cube	$\sqrt{3}/2$	0.866025	$\frac{1}{2}$	0.5
Octahedron	$\sqrt{2}/2$	0.707107	$\sqrt{6}/6$	0.408248
Dodecahedron	$\frac{1}{4}\left(\sqrt{15}+\sqrt{3}\right)$	1.401259	$\frac{1}{20}\sqrt{250+110\sqrt{5}}$	1.1113516
Icosahedron	$\frac{1}{4}\sqrt{10+2\sqrt{5}}$	0.951057	$\frac{1}{12}\sqrt{42+18\sqrt{5}}$	0.755761

Name	S/a^2		V/a^3	
Tetrahedron	$\sqrt{3}$	0.73205	$\sqrt{2}/12$	0.117851
Cube	6	6-	1	1-
Octahedron	$2\sqrt{3}$	3.46410	$\sqrt{2}/3$	0.471405
Dodecahedron	$3\sqrt{25+10\sqrt{5}}$	20.64573	$\frac{1}{4}\sqrt{15+7\sqrt{5}}$	7.663119
Icosahedron	$5\sqrt{3}$	8.66025	$\frac{5}{12}(3+\sqrt{5})$	2.181695

CYLINDERS

Given a line L and a curve C in a plane P, the cylinder with generator L and directrix C is the surface obtained by moving L parallel to itself, so that a point of L is always on C.

If L is parallel to the z-axis, the surface's implicit equation does not involve the variable z. Conversely, any implicit equation that does not involve one of the variables (or that can be brought to that form by a change of coordinates) represents a cylinder.

If C is a simple closed curve, we also apply the word cylinder to the solid enclosed by the surface generated in this way.

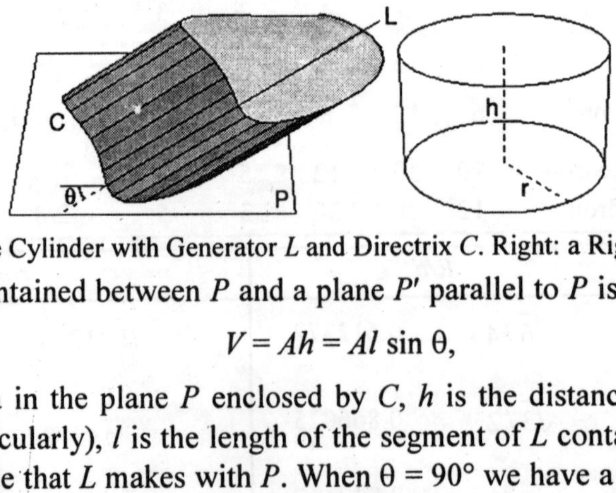

Fig. *Left:* an Oblique Cylinder with Generator L and Directrix C. Right: a Right Circular Cylinder.

The volume contained between P and a plane P' parallel to P is

$$V = Ah = Al \sin \theta,$$

where A is the area in the plane P enclosed by C, h is the distance between P and P' (measured perpendicularly), l is the length of the segment of L contained between P and P', and θ is the angle that L makes with P. When $\theta = 90°$ we have a right cylinder, and $h = l$. For a right cylinder, the lateral area between P and P' is hs, where s is the length (circumference) of C.

The most important particular case is the right circular cylinder (often simply called a cylinder). If r is the radius of the base and h is the altitude, the lateral area is

$$2\pi rh,$$

the total area is $2\pi r(r + h)$, and the volume is

$$\pi r^2 h.$$

The implicit equation of this surface can be written

$$x^2 + y^2 = r^2;$$

CONES

Given a curve C in a plane P and a point O not in P, the cone with vertex O and directrix C is the surface obtained as the union of all lines that join O with points of C.

If O is the origin and the surface is given implicity by an algebraic equation, that equation is homogeneous (all terms have the same total degree in the variables). Conversely, any homogeneous implicit equation (or one that can be made homogeneous by a change of coordinates) represents a cone.

If C is a simple closed curve, we also apply the word cone to the solid enclosed by the surface generated in this way.

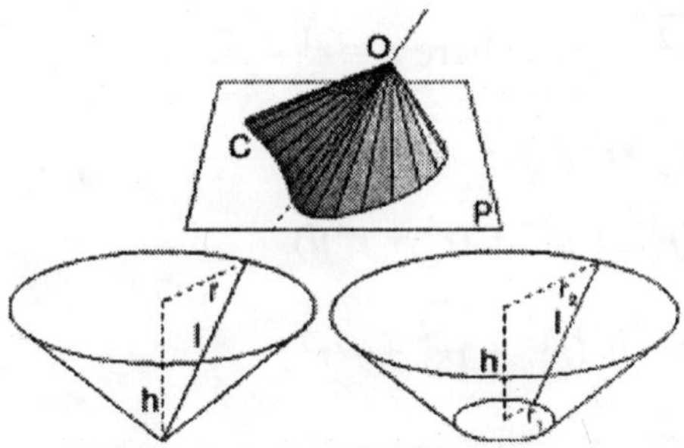

Fig. *Top:* A Cone with Vertex *O* and Directrix *C*. Bottom left: a right Circular Cone. Bottom right: *A* Frustum of the Latter.

The volume contained between P and the vertex O is

$$V = \frac{1}{3} Ah,$$

where A is the area in the plane P enclosed by C and h is the from O and P (measured perpendicularly).

The solid contained between P and a plane P' parallel to P (on the same side of the vertex) is called a frustum. Its volume is

$$V = \frac{1}{3} h(A + A' + \sqrt{AA'}),$$

where A and A' are the areas enclosed by the sections of the cone by P and P' (often called the bases of the frustum).

The most important particular case of a cone is the right circular cone (often simply called a cone).

If r is the radius of the base, h is the altitude, and l is the length between the vertex and a point on the base circle, we have the following relations:

l	$\sqrt{r^2 + h^2}$
Lateral area	$\pi rl = \pi r\sqrt{r^2 + h^2}$
Total area	$\pi r(l + r) = \pi r(r + \sqrt{r^2 + h^2})$
Volume	$\frac{1}{3}\pi r^2 h$

The implicit equation of this surface can be written

$$x^2 + y^2 = z^2;$$

For a frustum of right circular cone we have:

l	$\sqrt{r^2 + h^2}$, where $r = r^1 - y^2$
Lateral area	$\pi(r^1 + r^2)l\sqrt{r^2 + h^2}$
Total area	$\pi(r^{12} + r^{22} + (r^1 + r^2)l)$
Volume	$\frac{1}{3}\pi r^2 h\ (r^{12} + r^{22} + r^1 r^2)$

SPHERES

The set of points in space whose distance to a fixed point (the center) is a fixed positive number (the radius) is a sphere. A circle of radius r and center (x^0, y^0, z^0) has equation

$$(x - x^0) + (y - y)^2 + (z - z^0)^2 = r^2,$$

or

$$x^2 + y^2 + z^2 - 2xx^0 - 2yy^0 - 2zz^0 + x^{02} + y^{02} + z^{02} - r^2 = 0.$$

Conversely, an equation of the form

$$x^2 + y^2 + z^2 + 2dx + 2ey + 2fz + g = 0$$

defines a sphere if $d^2 + e^2 + f^2 > g$; the center is $(-d, -e, -f)$ and the radius is $\sqrt{d^2 + e^2 + f^2 - g}$.

Four points not on the same plane determine a unique sphere. If the points have coordinates (x^1, y^1, z^1), (x^2, y^2, z^2), (x^3, y^3, z^3) and (x^4, x^4, z^4), the equation of the sphere is

$$\begin{vmatrix} x^2+y^2+z^2 & x & y & z & 1 \\ x_2^1+y_1^2+z_2^2 & x_1 & y_1 & z_1 & 1 \\ x_2^2+y_2^2+z_2^2 & x_2 & y_2 & z_2 & 1 \\ x_3^3+y_3^2+z_3^2 & x_3 & y_3 & z_3 & 1 \\ x_4^2+y_4^2+z_4^2 & x_4 & y_4 & z_4 & 1 \end{vmatrix} = 0.$$

Given two points $P^1 = (x^1, y^1, z^1)$ and $P^2 = (x^2, y^2, z^2)$, there is a unique sphere whose diameter is P^1P^2; its equation is

$$(x - x^1)(x - x^2) + (y - y^1)(y - y^2) + (z - z^1)(z - z^2) = 0.$$

The area of a sphere of radius r is $4\pi r^2$, and the volume is $\frac{4}{3}\pi r^3$.

The area of a spherical polygon (that is, of a polygon on the sphere whose sides are arcs of great circles) is

$$S = \left(\sum_{i=1}^{n} \theta_i - (n-2)\pi\right) r^2,$$

where r is the radius of the sphere, n is the number of vertices, and the θ_i are the internal angles of the polygons in radians. In particular, the sum of the angles of a spherical triangle is always greater than $\pi = 180°$, and the excess is proportional to the area.

SPHERICAL CAP

Let the radius be r. The area of the curved region is $2\pi rh = \pi p^2$. The volume of the cap is $\frac{1}{3}\pi h^2(3r - h) = \frac{1}{8}\pi h(3a^2 + h^2)$.

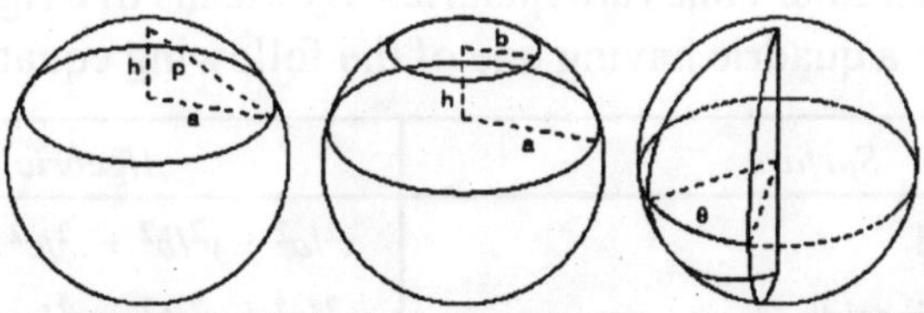

Fig. Left: a Spherical Cap. Middle: a Spherical Zone (of Two Bases). Right: a Spherical Segment.

Spherical Zone (of Two Bases)

Let the radius be r. The area of the curved region is $2\pi rh$. The volume of the zone is $\frac{1}{8}\pi h(3a^2 + 3b^2 + h^2)$.

Spherical Segment and Lune

Let the radius be r. The area of the curved region (lune) is $2r^2\theta$, the angle being measured in radians. The volume of the segment is $\frac{2}{3}r^3\theta$.

SURFACES OF REVOLUTION: THE TORUS

A surface of revolution is formed by the rotation of a planar curve C about an axis in the plane of the curve and not cutting the curve. The Pappus--Guldinus theorem says that:

1. The area of the surface of revolution on a curve C is equal to the product of the length of C and the length of the path traced by the centroid of C (which is 2pthe distance from this centroid to the axis of revolution).
2. The volume bounded by the surface of revolution on a simple closed curve C is equal to the product of the area bounded by C and the length of the path traced by the centroid of the area bounded by C.

When C is a circle, the surface obtained is a circular torus or torus of revolution. Let r be the radius of the revolving circle and let R be the distance from its center to the axis of rotation. The area of the torus is $4\pi Rr^2$, and its volume is $2\pi Rr^2$.

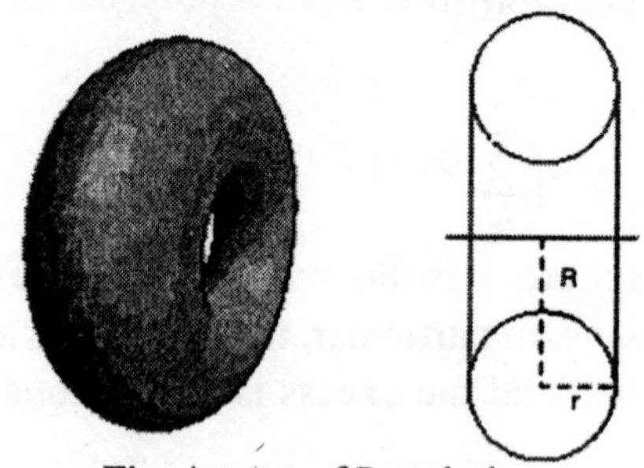

Fig. A torus of Revolution.

QUADRICS

A surface defined by an algebraic equation of degree two is called a quadric. Spheres, circular cylinders, and circular cones are quadrics. By means of a rigid motion, any quadric can be transformed into a quadric having one of the following equations (where $a,b,c \neq 0$):

S. No.	*Surfare*	*Algebric Equation*
1.	Real ellipsoid	$x^2/a^2 + y^2/b^2 + z^2/c^2 = 1$
2.	Imaginary ellipsoid	$x^2/a^2 + y^2/b^2 + z^2/c^2 = -1$
3.	Hyperboloid of one sheet	$x^2/a^2 + y^2/b^2 - z^2/c^2 = 1$
4.	Hyperboloid of two sheets	$x^2/a^2 + y^2/b^2 - z^2/c^2 = -1$
5.	Real quadric cone	$x^2/a^2 + y^2/b^2 - z^2/c^2 = 0$
6.	Imaginary quadric cone	$x^2/a^2 + y^2/b^2 + z^2/c^2 = 0$
7.	Elliptic paraboloid	$x^2/a^2 + y^2/b^2 + 2z = 0$
8.	Hyperbolic paraboloid	$x^2/a^2 - y^2/b^2 + 2z = 0$
9.	Real elliptic cylinder	$x^2/a^2 + y^2/b^2 = 1$
10.	Imaginary elliptic cylinder	$x^2/a^2 + y^2/b^2 = -1$

S. No.	*Surfare*	*Algebric Equation*
11.	Hyperbolic cylinder	$x^2/a^2 - y^2/b^2 = 1$
12.	Real intersecting planes	$x^2/a^2 - y^2/b^2 = 0$
13.	Imaginary intersecting planes	$x^2 + 2y = 0$
14.	Parabolic cylinder	$x^2 + 2y = 0$
15.	Real parallel planes	$x^2 = 1$
16.	Imaginary parallel planes	$x^2 = -1$
17.	Coincident planes	$x^2 = 0$

THE FIVE NONDEGENERATE REAL QUADRICS

Fig. The Ellipsoid (1).

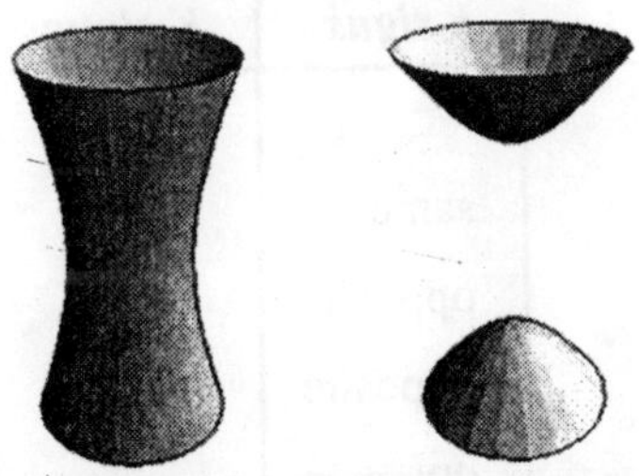

Fig. Left: Hyperboloid of One Sheet (3). Right: Hyperboloid of Two Sheets (4).

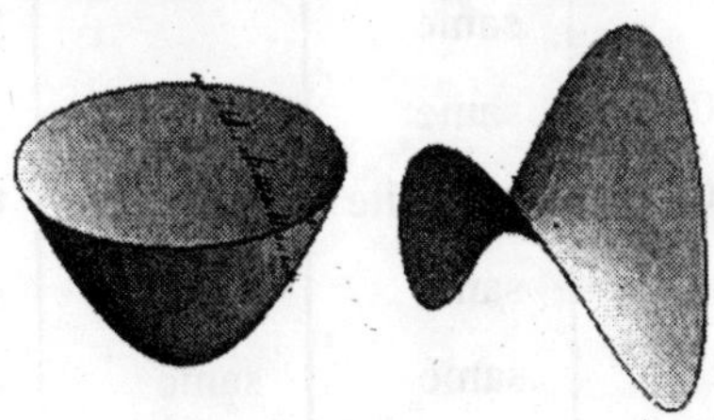

Fig. Left: Elliptic Paraboloid (7). Right: Hyperbolic Paraboloid (8).

Surfaces with equations (9) –(17) are cylinders over the planes curves of the same equation. Equations (2), (6), (10), (16), have no real solutions, so they do not describe surfaces in real three-dimensional space.

A surface with equation (5) can be regarded as a cone over a conic C (any ellipse, parabola or hyperbola can be taken as the directrix; there is a two-parameter family of essentially distinct cones over it, determined by the position of the vertex with respect to C).

The surfaces with equations (1) --(6) are central quadrics; in the form given, the center is at the origin. The quantities a, b, c are the semiaxes.

The volume of the ellipsoid with semiaxes *a, b, c* is $\frac{4}{3}\pi abc$. When two of the semiaxes are the same, we can also write the area of the ellipsoid in closed form. Suppose $b = c$, so the ellipsoid $x^2/a^2 + (y^2 + z^2)/b^2 = 1$ is the surface of revolution obtained by rotating the ellipse $x^2/a^2 + y^2/b^2 = 1$ around the x-axis. Its area is

$$2\pi b^2 + \frac{2\pi a^2 b}{\sqrt{a^2 - b^2}} \arcsin \frac{\sqrt{a^2 - b^2}}{a} \quad 2\pi b^2 + \frac{\pi a^2 b}{\sqrt{b^2 - a^2}} \log \frac{b + \sqrt{b^2 - a^2}}{b - \sqrt{b^2 - a^2}}.$$

The two quantities are equal, but only one avoids complex numbers, depending on whether $a>b$ or $a<b$. When $a>b$, we have a prolate spheroid, that is, an ellipse rotated around its major axis; when $a<b$ we have an oblate spheroid, which is an ellipse rotated around its minor axis.

Given a general quadratic equation in three variables,

$$ax^2 + by^2 + cz^2 + 2fyz + 2gzx + 2hxy + 2px + 2qy + 2rz + d = 0,$$

one can find out the type of conic it determines by consulting the following table:

ρ_2	ρ_4	Δ	*k signs*	*K signs*	*Type of quadric*
3	4	<0			real ellipsoid
3	4	>0	same		imaginary ellipsoid
3	4	>0	opposite		hyperboloid of one sheet
3	4	<0	opposite		hyperboloid of two sheets
3	3		opposite		real quadric cone
3	3		same		imaginary quadric cone
2	4	<0	same		elliptic paraboloid
2	4	>0	opposite		hyperbolic paraboloid
2	3		same	opposite	real elliptic cylinder
2	3		same	same	imaginary elliptic cylinder
2	3		opposite		hyperbolic cylinder
2	2		opposite		real intersecting planes
2	2		same		imaginary intersecting planes
1	3				parabolic cylinder
1	2			opposite	real parallel planes
1	2			same	imaginary parallel planes
1	1				coincident planes

The columns have the following meaning. Let

$$e = \begin{bmatrix} a & h & g \\ h & b & f \\ g & f & c \end{bmatrix} \text{ and } E = \begin{bmatrix} a & h & g & p \\ h & b & f & q \\ g & f & c & f \\ p & q & r & d \end{bmatrix};$$

let ρ^3 and r^4 be the ranks of *e* and *E*, and let Δ be the determinant of *E*. The column "*k* signs" refers to the nonzero eigenvalues of e, that is, the roots of

$$\begin{vmatrix} a-x & h & g \\ h & b-x & f \\ g & f & c-x \end{vmatrix} = 0'$$

if all nonzero eigenvalues have the same sign, choose "same", otherwise "opposite". Similarly, "*K* signs" refers to the sign of the nonzero eigenvalues of *E*.

Encyclopaedia of

SOLID GEOMETRY

Encyclopaedia of
SOLID GEOMETRY

Volume 1

R.K. Pandey

ANMOL PUBLICATIONS PVT. LTD.
NEW DELHI - 110 002 (INDIA)

ANMOL PUBLICATIONS PVT. LTD.
H.O.: 4374/4B, Ansari Road, Daryaganj,
New Delhi-110 002 (India)
Ph.: 23278000, 23261597
B.O.: No. 1015, Ist Main Road, BSK IIIrd Stage
IIIrd Phase, IIIrd Block,
Bangalore - 560 085 (India)
Visit us at: www.anmolpublications.com

Encyclopaedia of Solid Geometry

© Reserved

First Edition, 2009
ISBN 978-81-261-4010-7 (Set)

PRINTED IN INDIA

Printed at Mehra Offset Press, Delhi.

Contents

Preface

The Encyclopaedia of Solid Geometry is so presented that students will gain a knowledge of the Properties of Planes and Lines in space; Polyhedrons, Cones, Cylinders, and Spheres; further practice in the processes of logic which were begun in Plane Geometry; a realization that they must understand solid geometry if they wish to comprehend nature and art and the fascinating and diversified developments of modern means of Communication, Transportation, Construction, and Fabrication; and skills and attitudes that will be helpful as a part of the training essential in meeting his own personal problems of everyday living.

Author

List of Symbols

$\pm$	$\not\subset$	$\subset$	$\supset$	$\subseteq$	$\in$	$\notin$	∂
$\neq$	$\equiv$	$\approx$	$\angle$	∇	Σ	Π	Ω
ϕ	α	β	χ	ε	γ	η	λ
μ	ν	$\perp$	π	θ	ρ	σ	τ
$\Re$	ω	ξ	ψ	ζ	Υ	$\exists$	$\int$
$\iint$	$\iiint$	$\oint$	$\times$	$\sqrt{}$	$\therefore$	$\because$	Δ
$\geq$	$\leq$	$\Rightarrow$	$\Downarrow$	$\Leftrightarrow$	∞	$\propto$	$\mathbb{R}$
$\mathbb{Z}$	$\mathbb{N}$	$\mathbb{C}$	$\mathbb{Q}$				

Chapter 1

Introduction to Solid Geometry

In mathematics, solid geometry was the traditional name for the geometry of three-dimensional Euclidean space — for practical purposes the kind of space we live in. It was developed following the development of plane geometry. Stereometry deals with the measurements of volumes of various solid figures: cylinder, circular cone, truncated cone, sphere, prisms, blades, wine casks.

The Pythagoreans had dealt with the sphere and regular solids, but the pyramid, prism, cone and cylinder were not studied until the Platonists. Eudoxus established their measurement, proving the pyramid and cone to have one-third the volume of a prism and cylinder on the same base and of the same height, and was probably the discoverer of a proof that the volume of a sphere is proportional to the cube of its radius.

THE ORIGINS OF GEOMETRY

In the beginning *geometry* was a collection of rules for computing lengths, areas, and volumes. Many were crude approximations derived by trial and error. This body of knowledge developed and used in construction, navigation, and surveying by the Babylonians and Egyptians, was passed to the Greeks. The Greek historian Herodotus credits the Egyptians with having originated the subject, but there is much evidence that the Babylonians, the Hindu civilization, and the Chinese knew much of what was passed along to the Egyptians.

The Babylonians of 2,000 to 1,600 B.C knew much about navigation and astronomy, which required knowledge of *geometry*. They also considered the circumference of the circle to be three times the diameter. Of course, this would make $\pi = 3$ – a small problem. This value for π carried along to later times. The Roman architect Vitruvius took $\pi = 3$. Prior to this it seems that the Chinese mathematicians had taken the same value for π. This value for π was sanctified by the ancient Jewish civilization and sanctioned in the scriptures. In *I Kings* 7:23 we find:

He then made the sea of cast metal: it was round in shape, the diameter from rim to rim being ten cubits: it stood five cubits high, and it took a line thirty cubits long to go round it.—The New English Bible

Rabbi Nehemiah attempted to change the value of π to $\frac{22}{7}$ but was rejected. By 1,800 B.C. the Egyptians, according to the *Rhind* papyrus, had the approximation $\pi \sim \left(\frac{16}{9}\right)^2 \sim 3.1604$ It has been popularly expressed that in the 1920's the Indiana legislature passed a law mandating that $\pi = \frac{22}{7}$ or even 3. The bill appears to state that it is possible to square the circle. It apparently was tabled and never passed due to the fortuituous chance that a math professor from Purdue happened to be in the Senate that day for another purpose. We assume that it is still tabled, never to be brought up for debate again. In 1789 Johann Lambert proved that π is an irrational number, and in 1882 F. Lindemann proved that π is *transcendental*, *i.e.*, it is not the solution to any algebraic equation with rational mathfficients.

The ancient knowledge of *geometry* was passed on to the Greeks. They seemed to be blessed with an inclination toward speculative thinking and the leisure to pursue this inclination. They insisted that geometric statements be established by deductive reasoning rather than trial and error. This began with Thales of Milete.

He was familiar with the computations, right or wrong, handed down from Egyptian and Babylonian mathematics. In determining which of the computations were correct, he developed the first logical geometry. This orderly development of theorems by *proof* was the distinctive characteristic of Greek mathematics and was *new*.

This *new mathematics* of Thales was continued over the next two centuries by Pythagoras and his disciples. The Pythagoreans, as a religious sect, believed that the elevation of the soul and union with God were achieved by the study of *music* and *mathematics*. Nonetheless, they developed a large body of mathematics by using the deductive method.

Their foundation of plane geometry was brought to a conclusion around 400 B.C. in the *Elements* by the mathematician Hippocrates. This treatise has been lost, but many historians agree that it probably covered most of *Books I-IV* of Euclid's *Elements*, which appeared about a century later, *circa* 300 B.C.

Euclid was a disciple of the Platonic school. Around 300 B.C. he produced the definitive treatment of Greek geometry and number theory in his thirteen-volume *Elements*. In compiling this masterpiece Euclid built on the experience and achievements of his predecessors in preceding centuries: on the Pythagoreans for Books I-IV, VII, and IX, on Archytas for Book VIII, on Eudoxus for Books V, VI, and XII.

So completely did Euclid's work supersede earlier attempts at presenting geometry that few traces remain of these efforts. It's a pity that Euclid's heirs have not been able to collect royalties on his work, for he is the most widely read author in the history of mankind.

His approach to geometry has dominated the teaching of the subject for over two thousand years. Moreover, the axiomatic method used by Euclid is the prototype for all of what we now call *pure mathematics*. It is pure in the sense of *pure thought*: no physical experiments need be performed to verify that the statements are correct – only the reasoning in the demonstrations need be checked.

In this treatise, he organized a large body of known mathematics, including discoveries of his own, into the first formal system of mathematics.

This *formalness* was exhibited by the fact that the *Elements* began with an explicit statement of assumptions called *axioms* or *postulates*, together with definitions. The other statements—theorems lemmζ, corollaries – were then shown to follow logically from these axioms and definitions. *Books I-IV, VII,* and *IX* of the work dealt primarily with mathematics which we now classify as geometry, and the entire structure is what we now call *Euclidean geometry*.

Euclidean geometry was certainly conceived by its creators as an idealization of physical geometry. The entities of the mathematical system are concepts, suggested by, or abstracted from, physical experience but differing from physical entities as an idea of an object differs from the object itself. However, a remarkable correlation existed between the two systems.

The angle sum of a mathematical triangle was stated to be 180°, if one measured the angles of a physical triangle the angle sum did indeed seem to be 180°, and so it went for a multitude of other relations.

Because of this agreement between theory and practice, it is not surprising that many writers came to think of Euclid's axioms as *self evident truths*. Centuries later, the philosopher Immanuel Kant even took the position that the human mind is essentially *Euclidean* and can only conceive of space in Euclidean terms. Thus, almost from its inception, Euclidean geometry had something of the character of dogma.

EUCLID BASED HIS GEOMETRY ON FIVE FUNDAMENTAL ASSUMPTIONS

Postulate I

For every point P and for every point Q not equal to P there exists a unique line that passes through P and Q.

Postulate II

For every segment AB and for every segment CD there exists a unique point E such that B is between A and E and segment CD is congruent to segment BE.

Postulate III

For every point O and every point A not equal to O there exists a circle with center O and radius OA.

Postulate IV

All right angles are congruent to each other.

Before we study the Fifth Postulate, let me say a few words about his definitions. Euclid's methods are imperfect by *modern* standards. He attempted to define everything in terms of a more familiar notion, sometimes creating more confusion than he removed. As an example:

A *point* is that which has no part. A *line* is breadthless length. A straight line is a line which lies evenly with the points on itself. A *plane angle* is the inclination to one another of two lines which meet. When a straight line set upon a straight line makes adjacent angles equal to one another, each of the equal angles is a *right angle*.

Euclid did not define *length, distance, inclination, or set upon*. Once having made the above definitions, Euclid never used them. He used instead the *rules of interaction* between the defined objects as set forth in his five postulates and other postulates that he implicitly assumed but did not state.

Postulate V

If a straight line falling on two straight lines makes the interior angles on the same side less than two right angles, the two straight lines, if produced indefinitely, meet on that side on which the angles are less than two right angles.

No one seemed to like this Fifth Postulate, possibly not even Euclid himself—he did not use it. The reason that this statement seems out of place is that the first 4 postulates seem to follow from experience—try to draw more than one line through 2 different points. The Fifth Postulate is *unintuitive*. It does come from the study of parallel lines, though. Equivalent to this postulate is :

PLAYFAIR'S POSTULATE

Given a line and a point not on that line, there exists one and only one line through that point parallel to the given line.

Euclid's Fifth Postulate seemed to be too burdensome. It should follow from the other axioms. Since it is not intuitive, we should be able to prove it as a theorem. *We should be able to prove that it is* dependent *in this Axiom system*. If we have a set of axioms A_1, A_2,..,

A_n for our mathematical system and we can prove that Axiom A_n is derivable, or provable, from the other axioms, then is indeed redundant. In some sense we are looking for a *basis* for this *mathematical system*. Unlike vector spaces and linear algebra, there is not a unique number of elements in this *basis*, for it includes the axioms, definitions, and the rules of logic that you use.

Many people have tried to prove the Fifth Postulate. The first known attempt to *prove* Euclid V, as it became known, was by Posidionius(1st century B.C.). He proposed to replaced the definition of parallel lines (those that do not intersect) by defining them as coplanar lines that are everywhere equidistant from one another. It turns out that without Euclid V you cannot prove that such lines exist. It is true that such a statement that parallel lines are equidistant from one another is equivalent to Euclid V.

PTOLEMY USED THE FOLLOWING ASSUMPTION

For every line l and every point P not on l, there exists at most one line m through P such that m is parallel to l.

We will show at a later date that this statement is equivalent to Euclid V, and therefore this did not constitute a proof of Euclid V.

Proclus also attempted to prove Euclid V. His argument used a limiting process. He retained all of Euclid's definitions, all of his assumptions except Euclid V, and hence all of his propositions which did not depend on Euclid V. His plan was:

- To prove on this basis that a line which meets one of two parallels also meets the other.
- To deduce Euclid V from this proposition.

His handling of step (2) was correctly handled. The argument in step (1) runs substantially as follows. Let g and h be parallel lines and let another line k meet h in P. From Q, a point of k situated between g and h, drop a perpendicular to h. As Q recedes indefinitely far from P, its distance QR from h increases and exceeds any value, however great. In particular, QR will exceed the distance between g and h. For some position of Q, then, QR will equal the distance between g and h. When this occurs, k will meet g.

There are a number of assumptions here which go beyond those found in Euclid. I will mention only the following two:

- The distance from one of two intersecting lines to the other increases beyond all bounds as we recede from their common point,
- The distance between two parallels never exceeds some finite value.

The first of the two assumptions is not a grave error on the part of Proclus, for it can be proved as a theorem on the basis of what he assumed from Euclid. Unfortunately for Proclus, his second assumption is equivalent to Euclid V.

Nasiraddin, John Wallis, Legendre, Wolfgang Bolyai, Girolamo Saccheri, Johann Heinrich Lambert, and many others tried to prove Euclid V, and failed. In these failures there developed a goodly number of substitutes for Euclid V; *i.e.*, statements that were equivalent to the statement of Euclid V. The following is a list of some of these that are more common:

- Through a point not on a given line there passes not more than one parallel to the line.
- Two lines that are parallel to the same line are parallel to each other.
- A line that meets one of two parallels also meets the other.
- If two parallels are cut by a transversal, the alternate interior angles are equal.
- There exists a triangle whose angle-sum is two right angles.
- Parallel lines are equidistant from one another.
- There exist two parallel lines whose distance apart never exceeds some finite value.
- Similar triangles exist which are not congruent.
- Through any three noncollinear points there passes a circle.
- Through any point within any angle a line can be drawn which meets both sides of the angle.
- There exists a quadrilateral whose angle-sum is four right angles.
- Any two parallel lines have a common perpendicular.

It fell to three different mathematicians to independently show that Euclid V is not provable from the other axioms and what is derivable from them. These were Carl Friedrich Gauss, Jαnos Bolyai, and Nicolai Ivanovich Lobachevsky. Once these men broke the ice, the pieces of geometry began to fall into place.

More was learned about *non-Euclidean* geometries - - hyperbolic and elliptic or doubly elliptic. The elliptic geometry was studied by Riemann, gave rise to riemannian geometry and manifolds, which gave rise to differential geometry which gave rise to relativity theory, *et. al.*. This gives us something to anticipate as we learn more about geometry. We will spend our time studying *hyperbolic geometry*, for it lends itself to better study—not requiring major changes in the axiom system that we have chosen. We may have an opportunity to see that hyperbolic geometry is now lending itself to considerations in the latest research areas of mathematics.

SPHERICAL GEOMETRY

Whereas basic plane geometry is concerned with points and lines and their *interactions*, most of the early geometry of the Babylonians, Arabs, and Greeks was spherical geometry—the study of the Earth, idealized as a sphere. This early *science* was astronomy and the need to measure time accurately by the sun.

We have to come to some agreement on what *lines* and *line segments* on a sphere are going to be. A great circle on a sphere is the intersection of that sphere with a plane passing through the center of the sphere. Examples of great circles are the equator and the lines of constant longitude (such as the Greenwich Mean Time Line 0°).

These are good choices to play the role of lines on our sphere. For example, given any two non-antipodal points there is a unique great circle joining those two points. This is easy to see when you remember that three non-collinear points determine a plane. Take these two points and the center of the sphere as the three non-collinear points. The intersection of the plane determined by these points and the sphere is the great circle joining the two given points.

If great circles are to be lines, then we can measure the angle between two intersecting great circles as the angle formed by the intersection of the two defining planes with the plane tangent to the sphere at the point of intersection. With this definition of angle, we can form triangles on the sphere whose interior angle sum is greater than two right angles. In fact, we will show that the interior angle sum of *all* triangles on the sphere is greater than two right angles. We will be working with radian measure for most of the semester (to make many of the calculations neater). Thus, we are claiming that on a sphere the interior angle sums of any triangle is greater than π.

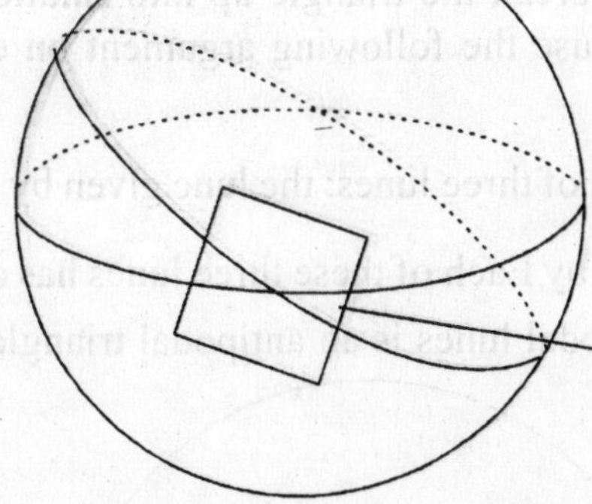

Fig. Angles in Spherical Geometry

In order to discuss how to measure the angle sums, we will first discuss the concept of the area of a triangle on the sphere. We will discuss this more fully later, but for the present time we will accept the following four properties of area for polygonal regions on the sphere:

- The sphere of radius R has area $4\pi R^2$;
- The area of a union of non-overlapping regions is the sum of their areas;
- The areas of congruent regions are equal;
- The ratio of the area enclosed by two great circles to the area of the whole sphere is the same as the ratio of the angle between them to 2π.

Theorem: *On a sphere of radius R, a triangle Δ ABC with interior angles α, β, and γ has area given by:*

arca (Δ *ABC*) = R^2 (α + β + γ − π)-

First, some terminology: a lune is one of the regions between two great circles and the antipodal points where they cross. Antipodal points or figures are those that correspond to opposite ends of a diameter of the sphere. A lune is given by the angle q at the center of the sphere between the two defining planes.

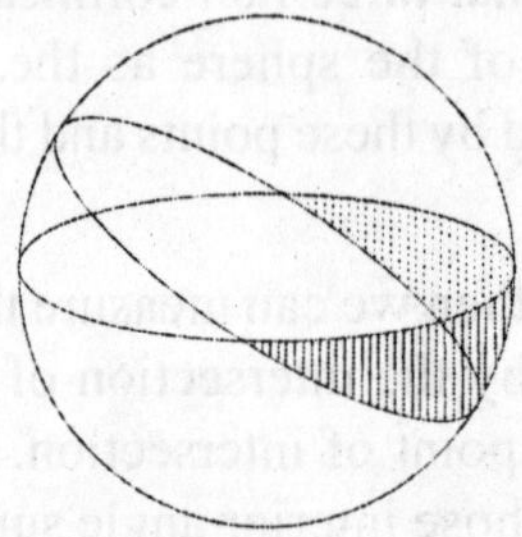

Fig. A Lune

Proof: From our assumptions above,

$$\text{Arca of the lunc} = \frac{\theta}{2\pi} \cdot 4\pi R^2 = 2\theta R^2$$

To make our proof easier, we may assume that the triangle lies in one hemisphere. If not you can use Assumption to break the triangle up into smaller triangles that do lie in one hemisphere. You can then use the following argument on each piece and then add them together.

A triangle is the intersection of three lunes: the lune given by $\overline{AB} \cap \overline{AC}$ the lune given by $\overline{AB} \cap \overline{AC}$ and the lune given by Each of these three lunes has an antipodal lune and the intersection of these three antipodal lunes is an antipodal triangle Δ *A'*A *B'*A *C'*A.

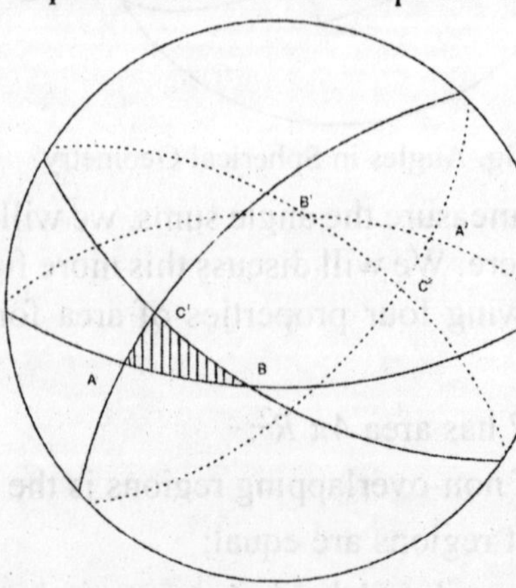

Fig. A Triangle and the Associated Lunes

We will denote the lunes by the angle that includes them, BCA, BAC, and ABC. Starting with the triangle and the lune ABC, add to them the lune ACB. This covers the triangle $\Delta A' B' C'$ three times. Now, add in the antipodal lune to lune CAB and the antipodal triangle $\Delta A' B' C'$. This covers the hemisphere determined by the line $\overline{BC}$ counting the triangle three times. So we get:

$$2\pi R^2 = 2\alpha R^2 + 2\gamma R^2 - 3. \text{ arca } (\Delta ABC) + \text{arca } (\Delta ABC)$$

Thus, we get our desired formula:

$$\text{arca } (\Delta ABC) = R^2 (\alpha + \beta + \gamma - \pi)$$

We call the value of $\alpha + \beta + \gamma - \pi$ the angle excess of the spherical triangle.

Note that since every triangle on the sphere has area, every triangle on the sphere has interior angle sum greater than If the radius of the sphere is very large and the triangle very small in area, then the angle excess will be almost 0.

Let Σ_R denote the sphere of radius R centered at the origin in R^3. In rectangular coordinates

$$\Sigma_R = \{(x, y, z) \mid x^2 + y^2 + z^2 = R^2\}$$

And in spherical coordinates

$$\Sigma R = \{(R \cos(\theta) \sin(\phi), R \sin(\theta) \sin(\phi), R \cos(\phi) \mid 0 \le \theta \le 2\pi, 0 \le \phi \le \pi\}.$$

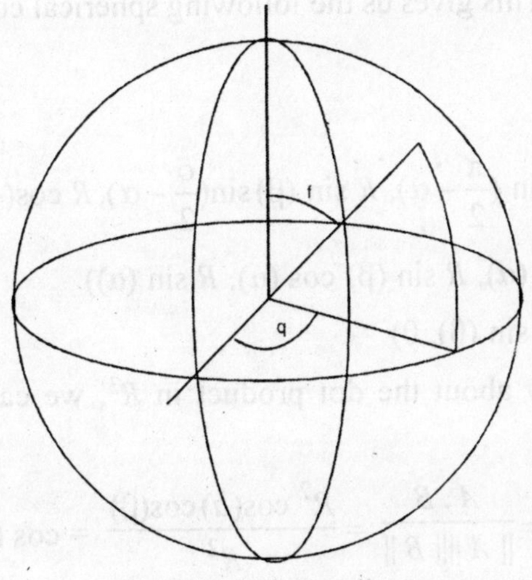

Fig. Spherical Coordinates

The lengthof a path between two points along a great circle is found by measuring the angle, in radians, made by the radii at the two endpoints and multiply it by the radius of the sphere, *i.e.* $R\theta$. Note that the length of a segment is unchanged when we rotate the sphere around some axis or reflect it through a great circle.

Spherical Pythagorean Theorem

Theorem: *For a right triangle Δ ABC on a sphere of radius R with a right angle at vertex C and sides of length a, b, and c, then*

$$\cos\left(\frac{C}{R}\right) = \cos\left(\frac{a}{R}\right)\cos\left(\frac{b}{R}\right)$$

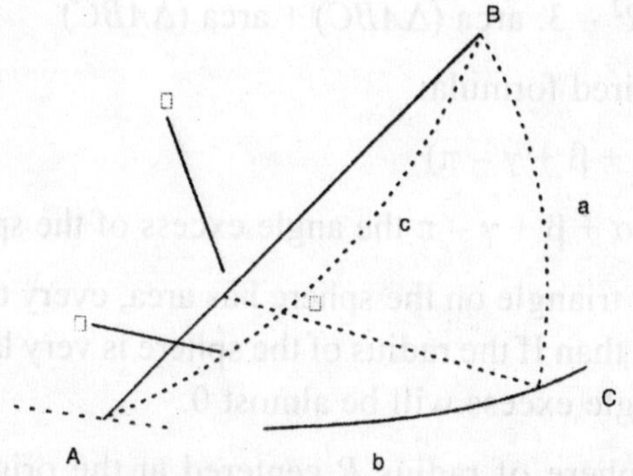

Fig. Spherical Right Triangle

Proof: Rotate the sphere so that the point A has coordinates $(R,0,0)$ and the point C lies in the xy-plane. This will make β and $\frac{\pi}{2}$ the angles determining the point B. Here β is the central angle determined by side AC, α is the angle determined by AC and γ is the angle determined by AB. This gives us the following spherical coordinates for the vertices of the triangle:

$$A = (R, 0, 0)$$

$$B = \left(R\cos(\beta)\sin\left(\frac{\pi}{2} - \alpha\right), R\sin(\beta)\sin\left(\frac{o}{2} - \alpha\right), R\cos\left(\frac{\pi}{2} - \alpha\right)\right)$$

$$= (R\cos(\beta)\cos(\alpha), R\sin(\beta)\cos(\alpha), R\sin(\alpha))$$

$$c = (R\cos(\beta), R\sin(\beta), 0)$$

Using what we know about the dot product in R^3, we can find the cosine of the angle γ.

$$\cos(\gamma) = \frac{A \,.\, B}{\| A \|\| B \|} = \frac{R^2\cos(a)\cos(\beta)}{R^2} = \cos(\alpha)\cos(\beta).$$

Now, in radian measure

$$\alpha = \frac{a}{R}\,\beta = \frac{b}{R}\,\gamma = \frac{c}{R}$$

And we are done.

Why is this called the spherical Pythagorean theorem? First, note that it does give a relationship of the hypotenuse of a right triangle with the two sides of this triangle. To see how it is related to the classical Pythagorean theorem, recall the Maclaurin series (Taylor series at $a = 0$) for the cosine function:

$$\cos(x) = 1 - \frac{x^2}{2!} + \frac{x^4}{4!} - \frac{x^6}{6!} + \ldots$$

Using this in the formula from the spherical Pythagorean theorem gives us:

$$1 - \frac{c^2}{2R^2} + \frac{c^4}{24R^4} - \ldots = \left(1 - \frac{a^2}{2R^2} + \frac{a^4}{24R^4} - \ldots\right)\left(1 - \frac{b^2}{2R^2} + \frac{b^4}{24R^4} - \ldots\right)$$

$$= 1 - \frac{a^2}{2R^2} - \frac{b^2}{2R^2} + \frac{a^2b^2}{4R^4} + \ldots$$

$$c2 + \frac{\text{terms in higher powers of } c}{R^2} = a2 + b2 + \frac{\text{terms in higher powers of } a \text{ and } b}{R^2}$$

Now, as R gets arbitrarily large while a, b, and c stay small, we get the classical Pythagorean theorem as a limiting case.

Spherical Sine Theorem

Theorem: *Let ΔABC be a spherical triangle on a sphere of radius R. Let a, b, and c denote the lengths of the sides in radians, and let $\angle A$, $\angle B$, and $\angle C$ denote the interior angles at each vertex. Then*

$$\frac{\sin\left(\frac{a}{R}\right)}{\sin(\angle A)} = \frac{\sin\left(\frac{b}{R}\right)}{\sin(\angle B)} = \frac{\sin\left(\frac{c}{R}\right)}{\sin(\angle C)}$$

Proof: We will first prove this for a right triangle and then use that result to prove the general case. In addition we can restrict our attention to triangles that lie entirely within a quarter of a hemisphere.

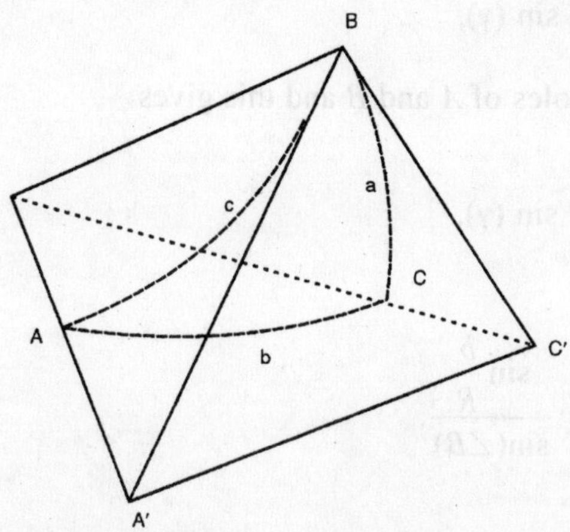

Let $\Delta\ ABC$ be a right triangle with right angle at C. Extend the radius OA to OA', where A' is the on the intersection of the tangent plane to the sphere at B with the line OA. Likewise, extend OC to OC'. This means that the triangles $\Delta\ OBC'$ and $\Delta\ OBA'$ are right triangles with right angle at B. Additionally$\angle A'B\,C' = \angle ABC = \angle B$.

Now, we have other right angles as well. First, planes $A'BC'$ and OBC are perpendicular, from above. Now, plane OCB is perpendiculare to plane OCA since $\angle ACB$ is a right angle. Look at the two planes $A'BC'$ and OAC. They intersect in the line $\overrightarrow{A'\,C'}$

Since $\overrightarrow{A'\,C'}$ lies in the plane OCA, and $\overrightarrow{O'\,C'}$ lies in the plane OCB, the two lines must be perpendicular. Thus, $\Delta\ O\ C'\ A'$ is a right triangle with right angle at C'. The classical Pythagorean theorem gives:

$$(O\,A')^2 = (BA')^2 + (OB)^2 = (BA')^2 + R^2$$

$$(OC')^2 = (BC')^2 + (OB)^2 = (BC')^2 + R^2$$

$$(OA')^2 = (A'C')^2 + (OC')^2.$$

Combining these equations gives us

$$(BA')^2 = (A'C')^2 + (BC')^2,$$

And so $\Delta\ A'B'C'$ is a triangle with right angle at C'. Let α, β and γ denote the angles at the center of the sphere determined by the arcs BC, AC, and AB, respectively. The definition of the sine function gives us:

$$\sin(\beta) = \frac{A'C'}{OA'} = \frac{A'C'}{BA'}\frac{BA'}{OA'} = \sin(\angle B)\sin(\gamma).$$

And so

$$\frac{\sin\frac{b}{R}}{\sin(\angle B)} = \sin(\gamma),$$

We can reverse the roles of A and B and this gives

$$\frac{\sin\frac{\alpha}{R}}{\sin(\angle A)} = \sin(\gamma),$$

And so

$$\frac{\sin\frac{\alpha}{R}}{\sin(\angle A)} = \frac{\sin\frac{b}{R}}{\sin(\angle B)}.$$

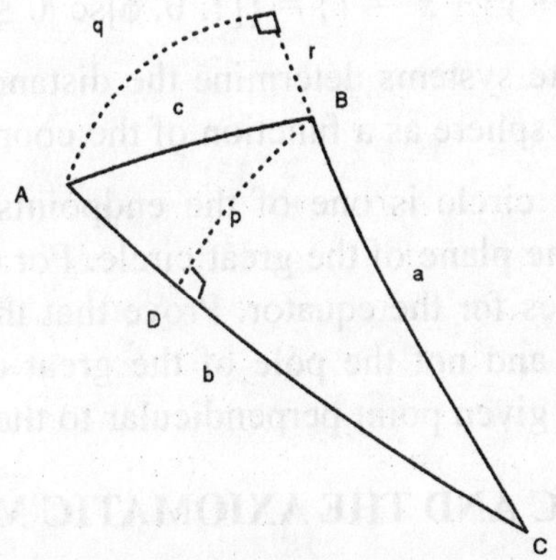

Altitudes

Now, if we have an arbitrary triangle, we can construct the spherical analogue of the altitude to reduce the relation for two sides to the case of a right triangle, which we just finished. For example in the following figure, inserting the altitudes from *A* and from *B*, we get

$$\sin\left(\frac{a}{R}\right)\sin(\angle C) = \sin\left(\frac{p}{R}\right) = \sin\left(\frac{c}{R}\right)\sin(\angle A).$$

From the other altitude we have

$$\sin\left(\frac{b}{R}\right)\sin(\angle C) = \sin\left(\frac{q}{R}\right) = \sin\left(\frac{c}{R}\right)\sin(\pi - \angle B) = \sin\left(\frac{c}{R}\right)\sin(\angle B),$$

And the theorem follows.

These theorems show that there are exactly six related pieces of information determining a spherical triangle — the three sides and the three angles. Given any three of these pieces you can determine the other three pieces. This means that similar triangles must be congruent! This is not immediate. We need one other piece which we are able to prove, but will not do so now.

Theorem: *Suppose that Δ ABC is a right triangle on the sphere of radius R with a right angle at C. Then*

$$\cos(A) = \sin(B).\cos\left(\frac{a}{R}\right)$$

Exercise

1. Prove that two great circles intersect.
2. Prove that two great circles bisect each other.
3. The sphere of radius one can be considered as

$$\Sigma_1 = \{(x, y, z) \,|x^2 + y^2 + z^2 = 1\} = \{[1, \theta, \phi]\text{sc} \,|0 \le \theta \le 2\pi, 0 \le \phi \le \pi\}$$

In these two coordinate systems determine the distance along great circles between two arbitrary points on the sphere as a function of the coordinates.

4. A *pole* of a great circle is one of the endpoints of a diameter of the sphere perpendicular to the plane of the great circle. For example, the North and South poles are both poles for the equator. Prove that through a given point, not on a given great circle and not the pole of the great circle, there is a unique great circle ghrough the given point perpendicular to the given great circle.

LOGIC AND THE AXIOMATIC METHOD

Mathematicians use a large number of methods to discover new results—trial and error, computation of special cases, inspired guessing, pulling results from thin air. The difference in this and an astrologer, for example, is that we have an accepted method, called the axiomatic method, for *proving* that these results are correct. Proofs give us assurance that results are correct. In many cases they also give more *general* results. For example, the Egyptians and Hindus knew by experiment that if a triangle has sides of lengths 3, 4, and 5, it is then a right triangle. But the Greeks proved that if a triangle has sides of lengths a, b, and c, and if $a^2 + b^2 = c^2$, then the triangle is a right triangle.

There is no amount of checking by experiment that could give this general result. Proofs give us insight into relationships among different things that we are studying, forcing us to organize our thoughts in a coherent way. If you gain nothing else from the course than this, you have still gained the greatest gift that mathematics has to offer. I wish to persuade you that a certain statement is true or false by *pure reasoning*. I could do this by showing you that the statement follows logically from some other statement that you may already believe. I may have to convince you that that statement is also true, and follows from another statement.

This process may continue until I reach a statement which you are willing to believe, one which does not need justification. That statement plays the role of an *axiom*. If no such statement exists, then I will be caught in an *infinite regress*, giving one proof after another *ad infinitum*. There are three requirements that must be met before we can agree that a proof is correct.

Requirement 1: There must be mutual understanding of the words and symbols used in the discourse.

Requirement 2: There must be acceptance of certain statements called *axioms* without justification.

Requirement 3: There must be agreement on how and when one statement *follows logically* from another, *i.e.*, agreement on certain rules of reasoning.

There should be no problem in reaching mutual understanding so long as we use terms familiar to both and use them consistently. If I use an unfamiliar term, you have the right to demand a *definition* of this term. Definitions cannot be given arbitrarily; they are subject to the rules of reasoning referred to in Requirement 2. Also, we cannot define every term that we use. In order to define one term we must use other terms, and to define these terms we must use still other terms, and so on. If we were not allowed to leave some terms *undefined*, we would get involved in infinite regress.

SETS

We need some basic information about sets in order to study the logic and the axiomatic method. This is not a formal study of sets, but consists only of basic definitions and notation.

Braces {and} are used to name or enumerate sets. The *roster* method for naming sets is simply to list all of the elements of a set between a pair of braces. For example the set of integers 1, 2, 3, and 4 could be named

$$\{1, 2, 3, 4\}.$$

This does not work well for sets containing a large number of elements, though it can be used. The more common method for this is known as the *set builder notation*. A property is specified which is held by all objects in a set. $P(x)$, read *P of x*, will denote a sentence referring to the variable x. For example,

- $x = 23$
- x is an odd integer.
- $1 \le x \le 4$.

The set of all objects x such that x satisfies $P(x)$ is denoted by

$$\{x \mid P(x)\}.$$

The set {1, 2, 3, 4} can be named

$$\{x \mid 1 \le x \le 4, x \in \mathbb{Z}\} = \{x \in \mathbb{Z} \mid 1 \le x \le 4\}.$$

From hence forth, the words *object, element*, and *member* mean the same thing when referring to sets. Sets will be denoted mainly by capital Roman letters and elements of the sets by small letters. The following have the same meaning:

- $a \in A$
- a is in set A
- a is a member of set A
- a is an element of set A

Likewise, $a \notin A$ means that a is not an element of set A.

A is a *subset* of B if every element of A is also an element of B. The following have the same meaning:

- $A \subset B$
- Every element of A is an element of B
- If $a \in A$, then $a \in B$
- A is included in B
- B contains A
- A is a subset of B

Note that a set is always a subset of itself.

If A and B are sets, then we say that $A = B$ if A and B represent the same set:

- $A = B$
- A and B are the same set
- A and B have the same members
- $A \subset B$ and $B \subset A$

The set which contains no elements is known as the *empty set*, and is denoted by θ. Note that for each set A, $\theta \subset A$.

The *intersection* of two sets A and B is the set of all elements common to both sets. The intersection is symbolized by $A \cap B$ or $\{x \mid x \in A \text{ and } x \in B\}$. The *union* of two sets A and B is the set of elements which are in A or B or both. The union is symbolized by $A \cap B$ or $\{x \mid x \in A \text{ and } x \in B\}$.

UNIVERSAL SETS AND COMPLIMENTS

When we are working in an area or on a certain problem, we always have a frame of reference in which we are working called a *universal set*. In our geometry course, it will be the set of points that lie on a plane. In calculus we consider the set of real numbers, the set of real functions, the set of differentiable functions, and the set of continuous functions as universal sets. The *complement* of a set A is defined to be the set of all elements of the universal set which are not in A, and is symbolized by $C\,A = A' = A^c$. Note that $A \cup A^c$ is always the universal set, while $A \cap A^c = \theta$

SENTENCES AND STATEMENTS

Logic and mathematical proof can be studies just like algebra. In fact, much of symbolic logic is just that. A declarative sentence which is true or false, but not both, is called a *statement*. The following are statements:

- Babe Ruth hit 714 home runs.
- Jack Nicklaus has won 20 major golf titles.
- $2 + 3 = 6$
- The 25,000th digit of π is 7.

The following are not statements:

- He is a golfer.
- Why is a duck?
- $x + 1 = 0$
- $x - y = a$

The sentence

He is a golfer.

Cannot be judged true or false because we do not know who *He* is. If the word *He* is replaced by *Bobby Jones* forming the sentence

Bobby Jones is a golfer.

The sentence becomes a true statement. Similarly, if x in the sentence

$$x + 1 = 0$$

Is replaced by 2, forming the sentence

$$2 + 1 = 0$$

The sentence then becomes a false statement.

The letter x is a variable in the sentence $x + 1 = 0$. A letter or other symbol that can represent various elements of a universal set is called a variable. We can make a sentence a statement by replacing its variables by elements of the universal set or by attaching phrases such as *For every* or *There exists* to the sentence. For example,

$$x < 3$$

Is not a statement, but each of the following is a statement:

- $1 < 3$
- $5 < 3$
- For every real number x, $x < 3$.
- There exists a real number x, such that $x < 3$.

Replacements for variables of a sentence are always chosen from some universal set. Any replacement which makes a sentence true is called a solution. The set of all solutions is called the *solution set* of the sentence.

SENTENCE CONNECTIVES

If P and Q are sentences, then the sentence *P and Q* is called the conjunction of P and Q, denoted by $P \wedge Q$. For any statement there are just two possible truth values, true (T) or false (F). If P and Q are both true, then $P \wedge Q$ is true. If one or both of P and Q are false, then $P \wedge Q$ is false. The truth table below defines the truth values of $P \wedge Q$ for all possible truth value combinations of P and Q.

If P and Q are sentences, then the sentence *P or Q* is called the disjunction of P and Q, denoted by $P \vee Q$. In mathematics we use an *inclusive* or. That is, $P \vee Q$ is true when P is true, or Q is true, or both are true. $P \vee Q$ is false only when P and Q are false. the truth table for $P \vee Q$ is thus defined below.

P	Q	$P \wedge Q$
T	T	T
T	F	F
F	T	F
F	F	F

P	Q	$P \vee Q$
T	T	T
T	F	T
F	T	T
F	F	F

P	$\sim P$
T	F
F	T

A negation, or denial, of a sentence is formed in many ways. For example the *negation* of

P: 2 is rational.

Is represented by each of the following:

- $\sim P$
- It is false that 2 is rational.
- 2 is not rational.
- 2 is irrational.

The truth table for negation is obvious.

You need to realize that there are other symbols, besides ~, for negations that are in common usage.

- $a \neq b$ means $\sim (a = b)$

- $a \geq b$ means $\sim (a < b)$
- $a \notin A$ means $\sim (a \in A)$

If P and Q are sentences, the sentence

If P, then Q

Is denoted by $P \rightarrow Q$ or $P \Rightarrow Q$. We construct a truth table for $P \Rightarrow Q$ just as for the the other connectives Λ, V, and ~. However, the definition is not at all obvious. Consider the sentence:

If I get an A in mathematics, then I will take the next course.

Suppose a fellow student says this. When is the sentence true and when is it false? Let P denote the statement

I get an A in mathematics

and let Q denote the statement

I will take the next course.

Consider the following four cases.

- P (true): He gets an A in mathematics.
  Q (true): He takes the next course.
- P (true): He gets an A in mathematics.
  Q (false): He does not take the next course.
- P (false): He does not get an A in mathematics.
  Q (true): He takes the next course.
- P (false): He does not get an A in mathematics.
  Q (false): He does not take the next course.

It is easy to see that (1) is true and that (2) is false.

You cannot claim that the original statement was false in (3) since he takes the next course even though he did not get an A.

Likewise, in (4) you cannot claim that the statement was false, since he did not get an A and he did not take the next course. The truth table for this sentence is as below:

P	Q	$P \Rightarrow Q$
T	T	T
T	F	F
F	T	T
F	F	T

The sentence $P \Rightarrow Q$ is called a conditional with P the antecedent and Q the consequent. In mathematics the conditional is encountered in many forms. The following have the same meaning:

- $P \Rightarrow Q$
- If P, then Q
- P implies Q
- Q if P, P only if Q
- Q provided P
- Q whenever P, Q when P
- P is a sufficient condition for Q
- Q is a necessary condition for P.

BICONDITIONALS AND COMBINATIONS OF CONNECTIVES

A sentence of the type

$$(P \Rightarrow Q) \wedge (Q \Rightarrow P)$$

Is called a biconditional, denoted

$$p \Leftrightarrow Q.$$

When P and Q are sentences, the truth table for $P \Leftrightarrow Q$ is:

P	Q	$P \Leftrightarrow Q$
T	T	T
T	F	F
F	T	F
F	F	T

In mathematics the biconditional is encountered in many forms. The following have the same meaning:

- $P \Leftrightarrow Q$
- P is equivalent to Q
- P if and only if Q
- Q if and only if P
- P iff Q
- If P, then Q and conversely
- If Q, then P and conversely
- P is a necessary and sufficient condition for Q
- Q is a necessary and sufficient condition for P.

Combinations of $\sim, \Rightarrow \Leftrightarrow \Lambda$, and V often occur. A facility at recognizing them is essential for mathematical reading and proof. Consider the following statement:

If p is prime, then if p is even p must be smaller than 7.

This breaks up into three statements:

- P: p is prime.
- Q: p is even.
- R: p must be smaller than 7.

We can then translate the original statement into

$$p \Rightarrow (Q \Rightarrow R).$$

If k is perpendicular to l and l is perpendicular to m, then k is parallel to m.

Let

- P: k is perpendicular to l
- Q: l is perpendicular to m
- R: k is parallel to m.

Then the sentence translates as

$$(P \Lambda Q) \Rightarrow R.$$

QUANTIFIERS

Sentences involving the phrases *For every..* and *There exists..* also play a very important role in the structure of mathematical sentences. The symbol $\forall$ called the universal quantifier , denotes phrases such as *For each*, *For every*, *For all*. A sentence such as

For every x, $P(x)$

can be translated symbolically into

$$\forall x\, p(x), \text{ or } \forall x, p(x).$$

The following sentences have the same meaning:

- $\forall x$, x is an integer $\Rightarrow x \in Q$,
- For every x, if x is an integer, then $x \in Q$,
- For all x, if x is an integer, then $x \in Q$,
- For each x, if x is an integer, then $x \in Q$,
- Every integer belongs to Q,
- Every integer is a rational number,
- If x is an integer, then $x \in Q$.

Note that in the last sentence the universal quantifier is understood and not written.

The symbol $\exists$ called the existential quantifier symbolizes phrases such as *There exists*, *There is at least one*, *For at least one*, and *Some*. A sentence such as

There exists an x such that $P(x)$

Translates symbolically to

$$\exists x\, P(x) \text{ or } \exists x,\, P(x).$$

The following have the same meaning:

- $\exists x$, x is a natural number.
- There exists an x such that x is a natural number.
- Some number is a natural number.
- There is at least one natural number.

Quantifiers often appear together. Consider the following examples.

$\forall x \forall y,\ x + y = 0$	For every x and for every y, $x + y = 0$.
$\forall x \exists y,\ x + y = 0$	forevery x there exists a y so that $x + y = 0$.
$\exists x \forall y,\ x + y = 0$	There exists an x such that for all y, $x + y = 0$.
$\exists x \exists y,\ x + y = 0$	There exists an x and there exists a y such that $x + y = 0$

The following sentence

For every x, if x is even, then there exists a y such that $x = 2y$.
Translates as

$$\forall x\, (x \text{ is even} \Rightarrow \exists y,\, x = 2y$$

These quantifiers refer to some universal set, which if not explicitly given, must be easily inferred from the context. We will be interested only in nonempty universal sets.

Definition: The sentence $\forall x\, P(x)$ is true if and only if the solution set of $P(x)$ equals the universal set. This sentence is false if the solution set is a proper subset of the universal set; *i.e.*, if there is at least one element of the universal set for which $P(x)$ is false.

Definition: The sentence $\exists x\, p(x)$ is true if the solution set of $P(x)$ is nonempty. This sentence is false if the solution set of $P(x)$ is empty; *i.e.*, if for every replacement of x by a member a of the universal set, $P(a)$ is false.

For more complicated mathematical sentences containing more quantifiers let us look at a few examples.

Example: Suppose $P(x,y)$ is a sentence with variables x and y. The sentence $\forall x \forall y$

$P(x, y)$ is true if and only if for every replacement of x and y by members a and b from the universal set, the statement $P(a,b)$ is true. The sentence is false if there is a replacement for x or a replacement for y for which the statement is false.

Example: The sentence $\exists x\ \forall y\ P(x, y)$ is true if there exists a replacement a for x such that $\forall y\ P(a, y)$ is true. This same a makes the sentence $P(a,b)$ true for every b in the universal set. Note that the sentence $\exists x \forall y, x > y$ is false. There is no replacement a for x which makes the sentence $\forall y, a > y$ true.

RULES OF REASONING

Mathematicians assume a certain class of sentences to be true before we ever prove any theorems in a mathematical system. We call these sentences *rules of reasoning*. They could be called reasoning axioms.

An important class of these rules of reasoning are known as **tautologies**. A tautology is a sentence which is true no matter what the truth value of its constituent parts.

Example: The sentence $P \Rightarrow (P\ V\ Q)$ is a tautology, where P and Q represent arbitrary mathematical sentences. We can show that this is a tautology from a truth table.

P	Q	PVQ	$P \Rightarrow (PVQ)$
T	T	T	T
T	F	T	T
F	T	F	T
F	F	F	T

The way in which we do this is to compute the truth values for $P\ V\ Q$ in the third column first, and then use columns one and three to compute the truth values in column four.

***Logic Axiom 1:** Every tautology is a rule of reasoning.*

The following are tautologies that we commonly use. You will find these listed in the *Rules of Logic* that you have been given.

1. $(P \Rightarrow Q) \Leftrightarrow (\sim Q \Rightarrow \sim P)$ *contrapositive*
2. $[P\ \Lambda (P \Rightarrow Q) \Rightarrow Q$ *Modus ponens*
3. $[(P \Rightarrow Q) \wedge (Q \Rightarrow R)] \Rightarrow (P \Rightarrow R)$ *Law of Syllogism*
4. $\sim (P\ \Lambda\ Q) \Leftrightarrow (\sim PV \sim Q)$
5. $\sim (P\ V\ Q) \Leftrightarrow (\sim P\ \Lambda \sim Q)$
6. $\sim (P \Rightarrow Q) \Leftrightarrow (P\ \Lambda \sim Q)$
7. $(P \Rightarrow Q) \Leftrightarrow (\sim P\ V\ Q)$

8. $(P \Lambda Q) \Rightarrow P$
9. $\sim\sim P \Leftrightarrow P$
10. $(P \Lambda Q) \Rightarrow (P V Q)$
11. $(P \Rightarrow \sim Q) \Rightarrow (Q \Rightarrow \sim P)$
12. $[\sim P \Rightarrow (R\Lambda \sim Q) \Lambda (R\Lambda \sim R)] \Rightarrow P$
13. $[(P\Lambda \sim Q) \wedge (R\Lambda \sim R)] \Rightarrow (P \Rightarrow Q)$
14. $P \Lambda \sim P$ *Law of the Excluded Middle*
15. $P \Leftrightarrow P$
16. $P \Rightarrow P$
17. $[P \Rightarrow (Q V R)] \Rightarrow [(P \Lambda \sim Q) \Rightarrow R]$
18. $[(P \Rightarrow S_1) \wedge (S_1 \Rightarrow S_2) \Lambda..\Lambda (S_{n-1} \Rightarrow S_n) \Lambda (S_n \Rightarrow R)] \Rightarrow (P \Rightarrow R)$ *Law of Syllogism*
19. $[(P \Rightarrow R) \wedge (Q \Rightarrow R)] \Rightarrow [(P V Q) \Rightarrow R]$ *Proof by Cases*
20. $(P \Lambda Q) \Leftrightarrow (Q \Lambda P)$
21. $(P V Q) \Leftrightarrow (Q V P)$
22. $[P \Rightarrow (R \Rightarrow Q)] \Leftrightarrow [(P \Lambda R) \Rightarrow Q]$
23. $[P \Lambda (Q \Lambda R)] \Leftrightarrow [(P \Lambda Q) \wedge R]$
24. $[P V (Q V R)] \Leftrightarrow [(P V Q) V R]$
25. $[P \Lambda (Q V R)] \Leftrightarrow [(P \Lambda Q) V (P \Lambda R)]$
26. $[P V (Q \Lambda R)] \Leftrightarrow [(P V Q) \wedge (P V R)]$
27. $[(P \Leftrightarrow Q1) \wedge . \wedge (Qn \Leftrightarrow Q)] \Rightarrow (P \Leftrightarrow Q)$

VALID ARGUMENTS

The tautologies in the preceding section are not all that there are. If you want to make a deduction based on a sentence, check its truth table. If it is a tautology, use it. Tautologies provide lots of reasoning theorems before we ever start deduction within a mathematical system.

There are actually two branches of formal logic: the statement calculus, involving statements and reasoning by tautology; and the predicate calculus, involving quantified sentences. All we are doing is taking a quick guided tour through informal logic, and so we will not study these areas in great detail.

However, from the predicate calculus we get another collection of reasoning sentences, some of which are listed below. These cannot be verified by tautology.

Logic Axiom 2: *Let U be a universal set. Each of the following is a rule of reasoning.*

1. $[\forall x, P(x) \Rightarrow Q(x)] \Rightarrow [\forall x P(x) \Rightarrow \forall x Q(x)]$,
2. $\forall x P(x) \Leftrightarrow [P(a)$, for any $a \in U]$,
3. $\exists x P(x) \Leftrightarrow [P(a)$, for somc $a \in U]$.

An argument is an assertion that from a certain set of sentences $S_1,..S_n$ (called *premises* or *assumptions*) one can deduce another sentence Q (called an *inference* or *conclusion*). Such an argument can be denoted by

$$S1,.., S_n \perp Q.$$

Arguments are either valid (correct) or invalid (incorrect).

Definition: 1 $S1,.., S_n \perp Q.$ is a Valid argumcnt if and only if $(S1 \wedge .. \wedge S_n) \Rightarrow Q$ is a rulc of rcasoning.

Logic Axiom: [Rule of Substitution] *Suppose $P \Leftrightarrow Q$. Then P and Q may be substituted for one another in any sentence.*

Logic Axiom: *Every sentence of the type $\sim \exists x P(x) \Leftrightarrow \forall x \sim P(x)$ is true.*

Logic Axiom: *Every sentence of the type*

$$\sim \forall x P(x) \Leftrightarrow \exists x \sim P(x) \text{ is true.}$$

To prove a sentence of the type $\forall x P(x)$ false, one could try to prove $\exists x \sim P(x)$ true. This is referred to as *providing a counterexample*.

Chapter 2

Geometry Concepts

Geometry is one of the oldest branches of mathematics. Applications of geometric constructions were made centuries before the mathematical principles on which the constructions were based were recorded. Geometry is a mathematical study of points, lines, planes, closed flat shapes, and solids. Using any one of these alone, or in combination with others, it is possible to describe, design, and construct every visible object. The purpose of this section is to provide a foundation of geometric principles and constructions on which many practical problems depend for solution.

TWO-DIMENSIONAL OBJECTS

Points

The *xy*-coordinate plane has two coordinate axes, the *x* and *y*-axis. They are perpendicular to each other. Non-perpendicular axes can be used; but, the computation cost is higher.

A point in the *xy*-plane is represented by two numbers, (x, y), where x and y are the coordinates of the x-axes and y-axes.

Lines

A line in the xy-plane has an equation as follows:

$$Ax + By + C = 0$$

It consists of three coefficients A, B and C. C is referred to as the constant term. If B is non-zero, the line equation can be rewritten as follows:

$$y = mx + b$$

where $m = -\frac{A}{B}$ and $b = -\frac{C}{B}$. This is the well-known slope-intercept form in which m and b are the slope and the intercept (i.e., the intersection point of the line and the y-axis). If B is zero, the line equation becomes $Ax + C = 0$, which is a line parallel to the y-axis and intersects the x-axis at point $\left(0, -\frac{C}{A}\right)$.

The line equation has three coefficients; but, there are only two independent ones. That is, given a line equation, dividing the equation by one of its non-zero coefficients will not change the line. For example, the line equation $4x + 5y + 7 = 0$ is equivalent to $x + 1.25y + 1.75 = 0$ and $0.8x + y + 1.4 = 0$. Dividing an equation by a non-zero constant is usually referred to as normalization. There is one normalization that is very important to us. It is used to compute the distance between the origin to a line.

Suppose the line equation is $Ax + By + C = 0$, the distance between the origin and the line is given as follows:

$$\text{Distance} = \frac{|C|}{\sqrt{A^2 + B^2}}$$

Thus, after normalizing the line equation by dividing it with the square root of the sum of the squares of A and B, the absolute value of the new constant term is the distance between the origin and the given line.

Parallel and Perpendicular Lines

Given two lines as follows:

$$Ax + By + C = 0$$
$$Ex + Fy + G = 0$$

they are parallel to each other if their slopes are equal. Since the slopes are $-\frac{A}{B}$ and $-\frac{E}{F}$ (if B and F are both non-zero), two lines are parallel to each other if and only if the following holds

$$AF = BE$$

Note that we can assume that B or F are both non-zero. Otherwise, if both are zero, the lines are parallel to the y-axis and hence are parallel to each other.

Two lines are perpendicular if and only if the product of their slopes is -1. With the line equations above, two lines are perpendicular to each other if and only if the following holds:

$$AE = -BF$$

THREE-DIMENSIONAL OBJECTS

Points

The coordinate system in space needs three coordinate axes, the x-, y- and z-axis. Therefore, a point in space has three components (x,y,z), where x, y and z are the coordinates of the x-, y- and z-axis.

Planes

A plane in space has an equation as follows:

$$Ax + By + Cz + D = 0$$

It consists of four coefficients *A*, *B*, *C* and *D*, where *D* is the constant term. Similar to the line case, the distance between the origin and the plane is given as

$$\text{Distance} = \frac{|D|}{\sqrt{A^2 + B^2 + C^2}}$$

The normal vector of a plane is its gradient. The gradient of an equation $f(x, y, z) = 0$ is defined as follows:

$$\nabla(f) = \left\langle \frac{\partial f}{\partial x}, \frac{\partial f}{\partial y}, \frac{\partial f}{\partial z} \right\rangle$$

For a plane, its normal vector is simply $< A, B, C >$.

Unfortunately, a line in space cannot be represented with a single equation. However, it can be considered as the intersection of two planes. To ease our discussion, we shall switch to using vectors, although this traditional notation is still very useful.

Vectors

Vectors have a very important advantage in geometric computing, because it is "coordinate free." The meaning of "coordinate free" will be clear in later discussions. All vectors will be in boldface like *a* and *A*.

A vector is similar to a point. If it is a vector in the plane (*resp.*, space), it has two (*resp.*, three) components. Thus, a vector in a *n*-dimensional space has *n* components. For our applications, we shall distinguish two types of vectors: position vectors and direction vectors. A position vector gives the position of a point. More precisely, a point is a vector. A direction vector gives a direction. Hence, it is not a point. In what follows, position vectors and direction vector are written with boldface upper case and lower case, respectively. For example, *A* and *a* are position and direction vectors, respectively. In many cases, such distinction is unnecessary.

As you have learned in linear algebra or in calculus, you can add and subtract vectors; but you can only multiply or divide a vector with constants.

The length of a vector is the square root of the sum of squares of all components. A unit-length vector is a vector whose length is one. A vector can be normalized by dividing its components with its length, converting the given vector to a unit-length one while keeping its direction the same. For example, if $a = < 3, 4, 5 >$, then the length of a, usually written as |a|, is SQRT(50) and the normalized a is $\left\langle \frac{3}{SQRT(50)}, \frac{4}{SQRT(50)}, \frac{5}{SQRT(50)} \right\rangle$.

Inner Product

Given two vectors *a* and *b*, its inner product, written as *a.b*, is the sum of the products of corresponding components. For example, if $a = < 1, 2, 3 >$ and $b = < 2, -1, 4 >$, then *a.b* $= 1*2 + 2*(-1) + 3*4 = 12$.

The geometric meaning of the inner product of *a* and *b* is the following:

$$a.b = |a|.|b|\cos(t)$$

More precisely, the inner product of *a* and *b* is equal to the product of the length of *a*, the length of *b*, and the cosine of the angle *t* between *a* and *b*. This is an important formula, because we know the following facts about *a* and *b* once the inner product becomes available:

- If *a.b* is zero, where *a* and *b* are non-zero vectors, then cos(*t*) must be zero and, as a result, *t* must be 90 degree. Therefore, *a* and *b* are perpendicular to each other.
- If *a.b* is equal to the product of lengths of *a* and *b*, the cosine of *t* is 1 and *t* is 0 degree. As a result, *a* and *b* are parallel to each other and point to the same direction.
- If *a.b* is equal to the negative product of lengths of *a* and *b*, the cosine of *t* is –1 and *a* and *b* are parallel to each other but point to opposite directions.

Lines

A line is defined by a based point *B* and a direction vector *d* which gives the direction of the line. Therefore, the vector equation of a line is

$$B + td$$

where *t* is a parameter. In many applications, the direction vector is of unit-length.

Planes

A plane, in its vector form, is specified by a based point *B* and its normal vector *n*. For an arbitrary point, or position vector, *X* on the plane, the direction vector from the base point *B* to *X*, *X* – *B*, must be perpendicular to the normal vector *n*. Therefore, we have (*X* – *B*).n must be zero. From $(X - B).n = 0$, we have the equation of a plane specified with a base point and its normal vector:

$$X.n - B.n = 0$$

Given the vector notation of lines and planes, it is very easy to compute the intersection point of a line and a plane. Let the given line be *A* + *td*. Let the plane be defined with a base point B and its normal vector n. Then, this plane has equation $X.n = B.n$. If the line intersects the plane, there must be a value of t such that the corresponding point lies on the plane. That is, there must be a t such that the point corresponding to this t would satisfy the plane equation. Since a point on the line is $A + td$, plugging $A + td$ into the plane equation yields

$$(A + td).n - B.n = 0$$

Rearranging the terms and solving for t yields

$$t = \frac{(B - A).n}{d.n}$$

Therefore, plugging this t into the line equation yields the intersection point.

In the above, if $d.n$ is zero, t cannot be solved and consequently no intersection point exists. The meaning of $d.n = 0$ is that d and n are perpendicular to each other. Since n is the normal vector of a plane and d is perpendicular to n, d must be parallel to the plane. If the line is parallel to the plane, no intersection point exists.

Cross Product

There is yet one more important concept about vector: the cross product of two vectors. Given two vectors a and b, their cross product, written as $a \times b$, is defined as follows:

$$a \times b = \left\langle \begin{vmatrix} a_2 & a_3 \\ b_2 & b_3 \end{vmatrix}, -\begin{vmatrix} a_2 & a_3 \\ b_1 & b_3 \end{vmatrix}, \begin{vmatrix} a_1 & a_2 \\ b_1 & b_2 \end{vmatrix} \right\rangle$$

where a = $< a_1, a_2, a_3 >$, $b = < b_1, b_2, b_3 >$, and || is a 2-by-2 determinant. In other words, the cross product is

$$a \times b = < a_2b_3 - a_3b_2, -(a_1b_3 - a_3b_1), a_1b_2 - a_2b_1 >$$

The cross product of a and b, $a \times b$, is perpendicular to both a and b and points to the direction based on the right-handed system in the order of a, b, $a \times b$. Therefore, $b \times a$ points to the opposite direction of $a \times b$. The length of $a \times b$ is $|a|\ |b| \sin(t)$, where t is the acute angle between a and b. Hence, if a and b are perpendicular to each other, the length of $a \times b$ is simply $|a|\ |b|$.

SIMPLE CURVES AND SURFACES

We have seen the simplest curves (lines) and surfaces (planes) in the previous page. Next to lines and planes, there are conics and quadric surfaces. Although conics and quadric surfaces have been around for about 2000 years, they are still the most popular objects in many computer aided design and modeling systems. We shall discuss conics, quadric surfaces and tori on this page only. Consulting your calculus and/or geometry books should be very helpful. Your linear algebra book should also cover some of these topics with a modern approach.

The following figures show to you three different ways of cutting a cone with a plane. The conic sections, from left to right, are an ellipse, a hyperbola and a parabola.

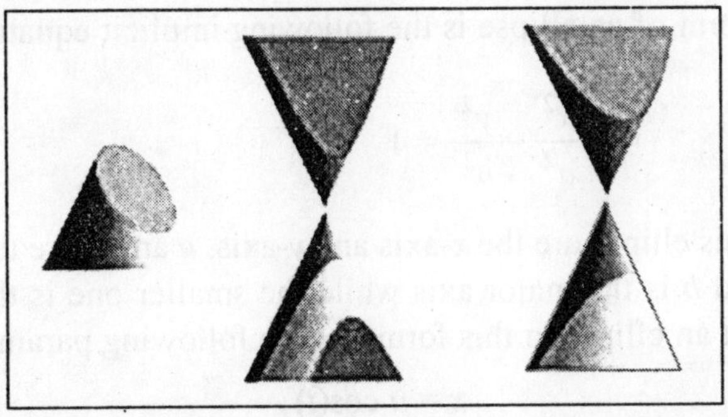

CURVES

Circles

The simplest non-linear curve is unquestionably the circle. A circle with center (a, b) and radius r has an equation as follows:

$$(x - a)^2 + (x - b)^2 = r^2$$

If the center is the origin, the above equation is simplified to

$$x^2 + y^2 = r^2$$

The above equations are referred to as the implicit form of the circle. The parametric form of a circle is

$$x = r \cos(t)$$

$$y = r \sin(t)$$

The following is the parametric form of a circle whose center is not the origin:

$$x = a + r \cos(t)$$

$$y = b + r \sin(t)$$

The above parametric form uses trigonometric functions. We shall discuss a parametric form of a circle without trigonometric functions later.

Conics in Normal Forms

A direct generalization of the circle is the so-called conic curves or simply conics. Greeks knew about conics very well. In fact, Apollonius of Perga (262 – 200 B.C.) wrote a book of several volumes about conics. Conics are the intersection curves of a plane and a circular cone (i.e., a cone whose base is a circle and whose axis is perpendicular to the base and through the center of the base circle).

There are three types of non-degenerate conics: ellipses, hyperbolas and parabolas.

Ellipses and hyperbolas are called central conics because they have a center of symmetry, while parabolas are non-central.

The normal form of an ellipse is the following implicit equation:

$$\frac{x^2}{a^2} - \frac{y^2}{b^2} = 1$$

The axes of this ellipse are the *x*-axis and *y*-axis, *a* and *b* are the axis lengths, and the larger one of *a* and *b* is the major axis while the smaller one is the minor axis. It is not difficult to see that an ellipse in this form has the following parametric form:

$$x = a\cos(t)$$

$$y = b\sin(t)$$

The center of an ellipse and hyperbola, in normal form, is the coordinate origin and the curve is symmetric about its center and its axes.

The normal form of a hyperbola is the following implicit equation:

$$x^2 = 4py$$

In this normal form, for any point (x, y) on a parabola, the value of y must be positive and the opening of this parabola is upward. The axis of this parabola is the y-axis. It is intersecting to note that the normal form of a parabola is already a parametric form. Or, if you like, you can rewrite it into the following:

$$x = t$$

$$y = \frac{t^2}{(4p)}$$

Conics in General Form

Conics are degree two curves because their most general form is the following degree two implicit polynomial:

$$Ax^2 + 2Bxy + Cy^2 + 2Dx + 2Ey + F = 0$$

In the above polynomial, the coefficients of xy, x and y are $2B$, $2D$ and $2E$, respectively. This polynomial has six coefficients; however, dividing it with a non-zero coefficient would reduce six to five. Thus, in general, five conditions can uniquely determine a conic. In linear algebra, you perhaps have learned the way of reducing the above polynomial to a normal form using eigenvalues and eigenvectors.

Frequently, we only want to know the curve type of a general second degree polynomial. In this case, as long as the second degree equation represents a conic rather than two intersecting or parallel lines, it can easily be done as follows:

- If $B^2 < A*C$, the general equation represents an ellipse.
- IF $B^2 = A*C$, the general equation represents a parabola.
- If $B^2 > A*C$, the general equation represents a hyperbola.

Expression $B^2 - A*C$ is called the discriminant of the general second degree polynomial. Based on the above, if the value of the discriminant is less than, equal to or greater than zero, the conic is an ellipse, a parabola, or a hyperbola.

Conics in Matrix Form

One nice thing of conics is that its general form can be rewritten compactly using matrices. First, each point $x = (x, y)$ is considered as a column vector whose third component is 1 and hence the transpose is a row vector, written as $x^T = [x, y, 1]$. Next, the six coefficients of the general second degree polynomial are used to construct a three-by-three symmetric matrix as follows:

$$x = \begin{bmatrix} x \\ y \\ 1 \end{bmatrix} \quad x^T = [x, y, 1] \quad Q = \begin{bmatrix} A & B & D \\ B & C & E \\ D & E & F \end{bmatrix}$$

It is not difficult to verify that the general second degree polynomial becomes

$$x^T Q x = 0$$

Now what you have learned from linear algebra can be applied to this matrix form.

SURFACES

Quadric Surfaces in Normal Forms

Quadric surfaces, or quadrics for short, consist of the following different types: ellipsoids, hyperboloids of one sheet, hyperboloids of two sheets, elliptic paraboloids, and hyperboloid paraboloids. The following are their normal forms in implicit forms and their shapes:

Ellipsoid

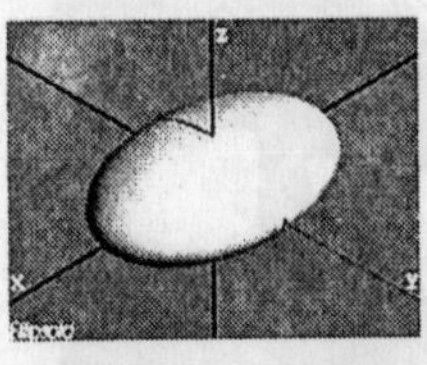

$$\frac{x^2}{a^2} + \frac{y^2}{b^2} + \frac{z^2}{c^2} = 1$$

Hyperboloid of One Sheet

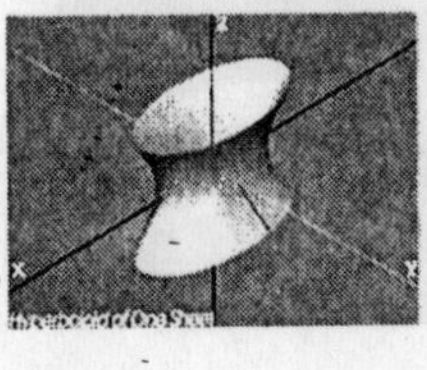

$$\frac{x^2}{a^2} + \frac{y^2}{b^2} - \frac{z^2}{c^2} = 1$$

Hyperboloid of Two Sheets

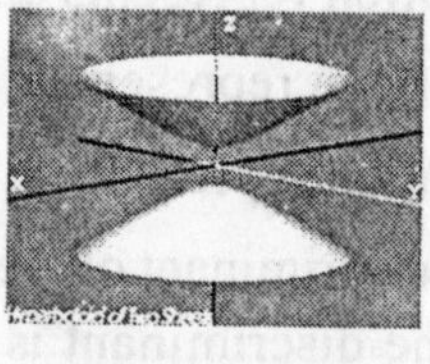

$$-\frac{x^2}{a^2}-\frac{y^2}{b^2}+\frac{z^2}{c^2}=1$$

Elliptic Paraboloid

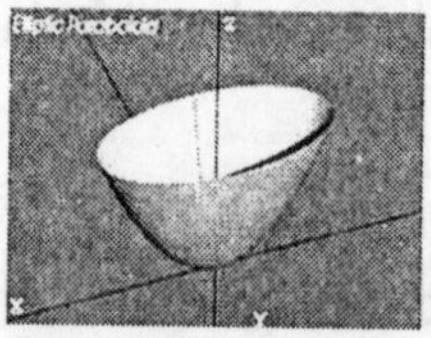

$$\frac{x^2}{a^2}+\frac{y^2}{b^2}=1$$

Hyperbolic Paraboloid

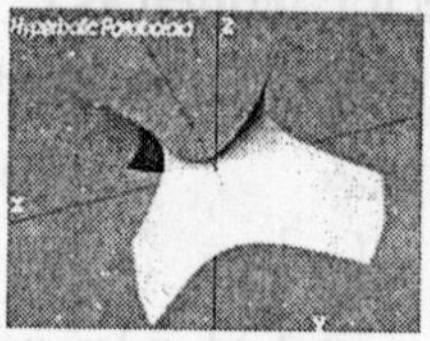

$$\frac{x^2}{a^2}+\frac{y^2}{b^2}=2cz$$

These five quadric surfaces are normally referred to as rank four quadrics. There are two types of rank three quadrics: cones and cylinders. Cylinders have three subtypes: elliptic cylinders, hyperbolic cylinders and parabolic cylinders as shown below:

Cone

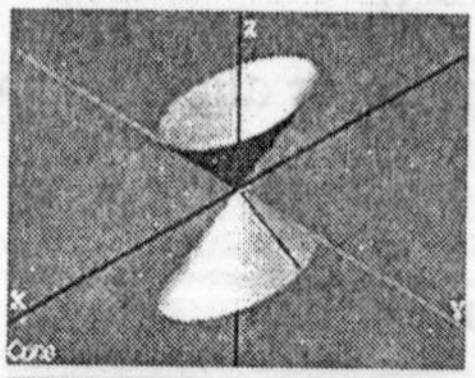

$$\frac{x^2}{a^2}+\frac{y^2}{b^2}+\frac{z^2}{c^2}=0$$

Elliptic Cylinder

$$\frac{x^2}{a^2}+\frac{y^2}{b^2}=1$$

Hyperbolic Cylinder

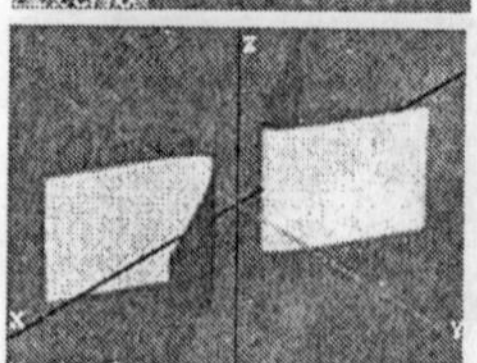

$$\frac{x^2}{a^2}-\frac{y^2}{b^2}=1$$

Parabolic Cylinder

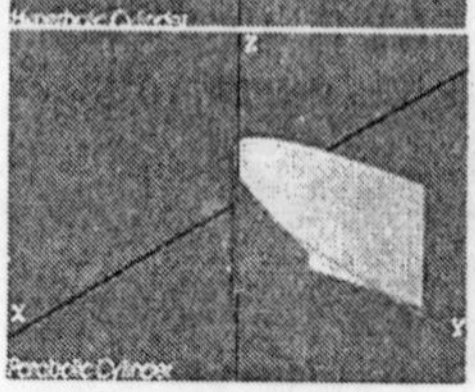

$$x^2=4py$$

QUADRIC SURFACES IN GENERAL FORM

The general form of quadric surfaces is the following:

$$Ax^2 + By^2 + Cz^2 + 2Dxy + 2Exz + 2Fyz + 2Gx + 2Hy + 2Iz + J = 0$$

It has ten coefficients; but, as mentioned in the discussion of conics, dividing the equation with one of its non-zero coefficients reduces the number of coefficients to nine. Please also note that except for the coefficients for x^2, y^2 and z^2 and the constant term, all coefficients has a multiplier of 2.

You might want to ask a question: is it possible to develop an algorithm for classifying the quadrics just like we did for conics? More precisely, if a general second degree polynomial is given, could we tell its type (i.e., ellipsoid, hyperboloid of one sheet, elliptic paraboloid, etc) by simply looking at their coefficients? The answer is always a "yes"; but the computation algorithm is quite complex. So, I would rather skip this algorithm. However, you can always use eigenvalues and eigenvectors to solve this problem.

QUADRIC SURFACES IN MATRIX FORM

The equation of a general quadric can also be put into matrix form:

$$x = \begin{bmatrix} x \\ y \\ z \\ 1 \end{bmatrix}, \quad x^T = [x, y, z, 1]$$

$$Q = \begin{bmatrix} A & D & E & G \\ D & B & F & H \\ E & F & C & I \\ G & H & I & J \end{bmatrix}$$

where (x, y, z) is the coordinates of a point. This form translates the general second polynomial of a quadric to the following matrix form:

$$x^T Qx = 0$$

Note that it is exactly identical to that of a conic. Therefore, matrices help to bring conics and quadrics into an identical form.

After knowing the matrix form of quadrics, we will be able to discuss the meaning of rank four and rank three quadrics. Consider the symmetric matrix Q that contains the coefficients of a general second degree polynomial. The rank of a matrix is the number of non-zero eigenvalues. Thus, rank four quadrics are those quadrics whose matrix Q are of rank four. It is easy to see (from their normal forms) that ellipsoids, hyperboloids and paraboloids are rank four quadrics, and cones and cylinders are rank three quadrics. If a general second degree polynomial factors into the product of two distinct degree one polynomials (i.e., planes), Q will have rank two.

TORI IN NORMAL FORM

A torus can be generated by rotating a circle, the minor circle, about a line, the rotation axis or axis of rotation. The moving circle has its center on another circle, the major circle. The radii of the major and minor circles are referred to as the major radius and minor radius, denoted by R and r, respectively.

If the rotation axis is the z-axis and the major circle lies on the xy-plane, the equation of the generated torus is the following:

$$(x^2 + y^2 + z^2 - (R^2 + r^2))^2 = 4R^2 (r^2 - z^2)$$

If R is greater than r, the result is a commonly seen torus as shown in the middle of the following figure.

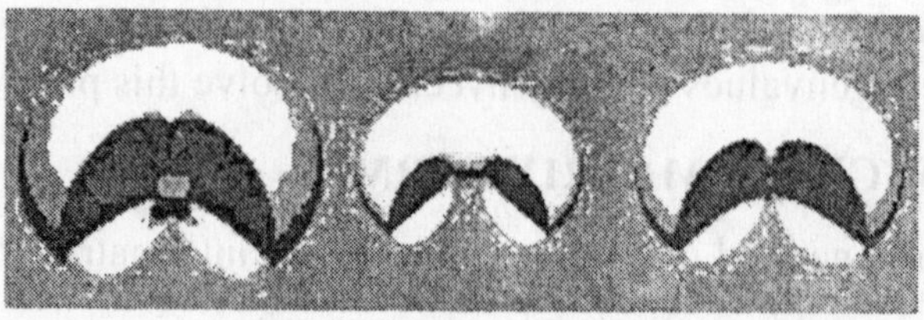

IF R is equal to r, then all moving circles are tangent to the rotation axis at the coordinate origin as shown in the right figure. If R is less than r, all moving circles intersect the rotation axis at two distinct points and the generated torus will have a olive like shape in the interior of the torus as shown in the left figure.

HOMOGENEOUS COORDINATES

One of the many purposes of using homogeneous coordinates is to capture the concept of infinity. In the Euclidean coordinate system, infinity is something that does not exist. Mathematicians have discovered that many geometric concepts and computations can be greatly simplified if the concept of infinity is used. This will become very clear when we move to curves and surfaces design. Without the use of homogeneous coordinates system, it would be difficult to design certain classes of very useful curves and surfaces in computer graphics and computer-aided design.

Let us consider two real numbers, a and w, and compute the value of $\frac{a}{w}$. Let us hold the value of a fixed and vary the value of w. As w getting smaller, the value of $\frac{a}{w}$ is getting larger. If w approaches zero, $\frac{a}{w}$ approaches to infinity! Thus, to capture the concept of infinity, we use two numbers a and w to represent a value v, $v = \frac{a}{w}$. If w is not zero, the value is exactly $\frac{a}{w}$. Otherwise, we identify the infinite value with $(a, 0)$. Therefore, the concept of infinity can be represented with a number pair like (a, w) or as a quotient a/w.

Let us apply this to the xy-coordinate plane. If we replace x and y with $\frac{x}{w}$ and $\frac{y}{w}$, a function $f(x, y) = 0$ becomes $f\left(\frac{x}{y}, \frac{y}{w}\right) = 0$. If function $f(x, y) = 0$ is a polynomial, multiplying it with w^n will clear all denominators, where n is the degree of the polynomial.

For example, suppose we have a line $Ax + By + C = 0$. Replacing x and y with $\frac{x}{w}$ and $\frac{y}{w}$ yields $A\left(\frac{x}{w}\right) + B\left(\frac{y}{w}\right) + C = 0$. Multiplying by w changes it to

$$Ax + By + Cw = 0.$$

Let the given equation be a second degree polynomial $Ax^2 + 2Bxy + Cy^2 + 2Dx + 2Ey + F = 0$. After replacing x and y with $\frac{x}{w}$ and y/w and multiplying the result with w^2, we have

$$Ax^2 + 2Bxy + Cy^2 + 2Dxw + 2Eyw + Fw^2 = 0$$

If you look at these two polynomials carefully, you will see that the degrees of all terms arc equal. In the case of a line, terms x, y and w are of degree one, while in the second degree polynomial, all terms (i.e., x^2, xy, y^2, xw, yw and w^2) are of degree two.

Given a polynomial of degree n, after introducing w, all terms are of degree n. Consequently, these polynomials are called homogeneous polynomials and the coordinates (x, y, w) the homogeneous coordinates.

Given a degree n polynomial in a homogeneous coordinate system, dividing the polynomial with wn and replacing $\frac{x}{w}$, $\frac{y}{w}$ with x and y, respectively, will convert the polynomial back to a conventional one. For example, if the given degree 3 homogeneous polynomial is the following:

$$x^3 + 3xy^2 - 5y^2w + 10w^3 = 0$$

the result is

$$x^3 + 3xy^2 - 5y^2 + 10 = 0$$

This works for three-dimension as well. One can replace a point (x, y, z) with $\left(\frac{x}{w}, \frac{y}{w}, \frac{z}{w}\right)$ and multiply the result by w raised to certain power. The resulting polynomial is a homogeneous one. Converting a degree n homogeneous polynomial in x, y, z and w back to the conventional form is exactly identical to the two-variable case.

AN IMPORTANT NOTES

Given a point (x, y, w) in homogeneous coordinates, what is its corresponding point in the *xy*-plane? From what we discussed for converting a homogeneous polynomial back to its conventional form, you might easily guess that the answer must be $\left(\frac{x}{w}, \frac{y}{w}\right)$. This is correct. Thus, a point (3, 4, 5) in homogeneous coordinates converts to point $\left(\frac{3}{5}, \frac{4}{5}\right) = (0.6, 0.8)$ in the *xy*-plane. Similarly, a point (x, y, z, w) in homogeneous coordinates converts to a point $\left(\frac{x}{w}, \frac{y}{w}, \frac{z}{w}\right)$ in space.

Conversely, what is the homogeneous coordinates of a point (x, y) in the *xy*-plane? It is simply $(x, y, 1)$! That is, let the *w* component be 1. In fact, this is only part of the story, because the answer is not unique. The homogeneous coordinates of a point (x, y) in the *xy*-plane is (xw, yw, w) for any non-zero *w*. Why is this true? Because (xw, yw, w) is converted back to (x, y). As a result, the following is important for you to memorize:

Converting from a homogeneous coordinates to a conventional one is unique; but, converting a conventional coordinates to a homogeneous one is not.

For example, a point (4, 2, 3) in space is convert to $(4w, 2w, 3w, w)$ for any non-zero *w*.

The Dimensionality of Homogeneous Coordinates

You perhaps have discovered that homogeneous coordinates need 3 and 4 components to represent a point in the *xy*-plane and a point in space, respectively. Therefore, a point in space (resp., the *xy*-plane) in homogeneous coordinates actually has four (resp., third) components. Adding a fourth (resp., third) component whose value is 1 to the coordinates of a point in space (resp., the *xy*-plane) converts it to its corresponding homogeneous coordinates.

Ideal Points or Points at Infinity

As mentioned at the very beginning of this page, homogeneous coordinates can easily capture the concept of infinity. Let a point (x, y) be fixed and converted to a homogeneous coordinate by multiplying with 1/w, (x/w,y/w,1/w). Let the value of w approach to zero, then (x/w,y/w) moves farther and farther away in the direction of (x, y). When w becomes zero, $\left(\frac{x}{w}, \frac{y}{w}\right)$ moves to infinity. Therefore, we would say, the homogeneous coordinate $(x, y, 0)$ is the ideal point or point at infinity in the direction of (x, y).

Let us take a look at an example. Let (3, 5) be a point in the xy-plane. Consider $\left(\frac{3}{w}, \frac{5}{w}\right)$. If w is not zero, this point lies on the line $y = \left(\frac{5}{3}\right) x$. Or, if you like the vector form,

$\left(\frac{3}{w}, \frac{5}{w}\right)$ is a point on the line $O + \left(\frac{1}{w}\right)d$, where the base point O is the coordinate origin and d is the direction vector <3, 5>. Therefore, as w approaches zero, the point moves to infinity on the line. This is why we say $(x, y, 0)$ is the ideal point or the point at infinity in the direction of (x, y).

The story is the same for points in space, where $(x, y, z, 0)$ is the ideal point or point at infinity in the direction of (x, y, z).

The concept of homogeneous coordinates and points at infinity in certain direction will become very important when we discuss representations of curves and surface.

A Simple Geometric Interpretation

Given a homogeneous coordinate (x, y, w) of a point in the xy-plane, let us consider (x, y, w) to be a point in space whose coordinate values are x, y and w for the x-, y- and w-axes, respectively. The line joining this point and the coordinate origin intersects the plane $w = 1$ at a point $\left(\frac{x}{w}, \frac{y}{w}, 1\right)$. Please verify this fact yourself. The following figure illustrates this concept.

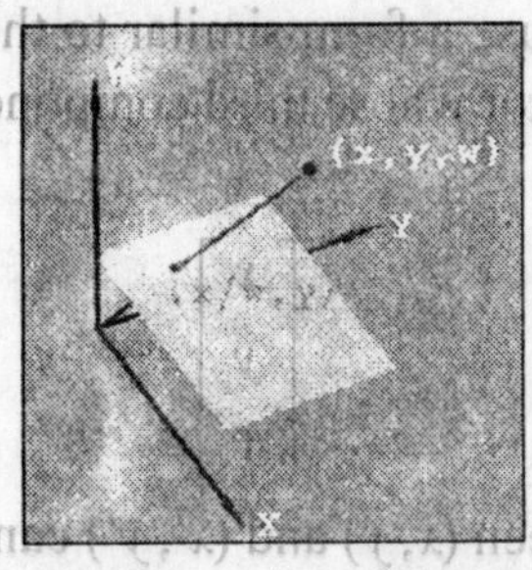

This transformation treats a two-dimensional homogeneous point as a point in three-dimensional space and projects (from the coordinate origin) this three-dimensional point to the plane $w = 1$. Therefore, as a homogeneous point moves on a curve defined by homogeneous polynomial $f(x, y, w)=0$, its corresponding point moves in three-dimensional space, which, in turn, is projected to the plane $w = 1$. Of course, $\left(\frac{x}{w}, \frac{y}{w}\right)$ moves on a curve in plane $w = 1$.

The above figure also shows clearly that while the conversion from the conventional Euclidean coordinates to homogeneous coordinates is unique, the opposite direction is not because all points on the line joining the origin and (x,y,w) will be projected to $\left(\frac{x}{w}, \frac{y}{w}, 1\right)$. This is also an important concept to be used in later lectures.

GEOMETRIC TRANSFORMATIONS

When talking about geometric transformations, we have to be very careful about the object being transformed. We have two alternatives, either the geometric objects are transformed or the coordinate system is transformed. These two are very closely related; but, the formulae that carry out the job are different. We only cover transforming geometric objects here.

We shall start with the traditional Euclidean transformations that do not change lengths and angle measures, followed by affine transformation. Finally, we shall talk about projective transformations.

EUCLIDEAN TRANSFORMATIONS

The Euclidean transformations are the most commonly used transformations. An Euclidean transformation is either a translation, a rotation, or a reflection. We shall discuss translations and rotations only.

TRANSLATIONS AND ROTATIONS ON THE *xy*-PLANE

We intend to translate a point in the xy-plane to a new place by adding a vector $<h, k>$. It is not difficult to see that between a point (x, y) and its new place (x', y'), we have $x' = x + h$ and $y' = y + k$. Let us use a form similar to the homogeneous coordinates. That is, a point becomes a column vector whose third component is 1. Thus, point (x, y) becomes the following:

$$\begin{bmatrix} x \\ y \\ 1 \end{bmatrix}$$

Then, the relationship between (x, y) and (x', y') can be put into a matrix form like the following:

$$\begin{bmatrix} x' \\ y' \\ 1 \end{bmatrix} = \begin{bmatrix} 1 & 0 & h \\ 0 & 1 & k \\ 0 & 0 & 1 \end{bmatrix} \cdot \begin{bmatrix} x \\ y \\ 1 \end{bmatrix}$$

$$\begin{bmatrix} x \\ y \\ 1 \end{bmatrix} = \begin{bmatrix} 1 & 0 & -h \\ 0 & 1 & -k \\ 0 & 0 & 1 \end{bmatrix} \cdot \begin{bmatrix} x' \\ y' \\ 1 \end{bmatrix}$$

Therefore, if a line has an equation $Ax + By + C = 0$, after plugging the formulae for x and y, the line has a new equation $Ax' + By' + (-Ah - Bk + C) = 0$.

If a point (x, y) is rotated an angle a about the coordinate origin to become a new point (x', y'), the relationships can be described as follows:

$$\begin{bmatrix} x' \\ y' \\ 1 \end{bmatrix} = \begin{bmatrix} \cos a & -\sin a & 0 \\ \sin a & \cos a & 0 \\ 0 & 0 & 1 \end{bmatrix} . \begin{bmatrix} x \\ y \\ 1 \end{bmatrix}$$

$$\begin{bmatrix} x \\ y \\ 1 \end{bmatrix} = \begin{bmatrix} \cos a & \sin a & 0 \\ -\sin a & \cos a & 0 \\ 0 & 0 & 1 \end{bmatrix} . \begin{bmatrix} x' \\ y' \\ 1 \end{bmatrix}$$

Thus, rotating a line $Ax + By + C = 0$ about the origin a degree brings it to a new equation:

$(A \cos a - B \sin a)x' + (A \sin a + B \cos a)y' + C = 0$

Translations and rotations can be combined into a single equation like the following:

$$\begin{bmatrix} x' \\ y' \\ 1 \end{bmatrix} = \begin{bmatrix} \cos a & -\sin a & k \\ \sin a & \cos a & k \\ 0 & 0 & 1 \end{bmatrix} . \begin{bmatrix} x \\ y \\ 1 \end{bmatrix}$$

$$\begin{bmatrix} x \\ y \\ 1 \end{bmatrix} = \begin{bmatrix} \cos a & \sin a & -h\cos a - k\sin a \\ \sin a & \cos a & h\sin a - k\cos a \\ 0 & 0 & 1 \end{bmatrix} . \begin{bmatrix} x' \\ y' \\ 1 \end{bmatrix}$$

The above means that rotates the point (x,y) an angle a about the coordinate origin and translates the rotated result in the direction of (h,k). However, if translation (h,k) is applied first followed by a rotation of angle a (about the coordinate origin), we will have the following:

$$\begin{bmatrix} x' \\ y' \\ 1 \end{bmatrix} = \begin{bmatrix} \cos a & -\sin a & h\cos a - k\sin a \\ \sin a & \cos a & h\sin a - k\cos a \\ 0 & 0 & 1 \end{bmatrix} . \begin{bmatrix} x \\ y \\ 1 \end{bmatrix}$$

$$\begin{bmatrix} x \\ y \\ 1 \end{bmatrix} = \begin{bmatrix} \cos a & \sin a & -h \\ -\sin a & \cos a & -k \\ 0 & 0 & 1 \end{bmatrix} . \begin{bmatrix} x' \\ y' \\ 1 \end{bmatrix}$$

Therefore, rotation and translation are not commutative!

In the above discussion, we always present two matrices, A and B, one for transforming x to x' (i.e., $x'=Ax$) and the other for transforming x' to x (i.e., $x = Bx'$). You can verify that the product of A and B is the identity matrix. In other words, A and B are inverse matrices of each other. Therefore, if we know one of them, the other is the inverse of the given one. For example, if you know A that transforms x to x', the matrix that transforms x' back to x is the inverse of A.

Let R be a transformation matrix sending x' to x : $x = Rx'$. Plugging this equation of x into a conic equation gives the following:

$$(Rx')^T Q(Rx') = 0$$

Rearranging terms yields

$$(x')^T(R^T QR)x' = 0$$

This is the new equation of the given conic after the specified transformation. Note that the new 3-by-3 symmetric matrix that represents the conic in a new position is the following:

$$R^T QR$$

Now you see the power of matrices in describing the concept of transformation.

TRANSLATIONS AND ROTATIONS IN SPACE

Translations in space is similar to the plane version:

$$\begin{bmatrix} x' \\ y' \\ z' \\ 1 \end{bmatrix} = \begin{bmatrix} 1 & 0 & 0 & p \\ 0 & 1 & 0 & p \\ 0 & 0 & 1 & r \\ 0 & 0 & 0 & 1 \end{bmatrix} \cdot \begin{bmatrix} x \\ y \\ z \\ 1 \end{bmatrix}$$

$$\begin{bmatrix} x \\ y \\ z \\ 1 \end{bmatrix} = \begin{bmatrix} 1 & 0 & 0 & -p \\ 0 & 1 & 0 & -p \\ 0 & 0 & 1 & -r \\ 0 & 0 & 0 & 1 \end{bmatrix} \cdot \begin{bmatrix} x' \\ y' \\ z' \\ 1 \end{bmatrix}$$

The above translates points by adding a vector <*p*, *q*, *r*>.

Rotations in space are more complex, because we can either rotate about the *x*-axis, the *y*-axis or the *z*-axis. When rotating about the *z*-axis, only coordinates of *x* and *y* will change and the *z*-coordinate will be the same. In effect, it is exactly a rotation about the origin in the *xy*-plane. Therefore, the rotation equation is

$$\begin{bmatrix} x' \\ y' \\ z' \\ 1 \end{bmatrix} = \begin{bmatrix} \cos a & -\sin a & 0 & 0 \\ \sin a & \cos a & 0 & 0 \\ 0 & 0 & 1 & 0 \\ 0 & 0 & 0 & 1 \end{bmatrix} \cdot \begin{bmatrix} x \\ y \\ z \\ 1 \end{bmatrix}$$

$$\begin{bmatrix} x \\ y \\ z \\ 1 \end{bmatrix} = \begin{bmatrix} \cos a & \sin a & 0 & 0 \\ -\sin a & \cos a & 0 & 0 \\ 0 & 0 & 1 & 0 \\ 0 & 0 & 0 & 1 \end{bmatrix} \cdot \begin{bmatrix} x' \\ y' \\ z' \\ 1 \end{bmatrix}$$

With this set of equations, letting a be 90 degree rotates (1, 0, 0) to (0, 1, 0) and (0, 1, 0) to (−1, 0, 0). Therefore, the x-axis rotates to the y-axis and the y-axis rotates to the negative direction of the original x-axis. This is the effect of rotating about the z-axis 90 degree.

Based on the same idea, rotating about the x-axis an angle a is the following:

$$\begin{bmatrix} x' \\ y' \\ z' \\ 1 \end{bmatrix} = \begin{bmatrix} 1 & 0 & 0 & 0 \\ 0 & \cos a & -\sin a & 0 \\ 0 & \sin a & \cos a & 0 \\ 0 & 0 & 0 & 1 \end{bmatrix} \cdot \begin{bmatrix} x \\ y \\ z \\ 1 \end{bmatrix}$$

$$\begin{bmatrix} x \\ y \\ z \\ 1 \end{bmatrix} = \begin{bmatrix} 1 & 0 & 0 & 0 \\ 0 & \cos a & \sin a & 0 \\ 0 & -\sin a & \cos a & 0 \\ 0 & 0 & 0 & 1 \end{bmatrix} \cdot \begin{bmatrix} x' \\ y' \\ z' \\ 1 \end{bmatrix}$$

Let us verify the above again with a being 90 degree. This rotates (0, 1, 0) to (0, 0, 1) and (0, 0, 1) to (0, −1, 0). Thus, the y-axis rotates to the z-axis and the z-axis rotates to the negative direction of the original y-axis.

But, rotating about the y-axis is different! It is because the way of measuring angles. In a right-handed system, if your right hand holds a coordinate axis with your thumb pointing in the positive direction, your other four fingers give the positive direction of angle measuring. More precisely, the positive direction for measuring angles is from the z-axis to x-axis. However, traditionally the angle measure is from the x-axis to the z-axis. As a result, rotating an angle a about the y-axis in the sense of a right-handed system is equivalent to rotating an angle $-a$ measuring from the x-axis to the z-axis. Therefore, the rotation equations are

$$\begin{bmatrix} x' \\ y' \\ z' \\ 1 \end{bmatrix} = \begin{bmatrix} \cos a & 0 & \sin a & 0 \\ 0 & 1 & 0 & 0 \\ -\sin a & 0 & \cos a & 0 \\ 0 & 0 & 0 & 1 \end{bmatrix} \cdot \begin{bmatrix} x \\ y \\ z \\ 1 \end{bmatrix}$$

$$\begin{bmatrix} x \\ y \\ z \\ 1 \end{bmatrix} = \begin{bmatrix} \cos a & 0 & -\sin a & 0 \\ 0 & 1 & 0 & 0 \\ \sin a & 0 & \cos a & 0 \\ 0 & 0 & 0 & 1 \end{bmatrix} \cdot \begin{bmatrix} x' \\ y' \\ z' \\ 1 \end{bmatrix}$$

Let us verify the above with rotating about the y-axis 90 degree. This rotates (1, 0, 0) to (0, 0, –1) and (0, 0, 1) to (1, 0, 0). Therefore, the x-axis rotates to the negative direction of the z-axis and the z-axis rotates to the original x-axis.

A rotation matrix and a translation matrix can be combined into a single matrix as follows, where the r's in the upper-left 3-by-3 matrix form a rotation and p, q and r form a translation vector. This matrix represents rotations followed by a translation.

$$x' = \begin{bmatrix} r_{11} & r_{12} & r_{13} & p \\ r_{21} & r_{22} & r_{23} & q \\ r_{31} & r_{32} & r_{33} & r \\ 0 & 0 & 0 & 1 \end{bmatrix} .x$$

You can apply this transformation to a plane and a quadric surface just as what we did for lines and conics earlier.

AFFINE TRANSFORMATIONS

Euclidean transformations preserve length and angle measure. Moreover, the shape of a geometric object will not change. That is, lines transform to lines, planes transform to planes, circles transform to circles, and ellipsoids transform to ellipsoids. Only the position and orientation of the object will change. Affine transformations are generalizations of Euclidean transformations. Under affine transformations, lines transforms to lines; but, circles become ellipses. Length and angle are not preserved. In this section, we shall discuss scaling, shear and general affine transformations.

Scaling

Scaling transformations stretch or shrink a given object and, as a result, change lengths and angles. So, scaling is not an Euclidean transformation. The meaning of scaling is making the new scale of a coordinate direction p times larger. In other words, the x coordinate is "enlarged" p times. This requirement satisfies $x¢ = px$ and therefore $x = \frac{x'}{p}$.

Scaling can be applied to all axes, each with a different scaling factor. For example, if the x-, y- and z-axis are scaled with scaling factors p, q and r, respectively, the transformation matrix is:

$$\begin{bmatrix} x' \\ y' \\ z' \\ 1 \end{bmatrix} = \begin{bmatrix} p & 0 & 0 & 0 \\ 0 & q & 0 & 0 \\ 0 & 0 & r & 0 \\ 0 & 0 & 0 & 1 \end{bmatrix} . \begin{bmatrix} x \\ y \\ z \\ 1 \end{bmatrix}$$

$$\begin{bmatrix} x \\ y \\ z \\ 1 \end{bmatrix} = \begin{bmatrix} \frac{1}{p} & 0 & 0 & 0 \\ 0 & \frac{1}{q} & 0 & 0 \\ 0 & 0 & \frac{1}{r} & 0 \\ 0 & 0 & 0 & 1 \end{bmatrix} . \begin{bmatrix} x' \\ y' \\ z' \\ 1 \end{bmatrix}$$

Shear

The effect of a shear transformation looks like "pushing" a geometric object in a direction parallel to a coordinate plane ($3D$) or a coordinate axis ($2D$). In the following, the red cylinder is the result of applying a shear transformation to the yellow cylinder:

How far a direction is pushed is determined by a shearing factor. On the xy-plane, one can push in the x-direction, positive or negative, and keep the y-direction unchanged. Or, one can push in the y-direction and keep the x-direction fixed. The following is a shear transformation in the x-direction with shearing factor a:

$$\begin{bmatrix} x' \\ y' \\ 1 \end{bmatrix} = \begin{bmatrix} 1 & a & 0 \\ 0 & 1 & 0 \\ 0 & 0 & 1 \end{bmatrix} . \begin{bmatrix} x \\ y \\ 1 \end{bmatrix}$$

$$\begin{bmatrix} x \\ y \\ 1 \end{bmatrix} = \begin{bmatrix} 1 & -a & 0 \\ 0 & 1 & 0 \\ 0 & 0 & 1 \end{bmatrix} . \begin{bmatrix} x' \\ y' \\ 1 \end{bmatrix}$$

The shear transformation in the y-direction with shearing factor b is the following:

$$\begin{bmatrix} x' \\ y' \\ 1 \end{bmatrix} = \begin{bmatrix} 1 & 0 & 0 \\ b & 1 & 0 \\ 0 & 0 & 1 \end{bmatrix} . \begin{bmatrix} x \\ y \\ 1 \end{bmatrix}$$

$$\begin{bmatrix} x \\ y \\ 1 \end{bmatrix} = \begin{bmatrix} 1 & 0 & 0 \\ -b & 1 & 0 \\ 0 & 0 & 1 \end{bmatrix} . \begin{bmatrix} x' \\ y' \\ 1 \end{bmatrix}$$

In space, one can push in two coordinate axis directions and keep the third one fixed. The following is the shear transformation in both x- and y-directions with shearing factors a and b, respectively, keeping the z-coordinate the same:

$$\begin{bmatrix} x' \\ y' \\ z' \\ 1 \end{bmatrix} = \begin{bmatrix} 1 & 0 & a & 0 \\ 0 & 1 & b & 0 \\ 0 & 0 & 1 & 0 \\ 0 & 0 & 0 & 1 \end{bmatrix} \cdot \begin{bmatrix} x \\ y \\ z \\ 1 \end{bmatrix}.$$

$$\begin{bmatrix} x \\ y \\ z \\ 1 \end{bmatrix} = \begin{bmatrix} 1 & 0 & -a & 0 \\ 0 & 1 & -b & 0 \\ 0 & 0 & 1 & 0 \\ 0 & 0 & 0 & 1 \end{bmatrix} \cdot \begin{bmatrix} x' \\ y' \\ z' \\ 1 \end{bmatrix}$$

Let us take a look at the effect of this shear transformation. Expanding the matrix equation gives the following:

$$\begin{aligned} x\text{¢} &= x + az \\ y\text{¢} &= y + bz \\ z\text{¢} &= z \end{aligned}$$

Thus, a point (x, y, z) in space is transformed to $(x + az, y + bz, z)$. Therefore, the z-coordinate does not change, while (x, y) is "pushed" in the direction of $(a, b, 0)$ with a factor z.

The following is the shear transformation in xz-direction:

$$\begin{bmatrix} x' \\ y' \\ z' \\ 1 \end{bmatrix} = \begin{bmatrix} 1 & a & 0 & 0 \\ 0 & 1 & 0 & 0 \\ 0 & c & 1 & 0 \\ 0 & 0 & 0 & 1 \end{bmatrix} \cdot \begin{bmatrix} x \\ y \\ z \\ 1 \end{bmatrix}$$

$$\begin{bmatrix} x \\ y \\ z \\ 1 \end{bmatrix} = \begin{bmatrix} 1 & -a & 0 & 0 \\ 0 & 1 & 0 & 0 \\ 0 & -c & 1 & 0 \\ 0 & 0 & 0 & 1 \end{bmatrix} \begin{bmatrix} x' \\ y' \\ z' \\ 1 \end{bmatrix}$$

The following is the shear transformation in yz-direction:

$$\begin{bmatrix} x' \\ y' \\ z' \\ 1 \end{bmatrix} = \begin{bmatrix} 1 & 0 & 0 & 0 \\ b & 1 & 0 & 0 \\ c & 0 & 1 & 0 \\ 0 & 0 & 0 & 1 \end{bmatrix} \begin{bmatrix} x \\ y \\ z \\ 1 \end{bmatrix}$$

$$\begin{bmatrix} x \\ y \\ z \\ 1 \end{bmatrix} = \begin{bmatrix} 1 & 0 & 0 & 0 \\ -b & 1 & 0 & 0 \\ -c & 0 & 1 & 0 \\ 0 & 0 & 0 & 1 \end{bmatrix} \cdot \begin{bmatrix} x' \\ y' \\ z' \\ 1 \end{bmatrix}$$

General Affine Transformations

The general affine transformation matrix has the following form:

$$\begin{bmatrix} x' \\ y' \\ z' \\ 1 \end{bmatrix} = \begin{bmatrix} a_{11} & a_{12} & a_{13} & a_{14} \\ a_{21} & a_{22} & a_{23} & a_{24} \\ a_{31} & a_{32} & a_{33} & a_{34} \\ 0 & 0 & 0 & 1 \end{bmatrix} \cdot \begin{bmatrix} x \\ y \\ z \\ 1 \end{bmatrix}$$

$$\begin{bmatrix} x \\ y \\ z \\ 1 \end{bmatrix} = \begin{bmatrix} a_{11} & a_{12} & a_{13} & a_{14} \\ a_{21} & a_{22} & a_{23} & a_{24} \\ a_{31} & a_{32} & a_{33} & a_{34} \\ 0 & 0 & 0 & 1 \end{bmatrix}^{-1} \begin{bmatrix} x' \\ y' \\ z' \\ 1 \end{bmatrix}$$

Comparing with all previous discussed matrices, rotations and translations included, you will see that all of them fit into this form and hence are affine transformations. Affine transformations do not alter the degree of a polynomial, parallel lines/planes are transformed to parallel lines/planes, and intersecting lines/plane are transformed to intersecting lines and planes. However, affine transformations do not preserve lengths and angle measures and as a result they will change the shape of a geometric object. The following shows the result of a affine transformation applied to a torus. A torus is described by a degree four polynomial. The red surface is still of degree four; but, its shape is changed by an affine transformation.

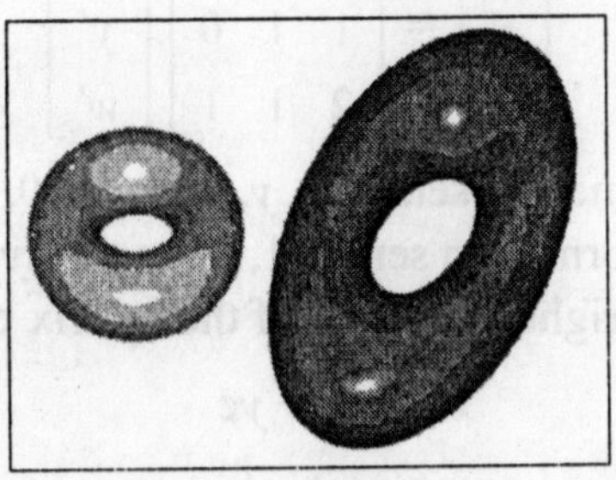

Note that the matrix form of an affine transformation is a 4-by-4 matrix with the fourth row 0, 0, 0 and 1. Moreover, if the inverse of an affine transformation exists, this affine transformation is referred to as non-singular; otherwise, it is singular. We do not use singular affine transformations in this course.

PROJECTIVE TRANSFORMATIONS

Projective transformations are the most general "linear" transformations and require the use of homogeneous coordinates. Given a point in space in homogeneous coordinate (*x*, *y*, *z*, *w*) and its image under a projective transform (*x¢*, *y¢*, *z¢*, *w¢*), a projective transform has the following form:

$$\begin{bmatrix} x' \\ y' \\ z' \\ w' \end{bmatrix} = \begin{bmatrix} p_{11} & p_{12} & p_{13} & p_{14} \\ p_{21} & p_{22} & p_{23} & p_{24} \\ p_{31} & p_{32} & p_{33} & p_{34} \\ p_{41} & p_{42} & p_{43} & p_{44} \end{bmatrix} \cdot \begin{bmatrix} x \\ y \\ z \\ w \end{bmatrix}$$

$$\begin{bmatrix} x \\ y \\ z \\ w \end{bmatrix} = \begin{bmatrix} p_{11} & p_{12} & p_{13} & p_{14} \\ p_{21} & p_{22} & p_{23} & p_{24} \\ p_{31} & p_{32} & p_{33} & p_{34} \\ p_{41} & p_{42} & p_{43} & p_{44} \end{bmatrix}^{-1} \cdot \begin{bmatrix} x' \\ y' \\ z' \\ w' \end{bmatrix}$$

In the above, the 4-by-4 matrices must be non-singular (i.e., invertible). Therefore, projective transformations are more general than affine transformations because the fourth row does not have to contain 0, 0, 0 and 1.

Projective transformation can bring finite points to infinity and points at infinity to finite range. Let us take a look at an example. Consider the following projective transformation:

$$\begin{bmatrix} x' \\ y' \\ w' \end{bmatrix} = \begin{bmatrix} 1 & -1 & 0 \\ -1 & 2 & 0 \\ -1 & 0 & 1 \end{bmatrix} \cdot \begin{bmatrix} x \\ y \\ w \end{bmatrix}$$

$$\begin{bmatrix} x \\ y \\ w \end{bmatrix} = \begin{bmatrix} 2 & 1 & 0 \\ 1 & 1 & 0 \\ 2 & 1 & 1 \end{bmatrix} \cdot \begin{bmatrix} x' \\ y' \\ w' \end{bmatrix}$$

Obviously, this transformation sends (*x*, *y*, *w*) = (1, 0, 1) to (*x¢*, *y¢*, *w¢*) = (1, –1, 0). That is, this projective transformation sends (1, 0) on the *xy*-plane to the point at infinity in direction <1, –1>. From the right-hand side of the matrix equation *x* = *Px¢* we have

x = 2*x¢* + *y¢*

y = *x¢* + *y¢*

w = 2*x¢* + *y¢* + *w¢*

Let us consider a circle $x^2 + y^2 = 1$. Plugging the above equations into the circle equation changes it to the following:

$$x^2 + 2xy + y^2 - 4xw - 2yw - w^2 = 0$$

Dividing the above by w^2 to convert it back to conventional form yields

$$x^2 + 2xy + y^2 - 4x - 2y - 1 = 0$$

This is a parabola! (Why?) Therefore, a circle that has no point at infinity is transformed to a parabola that does have point at infinity.

While projective transformations, like affine transformations, do not change the degree of a polynomial, two parallel (*i.e.*, intersecting) lines/planes can be transformed to two intersecting (*i.e.*, parallel) lines/planes. Please verify this fact yourself.

Although we do not use these facts and the concept of projective transformations immediately, it will be very helpful in later lectures.

MATRIX MULTIPLICATION AND TRANSFORMATIONS

We have introduced to you several transformations. We always show to you two forms, one from x to $x¢$ and the other the inverse from $x¢$ to x. In many cases, one may need several transformations to bring an object to its desired position. For example, one may need a transformation in matrix form $q = Ap$ bringing p to q, followed by a second transformation $r = Bq$ bringing q to r, followed by yet another transformation $s = Cr$ bringing r to s. The net effect of $p \to q \to r \to s$ can be summarized into a single transformation represented by the product of all involved matrices. Note that the first (resp., last) transformation matrix is the right-most (resp., left-most) in the multiplication sequence.

$$\begin{aligned} s = Cr &= C(Bq) = CBq \\ &= CB(Ap) = CBAp \end{aligned}$$

Therefore, to compute the net effect, we just compute CBA and use it as a single transformation, which brings p to s.

Let us take a look at an example. We want to perform the following transformations to an object:

- Scale in the x-direction using a scale factor 5 (i.e., making it five times larger).
- Followed by a rotation about z-axis 30 degree
- Followed by a shear transformation in x- and y-direction with shearing factor 2 and 3, respectively.
- Followed by a transformation moving the point in the direction of <2, 1, 2>.

Let the scaling, rotation, shearing and translation matrices be A, B, C and D, respectively. With previous discussion, we have the following, where matrix $H = DCBA$ is the net effect:

$$A = \begin{bmatrix} 5 & 0 & 0 & 0 \\ 0 & 1 & 0 & 0 \\ 0 & 0 & 1 & 0 \\ 0 & 0 & 0 & 1 \end{bmatrix}$$

$$B = \begin{bmatrix} \frac{\sqrt{3}}{2} & -\frac{1}{2} & 0 & 0 \\ \frac{1}{2} & \frac{\sqrt{3}}{2} & 0 & 0 \\ 0 & 0 & 1 & 0 \\ 0 & 0 & 0 & 1 \end{bmatrix}$$

$$C = \begin{bmatrix} 1 & 0 & 2 & 0 \\ 0 & 1 & 3 & 0 \\ 0 & 0 & 1 & 0 \\ 0 & 0 & 0 & 1 \end{bmatrix}$$

$$D = \begin{bmatrix} 1 & 0 & 0 & 2 \\ 0 & 1 & 0 & 1 \\ 0 & 0 & 1 & 2 \\ 0 & 0 & 0 & 1 \end{bmatrix}$$

$$H = DCBA = \begin{bmatrix} \frac{5\sqrt{3}}{2} & \frac{-1}{2} & 2 & 2 \\ \frac{5}{2} & \frac{\sqrt{3}}{2} & 3 & 1 \\ 0 & 0 & 1 & 2 \\ 0 & 0 & 0 & 1 \end{bmatrix}$$

Therefore, the net effect of transforming a point *x* of the initial object to the corresponding point x′ after the above four transformations is computed as *x¢* = *Hx* = *DCBAx*.

Chapter 3

Incidence Geometry

An incidence geometry is a mathematical structure composed of objects of various types and an incidence relation between them. The number of types of object used in the structure is called the rank of the incidence geometry. Incidence geometries can be modelled by vertex-colored graphs. In such a model, the colors correspond to the object types, and adjacent vertices of G correspond to incident objects. Incidence geometries of rank two are also known as incidence structures. In geometry, the relations of incidence are those such as 'lies on' between points and lines (as in 'point P lies on line L'), and 'intersects' (as in 'line L_1 intersects line L_2', in three-dimensional space). That is, they are the binary relations describing how subsets meet. The propositions of incidence stated in terms of them are statements such as 'any two lines in a plane meet'. This is true in a projective plane, though not true in Euclidean space of two dimensions where lines may be parallel.

Historically, projective geometry was introduced in order to make the propositions of incidence true (without exceptions such as are caused by parallels). From the point of view of synthetic geometry it was considered that projective geometry *should be* developed using such propositions as axioms. This turns out to make a major difference only for the projective plane (for reasons to do with Desargues' theorem). The modern approach is to define projective space starting from linear algebra and homogeneous co-ordinates. Then the propositions of incidence are derived from the following basic result on vector spaces: given subspaces U and V of a vector space W, the dimension of their intersection is at least dim U + dim V “ dim W. Bearing in mind that the dimension of the projective space P(W) associated to W is dim W “ 1, but that we require an intersection of subspaces of dimension at least 1 to register in projective space (the subspace {0} being common to all subspaces of W), we get the basic proposition of incidence in this form: linear subspaces L and M of projective space P meet provided dim L + dim M is at least dim P.

We are now ready to begin our study of geometries in earnest. We will study *neutral* geometry, based on the axioms of Hilbert. This means that we will study all that we can, almost, without the introduction of a *Parallel Postulate* of any sort. At the appropriate time we will add a *parallel postulate* and see where we will be led.

For an ease of notation, let $A * B * C$ denote the statement that the point B lies between the point A and the point C.

Definition: Let l be any line and let A and B be any points that do not lie on l. If $A = B$ or if the segment AB contains no point lying on l, we say that A and B are on the same side of$l$. If if $A \neq B$ and if AB contains a point of l, we say that A and B are on opposite sides of l.

Let us quickly review the incidence axioms.

Incidence Axiom 1: *For every point P and for every point Q not equal to P there exists a unique line l that passes through P and Q.*

Incidence Axiom 2: *For every line l there exists at least two distinct points incident with l*

Incidence Axiom 3: *There exist three distinct points with the property that no line is incident with all three of them.*

This does not seem like much, but already we can prove several easy properties that any set satisfying these three axioms must have.

Theorem: *If l and m are distinct lines that are not parallel, then l and m have a unique point in common.*

Let's be brave and give a formal and an informal proof of this theorem. Having done that, I think that you will see how an informal proof is really a *rigorous proof*, just not a formal proof.

Proof: First, the formal proof. We shall break the statement into its three constituent parts.

P: $l \neq m$

Q: $l \nparallel m$

R: l and m have a unique point in common.

We are to prove then that $(P \wedge Q) \Rightarrow R$.

S_1: $l \nparallel m$ and $l \neq m$

Assuming P and Q in an RCP proof

S_2: l and m have at least one point in common.

negation of the condition of being parallel

S_3: Assume l and m have more than one point in common.

Proof by Contradiction

S_4: From S_1 and S_3, l and m have at least

two points, A and B, in common

S_5: There exists a unique line through A and B

Axiom I – 1

S_6: Thus, $l = m$

S_4 *and* S_5

S_7: We have Q and ~ Q, a contradiction

S_1 *and* S_6

S_8: Thus l and m have a unique point in common

the other case of S_2

An informal proof of this result follows much the same line, but is easier to read.

Proof: Since l and m are not parallel and since $l \neq m$, they must have at least one point in common. Assume that they have more than one point in common. They then have at least two points in common. Axiom I-1 says that two points determine a unique line, so $l = m$, which is contrary to the hypothesis. Thus, l and m have a unique point in common.

Definition: Two or more lines are concurrent if they intersect in one common point.

Definition: Two or more points are collinear if they are all incident with the same line.

We have four other results to mention.

Theorem: *For every line there is at least one point not incident with it.*

Theorem: every point there is at least one line not incident with it.

Theorem: every point there exist at least two lines incident with it.

Theorem: exist three distinct lines which are not concurrent.

To introduce you to the concept of a *model for geometry*, let us look at a simple example of some mathematical object which satisfies the three *axioms of incidence*, based on our interpretation of the undefined concepts.

Example: Consider the set $U = \{A, B, C\}$ We shall interpret a *point* to be a singleton subset of U. Thus, $\{A\}$, $\{B\}$, and $\{C\}$ are points. We shall interpret a *line* to be a doubleton subset of U. The lines are then $\{A, B\}$, $\{A, C\}$, and $\{B, C\}$ We shall agree that a point is *incident* with a line if it is a subset of the *line*. Now, before we continue, we must verify that each of the *Incidence Axioms* is valid in this particular example.

Axiom I-1: If X and Y are any of the *points* of this geometry, then $P \to Q$ is the unique line which contains them, for there are only three possible lines.

Axiom I-2: If $P \to Q$ is any *line* in this geometry, then $\{X\}$ and $\{Y\}$ are two distinct points *incident* with it.

Axiom I-3: The points $\{A\}$, $\{B\}$ and $\{C\}$ are three distinct *points* which are not collinear.

Thus this is a model of a geometry which satisfies the *Incidence Axioms*. Such a geometry is called an *incidence geometry*. There are a number of different ways of visualizing this geometry.

Example: Again, consider the set $U = \{A, B, C\}$ We shall interpret a *point* to be a doubleton subset of U. The points are then $\{A, B\}$, $\{A, C\}$, and $\{B, C\}$ We shall interpret a *line* to be a singleton subset of U. Thus, $\{A\}$, $\{B\}$, and $\{C\}$ are lines. We shall agree that a point is *incident* with a line if it contains the *line* as a subset. Now, we must verify that each of the *Incidence Axioms* is valid in this example.

Axiom I-1: If X and Y are any of the *points* of this geometry, then $\{X\} \cap \{Y\}$ is the unique line which contains them. There are three possible points and each must contain two elements from U. Thus, the intersection in the set U will be nonempty.

Axiom I-2: If $\{X\}$ is any *line* in this geometry, then $P \to Q$ and $\{X, Z\}$ are two distinct points *incident* with it.

Axiom I-3: The points $\{A, B\}$, $\{B, C\}$, and $\{A, C\}$ are three distinct *points* which are not collinear, since their intersection is empty.

Thus this, too, is a model of an incidence geometry. It is sometimes referred to as the dual geometry to the previous example.

Note that since the three *incidence axioms* hold for each of these two examples, the five theorems must also hold. One further item to note about these geometries—there are no parallel lines. Given a line and a point not on that line in each of these two examples, there is no line through that point parallel to the given line. We say that these two models exhibit the elliptic parallel property. This property is not inherent in the *incidence axioms*, but is inherent in the examples. There are other examples of incidence geometries which do not exhibit the elliptic parallel property.

This implies that we cannot prove the Euclidean Parallel Postulate based only on the *incidence axioms*. In fact we cannot prove that parallel lines even exist, based solely on the incidence axioms. Furthermore, we cannot prove that they do not exist.

Example:

1. If we take $U = \{A, B, C, D\}$ and take the same interpretation for point and line as in Example 1, then we will have an incidence geometry which exhibits the Euclidean parallel property--unique parallels.

2. If we take $U = \{A, B, C, D, E\}$ and take the same interpretation for point and line as in Example 1, then we will have an incidence geometry which exhibits the hyperbolic parallel property--multiple parallels.

Definition: We say that two models for incidence geometry are isomorphic if there is a one-to-one correspondence between the points of the models, $P \leftrightarrow P'$ and the lines of the models $l \leftrightarrow l'$ which preserves the incidence relations; *i.e.*, P' is incident with l' if and only if P is incident with l.

Note that we will not be able to have a model with the elliptic parallel property isomorphic to a model with the hyperbolic or Euclidean parallel property, for incidence would not be preserved.

BETWEENNESS AXIOMS

You should review the *Betweenness Axioms* in our list of axioms. There are numerous Propositions proved in the text based on the *Betweenness Axioms*. We shall list most of them, but shall prove only a few.

Proposition: *For any two points A and B,*

- $\overrightarrow{AB} \cap \overrightarrow{BA} = \overline{AB}$.
- $\overrightarrow{AB} \cup \overrightarrow{BA} = \overleftrightarrow{AB}$

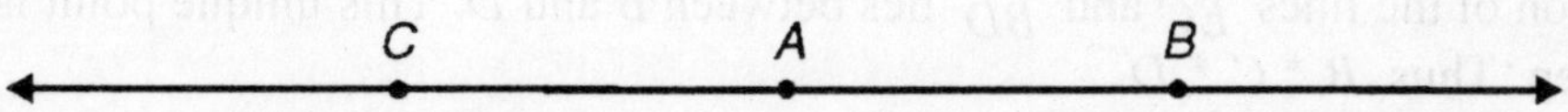

It seems clear that every point P lying on the line l through A, B, and C must either belong to ray $\overrightarrow{AB}$ or to an opposite ray $\overrightarrow{AC}$ This statement seems similar to the second assertion, but it is actually much more complicated. You will prove this assertion in your homework, with the addition of another axiom.

Let us call the assertion

$C * A * B$ and a point P is collinear with A, B, and C, implies that $P \; \Xi \; \overrightarrow{AB} \cup \overrightarrow{AC}$

As the *line separation property*. This is something that you will prove, but knowing that it can be proven, we shall use it as we need.

Recall the definitions of same side and opposite sides. Also, recall the last *Betweenness Axiom*. It is called the *plane separation axiom*. Note that it indirectly guarantees that our geometry will be only two-dimensional, for such an axiom will not hold in three-dimensional space. The points A and C may lie in the same plane with l yet the point B need not. In this case the conditions would be satisfied, yet the conclusion would not hold.

Definition: Let l be any line. A side of the line l containing A is the set of all points that are on the same side of l as A. This is also called the half-plane bounded by l containing A.

Proposition: *Every line bounds exactly two half-planes and these half-planes have no point in common.*

Proposition: *Given A * B * C and A * C * D, then B * C * D and A * B * D.*

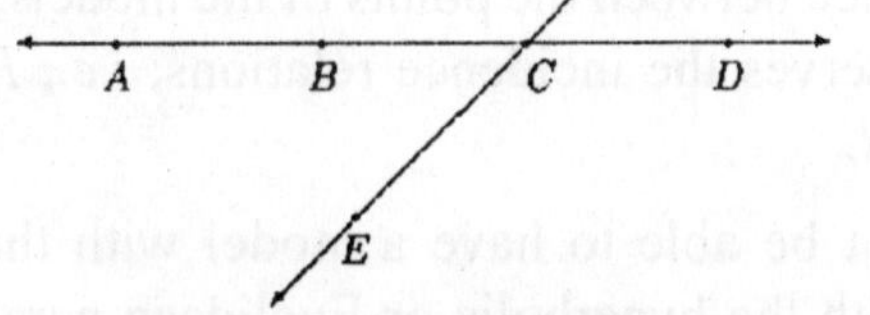

Fig. Distinct Point on the Line

Proof: *A*, *B*, *C*, and *D* are four distinct points on the line. From *Incidence Axiom* there is at least one point *E* not on the line $\overleftrightarrow{AD}$ Line $\overleftrightarrow{EC}$ exists by *Incidence Axiom*, and by hypothesis $\overleftrightarrow{AD} \cap \overleftrightarrow{EC} = \{C\}$Thus, *A* and *D* are on opposite sides of $\overleftrightarrow{EC}$. We claim that *A* and *B* lie on the same side of $\overleftrightarrow{EC}$.

Assume not; *i.e.*, assume that *A* and *B* lie on opposite sides of $\overleftrightarrow{EC}$. Then $\overleftrightarrow{EC}$ intersects $\overleftrightarrow{AB}$ in a point between *A* and *B*. By Propositionthat point must be *C*. Thus, we have that *A * C * B* and *A * B * C* which contradicts *Betweenness Axiom*. Thus, *A* and *B* lie on the same side of $\overleftrightarrow{EC}$ .Since *A* and *D* are on opposite sides of $\overleftrightarrow{EC}$, as we have proven earlier, by *Betweenness Axiom B* and *D* lie on opposite sides of $\overleftrightarrow{EC}$. Hence the unique point of intersection of the lines $\overleftrightarrow{EC}$ and $\overleftrightarrow{BD}$ lies between *B* and *D*. This unique point is *C*, by an earlier step. Thus, *B * C * D*

This proves Proposition.

Proposition: [Line Separation Property] If *C * A * B* and l is the line through *A*, *B*, *and C, then for every point P incident with P is incident with l, either ray* $\overrightarrow{AB}$ *or the opposite ray* $\overrightarrow{AC}$

Pasch's Theorem

Theorem: *If Δ ABC is any triangle and l is any line intersecting side AB in a point between A and B, then l also intersects either side AC or side BC. If C is not incident with l, then l does not intersect both AC and BC.*

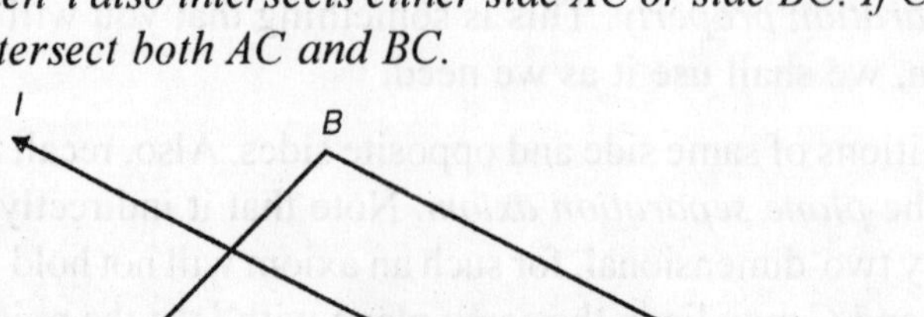

Fig. Pasch's Triangle

Proof: Either $C \in l$ or $C \notin l$ If $C \in l$ then we are done. Thus, let us assume that $C \notin l$ By hypothesis $A, B \notin l$ and $l \cap AB \neq \theta$ Thus, A and B are on opposite sides of l Then, since there are only two sides to any line, either C is on the same side of l as A or it is on the opposite side. If A and C are on the same side of l, then B and C are on opposite sides of l. Thus, $l \cap BC \neq \theta$ Likewise, if A and C are on opposite sides, then $l \cap AC \neq \theta$, and we are done.

Proposition: *Given $A * B * C$ then $AC = AB \cup BC$ and B is the only point common to the segments AB and BC.*

Proposition: *Given $A * B * C$ then B is the only point common to the rays $\overrightarrow{BA}$ and $\overrightarrow{BC}$ and $\overrightarrow{AB} = \overrightarrow{AC}$.*

Definition: Given an angle $\angle CAB$ we shall say that a point D is in the interior of the angle if D is on the same side of $\overleftrightarrow{AB}$ as C and on the same side of $\overleftrightarrow{AC}$ as B.

Note that this implies that the interior of an angle is the intersection of two half planes.

Proposition: *Let $\angle CAB$ be any angle and let D be any point incident with line $\overleftrightarrow{BC}$. D is in the interior of $\angle CAB$ if and only if $* D * C$*

Proposition: *If D is in the interior of angle $\angle CAB$, then*

- *So is every other point on the ray $\overrightarrow{AD}$ except A;*
- *No point on the opposite ray to $\overrightarrow{AD}$ is in the interior of $\angle CAB$;*
- *If $C * A * E$ then B is in the interior of $\angle DAE$.*

Definition: Ray $\overrightarrow{AD}$ is between rays $\overrightarrow{AC}$ and $\overrightarrow{AB}$ if $\overrightarrow{AB}$ and $\overrightarrow{AC}$ are not opposite rays *and* D is in the interior of $\angle CAB$

Crossbar Theorem

Theorem: *If $\overrightarrow{AD}$ is between $\overrightarrow{AB}$ and $\overrightarrow{AC}$ then $\overrightarrow{AD}$ intersects the segment BC.*

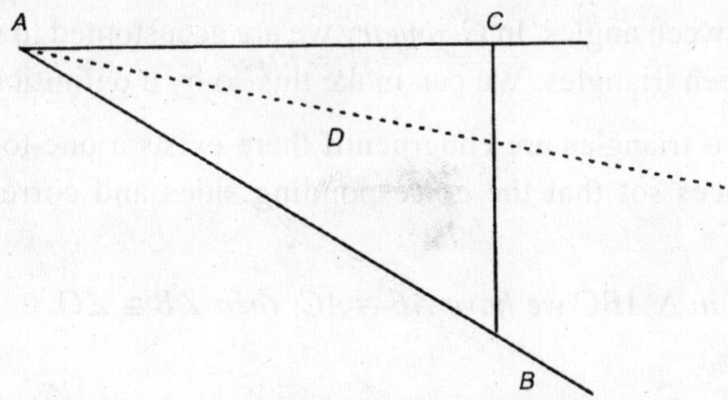

Fig. Crossbar Theorem

Proof: We are given a ray $\overrightarrow{AD}$ between rays $\overrightarrow{AB}$ and $\overrightarrow{AC}$. Let us use proof by contradiction and assume that $\overrightarrow{AD} \cap BC = \theta$

Let $\overrightarrow{AF}$ be the opposite ray to $\overrightarrow{AD}$ If $\overrightarrow{AF} \cap BC = P$, then $B * P * C$ and by Propositionwe have that P lies in the interior of $\angle CAB$.

However, this contradicts (*ii*) of Proposition, which says that no point on the opposite ray can be interior to the angle.

Thus, we have that $\overrightarrow{AF} \cap BC = \theta$ Now, this means that $\overleftrightarrow{AD}$ H $BC = \theta$ by Proposition, part (*ii*). It follows that B and C are on the same side of the line $\overleftrightarrow{AD}$.

Let E be a point on the line $\overleftrightarrow{AC}$ so that $C * A * E$. Then, C and E are on opposite sides of $\overleftrightarrow{AD}$ and by *Betweenness Axiom*, B and E are on opposite sides of $\overleftrightarrow{AD}$ Applying Proposition, part (*iii*) to this situation, we know that B is in the interior of $\angle DAE$, which means that B and E are on the same side of the line $\overleftrightarrow{AD}$. We now have a contradiction.

Thus, we have that $\overleftrightarrow{AD}$ H $BC \neq \theta$.

Lastly, we need to consider the incidence of rays and triangles.

Definition: The interiorof a triangle is the intersection of the interiors of the three angles. A point is in the exterior of the triangle if it is not in the interior or on any of the three sides.

Proposition:

- *If a ray* $\overrightarrow{PQ}$ *emanating from a point exterior to* ΔABC *intersects side AB in a point between A and B, then* $\overrightarrow{PQ}$ *also intersects AC or BC.*
- *If a ray emanates from an interior point of* ΔABC*, then it intersects on of the sides, and if it does not pass through a vertex, it intersects only one side.*

CONGRUENCE THEOREMS

According to our development, *congruence* has been a relationship between segments or a relationship between angles. In *Geometry* we are accustomed to seeing congruence as a relationship between triangles. We can make this so by a definition.

Definition: Two triangles are congruentif there exists a one-to-one correspondence between their vertices sot that the corresponding sides and corresponding angles are congruent.

Proposition: *If in* ΔABC *we have* $AB \cong AC$*, then* $\angle B \cong \angle C$.

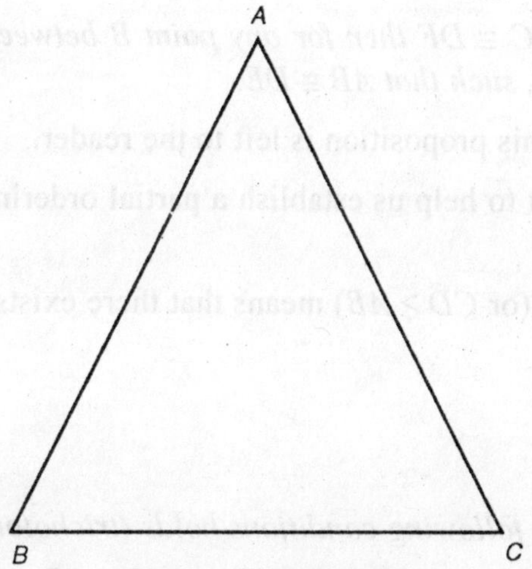

Fig. Isosceles Triangles

Proof: This simple proof is due to Pappus. Consider the correspondence of vertices $A \leftrightarrow A, B \leftrightarrow C$, and $C \leftrightarrow B$.

Under this correspondence, two sides and the included angle ΔABC of are congruent respectively to the corresponding sides and included angle of ΔACB Hence, by SAS the triangles are congruent. Therefore, the corresponding angles are congruent and $\angle B \cong \angle C$

Segment Subtraction

Proposition: *If $A * B * C$, $D * E * F$, $AB \cong DE$, and $AC \cong DF$ then $BC \cong EF$.*

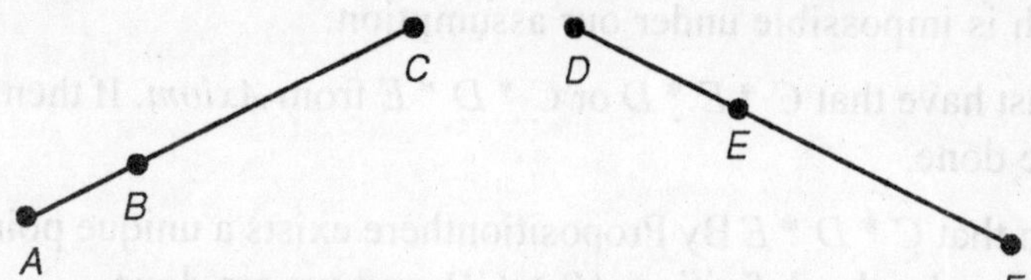

Fig. Segment Subtraction

Proof: From what we are given, let us assume that $BC \not\cong EF$ By *Axiom*, there exists a unique point G on ray $\overrightarrow{EF}$ so that $BC \cong EG$ By our hypothesis, we have that $G \neq F$ Since $AB \cong DC$ and $BC \cong EG$ by *Axiom* $AC \cong DG$.

But then by *Axiom* we have that $DF \cong DG$ It follows from *Axiom* that $F = G$. We have reached a contradiction. Thus, it must be that $BC \cong EF$ and we are done.

Proposition: *Given $AC \cong DF$ then for any point B between A and C, there exists a unique point E, $D * E * F$, such that $AB \cong DE$.*

Proof: The proof of this proposition is left to the reader.

We can use this result to help us establish a partial ordering on the line segments in the plane.

Definition: $AB < CD$ (or $CD > AB$) means that there exists a point E between C and D such that $AB \cong CE$

Segment Ordering

Proposition:

- *exactly one of the following conditions holds (trichotomy):*

  $AB < CD$, $AB \cong CD$, or $AB > CD$.

- *if $AB < CD$ and $CD \cong EF$ then $AB < EF$.*
- *if $AB > CD$ and $CD \cong EF$ then $AB > EF$.*
- *if $AB < CD$ and $CD < EF$, then $AB < EF$.*

Proof: We will prove the first item only. The proofs of the remaining three items will be left to you as homework.

We are given segments AB and CD. If $AB \cong CD$ then we are done. So, let us assume that $AB \not\cong CD$ By *Axiom* there is a unique point $E \in \overrightarrow{CD}$ so that $AB \cong CE$.$E \neq D$, else $AB \cong CE \cong CD$ which is impossible under our assumption.

Thus, we must have that $C * E * D$ or $C * D * E$ from *Axiom*. If then by definition $AB < CD$, and we are done.

Suppose then that $C * D * E$ By Propositionthere exists a unique point $F \in AB$ so that $AF \cong CD$ In this case by the definition $AB > CD$, and we are done.

Proposition: *Supplements of congruent angles are congruent.*

Proposition:

- *Vertical angles are congruent to each other.*
- *An angle congruent to a right angle is a right angle.*

Proposition: *For every line l and every point P there exists a line through P perpendicular to l*

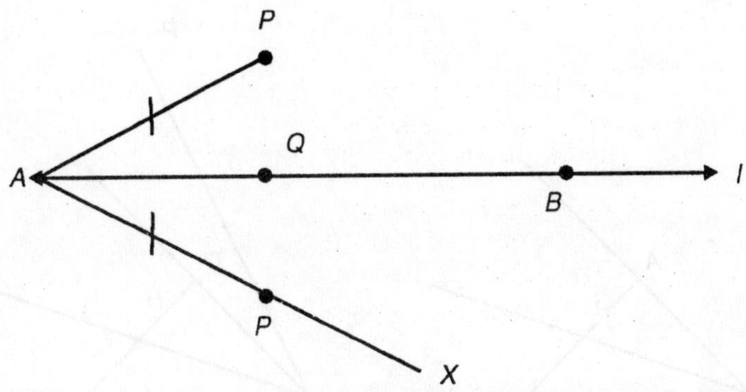

Proof: Either $P \in l$ or $P \notin l$ First, let us assume that $P \notin l$ Let $A, B \in l$ Such points exist by *Axiom*. The ray $\overrightarrow{AP}$ lies on one side of l. By *Axiom* there exists a ray $\overrightarrow{AX}$ on the opposite side of l from P so that

$$\angle XAB \cong \angle PAB.$$

By *Axiom* there is a point $P' \in \overrightarrow{AX}$ so that $AP \cong AP'$ Since P and P' are on opposite sides of l, the segment PP' intersects the line l. Let $\{Q\} = PP' \cap l$.

If $Q = A$, then $\angle PQB = \angle PAB \cong \angle P'AB = \angle P'QB$ Thus, $PP' \perp l$.

If $Q \neq A$ then $\Delta PAQ \cong \Delta P'AQ$ by *SAS*. Thus, from the definition of congruent triangles, $\angle PQA \cong \angle P'AQ$ and $PP' \perp l$.

Now, if $P \in l$ there is an $X \notin l$. From this point X apply the previous technique to construct a perpendicular line to l through X.

By *Axiom C-4* we can copy this angle on one side of l at P. From the Propositionthe other side of this angle is part of a line through P perpendicular to l.

Proposition: [ASA] *Given* ΔABC *and* ΔDEF *with* $\angle A \cong \angle D$, $\Pi C \cong \angle F$ *and* $AC \cong DF$ *Then* $\Delta ABC \cong \Delta DEF$.

Proposition: *If in* ΔABC *we have that* $\angle B \cong \angle C$ *then* $AB \cong AC$ *and* ΔABC *is isosceles.*

Proposition: [Angle Addition] *Given* $\overrightarrow{BG}$ *between* $\overrightarrow{BA}$ *and* $\overrightarrow{BC}$ $R \Rightarrow Q$ *between* $\overrightarrow{ED}$ *and* $\overrightarrow{EC}$, $\angle CBG \cong \angle FEH$ *and* $\angle GBA \cong \angle HED$ *Then* $\angle ABC \cong \angle DEF$.

Proposition: [Angle Subtraction] *Given* $\overrightarrow{BG}$ *between* $\overrightarrow{BA}$ *and* $\overrightarrow{BC}$, $R \Rightarrow Q$ *between* $\overrightarrow{ED}$ *and* $\overrightarrow{EF}$, $\angle CBG \cong \angle FEH$, *and* $\angle GBA \cong \angle HED$, *then* $\angle ABC \cong \angle DEF$.

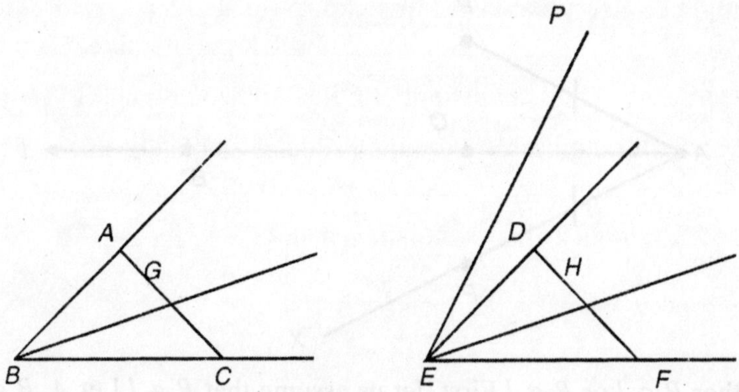

Fig. Angle subtraction

Proof: Since $\overrightarrow{BG}$ lies between $\overrightarrow{BA}$ and $\overrightarrow{BC}$ we may apply the *Crossbar Theorem* to find that $\overrightarrow{BG}$ intersects AC. Without loss of generality, we may assume that this point of intersection is, in fact, the point G. Then, we have that $A * G * C$ Assume that the points D, F, and H are chosen so that $AB \cong ED$, $BG \cong EH$, and $BC \cong EF$ This is nothing but a relabeling of the points.

Let us assume then that $\angle ABG \not\cong \angle DBG$ Then, there exists a unique ray, $\overrightarrow{EP}$ on the same side of $\overrightarrow{EH}$ as $\overrightarrow{ED}$ so that

$$\angle ABC \cong \angle ABG.$$

By our assumption, $\overrightarrow{EP} \neq \overrightarrow{ED}$ so by Proposition, $\angle ABC \cong \angle PEF$ Since $\angle ABC \cong \angle DEF$ the uniqueness of $\overrightarrow{EP}$ implies that $\overrightarrow{EP} = \overrightarrow{ED}$ a contradiction. Thus, $\angle ABG \cong \angle DEH$.

As with line segments, there is a natural method for defining an ordering on angles.

Definition: $\angle ABC < \angle DEF$ means that there exists a ray $\overrightarrow{EG}$ between $\overrightarrow{ED}$ and $\overrightarrow{EF}$ so that $\angle ABC \cong \angle GEF$

This gives us the following results, which are completely analogous to those for segments.

Ordering of Angles

Proposition:

- *Exactly one of the following conditions holds:*

  $\angle P < \angle Q$, $\angle P = \angle Q$, or $\angle P > \angle Q$.
- *If* $\angle P < \angle Q$ *and* $\angle Q \cong \angle R$ *then* $\angle P < \angle R$.
- *If* $\angle P > \angle Q$ *and* $\angle Q \cong \angle R$ *then* $\angle P > \angle R$.
- *If* $\angle P < \angle Q$ *and* $\angle Q < \angle R$ *then* $\angle P < \angle R$.

Proposition: *Given triangles ΔABC and ΔDEF If $AB \cong DE$, $BC \cong EF$ and $AC \cong DF$ then $\Delta ABC \cong \Delta DEF$.*

Euclid's Fourth Postulate

Proposition: *All right angles are congruent to each other.*

AXIOMS OF CONTINUITY

These axioms are the axioms which give us our correspondence between the real line and a Euclidean line. These are necessary to guarantee that our Euclidean plane is *complete*. The first axiom gives us some information about the relative sizes of segments as compared one to another. ARCHIMEDES' AXIOM. *If AB and CD are any segments, then there is a number n such that if segment CD is laid off n times on the ray $\overrightarrow{AB}$ emanating from A, then a point E is reached where n. $CD \cong AE$ and B is between A and E.* This is derived from the Archimedean Axiom in the real number system.

This should not be surprising, for we wish to have a one-to-one correspondence between each euclidean line and the set of *real numbers*. In the real line the Archimedean Postulate takes on the flavor: Archimedean Postulate: *Let M and e be any two positive numbers. Then there is a positive integer n such that ne > M.*

The main point for geometry is that if you choose any segment to be of unit length, then every segment has finite length with respect to this measure. Nothing can be too big, *n. CD > AB*; and nothing can be too small, *n. EF > CD* where CD was chosen as our unit length.

This is still not enough for the purposes of geometry, for the set of rational numbers Q satisfies this property, but causes trouble in another situation. Let us consider the set of rational points in the cartesian plane. Call this Q X Q Consider the ray passing from the origin through the point (1,1). This segment has length $\sqrt{2}$ Now, on the ray making up the positive real axis we are unable to find a point satisfying *Congruence Axiom 1.*

No point exists whose distance to the origin is $\sqrt{2}$ We require a stronger property. Dedekind's Axiom. *Suppose that the set of all points on a line l is the union $\Sigma_1 \cup \Sigma_2$ of two nonempty subsets such that no point of Σ_1 is between two points of Σ_2 and vice versa. Then there is a unique point O lying on l such that $P_1 * O * P_2$ if and only if $P_1 \in \Sigma_1$ and $P_2 \in \Sigma_2$ and $O \neq P_1, P_2$* In order to better understand this axiom, we need to study the concept in the real line.

Suppose that $a, b \in \mathbb{R}$ with $a < b$. Suppose that $S_1, S_2 \subset [a, b]$ satisfying

- $S_1 \cup S_2 = [a, b]$.
- $S1 \cap S2$.
- $a \in S_1$ and $b \in S_2$
- If $x \in S_1$ and $a < y < x$, then $y \in S_1$.

The fourth item above indicates that we have the property that if $X \in S_1$ then $[a, x] \subset S_1$ Let $z \in S_2$ CLAIM: If $z < w < b$, then $w \in S_2$ Assume not, so that $w \in S_1$ This means that $[a, w] \subset S_1$ This places $z \in S_1$, which then means that $z \in S_1 \cap S_2$ a contradiction to item. Thus, $w \in S_2$ and, in fact, $[z, b] \subset S_2$

There are three different situations possible here.

- $S_1 = \{a\}$ and $S_2 = (a, b]$.
- $S_1 = [a, b)$ and $S_2 = \{b\}$.
- If neither of the above two conditions occurs, then we classify numbers as S_1-numbers or S_2-numbers. We are willing to believe that there exists an element $c \in [a, b]$ so that if $x < c$ then $x \in S_2$ and if $x > c$ then The existence of such a real number is guaranteed by the *Least Upper Bound Axiom* for the real numbers. Since, S_1 is bounded above by b, it has a *least upper bound*. Does this least upper bound separate the sets S_1 and S_2 as above? Let c = lub S_1 and let $x \in S_1$ Then, by the definition of a least upper bound, $x < c$ and for any $c > 0$ $c - e \in S_1$ Thus, if $y < c$, then $y \in S_1$ If $y > c$ then y cannot be in S_1 for c is bigger than every element in S_1 Thus, $y \in S_2$

The sets S_1 and S_2 form a Dedekind cut of the set $[a,b]$, and c is called the Dedekind number of the cut. Our Dedekind's Axiom is a translation of this phenomenon to a line. Without Dedekind's Axiom there is no guarantee that there is a segment of length π or of length e, or of certain other *non-constructible* lengths, *e.g.* $\sqrt[3]{2}$ It is Dedekind's Axiom that allows us to make the correspondence of the line in our geometry and the real line. There are well-defined geometries that exist without the Dedekind Axiom, such as the geometry of the *surd plane*. They do not have all of the properties which we wish to have, or to which we are accustomed to having. It is with Dedekind's Axiom that we are able to introduce a coordinate system and do geometry analytically, in the fashion of Fermat and Descartes.

To see why we will want Dedekind's Axiom, consider the manner in which you construct the perpendicular to a given line at a given point. First, using the given point as a center, draw a circle of positive radius. The circle intersects the line in two points. At each of these points, you then construct a circle of larger radius and these two circles intersect in two points. Drawing the line between the two points of intersection gives a line perpendicular to the given line at the given point. There are two problems with this proof, and they are both very subtle. Why does the line intersect the circle at all? Why do the two circles then intersect? In the *surd plane* these are not necessarily true!

The first of these problems is addressed by the following principle, which follows from Dedekind's Axiom. We define a point A to be inside a circle centered at O with radius OR if $OA < OR$. A point B is outside the circle if $OB > OR$. Elementary Continuity Principle. *If one endpoint of a segment is inside a circle and the other outside, then the segment intersects the circle.* We shall prove this after we have developed some theorems in geometry that we shall need. The second problem above is that of the intersection of two circles. It is addressed by the following principle, which is again a corollary of Dedekind's Axiom. Circular Continuity Principle. *If a circle has one point inside and one point outside another circle γ then the two circles intersect in two points.* We shall prove this later also.

NEUTRAL GEOMETRY

How much of our "geometry" does not depend on a parallel axiom? How much is independent of this particular postulate? As you will see, a large portion of our knowledge does not depend on any parallel postulate. This study will then help us to see the true role of a parallel postulate in Geometry.

ALTERNATE INTERIOR ANGLES

Definition: Let L be a set of lines in the plane. A line k is transversalof L if

1. $k \notin L$, and
2. $k \cap m \neq \theta$ for all $m \in L$.

Let l be transversal to m and n at points A and B, respectively. We say that each of the angles of intersection of l and m and of l and n has a *transversal side* in and a *non-transversal side* not contained in l

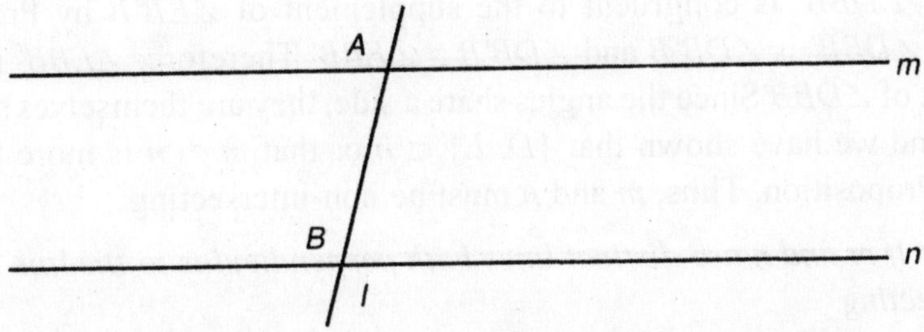

Definition: An angle of intersection of m and k and one of n and k are alternate interior angles if their transversal sides are opposite directed and intersecting, and if their non-transversal sides lie on opposite sides of l. Two of these angles are corresponding angles if their transversal sides have like directions and their non-transversal sides lie on the same side of l

Definition: If k and l are lines so that $k \cap l = \theta$ we shall call these lines non-intersecting.

We want to reserve the word parallel for later.

Alternate Interior Angle Theorem

Theorem: *If two lines cut by a transversal have a pair of congruent alternate interior angles, then the two lines are non-intersecting.*

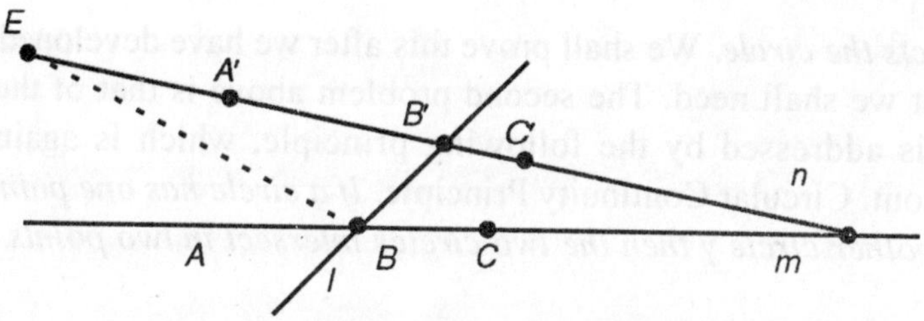

Fig. Alternate Interior Angles

Proof: Let *m* and *n* be two lines cut by the transversal *l*. Let the points of intersection be *B* and *B'*, respectively. Choose a point *A* on *m* on one side of *l*. and choose $A' \in n$ on the same side of *l* as *A*. Likewise, choose $OP \cong OA$ on the opposite side of *l* from *A*. Choose $\overrightarrow{AB} = \{C \in \overrightarrow{AB} \mid C \in AB$ or *B* lies between *A* and $C\}$ on the same side of *l* as *C*. Hence, it is on the opposite side of *l* from *A'*, by the *Plane Separation Axiom*.

We are given that $\angle A'\ B'B \cong \angle CBB'$ Assume that the lines *m* and *n* are not non-intersecting; *i.e.*, they have a nonempty intersection. Let us denote this point of intersection by *D*. *D* is on one side of *l*, so by changing the labeling, if necessary, we may assume that *D* lies on the same side of *l* as *C* and *C'*. By *Congruence Axiom 1* there is a unique point $E \in \overline{B'A'}$ so that $B'E \cong BD$ Since, $BB' \cong BB'$ (by Axiom C-2), we may apply the SAS Axiom to prove that

$$\Delta EBB' \cong \Delta DBB'.$$

From the definition of congruent triangles, it follows that $\angle DB'B \cong \angle EBB'$ Now, the supplement of $\angle DBB'$ is congruent to the supplement of $\angle EB'B$ by Proposition. The supplement of $\angle DBB'$ is $\angle DB'B$ and $\angle DB'B \cong \angle EBB'$ Therefore, $\angle EBB'$ is congruent to the supplement of $\angle DBB'$Since the angles share a side, they are themselves supplementary. Thus, $E \in n$ and we have shown that $\{D, E\} \subset n$ or that $m \cap n$ is more that one point, contradicting Proposition. Thus, *m* and *n* must be non-intersecting.

Corollary: *If m and n are distinct lines both perpendicular to the line l then m and n are non-intersecting.*

Proof: *l* Is the transversal to *m* and *n*. The alternate interior angles are right angles. By Propositionall right angles are congruent, so the *Alternate Interior Angle Theorem* applies. *m* and *n* are non-intersecting.

Corollary: *If P is a point not on l then the perpendicular dropped from P to is unique.*

Proof: Assume that *m* is a perpendicular to *l* through *P*, intersecting *l* at *Q*. If *n* is another perpendicular to *l* through *P* intersecting *l* at *R*, then *m* and *n* are two distinct lines perpendicular to *l* By the above corollary, they are non-intersecting, but each contains *P*. Thus, the second line cannot be distinct, and the perpendicular is unique.

The point at which this perpendicular intersects the line *l*, is called the *foot* of the perpendicular.

Corollary: *If l is any line and P is any point not on l there exists at least one line m through P which does not intersect l*

Proof: By Corollary there is a unique line, *m*, through *P* perpendicular to *l* By Propositionthere is a unique line, *n*, through *P* perpendicular to *m*. By Corollary 1 *l* and *n* are non-intersecting.

Note that while we have proved that there is a line through *P* which does not intersect *l* we have not (and cannot) proved that it is *unique*.

WEAK EXTERIOR ANGLE THEOREM

Let ΔABC be any triangle in the plane. This triangle gives us not just three segments, but in fact three lines.

Definition: An angle supplementary to an angle of a triangle is called an exterior angle of the triangle. The two angles of the triangle not adjacent to this exterior angle are called the remote interior angles.

Exterior Angle Theorem

Theorem: *An exterior angle of a triangle is greater than either remote interior angle.*

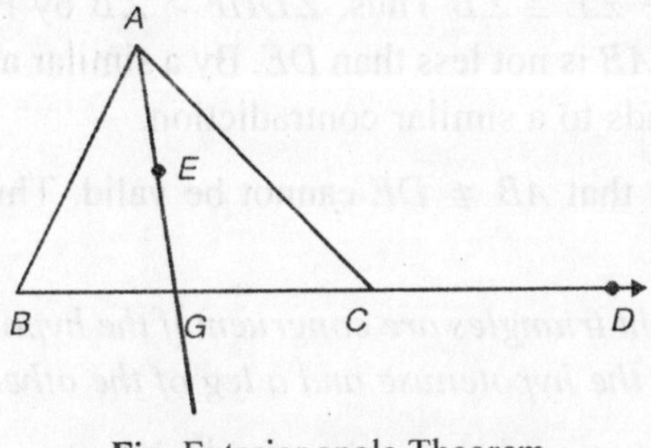

Fig. Exterior angle Theorem

Proof: We shall show that $\angle ACD > \angle A$ In a like manner, you can show that $\angle ACD > \angle B$ Then by using the same techniques, you can prove the same for the other two exterior angles.

By Proposition part *i)* either:

$$\angle A < \angle ACD \quad \angle A \cong \angle ACD \quad \angle A > \angle ACD$$

If $\angle A = \angle BAC \cong \angle ACD$, then by the *Alternate Interior Angle Theorem*, $\overleftrightarrow{AB}$ and $\overleftrightarrow{CD}$ are non-intersecting. This is impossible, since they both contain *B*.

Assume, then, that $\angle A > \angle ACD$ By the definition of this ordering on angles, there exists a ray $\overrightarrow{AE}$ between $\overrightarrow{AB}$ and so $\overrightarrow{AC}$ that

By the *Crossbar Theorem*, $\overrightarrow{AE}$ intersects *BC* in a point *G*. Again by the *Alternate Interior Angle Theorem* $\overleftrightarrow{AB}$ and $\overleftrightarrow{CD}$ are non-intersecting. This is a contradiction.

Thus, $\angle A < \angle ACD$

Proposition: [SAA Congruence] *In triangles* ΔABC *and* ΔDEF *given that* $AC \cong DF$, $\angle A \cong \angle D$ *and* $\angle B \cong \angle E$, *then* $\Delta ABC \cong \Delta DEF$.

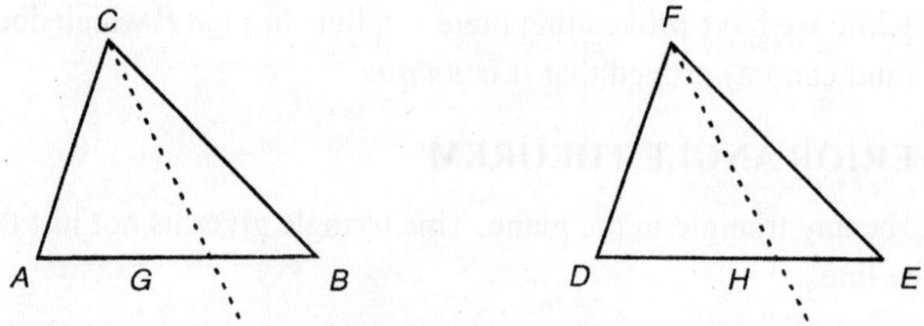

Fig. Side-Angle-Angle

Proof: If $AB \cong DE$ we are done by *Angle-Side-Angle*. Thus, let us assume that $AB \not\cong DE$ Then, by Propositionwe must have that either $AB < DE$ or $AB > DE$.

If $AB < DE$, then there is a point $H \in DE$ so that $AB \cong DH$ Then by the *SAS Axiom* $\Delta ABC \cong \Delta DHF$ Thus, $\angle B \cong \angle DHF$ But $\angle DHF$ is exterior to ΔFHE so by the *Exterior Angle Theorem* $\angle DHF > \angle E \cong \angle B$ Thus, $\angle DHF > \angle B$ by Proposition, and we have a contradiction. Therefore, *AB* is not less than *DE*. By a similar argument, we can show that assuming that $AB > DE$ leads to a similar contradiction.

Thus, our hypothesis that $AB \not\cong DE$ cannot be valid. Thus, $AB \cong DE$ and $\Delta ABC \cong \Delta DEF$ by *ASA*.

Proposition: *Two right triangles are congruent if the hypotenuse and a leg of one are congruent respectively to the hypotenuse and a leg of the other.*

Proposition: *Every segment has a unique midpoint.*

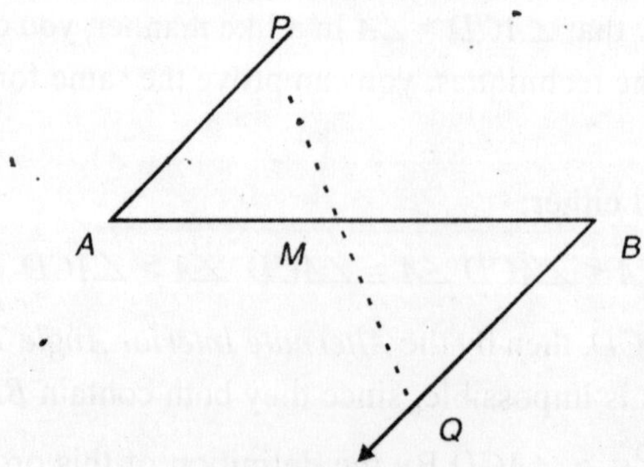

Fig. Segment Midpoint

Proof: Let AB be any segment in the plane, and let C be any point not on $\overleftrightarrow{AB}$ Such a point exists by Axiom. There exists a unique ray $\overrightarrow{BX}$ on the opposite side of $\overleftrightarrow{AB}$ from P such that $\angle PAB \cong \angle XBA$ by Axiom. There is a unique point $Q \in \overrightarrow{BX}$ so that $AP \cong BQ$ by Axiom. Q is on the opposite side of $\overleftrightarrow{AB}$ from P by Axiom. Since P and Q are on opposite sides of $\overleftrightarrow{AB}$ $PQ \cap \overleftrightarrow{AB} \neq \theta$ Let M denote this point of intersection. By Axiom, either $A * M * B$, $M * A * B$, $A * B * M$, $M = A$, or $M = B$.

We want to show that $A * M * B$, so let us assume $\sim A * M * B$ Since $\angle PAB \cong \angle QBA$ by construction, we have from the *Alternate Interior Angle Theorem* that $\overleftrightarrow{AP}$ and $\overleftrightarrow{BQ}$ are non-intersecting. If $M = A$ then $A, P, M \in \overleftrightarrow{AP}$ and $\overleftrightarrow{AP} = \overleftrightarrow{AB}$ which intersects $\overleftrightarrow{BQ}$ We can dispose of the case $M = B$ similarly.

Thus, assume that $M * A * B$ This will mean that the line $\overleftrightarrow{PA}$ will intersect side MB of ΔMBQ at a point between M and B.

Thus, by *Pasch's Theorem* it must intersect either MQ or BQ. It cannot intersect side BQ as $\overleftrightarrow{AP}$ and $\overleftrightarrow{BQ}$ are non-intersecting.

If $\overleftrightarrow{AP}$ intersects MQ then it must contain MQ for P, Q, and M are collinear. Thus, $M = A$ which we have already shown is impossible. Thus, we have shown that $M * A * B$ is not possible.

In the same manner, we can show that $A * B * M$ is impossible. Thus, we have that $A * M * B$. This means that $\angle AMP \cong \angle BMQ$ since they are vertical angles. By *Angle-Angle-Side* we have that $\Delta AMP \cong \Delta BMQ$ Thus, $AM \cong MB$ and M is the midpoint of AB.

Proposition:

- *Every angle has a unique bisector.*
- *Every segment has a unique perpendicular bisector.*

Proposition: *In a triangle ΔABC the greater angle lies opposite the greater side and the greater side lies opposite the greater angle; i.e., $AB > BC$ if and only if $\angle C > \angle A$*

Proposition: *Given ΔABC and $\Delta A'B'C'$ if $AB \cong A'B'$ and $BC \cong B'C'$ then $\angle B < \angle B'$ if and only if $AC < A'C'$.*

THEOREMS OF CONTINUITY

ELEMENTARY CONTINUITY PRINCIPLE

We will now take up the Axioms of Continuity. First, we will prove the two Continuity Principles and then discuss some of the different uses of the Continuity Axioms in our work. First, we shall need the famous *Triangle Inequality*. It is usually proved after we have given a measure to line segments, but that is not necessary.

Triangle Inequality

Proposition: *If A, B, and C are three noncollinear points, then AC<AB+BC, where the sum is segment addition.*

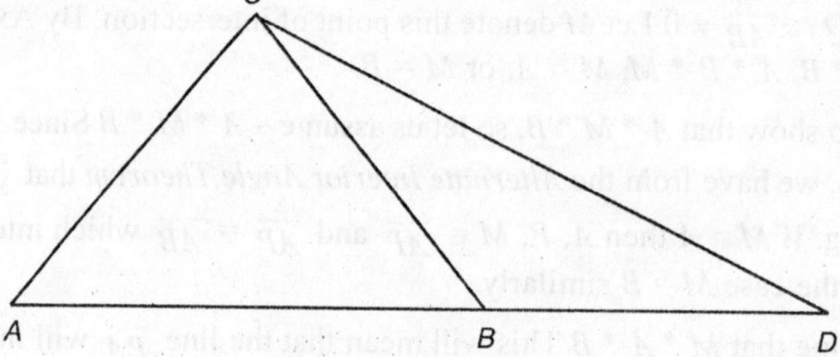

Fig. Triangle Inequality

Proof: There is a unique point D such that $A * B * D$ and $BD \cong BC$ by the first congruence axiom. Then $\angle BCD \cong \angle BDC$ by Proposition. Now, $AD \cong AB + BD$ and since $BD \cong BC$ we have that $AD \cong AB + BC$ By Proposition $\overrightarrow{CB}$ is between the rays $\overrightarrow{CA}$ and $\overrightarrow{CD}$ Then, by definition, $\angle ACD \cong \angle BCD$ By Proposition and what we have just shown, $\angle ACD > \angle ADC$ Thus, by Proposition $AD > AC$, and we are done.

Elementary Continuity Principle

Theorem: *If one endpoint of a segment is inside a circle and the other outside, then the segment intersects the circle.*

Proof: Let Γ be a circle centered at O with radius OR, so that $R \in \Gamma$ Let AB be the above segment. Let Int $(\Gamma) = \{A| \ OA < OR\}$ and Ext $(\Gamma) = \{B| \ OB > OR\}$. Let

$$\Sigma_1 = \{X \mid X \in AB \text{ and } OX < OR\}$$

$$\Sigma_2 = \{X \mid X \Xi AB \text{ and } OX \geq OR\}.$$

Note that neither of Σ_1 or Σ_2 is empty. We wish to show that these two sets form a Dedekind cut of the segment AB. There are then several things to show.

- If $Y \in AB$ then by Proposition we must have that $OX < OR$, $OX \cong OR$, or $OX > OR$ This puts X in one of the two sets. Therefore, $\Sigma_1 \cup \Sigma_2 = AB$
- Clearly, $\Sigma_1 \cap \Sigma_2 = \theta$
- We have that and
- Let and We need to show then that If then is exterior to so Thus, by Proposition$OC<OX$. Since $OX<OR$ by Proposition$OC<OR$ and

Now, assume that $OA \not\equiv OX$ Then, for the first case, assume that $OA < OX$. Hence, $\angle X < \angle A$ Again, $\angle A < \angle OCX$ which implies $OC < OX < OR$. Thus, $C \in \Sigma_1$ If $OA > OX$ reverse the labels in the previous argument and we again find that $C \in \Sigma_1$.

Thus, Σ_1 and Σ_2 form a Dedekind cut of AB. Let M be the cut point.

Claim: $OM \cong OR$

Assume $OM < OR$. This puts $M \in \Sigma_1$ and so $M \neq B$ Now, there is a segment which is less than the difference between OM and OR. Let M' be a point outside Γ and on AB so that MM' is congruent to that segment. Thus, we have that $M' \in \Sigma_2$ By the Triangle Inequality we have that

$$OM' < OM + MM'$$

$$OM + MM' < OR$$

$$OM' < OR$$

But, $M' \in \Sigma_2$ so that $OM' \geq OR$: a contradiction.

Assuming that $OM > OR$ we use the same technique to arrive at a contradiction. Thus, $OM \cong OR$ and $M \in \Gamma$.

MEASURE OF ANGLES AND SEGMENTS

To avoid some of the difficulties that we faced in the previous two proofs, and to facilitate matters at a later time, we will introduce a measure for angles and for segments.

The proof of the Theorem requires the axioms of continuity for the first time. The axioms of continuity are not needed if one merely wants to define the addition for congruence classes of segments and then prove the triangle inequality for these congruence classes. It is in order to prove several of our theorems that we need the measurement of angles and segments by real numbers, and for such measurement Archimedes's axiom is required. However, the fourth and eleventh parts of Theorem, the proofs for which require Dedekind's axiom, are never used in proofs in the text. It is possible to introduce coordinates without the continuity axioms, as in discussed in Appendix B of the text.

The notation $\angle A°$ will be used for the number of degrees in $\angle A$, and the length of segment AB will be denoted by $\overline{AB}$

Theorem: *There is a unique way of assigning a degree measure to each angle such that the following properties hold:*

- $\angle A°$ *is a real number such that* $0 < \angle A° < 180°$.
- $\angle A° = 90°$ *if and only if* $\angle A$ *is a right angle.*
- $\angle A° = \angle B°$ *if and only if* $\angle A \cong \angle B$
- *If* $\overrightarrow{AC}$ *is interior to* $P \notin l$ *then* $\angle DAB° = \angle DAC° + \angle CAB°$.
- *For every real number x between 0 and* 180, *there exists an angle* $\angle A$ *such that* $\angle A° = x°$

- *If $\angle B$ is supplementary to $\angle A$ then $\angle A^\circ + \angle B^\circ = 180^\circ$.*
- *if $\angle A^\circ > \angle B^\circ$ and only if $\angle A > \angle B$.*

Given a segment OI, called the unit segment. Then there is a unique way of assigning a length $\overline{AB}$ to each segment AB such that the following properties hold:

- *$\overline{AB}$ is a positive real number and $\overline{OI} = 1$.*
- *$\overline{AB} = \overline{CD}$ if and only if $AB \cong CD$.*
- *$A * B * C$ if and only if $\overline{AC} = \overline{AB} + \overline{BC}$.*
- *$\overline{AB} < \overline{CD}$ if and only if $AB < CD$.*
- *For every positive real number x, there exists a segment AB such that $\overline{AB} = x$.*

Definition: An angle $\angle A$ is acuteif $\angle A^\circ < 90^\circ$ and is obtuseif $\angle A^\circ > 90^\circ$.

Corollary: *The sum of the degree measures of any two angles of a triangle is less than* 180°.

This follows from the Exterior Angle Theorem.

Proof: We want to show that $\angle A^\circ + \angle B^\circ < 180^\circ$ From the Exterior Angle Theorem,

$$\angle A^\circ < \angle CBD^\circ$$

$$\angle A^\circ + \angle B^\circ < \angle CBD^\circ + \angle B^\circ = 180^\circ,$$

Since they are supplementary angles.

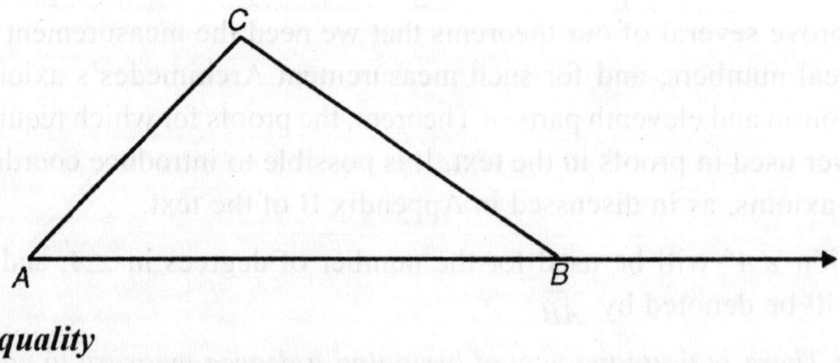

Triangle Inequality

Corollary: *If A, B, and C are three noncollinear points, then $\overline{AC} < \overline{AB} + \overline{BC}$.*

Theoremoffers an easier proof of this than the one that we gave.

SACCHERI-LEGENDRE THEOREM

This theorem gives us a setting for our later exploration into noneuclidean geometry.

Saccheri-Legendre Theorem

Theorem: *The sum of the degree measures of the three angles in any triangle is less than or equal to 180°;*

$$\angle A^\circ + \angle B^\circ + \angle C^\circ \le 180^\circ.$$

Proof: Let us assume not; *i.e.*, assume that we have a triangle ΔABC in which $\Pi A^\circ + \Pi B^\circ + \Pi C^\circ > 180^\circ$ So there is an $x \in \mathbb{R}^+$ so that

$$\angle A^\circ + \angle B^\circ + \angle C^\circ = 180^\circ + x.$$

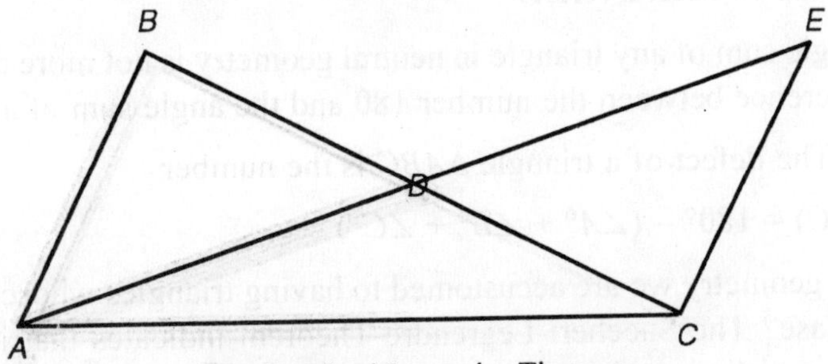

Fig. Saccheri-Legendre Theorem

Let D be the midpoint of BC and let E be the unique point on $\overrightarrow{AD}$ so that $DE \cong AD$ Then by SAS $\Delta BAD \cong \Delta CED$. This makes

$$\angle B^\circ = \angle DCE^\circ \; \angle E^\circ = \angle BAD^\circ.$$

Thus,

$$\begin{aligned}\angle A^\circ + \angle B^\circ + \angle C^\circ &= (\angle BAD^\circ + \angle EAC^\circ) + \angle B^\circ + \angle ACB^\circ \\ &= \angle E^\circ + \angle EAC^\circ + (\angle DEC^\circ + \angle ACD^\circ) \\ &= \angle E^\circ + \angle A^\circ + \angle C^\circ\end{aligned}$$

So, ΔABC and ΔACE have the same angle sum, even though they need not be congruent. Note that $\angle BAE^\circ + \angle CAE^\circ = \angle BAC^\circ$, hence

$$\angle CEA^\circ + \angle CAE^\circ = \angle BAC^\circ$$

It is impossible for both of the angles $\angle CEA$ and $\angle CAE$ to have angle measure greater than $1/2 \angle BAC^\circ$, so at least one of the angles has angle measure greater than or equal to $1/2 \angle BAC^\circ$.

Therefore, there is a triangle ΔACE so that the angle sum is $180^\circ + x$ but in which one angle has measure less than or equal to $1/2 \angle A^\circ$ Repeat this construction to get another triangle with angle sum $180^\circ + x$ but in which one angle has measure less than or equal to $1/4 \angle A^\circ$ Now there is an $n \in \mathbb{Z}^+$ so that

$$\frac{1}{2^n} \angle A^\circ \le x,$$

By the Archimedean property of the real numbers.

Thus, after a finite number of iterations of the above construction we obtain a triangle with angle sum $180^\circ + x$. in which one angle has measure less than or equal to x

Then the other two angles must sum to a number greater than 180°.

Corollary: *In ΔABC the sum of the degree measures of two angles is less than or equal to the degree measure of their remote exterior angle.*

THE DEFECT OF A TRIANGLE

Since the angle sum of any triangle in neutral geometry is not more than 180°, we can compute the difference between the number 180 and the angle sum of a given triangle.

Definition: The defect of a triangle ΔABC is the number

$$defect\ (\Delta ABC) = 180° - (\angle A° + \angle B° + \angle C°)$$

In euclidean geometry we are accustomed to having triangles whose defect is zero. Is this always the case? The Saccheri-Legrendre Theorem indicates that it may not be so. However, what we wish to see is that the *defectiveness* of triangles is preserved. That is, if we have one defective triangle, then all of the triangles are defective. By defective, we mean that the triangles have positive defect.

Additivity of Defect

Theorem: *Let ΔABC be any triangle and let D be a point between A and B. Then*

$$defect\ (\Delta ABC) = defect\ (\Delta ACD) + defect\ (\Delta BCD)$$

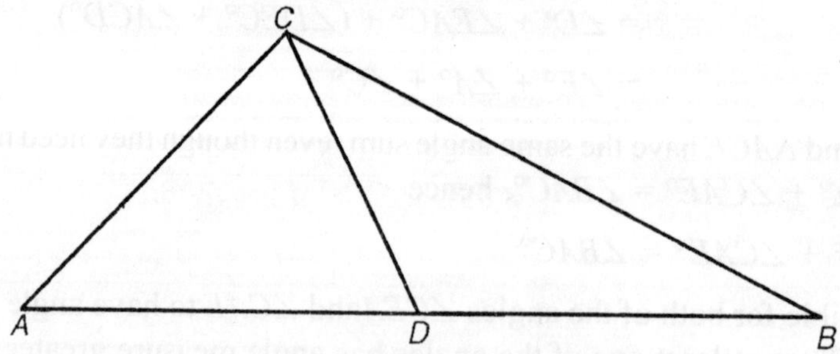

Proof: Since $\overrightarrow{CD}$ lies in $\angle ACB$, we know that

$$\angle ACB° = \angle ACD° + \angle BCD°,$$

And since $\angle ADC$ and $\angle BDC$ are supplementary angles $\angle ADC° + \angle BDC° = 180°$. Therefore,

$$\begin{aligned} defect\ (\Delta ABC) &= 180° - (\angle A° + \angle B° + \angle C°) \\ &= 180° - (\angle A° + \angle B° + \angle ACD° + \angle BCD°) \\ &= 180° + 180° - (\angle A° + \angle B°) \\ &\quad + \angle ACD° + \angle BCD° + \angle ACD° + \angle BDC°) \\ &= defect\ (\Delta\ ACD) + defect\ (\Delta BCD). \end{aligned}$$

Corollary: *Defect* (Δ*ABC*) = 0 *if and only if defect* (Δ*ACD*) = *defect* (Δ*BCD*) = 0.

A rectangleis a quadrilateral all of whose angles are right angles. We cannot prove the existence or non-existence of rectangles in Neutral Geometry. Nonetheless, the following result is extremely useful.

Theorem: *If there exists a triangle of defect 0, then a rectangle exists. If a rectangle exists, then every triangle has defect 0.*

Let me first outline the proof in five steps.

1. Construct a *right* triangle having defect 0.
2. From a right triangle of defect 0, construct a rectangle.
3. From one rectangle, construct arbitrarily large rectangles.
4. Prove that all *right* triangles have defect 0.
5. If every *right* triangle has defect 0, then *every* triangle has defect 0.

Having outlined the proof, each of the steps is relatively straightforward.

1. Construct a *right* triangle having defect 0.

Let us assume that we have a triangle Δ*ABC* so that *defect* (Δ*ABC*) = 0 We may assume that Δ*ABC* is not a right triangle, or we are done.

Now, at least two angles are acute since the angle sum of any two angles is always less than 180° Let us assume that ∠*A* and ∠*B* are acute. Also, let *D* be the foot of *C* on $\overleftrightarrow{AB}$. We need to know that *A* * *D* * *B*.

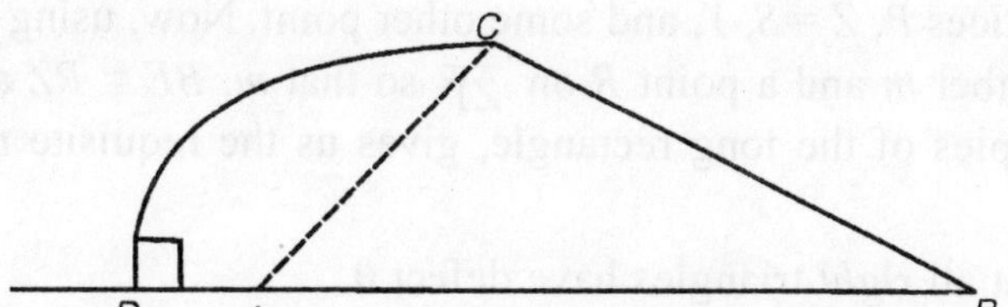

Fig. Right Triangle with Defect 0

Assume not; *i.e.*, assume that *D* * *A* * *B*. This means that ∠*CAB* is exterior to Δ*CAD* and, therefore, ∠*A*° > ∠*CAD*° = 90° This makes ∠*A* obtuse, a contradiction. Similarly, if *A* * *B* * *D* we can show that ∠*B* is obtuse. Thus, we must have that *A* * *D* * *B*.

This makes Δ*AFC* and Δ*BDC* right triangles.

By Corollary above, since Δ*ABC* has defect 0, each of them has defect 0, and we have two right triangles with defect 0.

2. From a right triangle of defect 0, construct a rectangie.

We now have a right triangle of defect 0. Take Δ*CBD* from Step, which has a right angle at *D*. There is a unique ray $\overrightarrow{CX}$ on the opposite side of $\overleftrightarrow{BC}$ from *D* so that

∠*DBC* ≅ ∠*BCX*,

from Axiom. From Axiom, there is a unique point E on $\overline{CX}$ such that $CE \cong BD$.

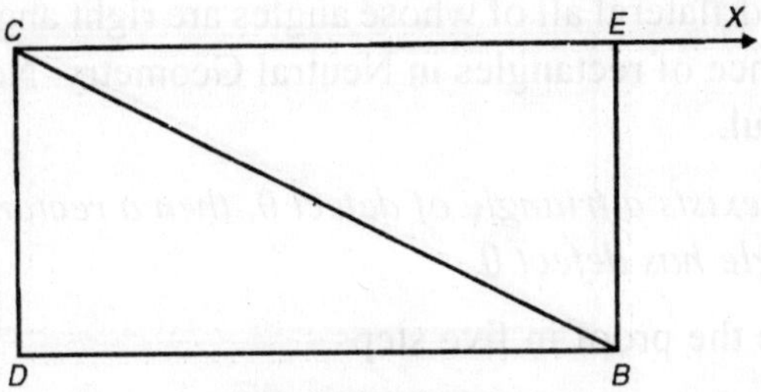

Fig. Rectangles

Thus, $\Delta CDB \cong \Delta BEC$ by *SAS*. Then $\angle BEC° = 90°$ and ΔBEC must also have defect 0. Now, clearly, Since *defect* $(\Delta DCB) = 0$

$\angle DBC° + \angle BCD° = 90°$

And, hence,

$$\angle ECB° + \angle BCD° = \angle ECD° = \angle 90°.$$

Likewise, $\angle EBD° = 90°$ and $\square CDBE$ is a rectangle.

3. From one rectangle, construct arbitrarily large rectangles.

Given any right triangle ΔXYZ we can construct a rectangle $\square PQRS$ so that $PS > XZ$ and $RS > YZ$. By applying Archimedes Axiom, we can find a number n so that we lay off segment BD in the above rectangle on the ray $\overline{ZX}$ to reach the point P so that $n.\ BD \cong PZ$ and $P * X * Z$ We make n copies of our rectangle sitting on $PZ = PS$. This gives us a rectangle with vertices P, $Z = S$, Y, and some other point. Now, using the same technique, we can find a number m and a point R on $\overline{ZY}$ so that $m.\ BE \cong RZ$ and $R * Y * Z$ Now, constructing m copies of the long rectangle, gives us the requisite rectangle containing ΔXYZ.

4. Prove that all *right* triangles have defect 0.

Let ΔXYZ be an arbitrary right triangle. We can embed it in a rectangle $\square PQRS$.

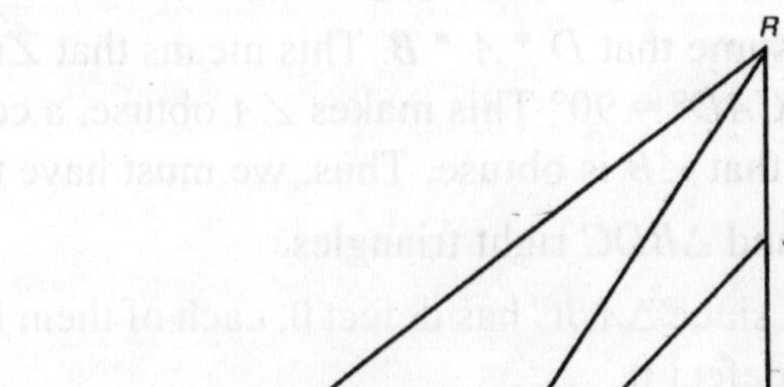

Fig. Defect 0 for all Right Triangles

Since $\Delta PQR \cong \Delta PSR$ we have that $\angle RPS° + \angle PRS° = 90°$ and then, ΔPRS has defect 0. Using Corollary we find *defect* $(\Delta RXY) = 0$ and thus, *defect* $(\Delta XYZ) = 0$ Therefore, each triangle has defect 0.

5. If every *right* triangle has defect 0, then *every* triangle has defect 0.

As in the first step, use the foot of a vertex to decompose the triangle into two right triangles, each of which has defect 0, from Step 4. Thus, the original triangle has defect 0.

Corollary: *If there is a triangle with positive defect, then all triangles have positive defect.*

THE WORK OF SACCHERI AND GAUSS

SACCHERI

Girolamo Saccheri was a Jesuit priest living from 1667 to 1733. Before he died he published a book entitled *Euclides ab omni nζvo vindicatus* (*Euclid Freed of Every Flaw*). It sat unnoticed for over a century and a half until rediscovered by the Italian mathematician Beltrami.

He wished to prove Euclid's Fifth Postulate from the other axioms. To do so he decided to use a *reductio ad absurdum* argument. He assumed the negation of the Parallel Postulate and tried to arrive at a contradiction. He studied a family of quadrilaterals that have come to be called *Saccheri quadrilaterals*. Let S be a convex quadrilateral in which two adjacent angles are right angles. The segment joining these two vertices is called the base. The side opposite the base is the summit and the other two sides are called the sides. If the sides are congruent to one another then this is called a Saccheri quadrilateral. The angles containing the summit are called the summit angles.

Theorem: *In a Saccheri quadrilateral*

- *The summit angles are congruent, and*
- *The line joining the midpoints of the base and the summit—called the altitude--is perpendicular to both.*

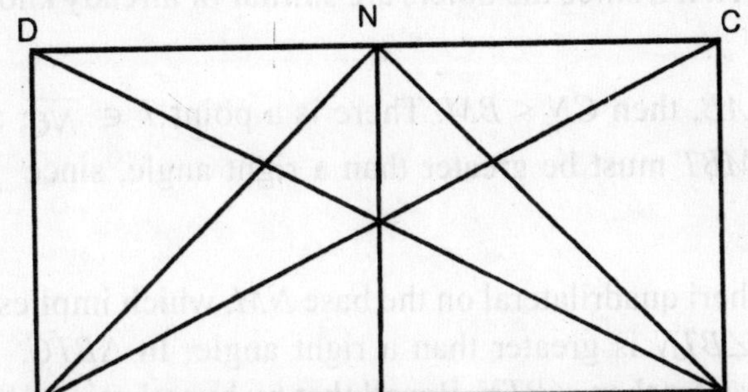

Proof: Let M be the midpoint of AB and let N be the midpoint of CD.

1. We are given that

$$\angle DAB^\circ = \angle ABC^\circ = 90^\circ.$$

Now, $AD \cong BC$ and $AB \cong AB$ so that by *SAS* $\Delta DAB \cong \Delta CAB$ which implies that $BD \cong AC$ Also, since $CD \cong CD$ then we may apply the *SSS* criterion to see that $\Delta CDB \cong \Delta DCA$ Then, it is clear that $\angle D \cong \angle C$

2. We need to show that $\overleftrightarrow{MN}$ is perpendicular to both $\overrightarrow{AB}$ and $\overrightarrow{CD}$ Now $DN \cong CN$, $AD \cong BC$ and $\angle D \cong \angle C$ Thus by *SAS* $\Delta ADN \cong \Delta BCN$ This means then that $AN \cong BN$ Also, $AM \cong BM$ and $MN \cong MN$ By *SSS* $\Delta ANM \cong \Delta BNM$ and it follows that $\angle AMN \cong \angle BMN$ They are supplementary angles, hence they must be right angles. Therefore $\overleftrightarrow{MN}$ is perpendicular to $\overrightarrow{AB}$.

' Using the analogous proof and triangles ΔDMN and ΔCMN we can show that $\overleftrightarrow{MN}$ is perpendicular to $\overrightarrow{CD}$.

Thus, we are done.

Saccheri considered all possible cases of such a quadrilateral. They are:

$\angle ADC$ is a right angle HRA, the hypothesis of the right angle.

$\angle ADC$ is an obtuse angle HOA, the hypothesis of the obtuse angle.

$\angle ADC$ is an acute angle HAA, the hypothesis of the acute angle.

We shall see that HRA is equivalent to Euclid's Postulate V, so we may take HOA or HAA as negations of Postulate V. The Three Musketeers Theorem implies that if one of HRA, HOA, or HAA holds for one quadrilateral, then it holds for all.

Theorem: *In a Saccheri quadrilateral ABCD on the base AB under the assumption HRA, HOA, or HAA we have $AB \sim CD$, $AB > CD$, or $AB < CD$, respectively, and the angle sum of a triangle is equal to, greater than, or less than two right angles, respectively.*

Proof: Let M and N denote the midpoints of AB andCD, respectively. We will work with the assumption HAA, since the others are similar or already known. We wish to show that $AB < CD$.

Suppose $CD < AB$, then $CN < BM$. There is a point $T \in \overrightarrow{NC}$ so that $NT \sim BM$ and $N * C * T$ Thus, $\angle MBT$ must be greater than a right angle, since $\overrightarrow{BC}$ is between $\overrightarrow{BM}$ and $\overrightarrow{BT}$.

NMBT is a Saccheri quadrilateral on the base *NM*, which implies that $\angle BTN \cong \angle TBN$. which implies that $\angle BTN$ is greater than a right angle. In ΔBTC. $\angle BCN$ is an exterior angle and hence is greater than $\angle BTN$. Recall that by hypothesis, $\angle BCN$ is an acute angle.

Thus, we have an acute angle greater than an angle that is greater than a right angle, our contradiction. Therefore, $AB < CD$.

Now, suppose we have a right triangle ΔABC with right angle at B. Construct AD perpendicular to AB, with $AD \sim BC$ Assuming HAA, $CD > AB$ and $\angle DAC > \angle ACB$ since the greater side subtends the greater angle. Now, $\angle CAB + \angle DAC$ is a right angle. Thus, the angle sum of ΔABC is less than two right angles.

Consider two parallel lines k and l and let $k = \overleftrightarrow{AB}$ and $l = \overleftrightarrow{CD}$ Let $X, Y \in \overleftrightarrow{CD}$ and construct lines perpendicular to $\overleftrightarrow{AB}$ through X and Y. Call these lines $\overleftrightarrow{XU}$ and $\overleftrightarrow{YV}$.

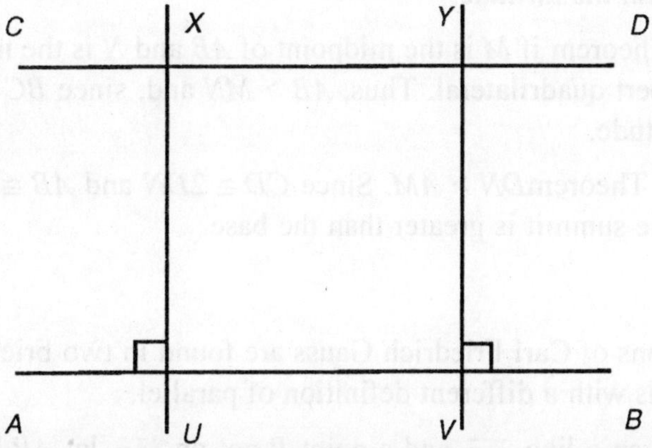

We are interested in the angles $\angle UXY$ and $\angle VYX$ If both angles are acute, then the two lines $\overleftrightarrow{AB}$ and $\overleftrightarrow{CD}$ have a common perpendicular between the segments UX and VY. If one of the angles is a right angle and the other is acute, then the two lines already have a common perpendicular. Suppose that $\angle UXY$ is acute and $\angle VYX$ is obtuse. If we move V away from U along $\overleftrightarrow{AB}$ then $\angle VYX$ may change to a right angle or remain obtuse.

Theorem: *As above and assuming HAA, if as V moves away from U along* $\overleftrightarrow{AB}$, $\angle VXY$ *remains obtuse, then* $\overleftrightarrow{CD}$ *is asymptotic to* $\overleftrightarrow{AB}$.

Saccheri now shows the following:

Theorem: *Given a point P not on a line l there are three classes of lines through P:*

- *Lines meeting l*
- *Lines with a common perpendicular to l and*
- *Lines withour a common perpendicular to l and hence asymptotic to l*

Saccheri now concludes "*the hypothesis of the acute angle is absolutely false; because it is repugnant to the nature of straight lines.*"

The flaw in Saccheri's work was observed by the Swiss mathematician J.H. Lambert in 1786. He himself contributed to the work in non-euclidean geometries with the following.

A convex quadrilateral three of whose angles are right angles is called a Lambert quadrilateral.

Under the HAA assumption the following are true.

Theorem: *The fourth angle of a Lambert quadrilateral is acute.*

Theorem: *The side adjacent to the acute angle of a Lambert quadrilateral is greater than its opposite side.*

Theorem: *In a Saccheri quadrilateral the summit is greater than the base and the sides are greater than the altitude.*

Proof: Using Theorem if M is the midpoint of AB and N is the midpoint of CD, then œ$AMND$ is a Lambert quadrilateral. Thus, $AB > MN$ and, since $BC \cong AB$ both sides are greater than the altitude.

Also, applying Theorem$DN > AM$. Since $CD \cong 2DN$ and $AB \cong 2AM$ it follows that $CD > AB$, so that the summit is greater than the base.

GAUSS

The contributions of Carl Friedrich Gauss are found in two brief unpublished letters and notes. He begins with a different definition of parallel.

Definition: Given a line $\overleftrightarrow{AB}$ and a point P not on $\overleftrightarrow{AB}$ let AP be the perpendicular from P to $\overleftrightarrow{AB}$ A line $\overrightarrow{PQ}$ is parallel to $\overleftrightarrow{AB}$ through P if for any line $\overrightarrow{PS}$ with S in the interior of $\angle APQ$ and $0 < \angle APS° < \angle APQ°$ it follows that $\overrightarrow{PS}$ intersects $\overleftrightarrow{AB}$.

So it would seem that as we consider the set of lines through the point P sweeping up from $\overrightarrow{AP}$ the parallel $\overrightarrow{PQ}$ is the first line not meeting $\overleftrightarrow{AB}$ Unless we assume Euclid's Postulate V, we do not know that this line is unique.

Thus it may depend on the side of $\overleftrightarrow{AP}$ we choose. Therefore, we speak of parallel lines *in a direction*. These lines will be called nonintersecting lines to distinguish them from parallel lines.

Theorem: *In a given direction, being parallel is an equivalence relation.*

Proof: We need to show three things:

1. First we must show that it is well-defined: that is, if $\overleftrightarrow{PQ}$ is parallel to $\overleftrightarrow{AB}$ and S is any point between P and Q, then $\overleftrightarrow{SQ}$ is parallel to $\overleftrightarrow{AB}$.
2. We must also show that the relation is reflexive; that is if $\overleftrightarrow{PQ}$ is parallel to $\overleftrightarrow{AB}$ in a given direction then $\overleftrightarrow{AB}$ is parallel to $\overleftrightarrow{PQ}$ in the same direction..

3. We must show that the relation is transitive; if $\overleftrightarrow{AB}$ is parallel to $\overleftrightarrow{PQ}$ in a given direction and $\overleftrightarrow{PQ}$ is parallel to $\overleftrightarrow{RS}$ in the same direction, then $\overleftrightarrow{AB}$ is parallel to $\overleftrightarrow{RS}$ in that direction.

To begin

1. Let $P * S * Q$ and consider any line through S, $\overrightarrow{SC}$ such that C is interior to $\angle APQ$ and $\angle BAP$ Construct $\overrightarrow{PC}$ By our assumption $\overrightarrow{PC}$ intersects $\overrightarrow{AB}$ Thus, $\overrightarrow{SC}$ has entered a triangle, so by Pasch's Theorem it must intersect $\overrightarrow{AB}$.
2. Let R be the foot of A in $\overrightarrow{PQ}$ Let S be a point in the interiors of $\angle BAR$ and $\angle ARQ$ and construct the line $\overrightarrow{AS}$.

Chapter 4

Conic Sections

In mathematics, a conic section (or just conic) is a curve obtained by intersecting a cone (more precisely, a circular conical surface) with a plane. The conic sections were named and studied as long ago as 200 BC, when Apollonius of Perga undertook a systematic study of their properties.

CONICS

Analytic geometry is roughly the same as plane geometry except that in analytic geometry, figures are studied in the context of the coordinate plane. Instead of focusing on the congruence of shapes like plane geometry, analytic geometry deals with the coordinates of shapes and formulas for their graphs in the coordinate plane. Much of analytic geometry focuses on the conics. A conic is a two-dimensional figure created by the intersection of a plane and a right circular cone. All conics can be written in terms of the following equation: $Ax^2 + Bxy + Cy^2 + Dx + Ey + F = 0$. The four conics we'll explore in this text are parabolas, ellipses, circles, and hyperbolas. The equations for each of these conics can be written in a standard form, from which a lot about the given conic can be told without having to graph it. We'll study the standard forms and graphs of these four conics, as well as the polar equations for the conics, which are useful for certain applications.

TYPES OF CONICS

The three types of conics are the hyperbola, ellipse, and parabola. The circle can be considered as a fourth type (as it was by Apollonius) or as a kind of ellipse. The circle and the ellipse arise when the intersection of cone and plane is a closed curve. The circle is obtained when the cutting plane is parallel to the plane of the generating circle of the cone. If the cutting plane is parallel to exactly one generating line of the cone, then the conic is unbounded and is called a parabola. In the remaining case, the figure is a hyperbola. In this case, the plane will intersect *both* halves (*nappes*) of the cone, producing two separate unbounded curves, though often one is ignored.

DEGENERATE CASES

There are multiple degenerate cases, in which the plane passes through the apex of the cone. The intersection in these cases can be a straight line (when the plane is tangential to the surface of the cone); a point (when the angle between the plane and the axis of the cone is larger than tangential); or a pair of intersecting lines (when the angle is smaller). Where the cone is a cylinder (the vertex is at infinity) cylindric sections are obtained. Although these yield mostly ellipses (or circles) as normal, a degenerate case of two parallel lines, known as a ribbon, can also be produced.

ECCENTRICITY

The four defining conditions above can be combined into one condition that depends on a fixed point F (the *focus*), a line L (the *directrix*) not containing F and a nonnegative real number e (the *eccentricity*). The corresponding conic section consists of all points whose distance to F equals e times their distance to L. For $0 < e < 1$ we obtain an ellipse, for $e = 1$ a parabola, and for $e > 1$ a hyperbola. For an ellipse and a hyperbola, two focus-directrix combinations can be taken, each giving the same full ellipse or hyperbola. The distance from the center to the directrix is a/e, where a is the semi-major axis of the ellipse, or the distance from the center to the tops of the hyperbola. The distance from the center to a focus is ae. In the case of a circle, the eccentricity $e = 0$, and one can imagine the directrix to be infinitely far removed from the center. However, the statement that the circle consists of all points whose distance is e times the distance to L is not useful, because we get zero times infinity. The eccentricity of a conic section is thus a measure of how far it deviates from being circular. For a given , the closer is to 1, the smaller is the semi-minor axis.

A particle moving under the influence of an inverse square force moves in an orbit that is a conic section; that is to say an ellipse, a parabola or a hyperbola. We shall prove this from dynamical principles in a later chapter. In this chapter we review the geometry of the conic sections. We start off, however, with a brief review of the geometry of the straight line.

The Straight Line

It might be thought that there is rather a limited amount that could be written about the geometry of a straight line. We can manage a few equations here, however, (there are 35 in this section on the Straight Line) and shall return for more on the subject.

Most readers will be familiar with the equation

$$y = mx + c$$

for a straight line. The slope (or gradient) of the line, which is the tangent of the angle that it makes with the x-axis, is m, and the intercept on the y-axis is c. There are various other forms that may be of use, such as

$$\frac{x}{x_0}+\frac{y}{y_0}=1$$

$$\frac{y-y_1}{x-x_1}=\frac{y_2-y_1}{x_2-x_1}$$

which can also be written

$$\begin{vmatrix} x & y & 1 \\ x_1 & y_1 & 1 \\ x_2 & y_2 & 1 \end{vmatrix}=0$$

$$x\cos\theta\,,+y\sin\theta=p$$

The four forms are illustrated in figure.

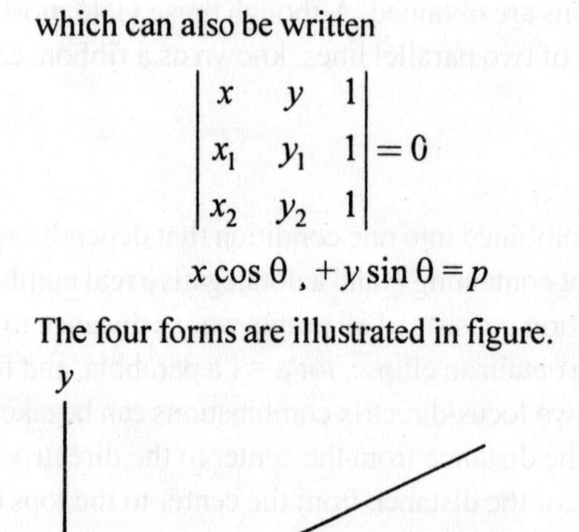

2.2.1

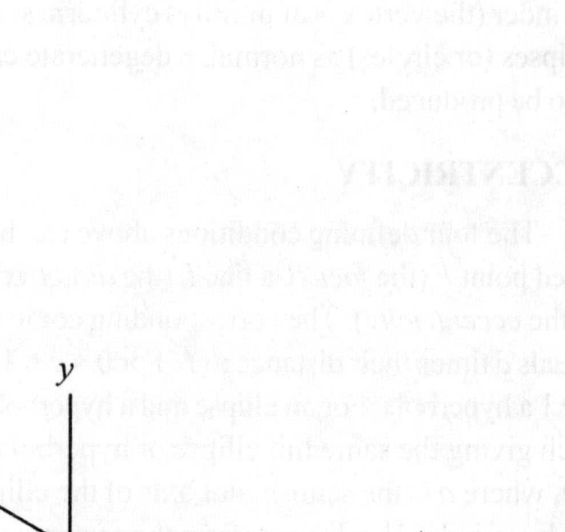

2.2.2

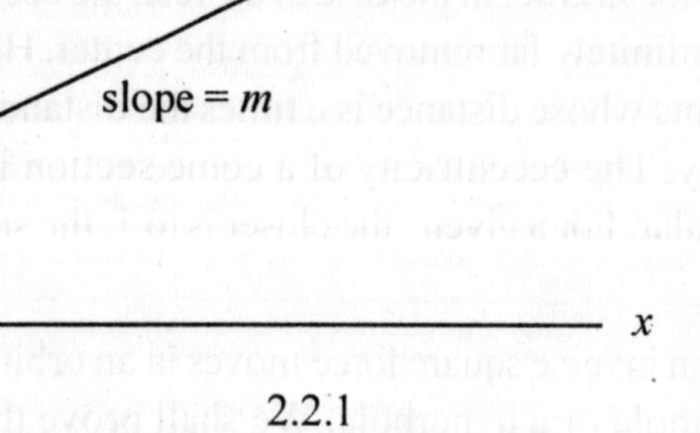

2.2.3 or 4

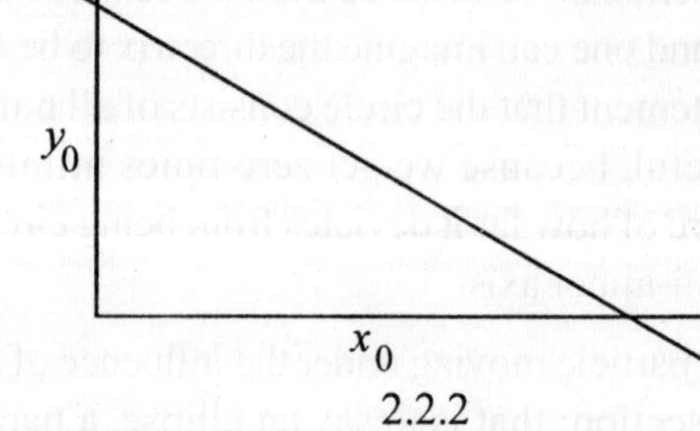

2.2.5

A straight line can also be written in the form

$$Ax+By+C=0.$$

If $C = 0$, the line passes through the origin. If $C \neq 0$, no information is lost, and some arithmetic and algebra are saved, if we divide the equation by C and re-write it in the form

$$ax + by = 1.$$

Let $P(x, y)$ be a point on the line and let $P_0(x_0, y_0)$ be a point in the plane not necessarily on the line. It is of interest to find the perpendicular distance between P_0 and the line. Let S be the square of the distance between P_0 and P. Then

$$S = (x - x_0)^2 + (y - y_0)^2$$

We can express this in terms of the single variable x by substitution for y from equation. Differentiation of S with respect to x will then show that S is least for

$$x = \frac{a + b(bx_0 - ay_0)}{a^2 + b^2}$$

The corresponding value for y, found from equations is

$$y = \frac{b + a(ay_0 - bx_0)}{a^2 + b^2}.$$

The point P described by equation is the closest point to P_0 on the line. The perpendicular distance of P from the line is $p = \sqrt{S}$ or

$$p = \frac{1 - ax_0 - by_0}{\sqrt{a^2 + b^2}}.$$

This is positive if P_0 is on the same side of the line as the origin, and negative if it is on the opposite side. If the perpendicular distances of two points from the line, as calculated from equation , are of opposite signs, they are on opposite sides of the line. If $p = 0$, or indeed if the numerator is zero, the point $P_0(x_0, y_0)$ is, of course, on the line.

Let $A(x_1, y_1)$, $B(x_2, y_2)$ and $C(x_3, y_3)$ be three points in the plane. What is the area of the triangle ABC?

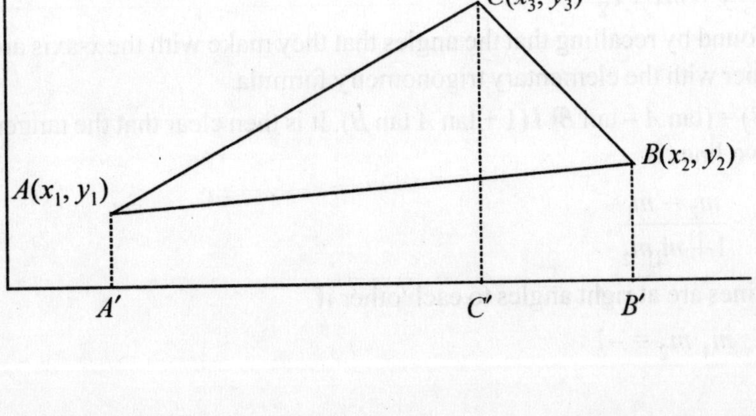

We see that

area of triangle ABC = area of trapezium $A'ACC'$

+ area of trapezium $C'CBB'$

– area of trapezium $A'ABB'$.

$$= \frac{1}{2}(x_3 - x_1)(y_3 + y_1) + \frac{1}{2}(x_2 - x_3)(y_2 + y_3) - \frac{1}{2}(x_2 - x_1)(y_2 + y_1)$$

$$= \frac{1}{2}[x_1(y_2 - y_3) + x_2(y_3 - y_1) + x_3(y_1 - y_2)]$$

$$= \frac{1}{2}\begin{vmatrix} x_1 & x_2 & x_3 \\ y_1 & y_2 & y_3 \\ 1 & 1 & 1 \end{vmatrix}$$

The reader might like to work through an alternative method, using results that we have obtained earlier. The same result will be obtained. In case the algebra proves a little tedious, it may be found easier to work through a numerical example, such as: calculate the area of the triangle ABC, where *A, B, C* are the points (2, 3), (7, 4), (5, 6) respectively. In the second method, we note that the area of a triangle is $\frac{1}{2}$ × base × height. Thus, if we can find the length of the side BC, and the perpendicular distance of A from BC, we can do it. The first is easy:

$$(BC)^2 = (x_3 - x_2)^2 + (y_3 - y_2)^2.$$

To find the second, we can easily write down the equation to the line BC from equation, and then re-write it in the other form. Then equation enables us to find the perpendicular distance of A from BC, and the rest is easy.

If the determinant in equation is zero, the area of the triangle is zero. This means that the three points are collinear.

The angle between two lines

$$y = mx + c_1$$

and $$y = mx + c_2$$

is easily found by recalling that the angles that they make with the x-axis are $\tan^{-1} m_1$ and $\tan^{-1} m_2$ together with the elementary trigonometry formula

$\tan(A - B) = (\tan A - \tan B) / (1 + \tan A \tan B)$. It is then clear that the tangent of the angle between the two lines is

$$\frac{m_2 - m_1}{1 + m_1 m_2}.$$

The two lines are at right angles to each other if

$$m_1 m_2 = -1$$

The line that bisects the angle between the lines is the locus of points that are equidistant from the two lines. For example, consider the two lines

$$-2x + 5y = 1$$

$$30x - 10y = 1$$

Making use of equation, we see that a point (x, y) is equidistant from these two lines if

$$\frac{1+2x-5y}{\sqrt{29}} = \pm\frac{1-30x+10y}{\sqrt{1000}}.$$

The significance of the ± will become apparent shortly. The + and – choices result, respectively, in

$$-8.568x + 8.079y = 1$$

and $$2.656x + 2.817y = 1.$$

An equation of the form

$$ax^2 + 2hxy + by^2 = 0$$

can be factored into two linear factors with no constant term, and it therefore represents two lines intersecting at the origin. It is left as an exercise to determine the angles that the two lines make with each other and with the x axis, and to show that the lines

$$x^2 + \left(\frac{a-b}{h}\right)xy - y^2 = 0$$

are the bisectors of 2.2.23 and are perpendicular to each other.

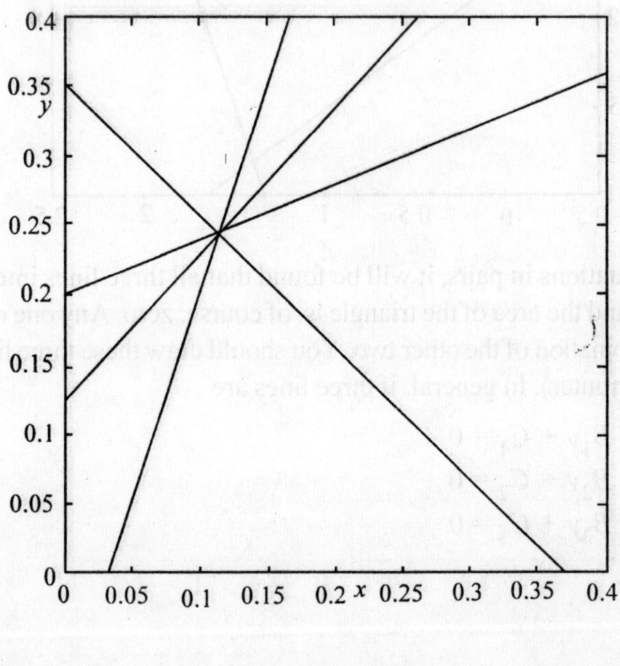

Given the equations to three straight lines, can we find the area of the triangle bounded by them? To find a general algebraic expression might be a bit tedious, though the reader might like to try it, but a numerical example is straightforward. For example, consider the lines

$$x - 5y + 12 = 0,$$
$$3x + 4y - 9 = 0,$$
$$3x - y - 3 = 0.$$

By solving the equations in pairs, it is soon found that they intersect at the points (–0.15789, 2.36842), (1.4, 1.2) and (1.92857, 2.78571). Application of equation gives the area as 1.544. Measure any side and the corresponding height with a ruler and see if the area is indeed about 1.54.

But now consider the three lines

$$x - 5y + 12 = 0,$$
$$3x + 4y - 9 = 0,$$
$$3x + 23y - 54 = 0.$$

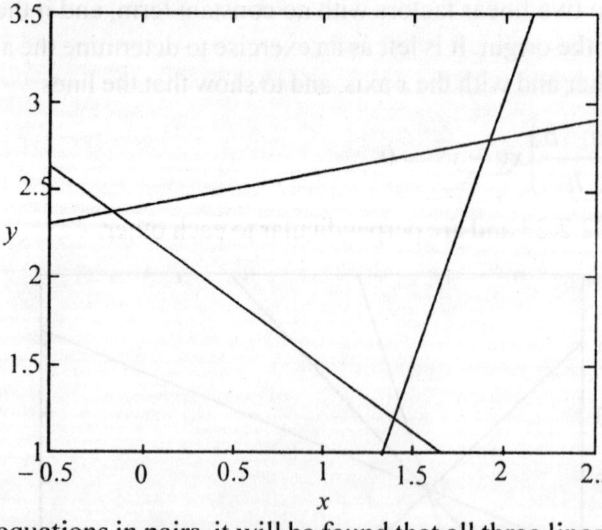

By solving the equations in pairs, it will be found that all three lines intersect at the same point (please do this), and the area of the triangle is, of course, zero. Any one of these equations is, in fact, a linear combination of the other two. You should draw these three lines accurately on graph paper (or by computer). In general, if three lines are

$$A_1x + B_1y + C_1 = 0$$
$$A_2x + B_2y + C_2 = 0$$
$$A_3x + B_3y + C_3 = 0$$

they will be concurrent at a single point if

$$\begin{vmatrix} A_1 & B_1 & C_1 \\ A_2 & B_2 & C_2 \\ A_3 & B_3 & C_3 \end{vmatrix} = 0.$$

Thus the determinant in equation provides a test of whether three points are collinear, and the determinant in equation provides a test of whether three lines are concurrent.

Finally - at least for the present chapter - there may be rare occasion to write the equation of a straight line in polar coordinates. It should be evident from figure that the equations

$$r = p\csc(\theta - \alpha) \text{ or } r = p\csc(\alpha - \theta)$$

describe a straight line passing at a distance p from the pole and making an angle ± with the initial line. If $p = 0$, the polar equation is merely $\theta = \alpha$.

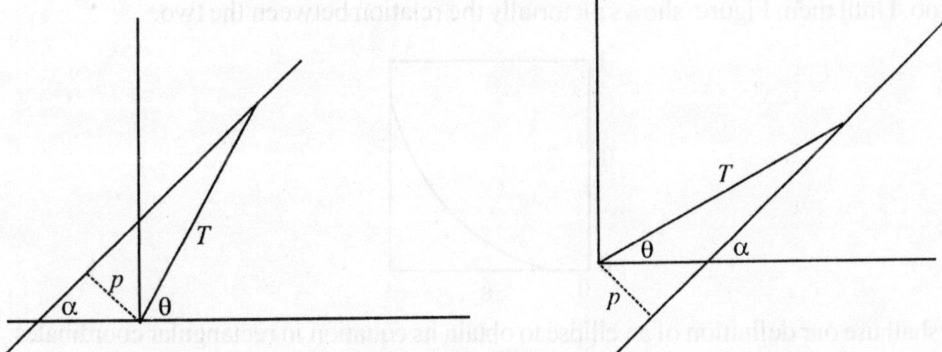

THE ELLIPSE

An ellipse is a figure that can be drawn by sticking two pins in a sheet of paper, tying a length of string to the pins, stretching the string taut with a pencil, and drawing the figure that results. During this process, the sum of the two distances from pencil to one pin and from pencil to the other pin remains constant and equal to the length of the string. This method of drawing an ellipse provides us with a formal definition, which we shall adopt in this chapter, of an ellipse, namely:

An ellipse is the locus of a point that moves such that the sum of its distances from two fixed points *called the foci is constant.*

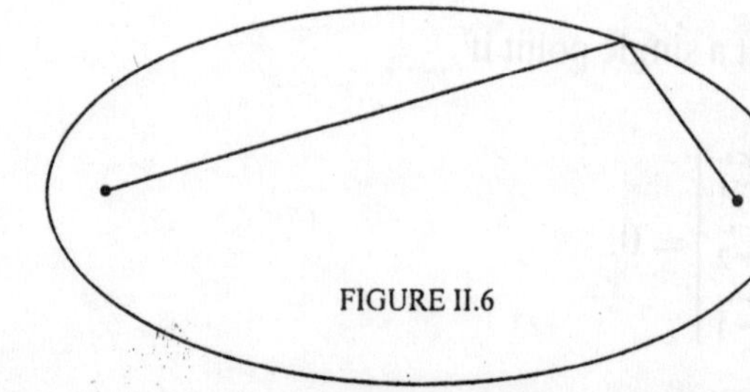

FIGURE II.6

We shall call the sum of these two distances (i.e the length of the string) $2a$.. The ratio of the distance between the foci to length of the string is called the *eccentricity e* of the ellipse, so that the distance between the foci is $2ae$, and e is a number between 0 and 1.

The longest axis of the ellipse is its major axis, and a little bit of thought will show that its length is equal to the length of the string; that is, $2a$. The shortest axis is the minor axis, and its length is usually denoted by $2b$. The eccentricity is related to the ratio b/a in a manner that we shall shortly discuss.

The ratio $\eta = (a - b)/a$ is called the *ellipticity* if the ellipse. It is merely an alternative measure of the noncircularity. It is related to the eccentricity, and we shall obtain that relation shortly, too. Until then, Figure shows pictorially the relation between the two.

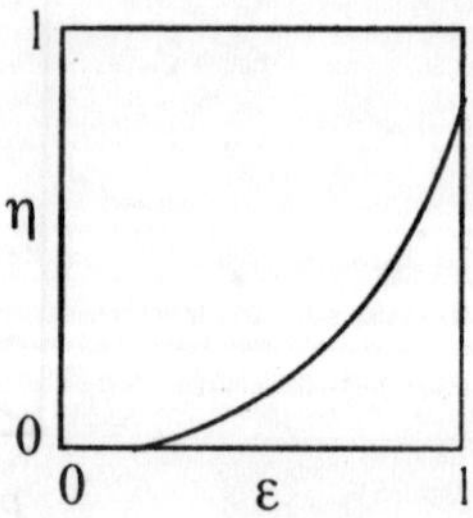

We shall use our definition of an ellipse to obtain its equation in rectangular coordinates. We shall place the two foci on the x-axis at coordinates $(-ae, 0)$ and $(ae, 0)$.

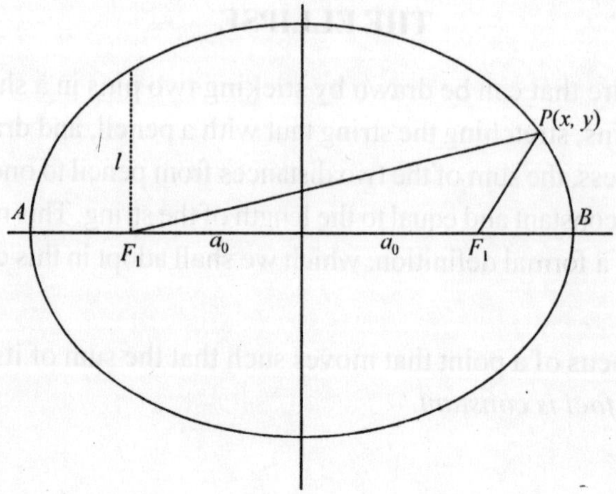

The definition requires that $PF_1 + PF_2 = 2a$. That is:

$$[(x+ae)^2+y^2]^{\frac{1}{2}}+[(x-ae)^2+y^2]^{\frac{1}{2}}=2a,$$

and this is the equation to the ellipse. The reader should be able, after a little bit of slightly awkward algebra, to show that this can be written more conveniently as

$$\frac{x^2}{a^2}+\frac{y^2}{a^2(1-e^2)}=1.$$

By putting $x = 0$, it is seen that the ellipse intersects the y-axis at $\pm a\sqrt{1-e^2}$ and therefore that $a\sqrt{1-e^2}$ is equal to the semi minor axis b. Thus we have the familiar equation to the ellipse

$$\frac{x^2}{a^2}+\frac{y^2}{b^2}=1$$

as well as the important relation between a, b and e:

$$b^2 = a^2 (1 - e^2)$$

The reader can also now derive the relation between ellipticity η and eccentricity e:

$$\eta=1-\sqrt{(1-e^2)}\,.$$

This can also be written

$$e^2=\sqrt{\eta(2-\eta)}$$

or $$e^2 + (\eta - 1)^2 = 1.$$

This shows, incidentally, that the graph of · versus e, which we have drawn in figure, is part of a circle of radius 1 centred at $e = 0$, $\eta = 1$.

You may find that ellipticity gives you a better idea than eccentricity of the noncircularity of an ellipse. For an exercise, you should draw in the positions of the foci of each of these ellipses, and decide whether eccentricity or ellipticity gives you a better idea of the "ex-centricity" of the foci. Note that the eccentricities of the orbits of Mars and Mercury are, respectively, about 0.1 and 0.2 (these are the most eccentric of the planetary orbits except for comet-like Pluto). and it is difficult for the eye to see that they depart at all from circles - although, when the foci are drawn, it is obvious that the foci are "ex-centric".

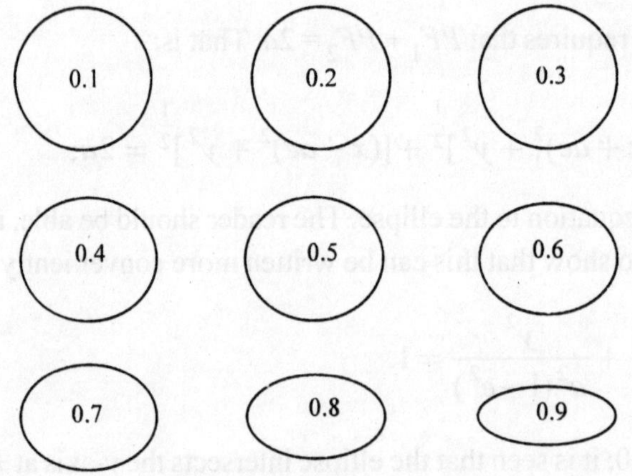

The number inside each ellipse is its eccentricity.

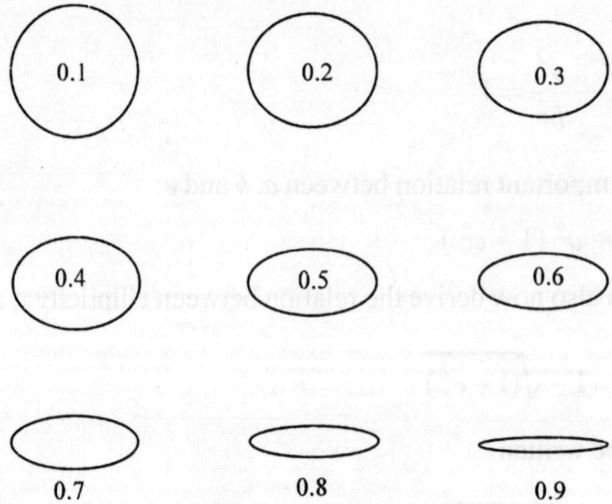

The figure inside or below each ellipse is its ellipticity.

In the theory of planetary orbits, the Sun will be at one focus. Let us suppose it to be at F_2. In that case the distance $F_2 B$ is the perihelion distance q, and is equal to

$$q = a(1 - e).$$

The distance $F_2 A$ is the aphelion distance Q (pronounced ap-helion by some and affelion by others – and both have defensible positions), and it is equal to

$$Q = a(1 + e).$$

A line parallel to the minor axis and passing through a focus is called a *latus rectum* (plural: *latera recta*). The length of a semi latus rectum is commonly denoted by l (sometimes by p). Its length is obtained by putting $x = ae$ in the equation to the ellipse, and it will be readily found that

$$l = a(1 - e^2).$$

The length of the semi latus rectum is an important quantity in orbit theory. It will be found, for example, that the energy of a planet is closely related to the semi major axis a of its orbit, while its angular momentum is closely related to the semi latus rectum.

The circle whose diameter is the major axis of the ellipse is called the *eccentric circle* or, preferably, the *auxiliary circle*. Its equation is

$$x^2 + y^2 = a^2.$$

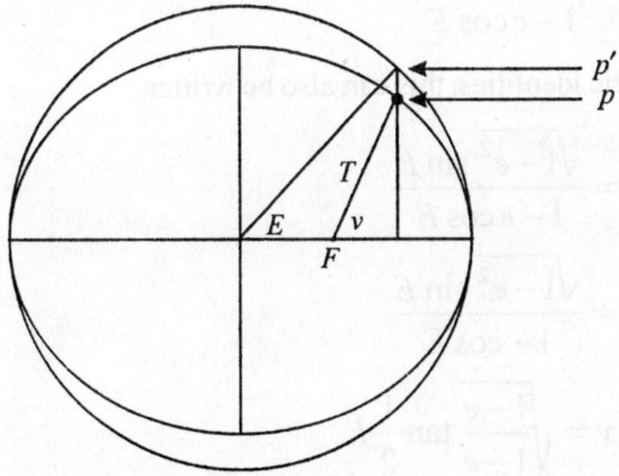

In orbit theory the angle v (denoted by f by some authors) is called the *true anomaly* of a planet in its orbit. The angle E is called the *eccentric anomaly*, and it is important to find a relation between them.

We first note that, if the eccentric anomaly is E, the abscissas of P' and of P are each $a \cos E$. The ordinate of P' is $a \sin E$. By putting $x = a \cos E$ in the equation to the ellipse, we immediately find that the ordinate of P is $b \sin E$. Several deductions follow. One is that any point whose abscissa and ordinate are of the form

$$x = a \cos E, y = b \sin E$$

is on an ellipse of semi major axis a and semi minor axis b. These two equations can be regarded as parametric equations to the ellipse. They can be used to describe an ellipse just as readily as

$$\frac{x^2}{a^2} + \frac{y^2}{b^2} = 1$$

and indeed this equation is the E-eliminant of the parametric equations.

The ratio PM/P'M for any line perpendicular to the major axis is b/a. Consequently the area of the ellipse is b/a times the area of the auxiliary circle; and since the area of the auxiliary circle is πa^2, it follows that the area of the ellipse is πab.

The distance r is called the *radius vector* (plural *radii vectores*), and from the theorem of Pythagoras its length is given by

$$r^2 = b^2 \sin^2 E + a^2 (\cos E - e)^2.$$

On substituting $1 - \cos^2 E$ for $\sin^2 E$ and $a^2 (1 - e^2)$ for b^2, we soon find that

$$r = a(1 - e \cos E)$$

It then follows immediately that the desired relation between v and E is

$$\cos v = \frac{\cos E - e}{1 - e\cos E}.$$

From trigonometric identities, this can also be written

$$\sin v = \frac{\sqrt{1-e^2}\sin E}{1 - e\cos E}$$

or

$$\tan v = \frac{\sqrt{1-e^2}\sin E}{1 - \cos E}$$

or

$$\tan\frac{1}{2}v = \sqrt{\frac{1-e}{1-e}}\tan\frac{1}{2}E.$$

There are a number of miscellaneous geometric properties of an ellipse, some, but not necessarily all, of which may prove to be of use in orbital calculations. We describe some of them in what follows.

Tangents to an ellipse

Find where the straight line $y = mx + c$ intersects the ellipse

$$\frac{x^2}{a^2} + \frac{y^2}{b^2} = 1.$$

The answer to this question is to be found by substituting $mx + c$ for y in the equation to the ellipse. After some rearrangement, a quadratic equation in x results:

$$(a^2m^2 + b^2)x^2 + 2a^2cmx + a^2(c^2 - b^2) = 0.$$

If this equation has two real roots, the roots are the x-coordinates of the two points where the line intersects the ellipse. If it has no real roots, the line misses the ellipse. If it has two coincident real roots, the line is tangent to the ellipse. The condition for this is found by setting the discriminant of the quadratic equation to zero, from which it is found that

$$c^2 = a^2m^2 + b^2.$$

Thus a straight line of the form

$$y = mx \pm \sqrt{a^2m^2 + b^2}$$

is tangent to the ellipse.

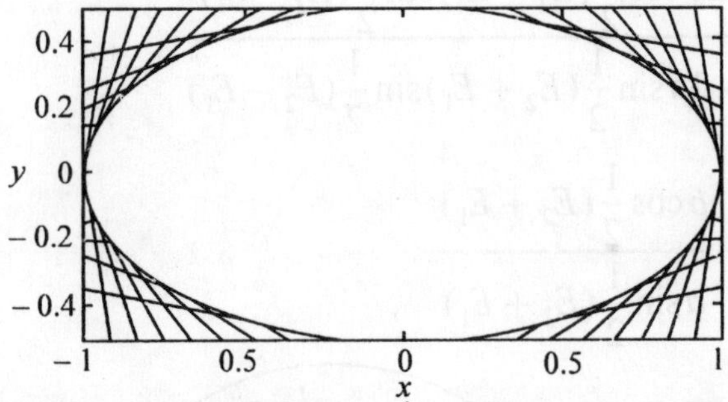

Fig. Lines, for $a = 2b$ and Slopes ($\tan^{-1} m$) of 0° to 180° in Steps of 10°

DIRECTOR CIRCLE

The equation we have just derived for a tangent to the ellipse can be rearranged to read

$$m^2 (a^2 - x^2) + 2mx + b^2 - y^2 = 0.$$

Now the product of the slopes of two lines that are at right angles to each other is −1. Therefore, if we replace m in the above equation by $-1/m$ we shall obtain another tangent to the ellipse, at right angles to the first one. The equation to this second tangent becomes (after multiplication throughout by m)

$$m^2 (b^2 - y^2) - 2mx + a^2 - x^2 = 0.$$

If we eliminate m from these two equations, we shall obtain an equation in x and y that describes the point where the two perpendicular tangents meet; that is, the equation that will describe a curve that is the locus of the point of intersection of two perpendicular tangents. It turns out that this curve is a circle of radius $\sqrt{a^2 + b^2}$, and it is called the *director circle*.

It is easier than it might first appear to eliminate m from the equations. We merely have to add the equations:

$$m^2 (a^2 + b^2 - x^2 - y^2) + a^2 + b^2 - x^2 - y^2 = 0.$$

For real m, this can only be if

$$x^2 + y^2 = a^2 + b^2,$$

which is the required locus of the director circle of radius $\sqrt{a^2 + b^2}$.

We shall now derive an equation to the line that is tangent to the ellipse at the point (x_1, y_1). Let $(x_1, y_1) = (a \cos E_1, b \sin E_1)$ and $(x_2, y_2) = (a \cos E_2, b \sin E_2)$ be two points on the ellipse.

The line joining these two points is

$$\frac{y - b\sin E_1}{x - a\cos E_1} = \frac{b(\sin E_2 - \sin E_1)}{a(\cos E_2 - \cos E_1)}$$

$$= \frac{2b\cos\frac{1}{2}(E_2+E_1)\sin\frac{1}{2}(E_2-E_1)}{-2a\sin\frac{1}{2}(E_2+E_1)\sin\frac{1}{2}(E_2-E_1)}$$

$$= -\frac{b\cos\frac{1}{2}(E_2+E_1)}{a\sin\frac{1}{2}(E_2+E_1)}.$$

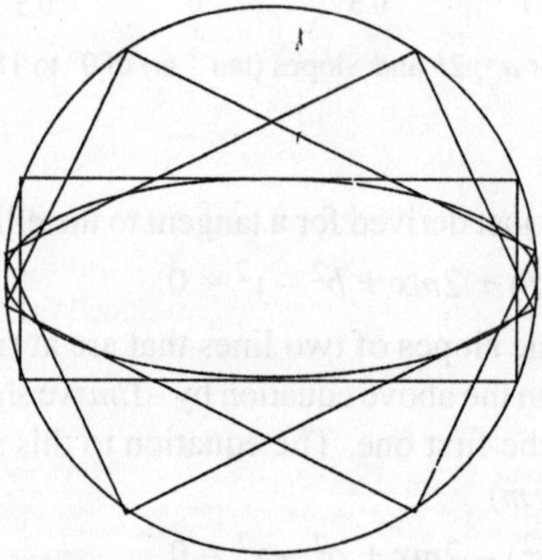

Now let E_2 approach E_1, eventually coinciding with it. The resulting equation

$$\frac{y-b\sin E}{x-a\cos E} = -\frac{b\cos E}{a\sin E},$$

in which we no longer distinguish between E_1 and E_2, is the equation of the straight line that is tangent to the ellipse at ($a\cos E$, $b\sin E$). This can be written

$$\frac{x\cos E}{a} + \frac{y\sin E}{b} = 1$$

or, in terms of (x_1, y_1),

$$\frac{x_1 x}{a^2} + \frac{y_1 y}{b^2} = 1,$$

which is the tangent to the ellipse at (x_1, y_1).

An interesting property of a tangent to an ellipse, the proof of which I leave to the reader, is that F_1P and F_2P make equal angles with the tangent at P. If the inside of the ellipse were a reflecting mirror and a point source of light were to be placed at F_1, it would be imaged at F_2. This has had an interesting medical application. A patient has a kidney stone. The patient is asked to lie in an elliptical bath, with the kidney stone at F_2. A small explosion is detonated at F_1; the explosive sound wave emanating from F_1 is focused as an implosion at F_2 and the kidney stone at F_2 is shattered. Don't try this at home.

DIRECTRICES

The two lines $x = \pm a/e$ are called the *directrices* (singular *directrix*) of the ellipse.

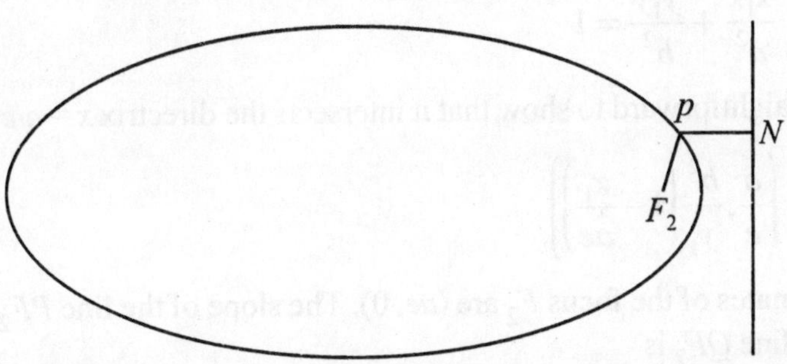

The ellipse has the property that, for any point P on the ellipse, the ratio of the distance PF_2 to a focus to the distance PN to a directrix is constant and is equal to the eccentricity of the ellipse. Indeed, this property is sometimes used as the definition of an ellipse, and all the equations and properties that we have so far derived can be deduced from such a definition. We, however, adopted a different definition, and the focus-directrix property must be derived. This is straightforward, for, (recalling that the abcissa of F_2 is ae) we see from figure that the square of the desired ratio is

$$\frac{(x-ae)^2+y^2}{(a/e-x)^2}.$$

On substitution of $b^2\left[1-\left(\frac{x}{a}\right)^2\right] = a^2(1-e^2)\left[1-\left(\frac{x}{a}\right)^2\right] = (1-e^2)(a^2-x^2)$

for y^2, the above expression is seen to reduce to e^2.

Another interesting property of the focus and directrix, although a property probably with not much application to orbit theory, is that if the tangent to an ellipse at a point P intersects the directrix at Q, then P and Q subtend a right angle at the focus.

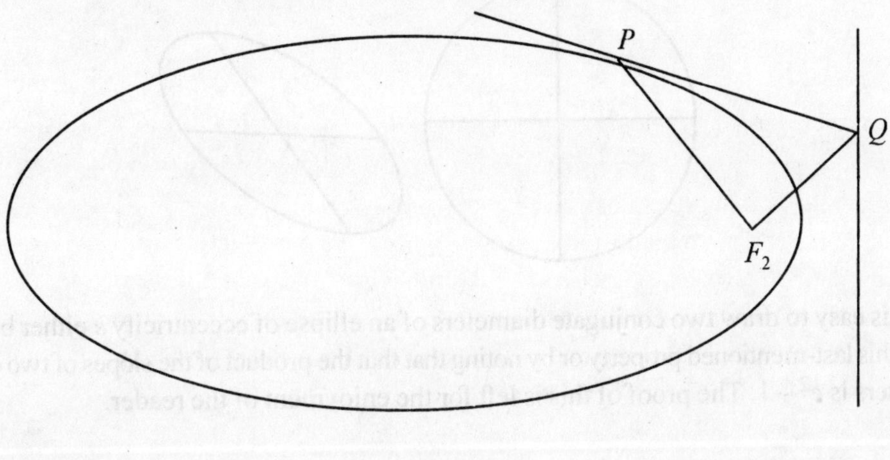

Thus the tangent at $P = (x_1, y_1)$ is

$$\frac{x_1 x}{a^2} + \frac{y_1 y}{b^2} = 1$$

and it is straightforward to show that it intersects the directrix $x = a/e$ at the point

$$\left(\frac{a}{e}, \frac{b^2}{y_1}\left(1 - \frac{x_1}{ae}\right)\right).$$

The coordinates of the focus F_2 are $(ae, 0)$. The slope of the line PF_2 is $(x_1 - ae)/y_1$ and the slope of the line QF_2 is

$$\frac{\frac{b^2}{y_1}\left(1 - \frac{x_1}{ae}\right)}{\frac{a}{e} - ae}.$$

It is easy to show that the product of these two slopes is “1, and hence that PF_2 and QF_2 are at right angles.

CONJUGATE DIAMETERS.

The left hand of figure shows a circle and two perpendicular diameters. The right hand figure shows what the circle would look like when viewed at some oblique angle. The circle has become an ellipse, and the diameters are no longer perpendicular. The diameters are called *conjugate diameters* of the ellipse. One is conjugate to the other, and the other is conjugate to the one. They have the property - or the definition - that each bisects all chords parallel to the other, because this property of bisection, which is obviously held by the perpendicular diameters of the circle, is unaltered in projection.

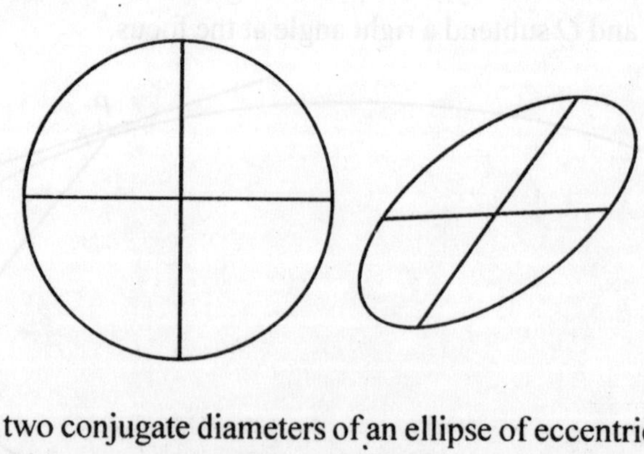

It is easy to draw two conjugate diameters of an ellipse of eccentricity e either by making use of this last-mentioned property or by noting that that the product of the slopes of two conjugate diameters is $e^2 - 1$. The proof of this is left for the enjoyment of the reader.

A LADDER PROBLEM.

No book on elementary applied mathematics is complete without a ladder problem. A ladder of length $a + b$ leans against a smooth vertical wall and a smooth horizontal floor. A particular rung is at a distance a from the top of the ladder and b from the bottom of the ladder. Show that, when the ladder slips, the rung describes an ellipse. (This result will suggest another way of drawing an ellipse.)

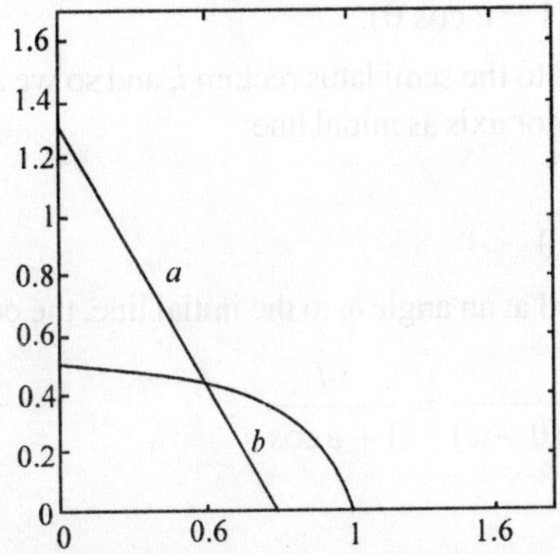

If you have not done this problem after one minute, here is a hint. Let the angle that the ladder makes with the floor at any time be E. That is the end of the hint.

The reader may be aware that some of the geometrical properties that we have discussed in the last few paragraphs are more of recreational interest and may not have much direct application in the theory of orbits. In the next subsection we return to properties and equations that are very relevant to orbital theory - perhaps the most important of all for the orbit computer to understand.

POLAR EQUATION TO THE ELLIPSE

We shall obtain the equation in polar coordinates to an ellipse whose focus is the pole of the polar coordinates and whose major axis is the initial line ($\theta = 0°$) of the polar coordinates. We have indicated the angle , of polar coordinates, and it may occur to the reader that we have previously used the symbol v for this angle and called it the true anomaly. Indeed at present, v and θ are identical, but a little later we shall distinguish between them.

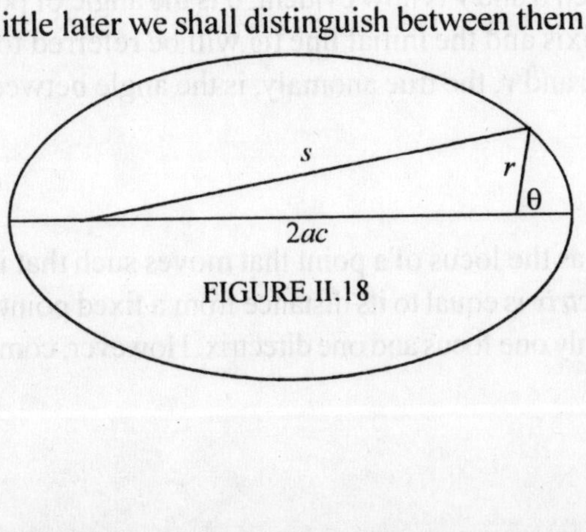

FIGURE II.18

From our definition of the ellipse, $s = 2a - r$, and so

$$s^2 = 4a^2e^2 + r^2 - 4ar + r^2.$$

From the cosine formulae for a plane triangle,

$$s^2 = 4a^2e^2 + r^2 + 4aer\cos\theta.$$

On equating these expressions we soon obtain

$$a(1 - e^2) = r(1 + e\cos\theta).$$

The left hand side is equal to the semi latus rectum l, and so we arrive at the polar equation to the ellipse, focus as pole, major axis as initial line:

$$r = \frac{l}{1 + e\cos\theta}.$$

If the major axis is inclined at an angle ω to the initial line, the equation becomes

$$r = \frac{l}{1 + e\cos(\theta - \omega)} = \frac{l}{1 + e\cos v}.$$

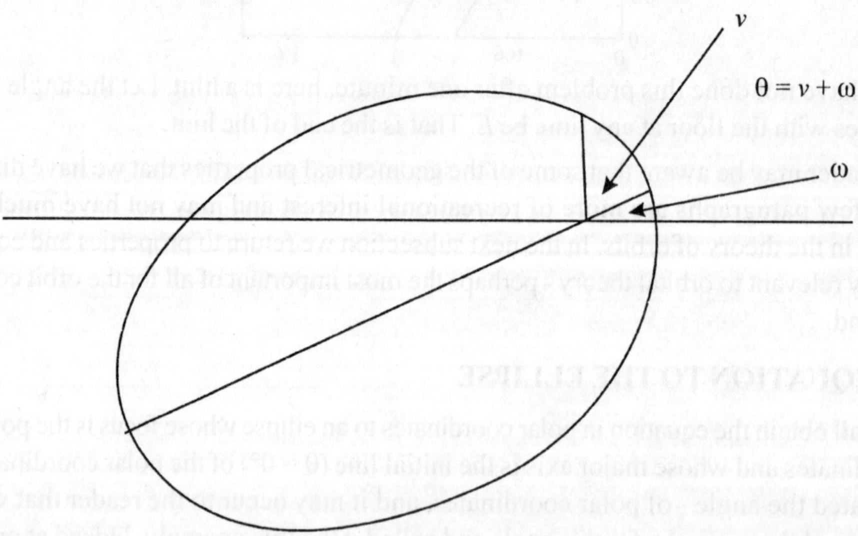

The distinction between θ and v is now evident. θ is the angle of polar coordinates, ω is the angle between the major axis and the initial line (ω will be referred to in orbital theory as the "argument of perihelion"), and v, the true anomaly, is the angle between the radius vector and the initial line.

THE PARABOLA

We define a parabola as the locus of a point that moves such that its distance from a fixed straight line called the *directrix* is equal to its distance from a fixed point called the *focus*. Unlike the ellipse, a parabola has only one focus and one directrix. However, comparison of this definition with the focus - directrix property of the ellipse (which can also be used to define the ellipse) shows that the parabola can be regarded as a limiting form of an ellipse with eccentricity equal to unity.

We shall find the equation to a parabola whose directrix is the line $y = -q$ and whose focus is the point $(q, 0)$. F is the focus and O is the origin of the coordinate system. The vertex of the parabola is at the origin. In an orbital context, for example, the orbit of a comet moving around the Sun in parabolic orbit, the Sun would be at the focus F, and the distance between vertex and focus would be the perihelion distance, for which the symbol q is traditionally used in orbit theory.

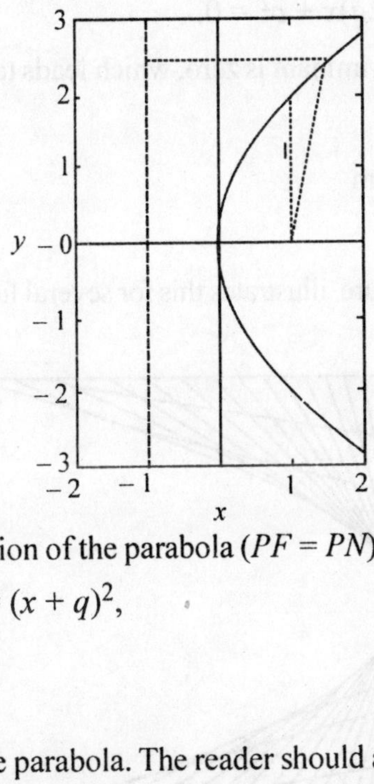

It is evident that the definition of the parabola ($PF = PN$) requires that

$$(x-q)^2 + y^2 = (x+q)^2,$$

from which

$$y^2 = 4qx,$$

which is the equation to the parabola. The reader should also try sketching the parabolas

$$y^2 = -4qx,\ x^2 = 4qy,\ x^2 = -4qy,\ (y-2)^2 = 4q(x-3).$$

The line parallel to the y-axis and passing through the focus is the *latus rectum*. Substitution of $x = q$ into $y^2 = 4ax$ shows that the latus rectum intersects the parabola at the two points $(q, \pm 2q)$, and that the length l of the semi latus rectum is $2q$.

The equations

$$x = qt^2,\ y = 2qt$$

are the parametric equations to the parabola, for $y^2 = 4qx$ results from the elimination of t between them. In other words, if t is any variable, then any point that satisfies these two equations lies on the parabola.

Most readers will know that if a particle is moving with constant speed in one direction and constant acceleration at right angles to that direction, as with a ball projected in a uniform gravitational field or an electron moving in a uniform electric field, the path is a parabola. In the constant speed direction the distance is proportional to the time, and in the constant acceleration direction, the distance is proportional to the square of the time, and hence the path is a parabola.

TANGENTS TO A PARABOLA.

Where does the straight line $y = mx + c$ intersect the parabola $y2 = 4qx$? The answer is found by substituting $mx + c$ for y to obtain, after rearrangement,

$$m^2x^2 + 2(mc - 2q)x + c^2 = 0.$$

The line is tangent if the discriminant is zero, which leads to

$$c = q/m.$$

Thus a straight line of the form

$$y = mx + q/m$$

is tangent to the parabola. Figure illustrates this for several lines, the slopes of each differing by 5o from the next.

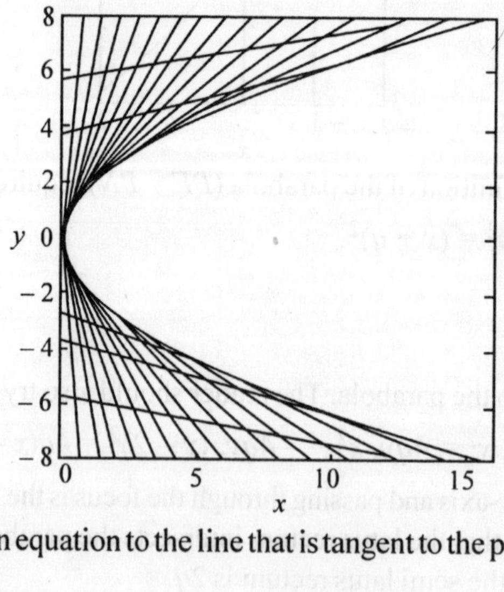

We shall now derive an equation to the line that is tangent to the parabola at the point (x_1, y_1).

Let (x_1, y_1) $(qt_1^2, 2qt_1)$ be a point on the parabola, and

Let (x_2, y_2) $(qt_2^2, 2qt_2)$ be another point on the parabola.

The line joining these two points is

$$\frac{y-2qt_1}{x-qt_1^2}=\frac{2q(t_2-t_1)}{q(t_2^2-t_1^2)}=\frac{2}{t_2+t_1}.$$

Now let t_2 approach t_1, eventually coinciding with it. Putting $t_1=t_2=t$ in the last equation results, after simplification, in

$$ty = x + qt^2,$$

being the equation to the tangent at $(qt^2, 2qt)$.

Multiply by $2q$:

$$2qty = 2q(x + qt^2)$$

and it is seen that the equation to the tangent at (x_1, y_1) is

$$y_1y = 2q\,(x_1 + x).$$

There are a number of interesting geometric properties, some of which are given here. For example, if a tangent to the parabola at a point P meets the directrix at Q, then, just as for the ellipse, P and Q subtend a right angle at the focus. The proof is similar to that given for the ellipse, and is left for the reader.

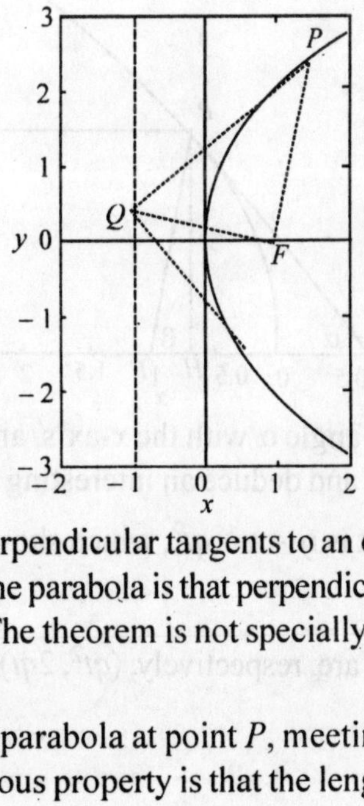

The reader will recall that perpendicular tangents to an ellipse meet on the director circle. The analogous theorem *vis-à-vis* the parabola is that perpendicular tangents meet on the directrix. This is also illustrated in figure. The theorem is not specially important in orbit theory, and the proof is also left to the reader.

Let PG be the normal to the parabola at point P, meeting the axis at G. We shall call the length GH the subnormal. A curious property is that the length of GH is always equal to l, the length of the semi latus rectum, irrespective of the position of P. This proof again is left to the reader.

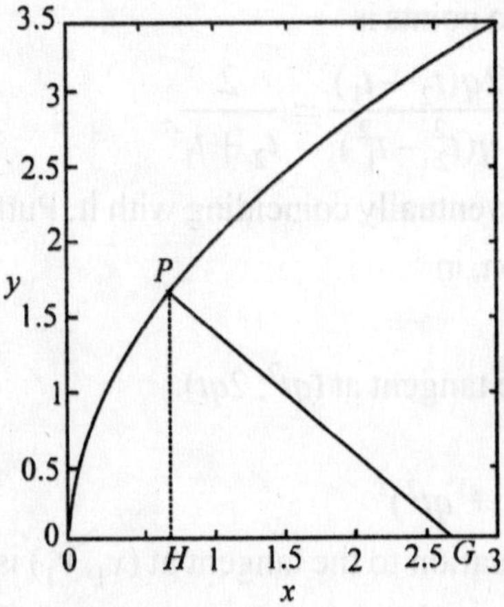

The following two geometrical properties, while not having immediate applications to orbit theory, certainly have applications to astronomy.

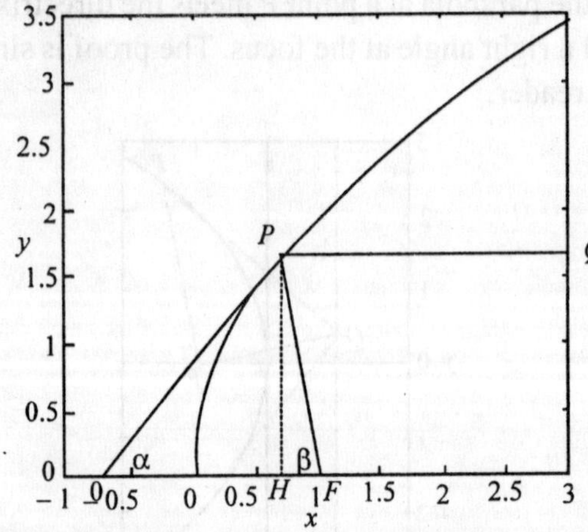

The tangent at P makes an angle α with the x-axis, and PF makes an angle β with the x-axis. We shall show that $\beta = 2\alpha$ and deduce an interesting consequence.

The equation to the tangent is $ty = x + qt^2$, which shows that

$$\tan \alpha = 1/t.$$

The coordinates of P and F are, respectively, $(qt^2, 2qt)$ and $(a, 0)$, and so, from the triangle PFH, we find.

$$\tan \beta = \frac{2t}{t^2 - 1}.$$

Let $\tau = 1/t$, then $\tan \alpha = \tau$ and $\tan \beta = 2\tau/(1 - \tau^2)$, which shows that $\beta = 2\alpha$.

This also shows that triangle JFP is isosceles, with the angles at J and P each being α. This can also be shown as follows.

From the equation $ty = x + qt^2$. we see that J is the point $(-qt^2, 0)$, so that $JF = q(t^2 + 1)$.

From triangle PFH, we see that

$$(PF)^2 = 4q^2t^2 + (t^2 - 1)^2 - q^2 (t^2 + 1)^2.$$

Therefore

$$PF = JF.$$

Either way, since the triangle JPF is isosceles, it follows that QP and PF make the same angle α to the tangent. If the parabola is a cross section of a telescopic mirror, any ray of light coming in parallel to the axis will be focussed at F, so that a paraboloidal mirror, used on-axis, does not suffer from spherical aberration. (This property holds, of course, only for light parallel to the axis of the paraboloid, so that a paraboloidal mirror, without some sort of correction, gives good images over only a narrow field of view.)

Now consider what happens when you stir a cup of tea. The surface takes up a shape that looks as though it might resemble the parabola $y = x^2/(4q)$:

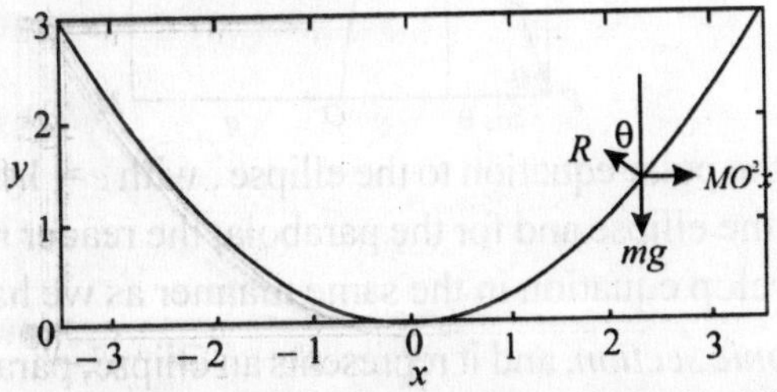

Suppose the liquid is circulating at angular speed ω. A tea leaf floating on the surface is in equilibrium (in the rotating reference frame) under three forces: its weight mg, the centrifugal force $m\omega^2 x$ and the normal reaction R. The normal to the surface makes an angle θ with the vertical (and the tangent makes an angle θ with the horizontal) given by

$$\tan\theta = \frac{\omega^2 x}{g}.$$

But the slope of the parabola $y = x^2/(4q)$ is $x/(2q)$, so that the surface is indeed a parabola with semi latus rectum $2q = g/\omega^2$.

This phenomenon has been used in Canada to make a successful large telescope (diameter 6 m) in which the mirror is a spinning disc of mercury that takes up a perfectly paraboloidal shape. Another example is the spin casting method that has been successfully used for the production of large, solid glass paraboloidal telescope mirrors. In this process, the furnace is rotated about a vertical axis while the molten glass cools and eventually solidifies into the required paraboloidal effect.

POLAR EQUATION TO THE PARABOLA

As with the ellipse, we choose the focus as pole and the axis of the parabola as initial line. We shall orient the parabola so that the vertex is toward the right.

We recall the focus-directrix property, $FP = PN$. Also, from the definition of the directrix, $FO = OM = q$, so that $FM = 2q = l$, the length of the semi latus rectum. It is therefore immediately evident from figure that $r \cos \theta + r = 2q = l$, so that the polar equation to the parabola is

$$r = \frac{l}{1+\cos\theta}.$$

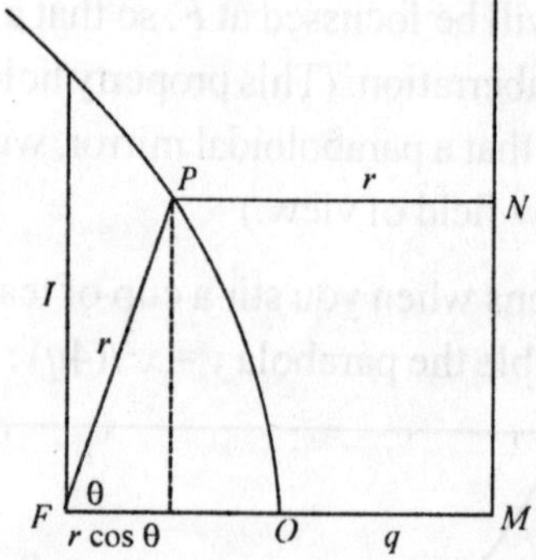

This is the same as the polar equation to the ellipse, with $e = 1$ for the parabola. I have given different derivations for the ellipse and for the parabola; the reader might like to interchange the two approaches and develop equation in the same manner as we have developed equation.

The equation to a *conic section*, and it represents an ellipse, parabola or hyperbola according to whether $e < 1$, $e = 1$ or $e > 1$.

THE HYPERBOLA

A hyperbola is the locus of a point that moves such that the difference between its distances from two fixed points called the foci is constant. We shall call the difference between these two distances $2a$ and the distance between the foci $2ae$, where e is the eccentricity of the hyperbola, and is a number greater than 1.

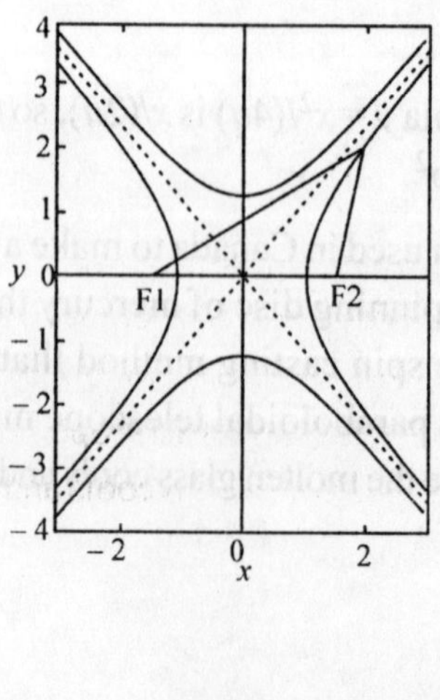

For example, in a Young's double-slit interference experiment, the *m*th bright fringe is located at a point on the screen such that the path difference for the rays from the two slits is *m* wavelengths. As the screen is moved forward or backwards, this relation continues to hold for the *m*th bright fringe, whose locus between the slits and the screen is therefore a hyperbola. The "Decca" system of radar navigation, first used at the D-Day landings in the Second World War and decommissioned only as late as 2000 on account of being rendered obsolete by the "GPS" (Global Positioning Satellite) system, depended on this property of the hyperbola. (Since writing this, part of the Decca system has been re-commissioned as a back-up in case of problems with GPS.)

Two radar transmitters some distance apart would simultaneously transmit radar pulses. A ship would receive the two signals separated by a short time interval, depending on the difference between the distances from the ship to the two transmitters. This placed the ship on a particular hyperbola. The ship would also listen in to another pair of transmitters, and this would place the ship on a second hyperbola. This then placed the ship at one of the four points where the two hyperbolas intersected. It would usually be obvious which of the four points was the correct one, but any ambiguity could be resolved by the signals from a third pair of transmitters.

In figure, the coordinates of F_1 and F_2 are, respectively, $(-ae, 0)$ and $(ae, 0)$.

The condition $PF_1 - PF_2 = 2a$ requires that

$$[(x + ae)^2 + y^2]^2 - [(x - ae)^2 + y^2]^{1/2} = 2a,$$

and this is the equation to the hyperbola. After some arrangement, this can be written

$$\frac{x^2}{a^2} - \frac{y^2}{a^2(e^2-1)} = 1,$$

which is a more familiar form for the equation to the hyperbola. Let us define a length b by

$$b^2 = a^2(e^2 - 1).$$

The equation then becomes

$$\frac{x^2}{a^2} - \frac{y^2}{b^2} = 1,$$

which is the most familiar form for the equation to a hyperbola.

Problem. When a meteor streaks across the sky, it can be tracked by radar. The radar instrumentation can determine the range (distance) of the meteoroid as a function of time. Show that, if the meteoroid is moving at constant speed (a questionable assumption, because it must be decelerating, but perhaps we can assume the decrease in speed is negligible during the course of the observation), and if the range r is plotted against the time, the graph will be a hyperbola. Show also that, if r^2 is plotted against t, the graph will be a parabola of the form

$$r^2 = at^2 + bt + c,$$

where $a = V^2$, $b = -V^2 t_0$, $c = V_2 t_0^{\ 2} + r_0^{\ 2}$, V = speed of the meteoroid, t_0 = time of closest approach, r_0 = distance of closest approach.

Radar observation of a meteor yields the following range-time data:

t (s)	r (km)
0.0	101.4 *
0.1	103.0
0.2	105.8
0.3	107.8
0.4	111.1
0.5	112.6
0.6	116.7
0.7	119.3
0.8	123.8 *
0.9	126.4
1.0	130.6
1.1	133.3
1.2	138.1
1.3	141.3 *

Assume that the velocity of the meteor is constant.

Determine:

(*i*) The time of closest approach (to 0.01 s)

(*ii*) The distance of closest approach (to 0.1 km)

(*iii*) The speed (to 1.0 km s^{-1})

If you wish, just use the three asterisked data to determine *a*, *b* and *c*. If you are more energetic, use *all* the data, and determine *a*, *b* and *c* by least squares, and the probable errors of V, t_0 and r_0.

The distance between the two vertices of the hyperbola is its transverse axis, and the length of the semi transverse axis is *a* – but what is the geometric meaning of the length *b*? This is discussed below in the next subsection (on the conjugate hyperbola) and again in a later section on the impact parameter. The lines perpendicular to the *x*-axis and passing through the foci are the two *latera recta*. Since the foci are at ($\pm ae$, 0), the points where the latera recta intersect the hyperbola can be found by putting $x = ae$ into the equation to the hyperbola, and it is then found that the length l of a semi latus rectum is

$$l = a(e^2 - 1).$$

The Conjugate Hyperbola

The equation

$$\frac{x^2}{a^2}-\frac{y^2}{b^2}=-1$$

is the equation to the conjugate hyperbola. The conjugate hyperbola is drawn dashed in figure, and it is seen that the geometric meaning of b is that it is the length of the semi transverse axis of the conjugate hyperbola. It is a simple matter to show that the eccentricity of the conjugate hyperbola is $e^2/(e^2-1)$.

The Asymptotes

The lines

$$y=\pm\frac{bx}{a}$$

are the *asymptotes* of the hyperbola. This can also be written

$$\frac{x^2}{a^2}-\frac{y^2}{b^2}=0.$$

Thus

$$\frac{x^2}{a^2}-\frac{y^2}{b^2}=c$$

is the hyperbola, the asymptotes, or the conjugate hyperbola, if $c=+1$, 0 or -1 respectively. The asymptotes are drawn as dotted lines in figure. The semi angle ψ between the asymptotes is given by

$$\tan\psi=b/a.$$

The Directrices

The lines $y=\pm a/e$ are the directrices, and, as with the ellipse (and with a similar proof), the hyperbola has the property that the ratio of the distance PF_2 to a focus to the distance PN to the directrix is constant and is equal to the eccentricity of the hyperbola. This ratio (i.e. the eccentricity) is less than one for the ellipse, equal to one for the parabola, and greater than one for the hyperbola. It is not a property that will be of great importance for our purposes, but is worth mentioning because it is a property that is sometimes used to define a hyperbola.

Parametric Equations to the Hyperbola

The reader will recall that the point $(a\cos E, b\sin E)$ is on the ellipse $(x^2/a^2)+(y^2/b^2)=1$ and that this is evident because this equation is the E-eliminant of $x=a\cos E$ and $y=b\sin E$. The angle E has a geometric interpretation as the eccentric anomaly. Likewise, recalling the relation $\cos h^2 \phi - \sin h^2 \phi = 1$, it will be evident that $(x^2 / a^2) - (y^2 / b^2) = 1$ can also be obtained as the ϕ–eliminant of the equations

$$x = a\cos h\ \phi,\ y = b \sin h\ \phi$$

These two equations are therefore the parametric equations to the hyperbola, and any point satisfying these two equations lies on the hyperbola. The variable ϕ is not an angle, and has no geometric interpretation analogous to the eccentric anomaly of an ellipse. The equations

$$x = a \sec E,\ y = b \tan E$$

can also be used as parametric equations to the hyperbola, on account of the trigonometric identity $1 + \tan^2 E = \sec^2 E$. In that case, the angle E does have a geometric interpretation (albeit not a particularly interesting one) in relation to the *auxiliary circle*, which is the circle of radius a centred at the origin. The meaning of the angle should be evident from figure, in which E is the eccentric angle corresponding to the point P.

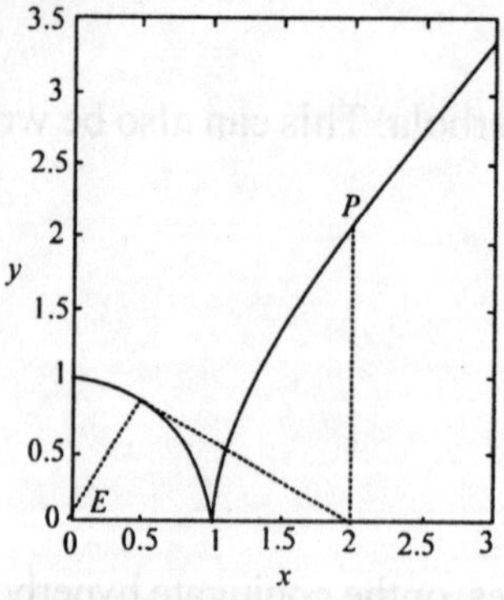

Impact Parameter

A particle travelling very fast under the action of an inverse square attractive force (such as an interstellar meteoroid or comet - if there are such things - passing by the Sun, or an electron in the vicinity of a positively charged atomic nucleus) will move in a hyperbolic path. We prove this in a later chapter, as well as discussing the necessary speed. We may imagine the particle initially approaching from a great distance along the asymptote at the bottom right hand corner of figure. As it approaches the focus, it no longer moves along the asymptote but along an arm of the hyperbola.

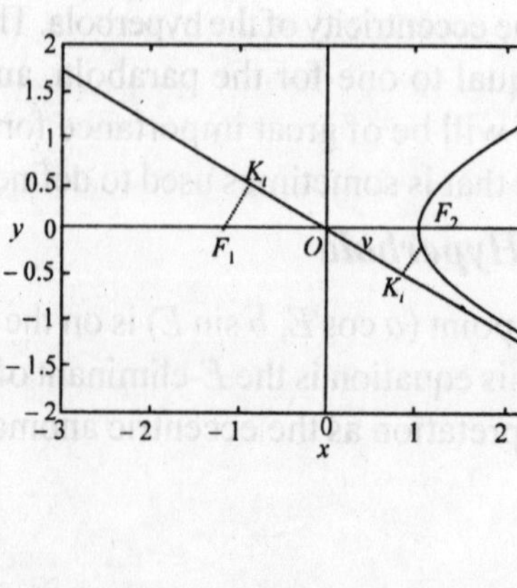

The distance K_2F_2, which is the distance by which the particle would have missed F_2 in the absence of an attractive force, is commonly called the *impact parameter*. Likewise, if the force had been a repulsive force (for example, suppose the moving particle were a positively charged particle and there were a centre of repulsion at F_1, F_1K_1 would be the impact parameter. Clearly, F_1K_1 and F_2K_2 are equal in length. The symbol that is often used in scattering theory, whether in celestial mechanics or in particle physics, is b - but is this b the same b that goes into the equation to the hyperbola and which is equal to the semi major axis of the conjugate hyperbola?

$OF_2 = ae$, and therefore $K_2F_2 = ae \sin \psi$. This, in conjunction with $\tan \psi = b/a$ and $b^2 = a^2(e^2 - 1)$, will soon show that the impact parameter is indeed the same b that we are familiar with, and that b is therefore a very suitable symbol to use for impact parameter.

Tangents to the Hyperbola

Using the same arguments as for the ellipse, the reader should easily find that lines of the form

$$y = mx \pm \sqrt{a^2m^2 - b^2}$$

are tangent to the hyperbola. This is illustrated in figure for a hyperbola with $b = a/2$, with tangents drawn with slopes 30° to 150° in steps of 5°. (The asymptotes have $\psi = 26° 34'$.)

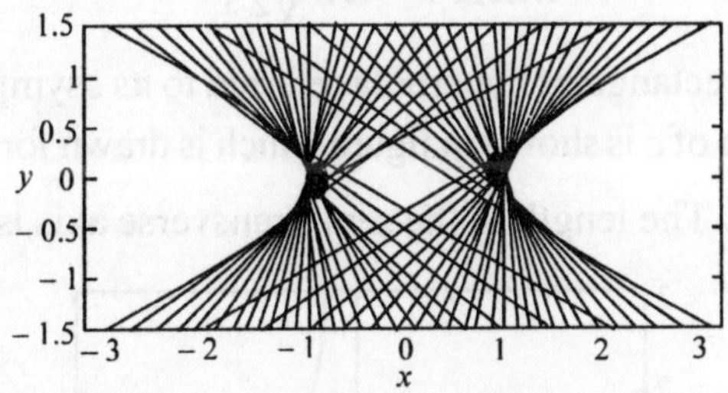

Likewise, from similar arguments used for the ellipse, the tangent to the hyperbola at the point (x, y) is found to be

$$\frac{x_1x}{a^2} - \frac{y_1y}{b^2} = 1.$$

Director Circle

As for the ellipse, and with a similar derivation, the locus of the points of intersection of perpendicular tangents is a circle, called the director circle, which is of radius $\sqrt{(a^2 - b^2)}$. This is not of particular importance for our purposes, but the reader who is interested might like to prove this by the same method as was done for the director circle of the ellipse, and might like to try drawing the circle and some tangents. If $b > a$, that is to say if È > 45o and the angle between the asymptotes is greater than 90o , the director circle is not real and it is of course not possible to draw perpendicular tangents.

Rectangular Hyperbola

If the angle between the asymptotes is 90°, the hyperbola is called a *rectangular hyperbola.* For such a hyperbola, $b = a$, the eccentricity is $\sqrt{2}$, the director circle is a point, namely the origin, and perpendicular tangents can be drawn only from the asymptotes.

The equation to a rectangular hyperbola is

$$x^2 - y^2 = a^2$$

and the asymptotes are at 45° to the x-axis.

Let Ox', Oy' be a set of axes at 45° to the x axis. (That is to say, they are the asymptotes of the rectangular hyperbola.) Then the equation to the rectangular hyperbola referred to its asymptotes as coordinate axes is found by the substitutions

$$\begin{pmatrix} x \\ y \end{pmatrix} = \begin{pmatrix} \cos 45° & \sin 45° \\ -\sin 45° & \cos 45° \end{pmatrix} \begin{pmatrix} x' \\ y' \end{pmatrix}$$

into $x^2 - y^2 = a^2$. This results in the equation

$$x'y' = \frac{1}{2}a^2 = c^2, \text{ where } c = a/\sqrt{2},$$

for the equation to the rectangular hyperbola referred to its asymptotes as coordinate axes. The geometric interpretation of c is shown in figure which is drawn for $c = 1$, and we have called the coordinate axes x and y. The length of the semi transverse axis is $c\sqrt{2}$.

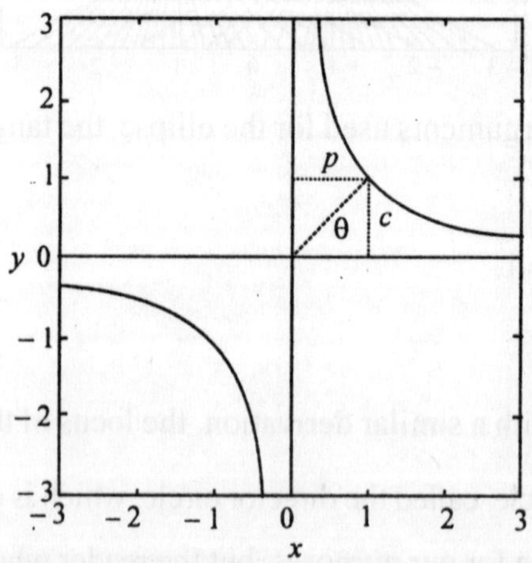

The simple equation $y = 1/x$ is a rectangular hyperbola and indeed it is this equation that is shown in figure.

It is left to the reader to show that the parametric equations to the rectangular hyperbola $xy = c^2$ (we have dropped the primes) are $x = ct$, $y = c/t$, that lines of the form $y = mx \pm 2c - m$ are tangent to $xy = c^2$, and that the tangent at (x_1, y_1) is $x_1y + y_1x = 2c$.

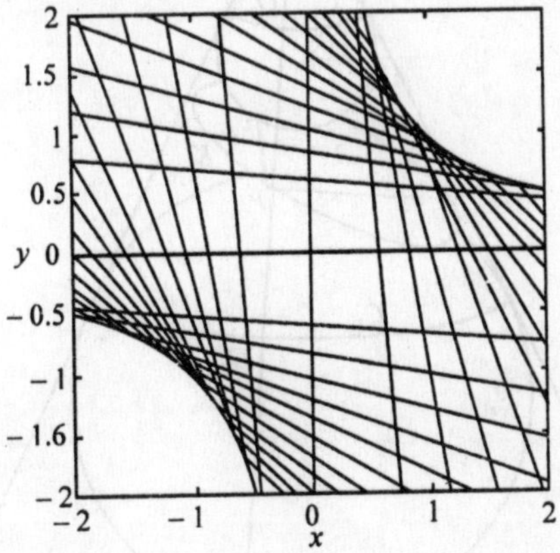

POLAR EQUATION TO THE HYPERBOLA

We found the polar equations to the ellipse and the parabola in different ways. Now go back and look at both methods and use either (or both) to show that the polar equation to the hyperbola (focus as pole) is

$$r = \frac{1}{1 + e\cos\theta}$$

This is the polar equation to any conic section - which one being determined solely by the value of e. You should also ask yourself what is represented by the equation

$$r = \frac{l}{1 - e\cos\theta}$$

Try sketching it for different values of e.

CONIC SECTIONS

We have so far defined an ellipse, a parabola and a hyperbola without any reference to a cone. Many readers will know that a plane section of a cone is either an ellipse, a parabola or a hyperbola, depending on whether the angle that the plane makes with the base of the cone is less than, equal to or greater than the angle that the generator of the cone makes with its base. However, given the definitions of the ellipse, parabola and hyperbola that we have given, proof is required that they are in fact conic sections. It might also be mentioned at this point that a plane section of a circular cylinder is also an ellipse. Also, of course, if the plane is parallel with the base of the cone, or perpendicular to the axis of the cylinder, the ellipse reduces to a circle.

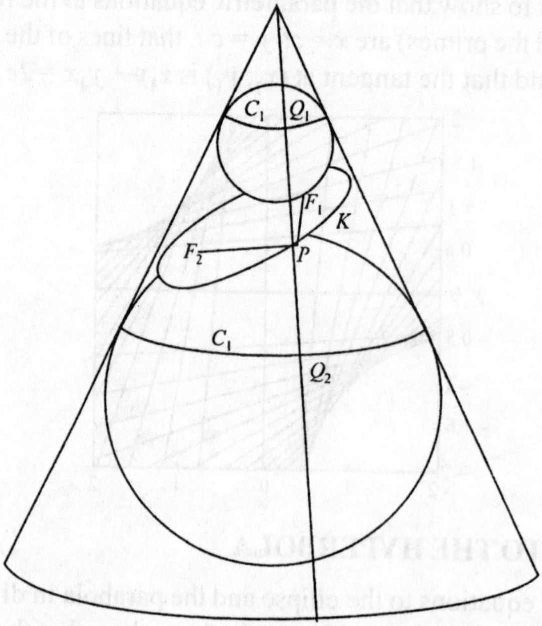

A simple and remarkable proof can be given in the classical Euclidean "Given. Required. Construction. Proof. Q.E.D." style.

Given: A cone and a plane such that the angle that the plane makes with the base of the cone is less than the angle that the generator of the cone makes with its base, and the plane cuts the cone in a closed curve K.

Required: To prove that K is an ellipse.

Construction: Construct a sphere above the plane, which touches the cone internally in the circle C_1 and the plane at the point F_1. Construct also a sphere below the plane, which touches the cone internally in the circle C2 and the plane at the point F_2.

Join a point P on the curve K to F_1 and to F_2.

Draw the generator that passes through the point P and which intersects C_1 at Q_1 and C_2 at Q_2.

Proof: $PF_1 = PQ_1$ (Tangents to a sphere from an external point.)

$PF_2 = PQ_2$ (Tangents to a sphere from an external point.)

$\therefore \quad PF_1 + PF_2 = PQ_1 + PQ_2 = Q_1 Q_2$

and Q_1Q_2 is independent of the position of P, since it is the distance between the circles C_1 and C_2 measured along a generator.

$\therefore \quad K$ is an ellipse and F_1 and F_2 are its foci.

A similar argument will show that a plane section of a cylinder is also an ellipse.

The reader can also devise drawings that will show that a plane section of a cone parallel to a generator is a parabola, and that a plane steeper than a generator cuts the cone in a hyperbola. The drawings are easiest to do with paper, pencil, compass and ruler, and will require some ingenuity. While I have seen the above proof for an ellipse in several books, I have not seen the corresponding proofs for a parabola and a hyperbola, but they can indeed be done, and the reader should find it an interesting challenge. If the reader can use a computer to make the drawings and can do better than my poor effort with figure, s/he is pretty good with a computer, which is a sign of a misspent youth.

The General Conic Section

The equation

$$\frac{x^2}{a^2}+\frac{y^2}{b^2}=1$$

represents an ellipse whose major axis is along the x axis and whose centre is at the origin of coordinates. But what if its centre is not at the origin, and if the major axis is at some skew angle to the x axis? What will be the equation that represents such an ellipse?

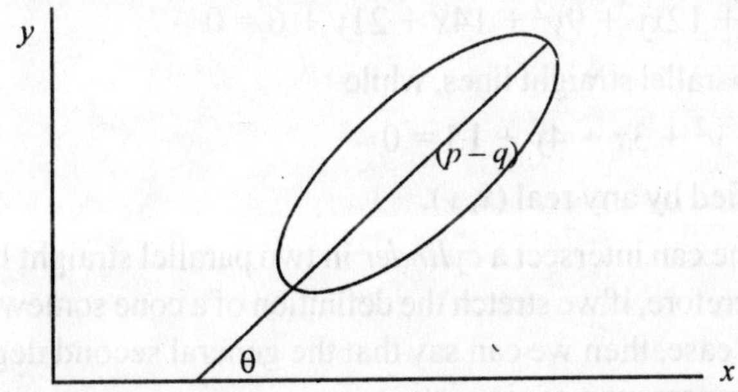

If the centre is translated from the origin to the point (p, q), the equation that represents the ellipse will be found by replacing x by $x-p$ and y by $y-q$. If the major axis is inclined at an angle, to the x axis, the equation that represents the ellipse will be found by replacing x by $x\cos\theta+y\sin\theta$ and y by $-x\sin\theta+y\cos\theta$. In any case, if the ellipse is translated or rotated or both, x and y will each be replaced by linear expressions in x and y, and the resulting equation will have at most terms in x^2, y^2, xy, x, y and a constant. The same is true of a parabola or a hyperbola. Thus, any of these three curves will be represented by an equation of the form

$$ax^2+2hxy+by^2+2gx+2fy+c=0.$$

(The coefficients a and b are not the semi major and semi minor axes.) The apparently random notation for the coefficients arises because these figures are plane sections of three-dimensional surfaces (the ellipsoid, paraboloid and hyperboloid) which are described by terms involving the coordinate z as well as x and y. The customary notation for these three-dimensional surfaces is very systematic, but when the terms in z are removed for the two-dimensional case, the apparently random notation a, b, c, f, g, h remains. In any case, the above equation can be divided through by the constant term without loss of generality, so that the equation to an ellipse, parabola or hyperbola can be written, if preferred, as

$$ax^2 + 2hxy + by^2 + 2gx + 2fy + 1 = 0.$$

Is the converse true? That is, does an equation of this form always necessarily represent an ellipse, parabola or hyperbola?

Not quite. For example,

$$6x^2 + xy - y^2 - 17x - y + 12 = 0$$

represents two straight lines (it can be factored into two linear terms - try it), while

$$2x^2 - 4xy + 4y^2 - 4x + 4 = 0$$

is satisfied only by a single point. (Find it.)

However, a plane section of a cone can be two lines or a single point, so perhaps we can now ask whether the general second degree equation must always represent a conic section. The answer is: close, but not quite.

For example,

$$4x^2 + 12xy + 9y^2 + 14x + 21y + 6 = 0$$

represents two parallel straight lines, while

$$x^2 + y^2 + 3x + 4y + 15 = 0$$

cannot be satisfied by any real (x, y).

However, a plane can intersect a *cylinder* in two parallel straight lines, or a single straight line, or not at all. Therefore, if we stretch the definition of a cone somewhat to include a cylinder as a special limiting case, then we can say that the general second degree equation in x and y does indeed always represent a conic section.

Is there any means by which one can tell by a glance at a particular second degree equation, for example

$$8x^2 + 10xy - 3y^2 - 2x - 4y - 2 = 0,$$

what type of conic section is represented? The answer is yes, and this one happens to be a hyperbola. The discrimination is made by examining the properties of the determinant

$$\Delta = \begin{vmatrix} a & h & g \\ h & b & f \\ g & f & c \end{vmatrix}$$

I have devised a table after the design of the dichotomous tables commonly used by taxonomists in biology, in which the user is confronted by a couplet (or sometimes triplet) of alternatives, and is then directed to the next part of the table. I shall spare the reader the derivation of the table; instead, I shall describe its use.

In the table, I have used the symbol $\overline{a}$ to mean the cofactor of a in the determinant, $\overline{h}$ the cofactor of h, $\overline{g}$ the cofactor of g, etc. Explicitly, these are

$$\overline{a} = bc - f^2,$$

$$\overline{b} = ca - g^2,$$

$$\overline{c} = ab - h^2,$$

$$\overline{f} = gh - af,$$

$$\overline{g} = hf - bg$$

and $\overline{h} = fg$ “ $ch.$

The first column labels the choices that the user is asked to make. At the beginning, there are two choices to make, 1 and 1' The second column says what these choices are, and the fourth column says where to go next. Thus, if the determinant is zero, go to 2; otherwise, go to 5. If there is an asterisk in column 4, you are finished. Column 3 says what sort of a conic section you have arrived at, and column 5 gives an example.

No matter what type the conic section is, the coordinates of its centre are $(\overline{g}/\overline{c}, \overline{f}/\overline{c})$ and the angle θ that its major or transverse axis makes with the x axis is given by

$$\tan 2\theta = \frac{2h}{a-b}.$$

Thus if x is first replaced with $x + \overline{g}/\overline{c}$ and y with $y + \overline{f}/\overline{c}$, and then the new x is replaced with $x \cos\theta - y \sin\theta$ and the new y with $x \sin\theta + y \cos\theta$, the equation will take the familiar form of a conic section with its major or transverse axis coincident with the x axis and its centre at the origin. Any of its properties, such as the eccentricity, can then be deduced from the familiar equations.

Key to the Conic Sections

1	$\Delta = \Delta$	2		
1'	$\Delta \neq \Delta$	5		
2	$\overline{c} > 0$	Point	*	$x^2 - 2xy + 2y^2 - 2x + 2 = 0$
2'	$\overline{c} = 0$	3		
2"	$\overline{c} < 0$	Two nonparallel straight lines 4		
3	$\overline{h} = 0$	Straight line	*	$4x^2 + 4xy + y^2 + 12x + 6y + 9 = 0$

3'	$\overline{h} \neq 0$	Two parallel straight line	*	$4x^2 + 12xy + 9y^2 + 14x + 21y + 6 = 0$
4	$a + b = 0$	Two perpendicular straight lines	*	$6x^2 + 5xy - 6y^2 + x + 8y - 2 = 0$
4'	$a + b \neq 0$	Two straight lines, neither parallel nor perpendicular	*	$6x^2 - xy - y^2 + 34x + 13y - 12 = 0$
5	$\overline{c} > 0$		6	
5'	$\overline{c} = 0$	Parabola	*	$9x^2 - 12xy + 4y^2 - 18x - 101y + 19 = 0$
5"	$\overline{c} < 0$		8	
6	$a\Delta > 0$	Nothing	*	$x^2 + y^2 + 3x + 4y + 15 = 0$
6'	$a\Delta < 0$		7	
7	$a = b$ and $h = 0$	Circle	*	$x^2 + y^2 - 6x - 8y + 9 = 0$
7'	Not so	Ellipse	*	$14x^2 - 4xy + 11y^2 - 44x - 58y + 71 = 0$
8	$a + b = 0$	Rectangular hyperbola	*	$7x^2 - 48xy - 7y^2 + 10x - 28y + 100 = 0$
8'	$a + b \neq 0$	Hyperbola (not Rectangular)	*	$8x^2 + 10xy - 3y^2 - 2x - 4y - 2 = 0$

FITTING A CONIC SECTION THROUGH FIVE POINTS.

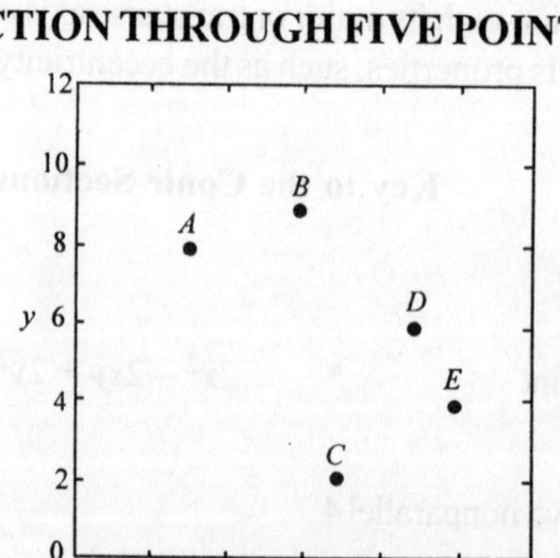

Fig. The Five Points $A(1, 8)$, $B(4, 9)$, $C(5, 2)$, $D(7, 6)$, $E(8, 4)$.

Problem: Draw a conic section through the five points.

The first thing to notice is that, since a conic section is of the form

$$ax^2 + 2hxy + by^2 + 2gx + 2fy + 1 = 0$$

(remember that there is no loss of generality by taking the constant term to be 1), five points are necessary and sufficient to define a conic section uniquely. One and only one conic section can be drawn through these five points. We merely have to determine the five coefficients. The most direct (but not the fastest or most efficient) way to do this is to substitute each of the (x, y) pairs into the equation in turn, thus obtaining five linear equations in the five coefficients.

There is a better way.

We write down the equations for the straight lines *AB, CD, AC* and *BD*. Let us call these equations $\alpha = 0$, $\beta = 0$, $\gamma = 0$ and $\delta = 0$ respectively.

Then $\alpha\beta = 0$ is the equation that represents the two straight lines *AB* and *CD*, and $\gamma\delta = 0$ is the equation that represents the two straight lines *AC* and *BD*. The equation $\alpha\beta + \lambda\gamma\delta = 0$, where λ is an arbitrary constant, is a second degree equation that represents any conic section that passes through the points *A, B, C* and *D*. By inserting the coordinates of *E* in this equation, we can find the value of that forces the equation to go through all five points. This model of unclarity will become clear on following an actual calculation for the five points of the present example.

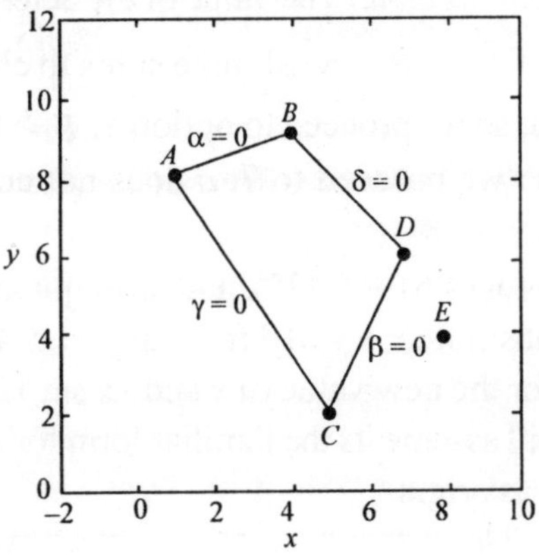

The four straight lines are

$\alpha = 0$: $\quad x - 3y + 23 = 0$

$\beta = 0$: $\quad 2x - y - 8 = 0$

$\alpha = 0$: $\quad 3x + y - 19 = 0$

$\alpha = 0$: $\quad x + y - 13 = 0$

The two ;pairs of lines are

$\alpha\beta = 0$: $\quad 2x^2 - 7xy + 3y^2 + 38x + y - 184 = 0$

$\gamma\delta = 0$: $\quad 3x^2 + 5xy + 2y^2 - 58x - 45y + 247 = 0$

and the family of conic sections that passes through A, B, C and D is

$$\alpha\beta + \lambda\gamma\delta = 0 \quad (2+3\lambda)x^2 - (7-5\lambda)xy + (3+2\lambda)y^2 + (38-58\lambda) + (1-45\lambda)y$$
$$-184+247\lambda = 0.$$

Now substitute $x = 8$, $y = 4$ to force the conic section to pass through the point E. This results in the value

$$\lambda = \frac{76}{13}.$$

The equation to the conic section passing through all five points is therefore

$$508x^2 + 578xy + 382y^2 - 7828x - 6814y + 32760 = 0$$

We can, if desired, divide this equation by 2 (since all coefficients are even), or by 32760 (to make the constant term equal to 1) but, to make the analysis that is to follow easier, I choose to leave the equation in the above form, so that the constants f, g and h remain integers.

The constants have the values

$$a = 508,\ b = 382,\ c = 32760,\ f = -3407,\ g = -3914,\ h = 289$$

and the cofactors have the values

$$\overline{a} = 906\,671,\ \overline{b} = 1\,322\,684,\ \overline{c} = 110\,535$$
$$\overline{f} = 599\,610,\ \overline{g} = 510\,525,\ \overline{h} = 867\,358$$

Let us consult the dichotomous table. The value of the determinant is $\Delta = a\overline{a} + b\overline{b} + c\overline{c}$ (or $h\overline{h} + b\overline{b} + f\overline{f}$, or $g\overline{g} + f\overline{f} + c\overline{c}$; try all three sums to check for arithmetic mistakes). It comes to $\Delta = -419\,939\,520$, so we proceed to option 5. $\overline{c} > 0$,so we proceed to option 6. a and Δ have opposite signs, so we proceed to 7. a does not equal b, nor is h equal to zero. Therefore we have an ellipse.

The centre of the ellipse is at (4.619, 5.425), and its major axis is inclined at an angle 128° 51' to the x-axis. If we now substitute $x + 4.619$ for x and $y + 5.425$ for y, and then substitute $x \cos 128°51' + y \sin 1280\ 51'$ for the new value of x and $-x \sin 128°\ 51' + y \cos 128°\ 51'$ for the new value of y, the equation will assume its the familiar form for an ellipse referred to its axes as coordinate axes and its centre as origin.

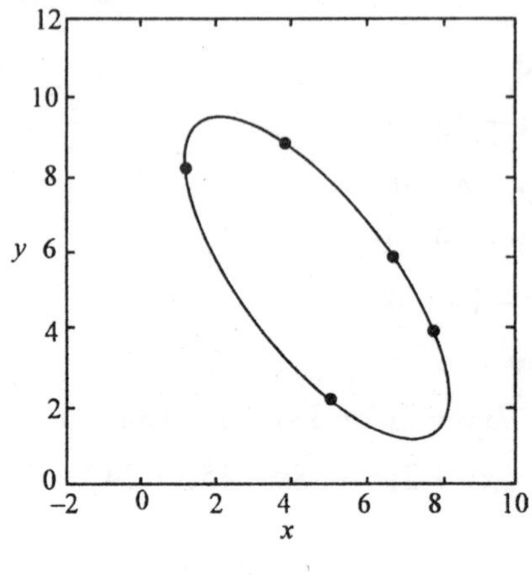

Fitting a Conic Section Through n Points

What is the best ellipse passing near to the following 16 points?

(1, 50) (11, 58) (20,63) (30,60)

(42, 59) (48, 52) (54, 46) (61, 42)

(61, 19) (45, 12) (35, 10) (25, 13)

(17, 17) (14, 22) (5, 29) (3, 43)

This is answered by substituting each point (x, y) in turn in the equation

$$ax^2 + 2hxy + by^2 + 2gx + 2fy + 1 = 0,$$

thus obtaining 16 equations in the coefficients a, h, b, g, f. (The constant term can be taken to be unity.) These are the equations of condition. The five normal equations can then be set up and solved to give those values for the coefficients that will result in the sum of the squares of the residuals being least, and it is in that sense that the "best" ellipse results. The details of the method are given in the chapter on numerical methods. The actual solution for the points given above is left as an exercise for the energetic.

It might be thought that we are now well on the way to doing some real orbital theory. After all, suppose that we have several positions of a planet in orbit around the Sun, or several positions of the secondary component of a visual binary star with respect to its primary component; we can now fit an ellipse through these positions. However, in a real orbital situation we have some additional information as well as an additional constraint. The additional information is that, for each position, we also have a time. The constraint is that the orbit that we deduce must obey Kepler's second law of planetary motion - namely, that the radius vector sweeps out equal areas in equal times. We shall have to await Part II before we get around actually to computing orbits.

Chapter 5

Straight Line

A line can be described as an ideal zero-width, infinitely long, perfectly straight curve (the term curve in mathematics includes "straight curves") containing an infinite number of points. In Euclidean geometry, exactly one line can be found that passes through any two points. The line provides the shortest connection between the points.

In two dimensions, two different lines can either be parallel, meaning they never meet, or may intersect at one and only one point. In three or more dimensions, lines may also be skew, meaning they don't meet, but also don't define a plane. Two distinct planes intersect in at most one line. Three or more points that lie on the same line are called *collinear*.

In geometry, a line segment is a part of a line that is bounded by two distinct end points, and contains every point on the line between its end points. Examples of line segments include the sides of a triangle or square. More generally, when the end points are both vertices of a polygon, the line segment is either an edge (of that polygon) if they are adjacent vertices, or otherwise a diagonal. When the end points both lie on a curve such as a circle, a line segment is called a chord (of that curve).

HOW CAN WE DRAW A STRAIGHT LINE

When using a compass to draw a circle, we are not starting with a model of a circle; instead we are using a fundamental property of circles that the points on a circle are a fixed distance from a center. Or we can say we use Euclid's definition of a circle . So now what about drawing a straight line: Is there a tool (serving the role of a compass) that will draw a straight line? One could say we can use a straightedge for constructing a straight line. Well, how do you know that your straightedge is straight? How can you check that something is straight? What does "straight" mean?

You can try to use Euclid's definition above. If you fold a piece of paper the crease will be straight – the edges of the paper needn't even be straight. This utilizes mirror symmetry to produce the straight line. Carpenters also use symmetry to determine straightness – they put two boards face to face, plane the edges until they look straight, and then turn one board over so the planed edges are touching. They then hold the boards up to the light. If the edges are not straight, there will be gaps between the boards where light will shine through. I

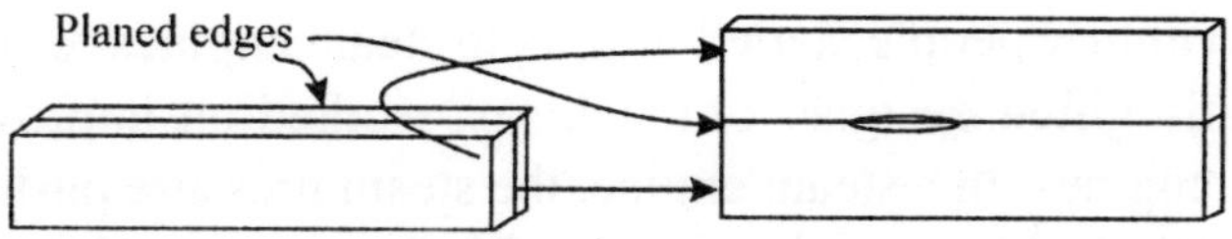

Fig. Carpenter's Method for Checking Straightness

When grinding an extremely accurate flat mirror, the following technique is sometimes used: Take three approximately flat pieces of glass and put pumice between the first and second pieces and grind them together. Then do the same for the second and the third pieces and then for the third and first pieces. Repeat many times and all three pieces of glass will become very accurately flat. Do you see why this works? What does this have to do with straightness?

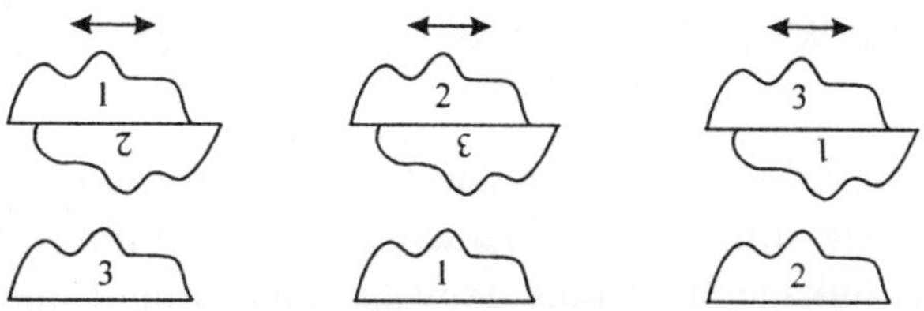

Fig. Grinding Flat Mirrors

We can also use the usual high school definition, "A straight line is the shortest distance between two points." This leads to producing a straight line by stretching a string.

This use of symmetry, stretching, and folding can also be extended to other surfaces. Sometimes we can get confused when reading in the literature that "straight line" is an undefined term or that straight lines on the sphere are "defined to be arcs of great circles." We find that putting "straight" in the context of the four historical strands helps clarify this: "Symmetry" comes mostly from the Art/Pattern Strand, "undefined term" comes from the axiomatics in the Building Structures Strand, and "shortest distance" comes mostly from the Navigation/Stargazing Strand.

But there is still left unanswered the question of whether there is a mechanism analogous to a compass that will draw an accurate straight line. We can find answers to this question in the history of mechanics, which leads us into the Motion/Machines Strand and to another meaning of "straight".

Turning circular motion into straight line motion has been a practical engineering problem since at least the 13th century. As we can see in some 13th century drawings of a sawmill linkages (rigid bars constrained to be near a plane and joined at their ends by rivets) were in use in the 13th century and probably were originated much earlier. Georgius Agricola's geological writings reflect firsthand observations not just of rocks and minerals, but of every aspect of mining technology and practice of the time. In the pictures of his work one can see link work that was widely used for converting the continuous rotation of a water wheel into a reciprocating motion suited to piston pumps. In 1588, Agostino Ramelli published his book on machines where linkages were widely used.

In the late 18th century people started turning to steam engines for power. James Watt (1736–1819), a highly gifted designer of machines, worked on improving the efficiency and power of steam engines. In a steam engine, the steam pressure pushes a piston down a straight cylinder. Watt's problem was how to turn this linear motion into the circular motion of a wheel (such as on steam locomotives). It took Watt several years to design the straight-line linkage that would change straight-line motion to circular one. Later, Watt told his son,

Though I am not over anxious after fame, yet I am more proud of the parallel motion than of any other mechanical invention I have ever made.

"Parallel motion" is a name Watt used for his linkage, which was included in an extensive patent of 1784. Watt's linkage was a good solution to the practical engineering problem.

But Watt's solution did not satisfy mathematicians, who knew that his linkage can draw only an approximate straight line. Mathematic ians continued to look for a planar straight-line linkage. A linkage that would draw an exact straight-line was not found until 1864–1871, when a French army officer, Charles Nicolas Peaucellier (1832–1913), and a Russian graduate student, Lipmann I. Lipkin (1851–1875), independently developed a linkage that draws an exact straight line.(There is not much known about Lipkin. Some sources mentioned that he was born in Lithuania and was a graduate student of Chebyshev in Saint

WHEN A LINE IS CALLED STRAIGHT?

In keeping with the spirit of the approach to geometry discussed in the Preface, we begin with a question that encourages you to explore deeply the concept of straightness. We ask you to build a notion of straightness from your experiences rather than accept a certain number of assumptions about straightness. Though it is difficult to formalize, straightness is a natural human concept.

(a) How can you check in a practical way if something is straight? How do you construct something straight — lay out fence posts in a straight line, or draw a straight line? Do this without assuming that you have a ruler, for then we will ask, "How can you check that the ruler is straight?"

At first, look for examples of straightness in your experiences. Go out and actually try walking along a straight line and then along a curved path; try drawing a straight line and then checking that a line already drawn is straight. As you look for properties of straight lines that distinguish them from non-straight lines, you will probably remember the following statement (which is often taken as a definition in high school geometry): *A straight line is the shortest distance between two points.* But can you ever measure the lengths of all the paths between two points? How do you find the shortest path? If the shortest path between two points is in fact a straight line, then is the converse true? Is a straight line between two points always the shortest path? We will return to these questions in later chapters.

A powerful approach to this problem is to think about lines in terms of symmetry. This will become increasingly important as we go on to other surfaces (spheres, cones, cylinders, and so forth). Two of the symmetries of lines are as follows:

- *Reflection symmetry in the line*, also called *bilateral symmetry* – reflecting (or mirroring) an object over the line.

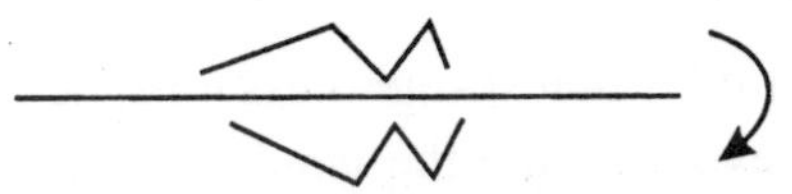

Fig. Reflection Symmetry of a Straight Line

- *Half-turn symmetry* – rotating 180° about any point on the line.

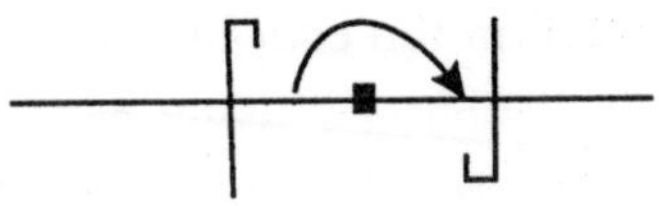

Fig. Half-turn Symmetry of a Straight Line

Although we are focusing on a symmetry *of the line* in each of these examples, notice that the symmetry is not a property of the line by itself but includes the line and the space around the line. The symmetries preserve the local environment of the line. Notice how in reflection and half-turn symmetry the line and its local environment are both part of the symmetry action and how the relationship between them is integral to the action. In fact, reflection in the line does not move the line at all but exhibits a way in which the spaces on the two sides of the line are the same.

DEFINITIONS. An ***isometry*** is a transformation that preserves distances and angle measures. A ***symmetry of a figure*** is an isometry of a region of space that takes the figure (or the portion of it in the region) onto itself.

(b) *What symmetries does a straight line have*?

Try to think of other symmetries of a line as well (there are quite a few). Some symmetries hold only for straight lines, while some work for other curves as well. Try to determine which symmetries are specific to straight lines and why. Also think of practical applications of these symmetries for constructing a straight line or for determining if a line is straight.

(c) *What is in common among the different notions of straightness? Can you write a definition of "straight line"?*

Look for things that you call "straight." Where do you see straight lines? Why do you say they are straight? Look for both physical lines and nonphysical uses of the word "straight". What symmetries does a straight line have? How do they fit with the examples that you have found and those mentioned above? Can we use any of the symmetries of a line to define straightness? The intersection of two (flat) planes is a straight line – why does this work? Does it help us understand "straightness"?

Imagine (or actually try!) walking while pulling a long thread with a small stone attached. When will the stone follow along your path? Why? This property is used to pick up a fallen water skier. The boat travels by the skier along a straight line and thus the tow rope follows the path of the boat. Then the boat turns in an arc in front of the skier. Because the boat is no longer following a straight path, the tow rope moves in toward the fallen skier. What is happening?

Another idea to keep in mind is that straightness must be thought of as a local property. Part of a line can be straight even though the whole line may not be. For example, if we agree that this line is straight,

and then we add a squiggly part on the end, like this,

would we now say that the original part of the line is not straight, even though it hasn't changed, only been added to? Also note that we are not making any distinction here between "line" and "line segment." The more generic term "line" generally works well to refer to any and all lines and line segments, both straight and non-straight.

You are likely to bring up many ideas of straightness. It is necessary then to think about what is common among all of these straight phenomena. Think about and formulate some answers for these questions before you read any further. Do not take anything for granted unless you see why it is true. No answers are predetermined. You may come up with something that we have never imagined. Consequently, it is important that you persist in following your own ideas.

You should not read further until you have expressed your thinking and ideas through writing or talking to someone else.

THE SYMMETRIES OF A LINE

Reflection-in-the-line Symmetry

It is most useful to think of reflection as a "mirror" action with the line as an axis rather than as a "flip-over" action that involves an action in 3-space. In this way we can extend the notion of reflection symmetry to a sphere (the flip-over action is not possible on a sphere). Notice that this symmetry cannot be used as a definition for straightness because we use straightness to define reflection symmetry. This same comment applies to most of the other symmetries discussed below.

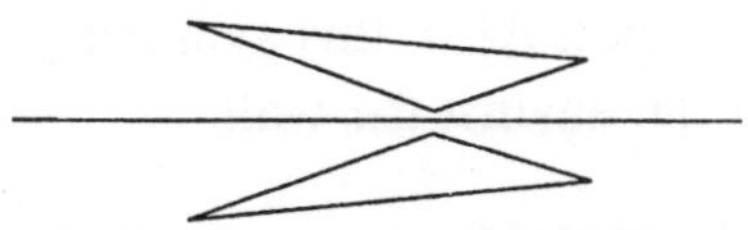

Fig. Reflection-in-the-line Symmetry

The light gray triangle is the image of the dark gray triangle under the action of the symmetry on the space around the line.

- *Practical application*: We can produce a straight line by folding a piece of paper because this action forces symmetry along the crease. Above we showed a carpenter's example.

Reflection-perpendicular to the Line Symmetry

A reflection through *any* axis perpendicular to the line will take the line onto itself. Note that circle s also have this symmetry about any diameter.

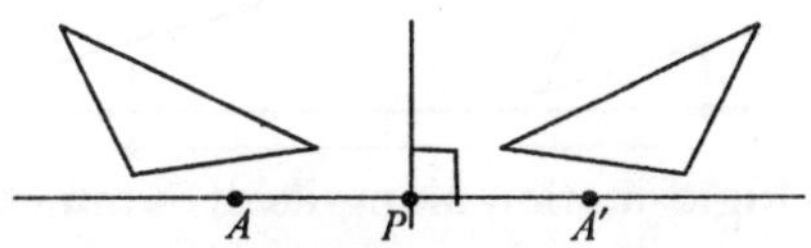

Fig. Reflection-perpendicular-to-the-line Symmetry

- *Practical applications*: You call tell if a straight segment is perpendicular to a mirror by seeing if it looks straight with its reflection. Also, a straight line can be folded onto itself.

Half-turn Symmetry

A rotation through half of a full revolution about any point P on the line takes the part of the line before P onto the part of the line after P and vice versa. Note that some non-straight lines, such as the letter Z, also have half-turn symmetry — but not about *every* point.

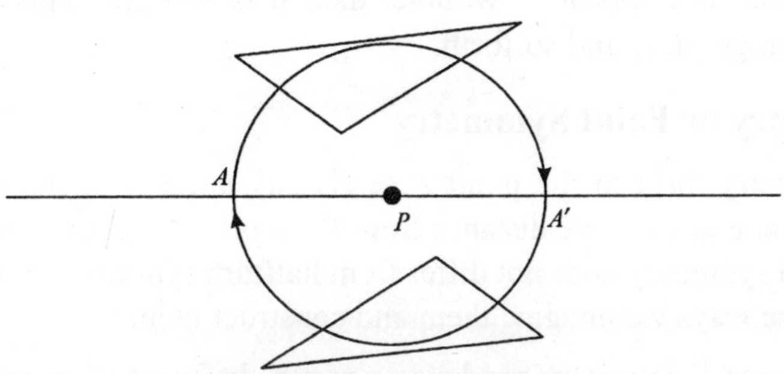

Fig. Half-turn Symmetry

- *Practical applications*: Half-turn symmetry exists for the slot on a screw and the tip of the screwdriver (unless you are using Phillips-head screws and screwdrivers, which also have quarter-turn symmetry) and thus we can more easily put the tip of the screwdriver into the slot. Also, this symmetry is involved when a door (in a straight wall) opens up flat against the wall.

Rigid-motion-along-itself Symmetry

For straight lines in the plane, we call this *translation symmetry*. Any portion of a straight line may be moved along the line without leaving the line. This property of being able to move rigidly along itself is not unique to straight lines; circles (rotation symmetry) and circular helixes (screw symmetry) have this property as well.

- *Practical applications*: Slide joints such as in trombones, drawers, nuts and bolts, and so forth, all utilize this symmetry.

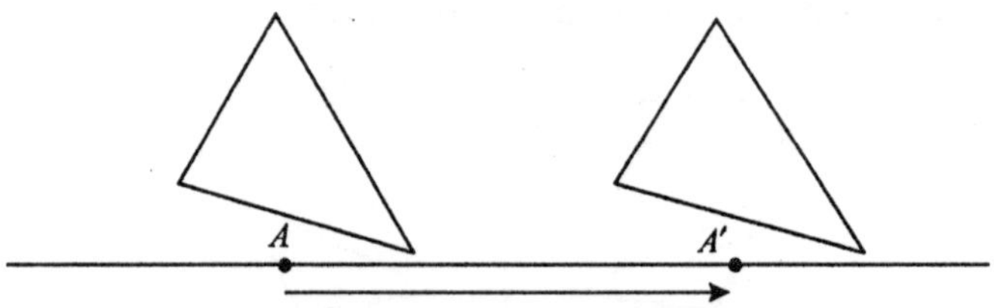

Fig. Rigid-motion-along-itself Symmetry

3-dimensional-rotation symmetry: In a 3-dimensional space, rotate the line around itself through *any* angle using itself as an axis.

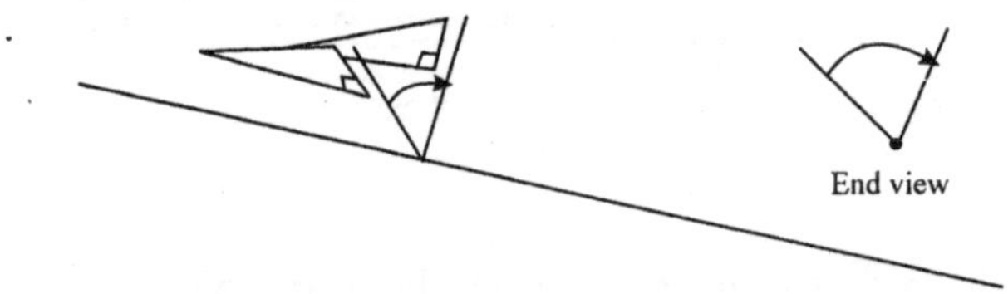

Fig. Dimensional-rotation Symmetry

- *Practical applications*: This symmetry can be used to check the straightness of any long thin object such as a stick by twirling the stick with itself as the axis. If the stick does not appear to wobble, then it is straight. This is used for pool cues, axles, hinge pins, and so forth.

Central Symmetry or Point Symmetry

Central symmetry through the point P sends any point A to the point on the line determined by A and P at the same distance from P but on the opposite side from P. In two dimensions central symmetry does not differ from halfturn symmetry in its end result, but they do differ in the ways we imagine them and construct them.

- In 3-space, central symmetry produces a result different than any single rotation or reflection (though we can check that it does give the same result as the composition of three reflections through mutually perpendicular planes). To experience central symmetry in 3-space, hold your hands in front of you with the palms facing each other and your left thumb up and your right thumb down. Your two hands now have approximate central symmetry about a point midway between the center of the palms; and this symmetry cannot be produced by any reflection or rotation.

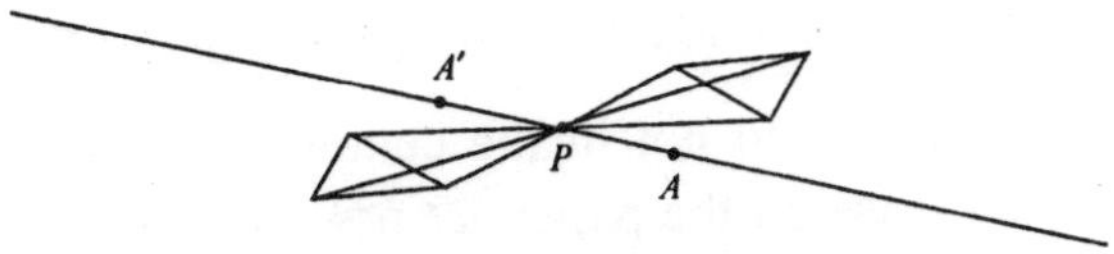

Fig. Central Symmetry

Similarity or Self-similarity "Quasi-symmetry"

Any segment of a straight line (and its environs) is similar to (that is, can be magnified or shrunk to become the same as) any other segment. This is not a symmetry because it does not preserve distances but it could be called a "quasi-symmetry" because it *does* preserve the measure of angles.

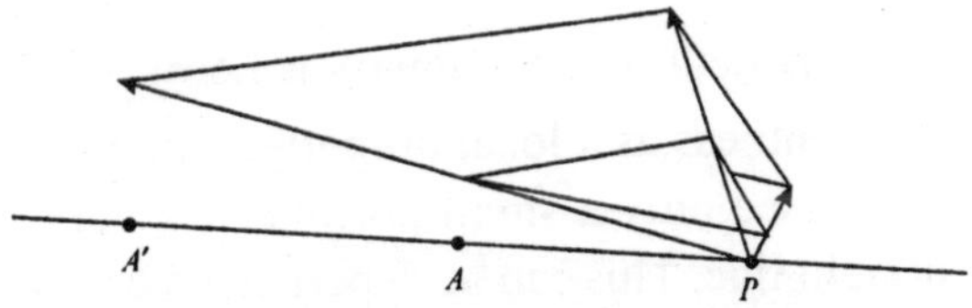

Fig. Similarity "Quasi-symmetry"

- Logarithmic spirals such as the chambered nautilus have selfsimilarity as do many fractals.

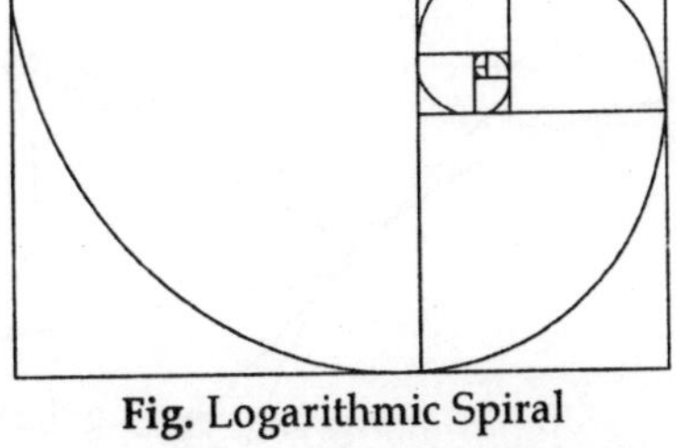

Fig. Logarithmic Spiral

Clearly, other objects besides lines have some of the symmetries mentioned here. It is important for you to construct your own such examples and attempt to find an object that has all of the symmetries but is not a line. This will help you to experience that straightness and the seven symmetries discussed here are intimately related. You should come to the conclusion that while other curves and figures have some of these symmetries, only straight lines have all of them.

LOCAL (AND INFINITESIMAL) STRAIGHTNESS

Previously, you saw how a straight line has reflection-in-the-line symmetry and half-turn symmetry: One side of the line is the same as the other. But, as pointed out above, straightness is a local property in that whether a segment of a line is straight depends only on what is near the segment and does not depend on anything happening away from the line. Thus each of the symmetries must be able to be thought of (and experienced) as applying only locally. This will become particularly important later when we investigate straightness on the cone and cylinder. For now, it can be experienced in the following way:

When a piece of paper is folded not in the center, the crease is still straight even though the two sides of the crease on the paper are not the same.

So what is the role of the sides when we are checking for straightness using reflection symmetry? Think about what is important near the crease in order to have reflection symmetry.

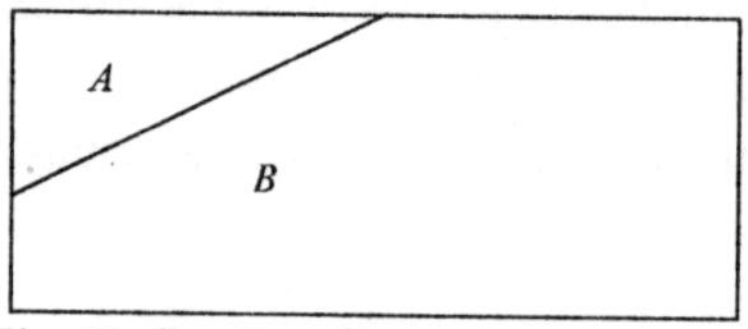

Fig. Reflection Symmetry is Local

When we talk about straightness as a local property, you may bring out some notions of scale . For example, if you see only a small portion of a very large circle , it will be indistinguishable from a straight line. This can be experienced easily on many of the modern graphing programs for computers. Also, a microscope with a zoom lens will provide an experience of zooming. If a curve is smooth (or differentiable), then if we "zoom in" on any point of the curve, eventually the curve will be indistinguishable from a straight line segment.

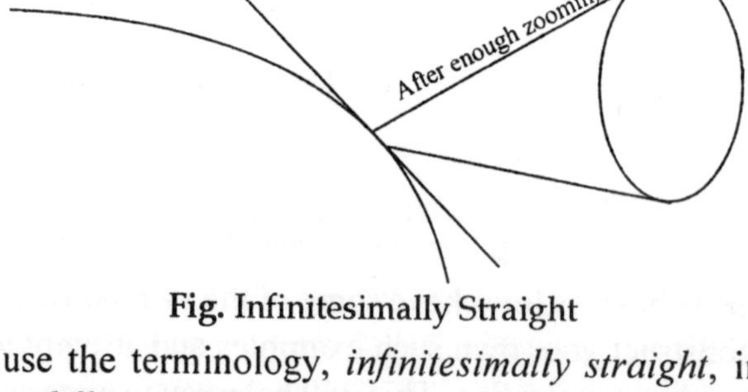

Fig. Infinitesimally Straight

We sometimes use the terminology, *infinitesimally straight*, in place of the more standard terminology, *differentiable*. We say that a curve is *infinitesimally straight* at a point p if there is a straight line l such that if we zoom in enough on p, the line and the curve become indistinguishable. When the curve is parametrized by arc length this is equivalent to the curve having a well-defined velocity vector at each point.

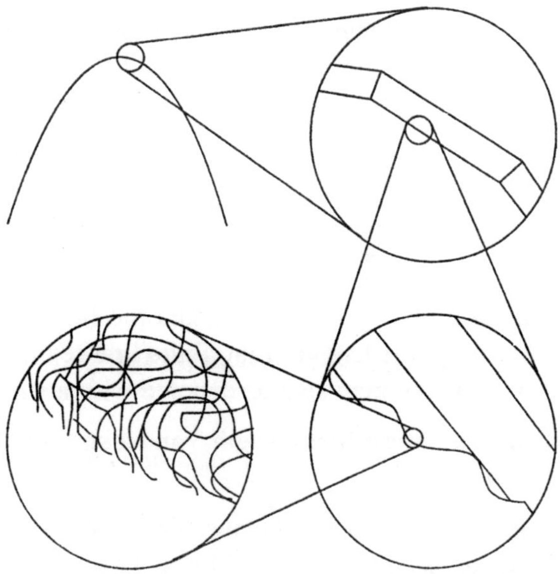

Fig. Straightness and Smoothness Depend on the Scale

In contrast, we can say that a curve is *locally straight at a point* if that point has a neighborhood that is straight. In the physical world the usual use of both *smooth* and *locally straight* is dependent on the scale at which they are viewed. For example, we may look at an arch made out of wood – at a distance it appears as a smooth curve; then as we move in closer we see that the curve is made by many short straight pieces of finished (planed) boards, but when we are close enough to touch it, we see that its surface is made up of smooth waves or ripples, and under a microscope we see the non-smoothness of numerous twisting fibers.

This is equivalent to the usual definitions of being differentiable at p. For example, if $t(x) = f(p) + f'(p)(x - p)$ is the equation of the line tangent to the curve $(x,f(x))$ at the point $(p,f(p))$, then, given $\varepsilon > 0$ (the distance of indistinguishability), there is a $\delta > 0$ (the radius of the zoom window) such that, for $|x - p| < \delta$ (for x within the zoom window), $|f(x) - t(x)| < \varepsilon$ [$f(x)$ is indistinguishable from $t(x)$]. This last inequality may look more familiar in the form

$$f(x) - t(x) = f(x) - f(p) - f'(p)(x - p) = \{[f(x) - f(p)]/(x - p) - f'(p)\}(x - p) < \varepsilon.$$

In general, the value of δ might depend on p as well as on ε. Often the term *smooth* is used to mean continuously differentiable, which the interested reader can check is equivalent (on closed finite intervals) to, for each $\varepsilon > 0$, there being one $\delta > 0$ that works for all p.

STRAIGHT LINE

A line is a straight one-dimensional figure having no thickness and extending infinitely in both directions. A line is sometimes called a straight line or, more archaically, a right line, to emphasize that it has no "wiggles" anywhere along its length. While lines are intrinsically one-dimensional objects, they may be embedded in higher dimensional spaces.

Harary called an edge of a graph a "line."

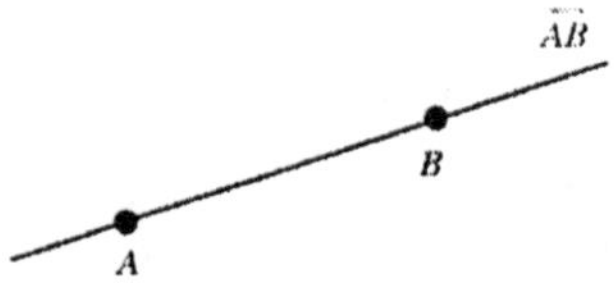

A line is uniquely determined by two points, and the line passing through points A and B is denoted AB . Similarly, the length of the finite line segment terminating at these points may be denoted AB . A line may also be denoted with a single lower-case letter.

Euclid defined a line as a "breadthless length," and a straight line as a line that "lies evenly with the points on itself".

Consider first lines in a two-dimensional plane. Two lines lying in the same plane that do not intersect one another are said to be parallel lines. Two lines lying in different planes that do not intersect one another are said to be skew lines.

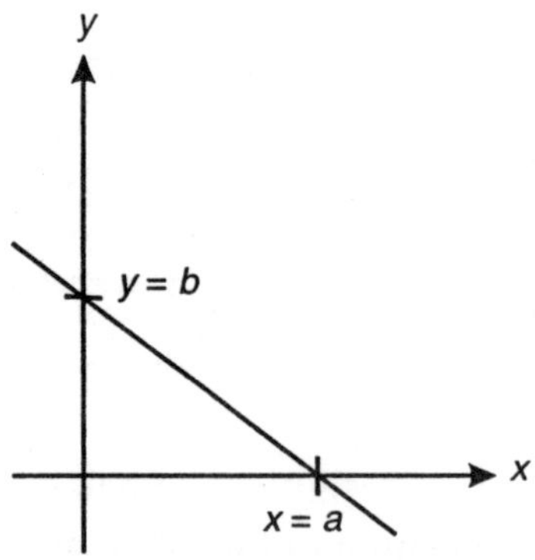

The line with x-intercept a and y-intercept b is given by the *intercept form*

$$\frac{x}{a}+\frac{y}{b}=1$$

The line through (x_1, x_1) with slope m is given by the *point-slope form*

$$y-y_1=m(x-x_1)$$

The line with y -intercept b and slope m is given by the *slope-intercept form*

$$y=mx+b$$

The line through (x_1, y_1) and (x_2, y_2) is given by the *two point form*

$$y-y_1=\frac{y_2-y_1}{x_2-x_1}(x-x_1).$$

A parametric form is given by

$x = x_0 + at$

$y = y_0 + bt$

Other forms are

$a(x - x_1) + b\,(y - y_1) = 0$

$ax + by + c = 0$

$$\begin{vmatrix} x & y & 1 \\ x_1 & y_1 & 1 \\ x_2 & y_2 & 1 \end{vmatrix} = 0.$$

A line in two dimensions can also be represented as a vector. The vector along the line

$ax + by = 0$

is given by

$$t\begin{bmatrix} -b \\ \hline a \end{bmatrix},$$

where $t \in \mathbb{R}$. Similarly, vectors of the form

$$t\begin{bmatrix} b \\ \hline a \end{bmatrix}$$

are perpendicular to the line.

Three points lie on a line if

$$\begin{vmatrix} x_1 & y_1 & 1 \\ x_2 & y_2 & 1 \\ x_3 & y_3 & 1 \end{vmatrix} = 0$$

The angle between lines

$A_1 x + B_1 y + C_1 = 0$

$A_2 x + B_2 y + C_2 = 0$

is

$$\tan\theta = \frac{A_1 B_2 - A_2 B_1}{A_1 A_2 + B_1 B_2}.$$

The line joining points with trilinear coordinates $\alpha_2 : \beta_1 : \gamma_1$ and $\alpha_1 : \beta_2 : \gamma_2$ is the set of point $\alpha : \beta : \gamma$ satisfying

$$\begin{vmatrix} \alpha & \beta & y \\ \alpha_1 & \beta_1 & y_1 \\ \alpha_2 & \beta_2 & y_2 \end{vmatrix} = 0$$

$$(\beta_1 y_2 - y_1\beta_2)\alpha + (y_1\alpha_2 - \alpha_1 y_2)\beta + (\alpha_1\beta_2 - \beta_1\alpha_2)\gamma = 0$$

The line through P_1 in the direction (a_1, b_1, c_1) and the line through P_2 in direction (a_2, b_2, c_2) intersect iff

$$\begin{vmatrix} x_2 - x_1 & y_2 - y_1 & z_2 - z_1 \\ a_1 & b_1 & c_1 \\ a_2 & b_2 & c_2 \end{vmatrix} = 0$$

The line through a point $\alpha' : \beta' : \gamma'$ parallel to

$l\alpha + m\beta + n\gamma = 0$

is

$$\begin{vmatrix} \alpha & \beta & y \\ \alpha' & \beta' & y' \\ bn - cm & cl - an & am - bl \end{vmatrix} = 0$$

The lines

$$l\alpha + m\beta + n\gamma = 0$$

$$l' + m'\beta + n'\gamma = 0$$

are parallel if

$$a(mn' - nm') + b(nl' - ln') + c(lm' - ml') = 0$$

for all (a, b, c), and perpendicular if

$$(ll' + mm' + nn') - (mn' + m'n)\cos A - (nl' + n'l)\cos B - (lm' + l'm)\cos c = 0$$

for all (a, b, c).

The line through a point $\alpha' : \beta' : \gamma$ perpendicular to (□) is given by

$$\begin{vmatrix} \alpha & \beta & \gamma \\ \alpha' & \beta' & \gamma' \\ l - m\cos C - n\cos B & m - n\cos A - l\cos C & n - l\cos B - m\cos A \end{vmatrix} = 0$$

In three-dimensional space, the line passing through the point (x_0, y_0, z_0) and parallel to the nonzero vector $V = (a, b, c)$ has parametric equations

$$x = x_0 + at$$

$$y = y_0 + bt$$

$$z = z_0 + ct,$$

written concisely as

$$x = x_0 + vt$$

Similarly, the line in three dimensions passing through (x_1, y_1) and (x_2, y_2) has parametric vector equation

$$x = x_1 + (x_2 - x_1)t,$$

where this parametrization corresponds to $x(t = 0) = x_1$ and $x(t = 1) = x_2$

EQUATION OF A STRAIGHT LINE

The general equation of a straight line is given in the form:

$$y = mx + b$$

What Does it Stand for?

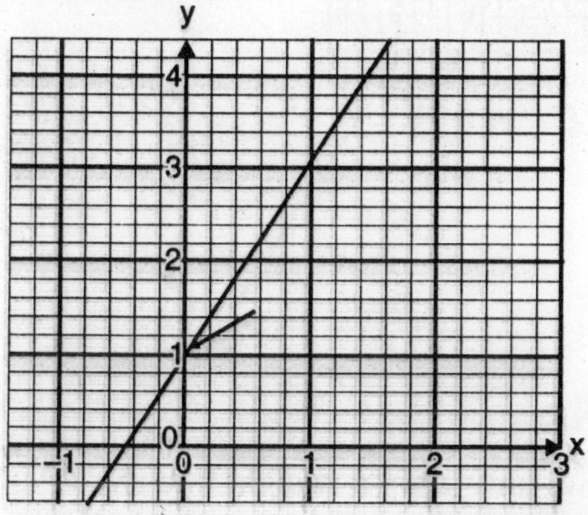

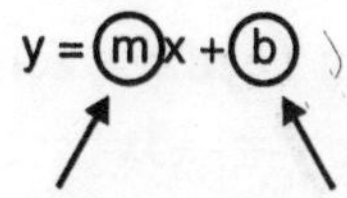

y = how far up

x = how far along

m = Gradient or Slope (how steep the line is)

b = the Y Intercept (where the line crosses the Y axis)

Knowing this we can work out the equation of a straight line:

Example:

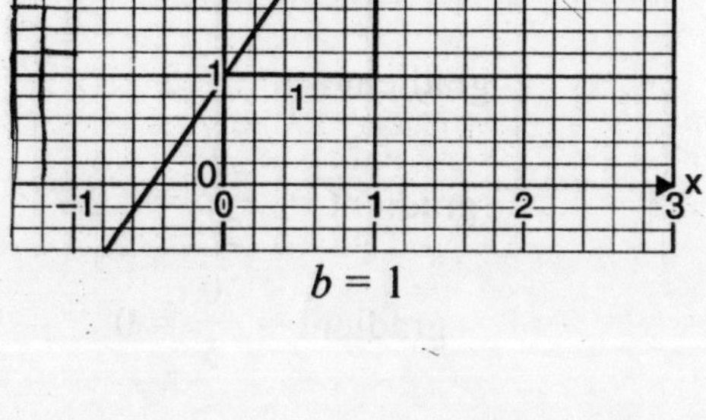

$b = 1$

Thereforey $= 2x + 1$

Example:

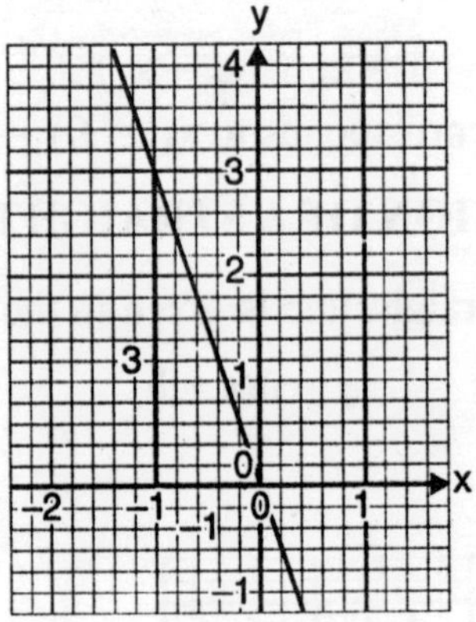

m = 3

b = 0

This gives us y = -3*x* + 0

We do not need the zero!

Therefore y = – 3*x*

GRADIENT OF A STRAIGHT LINE

The gradient of a straight line shows how steep a straight line is.

The method to calculate the gradient is:

Gradient = Change in Y

Change in X

The gradient of the above line = 3/3 = 1

The gradient of the above line is equal to 1

Different Types

There are different types of gradients of a straight line. Below are listed most of the ones that you will come up against.

Remember, starting from the left end of the line going across to the right is positive, but going across to the left is negative. And up is positive, but down is negative:

$$\text{gradient} = \frac{3}{5} = 0.6$$

$$\text{gradient} = \frac{4}{2} = 2$$

$$\text{gradient} = \frac{-4}{2} = -2$$

$$\text{gradient} = \frac{0}{5} = 0$$

Y Intercept of a Straight Line

The Y intercept of a straight line is simply where the line crosses the Y axis.

Example: In the above diagram the line crosses the Y axis at 1.

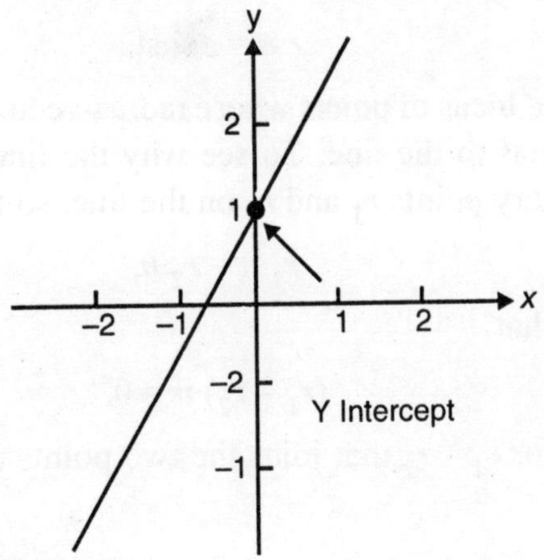

The Y intercept is equal to 1.

$$y = y_1 + [(y_2 - y_1) / (x_2 - x_1)]\cdot(x - x^1),$$

where x_1 and x_2 are assumed to be different. In case they are equal, the equation is simplified to

$$x = x_1$$

and does not require a second point.

Equation (1) can also be written as

$$y - y_1 = [(y_2 - y_1) / (x_2 - x_1)]\cdot(x - x_1),$$

or even as

$$(x_2 - x_1)\cdot(y - y_1) = (y_2 - y_1)\cdot(x - x_1),$$

where one does not have to worry whether $x_1 = x_2$ or not. However, the simplest for me to remember is this

$$(y - y_1)/(y_2 - y_1) = (x - x_1)/(x_2 - x_1)$$

which is not as universal is the one before.

General Equation

A straight line is defined by a linear equation whose general form is

$$Ax + By + C = 0,$$

where A, B are not both 0.

The coefficients A and B in the general equation are the components of vector $n = (A, B)$ normal to the line. The pair $r = (x, y)$ can be looked at in two ways: as a point or as a radius-vector joining the origin to that point. The latter interpretation shows that a straight line is the locus of points r with the property

$$r \cdot n = \text{const.}$$

That is a straight line is a locus of points whose radius-vector has a fixed scalar product with a given vector n, normal to the line. To see why the line is normal to n, take two distinct but otherwise arbitrary points r_1 and r_2 on the line, so that

$$r_1 \cdot n = r_2 \cdot n.$$

But then we conclude that

$$(r_1 - r_2) \cdot n = 0.$$

In other words the vector $r_1 - r_2$ that joins the two points and thus lies on the line is perpendicular to n.

Normalized Equation

The norm $\|n\|$ of a vector $n = (A, B)$ is defined via $\|n\|^2 = A^2 + B^2$ and has the property that, for any non-trivial vector n, $n/\|n\|$ is a unit vector, i.e., $\| n/\|n\| \| = 1$.

Note that the line defined by a general equation would not change if the equation were to be multiplied by a non-zero coefficient. This property can be used to keep the coefficient A non-negative. It can also be used to normalize the equation by dividing it by $\|n\|$. As a result, in a normalized equation

$$Ax + By + C = 0,$$

$$A_2 + B_2 = 1.$$

(In the applet, the coefficients of the normalized equation are rounded to up to 6 digits, for which reason the above identity may only hold approximately.)

The normalized equation is conveniently used in determining the distance from a point to a line.

Intercept-intercept

Assume a straight line intersects x-axis at $(a, 0)$ and y-axis at $(0, b)$. Then it is defined by the equation

$$x/a + y/b = 1,$$

which also can be written as

$$xb + ya = ab.$$

The latter form is somewhat more general as it allows either a or b to be 0. a and b are defined as *x-intercept* and *y-intercept* of the linear function. These are signed distances from the points of intersection of the line with the axes.

Point-slope

The equation of a straight line through point (a, b) with a given slope of m is

$$y = m(x - a) + b.$$

As a particular case, we have

Slope-intercept Equation

The equation of *a* line with a given slope m and the y-intercept b is

$$y = mx + b.$$

Parametric Equation

A line through point $r_0 = (a, b)$ parallel to vector $u = (u, v)$ is given by

$(x, y) = (a, b) + t\cdot(u, v)$, where t is any real number. In the vector form, we have

$$r = r_0 + t\cdot u,$$

where $r = (x, y)$.

Implicit Equation

A line through point $r_0 = (a, b)$ perpendicular to vector $n = (m, n)$ is given by

$$m(x - a) + n(y - b) = 0,$$

or if we take $r = (x, y)$, a generic point on the line, we see that

$$n\cdot(r - r_0) = 0,$$

where dots indicates the scalar product of two vectors.

Chapter 6

The Ellipse

An ellipse can be defined in several different ways. For example, it may be defined as a "section" (i.e., a planar slice) of a cone, from which we get the name "conic sections". This is how the ellipse (and hyperbola and parabola) were originally defined by Menaechmus and later by Apollonius. Alternatively, an ellipse can be defined as a plane section of a cylinder (not surprisingly, since a cylinder is a limiting case of a cone). Both of these definitions involve constructions outside the plane of the conic. The most common modern definitions are those expressed entirely within the plane. For example, we can define an ellipse as the locus of points for which the ratio of the distance to a fixed point (the focus) and the distance to a fixed line (the directrix) is a constant less than 1. Similarly, we can define an ellipse as the locus of points for which the sum of the distances to two fixed points (the foci) is a constant.

Another characteristic property of an ellipse is that the rays from the foci to any point on the ellipse make equal angles with the tangent to the ellipse at that point. It follows that rays of light emanating from one focus of an ellipse will all be reflected off the interior surface of the ellipse and re-converge on the other focus. Indeed if we imagine an ellipsoid (the surface generated by rotating an ellipse about the axis connecting its foci), all the light emanating from a point-like source of light at one focus would converge on the other focus. Moreover, due to the fact that the sum of the distances from each point to the two foci is constant, it follows that the light would be in phase, because the optical path length of each of the rays would be identical. Presumably the light rays would pass right through the other focus (if nothing is there to intercept them), reflect off the ellipse again, and re-converge on the source, still perfectly in phase.

If we imagine ourselves located at one focus of a large ellipsoid with mirrored interior surface, a small ball located at the other focus would be visible to us at apparently the same distance in all directions. In effect, we would perceive ourselves to be at the center of a sphere whose interior surface was the exterior surface of the small ball at the other focus. In this sense, each focus of an ellipsoid totally "encloses" the other. It's interesting to consider this scenario in terms of the retarded and advanced wave solutions of Maxwell's equations, which are associated with outward and inward going wavefronts, respectively. In this elliptical chamber, each wavefront going from one focus to the other is both outward and inward.

Considering how many different properties characterize an ellipse, rather than arbitrarily assuming one property as *the* defining property and then proving the others, it makes more sense to treat all the properties on an equal footing, and prove that they all go together. This can be efficiently done by examining a slightly different construction from all those noted above. Let us construct a curve by first slicing a right cone with a plane (in the manner of the ancient Greeks), and then projecting the intersection of these surfaces perpendicularly onto a plane normal to the axis of the cone. Note that the projection onto the normal plane distinguishes this construction from the original "conic section" definition of Menaechmus. This projective construction is illustrated in the figure below.

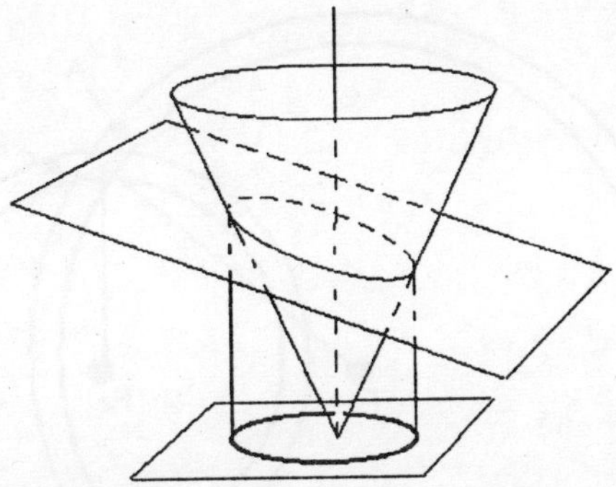

By symmetry, the locus of intersection between the cone and the plane is also the locus on intersection of two right cones with parallel axes offset from each other, as shown in the figure below.

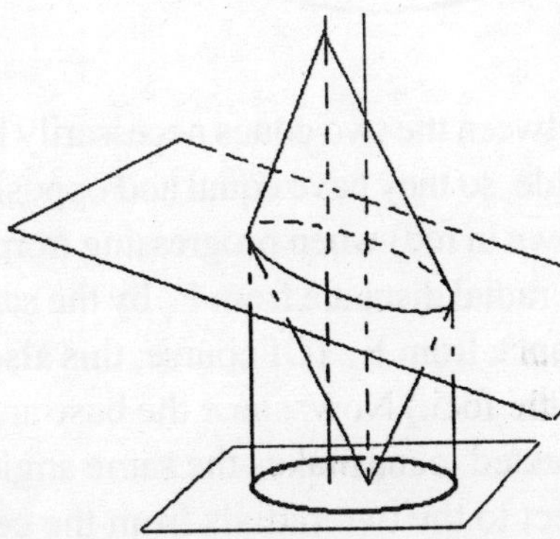

We now select an arbitrary point on the locus of intersection, and consider a line drawn from the lower cone vertex to that point, and another line from that point to the upper cone vertex. Since both the cones are right cones with the same expansion angle α and parallel axes, it's clear that the combined vertical height of those two line segments is :

$$H = \sin(a)L_1 + \sin(a)L_2,$$

likewise the sum of the projections of those line segments into the horizontal plane is:

$$\cos(a)L_1 + \cos(a)L_2.$$

Hence, from the fact that H is the same, regardless of what point we select on the locus of intersection, it follows that the sum of projections on the horizontal plane is a constant. Thus the locus projected onto the horizontal plane consists of point such that the sum of the distances to the two axes is constant. In other words, it is an ellipse, and the points at which the cone axes intersect the horizontal plane are the foci.

We can also see immediately that this locus has the "reflection" property. Viewing a region around the locus of intersection from directly above, we have the situation shown below.

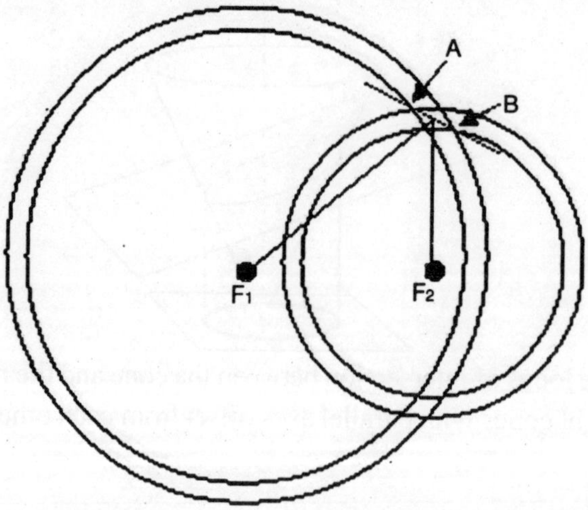

The locus of intersection between the two cones necessarily lies on both of the cones, and they have the same expansion angle, so they have equal and opposite changes in radial distance with height. Hence the locus (shown in red) when progressing from point A to an incrementally nearby point B must increase the radial distance from F_1 by the same incremental amount that it decreases the incremental distance from F_2. (Of course, this also follows immediately from the constant sum of distances to the foci.) Now, since the base angles of an isosceles triangle are equal, it follows that the projected locus makes the same angle with respect to both of the circles, and therefore with respect to the two radials from the centers of those circles. This shows how closely related the "constant sum" property is to the "equal angle" property.

The algebraic approach to the ellipse is less elegant, but does reveal some interesting relationships. To express the conditions of an ellipse in algebraic form, consider a side-view of a sliced cone as shown below. (Both foci lie on the x axis, but only one is shown.)

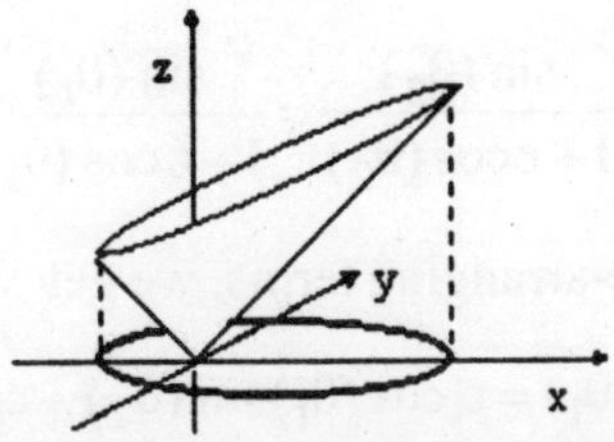

Suppose a ray of length R emanating from the focus at the origin makes an angle α with the xy plane, and let θ denote the angle of the projected ray (on the xy plane) relative to the positive x axis. Then the coordinates of the endpoint of the ray are

$$x = R\cos(\alpha)\cos(\theta) \quad y = R\cos(\alpha)\sin(\theta) \quad z = R\sin(\alpha).$$

For points on the sliced rim, the height z is a linear function of x, so we can write z in the form $z = k_0 + k_1 x$ for some constants k_0 and k_1. Thus we have

$$R\sin(\alpha) = k_0 - k_1 R\cos(\alpha)\cos(\theta).$$

We also know that the length of the projected ray on the xy plane is $r = R\cos(\alpha)$, so we can solve the above equation for R and insert into the equation for r to give the polar form of the equation for an ellipse with the pole at one focus:

$$r = \frac{k_0\cos(\alpha)}{\sin(\alpha) + k_1\cos(\alpha)\cos(\theta)} = \frac{h}{1+\varepsilon\cos(\theta)}$$

where

$$h = \frac{k_0\cos(\alpha)}{\sin(\alpha)} \qquad \varepsilon = \frac{k_1\cos(\alpha)}{\sin(\alpha)}.$$

From this we can easily see that h is the "semi-latus rectum" and ε is the eccentricity, and how these relate to the major and minor radii of the ellipse. Now we consider a general ellipse in standard position, with rays from the foci to an arbitrary point as shown below.

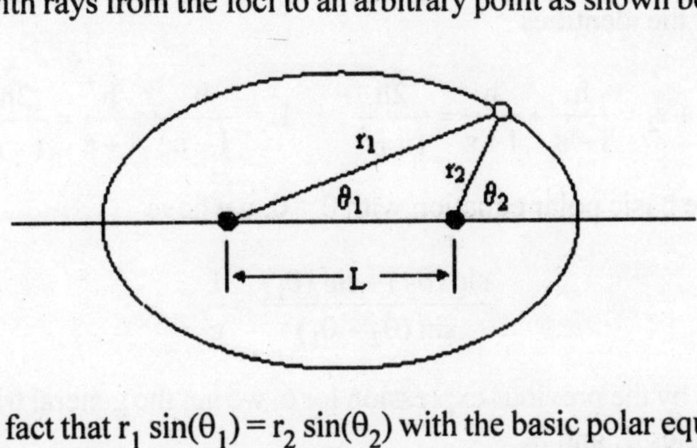

Combining the fact that $r_1\sin(\theta_1) = r_2\sin(\theta_2)$ with the basic polar equations, with the sign in the denominator reversed for the opposite foci, we have the relation between the angles

$$\frac{\sin(\theta_2)}{1+\varepsilon\cos(\theta_2)} = \frac{\sin(\theta_1)}{1-\varepsilon\cos(\theta_1)}.$$

Clearing the fractions and re-arranging terms, we get

$$\sin(\theta_2) - \sin(\theta_1) = \varepsilon[\cos(\theta_1)\sin(\theta_2) + \cos(\theta_2)\sin(\theta_1)].$$

Thus recognizing that the quantity in the square brackets is $\sin(\theta_1 + \theta_2)$, we find that the eccentricity also represents this invariant function of the ray angles

$$\frac{\sin(\theta_2) - \sin(\theta_1)}{\sin(\theta_1 + \theta_2)} = \varepsilon.$$

A similar identity involving the sines of the rays angles can be inferred directly from the relations

$$r_1 \sin(\theta_1) - r_2 \sin(\theta_2) = 0$$
$$r_1 \cos(\theta_1) - r_2 \cos(\theta_2) = L.$$

Solving these equations for r_1 and r_2, we get

$$r_1 = \frac{L\sin(\theta_2)}{\sin(\theta_1 - \theta_1)} \quad r_2 = \frac{L\sin(\theta_1)}{\sin(\theta_2 - \theta_1)}.$$

Therefore, the constant sum of the two radii divided by the distance between the foci is another invariant function of the angles

$$\frac{r_1 + r_2}{L} = \frac{\sin(\theta_2)\sin(\theta_1)}{\sin(\theta_2 - \theta_1)}.$$

Now, in view of the identities

$$r_1 + r_2 = \frac{h}{1-\varepsilon} + \frac{h}{1+\varepsilon} = \frac{2h}{1-\varepsilon^2} \qquad L = \frac{h}{1-\varepsilon} + \frac{h}{1+\varepsilon} = \frac{2h\varepsilon}{1-\varepsilon^2}$$

which come from the basic polar equation with $\theta = 0$, we have

$$\frac{\sin(\theta_2) + \sin(\theta_1)}{\sin(\theta_2 - \theta_1)} = \frac{1}{\varepsilon}.$$

Multiplying this by the previous expression for e, we get the general trigonometric identity, valid for any two angles θ_1 and θ_2,

$$\sin(\theta_2)^2 - \sin(\theta_1)^2 = \sin(\theta_2 + \theta_1)\sin(\theta_2 + \theta_1).$$

We can also express, in algebraic terms, the previous geometrical proof of the reflection property. Taking the differential of equations, squaring, and summing the results, we get the metric line element for the base plane in terms of the two systems of polar coordinates

$$(ds)^2 = (dr_1)^2 + r_1^2(d\theta_1)^2 = (dr_2)^2 + r_2^2(d\theta_2)^2.$$

Since, for an ellipse, we have $r_1 + r_2 =$ constant, it follows that $dr_1 = -dr_2$, and hence

$$r_1\, d\theta_1 = r_2\, d\theta_2$$

These quantities represent the projections of an incremental segment ds of the ellipse onto directions normal to the rays from the two foci, as shown below.

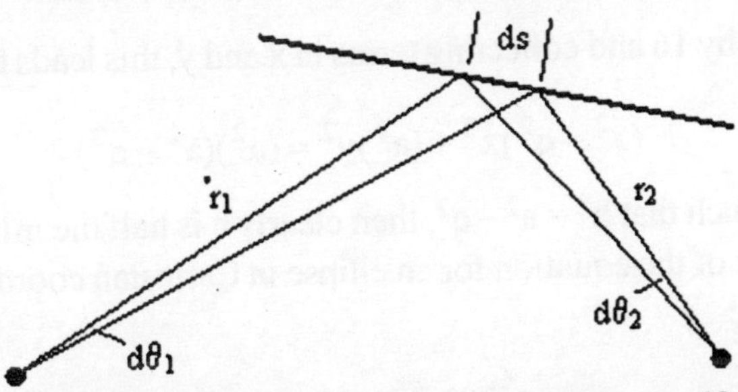

The small triangle with base ds is isosceles because $r_1 d\theta_1 = r_2 d\theta_2$, and so the base angle are equal, which (as before) implies that the radial rays make equal angles with the ellipse. Note that the areas swept out by the rays are half of $r^2 d\theta$. Also, if we divide through the equation by dt we get the kinematic relation $r_1\omega_1 = r_2\omega_2$ where ω denote angular speed. Hence the areas swept out by the two rays per unit time are related by

$$r_2\, \dot{A}_1 = r_1\, \dot{A}_2.$$

We haven't yet shown that the definition of an ellipse as the projected locus of the intersection of a plane and a right cone agrees with the original definition of the ancient Greeks, which regarded the intersection itself (in the slicing plane) as the locus. To show this, it's convenient to give the general form of the equation of an ellipse in Cartesian coordinates. Placing the two foci on the x axis at the positions –q and q, and letting "a" denote half the major diameter, the constant sum of distances to an arbitrary point with coordinates x,y on the ellipse is

$$2a = \sqrt{(x+q)^2 + y^2} + \sqrt{(x-q)^2 + y^2}.$$

Letting u and v denote the quantities under the square root signs, we can square both sides and re-arrange terms to give

$$4a^2 - u - v = 2\sqrt{uv}.$$

Squaring again, we get

$$16a^4 - 8a^2(u + v) + (u + v)^2 = 4uv$$

which can be written as

$$16a^4 - 8a^2(u + v) + (u + v)^2 = 0.$$

Re-inserting the expressions for u and v, we get

$$16a^4 - 8a^2(2x^2 + 2y^2 + 2q^2) + (4qx)^2 = 0.$$

Dividing through by 16 and collecting terms in x and y, this leads to

$$(a^2 - q^2)x^2 + (a^2)y^2 = (a^2)(a^2 - q^2).$$

If we define "b" such that $b^2 = a^2 - q^2$, then clearly b is half the minor diameter. Hence we have the standard form of the equation for an ellipse in Cartesian coordinates with the origin at the center of the ellipse

$$\frac{x^2}{a^2} + \frac{y^2}{b^2} = 1.$$

Now we can see that projecting the conic section onto the base plane simply re-scales the lengths in the x direction, leaving the y direction unaffected, so we still have a locus of the same form, with a different value of "a". The above derivation is reversible, so the resulting locus still satisfies the condition that the sum of the rays from the two foci is constant, from which all the other attributes of the ellipse follow.

A ROTATING DISK IN TRANSLATION

If a circular disk of radius R is rotating with an angular speed ω in equilibrium about its center, while the center is translating uniformly in the plane of rotation with velocity v (in terms of a fixed system of inertial coordinates), then the portions of the rim at right angles to the direction of translation are moving with speed v increased or decreased by approximately ωR. With respect to the rest frame inertial coordinates of the disk's center, the disk is perfectly circular (by definition), but, with respect to the "stationary" coordinates in terms of which the disk's center is moving at the speed v, the spatial extent of the disk in an "instant" is shortened into an ellipse, with the shorter axis pointing in the direction of travel. This may seem to imply that the transformation is symmetrical about the axis of translation, contradicting the fact that the opposite edges of the disk are moving at different speeds with respect to the "stationary" coordinates. However, the result of the Lorentz transformation applied to a spinning disk is not actually symmetrical about the axis of translation.

In terms of the inertial coordinates of the disk's center, a point on the disk at the location r,q (fixed with respect to the disk) and the time t is

$$t(r,\theta,t) = t \qquad x(r,\theta,t) = r\cos(\omega t + \theta) \qquad y(r,\theta,t) = r\sin(\omega t + \theta).$$

If we now consider a system of coordinates T,X,Y with respect to which the disk's center is moving uniformly in the negative x direction with speed v, then the Lorentz transformation gives

$$T(r,\theta,t) = \frac{t - vx}{\sqrt{1-v^2}} = \frac{t - vr\cos(\omega t + \theta)}{\sqrt{1-v^2}}$$

$$X(r,\theta,t) = \frac{x - vt}{\sqrt{1-v^2}} = \frac{r\cos(\omega t + \theta) - vt}{\sqrt{1-v^2}}$$

$$Y(r,\theta,t) = r\sin(\omega t + \theta)$$

Our objective to determine the configuration of the points of the disk with respect to these coordinates at a given "instant". From the equation for T(r,θ ,t) we see that at the instant T = 0 we must have

$$t = vr\cos(\omega t + \theta)$$

The values of v and w are fixed, so for any gives values of r and θ we can solve this "Kepler equation" for t(r, θ). By definition this value of t gives T = 0. We can then insert this value of t into the equations for X(r,θ, t) and Y(r,θ, t) to give the coordinates of the point [r, θ] on the disk. We can also substitute vrcos(wt + θ) for t in the expression for X. This gives the result

$$X(r,\theta)_{T=0} = r\cos(\omega\, t(r,\theta) + \theta)\sqrt{1-v^2}$$
$$Y(r,\theta)_{T=0} = r\sin(\omega\, t)(r,\theta) + \theta)$$

Obviously the spatial outline of the disk is an ellipse with respect to these coordinates, and this shape is symmetrical about the axis of translation. However, the spatial mapping of the points of the disk is not symmetrical, because the rotation of the disk causes the lower part to have a lower velocity (with respect to the "stationary" coordinates) than the upper part. If we plot lines on the disk that represent in the x,y,t coordinates purely radial loci of constant θ, we find that these loci map to more complicated curved with respect to the X,Y,T coordinates, as illustrated below.

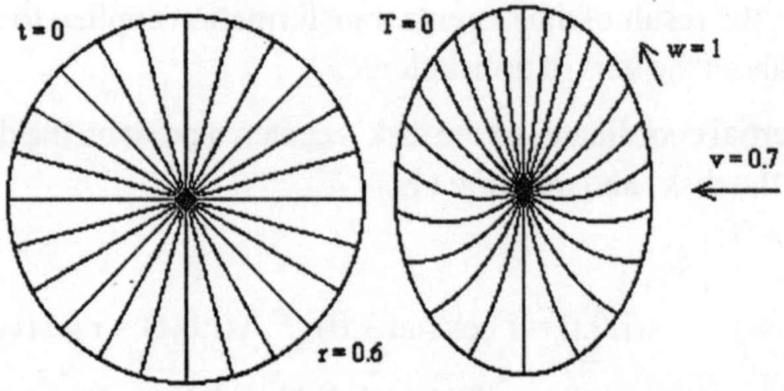

A UNIQUE INTERSECTION PATTERN FOR PLANE CURVES

As discussed in the note on Cramer's Paradox, any two algebraic plane curves of degree n intersect in exactly n^2 complex points (counting multiplicities and points at infinity), and each such curve is uniquely determined by a set of n(n+3)/2 general points. The apparent paradox (for n greater than 2) is resolved by the fact that of the n^2 points of intersection between two curves of degree n there are only n(n+3)/2-1 degrees of freedom. Thus the n^2 points can be partitioned into n(n+3)/2-1 independent points and (n-1)(n-2)/2 dependent points. For the particular case n = 10 the 10^2 intersection points consist of 8^2 independent points and 6^2 dependent points, so we can imagine taking the 64 points of intersection of a pair of octics as the independent intersection points for a family of decatics, such that the 36 dependent points might be the intersection points of two sextics. This possibility arises because with n = 10 the quantities n(n+3)/2-1 and (n-1)(n-2)/2 are both non-zero squares. No other value of n satisfies this condition, so the case n = 10 is unique.

To prove uniqueness, we can solve each of the conditions

$$\frac{n(n+3)}{2} - 1 = x^3 \qquad \frac{(n-1)(n-2)}{2} = y^2$$

for n, which gives n = (-3 ± r)/2 and n = (3 ± s)/2 where r and s are integers such that

$$r^2 - 8x^2 = 17 \qquad s^2 - 8y^2 = 1$$

The Pell equation on the left has two sequences of solutions, r = 5, 23, 133, 775, etc., and r = 7, 37, 215, 1253, etc. Each of these sequences satisfies the recurrence $r_k - 6r_{k-1} + r_{k-2} = 0$, so we can give the values of r_k as explicit functions of k, leading to the explicit expressions for the corresponding positive values of n shown below.

$$n_k = \frac{1}{2}\left[-3 + \frac{5+\sqrt{8}}{2}\left(3+\sqrt{8}\right)^k + \frac{5-\sqrt{8}}{2}\left(3-\sqrt{8}\right)^k\right]$$

$$n_k{}^f = \frac{1}{2}\left[-3 + \frac{7+2\sqrt{8}}{2}\left(3+\sqrt{8}\right)^k + \frac{7-2\sqrt{8}}{2}\left(3-\sqrt{8}\right)^k\right].$$

The other Pell equation has just a single sequence of solutions, s = 1, 3, 17, 99, etc., and again these values satisfy the recurrence $s_j - 6s_{j-1} + s_{j-2} = 0$, so we can give the corresponding values of n explicitly as

$$n_j" = \frac{1}{2}\left[3 + \frac{1}{2}\left(3+\sqrt{8}\right)^j + \frac{1}{2}\left(3-\sqrt{8}\right)^j\right].$$

(We omit the trivial solutions n = 1 and n = 0 given by the other sign.) It's easy to verify that the only common value of n_k (or n_k') and n_j" for small indices k and j is $n_1 = n_2" = 10$. For larger indices equality is ruled out by placing bounds on the difference between nk (or n_k') and n_j". We have

$$n_k - n_j" = \frac{1}{2}\left[3 + \frac{5+\sqrt{8}}{2}\left(3+\sqrt{8}\right)^k + \frac{5-\sqrt{8}}{2}\left(3-\sqrt{8}\right)^k\right]$$
$$-\frac{1}{2}\left[-3 + \frac{1}{2}\left(3+\sqrt{8}\right)^j + \frac{1}{2}\left(3-\sqrt{8}\right)^j\right]$$

which vanishes if and only if

$$\left(5+\sqrt{8}\right)\left(3+\sqrt{8}\right)^k + \left(5-\sqrt{8}\right)\left(3-\sqrt{8}\right)^k - \left(3+\sqrt{8}\right)^j - \left(3-\sqrt{8}\right)^j = 12.$$

The smaller root of the characteristic equation (3 minus the square root of 8) is much less than 1, so as the indices k and j increase, these terms become negligible, so we need to satisfy the approximate equality

$$\left(5+\sqrt{8}\right)\left(3+\sqrt{8}\right)^k - \left(3+\sqrt{8}\right)^j \approx 12.$$

Clearly we must have j > k, so this can be written as

$$\left(3+\sqrt{8}\right)^k + \left[\left(5-\sqrt{8}\right) - \left(3+\sqrt{8}\right)^{j-k}\right] \approx 12.$$

The minimum magnitude of the quantity in square brackets for integer values of j-k is 2, which occurs with j-k = 1. Thus in order to achieve equality the leading factor cannot exceed approximately 6. The argument of this factor is 5.828..., so the only possible solution occurs with k = 1. For any higher values of k, equality is ruled out. This proves that the only solution is the one with k = 1 and j = 2, which gives n = 10. By similar reasoning we can rule out equality between n_k' and n_j" for any indices k and j.

SERIES SOLUTION FOR RELATIVISTIC ORBITS

Solving the field equations of general relativity for the orbit of a small test particle in Schwarzschild spacetime leads to the non-linear differential equation

$$u + u = \frac{m}{h^2} + 3mu^2$$

where m is a measure of the mass of the gravitating body, h is a measure of the angular momentum, and u = 1/r. The parameters m and h are both normalized to have units of length. The second derivative of u in the above equation is with respect to the angular position q in the orbit. The second term on the right side is what distinguishes this equation from the Newtonian equation of motion. An approximate solution of (1) for planetary orbits is given in Anomalous Precesion, but it's also interesting to consider the power-series solution to this equation. Let u be represented by the infinite series

$$u(\theta) = c_0 + c_1\theta + c_2\theta + c_3\theta + \ldots$$

where c_0, c_1, ... are real constants. Inserting into (1) and collecting terms by powers of θ, and setting each coefficient to zero, we get the following conditions on the coefficients:

$$c_2 = \frac{1}{2}[-c_0 + 3m(c_0c_0)] + \frac{m}{2h^2}$$

$$c_3 = \frac{1}{6}[-c_1 + 3m(c_0c_1 + c_1c_0)]$$

$$c_4 = \frac{1}{12}[-c_2 + 3m(c_0c_2) + c_1c_1 + c_2c_0)]$$

$$c_n = \frac{1}{n(n-1)}\left[-c_{n-2} + 3m\sum_{j=0}^{n-2} c_jc_{n-2-j}\right]$$

If the term of (1) involving u^2 was absent, the convolutions in the above equations would also be absent, and we would have simply $c_n = -c_{n-2}/(n(n-1))$. This is the Newtonian case, and with the initial conditions $c_0 = u_0$ and $c_1 = 0$ we have

$$u(\theta) = \frac{m}{h^2} + \left(u_0 - \frac{m}{h^2}\right)\left[1 - \frac{1}{2!}\theta^2 + \frac{1}{4!}\theta^4 - \frac{1}{6!}\theta^6 + \ldots\right]$$

$$= \frac{m}{h^2}\left(u_0 - \frac{m}{h^2}\right)\cos(\theta).$$

Thus the bound Newtonian orbits satisfy the equation

$$u(\theta) = \frac{h^2/m}{1 + \left(\frac{h^2}{r_0 m} - 1\right)\cos(\theta)}$$

which is the equation of a conic section with the gravitating body at one focus. The ratio $p = h^2/m$ is the semi-latus rectum, and r_0 is the perihelion radius. Letting a and b denote the major and minor radii and e the linear eccentricity of the ellipse, we have the elementary relations $e^2 = a^2 - b^2$, $p = b^2/a$, and $r_0 = a - e$, so the quantity in parentheses in the above equation can be written as

$$\frac{h^2}{r_0 m} - 1 = \frac{p}{a-e} - 1 = \frac{\frac{b^2}{a} - a + e}{a-e} = \frac{b^2 - a^2 + ae}{a^2 - ae} = \frac{-e^2 + ae}{a^2 - ae} = \frac{e}{a}.$$

Thus the quantity in parentheses is the numerical eccentricity $\varepsilon = e/a$. The orbit is an ellipse, a hyperbola, or a parabola respectively depending on whether $|\varepsilon|$ is less than, greater than, or equal to 1.

To account for the relativistic correction term involving u^2 in equation we must include the convolution terms in the series coefficients. We again find that, with the initial condition $c_1 = 0$, all the subsequent coefficients of odd powers of θ are also zero, so $u(\theta)$ is an even function (meaning it is the same in both directions of θ).

For the planet Mercury the orbital eccentricity is $e = 0.026$, the semi-major axis is $a = (5.79)10^7$ km, and the normalized mass of the gravitating body (the sun) is $m = 1.475$ km. From these we can compute the perihelion distance $r_0 = (1-e)a = (4.59)10^7$ km, and we also have

$$\frac{h^2}{m} = \left(1 - \varepsilon^2\right)a = (5.54)10^7 \text{ km}.$$

Setting $c_0 = 1/r_0$, we can evaluate the power series coefficients of $u(\theta)$ to give

$$c_0 = (2.1751)10^{-8} \quad c_2 = (1.86)10^{-9} \quad c_4 = (1.5)10^{-10}$$

and so on. Based on the first 60 terms (which is 30 non-zero terms, since all the odd-ordered terms are zero), the function $u(\theta)$ and its derivatives are plotted below.

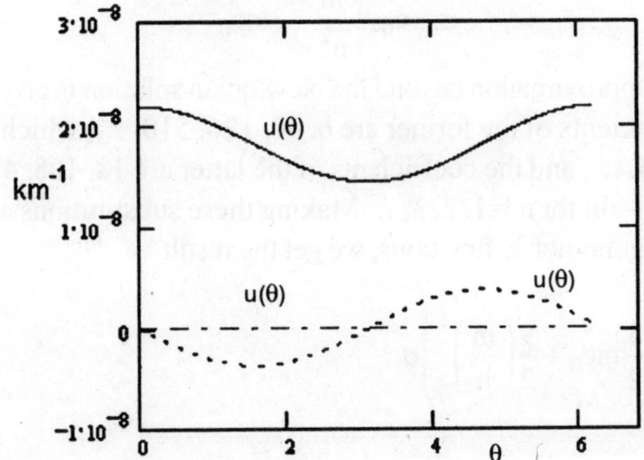

The first derivative is zero at each perihelion and apihelion, and the first apihelion occurs at $\theta = 3.14159290423\ldots$, which exceeds π by $(2.506)10^{-7}$ radians. Since $u(\theta)$ is an even function, we know the increment is the same for the apihelion near $-\pi$, so the total precession per revolution is twice this amount, which is $(5.012)10^{-7}$ radians per revolution, equal to 0.1034 arc seconds per revolution. Mercury completes 414 revolutions per century, so it's relativistic precession is 42.9 arc seconds per centuy, in close agreement with the observational value.

Algebraically the first several coefficients of the power series of $u(\theta)$ are

$$c_0$$

$$c_2 = -\frac{\sigma}{2!}$$

$$c_4 = -\frac{\sigma}{4!}(1 - 6mc_0)$$

$$c_6 = -\frac{\sigma}{6!}\left(1 - 30mc_0 + 18\left(\frac{m}{h}\right)^2 + 90m^2{c_0}^2\right)$$

$$c_8 = -\frac{\sigma}{8!}\left(1 - 126mc_0 + 108\left(\frac{m}{h}\right)^2 + 1080m^2{c_0}^2 - 648mc_0\left(\frac{m}{h}\right)^2 - 2160m^3{c_0}^3\right)$$

$$c_{10} = -\frac{\sigma}{10!}\left(1 - 510mc_0 + 486\left(\frac{m}{h}\right)^2 + 1053m^2{c_0}^2 - 11880mc_0\left(\frac{m}{h}\right)^2 \cdots\right)$$

where

$$\sigma = c_0 - \frac{m}{h^2} - 3m{c_0}^2$$

The first order of approximation beyond the Newtonian solution is given by the terms in mc_0 and $(m/h)^2$. The coefficients of the former are 6, 30, 126, 510, …, which can be expressed as $2(4^n - 1)$ for $n = 1, 2, 3, \ldots$, and the coefficients of the latter are 18, 108, 486, …, which can be expressed as $8(4^n - 1) - 6n$ for $n = 1, 2, 3, \ldots$ Making these substitutions and consolidating the resulting series into trigonometric functions, we get the result

$$u(\theta) \approx c_0 - \left(1 + \frac{3}{2}mc_0 + \frac{9}{2}\left(\frac{m}{h}\right)^2\right)\sigma$$

$$+\left[\left(1+2mc_0+4\left(\frac{m}{h}\right)^2\right)\cos(\theta)-\left(\frac{1}{2}mc_0-\frac{1}{2}\left(\frac{m}{h}\right)^2\right)\cos(2\theta)+3\left(\frac{m}{h}\right)^2\theta\sin(\theta)\right]\sigma$$

Naturally we still have $u(0) = c_0$., since the contributions of the other terms cancel out at this special value. Now, since c_0 and m/h^2 are many orders of magnitude smaller than 1 (for typical astronomical situations), the constant term is closely approximated by simply m/h^2. Also, the term involving $\cos(2\theta)$ represents just a very slight periodic oscillation affecting the shape of the orbit, with no secular effect. Thus to determine the secular effects we can examine the close approximation

$$u(\theta) \approx \frac{m}{h^2}+\left(c_0-\frac{m}{h^2}\right)\left[\cos(\theta)+3\left(\frac{m}{h}\right)^2\theta\sin(\theta)\right]$$

The quantity $3(m/h)^2\theta$ is very small for the first few revolutions, so for this part of the curve the quantity in the square brackets can be closely approximated by

$$\cos\left(3\left(\frac{m}{h}\right)^2\theta\right)\cos(\theta)+\sin\left(3\left(\frac{m}{h}\right)^2\theta\right)\sin(\theta)=\cos\left(\theta-3\left(\frac{m}{h}\right)^2\theta\right).$$

Consequently the orbit precesses by $6\pi(m/h)^2$ per revolution. The term involving $\theta\sin(\theta)$ appears naturally from the conditions on the power series coefficients, but it is often introduced simply as a trial solution, and/or justified as a resonance term. Of course, for sufficiently large angles θ, the approximation for $u(\theta)$ increases without limit, but m/h is so small that the above expressions are valid over many revolutions.

Incidentally, for some special cases we can find an exact closed-form solution to equation. Notice that it can be written in the form

$$\left(u-\frac{1}{6m}\right)^2-\frac{1}{3m}\ddot{u}=\frac{1}{36m^2}-\frac{1}{3h^2}$$

In the special case when $h^2 = 12m^2$. the right hand side vanishes, and we have the equation

$$3my^2=\ddot{y} \quad \text{where } y=u-\frac{1}{6m}$$

This has the exact solution $y = 2/(m\theta^2)$, so we have

$$r(\theta)=\frac{6m}{1+\dfrac{12}{\theta^2}}.$$

Hence the orbit spends infinitely much time arbitrarily close to a circular orbit at radius 6m for large negative values of θ, but then as θ approaches zero the orbit spirals down to the center of the field, as shown in the left-hand figure below.

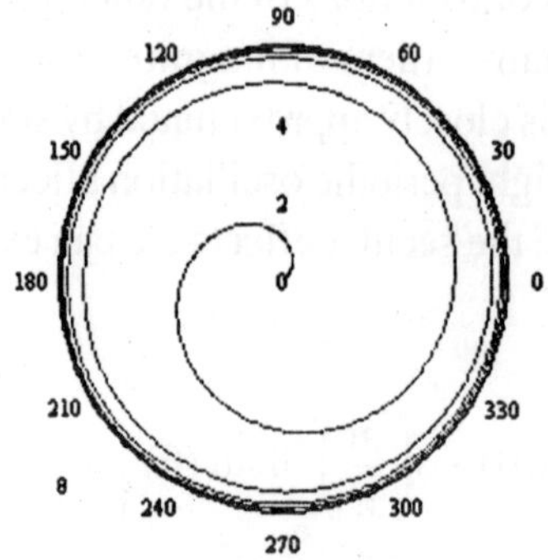

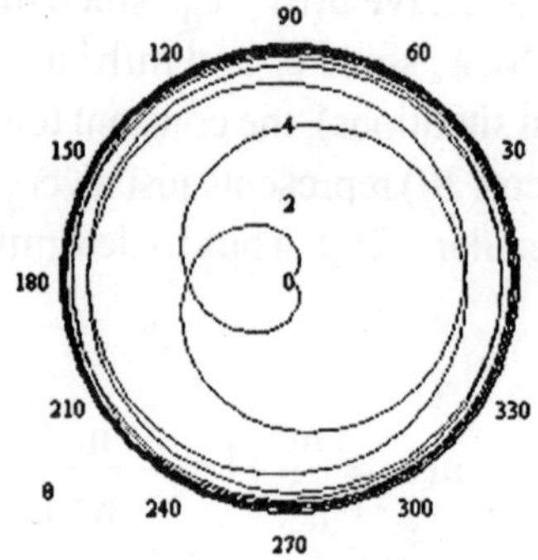

Assuming it continues through the singularity as θ passes through zero, the orbit then spirals back up and re-converges on a circular orbit of radius 6m, as illustrated in the right-hand figure above.

HARMONIA MENSURARUM

The second edition of Isaac Newton's Principia was published by Richard Bentley, the famous classical scholar and Master of Trinity College. Bentley was not universally well-liked (to put it mildly), and was certainly no scientist or mathematician, so it may seem odd that Newton chose him as his publisher, especially considering how many better-qualified men (such as David Gregory) would have jumped at the chance. True, Bentley had delivered a series of lectures in 1692 using Newton's theories to "prove" the existence of an intelligent creator-exactly the kind of thinking that appealed to Newton-but it was still an odd partnership. When he was later asked why he had chosen Bentley as publisher, Newton answered (probably in jest) that Bentley "was covetous & loved money & therefore I lett him that he might get money". That may be, but it's also possible that Newton (then 67) was not eager to engage a publisher who would subject the book to detailed scrutiny. He seems to have hoped to correct a few typos and ship it off to the printers, and with Bentley as editor this is probably what would have happened.

Whatever Newton's intentions may have been, events took a different turn when Bentley enlisted the bright young (27) mathematician Roger Cotes (1682-1716) to do the actual editing of Newton's great work. Cotes was the newly appointed professor of astronomy at Trinity College, a position he had gained with Bentley's support. It soon became clear to Newton that this new edition was not going to be an easy ride. Shortly after the project began, Cotes wrote to inform Newton of two errors he had found in a table of integrals referenced in the corollary of Proposition XCI. In reply, Newton tried to dampen his enthusiasm

I thank you for your letter and the corrections... I would not have you be at the trouble of examining all the demonstrations in the Principia. It's impossible to print the book without some faults, and if you print by the copy sent you, correcting only such faults as occur in reading over the sheets to correct them as they are printed off, you will have labor more than it's fit to give you. Apparently a cursory approach to the new edition wasn't what Cotes had expected. He showed the letter to Bentley, and asked him to clarify the scope of the task. Bentley wrote back to Newton: You need not be so shy of giving Mr. Cotes too much trouble. He has more esteem for you and more obligations to you than to think that trouble too grievous. But however he does it at my orders, to whom he owes more than that. And so pray you, be easy as to that. We will take care that no little slip in a calculation shall pass this fine edition.

This zeal may have been a bit threatening, from Newton's point of view, but eventually he warmed to the task, and Cotes succeeded in getting Newton to thoroughly re-work much of Books II and III, including the section on the resistance of fluids to motion and the extremely difficult sections on the lunar motion (which Newton said made his head ache). Newton also added the famous General Scholium to the end of Book III. Work on the Principia occupied Cotes for nearly four years. Interestingly, the notions of "action at a distance" and gravity as an innate property of matter, commonly attributed to Newton, were actually first clearly articulated by Cotes in his preface to the second edition.

Cotes died at the age of 34, prompting Newton to make the famous remark that "if he had lived, we might have known something". Although Cotes published almost no original work during his life, some of his writings were published posthumously by his uncle. Among the results that tend to support Newton's comment, we find that Cotes in 1714 was the first to note the fundamental identity

$$\ln[\cos(\theta) + i\sin(\theta)] = i\theta$$

where $i = \sqrt{-1}$. A generation later the great Swiss mathematician Euler deduced the same result and expressed it in the more familiar form

$$e^{i\theta} = \cos(\theta) + i\sin(\theta)$$

This identity can be seen as an expression of the correspondence between circular and hyperbolic measures, between exponential and trigonometric measures, and between orthogonal and polar measures, not to mention between real and complex measures, all of which seemed to be within Cotes' grasp. He named the treatise containing this result *Harmonia Mensurarum*, meaning "harmony of measures".

Another interesting and "harmonious" result found by Cotes concerns the points of intersection between a pencil of straight lines and an algebraic curve. Recall that an algebraic curve is defined by a function $f(x,y) = 0$ where f is a polynomial in the variables x and y. This means that $f(x,y)$ is the sum of terms of the form $c_{m,n}\, x^m y^n$. The largest value of m+n for any term with non-zero coefficient is called the degree of f. An algebraic curve of the first degree is a straight line, whereas the algebraic curves of degree 2 are the conics, i.e., circles, ellipses, hyperbolas, parabolas. We can have algebraic curves of any degree, including cubics, quartics, etc.

Given an arbitrary algebraic curve of degree n, and an arbitrary point O on the plane, suppose a straight line through the point O intersects the curve at n points, labeled $P_1, P_2, \ldots P_n$, and let r_j denote the distance from O to the point P_j. Now mark the point Q on the line at the distance r_q from O, where r_q is the harmonic mean of $r_1, r_2, \ldots, r_n$. If we repeat this process for many different lines through O (each intersecting the curve at n points), Cotes stated that the constructed "Q" points will all lie on a straight line.

Oddly enough, since this theorem involves the harmonic mean, the historian of mathematics W.W. Rouse Ball evidently concluded that this was the theorem from which Cotes' *Harmonia Mensurarum* took its name, and this claim is repeated in Boyer's history. However, Hollingdale notes that this result actually was part of a different treatise, called *On the Nature of Curves*.

In any case, it's a nice result, and at first glance seems as if it might be quite difficult to prove-even assuming it's true, which may not be intuitively obvious. We can easily imagine plane curves for which the theorem does not hold, but they are not algebraic curves. (It turns out that, in effect, we could almost use Cotes' theorem to define algebraic curves.) Another potential objection is that some of the lines through the point O may intersect an algebraic curve of degree n in n points, whereas other lines through O may not, so it may seem that the theorem must break down at some point.

However, this theorem is actually quite easy to prove, provided we accept the fundamental theorem of algebra, which assures us that a line always intersects an algebraic curve of degree n in exactly n points (not necessarily distinct), with the understanding that the points of intersection may have complex coefficients. To prove the theorem, we can, without loss of generality, set the point O at the origin of x,y coordinates, in terms of which our algebraic curve is defined by a polynomial of the form

$$f(x,y) = c_{0,0} + c_{1,0}x + c_{0,1}y + c_{2,0}x^2 + c_{1,1}xy + c_{0,2}y^2 + \ldots = 0.$$

Also, a straight line through the point O has the equation $y = \lambda x$ for some constant λ. To find the points of intersection we substitute for y in the above polynomial to arrive at the single-variable polynomial of degree n

$$c_{0,0} + \left(c_{1,0}x + c_{0,1}\lambda\right)x + \left(c_{2,0} + c_{1,1}\lambda + c_{0,2}\lambda^2\right)x^2 + \ldots = 0.$$

Since this polynomial is of degree n, it has n (possibly complex) roots, and these correspond to the x coordinates of the n points of intersection. Now, recall from the basic theory of equations that for the general polynomial of degree n

$$C_n x^n - C_{n-1}\, x^{n-1} + \ldots + (-1)^n C_0 = 0$$

the value of C_j/C_n equals the sum of all products of the roots taken j at a time. For example, C_{n-1}/C_n is just the sum of the roots, whereas C_0/C_n is just the product of the roots. Likewise, C_1/C_n is the sum of all products of n−1 roots. It follows that C_1/C_0 is the (negative) sum of the reciprocals of the roots.

The distances from O of points on the line are all directly proportional to the x coordinates of those points, so we can simply work with the x coordinates of the intersections points and compute the harmonic mean to give the x coordinate of the "Q" point. Thus we have

$$\frac{1}{x_Q}=\frac{1}{n}\left(\frac{1}{x_1}+\frac{1}{x_2}+\ldots+\frac{1}{x_n}\right)=-\frac{1}{n}\frac{c_{1,0}+c_{0,1}\lambda}{c_{0,0}}.$$

The point Q is also on the line $y = \lambda x$, so we have $y_Q = \lambda x_Q$. Substituting y_Q/x_Q for λ in the above expression and re-arranging gives

$$nc_{0,0}+c_{1,0}x_Q+c_{0,1}y_Q=0.$$

Since n and the coefficients $c_{i,j}$ are constants for a given algebraic curve, we see that the coordinates of the Q points all satisfy this same equation of a line, regardless of the value of k, so this proves Cotes' theorem. Notice that this works for all lines through the point O, even those that don't have a real point of intersection with the curve, because we are allowing the coordinates of the intersection points to be complex.

As a simple illustration, consider the algebraic curve of degree 2 defined by the polynomial

$$f(x,y)=4x^2-6xy+10y^2+13x-21y+8=0$$

This gives an ellipse as shown in the figure below.

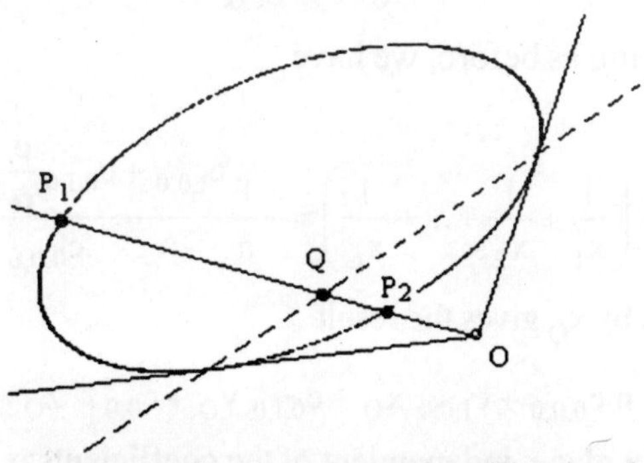

For any line through the point O that intersects the ellipse at the points P_1 and P_2 we can find the point Q such that

$$\frac{1}{OQ}=\frac{1}{2}\left(\frac{1}{OP_1}+\frac{1}{OP_2}\right)$$

The dashed line in the figure represents the locus of such points Q for all possible straight lines through the point O. This locus extends outside the ellipse, into the region where there are no real intersections between the ellipse and a line through O, but in this region there are complex points of intersection, and the distance OQ computed from harmonic mean of those distances is purely real-valued, and places the point Q on the dashed line.

Incidentally, the writings of Cotes on the subject of Newton's fluents and fluxions, as well as the measures of curves and integration formulae, were evidently known to Edgar Allan Poe, who included among his marginal notes the whimsical comment

I think I will set about a lyric on the Quadrature of Curves—or the Arithmetic of Infinites. Cotes, however, supplies me a ready-made title, in his "Harmonia Mensurarum," and there is no reason why I should not be fluent, at least, upon the fluents of fractional expressions.

The theorem can immediately be generalized to higher dimensions. For example, given any algebraic surface defined by the polynomial function $f(x,y,z) = 0$ of degree n, we can construct a straight line from the origin O intersecting the surface at n points. If we denote by Q the point on this line whose distance from the origin is the harmonic mean of the distances to the n points of intersection with the surface, the theorem asserts that the locus of such Q points is a flat plane. This can be demonstrated by essentially the same presented above. The equation of the line can be expressed in parametric form by the equations

$$x(t) = \alpha t \qquad y(t) = \beta t \qquad z(t) = \gamma t$$

for constants α, β, γ, so we have $y = (\beta/\alpha)x$, $z = (\gamma/\alpha)x$. Substituting these expressions for x and y into the polynomial $f(x,y,z) = 0$ gives a polynomial in x of degree n, and the lowest order terms are

$$c_{0,0,0} + \left(c_{1,0,0} + \frac{\beta}{\alpha}c_{0,1,0} + \frac{\gamma}{\alpha}c_{0,0,1}\right)x + ... = 0.$$

By the same reasoning as before, we have

$$\frac{1}{x_Q} = \frac{1}{n}\left(\frac{1}{x_1} + \frac{1}{x_2} + ... + \frac{1}{x_n}\right) = -\frac{1}{n}\frac{c_{1,0,0} + c_{0,1,0}\frac{\beta}{\alpha} + c_{0,0,1}\frac{\gamma}{\alpha}}{c_{0,0,0}}.$$

Multiplying through by x_Q gives the result

$$n\, c_{0,0,0} + c_{1,0,0}\, x_Q + c_{0,1,0}\, y_Q + c_{0,0,1}\, z_Q = 0$$

which is the equation of a plane, independent of the coefficients α, β, γ of the line. Hence the locus of "Q" points is a plane.

In general, the locus of points whose distance along a line from a fixed point O in d-dimensional space is the harmonic mean of the distances along the line from the point O to all the points of intersection of the line with a fixed algebraic surface in the d coordinates is a flat d–1 dimensional space.

This theorem can also be extended in other ways. For example, given an algebraic plane curve of degree n, consider a pencil of lines through an arbitrary point O. Each line intersects with the curve in n points (possibly with complex coordinates). For any given line through the point O, let $r_1, r_2, ..., r_n$ denote the distances from O to the points of intersection with the curve, and define the point Q on this line at a distance r_Q from the point O by the equation

$$\frac{1}{r_Q^{\;2}} = \frac{2}{n(n-1)} \sum_{i<j} \frac{1}{r_i r_j}$$

In other words, $1/r_Q^{\;2}$ is the average of all products of two distinct inverses of the distances from O to the curve. In this case it's easy to show that the locus of the "Q" points for all the lines through O is a central conic around O. Of course, if the original given curve itself is of degree n = 2 (meaning that it is a conic), then each line through O intersections with the curve at just two points, at the distances r_1 and r_2, and in this case r_Q is just the algebraic mean of those two distances. The figure below shows an ellipse in white and the locus of points in green such that the distance from the origin along any line through the origin is the geometric mean of the two distances to the ellipse along that line.

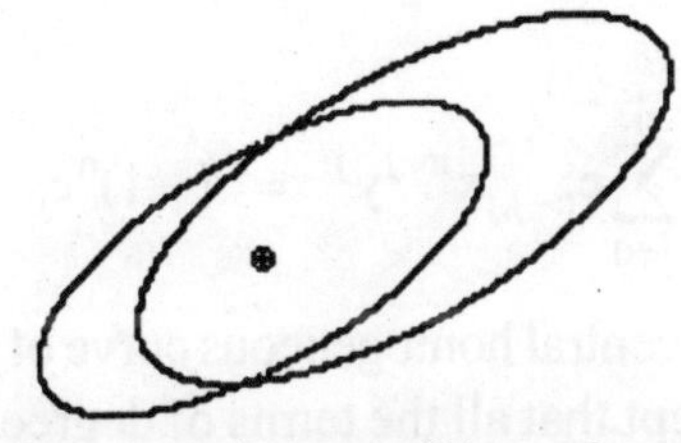

If the origin point is outside the given (white) ellipse, the result is still applicable, but in such a case the coordinates of the intersection points between the ray from the origin and the white ellipse are complex for rays outside the tangent rays. Nevertheless, there are still two intersection points, and the geometric mean of the distances is real, following the same ellipse as shown in red below.

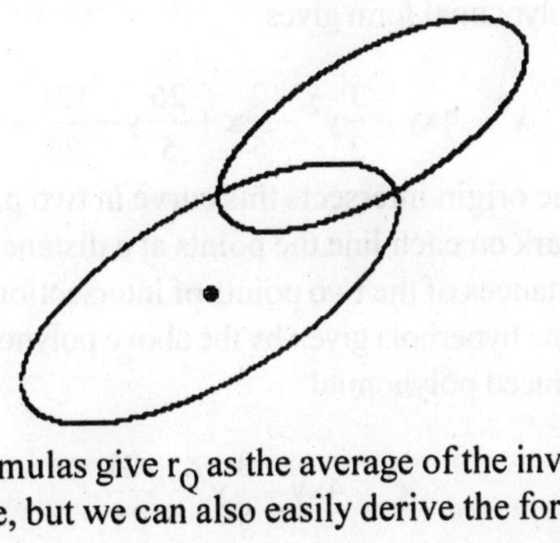

For n > 2 the above formulas give r_Q as the average of the inverses of all products of two distances to the given curve, but we can also easily derive the formulas for r_Q defined as the geometric mean of all the distances $r_1, r_2,.., r_n$ from the point O to the curve for all values of n. (This is equivalent to the previous case when n = 2, but not for n > 2.) For the general plane polynomial f(x,y) of degree n we can set $y = \lambda x$ to give the polynomial of x values of the points of intersection between the curve and a line through the origin with slope λ. The result is

$$c_{0,0} + (c_{1,0} + c_{0,1}\lambda)x + (c_{2,0} + c_{1,1}\lambda + c_{0,2}\lambda^2)x^2 + ... + \left(\sum_{j=0}^{n} c_{n-j,j}\lambda^j\right)x^n = 0.$$

To set r_Q equal to the geometric means of the n distances, we must set x_Q to the geometric mean of the roots of this equation, so we have

$$x_Q{}^n = \frac{(-1)^n c_{0,0}}{\sum_{j=0}^{n} c_{n-j,j}\lambda^j}$$

where $\lambda = y_Q/x_Q$. Multiplying through by the denominator on the right side, we have the locus of "Q" points:

$$\sum_{j=0}^{n} c_{n-j,j} x^{n-j} y^j = (-1)^n c_{0,0}.$$

Consequently the locus is a central homogeneous curve of degree n. Moreover, it is identical to the original plane curve, except that all the terms of degree 1 through n–1 are deleted.

As a simple illustration, suppose the given plane curve is the hyperbola

$$(x-1)^2 - 4(x-1)\left(y - \frac{2}{5}\right) - \frac{3}{2}\left(y - \frac{2}{5}\right)^2 = 4.$$

Expanding this into polynomial form gives

$$x^2 - 4xy - \frac{3}{2}y^2 - \frac{2}{5}x + \frac{26}{5}y - \frac{121}{25} = 0.$$

Every line through the origin intersects this curve in two points (possibly with complex coordinates), and if we mark on each line the points at a distance from the origin equal to the geometric mean of the distances of the two points of intersection with the hyperbola, then the locus of all these points is the hyperbola given by the above polynomial with the terms of the first degree deleted, i.e., the reduced polynomial

$$x^2 - 4xy + \frac{3}{5}y^2 - \frac{121}{25}.$$

A plot of these curves is shown below.

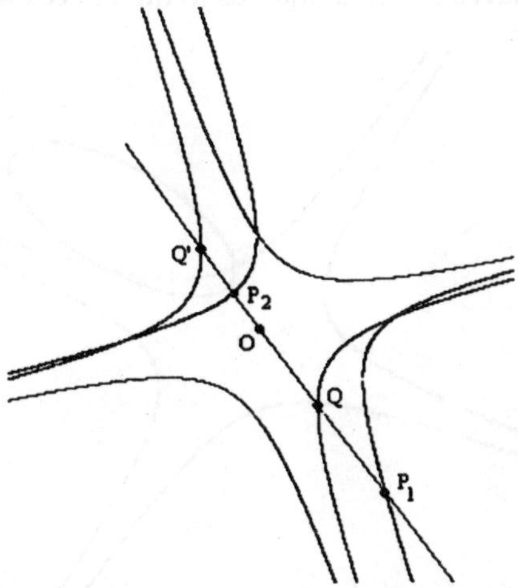

The white hyperbola is the original plane curve, and the point marked "O" is the origin. A sample line drawn through the origin is shown in yellow, intersecting the white hyperbola at the points P_1 and P_2. We then mark the points Q and Q' on this line at a distance given by the geometric mean of the distances from the origin to the points P_1 and P_2. As we rotate the line through O, these Q and Q' values generate the hyperbola shown in green and red. The red portions are where the intersection points have complex coordinates. Notice that we generate two conjugate hyperbolas by setting the reduced polynomial to +121/25 or −121/25.

In this same way we can characterize the locus of points generated by *any* symmetrical function of the distances from a point to a given curve. As an example, suppose we define $1/r_Q{}^n$ as the means of the *inverse squares* of the distances along a line from the origin to the n points of intersection with the given curve. Since we have

$$\sum_i \frac{1}{x_i} = -\frac{\left(c_{1,0} + c_{0,1}\lambda\right)}{c_{0,0}} \qquad \sum_{i<j} \frac{1}{x_i x_j} = -\frac{\left(c_{2,0} + c_{1,1}\lambda + c_{0,2}\lambda^2\right)}{c_{0,0}}$$

it follows that

$$\sum_i \frac{1}{x_i^2} = \left(\frac{c_{1,0} + c_{0,1}\lambda}{c_{0,0}}\right) - 2\frac{\left(c_{2,0} + c_{1,1}\lambda + c_{0,2}\lambda^2\right)}{c_{0,0}}.$$

Equating this to $n/x_Q{}^n$ and clearing fractions (with $\lambda = y_Q/x_Q$) gives the conic locus

$$\left(c_{10}{}^2 - 2c_{00}c_{20}\right)x_Q{}^2 + 2\left(c_{10}c_{01} - c_{00}c_{11}\right)x_Q y_Q + \left(c_{10}{}^2 - 2c_{00}c_{20}\right)y_Q{}^2 = nc_{00}{}^2.$$

In this case the Q locus need not have the same discriminant as the original curve. As an example, the figure below shows the locus of points (green) whose inverse-square distances are the mean of the inverse squares of the distances to the curve (white).

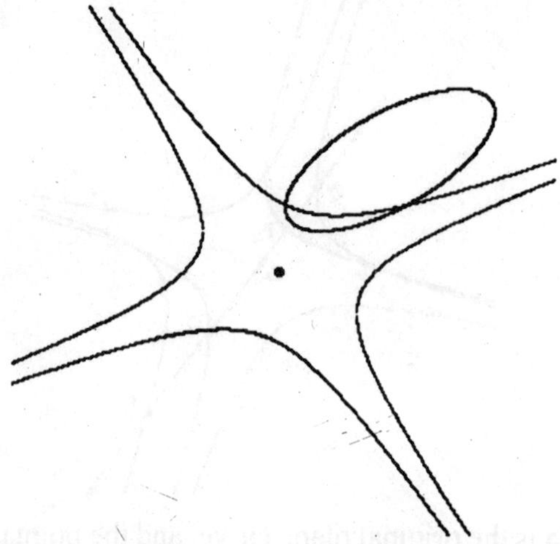

The red lines are generated by the same condition, applied to the complex points of intersection.

Chapter 7

Parametric Curves

We have thus far encountered curves as the graphs of functions $f(x)$. Such a graph consists of all points $(x, f(x))$ in the plane with x in the domain of f. In this chapter, we explore two further ways of describing curves that may be more convenient under certain circumstances.

A parametric curve is described by a pair of functions $x(t)$ and $y(t)$ and consists of the set of points $(x(t), y(t))$, where t varies over a specified domain. These functions describe how the x- and y-coordinates depend upon a third variable (t, often thought of as representing time), allowing somewhat more freedom than when y is forced to depend upon x directly. Parametric curves can be used, for example, to describe the motion of an object in a plane over a period of time. After a more thorough introduction, we will show how to calculate the velocity and acceleration of an object whose motion is given by parametric functions. We will then investigate how to find the length of a parametric curve.

A polar curve is described by a function $r(\theta)$ and consists of the points $[r(\theta), \theta]$ at a counterclockwise angle of è radians from the positive x- axis and a distance r from the origin, where θ varies over the domain of the function $r(\theta)$. A polar curve is therefore essentially the graph of a function, but in polár coordinates rather than the usual Cartesian coordinates. Polar curves can be used to give relatively simple expressions for intricate and interesting curves. Our focus with polar curves will be to find the area enclosed by a polar curve between two values of θ.

Each face is surrounded by edges, which could be line segments or curve segments, and the face itself is part of a surface (*i.e.*, a surface patch). In this unit, we shall discuss the general concept of curve segments in parametric form.

A parametric curve in space has the following form:

$$f: [0,1] \rightarrow (f(u), g(u), h(u))$$

where $f()$, $g()$ and $h()$ are three real-valued functions. Thus, $f(u)$ maps a real value u in the closed interval [0, 1] to a point in space. The domain of these real functions and vector-valued function $f()$ does not have to be [0, 1]. It can be any closed interval; but, for simplicity, we restrict the domain to [0, 1]. Thus, for each u in [0, 1], there corresponds to a point $(f(u), g(u), h(u))$ in space. In this course, functions $f()$, $g()$ and $h()$ are always polynomials.

Note that if function $h()$ is removed from the definition of $f()$, $f()$ has two coordinate components and becomes a curve in the coordinate plane.

Examples:

1. We have seen that a straight line is defined as $B + td$, where B is a base point and d is a direction vector. Thus, if $f()$ is defined as

$$f(u) = b_1 + ud_1$$
$$g(u) = b_2 + ud_2$$
$$h(u) = b_3 + ud_3$$

   where $B = < b_1, b_2, b_3 >$, $d = < d_1, d_2, d_3 >$, and $f()$ is a parametric curve that maps [0, 1] to the line segment between B and $B + d$, inclusive.
2. A circle has the following non-polynomial form:

$$x(u) = r \cos(2*PI*u) + p$$
$$y(u) = r \sin(2*PI*u) + q$$

   It has center (p, q) and radius r. Since the parameter u is in [0, 1], the value of $2*PI*u$ is in $[0, 2*PI]$ (*i.e.*, from 0 degree to 360 degree).

   Let us eliminate u. First, change the above equations to the following. For convenience, we drop (u) from $x(u)$ and $y(u)$.

$$x - p = r\cos(2*PI*u)$$
$$y - q = r \sin(2*PI*u)$$

   Then, squaring both equations and adding them together yields

$$(x - p)^2 + (y - q)^2 = r^2$$

   Thus, it shows that the given parametric curve is indeed a circle with center at (p, q) and radius r.
3. A space cubic curve has the following form:

$$f(u) = u$$
$$g(u) = u^2$$
$$h(u) = u^3$$

   The following figure shows this curve in the range of [–1, 1]. It is contained in the box (in white) defined by (–1, 0, –1) and (1, 1, 1)

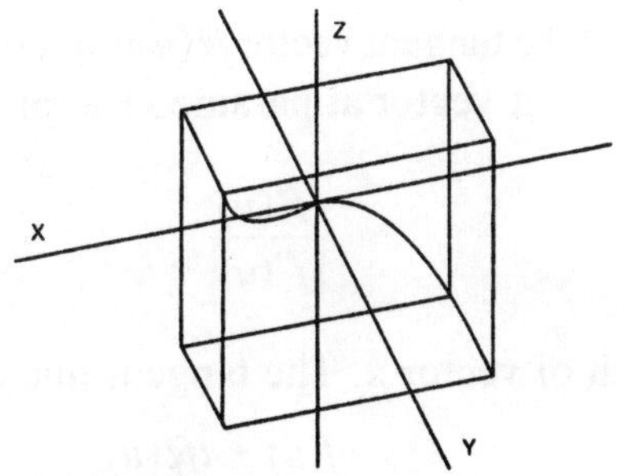

4. The circular helix is defined as follows:

   $$f(u) = (\ a\cos(u),\ a\sin(u),\ bu)$$

   The following figure shows the curve in [0, 4*PI]. The initial point is (a, 0, 0) and the endpoint is (a, 0, b*4*PI). Note that this curve lies on the cylinder of radius a and axis the z-axis.

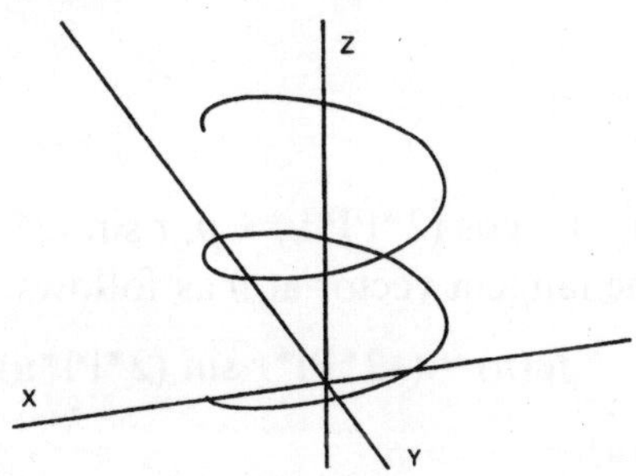

TANGENT VECTOR AND TANGENT LINE

Consider a fixed point X and a moving point P on a curve. As point P moves toward X, the vector from X to P approaches the tangent vector at X. The line that contains the tangent vector is the tangent line.

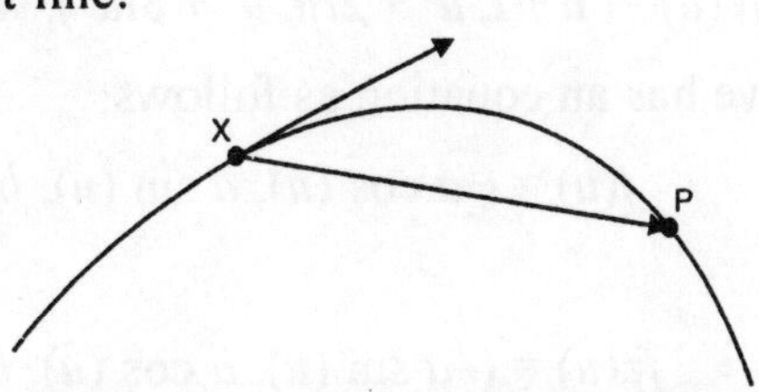

Computing the tangent vector at a point is very simple. Recall from your calculus knowledge that the derivative of parametric curve *f*(*u*)is the following:

f′(*u*) = (*f*¢(*u*), *g*¢(*u*), *h*¢(*u*))

where *f*¢(*u*) = $\dfrac{df}{du}$, *g*¢(*u*)= $\dfrac{dg}{du}$

and *h*¢(*u*) = $\dfrac{dh}{du}$.

In general, the length of the tangent vector *f¢*(*u*)is not one, and normalization is required. That is, the unit-length tangent vector at parameter *u*, or at point *f*(*u*), is

$$\frac{f'(u)}{|f'(u)|}$$

where $|x|$ is the length of vector x. The tangent line at *f*(*u*)is either

f(*u*) + *tf¢*(*u*)

or if unit-length direction vector is preferred

$$f(u)+t\left(\frac{f'(u)}{|f'(u)|}\right)$$

where *t* is a parameter.

EXAMPLES

"Consider the circle *f*(*u*) = (*r* cos (2*PI**u*) + *p*, r sin (2*PI**u*) + *q*), where *u* is in the range of 0 and 1. We have the tangent vector at *u* as follows:

f¢(*u*) = (–2*PI**r* sin (2*PI**u*), 2*PI**r* cos (2*PI**u*))

and the tangent line at *f*(*u*):

f(*u*)+ *tf¢*(*u*) = (*r* cos (2*PI**u*) + *p*, *r* sin(2*PI**u*) + *q*)
+ *t*(–2*PI* *r* sin(2*PI**u*), 2*PI**r* cos (2*PI**u*)

"Consider a space cubic curve *f*(*u*)= (u, u^2, u^3). We have tangent vector *f¢*(*u*)= (1, 2*u*, $3u^2$) and tangent line *f*(*u*)+ *tf¢*(*u*)= ($u + t$, $u^2 + 2tu$, $u^3 + 3tu^2$), where *t* is the line parameter.

"The circular helix curve has an equation as follows:

f(*u*) = (*a* cos (*u*), *a* sin (*u*), *bu*)

It has tangent vector

f¢(*u*) = (–*a* sin (*u*), *a* cos (*u*), *b*)

and tangent line

f(*u*) + *tf¢*(*u*) = (*a*(cos(*u*) – *t* sin(*u*)), *a*(sin(*u*) + *t* cos(*u*)), *b*(*t* + *u*))

NORMAL VECTOR AND CURVATURE

Consider a fixed point *f*(*u*)and two moving points *P* and *Q* on a parametric curve. These three points determine a plane. As *P* and *Q* moves toward *f*(*u*), this plane approaches a limiting position. This is the osculating plane at *f*(*u*). Obviously, the osculating plane at *f*(*u*)contains the tangent line at *f*(*u*). It can be shown that the osculating plane is the plane that passes through $f(u)$and contains both $f¢(u)$and $f''(u)$. More precisely, any point on this plane has an equation as follows, where p and q are parameters:

$$f(u) + pf'(u) + qf''(u)$$

The binormal vector $b(u)$is the unit-length vector of the cross-product of f′(*u*)and f″(*u*):

$$b(u) = \frac{(f'(u) \times f''(u)}{|(f'(u) \times f''(u))|}$$

Thus, the binormal vector $b(u)$is perpendicular to both $f¢(u)$and $f²(u)$and hence perpendicular to the osculating plane. The line $f(u) + tb(u)$is the binormal line at $f(u)$.

The normal vector is the vector perpendicular to both tangent and binormal vectors with its direction determined by the right-handed system. That is, the unit-length normal vector $n(u)$is defined to be

$$n(u) = \frac{b(u) \times f'(u)}{|(b(u) \times f'(u)|}$$

The line $f(u) + tn(u)$ is the normal line at $f(u)$. Therefore, tangent vector $f¢(u)$, normal vector $n(u)$and binormal vector $b(u)$form a coordinate system with origin $f(u)$. The tangent line, binormal line and normal line are the three coordinate axes with positive directions given by the tangent vector, binormal vector and normal vector, respectively. These three vectors are usually referred to as the moving triad or triad at point $f(u)$. Moving triad is also called moving trihedron. The following figure shows their relationship. Note that the tangent, normal and $f''(u)$vectors are on the same plane.

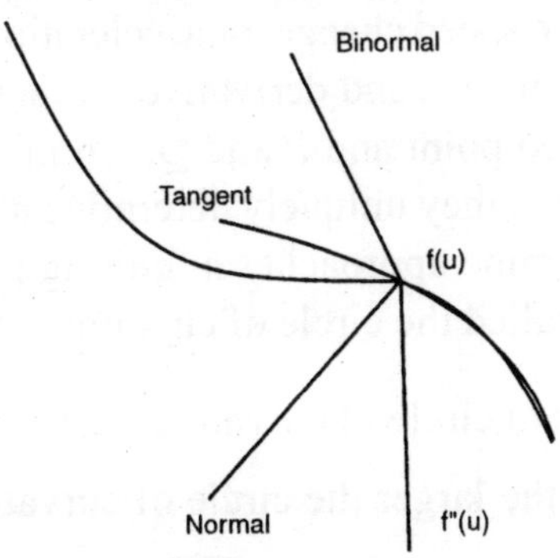

An Example

Let us compute the tangent, binormal and normal vectors of the circular helix curve:

$$f(u) = (a\cos(u), a\sin(u), bu)$$

The first and second derivatives are as follows:

$$f¢(u) = (-a \sin(u), a \cos(u), b)$$
$$f^2(u) = (-a \cos(u), -a \sin(u), 0)$$

The non-unit-length binormal vector is the cross-product of f′(*u*)and f″(*u*), in this order:

$$b(u) = f'(u) \times f'(u)$$
$$= (ab \sin(u), -ab \cos(u), a^2)$$

The non-unit-length normal vector is the cross-product of the binormal vector and the tangent vector, in this order:

$$n(u) = b(u) \times f¢(u)$$
$$= (-a(a^2 + b^2) \cos(u), -a(a^2 + b^2) \sin(u), 0)$$

If you compare $n(u)$and $f''(u)$, you will see that these vectors are parallel to each other (i.e., coefficients are proportional) and are both parallel to the *xy*-plane. As a result, after normalizing all involving vectors, the normal and the second derivative vectors are identical. This is shown by the following figure. It is computed at $u = 1$.

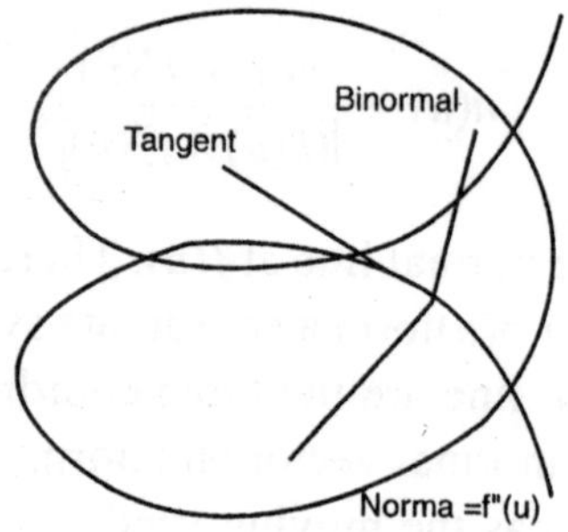

Curvature

As we all know, the tangent vector measures the change of "distance" and thus gives the speed of a moving point. The speed change, or acceleration, is measured by the derivative of the tangent vector, which is the second derivative. In fact, there is one more interesting interpretation. Take *X* as a fixed point and *P* and *Q* two moving points. As long as not all of these three points lie on a line, they uniquely determine a circle. As both *P* and *Q* moves toward *X*, the circle they determine approaches a limiting position as shown with dot-line below. This limiting circle is called the circle of curvature at *X* and its center and radius, *O* and *r*, are the center and radius of circle of curvature, respectively. More importantly, $\frac{1}{r}$ is the curvature at *X*. Therefore, the larger the circle of curvature, the smaller the curvature.

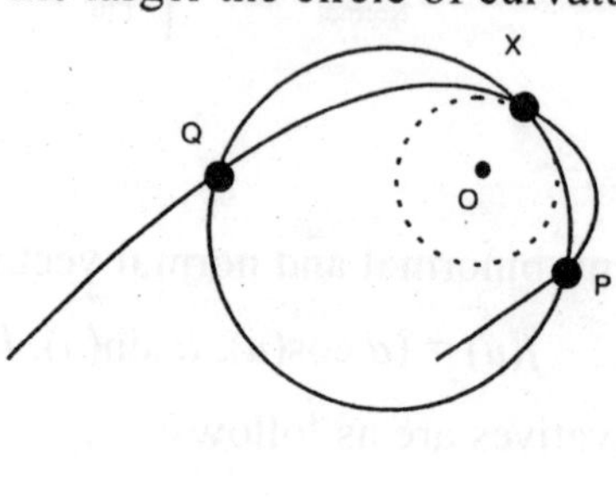

From the definition of osculating plane, we know that this circle of curvature must be on the osculating plane. Since the circle of curvature is tangent to the curve and hence the tangent line, the center of curvature lies on the normal line.

The curvature at u, $k(u)$, can be computed as follows:

$$k(u) = \frac{|f'(u) \times f''(u)|}{|f'(u)|^3}$$

Let us continue with the previous example of the circular helix curve. From the computations of $f\text{¢}(u)$and $f^2(u)$, we have

$$f\text{¢}(u) = (-a \sin(u), a \cos(u), b)$$

$$f\text{¢}(u) \times f^2(u) = (ab \sin(u), -ab \cos(u), a^2)$$

$$|f'(u)| = \text{SQRT } (a^2 + b^2)$$

$$|f'(u) \times f''(u)| = a \text{ SQRT } (a^2 + b^2)$$

$$k(u) = \frac{a}{(a^2 + b^2)}$$

Therefore, the curvature at any point on the curve is a constant $\dfrac{a}{(a^2 + b^2)}$.

Since the radius of the circle of curvature is $\dfrac{1}{k}$, we see that the center of circle of curvature is located at a distance of $\dfrac{(a^2 + b^2)}{a}$ from $f(u)$in the normal direction $n(u)$.

More Examples

- Consider a straight line:

$$f(u) = (a + up, b + uq, c + ur)$$

We have the following:

$$f'(u) = (p, q, \text{r})$$

$$|f'(u)| = \text{SQRT } (p^2 + q^2 + r^2)$$

$$f^2(u) = (0, 0, 0)$$

$$f\text{¢}(u) \times f^2(u) = (0, 0, 0)$$

$$k(u) = 0$$

Therefore, the curvature of a straight line is zero everywhere. Note that none of the binormal and normal vectors is well-defined because $f''(u)$is a zero vector.

- Consider a circle on the xy-plane:

$$f(u) = (r \cos(u)+ p, r \sin(u)+ q, 0)$$

Since it is on the *xy*-plane, the third coordinate function is always 0. From the given circle's equation we have the following:

$$f'(u) = (-r \sin(u), r \cos(u), 0)$$

$$f''(u) = (-r \cos(u), -r \sin(u), 0)$$

$$f'(u)\times f''(u) = (0, 0, r2)$$

$$| f¢(u) | = r$$

$$| f¢(u) \times f²(u) | = r^2$$

$$b(u) = \frac{(f'(u)\times f''(u))}{| f'(u)\times(f''(u)|} = (0, 0, 1)$$

$$n(u) = \frac{(b(u)\times f'(u))}{| b(u)\times(f'(u)|} = (-\cos(u), -\sin(u), 0)$$

$$k(u) = \frac{1}{r}$$

Thus, the unit-length tangent vector is $(-\sin(u), \cos(u), 0)$, the binormal vector is (0, 0, 1), and the normal vector is $(-\cos(u), \sin(u), 0)$. The binormal vector is always perpendicular to the xy-plane while both the tangent and normal vectors lie on the *xy*-plane. The curvature of a circle is a constant $\frac{1}{r}$. As a result, the radius of the circle of curvature is r and the circle of curvature is the given circle itself.

- Consider the space cubic defines as follows:

$$f(u) = (u, u^2, u^3)$$

The following are the computation of curvature $k(u)$:

$$f¢(u) = (1, 2u, 3u^2)$$

$$| f'(u) | = \text{SQRT} (1 + 4u^2 + 9u^4)$$

$$f''(u) = (0, 2, 6u)$$

$$f'(u)\times f''(u) = (6u^2, -6u, 2)$$

$$| f¢(u) \times f²(u) | = 2\ \text{SQRT} (1 + 9u^2 + 9u^4)$$

$$k(u) = \frac{2SQRT(1+9u^2+9u^4)}{SQRT(1+4u^2+9u^4))^3}$$

MOVING TRIAD

The moving triad is not only a mathematical concept. It also provides vital information of the characteristics of a moving object. For example, if you are on a airplane or a roller-coaster, as you are "flying" you are following a curve. Therefore, you are moving in the direction of the tangent vector, your "up" vector is in the direction of the binormal vector and the rate of turning and turning direction are given by the curvature and the direction of the normal vector, respectively. That is, while you are rolling or tumbling, this triad always provides you with the forward, up and turning directions. Therefore, to correctly animate a moving object, you need to know the geometric characteristics of the moving triad.

There is one more parameter, the torsion, that describing the way of twisting of a space curve. A planar curve does not twist and hence its torsion is zero. Since we will not use torsion in this course, it will not be introduced here.

CONTINUITY ISSUES

As mentioned in the discussion of B-rep, the edges and faces of a solid can be curve segments and surface patches rather than line segments and polygons. However, this may create some problems. In the figure shown below, we have three curvilinear patches of a B-rep joining together. Two boundary curve segments are shown in yellow and white meeting at a vertex *X*. Let these two curves be described as $f(u)$and g(v), where u and v are values in intervals $[a, b]$ and [m,n], respectively. The problem is: how we can make sure that these curves join together in a "smooth" way.

Consider the "right-end" of curve $f(b)$ and the "left-end" of curve g(m). If $f(b)$ and $g(m)$ are equal as shown in the above figure, we shall say curves $f()$ and $g()$ are C^0 continuous at $f(b)$=g(m). If for all $i = k$, the i-th derivatives at $f(b)$ and g(m) are equal, we shall say that the curves are C^k continuous at point $f(b) = g(m)$. Intuitively, two curves are C^0 continuous at the joining point if we can go from one curve to the other without crossing a gap because these two curves connect to each other. Two curves are C^1 continuous at the joining point if the speed (*i.e.*, first derivative) does not change when crossing one curve to the other.

Similarly, two curves are C_2 continuous at the joining point if, in addition to the speed, the acceleration (*i.e.*, second derivative) is also the same when crossing one curve to the other. Therefore, C1 continuous is "smoother" than C^0 continuous at the joining point, C^2 continuous is "smoother" than C1 continuous at the joining point, and so on. Moreover, if the curvatures of the curves at the joining point are equal, we will say they are curvature continuous at the joining point. Intuitively, two curves are curvature continuous if the turning rate is the same at the joining point; however, the second derivatives may not be the same at the joining point. In other words, curvature continuous does not guarantee C^2 continuous; but, C^2 continuous does imply curvature continuous.

It is clear from the definition that if two curve segments are C^k continuous at $f(b) = g(m)$, they are also Ci continuous for all i less than or equal to k. On the other hand, if the k-th derivatives of two curve segments at their joining point are not equal, they cannot be C^i continuous for any i larger than or equal to k.

Let us take a look at a simple example. The following curve consists of two parabolas:

$$f(u) = (u, -u^2, 0)$$

$$g(v) = (v, v^2, 0)$$

where $f()$, the left curve, and $g()$, the right curve, have domains [–1, 0] and [0, 1], respectively. It is easy to check that $f(0) = g(0)$ is the coordinate origin. Let us check to see if these two curves are C^2 continuous there.

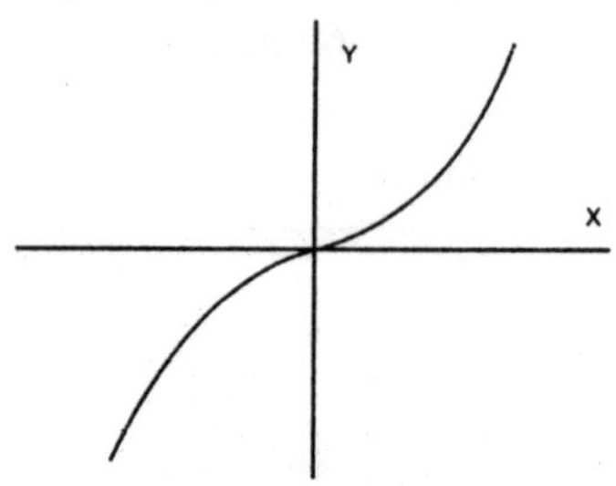

The following are necessary information for this checking:

$$f¢(u) = (1, -2u, 0)$$

$$f^2(u) = (0, -2, 0)$$

$$g¢(u) = (1, 2v, 0)$$

$$g^2(v) = (0, 2, 0)$$

Since $f¢(0) = g¢(0) = (1, 0, 0)$, these two curves are C^1 continuous at the origin. However, since $f^2(0) = (0, -2, 0)$ is not equal to $g^2(v) = (0, 2, 0)$, they are not C^2 continuous at the origin. In fact, $f''(u)$points to the south while $g''(u)$points to north, and both are constant (*i.e.*, do not change with u and v). Therefore, once a moving point crosses the origin from the green segment to the yellow segment, the second derivative changes its direction abruptly. Therefore, the curve segments are not C^2 continuous at the origin.

Let us computes their curvatures:

Curvature of $$f(u) = \frac{2}{(1+4u^2)^{1.5}}$$

Curvature of $$g(v) = \frac{2}{(1+4v^2)^{1.5}}$$

They have the same forms, although u is in [–1, 0] and v is in [0, 1]. Therefore, the two curve segments are curvature continuous at the joining point (*i.e.*, $f(0) = g(0)$)! Thus, we see that two curve segments may be C^1 continuous and even curvature continuous, but not C^2 continuous.

Problems with Parametric Representations

C^k continuity looks a promising way of specifying how smooth two curves are joint together. However, it does have a problem. Consider the following two line segments:

$$f(u) = A + u(B - A)$$

$$g(v) = B + v(C - B)$$

where A, B and C are three collinear points as shown in the figure below.

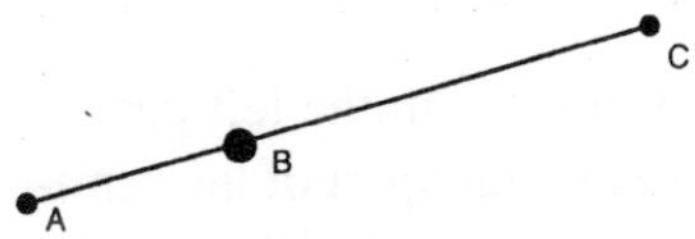

As u (resp., v) changes from 0 to 1, $f(u)$ (resp., g(v)) runs from A to B (resp., B to C). Line segments $f(u)$and $g(v)$are obviously C^0 continuous at the joining point B. Is it C^1 continuous?

$$f'(u) = B - A$$

$$g'(v) = C - B$$

Thus, $f'(u) = B - A$ is, in general, not equal to $g'(v) = C - B$ and consequently these two line segments are not C^1 continuous at the joining point B!

Is it strange? Yes, in fact, this is a problem of parameterization. If we replace the direction vectors $B - A$ and $C - B$ with unit-length vectors and change the domain of parameters u and v, this problem will disappear. That is, the above equations are changed to the following:

$$F(u) = \frac{A + u(B-A)}{|B-A|}$$

$$G(v) = \frac{B + v(C-B)}{|C-B|}$$

where u is in the range of 0 and $|B - A|$ and v is in the range of 0 and $|C - B|$. Now, since we have $F¢(u) = G¢(v)$ = the unit-length vector in the direction of A to C, the line segments are C^1 continuous. Thus, reparameterizing the curve segments may overcome the problem.

Let us look at another example, where *PI* is for 3.1415926 and *u* and *v* are in the range of [0, 1].

$$f(u) = \left(-\cos\left(u^2 \frac{PI}{2}\right), \sin\left(u^2 \frac{PI}{2}\right), 0\right)$$

$$g(v) = \left(\sin\left(v^2 \frac{PI}{2}\right), \cos\left(v^2 \frac{PI}{2}\right), 0\right)$$

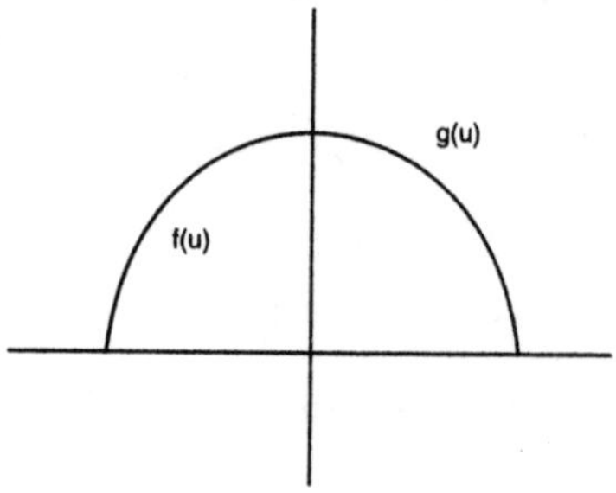

As *u* moves from 0 to 1, $f(u)$traces out the left part of the semi-circle. Similarly, as *v* moves from 0 to 1, $g(u)$traces out the right part of the semi-circle. These two have a joining point, shown in red, at (0, 1, 0) = $f(1) = g(0)$. We have the following:

$$f¢(u) = \left(PI\, u \sin\left(u^2 \frac{PI}{2}\right), PI\, u \cos\left(u^2 \frac{PI}{2}\right), 0\right)$$

$$f^2(u) = \left(PI^2 u^2 \cos\left(u^2 \frac{PI}{2}\right), -PI^2 u^2 \sin\left(u^2 \frac{PI}{2}\right), 0\right)$$

$$f'(u) \times f''(u) = (0, 0, -PI^3 u^3)$$

$$|f'(u)| = \text{PI}\, u$$

$$|f'(u) \times f''(u)| = PI^3 u^3$$

$$k(u) = 1$$

$$g¢(v) = \left(PI\, v \cos\left(v^2 \frac{PI}{2}\right), -PI\, v \sin\left(v^2 \frac{PI}{2}\right), 0\right)$$

$$g''(v) = \left(PI^2 v^2 \cos\left(v^2 \frac{PI}{2}\right), -PI^2 v^2 \sin\left(v^2 \frac{PI}{2}\right), 0\right)$$

$$g¢(v) \times g^2(v) = (0, 0, -PI^3\, u^3)$$

$$|\, g¢(v)\, | = PI\, v$$

$$|\, g¢(v) \times g^2(v)\, | = PI^3\, v^3$$

$$k(v) = 1$$

Note that both $g¢(0)$ and $g^2(0)$ are zero vectors and are not well-defined. As a result, we cannot talk about the continuity property at the joining point at all; but, from the figure, it "looks" like these two curves do have some continuity property because, at least, they have a common tangent line.

Let us reparameterize these curves (*i.e.*, change their parametric equations without changing their shapes). Let $u^2 = p$ in $f(u)$ and let $v^2 = q$ in $g(v)$. The new equations are:

$$f(p) = \left(-\cos\left(p\frac{PI}{2}\right), \sin\left(p\frac{PI}{2}\right), 0\right)$$

$$g(q) = \left(\sin\left(q\frac{PI}{2}\right), \cos\left(q\frac{PI}{2}\right), 0\right)$$

Their derivatives are

$$f'(p) = \left(\left(\frac{PI}{2}\right)\sin\left(p\frac{PI}{2}\right), \left(\frac{PI}{2}\right)\cos\left(p\frac{PI}{2}\right), 0\right)$$

$$f''(p) = \left(\left(\frac{PI}{2}\right)^2 \cos\left(p\frac{PI}{2}\right), -\left(\frac{PI}{2}\right)^2 \sin\left(p\frac{PI}{2}\right), 0\right)$$

$$g'(q) = \left(\left(\frac{PI}{2}\right)\cos\left(q\frac{PI}{2}\right), -\left(\frac{PI}{2}\right)\sin\left(q\frac{PI}{2}\right), 0\right)$$

$$g''(q) = \left(-\left(\frac{PI}{2}\right)^2 \sin\left(q\frac{PI}{2}\right), -\left(\frac{PI}{2}\right)^2 \cos\left(q\frac{PI}{2}\right), 0\right)$$

$$f'(p) \times f''(p) = g'(q) \times g''(q) = \left(0, 0, -\left(\frac{PI}{2}\right)^3\right)$$

$$|\, f'(p) \times f''(p)\, | = |\, g'(q) \times g''(q)| = \left(\frac{PI}{2}\right)^3$$

$$|f'(p)| = |g'(q)| = \frac{PI}{2}$$

$$k(p) = k(q) = 1$$

Therefore, after changing variables, both $f'(1)$ and $g'(0)$ are equal to $\left(\frac{PI}{2}, 0, 0\right)$ and hence they are C1 continuous. Moreover, both $f''(1)$ and $g''(0)$ are equal to $\left(0, -\left(\frac{PI}{2}\right)2, 0\right)$ and hence they are C^2 continuous! They are also curvature continuous because they have curvature 1 everywhere (remember they are circles). Thus, reparameterization or change of variables have a dramatic impact on continuity.

Arc Length Parameterization

Because different parametric forms of the same curve may induce different continuity results, one must ask the question: Is there a parameterization that we can trust for continuity discussion? The answer is yes, since mathematicians have solved this problem long time ago. The trick is to use arc length as a parameter.

Let a curve segment have length s. One can parameterize this curve such that $f(u)$is the point having a distance u from the initial point $f(0)$, where u is in the range of 0 and s. With this arc length parameterization, as u moves from 0 to s, $f(u)$moves on the curve from $f(0)$ to $f(s)$ in the same speed. Therefore, the tangent vector which measures speed is of unit-length. Not only this, many formulas shown on this and previous pages can be simplified.

Why don't we use arc length parameterization to simplify our computation? The answer is quite simple. While arc length parameterization is simple and elegant in theory, it is tedious in computation and impractical. One can easily design a curve in some handy parameterization; but, reparameterizing it with arc length sometime is extremely difficult. That is, finding arc length is not an easy task, since it requires to integrate a function involving the use of square root.

Therefore, even though non-arc length parameterization could cause problems, we will not use the arc length parameterization.

Geometric Continuity

Many C^1 continuous curves are curvature continue but not C^2 at the joining point and some of them may not even twice differentiable. These curves look smooth at the joining point and also look smooth when moving from a segment to the other. Moreover, as mentioned earlier, after a change of variables, some of them could become C^2 at the joining point. But, a reparameterization that can make this to happen could be difficult to find. therefore, we intend to relax the requirement of C^2 to the following:

Two curve segments are said Gk geometric continuous at the joining point if and only if all i-th derivatives, *i* less than or equal to *k*, computed with arc length parameters agree at the joining point.

Oops, we use arc length in the definition! But, don't worry, since arc length parameterization is not required as shown by the following equivalent definition:

Two curve segments are said G^k geometric continuous at the joining point if and only if there exists two parameterizations, one for each curve segment, such that all *i*-th derivatives, *i* less than or equal to *k*, computed with these new parameterizations agree at the joining point.

This is better but not good enough because we really do not know how to find such parameterizations. Fortunately, the cases of $k = 1$ and $k = 2$ are very simple. Let us start with the case of $k = 1$.

Two C^0 curve segments are said G^1 geometric continuous at the joining point if and only if vectors $f'(u)$and $g'(v)$are in the same direction at the joining point. Note that $f¢(u)$and $g¢(v)$are evaluated at the joining point.

Since the tangent vectors at the joining point have the same direction, both curves have the same tangent line at the joining point. However, the converse does not hold. More precisely, two curve segments having the same tangent line does not imply they are G^1 at the joining point. In the following figure, curves $f(u)$and $g(v)$have a common point at which the tangent lines are the same. However, the tangent vectors point to the opposite directions, and, as a result, they are not G^1 at the joining point. Based on this definition, all previously discussed example curves are G^1 at the joining points.

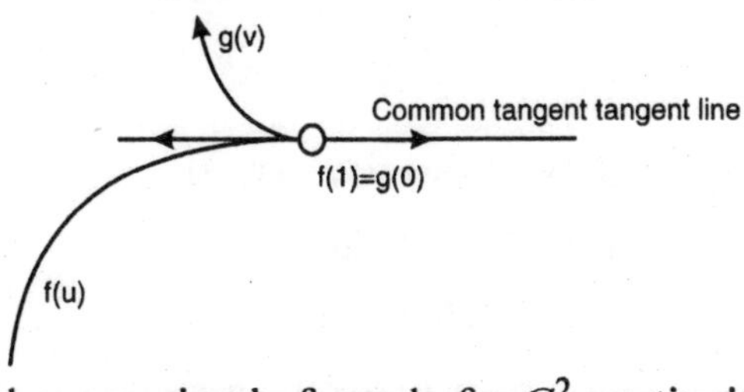

G. Neilson has found a very simple formula for G^2 continuity as follows:

Two C^1 curve segments are said G^2 geometric continuous at the joining point if and only if vector $f''(u) - g''(v)$ is parallel to the tangent vector at the joining point. Note that $f^2(u)$and $g^2(v)$are evaluated at the joining point.

The beauty of this characterization for G^2 continuity is that it works for any parameterization. Before using this characterization, you should check for C^1 condition first.

Let us use an example to illustrate Nielson's result. Consider the following two parabola segments with joining point at (0, 1, 0):

$$f(u) = (-1 + u^2, 2u - u^2, 0)$$
$$g(v) = (2u - u^2, 1 - u^2, 0)$$

Both segments have domain [0, 1]. The joining point is

$$f(1) = g(0) = (\ 0,\ 1,\ 0\ ).$$

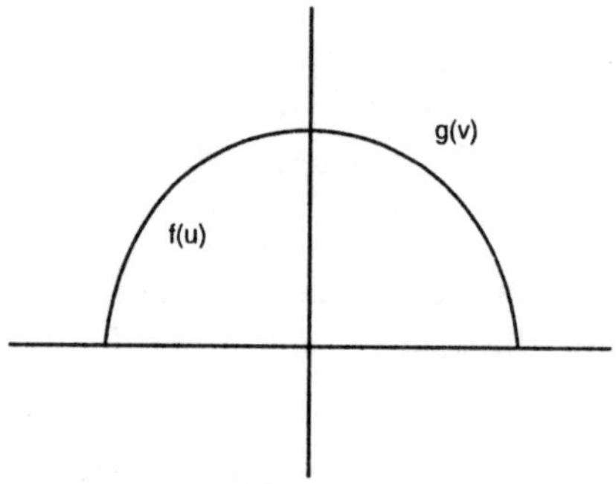

The following are essential calculations:

$$f¢(u) = (\ 2u,\ 2 - 2u,\ 0\ )$$
$$f^2(u) = (\ 2,\ -2,\ 0\ )$$
$$f^2(u) \times f^2(u) = (0, 0, -4)$$
$$|\, f¢(u) \,| = 2\ SQRT(1 - 2u + 2u^2)$$
$$|\, f¢(u) \times f^2(u) \,| = 4$$

$$k(u) = \frac{1}{\left(2(1 - 2u + 2u^2)\right)^{1.5}}$$

$$g'(v) = (2 - 2v, -2v, 0)$$
$$g''(v) = (-2, -2, 0)$$
$$g'(v) \times g''(v) = (0, 0, -4)$$
$$|\, g'(v) \,| = 2\ SQRT(1 - 2v + 2v2\ )$$
$$|\, g'(v) \times g''(v) \,| = 4$$

$$k(v) = \frac{1}{\left(2(1 - 2v + 2v^2)^{1.5}\right)}$$

From $f¢(1) = g¢(0) = (\ 2, 0, 0\ )$, we know that these two curve segments are C1 at the joining point. Since f″(1) = (2, -2, 0) is not equal to g″(0) = (-2, -2, 0), they are not C2. But, since their curvature functions are the same (i.e., k*f*(*u*)= kg(v)=1), these two curves are curvature continuous at the joining point.

Let us check for G^2. Since the curves are C^1, checking for G^2 is meaningful. Since $f''(1) - g''(0) = (4, 0, 0)$ is parallel to the tangent vector $(2, 0, 0)$ at the joining point, these two curve segments are G2 continuous at the joining point $(0, 1, 0)$.

By the definition of G^k, you could find two parameterizations, not necessarily of arc length type, so that the reparameterized segments are of C^2.

RATIONAL CURVES

Parametric representations using polynomials are simply not powerful enough, because many curves (e.g., circles, ellipses and hyperbolas) can not be obtained this way. One way to overcome this shortcoming is to use homogeneous coordinates. For example, a curve in space (resp., plane) is represented with four (resp., three) functions rather than three (resp., two) as follows:

Space curve: $F(u) = (x(u), y(u), z(u), w(u))$

Plane curve: $F(u) = (x(u), y(u), w(u))$

where u is a parameter in some closed interval $[a, b]$. Converting this curve to their conventional form yields the following:

Space curve: $$f(u) = \left(\frac{x(u)}{w(u)}, \frac{y(u)}{w(u)}, \frac{z(u)}{w(u)} \right)$$

Plane curve: $$f(u) = \left(\frac{x(u)}{w(u)}, \frac{y(u)}{w(u)} \right)$$

Obviously, if $w(u) = 1$, a constant function, the homogeneous forms reduce to conventional forms.

A parametric curve in homogeneous form is referred to as a rational curve. To make a distinction, we shall call a curve in polynomial form a polynomial curve.

An Example

Let us use an example to illustrate the power of the rational form. Consider a second degree parametric form:

$$x = f(u) = au^2 + bu + c$$
$$y = g(u) = pu^2 + qu + r$$

Let the domain of u be the real line. Now, let u run from negative infinity to positive infinity. The values of x and y will both go to infinity, positive or negative. In other words, the curve described by the above parametric form contains at least one point at infinity. Since all points of a circle are in finite range, it is impossible to obtain a circle from the above parametric form. So, this shows a weakness of parametric form.

We can also reach this conclusion through calculation. Let us start with the following form:

$$x = au^2 + bu + c$$
$$y = pu^2 + qu + r$$

Assume a and p are both non-zero. Dividing the first and second equation by a and p, respectively, we have

$$\frac{x}{a} = u^2 + \frac{b}{a}u + \frac{c}{a}$$
$$\frac{u}{p} = u^2 + \frac{q}{p}u + \frac{r}{p}$$

Now subtract the second from the first to eliminate the u2 term. Then, solving for u yields:

$$u = \frac{p(x-c) - a(y-r)}{bp - aq}$$

Finally, plugging this u back into the first equation (of the parametric form) gives

$$x = a\left[\frac{p(x-c)-a(y-r)}{bp-aq}\right]^2 + b\left[\frac{p(x-c)-a(y-r)}{bp-aq}\right] + c$$

Clearing the denominators and rearranging terms gives

$$Ax^2 - 2Bxy + Cy^2 + Dx + Ey + F = 0$$

where

$$A = ap^2$$
$$B = a^2p$$
$$C = a^3$$

Recall that to determine the type of this quadric curve, we do not need the values of D, E and F. Since B2-A×C=0, this curve is a parabola.

Rational Forms of Common Curves

Let us find the rational form of a circle.

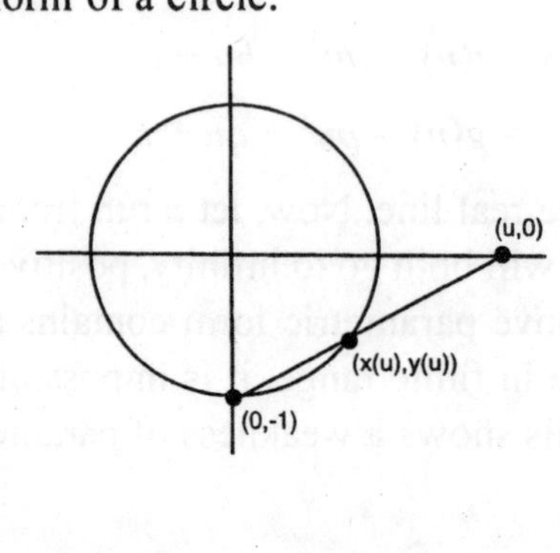

The above shows a circle with center at the origin and radius 1. Let u be a parameter for this circle. For convenience, we use the x-axis for the value of u, and, as a result, for any particular value of u there is a point $(u, 0)$ on the x-axis. Join a line between $(u, 0)$ and the south pole of the circle $(0, -1)$. Since a line intersects a circle in two points, let the other point, which is not the south pole, be $(x(u), y(u))$. As u moves on the x-axis, its corresponding point moves on the circle. Any finite u has such a corresponding point. The infinite u corresponds to the south pole. On the other hand, any point on the circle that is not the south pole corresponds to a u on the x-axis.

The line joining the south pole and $(u, 0)$ is $x = uy + u$. The circle equation is $x^2 + y^2 = 1$. Plugging the line equation into the circle equation and solving for y yields two roots for y. One of them must be $y = -1$, since this line passes through the south pole. The other root is $y = \frac{(1-y^2)}{(1+y^2)}$. Plugging this y into the line equation yields $x = \frac{2u}{(1+u^2)}$. Therefore, for each u, the corresponding point on the circle is

$$x = \frac{2u}{1+u^2}$$

$$y = \frac{1-u^2}{1+u^2}$$

As a result, a circle has a rational parameterization. When u moves to infinity, x approaches 0 while y approaches -1 (use L'Hospital's rule).

In fact, this computation tells us more. This circle has a trigonometric parametric form $(\cos(t), \sin(t))$, where t is in the range of 0 and $2PI$. Therefore, a point on the circle can have two different representations, but their values are equal. That is, $(\cos(t), \sin(t)) = \left(\frac{(2u)}{(1+u^2)}, \frac{(1-u^2)}{(1+u^2)}\right)$ for some u. Hence, trigonometric functions cos(t) and sin(t) can be parameterized as follows:

$$\cos(t) = \frac{2u}{1+u^2}$$

$$\sin(t) = \frac{1-u^2}{1+u^2}$$

With a parameterization for $\cos(t)$ and $\sin(t)$ in hand, you can easily find a rational parameterization for an ellipse and a hyperbola. Let us do the hyperbola part and leave the ellipse as an exercise. A hyperbola in its normal form has an equation as follows:

$$\frac{x^2}{a^2} + \frac{y^2}{b^2} = 1$$

where a and b are the semi-major axis and semi-minor axis lengths. It is easy to verify that the following is a correct parameterization for the hyperbola:

$$x = a \sec(t)$$

$$y = b \tan(t)$$

Since tan(t) = sin(t) / cos(t) and sec(t) = 1 / cos(t), plugging the rational parameterization for sin(t) and cos(t) into the above equations will convert them from a trigonometric parametric form to a rational one as follows:

$$x = a\frac{1+u^2}{2u}$$

$$y = b\frac{1-u^2}{2u}$$

Uniformization Theorems

You have seen the way of converting a polynomial parametric form to a polynomial $p(x, y) = 0$ in a previous example. Rational forms can also be converted to polynomials. The trick is by eliminating the parameter u from all parametric equations. It is tedious; but it can be done. In fact, with the elimination theory, one can convert a polynomial or rational parametric form to an equivalent implicit polynomial in two variables. This is called implicitization.

The question is: given a curve in an implicit polynomial form, is it possible to find a parametric form, polynomial or rational, that describes the same The Ellipse *182182* Encyclopaedia of Solid GeometryThe Ellipse *182182* Encyclopaedia of Solid Geometry-curve? Unfortunately, the answer is negative. That is, there exists curves which do not have any polynomial or rational parametric forms (*e.g.*, the elliptic curve $y^2 = x^3 + ax + c$). This is the result of uniformization theorems.

Due to this fact, implicit polynomials can describe more curves than polynomial and rational parametric forms. Why don't we use implicit polynomials? The reason is simple. Parametric curves are easy to use. While implicit polynomial curves are powerful and can describe more curves, many of its important properties are difficult to compute and implicit curves are in general difficult to use. We shall cover implicit curves and surfaces later.

Chapter 8

3-D Co-ordinate Geometry

In mathematics the dimension of a space is roughly defined as the minimum number of coordinates needed to specify every point within it. Dimensions can be thought of as the axes in a Cartesian coordinate system, which in a three-dimensional system run left-right, up-down and forward-backward. A set of three coordinates on these axes, or any other three-dimensional coordinate system, specifies the position of a particular point in space. In the physical world, according to the theory of relativity the fourth dimension is time, which runs before-after. An event's position in space and time is therefore specified if four coordinates are given.

THREE-DIMENSIONAL SPACE

Three-dimensional space is a geometric model of the physical universe in which we live. The three dimensions are commonly called length, width, and depth (or height), although any three mutually perpendicular directions can serve as the three dimensions. A three-dimensional picture, as one sees with a stereopticon or a View Master, tricks the human eye into experiencing the illusion of realistic depth, while the real depth is a mere fraction of a millimeter. In physics, our three-dimensional space is viewed as embedded in 4-dimensional space-time, called Minkowski space. The idea behind space-time is that time is hyperbolic-orthogonal to each of the three spatial dimensions.

In mathematics, Cartesian geometry (analytic geometry) describes every point in three-dimensional space by means of three coordinates. Three Coordinate axes are given, each perpendicular to the other two at the origin, the point at which they cross. They are usually labeled x, y, and z. Relative to these axes, the position of any point in three-dimensional space is given by an ordered triple of real numbers, each number giving the distance of that point from the origin measured along the given axis, which is equal to the distance of that point from the plane determined by the other two axes. Other popular methods of describing the location of a point in three-dimensional space include cylindrical coordinates and spherical coordinates, though there are an infinite number of possible methods. See Euclidean space.

Another mathematical way of viewing three-dimensional space is found in linear algebra, where the idea of independence is crucial. Space has three dimensions because the length of a box is independent of its width or breadth. In the technical language of linear algebra, space is three dimensional because every point in space can be described by a linear combination of three independent vectors. In this view, space-time is four dimensional because the location of a point in time is independent of its location in space. Three-dimensional space has a number of properties that distinguish it from spaces of other dimensions. It is, for example, the smallest dimension in which it is possible to tie a knot in a piece of string. Many of the laws of physics, such as the various inverse square laws, depend on dimension three . The understanding of three-dimensional space in humans is thought to be learned during infancy using unconscious inference, and is closely related to hand-eye coordination. The visual ability to perceive the world in three dimensions is called depth perception.

Various geometrical figures in three dimensions in three-dimensional space can be described relative to a set of mutually orthogonal axes Ox, Oy, Oz, and a point can be represented by a set of rectangular coordinates (x, y, z). The point can also be represented by cylindrical coordinates (ρ, ϕ, z) or spherical coordinates (r, θ, ϕ). In this chapter, we are concerned mostly with (x, y, z). The rectangular axes are usually chosen so that when you look down the z-axis towards the xy-plane, the y-axis is 90° counter clockwise from the x-axis. Such a set is called a right-handed set. A left-handed set is possible, and may be useful under some circumstances, but, unless stated otherwise, it is assumed that the axes chosen in this chapter are right-handed.

An equation connecting x, y and z, such as

$$f(x,y,z) = 0$$

or $$z = z(x, y)$$

describes a two-dimensional surface in three-dimensional space. A line (which need be neither straight nor two-dimensional) can be described as the intersection of two surfaces, and hence a line or curve in three-dimensional coordinate geometry is described by two equations, such as

$$f(x, y, z) = 0$$

and $$g(x, y, z) = 0.$$

In two-dimensional geometry, a single equation describes some sort of a plane curve. For example,

$$y^2 = 4qx$$

describes a parabola. But a plane curve can also be described in parametric form by two equations. Thus, a parabola can also be described by

$$x = qt^2$$

and $$y = 2qt$$

Similarly, in three-dimensional geometry, a line or curve can be described by three equations in parametric form. For example, the three equations

$$x = a \cos t$$

$$y = a \sin t$$

$$z = ct$$

describe a curve in three-space. Think of the parameter t as time, and see if you can imagine what sort of a curve this is.

We shall be concerned in this chapter mainly with six types of surface: the plane, the ellipsoid, the paraboloid, the hyperboloid, the cylinder and the cone.

Planes and Straight Lines

The geometry of the plane and the straight line is, of course, rather simple, so that we can dispose of them in this brief introductory section in a mere 57 equations.

The equation

$$Ax + By + Cz + D = 0$$

represents a plane. If $D \neq 0$ it is often convenient, and saves algebra and computation with no loss of information, to divide the equation through by D and re-write it in the form

$$ax + by + cz = 1.$$

The coefficients need not by any means all be positive. If $D = 0$, the plane passes through the origin of coordinates, and it may be convenient to divide the equation by C and hence to rewrite it in the form

$$ax + by + z = 0.$$

The plane represented by equation intersects the yz–, zx– and xy–planes in the straight lines

$$by + cz = 1$$

$$cz + ax = 1$$

$$ax + by = 1$$

and it intersects the x-, y- and z-axes at

$$x = x_0 = 1/a$$
$$y = y_0 = 1/b$$
$$z = z_0 = 1/c$$

The geometry can be seen in figure.

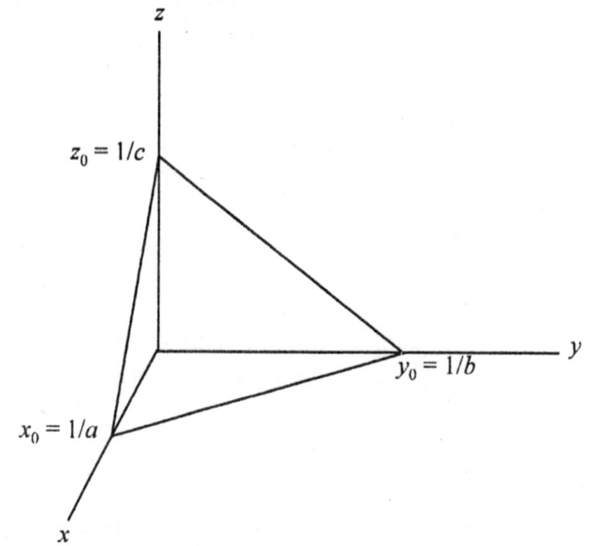

Another way of writing the equation to the plane would be

$$\frac{x}{x_0} + \frac{y}{y_0} + \frac{z}{z_0} = 1.$$

In this form, x_0, y_0 and z_0 are the intercepts on the x-, y- and z-axes.

Let $P(x, y, z)$ be a point on the plane 4.2.2. The square S of the distance of that point from the origin is

$$S = x^2 + y^2 + z^2.$$

But since (x, y, z) is on the plane, we can write S in terms of x and y alone, by substituting for z from equation :

$$S = x^2 + y^2 + \frac{(1 - ax - by)^2}{c^2}.$$

The distance to the origin is least for a point on the plane such that $\frac{\partial S}{\partial x}$ and $\frac{\partial S}{\partial y}$ are both zero. After some algebra, we find that the point on the plane that is closest to the origin is the point

$$\left(\frac{a}{a^2+b^2+c^2}, \frac{b}{a^2+b^2+c^2}, \frac{c}{a^2+b^2+c^2}\right)$$

Consequently the perpendicular distance of the origin from the plane is the orthogonal sum of these coordinates, which is

$$\frac{1}{\sqrt{a^2+b^2+c^2}}$$

Substitution of x_0 for $1/a$, y_0 for $1/b$ and z_0 for $1/c$ will make these results seem rather obvious.

Further, *OP* is normal to the plane, and the *direction cosines* of *OP*, i.e. of the normal to the plane, are

$$\frac{a}{\sqrt{a^2+b^2+c^2}}, \frac{b}{\sqrt{a^2+b^2+c^2}}, \frac{c}{\sqrt{a^2+b^2+c^2}}$$

The coefficients a, b, c are direction ratios of the normal to the plane; that is to say, they are numbers that are proportional to the direction cosines.

Example: Consider the plane

$$0.50x + 0.25y + 0.20z = 1$$

The plane intersects the x-, y- and z-axes at (2, 0, 0), 0, 4, 0) and (0, 0, 5). The point on the plane that is closest to the origin is (1.4184, 0.7092, 0.5674). The perpendicular distance of the origin from the plane is 1.6843. The direction cosines of the normal to the plane are (0.8422, 0.4211, 0.3369).

Now, let P(x, y, z) be a point in the plane and let P0 (x_0, y_0, z_0) be some point not necessarily in the plane. It is of interest to find the perpendicular distance between P_0 and the plane. Let S be the square of the distance between P_0 and P. Then

$$S = (x - x_0)^2 + (y - y_0)^2 + (z - z_0)^2$$

We can express this in terms of x and y alone by substitution for z from equation. The distance of P from P_0 is least when $\frac{\partial S}{\partial x}$ and $\frac{\partial S}{\partial y}$ are simultaneously zero. These two conditions result in

$$(a^2 + c^2)x = a + c^2 x_0 - acz_0 - aby$$

$$(b^2 + c^2)y = b + c^2 y_0 - bcz_0 - abx$$

These, combined with equation, result in

$$x = \frac{(b^2+c^2)x_0 + a(1 - by_0 - cz_0)}{a^2+b^2+c^2}$$

$$y = \frac{(c^2 + a^2)y_0 + b(1 - cz_0 - ax_0)}{a^2 + b^2 + c^2}$$

$$z = \frac{(a^2 + b^2)z_0 + c(1 - ax_0 - by_0)}{a^2 + b^2 + c^2}$$

These are the coordinates of the point P in the plane that is nearest to P_0. The perpendicular distance between P and P_0 is $p = \sqrt{S}$, or

$$p = \frac{1 - ax_0 - by_0 - cz_0}{\sqrt{a^2 + b^2 + c^2}}.$$

This is positive if P_0 is on the same side of the plane as the origin, and negative if it is on the opposite side. If the perpendicular distances of two points from the plane, as calculated from equation, are of opposite signs, they are on opposite sides of the plane. If $p = 0$, or indeed if the numerator of equation is zero, the point P_0 (x_0, y_0, z_0) is, of course, in the plane.

An equation for the plane containing three specified points can be found as follows. Let (x_1, y_1), (x_2, y_2), (x_3, y_3) be the three specified points, and let (x, y) be any point in the plane that contains these three points. Each of these points must satisfy an equation of the form. That is,

$$xA + yB + zC + D = 0$$
$$x_1A + y_1B + z_1C + D = 0$$
$$x_2A + y_2B + z_2C + D = 0$$
$$x_3A + y_3B + z_3C + D = 0$$

In these equations, we are treating A, B, C, D as unknowns, and the $x, y, z, x_1, y_1 \ldots$ as coefficients. We have four linear equations in four unknowns, and no constant term. From the theory of equations, these are consistent only if each is a linear combination of the other three. This is satisfied only if the determinant of the coefficients is zero:

$$\begin{vmatrix} x & y & z & 1 \\ x_1 & y_1 & z_1 & 1 \\ x_2 & y_2 & z_2 & 1 \\ x_3 & y_3 & z_3 & 1 \end{vmatrix} = 0$$

and this is the equation to the required plane containing the three points. The reader will notice the similarity of this equation to equation for a line passing between two points in twodimensional geometry. The reader might like to repeat the argument, but requiring instead the four points to satisfy an equation of the form. There will then be four linear equations in three unknowns. Otherwise the argument is the same.

We now move on to the question of finding the area of a triangle whose vertices are given. It is straightforward to do this with a numerical example, and the reader is now encouraged to write a computer program, in whatever language is most familiar, to carry out the following tasks. Read as data the *x-y-z* coordinates of three points *A, B, C*. Calculate the lengths of the sides *a, b, c, a* being opposite to A, etc. Calculate the three angles at the vertices of the triangle, in degrees and minutes, and check for correctness by verifying that their sum is 180°. If an angle is obtuse, make sure that the computer displays its value as a positive angle between 90° and 180^0. Finally, calculate the area of the triangle.

The data for several triangles could be written into a data file, which your program reads, and then writes the answers into an output file. Alternatively, you can type the coordinates of the vertices of one triangle and ask the computer to read the data from the monitor screen, and then to write the answers on the screen followed by a message such as "Do you want to try another triangle (1) or quit (2)?". Your program should also be arranged so that it writes an appropriate message if the three points happen to be collinear.

It should be easy to calculate the sides. The angles can then be calculated from equation and the area from each of the four equations. They should all yield the correct answer, of course, but the redundant calculations serve as an important check on the correctness of your programming, as also does your check that the three angles add to 180°. Where there are two of more ways of performing a calculation, a careful calculator will do all of them as a check against mistakes, whether the calculation is done by hand or by computer.

Example: If the coordinates of the vertices are

A(7, 4, 3), B(11, 6, 2), C(9, 2, 4)

the sides are $a = 4.583$, $b = 3.000$, $c = 4.899$,

and the angles are $A = 36° 42'$, $B = 77° 24'$, $C = 65° 54'$,

which add up to 180°. The area is 6.708.

Example: If the coordinates of the vertices are

A(6, 4, 9), B(2, 6, 17), C(8, 3, 5)

the area of the triangle is zero and the points are collinear.

The foregoing showed that it was not difficult to calculate numerically the area of the triangle from the coordinates of its vertices. Is it easy to find a simple explicit algebraic formula for the area in terms of (x_1, y_1, z_1), (x_2, y_2, z_2) and (x_3, y_3, z_3)? On referring to figure, we can proceed as follows.

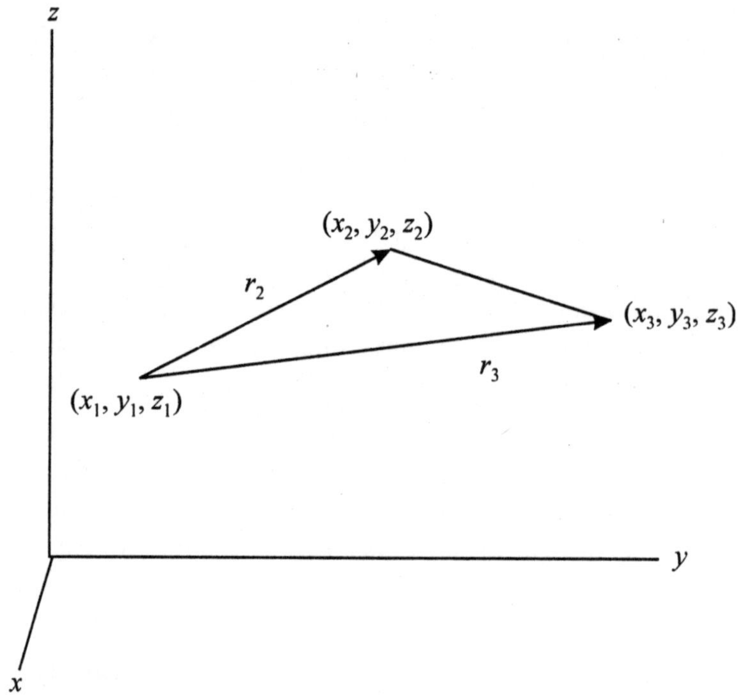

The vectors r_2 and r_3 can be written

$$r_2 = (x_2 - x_1)i + (y_2 - y_1)j + (z_2 - z_1)k$$
$$r_3 = (x_3 - x_1)i + (y_3 - y_1)j + (z_3 - z_1)k$$

where i, j, k are unit vectors parallel to the x-, y- and z- axes.

The cross product of r_2 and r_3 gives the (vector) area of the parallelogram of which they form two sides. The area A of the triangle is half of this, so that

$$\begin{aligned} 2A &= r_2 \times r_3 \\ &= [(y_2 - y_1)(z_3 - z_1) - (y_3 - y_1)(z_2 - z_1)]i \\ &+ [(z_2 - z_1)(x_3 - x_1) - (z_3 - z_1)(x_2 - x_1)]j \\ &+ [(x_2 - x_1)(y_3 - y_1) - (x_3 - x_1)(y_2 - y_1)]k \end{aligned}$$

The magnitude of this vector can be found in the usual way, to obtain

$$\begin{aligned} 4A^2 &= [(y_1 (z_2 - z_3) + y_2 (z_3 - z_1) + y_3 (z_1 - z_2)]^2 \\ &+ [(z_1 (x_2 - x_3) + z_2 (x_3 - x_1) + z_3 (x_1 - x_2)]^2 \\ &+ [(x_1(y_2 - y_3) + x_2 (y_3 - y_1) + x_3 (y_1 - y_2)]^2. \end{aligned}$$

The reader can verify that, if $z_1 = z_2 = z_3$, this reduces to equation for the area of a triangle the xy-plane. Equation can also be written

$$4A^2 = \begin{vmatrix} y_1 & y_2 & y_3 \\ z_1 & z_2 & z_3 \\ 1 & 1 & 1 \end{vmatrix}^2 + \begin{vmatrix} z_1 & z_2 & z_3 \\ x_1 & x_2 & x_3 \\ 1 & 1 & 1 \end{vmatrix}^2 + \begin{vmatrix} x_1 & x_2 & x_3 \\ y_1 & y_2 & y_3 \\ 1 & 1 & 1 \end{vmatrix}^2$$

This gives the area explicitly in terms of the coordinates of the vertices. If it is zero, the points are collinear.

The volume of a tetrahedron is $\frac{1}{6}$ × base × height. By combining equation for the area of a triangle with equation for the perpendicular distance of a point form a plane, we can determine that the volume of the tetrahedron whose vertices are

$$(x_0, y_0, z_0), (x_1, y_1, z_1), (x_2, y_2, z_2), (x_3, y_3, z_3)$$

is $$\frac{1}{6}\begin{vmatrix} x_1 & y_1 & z_1 & 1 \\ x_2 & y_2 & z_2 & 1 \\ x_3 & y_3 & z_3 & 1 \end{vmatrix}$$

If this determinant is zero, the four points are coplanar.

In three-dimensional coordinate geometry, a straight line is described by two equations, being the intersection of two planes:

$$a_1x + b_1x + c_1z = 1$$

$$a_2x + b_2x + c_2z = 1$$

If $a_1/a_2 = b_1/b_2 = c_1/c_2$, the normals to the two planes have the same direction ratios, so the planes are parallel and do not intersect. Otherwise the normals to the two planes have different direction ratios (a_1, b_1, c_1), (a_2, b_2, c_2), and, since the line of intersection of the planes is at right angles to both normals, the direction ratios of the line are found from the cross product of vectors normal to the planes. The direction ratios of the line of intersection are therefore

$$(b_1c_2 - b_2c_1, c_1a_2 - c_2a_1, a_1b_2 - a_2b_1)$$

The line crosses the *yz-*, *zx-* znd *xy-* planes at

$$y = \frac{c_1 - c_2}{b_1c_2 - b_2c_1} \qquad z = \frac{b_1 - b_2}{b_1c_2 - b_2c_1}$$

$$z = \frac{a_1 - a_2}{c_1a_2 - c_2a_1} \qquad x = \frac{c_1 - c_2}{c_1a_2 - c_2a_1}$$

$$x = \frac{b_1 - b_2}{a_1b_2 - a_2b_2} \qquad y = \frac{a_1 - a_2}{a_1b_2 - a_2b_1}$$

An example of computing a straight line from the intersection of two planes occurs in meteor astronomy. We can assume a flat Earth, which is tantamount to supposing that the height of a meteor is negligible compared with the radius of Earth, and the height of an observer above sea level is negligible compared with the height of the meteor. Since the heights of meteors are typically a few tens of km, both of these approximations are reasonable, at least for noninstrumental eyewitness accounts.

We suppose that, relative to an arbitrary origin O on the surface of Earth, a witness A is 15 km east and 5 km north of the origin. He sees a fireball start at an angle $\theta = 25°.5$ from his zenith and at an azimuth $\phi = 54°.5$ counterclockwise from his east, and it finishes at $\theta = 36°.7$, $\phi = 16°.7$.

Show that the plane containing the witness and the meteor is

$$0.0363x + 0.0911y - 0.0454z = 1$$

A second witness, 30 km east and 15 km north of O, estimates the zenith distance and azimuth of two points on the meteor track to be $\theta = 29°.6$, $\phi = 202°.9$ and $\theta = 33°.6$, $\phi = 242°.9$

Show that the plane containing this second witness and the meteor is

$$0.0257x + 0.0153y + 0.0168z = 1$$

These two equations describe the path of the fireball through the air. Show that, if the meteoroid carries on moving in a straight line, it will strike the ground as a meteorite 42.4 km east and 6.0 km south of the origin O.

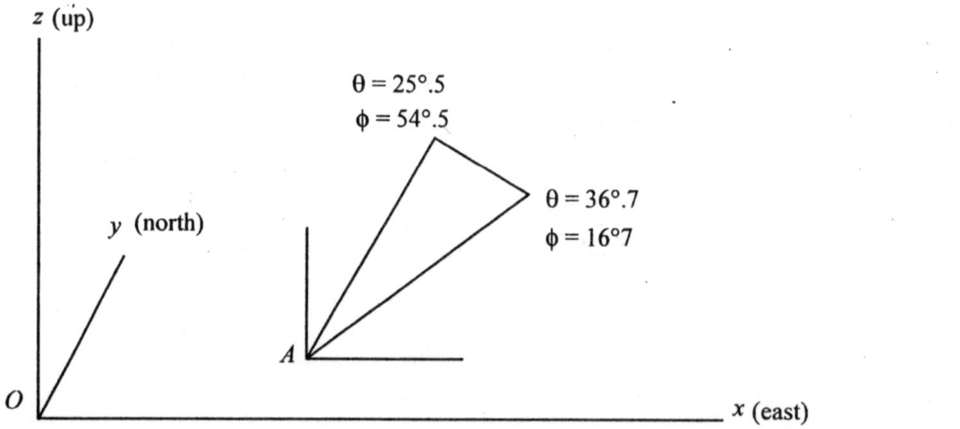

As we have just discussed, two nonparallel planes intersect in a straight line. Usually, three nonparallel planes intersect at a single unique point; for, if L is a line formed from the intersection of planes P_1 and P_2, L will usually intersect the plane P_3 at a point.

Example: The planes

$$2x + 3y + 4z - 9 = 0$$

$$x + y - 8z + 6 = 0$$
$$5x + 6y - 12z + 1 = 0$$

intersect at (1, 1, 1).

It will be recalled from the theory of linear equations that three equations

$$A_1x + B_1y + C_1z + D_1 = 0$$
$$A_2x + B_2y + C_2z + D_2 = 0$$
$$A_3x + B_3y + C_3z + D_3 = 0$$

have a unique solution cnly if

$$\Delta = \begin{vmatrix} A_1 & B_1 & C_1 \\ A_2 & B_2 & C_2 \\ A_3 & B_2 & C_3 \end{vmatrix} \neq 0$$

and, in the geometrical interpretation, this is the condition that three planes meet in a single point.

Consider, however, the three planes

$$2x + 3y + 4z - 9 = 0$$
$$x + y - 8z + 6 = 0$$
$$5x + 6y - 20z + 12 = 0$$

The direction ratios of the three lines found by combining the planes in pairs are

$$(-28, 20, -1)\ (-84, 60, -3)\ (28, -20, 1)$$

It will be observed that each is a multiple of either of the others, and the direction cosines of each of the three lines are identical apart from sign: ($\mp$ 0.813, $\pm$0.581, $\mp$ 0.029).

The three lines are, in fact, parallel, and the three planes enclose a prism. A condition for this is that

$$\Delta = 0.$$

But consider now the planes

$$2x + 3y + 4z - 9 = 0$$
$$x + y - 8z + 6 = 0$$
$$5x + 6y - 20z + 9 = 0$$

Not only does $\Delta = 0$, but also

$$\Delta' = \begin{vmatrix} A_1 & B_1 & D_1 \\ A_2 & B_2 & D_2 \\ A_3 & B_3 & D_3 \end{vmatrix} = 0$$

The three lines obtained by combining equations in pairs are in fact identical, and the three planes meet in a single line. Each of equations is a linear combination of the other two.

In summary, three nonparallel planes meet in a single line if $\Delta \neq 0$. They meet in a single point if $\Delta = \Delta' = 0$. They enclose a prism if $\Delta = 0$, $\Delta' \neq 0$.

The Ellipsoid

Consider the equation

$$\frac{x^2}{a^2}+\frac{z^2}{c^2}=1,$$

with $a > c$, in the *xz*-plane. The length of the semi major axis is a and the length of the semi minor axis is c. If this figure is rotated through 360° about its minor (z-) axis, the three- dimensional figure so obtained is called an *oblate spheroid*. The figure of the Earth is not exactly spherical; it approximates to a very slightly oblate spheroid, the ellipticity $(c - a)/a$ being only 0.00335. (The actual figure of the Earth, mean sea level, is often referred to as the *geoid*.)

The equation to the oblate spheroid referred to above is

$$\frac{x^2}{a^2}+\frac{y^2}{a^2}+\frac{z^2}{c^2}=1.$$

If the ellipse is rotated through 360° about its major (x-) axis, the figure so obtained is called a *prolate* spheroid. A rugby football (or, to a lesser extent, a North American football, which is a bit too pointed) is a good approximation to a prolate spheroid.

The equation to the prolate spheroid just described is

$$\frac{x^2}{a^2}+\frac{y^2}{c^2}+\frac{z^2}{c^2}=1.$$

Either type of spheroid can be referred to as an "ellipsoid of revolution".

The figure described by the equation

$$\frac{x^2}{a^2}+\frac{y^2}{b^2}+\frac{z^2}{c^2}=1$$

is a *tri-axial ellipsoid*. Unless stated otherwise, I shall adopt the convention $a > b > c$, and choose the coordinate axes such that the major, intermediate and minor axes are along the x-, y- and z-axes respectively. A tri-axial ellipsoid is not an ellipsoid of revolution; it cannot be obtained by rotating an ellipse about an axis.

The special case $a = b = c$:

$$x^2 + y^2 + z^2 = a^2$$

is, of course, a sphere.

Figure shows the cross-section of a tri-axial ellipse in the *yz*- plane (a), the *xz*-plane (b) and (twice - (c), (d)) the *xy*-plane. If you imagine your eye wandering in the *xz*-plane from the *x*-axis (a) to the *z*-axis (c), you will be convinced that there is a direction in the *xz*-plane from which the

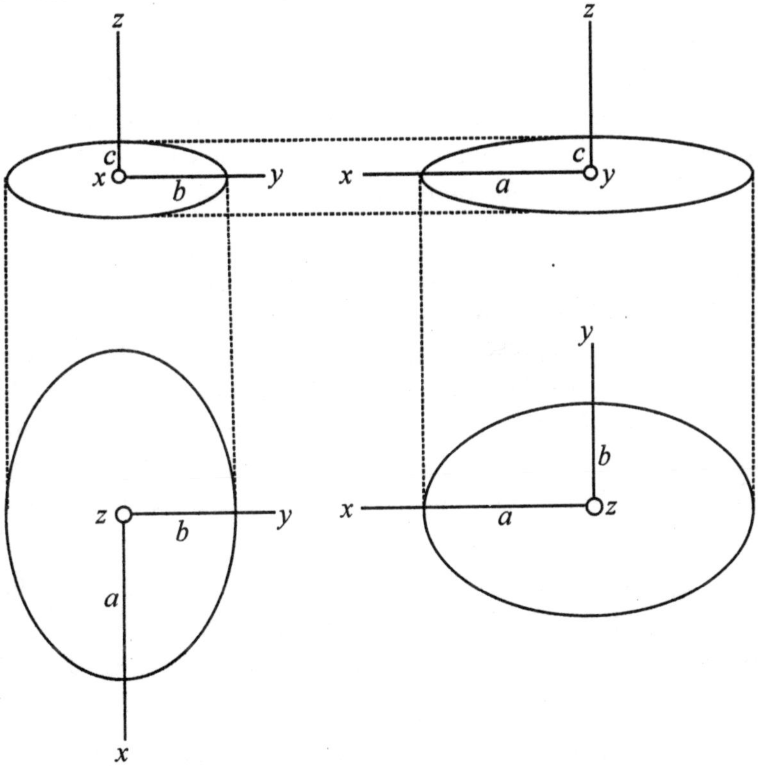

cross-section of the ellipse is a circle. There are actually two such directions, symmetrically situated on either side of the *z*-axis, but there are no such directions in either the *xy*- or the *yz*-planes from which the cross-section of the ellipsoid appears as a circle. Expressed otherwise, there are two planes that intersect the ellipsoid in a circle. This fact is of some importance in the description of the propagation of light in a bi-axial crystal, in which one of the wavefronts is a tri-axial ellipsoid.

Let us refer the ellipsoid to a set of axes $Ox'y'z'$ such that the angles z'Oz and $x'Ox$ are each θ, and the y'- and y-axes are identical. The equation of the ellipsoid referred to the new axes is (by making use of the usual formulas for the rotation of axes)

$$\frac{(z'\sin\theta + x'\cos\theta)^2}{a^2} + \frac{y^2}{b^2} + \frac{(z'\cos\theta - x'\sin\theta)^2}{c^2} = 1.$$

The cross-section of the ellipsoid in the $x'y'$-plane (i.e. normal to the z'-axis) is found by putting $z' = 0$:

$$\frac{(x'\cos\theta)^2}{a^2}+\frac{y^2}{b^2}+\frac{(x'\sin\theta)^2}{c^2}=1.$$

This is a circle if the coefficients of *x'* and *y'* are equal. Thus it is a circle if

$$\cos^2\theta=\frac{a^2(b^2-c^2}{b^2(a^2-c^2)}.$$

Thus, a plane whose normal is in the *xz*-plane (i.e. between the major and minor axis) and inclined at an angle θ to the minor (*z*-) axis, cuts the tri-axial ellipsoid in a circle. As viewed from either of these directions, the cross-section of the ellipsoid is a circle of radius *b*.

As an asteroid tumbles over and over, its brightness varies, for several reasons, such as its changing phase angle, the directional reflective properties of its regolith, and, of course, the cross-sectional area presented to the observer. The number of factors that affect the light-curve of a rotating asteroid is, in fact, so large that it is doubtful if it is possible, from the light-curve alone, to deduce with much credibility or accuracy the true shape of the asteroid. However, it is obviously of some interest for a start in any such investigation to be able to calculate the cross-sectional area of the ellipsoid as seen from some direction (θ, ϕ).

Let us erect a set of coordinate axes O*x'y'z'* such that O*z'* is in the direction (θ, ϕ), first by a rotation through ϕ about O*z* to form intermediate axes $Ox_1y_1z_1$, followed by a rotation through θ about Oy_1. The (*x', y', z'*) coordinates are related to the (*x, y, z*) coordinates by

$$\begin{pmatrix}x\\y\\z\end{pmatrix}=\begin{pmatrix}\cos\phi & -\sin\phi & 0\\ \sin\phi & \cos\phi & 0\\ 0 & 0 & 1\end{pmatrix}\begin{pmatrix}\cos\theta & 0 & \sin\theta\\ 0 & 1 & 0\\ -\sin\theta & 0 & \cos\theta\end{pmatrix}\begin{pmatrix}x'\\y'\\z'\end{pmatrix}$$

If we substitute for *x, y, z* in equation from equation , we obtain the equation to the ellipsoid referred to the O*x'y'z'* coordinate systems. And if we put *z'* = 0, we see the elliptical crosssection of the ellipsoid in the plane normal to O*z'*. This will be of the form

$$Ax'2 + 2Hx'y' + By'2 = 1,$$

where

$$A=\cos^2\theta\left(\frac{\cos^2\phi}{a^2}+\frac{\sin^2\phi}{b^2}\right)+\frac{\sin^2\theta}{c^2},$$

$$2H=2\cos^2\theta\sin\phi\cos\phi\left(\frac{1}{b^2}-\frac{1}{a^2}\right),$$

$$B = \frac{\sin^2\phi}{a^2} + \frac{\cos^2\phi}{b^2}.$$

This is an ellipse whose axes are inclined at an angle ψ from Ox' given by

$$\tan 2\psi = \frac{2H}{A-B}.$$

By replacing x' and y' by x'' and y''. where

$$\begin{pmatrix} x' \\ y' \end{pmatrix} = \begin{pmatrix} \cos\psi & -\sin\psi \\ \sin\psi & \cos\psi \end{pmatrix} \begin{pmatrix} x'' \\ y'' \end{pmatrix}$$

we shall be able to describe the ellipse in a coordinate system $Ox''y''$ whose axes are along the axes of the ellipse, and the equation will be of the form

$$\frac{x''^2}{a''^2} + \frac{y''^2}{b''^2} = 1$$

and the area of the cross-section is $\pi a''b''$.

For example, suppose the semi axes of the ellipsoid are $a = 3$, $b = 2$, $y = 1$, and we look at it from the direction $\theta = 60°$, $\phi = 45°$. Following equations we obtain for the equation of the elliptical cross-section referred to the system $Ox'y'z'$

$$0.79513\dot{8}\, x'^2 + 0.069\dot{4}\, x'y' + 0.180\dot{5}\, y'2 = 1.$$

From equation we find $\psi = 3°.22338$. Equation then transforms equation to

$$0.797094x''^2 + 0.178600y''^2 = 1$$

or

$$\frac{x''^2}{(1.1201)^2} + \frac{y''^2}{(2.3662)^2} = 1.$$

The area is $\pi \times 1.1202 \times 2.3663 = 8.362$.

The Paraboloid

The equation $x = 4qz = 2lz$ is a parabola in the xz-plane. The distance between vertex and focus is q, and the length of the semi latus rectum $l = 2q$. The equation can also be written

$$\frac{x^2}{a^2} = \frac{z}{h}$$

Here a and h are distances such that $x = a$ when $z = h$, and the length of the semi latus rectum is $l = a^2 /(2h)$.

If this parabola is rotated through 360o about the z-axis, the figure swept out is a *paraboloid of revolution*, or *circular paraboloid*. Many telescope mirrors are of this shape. The equation to the circular paraboloid is

$$\frac{x^2}{a^2} + \frac{y^2}{a^2} = \frac{z}{h}.$$

The cross-section at $z = h$ is a circle of radius a.

The equation

$$\frac{x^2}{a^2} + \frac{y^2}{a^2} = \frac{z}{h},$$

in which we shall choose the x- and y-axes such that $a > b$, is an elliptic paraboloid and, if $a \neq b$, is not formed by rotation of a parabola. At $z = h$, the cross section is an ellipse of semi major and minor axes equal to a and b respectively. The section in the plane $y = 0$ is a parabola of semi latus rectum $a^2 /(2h)$. The section in the plane $x = 0$ is a parabola of semi latus rectum $b^2 /(2h)$. The elliptic paraboloid lies entirely above the xy-plane.

The equation

$$\frac{x^2}{a^2} - \frac{y^2}{b^2} = \frac{z}{h}$$

is a hyperbolic paraboloid, and its shape is not quite so easily visualized. Unlike the elliptic paraboloid, it extends above and below the plane. It is a saddle-shaped surface, with the saddle point at the origin. The section in the plane $y = 0$ is the "nose down" parabola $x^2 = a^2 z/h$ extending above the xy-plane. The section in the plane $x = 0$ is the "nose up" parabola $y^2 = -b^2 z/h$ extending below the xy-plane. The section in the plane $z = h$ is the hyperbola

$$\frac{x^2}{a^2} - \frac{y^2}{b^2} = 1.$$

The section with the plane $z = -h$ is the conjugate hyperbola

$$\frac{x^2}{a^2} - \frac{y^2}{b^2} = -1.$$

The section with the plane $z = 0$ is the asymptotes

$$\frac{x^2}{a^2} - \frac{y^2}{b^2} = 0.$$

The surface for $a = 3$, $b = 2$, $h = 1$ is drawn in figure.

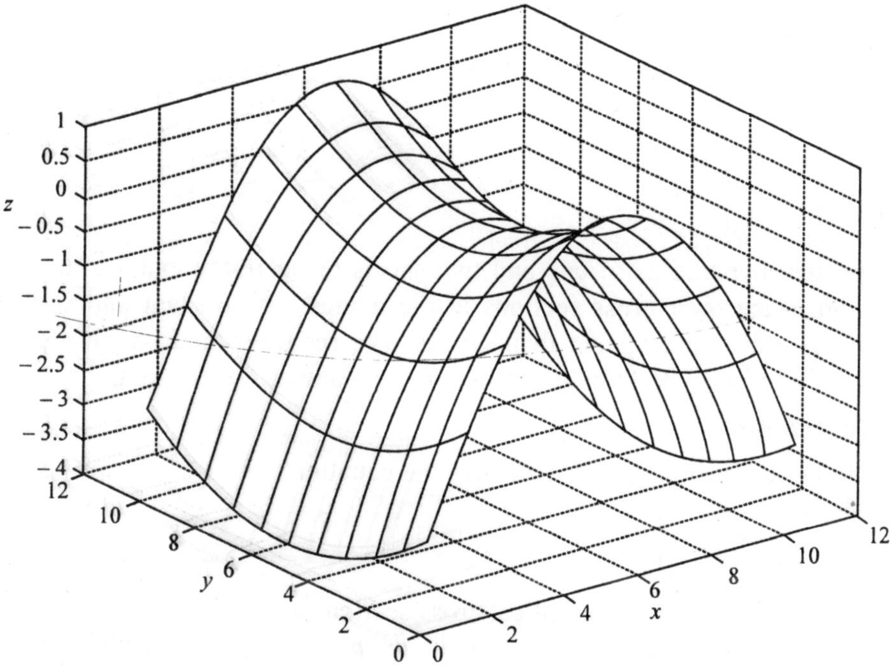

The Hyperboloid

The equation

$$\frac{x^2}{a^2} - \frac{z^2}{c^2} = 1$$

is a hyperbola, and a is the semi transverse axis.

If this figure is rotated about the z-axis through 360°, the surface swept out is a *circular hyperboloid* (or *hyperboloid of revolution*) of one sheet. Its equation is

$$\frac{x^2}{a^2} + \frac{y^2}{a^2} - \frac{z^2}{c^2} = 1.$$

Imagine two horizontal rings, one underneath the other. The upper one is fixed. The lower one is suspended from the upper one by a large number of vertical strings attached to points equally spaced around the circumference of each ring. Now twist the lower one through a few degrees about a vertical axis, so that the strings are no longer quite vertical, and the lower ring rises slightly. These strings are generators of a circular hyperboloid of one sheet.

If this figure is rotated about the z-axis through 360°, the surface swept out is a *circular hyperboloid* (or *hyperboloid of revolution*) *of two sheets*. Its equation is

$$\frac{x^2}{a^2} - \frac{y^2}{c^2} - \frac{z^2}{c^2} = 1.$$

The equations

$$\frac{x^2}{a^2} + \frac{y^2}{c^2} - \frac{z^2}{c^2} = 1$$

and

$$\frac{x^2}{a^2} - \frac{y^2}{c^2} - \frac{z^2}{c^2} = 1$$

represent hyperbolas of one and two sheets respectively, but are not hyperbolas of revolution, since their cross sections in the planes z = constant and x = constant > a respectively are ellipses rather than circles. The reader should imagine what the cross-sections of all four hyperboloids are like in the planes $x = 0$, $y = 0$ and $z = 0$.

The Cylinder

In three-dimensional solid geometry the equation

$$\frac{x^2}{a^2} + \frac{y^3}{b^2} = 1$$

represents a cylinder of elliptical cross-section, whose axis coincides with the z-axis. The equation to a cylinder with an axis in another position and with another orientation can be obtained by the usual processes of translation and rotation of axes.

The Cone

The equation

$$x^2 + y^2 = a^2z^2$$

represents a circular cone whose vertex in at the origin and whose axis coincides with the z-axis. The semi-vertical angle α of the cone is given by

$$\alpha = \tan^{-1} a.$$

In this context, the word "vertical" has nothing to do with "upright"; it merely means "of or at the vertex". A little knowledge of Latin is useful even today!

The equation

$$\frac{x^2}{a^2} + \frac{y^3}{b^2} = z^2$$

is a cone with elliptical cross-section.

If the vertex of the cone remains at the origin, but the vertex is in some arbitrary direction (described, for example, by the direction cosines, or by spherical angles θ and ϕ) the equation can be derived by a rotation of coordinate axes. This will introduce terms in *yz, zx* and *xy*, but it will not produce any terms in *x, y* or z, nor will it introduce a constant. Therefore the equation to such a cone will have only second degree terms in it. The equation will be of the form

$$ax^2 + by^2 + cz^2 + 2fyz + 2gzx + 2hxy = 0$$

With one proviso, the converse is usually true, namely that equation represents a cone with vertex at the origin. The one proviso is that if

$$\begin{vmatrix} a & h & g \\ h & b & f \\ g & f & c \end{vmatrix} = 0$$

equation will factorize into two linear expressions, and will represent two planes, which intersect in a line that contains the origin - and this could be regarded as a special case of a cone, with zero vertical angle. If, however, the two linear factors are identical, the two planes are coincident.

The equation to a cone of semi vertical angle α whose vertex is at the origin and whose axis has direction cosines (l, m, n) can be found as follows. The equation

$$lx + my + nz + h = 0$$

represents a plane that is perpendicular to the axis of the cone and is at a distance h from the origin (i.e. from the vertex of the cone). Let P (x, y, z) be a point in the plane and also on the surface of the cone, at a distance r from the origin. The semi vertical angle of the cone is then given by

$$\cos \alpha = h / r.$$

But $$r^2 = x^2 + y^2 + z^2$$

and, from equation,

$$h^2 = (lx + my + nz)^2$$

Thus $$(x^2 + y^2 + z^2)\cos^2 \alpha = (lx + my + nz)^2$$

is the required equation to the cone.

It might be thought that the equation to an inclined cone is unlikely to find much application in astronomy. Here is an application.

A fireball (i.e. a bright meteor, potentially capable of depositing a meteorite) moves down through the atmosphere with speed V along a straight line trajectory with direction cosines (l, m, n) referred to some coordinate system whose *xy*-plane is on the surface of Earth (assumed flat). If v is the speed of sound, and $V > v$, the meteoroid will generate a conical shock front of semi vertical angle α given by

$$\sin \alpha = v / V.$$

At a time t before impact, the coordinates of the vertex will be (lVt, mVt, nVt), and the equation to the conical shock front will then be

$$[(x + lVt)^2 + (y + mVt)^2 + (z + nVt)^2] \cos^2 a = [l(x + lVt) + m(y + mVt) + n(z + nVt)]^2$$

Part of this shock front (at time t before impact) has already reached ground level, and it intersects the ground in a conic section given by putting $z = 0$ in equation:

$$(\cos^2 \alpha - l^2)x^2 + (\cos^2 \alpha - m^2)y^2 - 2lmxy - 2Vt \sin^2 \alpha\,(lx + my) - V^2t^2 \sin^2 \alpha = 0$$

and witnesses on the ground on any point on this conic section will hear the shock front at the same time.

The General Second Degree Equation in Three Dimensions

The general second degree equation in three dimensions is

$$ax^2 + by^2 + cz + 2fyz + 2gzx + 2hxy + 2ux + 2vy + 2wz + d = 0$$

This may represent a plane or pair of planes (which, if not parallel, define a straight line), or an ellipsoid, paraboloid, hyperboloid, cylinder or cone. The equation could, if convenient, be divided through by d (or any of the other constants), and there are in reality only nine independent constants. Therefore nine points in space are sufficient to determine the second degree surface on this they lie.

If d is zero, the surface contains the origin. If u, v and w are all zero, and the surface is an ellipsoid, hyperbolic paraboloid or a hyperboloid, the origin is at the centre of the figure. If the figure is an elliptic paraboloid, the origin is at the vertex. If u, v, w and d are all zero, the surface is a cone with the proviso mentioned. If a, b, c, f, g, h are all zero, the surface is a plane.

Let us consider a particular example:

$$3x^2 - 4y^2 + 6z^2 + 8yz - 2zx + 4xy + 14x - 10y - 4z + 5 = 0$$

What sort of a surface is this ?

We need to do two things. First we need to rotate the coordinate axes so that they are parallel to the figure axes. The equation referred to the figure axes will have no terms in yz, zx or xy. Then we need to translate the axes so that the origin is at the centre of the figure (or at the vertex, if it is an elliptical paraboloid).

Mathematically, we need to find the eigenvectors of the matrix

$$\begin{vmatrix} a & h & g \\ h & b & f \\ g & f & c \end{vmatrix} = \begin{vmatrix} 3 & 2 & -1 \\ 2 & -4 & 4 \\ 1 & 4 & 6 \end{vmatrix}$$

Some readers will readily know how to do this. Others may not, and may not even be quite certain what an eigenvector is. In any case, the eigenvectors are found to be

$$\begin{pmatrix} l_{11} \\ l_{12} \\ l_{13} \end{pmatrix} = \begin{pmatrix} -0.0695481 \\ +0.3188310 \\ +0.9452565 \end{pmatrix} \quad \begin{pmatrix} l_{12} \\ l_{22} \\ l_{32} \end{pmatrix} = \begin{pmatrix} -0.2406405 \\ +0.9142071 \\ +0.3260635 \end{pmatrix} \quad \begin{pmatrix} l_{13} \\ l_{23} \\ l_{33} \end{pmatrix} = \begin{pmatrix} -0.9681194 \\ -0.2501441 \\ +0.0131423 \end{pmatrix}$$

with corresponding eigenvalues 7.422 7590, 5.953 0969, 3.530 3380.

The elements of the eigenvectors are the direction cosines of the present coordinate axis with respect to the figure axes. To express the equation to the surface relative to coordinate axes that are parallel to the figure axes, we replace

x by $l_{11}x + l_{12}y + l_{13}z$

y by $l_{21}x + l_{22}y + l_{23}z$

z by $l_{31}x + l_{32}y + l_{33}z$

This will make the terms in yz, zx and xy vanish; this should be checked numerically, particularly as it is easy to rotate the axes in the wrong sense. When the substitutions are made, the equation is found to be

$$7.4227590x^2 - 5.9530969y^2 + 3.5303380z^2 - 7.9430994x - 11.2067840y$$
$$- 11.1047998z + 5 = 0.$$

Notice that there are now no terms in yz, zx or xy.

Now we need to translate the origin of coordinates to the centre of the figure (or to the vertex if it is an elliptic paraboloid). It will readily be seen that this can be done by substituting

$x - \alpha$ for x

$y - \beta$ for y

$z - \gamma$ for z

where α = (coefficient of x)/(twice the coefficient of x^2) = – 0.535 050 336

β = (coefficient of y)/(twice the coefficient of y^2) = + 0.941 256 643

γ = (coefficient of z)/(twice the coefficient of z^2) = – 1.572 767 224.

The equation then becomes

$$7.422\,7590\, x^2 - 5.953\,0969\, y^2 + 3.530\,3380\, z^2 - 0.583\,3816 = 0$$

or

$$\frac{x^2}{(0.280346)^2} - \frac{y^2}{(0.313044)^2} + \frac{z^2}{(0.406507)^2} = 1.$$

The surface is a hyperboloid of one sheet, elliptical in any y = constant cross-section.

The surfaces described by second-degree equations in three dimensions - ellipsoids, paraboloids, hyperboloids, cones and cylinders - are generally called quadric surfaces. The surface described by the equation

$$\left[(x^2+y^2)^{\frac{1}{2}}-a\right]^2+z^2=b^2, b<a$$

is not one of the quadric surfaces. If the square root is isolated and squared, the resulting equation will contain terms of degree four. The surface is a fairly familiar one, and the reader should try to imagine what it is. Failing that, if your computer skills are up to it, you might try to draw the surface in three-dimensional space. The only hint I give is to suggest that you put $y = 0$ in equation to see what the section is in the z-x plane.

Matrices

This book has assumed a knowledge of matrices, which may or may not be justified. In this section, we do not attempt a thorough treatment of the subject, which must be sought elsewhere, but the remarks may be of use both to novices and to the more experienced.

However you may have the opportunity of learning about the manipulation of matrices, it is suggested that you should aim to understand and know how to carry out at least the following operations on matrices whose elements are real numbers:

Multiply a vector by a matrix

Multiply a matrix by a matrix

Calculate the determinant of a square matrix

Invert a square matrix

Diagonalize a symmetric matrix

Test a matrix for orthogonality

Numerous other aspects of matrix manipulation are possible, and the subject expands greatly if we allow the elements to be complex numbers. These six, however, are particularly useful for many applications. It might be mentioned, however, that solving simultaneous linear equations by Kramers' Rule or by inverting a matrix are very inefficient ways of solving such equations, and that is not the main purpose of acquiring the above skills.

Most, or doubtless all, of the above operations are available in many modern mathematical computer packages. This is not what I mean, however, by "understand and know how to carry out". The student should carry out at least once by hand calculator, step by step, each of the above operations, and, at each step, try to understand not only the algebraic and arithmetic steps, but also try to visualize the geometric interpretation, particularly when rotating axes and calculating eigenvectors. After doing a hand calculation, you should then write a series of short computer programs (rather than one vast, all-encompassing matrix package) for each operation, so that when, in future, you need to do any of these things, you can instantly obtain the answer without having to go through tedious calculations. For example, in the previous section, when I needed the eigenvectors of the matrix, I was able to generate them with a single word "EIGEN" on a computer; a considerable amount of arithmetic was actually performed by the computer.

On the question of testing a matrix for orthogonality, the usual application in mechanical and geometrical problems is to test a matrix of direction cosines. The tests can not only detect mistakes, but it can locate them and even suggest what the correct element should be. Tests for orthogonality are as follows. The student should try to think of the geometric interpretation of each.

The sum of the squares of the elements in any row or any column is unity. (This test does not guard against mistakes in signs of the elements.)

The sum of the products of corresponding elements in any two rows or in any two columns is zero.

Every element is equal to plus or minus its cofactor.

The determinant of the matrix is plus or minus one.

A minus sign in the last two tests indicates that the two sets of axes differ in chirality (handedness). This usually does not matter, and can easily be dealt with by reversing the signs of the elements of one eigenvector and of its corresponding eigenvalue.

CONIC SECTIONS AND THEIR APPLICATIONSN

PLANE SECTIONS OF A CONE

In this chapter I will discuss what the intersection of a plane with a right circular cone looks like. A number of topics discussed will be of use in drawing figures associated with such intersections.

I begin with a few elementary topics which arise. Many of these are elementary and perhaps tedious, but if you try to draw the various figures accurately you will need them (or possibly simple variations).

SOLVING QUADRATIC EQUATIONS NUMERICALLY

The first topic involves solving quadratic equations in the presence of limited precision in calculations. Suppose we know that

$$ax^2 + bx + c = 0$$

has two real roots, which we want to compute. The quadratic formula gives

$$\lambda = \frac{-b \pm \sqrt{b^2 - 4ac}}{2a} = \left(\frac{b}{2a}\right)\left(-1 \pm \sqrt{1 - \frac{4ac}{b^2}}\right)$$

If $4ac/b^2$ is small then one of these involves the subtraction of two positive numbers about the same size, and a large rounding error will result. This is called a **cancellation** error. To avoid it, we first observe that since

$$ax^2 + bx + c = a(x^2 + (b = a)x + (c = a)) = a(x - \lambda_1)(x - \lambda_2)$$

the product of the two roots is c/a. Therefore we set in sequence

$$B = b/2a$$
$$D = 4ac/b^2$$
$$= c/aB^2$$
$$\lambda = -1 - \sqrt{1 - D}$$
$$\lambda_1 = B\ell$$
$$\lambda_2 = BD/\ell$$

PLANES AND PROJECTIONS

I recall here some elementary formulas involving projections in 3D. The basic formula everything else depends on solves the following problem: Given vectors u and v in 3D, what is the perpendicular projection of u onto the line along v? The answer is

$$u_0 = \left(\frac{u.v}{v.v}\right)v$$

as is usually explained when dot products are introduced.

Suppose now we are given a plane

$$Ax + By + Cz = D.$$

Recall that the geometrical significance of the coefficient vector = (A, B, C) is that it is perpendicular to the plane. A unique multiple of will actually lie on the plane. It is 0 = $t(A, B, C)$ where

$$A(tA) + B(tB) + C(tC) = D,\ t = \frac{D}{A^2 + B^2 + C^2} = \frac{D}{\alpha.\alpha}.$$

Suppose $v = (x, y, z)$ given. What is the perpendicular projection $v?$ of v onto the plane? Let v_0 be the projection of the vector v onto the line along . Then the picture shows that

$$v_\perp = v - (v_0 - \alpha_0) = v - \frac{v.\alpha - D}{\alpha.\alpha}\alpha\ .$$

The distance from v to the plane is therefore

$$\frac{v.\alpha - D}{\|\alpha\|}.$$

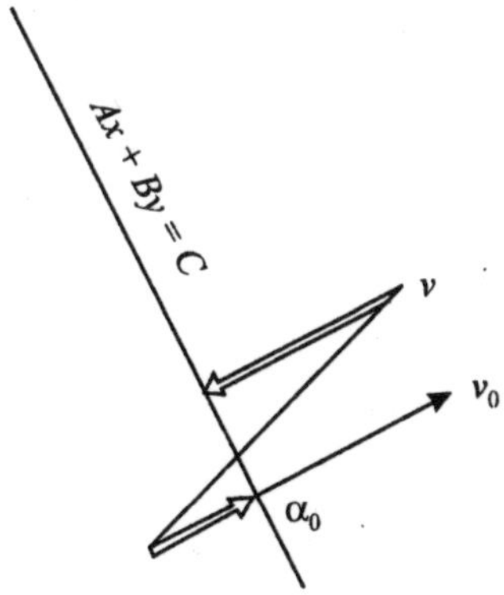

ANGULAR BISECTORS

A circle is said to be inscribed in a triangle if it lies completely inside the triangle and touches all three sides.

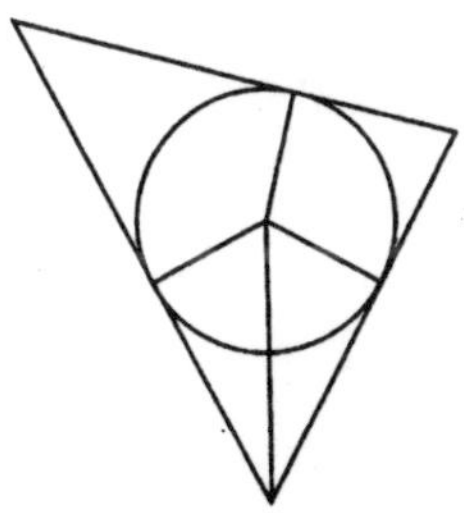

In this section we want to find (1) the centre of this circle, (2) its radius, and (3) the points of contact with the sides. We will assume that the triangle is oriented so that the origin is one vertex and the positive *y*axis bisects one of its angles. Let θ be half this vertex angle. We shall also assume that the side opposite the origin is given by an equation $Ax + By = C$.

The centre of the circle lies on the *y*-axis. Let it be $P = (0, y)$. Its distance from the left and right sides is thus guaranteed to be the same. This common distance is $y \sin \theta$. Its distance from the other side is

$$\frac{|By - C|}{\sqrt{A^2 + B^2}}$$

so we must solve the equation

$$y \sin \theta = \frac{|By - C|}{\sqrt{A^2 + B^2}}.$$

We do this by squaring both sides and rearranging to get

$$y^2 \sin^2 (A^2 + B^2) = (By - C)^2$$

leading to the quadratic equation

$$(B^2 - (A^2 + B^2) \sin^2 \theta)\, y^2 - 2BCy + C^2 = 0.$$

Exercise: The number y is one of the two roots of this equation. What is the other root? Which is the larger root? What does it mean if one root is negative?

The coordinates of the two contact points on the left and right sides are $(\pm y \sin \theta \cos \theta,\ y - y \sin^2 \theta)$. According to the discussion of projections, the third contact point is

$$(0, y) - \left(\frac{By - C}{A^2 + B^2} \right)(A, B)$$

DRAWING CONES

We start with a standard cone whose vertex angle is 2θ.

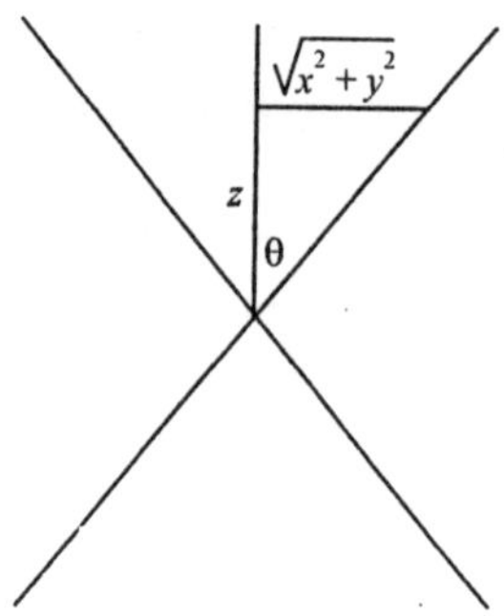

Its equation is

$$x^2 + y^2 = z^2 \tan^2 \theta.$$

We shall want to draw it in projection after a rigid transformation has been applied to it. There are several ways to draw a cone in 3D, but in these notes I shall draw only its outline, which consists of a pair of straight lines. The first step in drawing the transformed cone is to find the transformed axis. This is simple enough to calculate. By means of a 2D rotation, we can in effect assume it to be the y-axis.

The only extra piece of information needed to calculate the cone's outline is the angle its axis makes with respect to the (x, y) plane. Call it . Here is the picture we get for different values of α in effect rotating the top of the cone towards us.

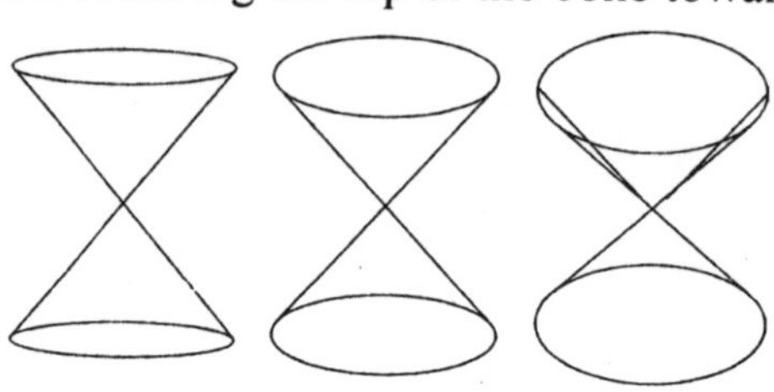

The point is that the front of the cone obscures its middle, even in projection. In effect we confront the following problem: Given an ellipse of axes a and b with centre at $(x, 0)$, what are the tangents from the origin to this ellipse? This is easily answered by scaling the ellipse vertically so as to turn it into a circle. The angle φ has

$$\sin \varphi = a/x$$

where $\quad r = a/x.$

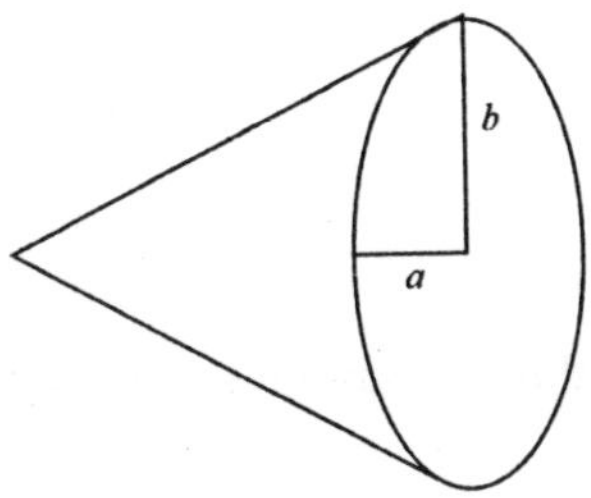

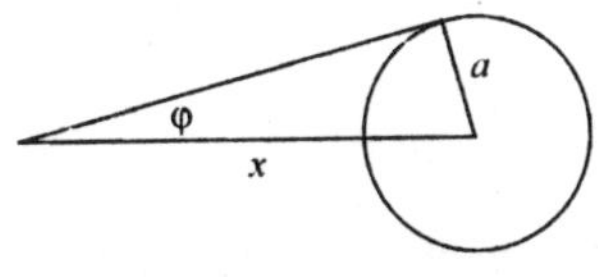

Note that this same angle φ is that made by the radius to the point of contact with the vertical, and is thus essentially the angular parameter of the corresponding point on the original ellipse The exact point of contact on the original ellipse is therefore

$$(x - a \sin \varphi, b \cos \varphi).$$

CONIC INTERSECTIONS

The intersection of any cone and a plane is always an ellipse, a parabola, or an hyperbola. It is simple to parametrize it, and not too difficult to tell exactly what its location and dimensions are (when the cone is rightcircular). Namely, we can essentially parametrize it as we do a circle. Recall that as t varies from 0 to 2π the point $(\cos t, \sin t)$ runs around the unit circle in the (x, y) plane. Therefore the point

$$(\tau \cos t, \tau \sin t, 1)$$

runs around the intersection of our cone with the plane $z = 1$, where $\tau = \tan \theta$. If the intersection is an ellipse, there will be for every point on this circle a unique point on the same line through the origin and also on the plane, of the form

$$k(\tau \cos t, \tau \sin t, 1).$$

If the plane is $Ax + By + Cz = D$ then we must have

$$k(A\tau \cos t + B\tau \sin t + C) = D, \frac{D}{A\tau\cos t + b\tau\sin t + C}.$$

If the intersection is not an ellipse then this parametrization will fail when the denominator is 0. This makes sense, since a circle cannot possibly parametrize an hyperbola or parabola. It is instructive to look carefully at the denominator to see exactly when it vanishes. If

$$A\tau \cos t + B\tau \sin t + C = 0$$

this means that the point ($\tau \cos t$, $\tau \sin t$) lies on the planar line

$$Ax + By + C = 0.$$

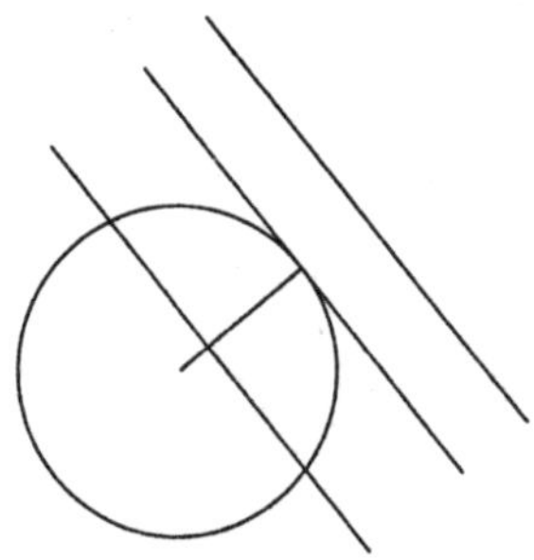

There are three possibilities as to exactly what the intersection of a circle and a straight line can be::

- There is no intersection.
- The line is tangent to the circle, and there is one point of intersection.
- There are two points of intersection.

How do we tell which case occurs? Recall that the geometrical significance of the coefficients A, B, C is that the direction perpendicular to the line is (A, B) and the signed distance from the origin to the line is

$$\frac{-C}{\sqrt{A^2+B^2}}.$$

The line will therefore intersect the circle of radius τ only when

$$\frac{|C|}{\sqrt{A^2+B^2}} \leq \tau$$

or

$$C^2 \text{ £ } \tau^2(A^2 + B^2).$$

It will intersect in exactly one point when this is an exact equality. Furthermore we can see how to find the intersection geometrically. The two points of intersection must be symmetrically placed with respect to the perpendicular direction (A, B). Multiplying (A, B) by $\tau = C$, we may as well assume that the direction vector (A, B) just touches the circle. The angle between this direction and the angle towards the intersection points satisfies

$$\cos\alpha = \frac{-C}{\sqrt{A^2+B^2}}.$$

We can now see that if the intersection is an ellipse we parametrize it by the whole circle; if it is a parabola we parametrize it by the circle minus a single point; and if an hyperbola we parametrize each piece of the hyperbola by one of the arcs we get by removing the two points of intersection with the line $Ax + By + C = 0$.

FOCAL POINTS

Certain points called the focal points play an important role in the theory of conic sections. It was only with Kepler's discovery of his first law of planetary motion that they became of general interest, but they were in fact discovered first for purely mathematical reasons by the Greeks, and in effect allowed the Greeks to formulate the algebraic equations describing conic sections long before coordinate systems were introduced in full generality.

We shall look only at ellipses. Here the fundamental fact is that one can locate two points F_1 and F_2 inside the ellipse with the following simple property:

If a line is drawn from one focus to the periphery of the ellipse and then to the other focus, the total length of the path is independent of which point on the periphery we move to.

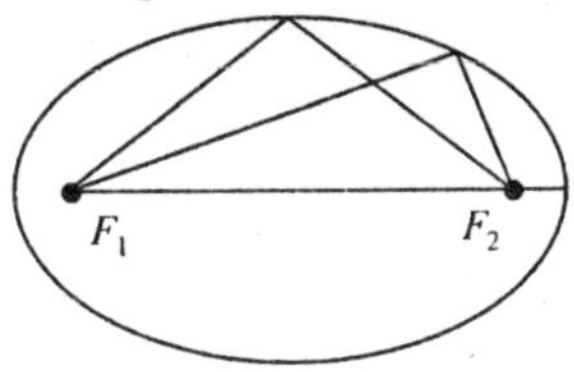

Before we begin to think about why this is true, we must locate the points $F1$ and $F2$. We can do this if we apply the characteristic property to just two paths. If a and b are the semimajor axes of the ellipse, then its equation is

$$\left(\frac{x}{a}\right)^2 + \left(\frac{y}{b}\right)^2 = 1.$$

If $F_1 = (-f, 0)$ then $F_2 = (f, 0)$. We first consider the path from F_2 to the right hand end extremity of the ellipse, then back to F_1. Its length is $(a - f) + a + f = 2a$. On the other and we consider the path from F_2 to the top point $(0, b)$ and then to F_1. Its length must also be $2a$, and by Pythagoras' Theorem

$$f = \sqrt{a^2 - b^2}.$$

We shall see two proofs of the assertion above about the equality of path lengths. The first is algebraic. Choose a point $P = (x, y)$ on the ellipse. We can calculate the distances from F_1 to P and F_2 to P and add them them up to see whether or not we get $2a$. I leave the details as an exercise – at one point a bit of tricky algebra is required. This proof is straightforward, but not very illuminating.

DANDELIN'S SPHERES

The proof of the focai property of ellipses that is in many ways the most satisfying was not discovered until the early nineteenth century, although the property itself had been known for more than 2000 years before that. The discoverer was a Belgian mathematician named Dandelin, who is relatively unknown. His proof starts with the construction of the ellipse as a conic section. So we choose a right circular cone and slice it with a plane which is horizontal enough to produce an ellipse. Dandelin's proof then inserts into the cone, above and below the plane, the largest spheres which will fit and at the same time touch the plane. We may choose our coordinates so that the axis of the cone is the zaxis, and the equation of the plane is

$$Ax + Cz = D.$$

Exercise: If the original equation is $Ax + By + Cz = D$, what change of coordinates do we have to make to get the new $B = 0$? What is the value of the new A?

If we then look sideways at the configuration (projecting along the yaxis), we see this picture:

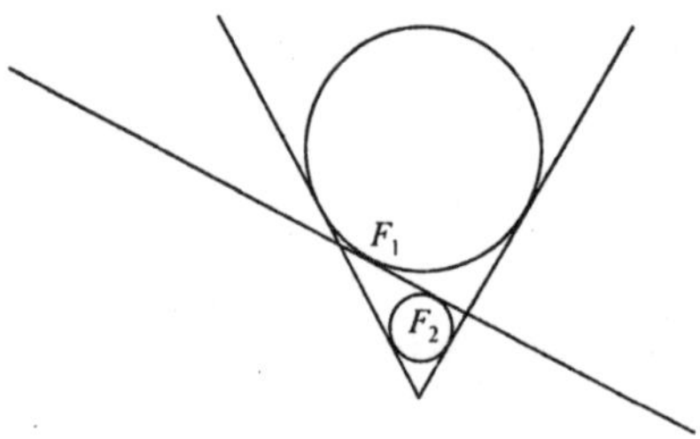

Each sphere will touch the plane at a single point. The points of contact of the spheres with the plane are defined in Dandelin's proof to be the focal points of the ellipse. Of course they have to be inside the cone and hence inside the ellipse. It is more or less evident from a symmetry argument that the curve of intersection has an axis of symmetry and that the focal points lie on it.

Here is the threedimensional view of Dandelin's construction of the foci:

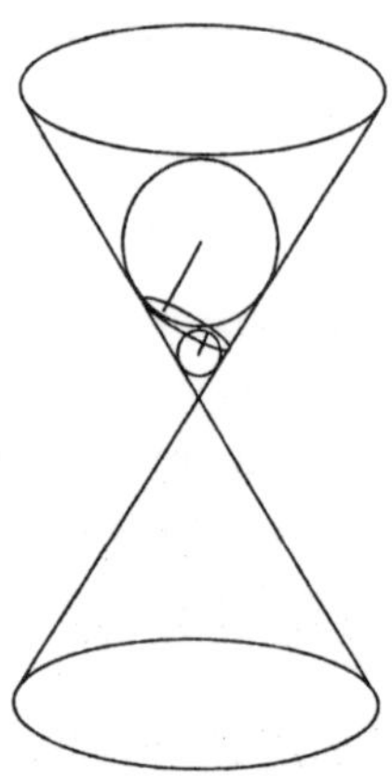

And here is the diagram to be used in the proof of the focal property:

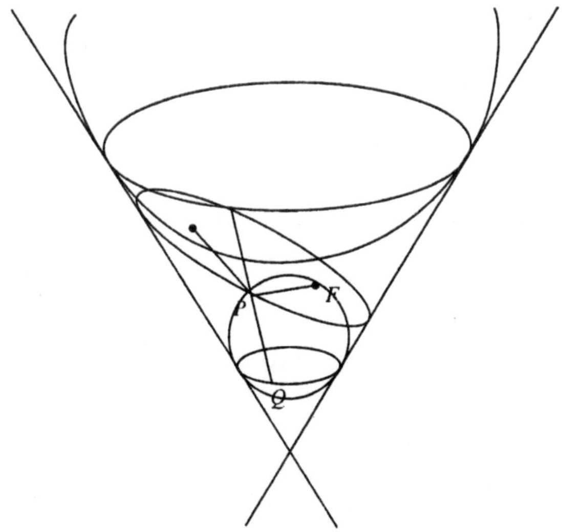

The proof now runs like this: Pick one of the spheres, say the smaller one. Let F be the focus touching that sphere. and let P be a point on the intersection curve. Draw a line segment from F to P, and and another along the surface of the cone from P to the circle of contact with the sphere, say hitting at Q. The two lines PF and PQ are both tangent to the sphere, and must be the same length. Therefore the distance from F to P is the same as the distance from P to the circle of contact. Repeat this argument for the other sphere. We see that the total distance from F_1 to P to F_2 is the same as the distance along the cone between the two circles of contact, and therefore does not depend on P.

THE DIRECTRIX

If E is an ellipse with semi-major axes a and b, the eccentricity of the curve is defined to be the ratio

$$e = \frac{\sqrt{a^2 - b^2}}{a}.$$

This is also the ratio of the focal distance f to a. If $e = 0$ the curve is a circle, and the closer e gets to 1 the flatter the ellipse is. The eccentricity can also be defined for other conic sections.

For each conic section and choice of focus F there exists a unique line ℓ, called the directrix determined by these data, with the following property: if P is any point on the curve then

$$\frac{d(P,F)}{d(P,\ell)} = e.$$

It is simple enough to figure out where the directrix of an ellipse has to be if it is to have this property, and it is straightforward to calculate that this line does indeed have this property.

Exercise: For the ellipse

$$\left(\frac{x}{a}\right)^2+\left(\frac{y}{b}\right)^2=1$$

tell where its directrix has to be. Prove the focal-directrix property for it.

The calculation is not too instructive. More satisfying is that we can use Dandelin's construction to give a geometrical proof of it.

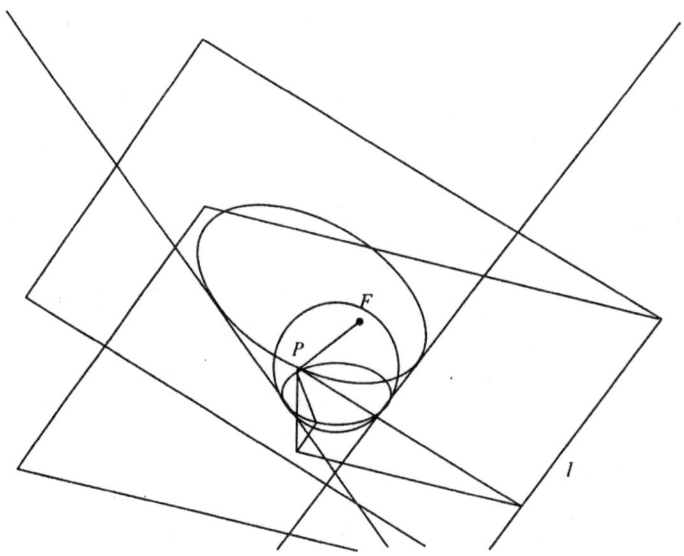

Exercise: Use the picture above to explain why $d(P, F)/d(P, \ell)$ is independent of P.

Chapter 9

Co-ordinate System in the Plane

In mathematics, a plane is a two-dimensional manifold or surface that is perfectly flat. Informally, it can be thought of as an infinitely vast and infinitesimally thin sheet oriented in some space. Formally, it is an affine space of dimension two.

When working in two-dimensional Euclidean space, the definite article is used, the plane, to refer to the whole space. Many fundamental tasks in geometry, trigonometry, and graphing are performed in two-dimensional space, or in other words, in the plane. A lot of mathematics can be and has been performed in the plane, notably in the areas of geometry, trigonometry, graph theory and graphing. All two-dimensional figures are assumed to be on a plane, even on *the* plane, unless otherwise specified.

When we talk about "The point with coordinates (x, y)" or "The curve with equation $f(x,y)$", we will always have in mind cartesian coordinates. If a formula involves another type of coordinates, this will be stated explicitly.

SUBSTITUTIONS AND TRANSFORMATIONS

Formulas for changes in coordinate systems can lead to confusion because (for example) moving the coordinate axes up has the same effect on equations as moving objects down while the axes stay fixed. To avoid confusion, we will carefully distinguish between transformations of the plane and substitutions, as explained below. Similar considerations will apply to transformations and substitutions in three dimensions.

SUBSTITUTIONS

A substitution, or change of coordinates, relates the coordinates of a point in one coordinate system to those of *the same point in a different coordinate system*. Usually one coordinate system has the superscript ' and the other does not, and we write

$$\begin{cases} x = F_x(x',y'), \\ y = F_y(x',y') \end{cases} \text{ or } (x, y) = F(x',y').$$

This means: given the equation of an object in the unprimed coordinate system, one obtains the equation of the same object in the primed coordinate system by substituting $Fx(x',y')$ for x and $Fy(x',y')$ for y in the equation. For instance, suppose the primed coordinate system is obtained from the unprimed system by moving the axes up a distance d. Then $x = x'$ and $y = y' + d$. The circle with equation $x^2 + y^2 = 1$ in the unprimed system has equation $x'^2 + (y'+d)^2 = 1$ in the primed system. Thus transforming an implicit equation in (x,y) into one in (x',y') is immediate. diate.

The point $P = (a,b)$ in the unprimed system, with equation $x = a$, $y = b$, has equation $Fx(x',y') = a$, $Fy(x',y') = b$ in the new system. To get the primed coordinates explicitly one must solve for x' and y' (in the example just given we have $x' = a$, $y' + d = b$, which yields $x' = a$, $y' = b-d$). Therefore if possible we give the inverse equations

$$\begin{cases} x' = G_{x'}(x, y'), \\ y' = G_{y'}(x, y') \end{cases} \text{ or } (x¢, y¢) = G(x, y),$$

which are equivalent to (1) if $G(F(x,y)) = (x,y)$ and $F(G(x,y)) = (x,y)$. Then to go from the unprimed to the unprimed system one merely plugs the known values of x and y into these equations. This is also the best strategy when dealing with a curve expressed parametrically, that is: $x = x(t)$, $y = y(t)$.

TRANSFORMATIONS

A transformation associates to each point (x,y) *a different point in the same coordinate system*; we denote this by

$$(x, y) \mapsto F(x, y),$$

where F is a map from the plane to itself (a two-component function of two variables). For example, translating down by a distance d is accomplished by $(x,y) \mapsto (x, y-d)$. Thus the action of the transformation on a point whose coordinates are known (or on a curve expressed parametrically) can be immediately computed.

If, on the other hand, we have an object (say a curve) defined *implicitly* by the equation $C(x,y) = 0$, finding the equation of the transformed object requires using the *inverse transformation*

$$(x,y) \mapsto G(x,y)$$

defined by $G(F(x,y)) = (x,y)$ and $F(G(x,y)) = (x,y)$. The equation of the transformed object is $C(G(x,y)) = 0$. For instance, if C is the circle with equation $x^2 + y^2 = 1$ and we are translating down by a distance d, the inverse transformation is

$$(x,y) \mapsto (x, y+d)$$

(translating up), and the equation of the translated circle is $x^2 + (y + d)^2 = 1$.

Using Transformations to Perform Changes of Coordinates

Usually, we will not give formulas of the form (1) for changes between two coordinate systems of the same type, because they can be immediately derived from the corresponding formulas for transformations, which are given here. We give two examples for clarity.

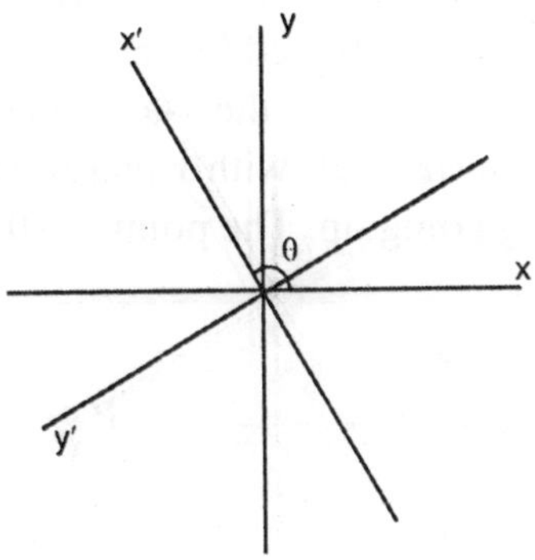

Fig. Change of Coordinates by a Rotation.

Let the two cartesian coordinate systems (x,y) and $(x¢,y¢)$ be related as follows: they have the same origin, and the positive x'-axis is obtained from the positive x-axis by a (counterclockwise) rotation through an angle q. If a point has coordinates (x,y) in the unprimed system, its coordinates (x',y') in the primed system are the same as the coordinates in the unprimed system of a point that undergoes the inverse rotation, that is, a rotation by an angle $a = -q$. This transformation acts as follows:

$$(x, y) \mapsto \begin{pmatrix} \cos\theta & \sin\theta \\ -\sin\theta & \cos\theta \end{pmatrix}(x, y) = (x \cos q + y \sin q, - x \sin q + y \cos q).$$

Therefore the right-hand side of (3) is (x',y'), and the desired substitution is

$$x' = x \cos \theta + y \sin \theta ,$$
$$y' = -x \sin \theta + y \cos \theta .$$

Switching the roles of the primed and unprimed systems we get the equivalent substitution

$$x' = x \cos \theta + -y \sin \theta ,$$
$$y' = x \sin \theta + y \cos \theta$$

(since the x-axis is obtained from the x'-axis by a rotation through an angle $- \theta$).

Similarly, let the two cartesian coordinate systems (x,y) and (x',y') differ by a translation: x is parallel to x' and y to y', and the origin of the second system coincides with the point (x^0, y^0) of the first system. The coordinates (x,y) and (x',y') of a point are related by

$$x = x' + x_0, x' = x - x_0,$$
$$y = y' + y_0, y' = y - y_0.$$

CARTESIAN COORDINATES IN THE PLANE

In cartesian coordinates (or rectangular coordinates), the ``address" of a point P is given by two real numbers indicating the positions of the perpendicular projections from the point to two fixed, perpendicular, graduated lines, called the axes. If one coordinate is denoted x and the other y, the axes are called the x-axis and the y-axis, and we write $P = (x,y)$. Usually the x-axis is drawn horizontal, with x increasing to the right, and the y-axis is drawn vertical, with y increasing going up. The point $x = 0$, $y = 0$ is the origin, where the axes intersect.

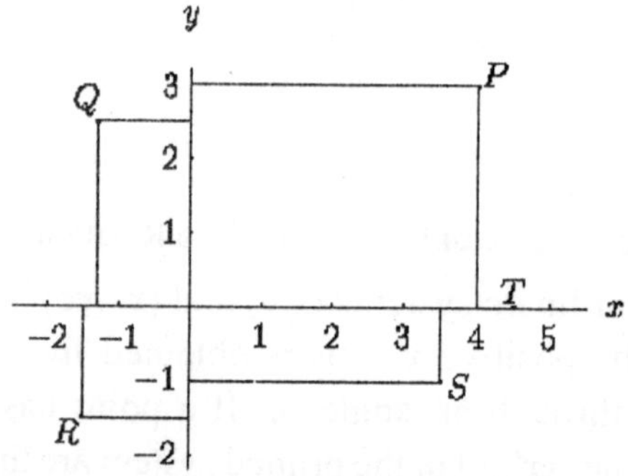

Fig. In Cartesian Coordinates, P =(4,3), Q=(-1.3,2.5), R =(-1.5,-1.5), S=(3.5,-1), and T=(4.5,0). The axes Divide the Plane into four Quadrants: P is in the First Quadrant, Q in the Second, R in the Third, and S in the Fourth. T is on the Positive x-axis.

POLAR COORDINATES IN THE PLANE

In polar coordinates a point P is also characterized by two numbers: the distance $r \geq 0$ to a fixed pole or origin O, and the angle the ray OP makes with a fixed ray originating at O, which is generally drawn pointing to the right (this is called the initial ray). The angle q is only defined up to a multiple of 360° or 2 p. In addition, it is sometimes convenient to relax the condition $r \geq 0$ and allow r to be a signed distance, so (r,θ) and $(-r, \theta+180°)$ represent the same point.

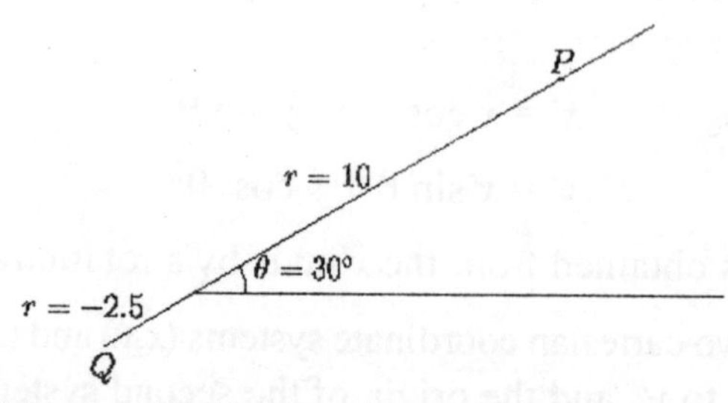

Fig. Among the Possible sets of Polar Coordinates for P are: (10, 30°), (10, 390°) and (10,-330°). Among the Sets of Polar Coordinates for Q are: (2.5, 210°) and (-2.5, 30°).

RELATIONS BETWEEN CARTESIAN AND POLAR COORDINATES

Consider a system of polar coordinates and a system of cartesian coordinates with the same origin. Assume the initial ray of the polar coordinate system coincides with the positive x-axis, and that the ray $\theta = 90°$ coincides with the positive y-axis. Then the polar coordinates (r, θ) and the cartesian coordinates (x,y) of the same point are related as follows:

$$\begin{cases} x = r\cos\theta, \\ y = r\sin\theta, \end{cases} \quad \begin{cases} r = r\sqrt{x^2+y^2}, \\ \theta = a.rcta.n\dfrac{y}{x}, \end{cases} \quad \begin{cases} \sin\theta = \dfrac{y}{\sqrt{x^2+y^2}}, \\ \cos\theta = \dfrac{x}{\sqrt{x^2+y^2}}, \end{cases}$$

HOMOGENEOUS COORDINATES IN THE PLANE

A triple of real numbers $(x{:}y{:}t)$, with $t \neq 0$, is a set of homogeneous coordinates for the point P with cartesian coordinates $(x/t, y/t)$. Thus the same point has many sets of homogeneous coordinates: $(x{:}y{:}t)$ and $(x'{:}y'{:}t')$ represent the same point if and only if there is some real number a such that $x' = a\,x$, $y' = y$, $t' = t$. If P has cartesian coordinates (x^0, y^0), one set of homogeneous coordinates for P is $(x^0, y^0, 1)$.

When we think of the same triple of numbers as the cartesian coordinates of a point in three-dimensional space, we write it (x,y,t) instead of $(x{:}y{:}t)$. The connection between the point in space with cartesian coordinates (x,y,t) and the point in the plane with homogeneous coordinates $(x{:}y{:}t)$ becomes apparent when we consider the plane $t = 1$ in space, with cartesian coordinates given by the first two coordinates x, y of space. The point (x,y,t) in space can be connected to the origin by a line L that intersects the plane $t = 1$ in the point with cartesian coordinates $(x/t, y/t)$, or homogeneous coordinates $(x{:}y{:}t)$.

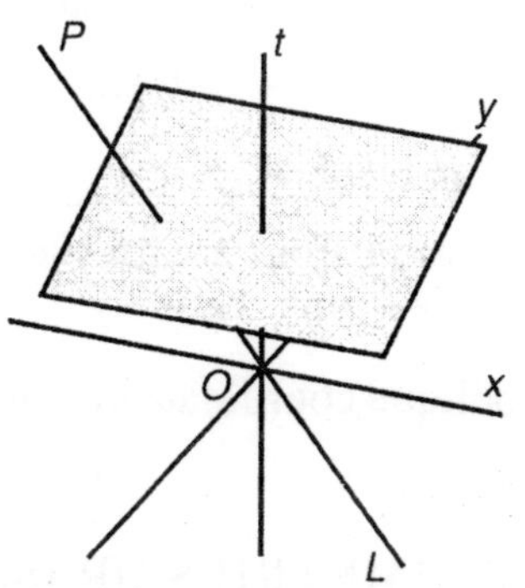

Fig. The Point P with Spatial Coordinates (x, y, t) Projects to the Point Q with Spatial Coordinates $(x/t, y/t, 1)$. The Plane Cartesian Coordinates of Q are $(x/t, y/t)$, and $(x{:}y{:}t)$ is One Set of Homogeneous Coordinates for Q. Any Point on the Line L (Except for the Origin O) Would also Project to P'.

Projective coordinates are useful for several reasons, one the most important being that they allow one to unify all symmetries of the plane (as well as other transformations) under a single umbrella. All these transformations can be regarded as linear maps in the space of triples $(x:y:t)$, and so can be expressed in terms of matrix multiplications.

If we consider triples $(x:y:t)$ such that at least one of x, y, t is nonzero, we can name not only the points in the plane but also points "at infinity". Thus $(x:y:t)$ represents the point at infinity in the direction of the line with slope y/x.

OBLIQUE COORDINATES IN THE PLANE

The following generalization of cartesian coordinates is sometimes useful. Consider two axes (graduated lines), intersecting at the origin but not necessarily perperdicularly. Let the angle between them be w. In this system of oblique coordinates, a point P is given by two real numbers indicating the positions of the projections from the point to each axis, in the direction of the other axis. The first axis (x-axis) is generally drawn horizontally. The case $w = 90°$ yields a cartesian coordinate system.

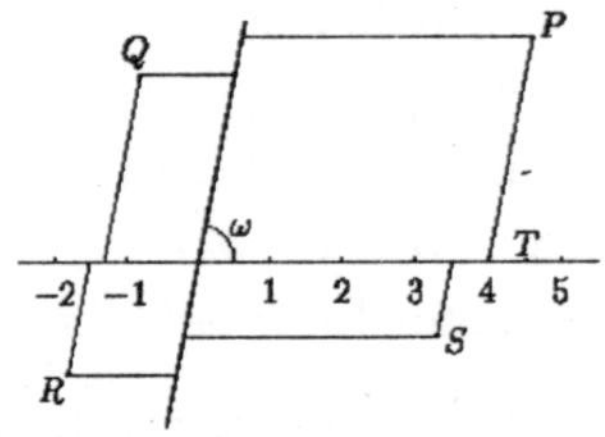

Fig. In Oblique Coordinates, $P = (4,3)$, $Q = (-1.3,2.5)$, $R = (-1.5,-1.5)$, $S = (3.5,-1)$, and $T = (4.5,0)$.

RELATIONS BETWEEN TWO OBLIQUE COORDINATE SYSTEMS

Let the two oblique coordinate systems (x,y) and (x',y'), with angles w and w', share the same origin, and suppose the x'-axis makes an angle with the x-axis. The coordinates (x,y) and (x',y') of a point in the two systems are related by

$$x = (x' \sin(w-\theta) + y' \sin(w - w' - \theta))/ \sin w,$$

$$y = (x' \sin \theta + y' \sin(w' + \theta))/ \sin w.$$

This formula also covers passing from a cartesian system to an oblique system and vice versa, by taking $w = 90°$ or $w' = 90°$.

The relation between two oblique coordinate systems that differ by a translation is the same as for cartesian systems.

PLANE SYMMETRIES OR ISOMETRIES

A transformation of the plane (invertible map of the plane to itself) that preserves distances is called an isometry of the plane. Every isometry of the plane is of one of the following types:

- The identity (which leaves every point fixed);
- A translation by a vector *v*;
- A rotation through an angle around a point *P*;
- A reflection in a line *L*;
- A glide-reflection in a line *L* with displacement *d*.

Although the identity is a particular case of a translation and of a rotation, and reflections are particular cases of glide-reflections, it is more intuitive to consider each case separately.

FORMULAS FOR SYMMETRIES IN CARTESIAN COORDINATES

In the formulas below, a multiplication between a matrix and a pair of coordinates should be carried out regarding the pair as a column vector (or a matrix with two rows and one column). Thus $\begin{pmatrix} a\,b \\ c\,d \end{pmatrix}(x,y) = (ax + by,\ cx + dy)$.

Translation by (x^0, y^0):

$$(x,y) \mapsto (x + x^0,\ y + y^0)$$

Rotation through a (counterclockwise) around the origin:

$$(x,y) \mapsto \begin{pmatrix} \cos a & -\sin a \\ \sin a & \cos a \end{pmatrix}(x,y)$$

Rotation through a (counterclockwise) around an arbitrary point (x^0, y^0):

$$(x,y) \mapsto (x_0, y_0) + \begin{pmatrix} \cos a & -\sin a \\ \sin a & \cos a \end{pmatrix}(x - x_0 y - y_0)$$

Reflection in the x-axis:

$$(x,y) \mapsto (x,-y)$$

Reflection in the y-axis:

$$(x,y) \mapsto (-x,y)$$

Reflection in the xy-diagonal:

$$(x,y) \mapsto (y,x)$$

Reflection in a line with equation $ax + by + c = 0$:

$$(x,y) \mapsto \frac{1}{a^2 + b^2}\left(\begin{pmatrix} b^2 - a^2 & -2ab \\ -2ab & a^2 - b^2 \end{pmatrix}(x,y) - (2ac, 2bc)\right)$$

Reflection in a line going through (x^0 ,y^0) and making an angle a with the x-axis:

$$(x,y)\mapsto(x_0,y_0)+\begin{pmatrix}\cos 2a & \sin 2a\\ \sin 2a & -\cos 2a\end{pmatrix}(x-x_0,y-y_0)$$

Glide-reflection in a line L with displacement d: Apply first a reflection in L, then a translation by a vector of length d in the direction of L, that is, by the vector

$$\frac{1}{a^2+b^2}(\pm ad,\mp bd)$$

if L has equation $ax + by + c = 0$.

FORMULAS FOR SYMMETRIES IN HOMOGENEOUS COORDINATES

All isometries of the plane can be expressed in homogeneous coordinates in terms of multiplication by a matrix. This fact is useful in the implementation of these transformations on the computer, for example. It also means that the successive application of transformations reduces to matrix multiplication. The corresponding matrices are as follows:

Translation by (x^0 ,y^0):

$$T(x_0, y_0)=\begin{bmatrix}1\,0\,x_0\\ 0\,1\,y_0\\ 0\,0\;1\end{bmatrix}$$

Rotation through a around the origin:

$$R_a=\begin{bmatrix}\cos a & -\sin a & 0\\ \sin a & \cos a & 0\\ 0 & 0 & 1\end{bmatrix}$$

Reflection in a line going through the origin and making an angle a with the x-axis:

$$M_a=\begin{bmatrix}\cos 2a & \sin 2a & 0\\ \sin 2a & -\cos 2a & 0\\ 0 & 0 & 1\end{bmatrix}$$

From this one can deduce all other transformations: for example, to find the matrix for a rotation through a around an arbitrary point P = (x^0 ,y^0), we apply a translation by -(x^0 ,y^0) to move P to the origin, then a rotation through around the origin, then a translation by (x^0 ,y^0):

$$T_{(x0,y0)}R_aT_{-(x0,y0)}=\begin{bmatrix}\cos a & -\sin a & x_0-x_0\cos a+y_0\sin a\\ \sin a & \cos & y_0-\cos a-x_0\sin a\\ 0 & 0 & 1\end{bmatrix}$$

(notice the order of the multiplication).

FORMULAS FOR SYMMETRIES IN POLAR COORDINATES

Rotation around the origin through an angle a :

$$(r, q) \mapsto (r, q + a).$$

Reflection in a line through the origin and making an angle a with the positive x-axis:

$$(r, q) \mapsto (r, 2a - q).$$

OTHER TRANSFORMATIONS OF THE PLANE

SIMILARITIES

A transformation of the plane that preserves shapes is called a similarity. Every similarity of the plane is obtained by composing a proportional scaling transformation (also known as a homothety) with an isometry. A proportional scaling transformation centered at the origin has the form

$$(x, y) \mapsto (ax, ay),$$

where $a \neq 0$ is the scaling factor (a real number). The corresponding matrix in homogeneous coordinates is

$$H_a = \begin{bmatrix} a & 0 & 0 \\ 0 & a & 0 \\ 0 & 0 & 1 \end{bmatrix}.$$

In polar coordinates, the transformation is

$$(r, \theta) \mapsto (ar, \theta)$$

TRANSFORMATIONS

A transformation that preserves lines and parallelism (maps parallel lines to parallel lines) is an affine transformation. There are two important particular cases of such transformations:

A nonproportional scaling transformation centered at the origin has the form

$$(x, y) \mapsto (ax, by),$$

where $a, b \neq 0$ are the scaling factors (real numbers). The corresponding matrix in homogeneous coordinates is

$$H_{a,b} = \begin{bmatrix} a & 0 & 0 \\ 0 & b & 0 \\ 0 & 0 & 1 \end{bmatrix}.$$

A shear preserving horizontal lines has the form

$$(x, y) \mapsto (x + ry, y),$$

where r is the shearing factor. The corresponding matrix in homogeneous coordinates is

$$S_r = \begin{bmatrix} 1 & r & 0 \\ 0 & b & 0 \\ 0 & 0 & 1 \end{bmatrix}.$$

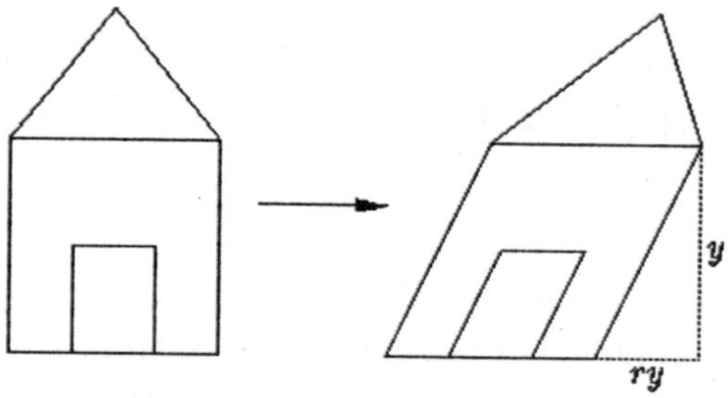

Fig. A Shear with Factor $r = ½$.

Every affine transformation is obtained by composing a scaling transformation with an isometry, or a shear with a homothety and an isometry.

PROJECTIVE TRANSFORMATIONS

A transformation that maps lines to lines (but does not necessarily preserve parallelism) is a projective transformation. Any plane projective transformation can be expressed by an invertible 3×3 matrix in homogeneous coordinates; conversely, any invertible 3×3 matrix defines a projective transformation of the plane. Projective transformations (if not affine) are not defined on all of the plane, but only on the complement of a line (the missing line is ``mapped to infinity").

A common example of a projective transformation is given by a perspective transformation. Strictly speaking this gives a transformation from one plane to another, but if we identify the two planes by (for example) fixing a cartesian system in each, we get a projective transformation from the plane to itself.

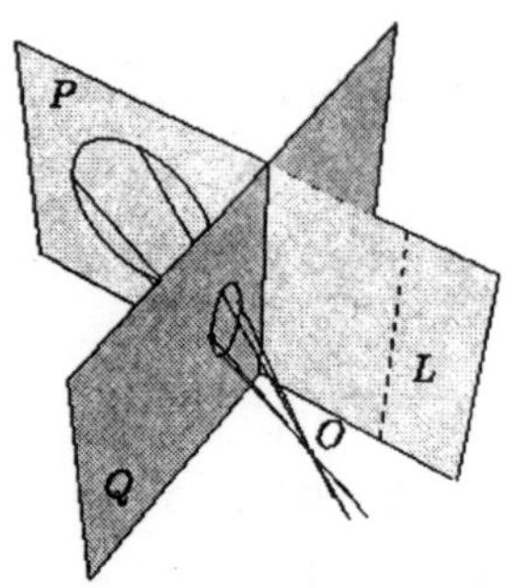

Fig: A Perspective Transformation with center O, Mapping the plane P to the Plane Q. The Transformation is not defined on the line L, where P intersects the plane Parallel to Q and going Throught O.

LINES

The (cartesian) equation of a straight line is linear in the coordinates x and y, that is, of the form $ax + by + c = 0$.

The slope of this line is $-a/b$, the intersection with the x-axis (or x-intercept) is $x = -c/a$, and the intersection with the y-axis (or y-intercept) is $y = -c/b$. If $a = 0$ the line is parallel to the x-axis, and if $b = 0$ the line is parallel to the y-axis. (In an oblique coordinate system everything in the preceding paragraph remains true, except for the value of the slope.)

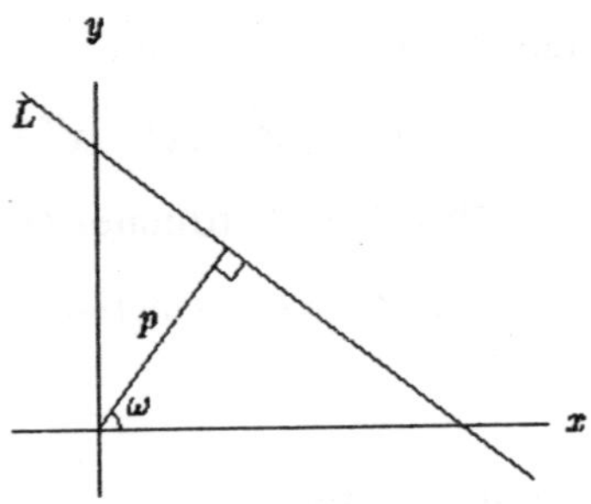

Fig: The Normal form of L is $x \cos w + y \sin w = p$.

When $a + b = 1$ and $c \leq 0$ in the equation $ax + by + c = 0$, the equation is said to be in normal form. In this case c is the distance of the line to the origin, and

$$w = \arcsin a = \arccos b$$

is the angle that the perpendicular dropped to the line from the origin makes with the positive x-axis.

To reduce an arbitrary equation $ax + by + c = 0$ to normal form, divide by $\pm\sqrt{a^2 + b^2}$, where the sign of the radical is chosen opposite the sign of c when $c \neq 0$ and the same as the sign of b when $c = 0$.

Lines with prescribed properties

Line of slope m intersecting the x-axis at $x = x^0$:

$$y = m\,(x - x^0)$$

Line of slope m intersecting the y-axis at $y = y$:

$$y = mx + y^0$$

Line intersecting the x-axis at $x = x$ and the y-axis at $y = y^0$:

$$x/x^0 + yy^0 = 1$$

(This formula remains true in oblique coordinates.)

Line of slope m going though (x^0, y^0):

$$y - y^0 = m(x - x^0)$$

Line going through points (x^0, y^0) and (x^1, y^1):

$$\frac{y - y_1}{x - x_1} = \frac{y_0 - y_1}{x_0 - x_1} \quad \text{or} \quad \begin{vmatrix} x & y & 1 \\ x_0 & y_0 & 1 \\ x_1 & y_1 & 1 \end{vmatrix} = 0$$

(These formulas remain true in oblique coordinates.)

Slope of line going through points (x^0, y^0) and (x^1, y^1):

$$(y^1 - y^0)/(x^1 - x^0)$$

Line going through points with polar coordinates (r^0, θ_0) and (r^1, θ^1):

$$r\,(r^0 \sin(\theta - \theta^0) - r^1 \sin(\theta - \theta^1)) = r^0 r^1 \sin(\theta^1 - \theta^0).$$

DISTANCES

The distance between two points in the plane is the length of the line segment joining the two points. If the points have cartesian coordinates (x^0, y^0) and (x^1, y^1), this distance is

$$\sqrt{(x_1 - x_0)^2 + (y_1 - y_0)^2}\,.$$

If the points have polar coordinates (r^0, q^0) and (r^1, q^1), this distance is

$$\sqrt{r_0^2 + r_1^2 - 2r_0 r_1 \cos(\theta_0 - \theta_1)}$$

If the points have oblique coordinates (x^0, y^0) and (x^1, y^1), this distance is

$$\sqrt{(x_1 - x_0)^2 + (y_1 - y_0)^2 + 2(x_1 - x_0)\cos w,}$$

where is the angle between the axes.

The point k% of the way from $P^0 = (x^0, y^0)$ to $P = (x^1, y^1)$ is

$$\left(\frac{kx + (100 - k)x}{100}, \frac{ky + (100 - k)y}{100}\right).$$

(The same formula works also in oblique coordinates.) This point divides the segment $P^0 P^1$ in the ratio k:(100-k). As a particular case, the midpoint of $P^0 P^1$ is given by

$$(½(x^1 + x^2), ½(y^1 + y^2)).$$

The distance from the point (x^0, y^0) to the line $ax + by + c = 0$ is

$$\frac{ax_0 + by_0 + c}{\sqrt{a^2 + b^2}}$$

ANGLES

The angle between two lines $a^0 x + b^0 y + c^0 = 0$ and $a^1 x + b^1 y + c = 0$ is

$$\arctan \frac{b_1}{a_1} - \arctan \frac{b_0}{a_0} = \arctan \frac{a_0 b_1 - a_1 b_0}{a_0 a_1 + b_0 b_1}.$$

In particular, the two lines are parallel when $a^0 b^0 = a^1 b^0$, and perpendicular when $a^0 a^1 = - b^0 b^1$.

The angle between two lines of slopes m^0 and m^1 is

$$\arctan((m^1 - m^0)/(1 + m^0 m^1)).$$

In particular, the two lines are parallel when $m^0 = m^1$
and perpendicular when $m^0 m^1 = -1$.

CONCURRENCE AND COLLINEARITY

Three lines $a^0 x + b^0 y + c^0 = 0$, $a^1 x + b^1 y + c^1 = 0$, and $a^2 x + b^2 y + c^2 = 0$ are concurrent if and only if

$$\begin{vmatrix} a_0 & b_0 & c_0 \\ a_1 & b_1 & c_1 \\ a_2 & b_2 & c_2 \end{vmatrix} = 0$$

(This remains true in oblique coordinates.)

Three points (x^0, y^0), (x^1, y^1) and (x^2, y^2) are collinear if and only if

$$\begin{vmatrix} x_0 & y_0 & 1 \\ x_1 & y_1 & 1 \\ x_2 & x_2 & 1 \end{vmatrix} = 0$$

(This remains true in oblique coordinates.)

Three points with polar coordinates (r^0, θ^0), (r^1, θ^1) and (r^2, θ^2) are collinear if and only if

$$r^1 r^2 \sin(\theta^2 - \theta^1) + r^0 r^1 \sin(\theta^1 - \theta^0) + r^2 r^0 \sin(\theta^0 - \theta^2) = 0.$$

POLYGONS

Given $k \geq 3$ points $A_1,\ldots,A_k$ in the plane, in a certain order, we obtain a k-sided polygon or k-gon by connecting each point to the next, and the last to the first, with a line segment. The points A_i are the vertices and the segments $Ai\ A_{i+1}$ are the sides or edges of the polygon. When $k = 3$ we have a triangle, when $k = 4$ we have a quadrangle or quadrilateral, and so on. Here we will assume that all polygons are simple: this means that no consecutive edges are on the same line, and no two edges intersect (except that consecutive edges intersect at the common vertex).

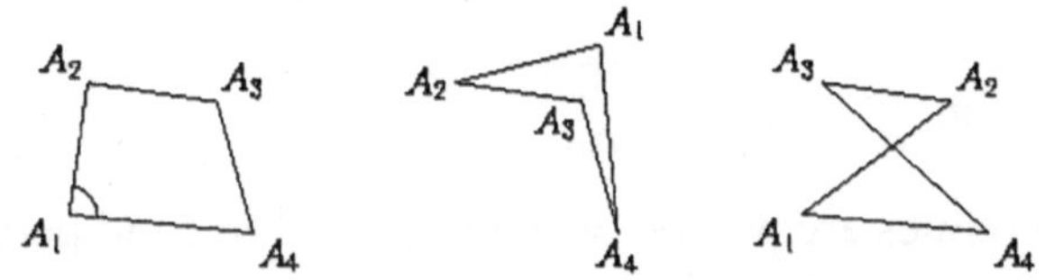

Fig. Two Simple Quadrilaterals (left and middle) and One that is not Simple (right). We will not Consider Non-simple Polygons

When we refer to the angle at a vertex A_k we have in mind the interior angle. We denote this angle by the same symbol as the vertex. The complement of A_k is the exterior angle at that vertex; geometrically, it is the angle between one side and the extension of the adjacent side. *In any k-gon, the sum of the angles equals 2(k-2) right angles, or* 2(*k*-2)×90°: for example, the sum of the angles of a triangle is 180°.

The area of a polygon whose vertices A_i have coordinates (x_i, y_i), for $1 \le i \le k$, is the absolute value of

$$\text{area} = \frac{1}{2}(x_1y_2 - x_2y_1) + \cdots + \frac{1}{2}(x_{k-1}y_k - x_ky_{k-1}) + \frac{1}{2}(x_ky_1 - x_1y_k),$$

$$= \frac{1}{2}\sum_{i=1}^{k}(x_1y_{i+1} - x_{i+1}y_i)$$

where in the sumation we take $x_{k+1} = x_1$ and $y_{k+1} = y_1$. In particular, for a triangle we have

$$\text{area} = \frac{1}{2}(x_1y_2 - x_2y_2 + x_2y_3 - x_3y_2 + x_3y_1 - x_1y_3) = \frac{1}{2}\begin{vmatrix} x_1 & y_1 & 1 \\ x_2 & y_2 & 1 \\ x_3 & y_3 & 1 \end{vmatrix}.$$

In oblique coordinates with angle *w* between the axes, the area is as given above, times *sin w*.

If the vertices have polar coordinates (r_i, θ_i), for $1 \le i \le k$, the area is the absolute value of

$$\text{area} = \frac{1}{2}\sum_{i=1}^{k} r_ir_{i+1}\sin(\theta_{i+1} - \theta_i),$$

where we take $r_{k+1} = r_1$ and $\theta_{k+1} = \theta_1$.

Formulas for specific polygons in terms of side lengths, angles, etc. are given in the next sections.

TRIANGLES

Because the angles of a triangle add up to 180°, at least two of them must be acute (less than 90°). In an acute triangle all angles are acute. A right triangle has one right angle, and an obtuse triangle has one obtuse angle.

The altitude corresponding to a side is the perpendicular dropped to the line containing that side from the opposite vertex. The bisector of a vertex is the line that divides the angle at that vertex into two equal parts. The median is the segment joining a vertex to the midpoint of the opposite side.

Every triangle also has an inscribed circle tangent to its sides and interior to the triangle (in other words, any three nonconcurrent lines determine a circle). The center of this circle is the point of intersection of the bisectors. We denote the radius of the inscribed circle by r.

Every triangle has a circumscribed circle going through its vertices; in other words, any three noncollinear points determine a circle. The point of intersection of the medians is the center of mass of the triangle (considered as an area in the plane). We denote the radius of the circumscribed circle by R.

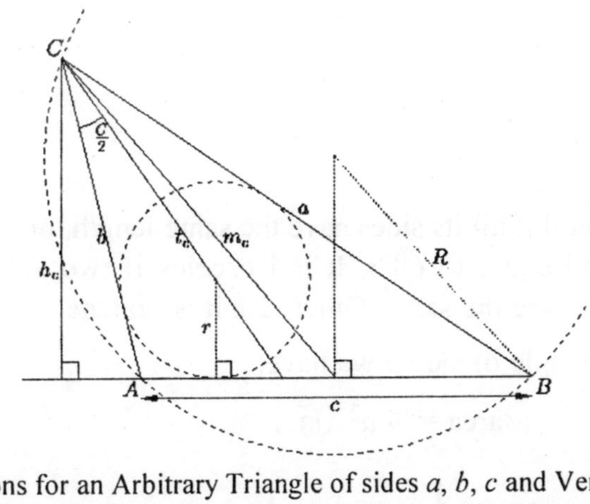

Fig. Notations for an Arbitrary Triangle of sides a, b, c and Vertices A, B, C. The Altitude Corresponding to C is h_c , the Median is m_c, the Bisector is t_c. The Radius of the Circumscribed Circle is R, that of the Inscribed Circle is r.

Introduce the following notations for an arbitrary triangle of vertices A, B, C and sides a, b, c. Let h_c, t_c and m_c be the lengths of the altitude, bisector and median originating in vertex C, let r and R be as usual the radii of the inscribed and circumscribed circles, and let $s = \frac{1}{2}(a + b + c)$. Then:

$$A + B + C = 180°,$$

$$c^2 = a^2 + b^2 - 2ab \cos C \text{ (law of cosines)},$$

$$a = b \cos C + c \cos B,$$

$$\frac{a}{\sin A} = \frac{b}{\sin B} = \frac{c}{\sin C} \text{ (law of sines),}$$

$$\text{area} = \frac{1}{2}h_c c = \frac{1}{2}ab\sin C = \frac{c^2 \sin A \sin B}{2\sin C} = rs = \frac{abc}{4R}$$

$$= \sqrt{s(s-a)(s-b)(c-c)} \text{ (Heron),}$$

$$r = c \sin \frac{1}{2}A \sin\frac{1}{2}B \sec\frac{1}{2}C = \frac{ab\sin C}{2s} = (s-c)\tan\frac{1}{2}C$$

$$= \left(\frac{1}{h_a} + \frac{1}{h_b} + \frac{1}{h_c}\right)^{-1},$$

$$R = \frac{c}{2\sin C} = \frac{abc}{4a.rea},$$

$$h_c = a \sin B = b \sin A = \frac{2a.rea}{c},$$

$$t_c = \frac{2ab}{a+b}, \cos\frac{1}{2}C = \sqrt{ab\left(1-\frac{c^2}{(a+b)^2}\right)},$$

$$m_c = \sqrt{\frac{1}{2}a^2 + \frac{1}{2}b^2 - \frac{1}{4}c^2}.$$

A triangle is equilateral if all its sides have the same length, or, equivalently, if all its angles are the same (and equal to 60°). It is isosceles if two sides are the same, or, equivalently, if two angles are the same. Otherwise it is scalene.

For an equilateral triangle of side a we have:

$$\text{area} = \tfrac{1}{4} a^2 \sqrt{3},$$

$$r = \frac{1}{8} a \sqrt{3},$$

$$R = 1/3\ a \sqrt{3},$$

$$h = \tfrac{1}{2} a \sqrt{3},$$

where h is any altitude. The altitude, the bisector and the median for each vertex coincide.

For an isosceles triangle, the altitude for the unequal side is also the corresponding bisector and median, but this is not true for the other two altitudes. Many formulas for an isosceles triangle of sides a, a, c can be immediately derived from those for a right triangle of legs a, $\tfrac{1}{2} c$.

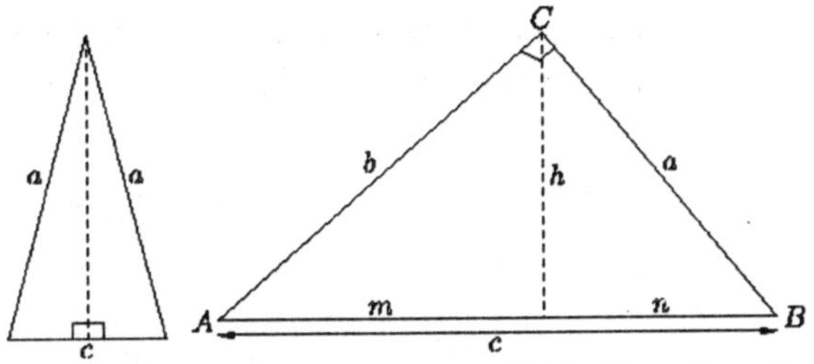

Fig. Left: An Isosceles Triangle can be Divided into Two Congruent Right Triangles. Right: Notations for a Right Triangle.

For a right triangle the hypothenuse is the longest side opposite the right angle; the legs are the two shorter sides, adjacent to the right angle. The altitude for each leg equals the other leg., h denotes the altitude for the hypothenuse, while m and n denote the segments into which this altitude divides the hypothenuse.

The following formulas apply for a right triangle:

$$A + B = 90°$$

$$r = ab/(a + b + c)$$

$$a = c \sin A = c \cos B$$

$$mc = b^2$$

$$\text{area} = ½\, ab$$

$$c^2 = a^2 + b^2 \text{ (Pythagoras)}$$

$$R = ½\, c$$

$$b = c \sin B = c \cos A$$

$$nc = a^2$$

$$hc = ab$$

The hypothenuse is a diameter of the circumscribed circle. The median joining the midpoint of the hypothenuse (the center of the circumscribed circle) to the right angle makes angles $2A$ and $2B$ with the hypothenuse.

Additional facts about triangles:

- In any triangle, the longest side is opposite the largest angle, and the shortest side is opposite the smallest angle. This follows from the law of sines.
- In a triangle ABC, let D, E and F be points on the lines BC, CA and AB, respectively. Then the lines AD, BE and CF are concurrent if and only if the signed distances BD, CE, ... satisfy

$$BD \,.\, CE \,.\, AF = DC \,.\, EA \,.\, FB \,.$$

This is so in three important particular cases: when the three lines are the medians, when they are the bisectors, and when they are the altitudes.

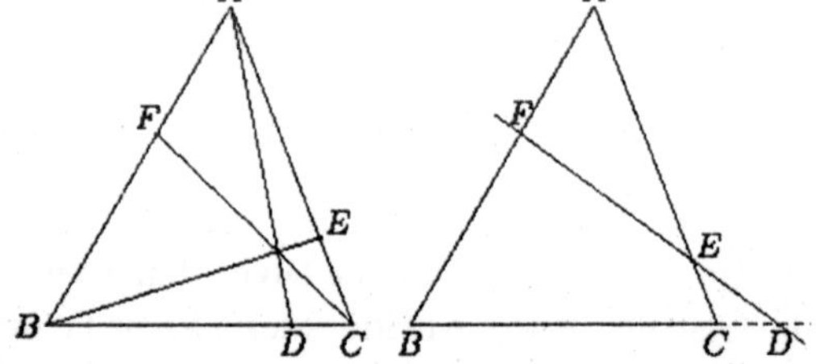

Fig. Left: Ceva's Theorem. Right: Menelaus's Theorem.

- In a triangle ABC, let D, E and F be points on the lines BC, CA and AB, respectively. Then D, E and F are collinear if and only if the signed distances BD, CE, ... satisfy

$$BD \, . \, CE \, . \, AF = - DC \, . \, EA \, . \, FB \, .$$

- Each side of a triangle is less than the sum of the other two. For any three lengths such that each is less than the sum of the other two, there is a triangle with these side lengths.

QUADRILATERALS

The following formulas give the area of a general quadrilateral.

$$\text{aeea} = \frac{1}{2} pq \sin\theta = \frac{1}{4}(b^2 + d^2 - a^2 - c^2)\tan\theta$$

$$= \frac{1}{4}\sqrt{4p^2q^2 - (b^2 + d^2 - a^2 - c^2)^2}$$

$$= \sqrt{(s-a)(s-b)(s-c)(s-d) - abcd\cos\frac{1}{2}(A+C)}.$$

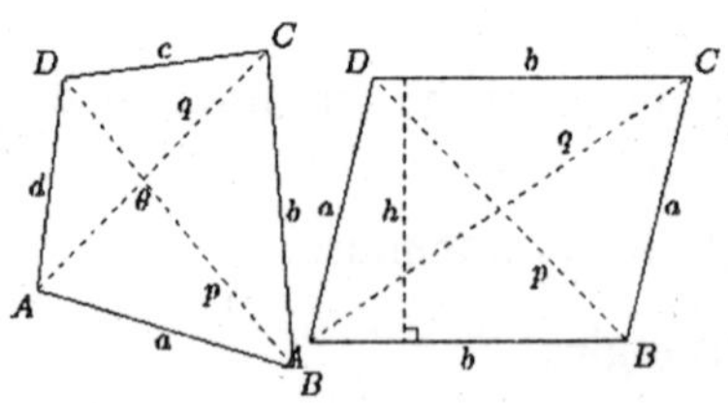

Fig. Left: Notation for a General Quadrilateral; in Addition $s = ½ (a + b + c + d)$. Right: a Parallelogram.

Often, however, it is easiest to compute the area by dividing the quadrilateral into triangles. One can also divide into triangles to compute one side given the other sides and angles, etc.

More formulas can be given for special cases of quadrilaterals. In a parallelogram opposite sides are parallel . It follows that opposite sides have the same length and that two consecutive angles add up to 180°. In the notation of the figure, we have

$$A = C,\ B = D,$$

$$h = a \sin A = a \sin B,$$

$$p = \sqrt{a^2 + b^2 - 2ab \cos A},$$

$$A + B = 180°,$$

$$\text{area} = bh,$$

$$q = \sqrt{a^2 + b^2 - 2ab \cos B}.$$

(All this follows from the triangle formulas applied to the triangles *ABD* and *ABC*.)

Two particular cases of parallelograms are:

- The rectangle ▭, where all angles equal 90°. The diagonals of a rectangle have the same length. The general formulas for parallelograms reduce to

$$h = a$$

$$\text{area} = ab$$

$$p = q = \sqrt{a^2 + b^2}.$$

- The rhombus or diamond ◇, where adjacent sides have the same length ($a = b$). The diagonals of a rhombus are perpendicular. In addition to the general formulas for parallelograms we have area = ½ pq and $p^2 + q^2 = 4a^2$.

The square or regular quadrilateral is both a rectangle and a rhombus.

A quadrilateral is a trapezoid if two sides are parallel. In the notation of the figure below we have

$$A + D = B + C = 180°,$$

$$\text{area} = \tfrac{1}{2}(AB + CD)h.$$

A quadrilateral is cyclic if it can be inscribed in a circle, that is, if its four vertices belong to a single, circumscribed, circle. This is possible if and only if the sum of opposite angles is 180°. If R is the radius of the circumscribed circle:

$$\text{area} = \sqrt{(s-a)(s-b)(s-c)(s-d)} = \frac{1}{2}(ac + bd)\sin\theta$$

$$= \frac{\sqrt{(ac+bd)(ad+bc)(ab+cd)}}{4R} \text{ (Brahnnagupta)},$$

$$p = \sqrt{\frac{(ac+bd)(ad+cd)}{(ad+bc)}},$$

$$R = \frac{1}{4}\sqrt{\frac{(ac+bd)(ad+bc)(ab+cd)}{(s-a)(s-b)(s-c)(s-d)}},$$

$$\sin\theta = \frac{2area}{ac+bd},$$

$pq = ac + bd$ (Ptolenny).

A quadrilateral is circumscribable if it has an inscribed circle (that is, a circle tangent to all four sides). Its area is rs, where r is the radius of the inscribed circle and s is as above.

For a quadrilateral that is both cyclic and circumscribable we have the following additional equalities, where m is the distance between the centers of the inscribed and circumscribed circles:

$$a + c = b + d,$$

$$R = \frac{1}{4}\sqrt{\frac{(ac+bd)(ad+bc)(ab+cd)}{abcd}},$$

$$\text{area} = \sqrt{abcd} = rs,$$

$$\frac{1}{r^2} = \frac{1}{(R-m)^2} + \frac{1}{(R+m)^2}$$

Additional fact about quadrilaterals:

- The diagonals of a quadrilateral with consecutive sides a, b, c, d are perpendicular if and only if $a^2 + c^2 = b^2 + d^2$.

REGULAR POLYGONS

A polygon is regular if all its sides are equal and all its angles are equal. Either condition implies the other in the case of a triangle, but not in general. (A rhombus has equal sides but not necessarily equal angles, and a rectangle has equal angles but not necessarily equal sides.)

For a k-sided regular polygon of side a, let θ be the angle at any vertex, r and R the radii of the inscribed and circumscribed circles (r is called the apothem). As usual, let $s = \frac{1}{2} ka$ be the half-perimeter. Then:

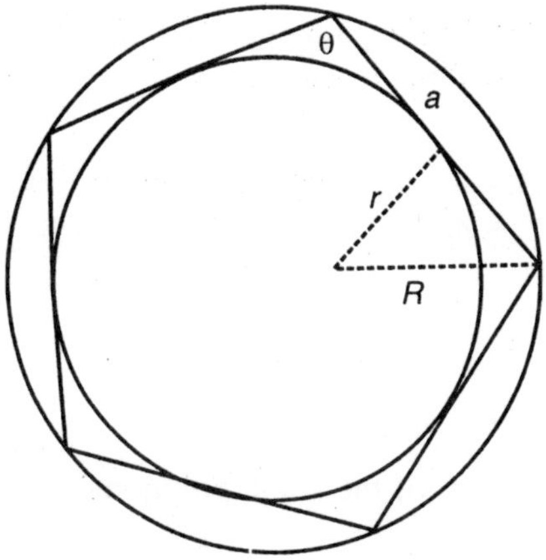

$$\theta = \left(\frac{k-2}{k}\right)180°,$$

$$a = 2r \tan \frac{180°}{k} = 2R \sin \frac{180°}{k},$$

$$\text{area} = \frac{1}{4}ks^2 \cot\frac{180°}{k} = kr^2 \tan\frac{180°}{k} = \frac{1}{2}kR^2 \sin\frac{360°}{k},$$

$$r = \frac{1}{2}a\cot\frac{180°}{k},$$

$$R = \frac{1}{2}a\csc\frac{180°}{k},$$

Name	k	Area	r	R
Equilateral Triangle	3	$0.43301\ a^2$	$0.28868\ a$	$0.57735\ a$
Square	4	a^2	$0.50000\ a$	$0.70711\ a$
Regular Pentagon	5	$1.72048\ a^2$	$0.68819\ a$	$0.85065\ a$
Regular Hexagon	6	$2.59808\ a^2$	$0.86603\ a$	a
Regular Heptagon	7	$3.63391\ a^2$	$1.03826\ a$	$1.15238\ a$
Regular Octagon	8	$4.82843\ a^2$	$1.20711\ a$	$1.30656\ a$
Regular Nonagon	9	$6.18182\ a^2$	$1.37374\ a$	$1.46190\ a$
Regular Decagon	10	$7.69421\ a^2$	$1.53884\ a$	$1.61803\ a$
Regular Undecagon	11	$9.36564\ a^2$	$1.70284\ a$	$1.77473\ a$
Regular Dodecagon	12	$11.19625a^2$	$1.86603\ a$	$1.93185\ a$

If a_k denotes the side of a k-sided regular polygon inscribed in a circle of radius R, we have

$$a_{2k} = \sqrt{2R^2 - R\sqrt{4R^2 - a_k{}^2}}.$$

If A_k denotes the side of a k-sided regular polygon circumscribed about the same circle, we have

$$A_{2k} = \frac{2RA_k}{2R + \sqrt{4R^2 + A_k^{\ 2}}},$$

In particular,

$$A_{2k} = \frac{a_k A_k}{a_k + A_k}, \; a_{2k} = \sqrt{\frac{a_k A_{2k}}{2}},$$

The areas s_k, s_{2k}, S_{2k} and of the same polygons satisfy

$$s_{2k} = \sqrt{s_k S_k,} \; S_{2k} = \frac{2s_k S_{k,}}{s_{2k} + S_k}.$$

CIRCLES

The set of points whose distance to a fixed point (the center) is a fixed positive number (the radius) is a circle. A circle of radius r and center (x^0, y^0) has equation

$$(x-x^0)^2 + (y - y^0)^2 = r^2,$$

or

$$x^2 + y^2 - 2xx^0 - 2yy^0 + x^{02} + y^{02} - r^2 = 0.$$

Conversely, an equation of the form

$$x^2 + y^2 + 2dx + 2ey + f = 0$$

defines a circle if $d^2 + e^2 > f$; the center is $(-d, -e)$ and the radius is $\sqrt{d^2 + e^2 - f}$.

Three points not on the same line determine a unique circle. If the points have coordinates (x^1, y^1), (x^2, y^2) and (x^3, y^3), the equation of the circle is

$$\begin{vmatrix} x^2 + y^2 & x & y & 1 \\ x_1^{\ 2} + y_1^{\ 2} & x_1 & y_1 & 1 \\ x_2^{\ 2} + y_2^{\ 2} & x_2 & y_2 & 1 \\ x_3^{\ 2} + y_3^{\ 2} & x_3 & y_3 & 1 \end{vmatrix} = 0$$

A chord of a circle is a line segment between two points. A diameter is a chord that goes through the center, or the length of such a chord (therefore the diameter is twice the radius). Given two points $P^1 = (x^1, y^1)$ and $P^2 = (x^2, y^2)$, there is a unique circle whose diameter is $P^1 P^2$; its equation is

$$(x - x^1)(x - x^2) + (y - y^1)(y - y^2) = 0.$$

The length or circumference of a circle of radius *r* is 2p *r*, and the area is p r^2. The length of the arc of circle subtended by ăn angle q, is *r*q. Other relations between the radius, the arc length, the chord, and the areas of the corresponding sector and segment are:

$$d = R\cos\frac{1}{2}\theta = \frac{1}{2}c\cot\frac{1}{2}\theta = \frac{1}{2}\sqrt{4R^2 - c^2},$$

$$c = 2R\sin\frac{1}{2}\theta = 2d\tan\frac{1}{2}\theta = 2\sqrt{R^2 - d^2}, = \sqrt{4h(2R - h)},$$

$$\theta = \frac{s}{R} = 2\text{ arc cos }\frac{d}{R} = 2\text{ arctan }\frac{c}{2d} = 2\text{ arc sin }\frac{c}{2R},$$

$$\text{area of sector} = \frac{1}{2}Rs = \frac{1}{2}R^2\theta,$$

$$\text{area of segment} = \frac{1}{2}R^2(\theta - \sin\theta) = \frac{1}{2}(R_s - cd) = R^2\text{ arc cos }\frac{d}{R} - d\sqrt{R^2 - d^2}$$

$$= R^2\text{ arc cos }\frac{R-h}{R} - (R-h)\sqrt{2Rh - h^2}.$$

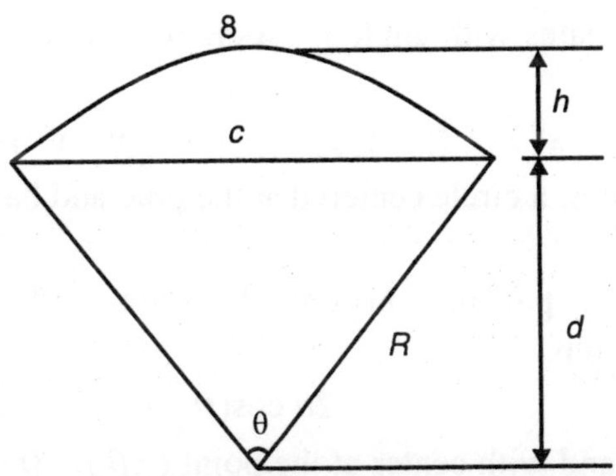

Fig: The arc of Circle Subtendend by the Angle *q* is *s*; the Chord is *c*; the Sector is the Whole Slice of the pie; the Segment is the Cap Bounded by the arc and the Chord (That is, the Slice Minus the Triangle).

OTHER PROPERTIES OF CIRCLES

- If a line intersects a circle of center *O* at points *A* and *B*, the segments *OA* and *OB* make equal angles with the line. In particular, a tangent line is perpendicular to the radius that goes through the point of tangency.
- Given a fixed circle and a fixed point *P* in the plane, and a line through *P* that intersects the circle at *A* and *B* (with *A* = *B* for a tangent). Then the product of the distances $PA \times PB$ is the same for all such lines. It is called the power of *P* with respect to the circle.

- If the central angle *AOB* equals q, the angle *ACB*, where *C* is any point on the circle, equals ½ q or 180°-½ q. Conversely, given a segment *AB*, the set of points that ``sees" *AB* under a fixed angle is an arc of a circle. In particular, the set of points that see *AB* under a right angle is a circle with diameter *AB*.

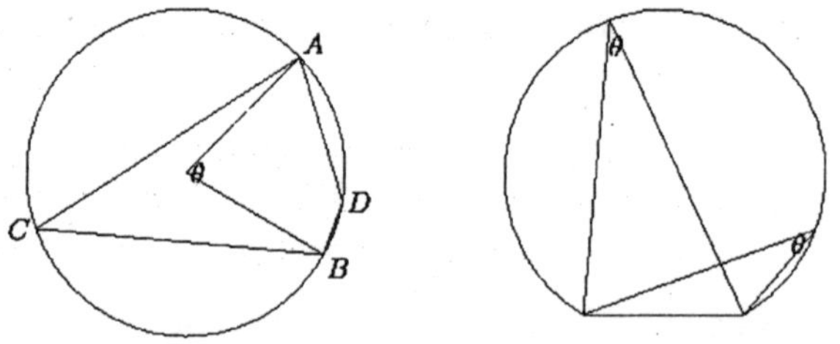

Fig. Left: The angle *ACB* Equals ½ q for any *C* in the Long arc *AB*; and *ADB* Equals 180°-½ q for any *D* in the Short arc *AB*. Right: The Locus of Points from which the Segment *AB* Subtends a Fixed Angle q is an arc of Circle.

- Let P^1 , P^2 , P^3 , P^4 be points in the plane, and let d_{ij}, for $1 \leq i,j \leq 4$, be the distance between P_i and P_j . A necessary and sufficient condition for the points to all lie on the same circle (or line) is that one of the following equalities be satisfied:

$$\pm d^{12}\ d^{34} \pm d^{13}\ d^{24} \pm d^{14}\ d^{23} = 0.$$

This is equivalent to Ptolemy's formula for cyclic quadrilaterals.

- In oblique coordinates with angle w , a circle of center (x^0 ,y^0) and radius r has equation

$$(x - x^0)^2 + (y - y^0) + 2(x - x^0)(y - y^0) \cos w = r^2 .$$

- In polar coordinates, a circle centered at the pole and having radius a has equation $r = a$.

A circle of radius a, passing through the pole, and with center at the point $(r,\theta) = (a, \theta^0)$ has equation

$$r = 2a \cos(\theta - \theta^0).$$

A circle of radius a and with center at the point $(r,\theta) = (r^0 ,\theta^0)$ has equation

$$r - 2r^0\, r \cos(\theta - \theta^0) + r^{02} - a^2 = 0.$$

CONICS

A conic (or conic section) is a plane curve that can be obtained by intersecting a cone with a plane that does not go through the vertex of the cone. There are three possibilities, depending on the relative position of the cone and the plane. If no line of the cone is parallel to the plane, the intersection is a closed curve, called an ellipse. If one line of the cone is parallel to the plane, the intersection is an open curve whose two ends are asymptotically parallel; this is called a parabola. Finally, there may be two lines in the cone parallel to the plane; the curve in this case has two open pieces, and is called a hyperbola.

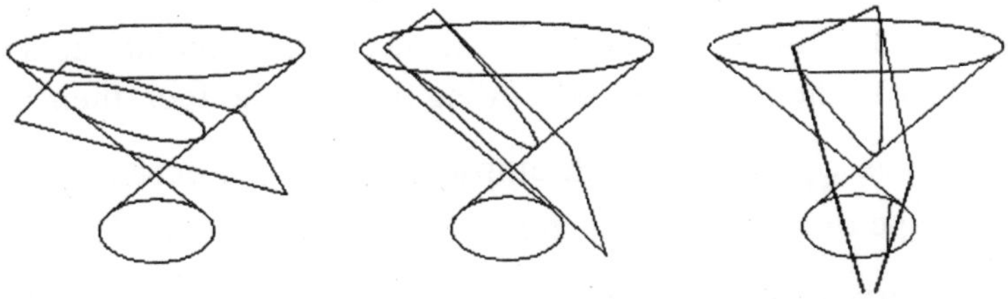

Fig: Left: A Section of a cone by a plane can yield an ellipse (Left), a Parabola (Middle) or a Hyperbola (Right).

ALTERNATIVE CHARACTERIZATION

Assume given a point F in the plane, a line d not going through F, and a positive real number e. The set of points P such that the distance PF is e times the distance from P to d (measured along a perpendicular) is a conic. We call F the focus, d the directrix and e the eccentricity of the conic. If $e < 1$ we have an ellipse, if $e = 1$ a parabola, and if $e > 1$ a hyperbola. This construction gives all conics except the circle, which is a particular case of the ellipse according to the earlier definition (we can recover it by taking the limit $e \to 0$).

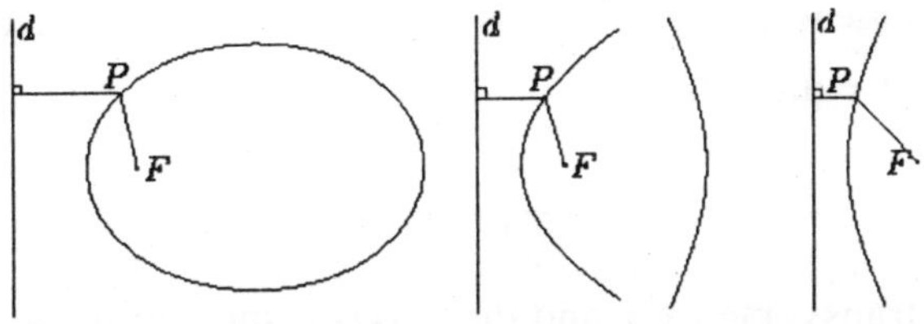

Fig: Left: Definition of conics by means of the ratio (Eccentricity) between the Distance to a Point and the Distance to a line. On the left, e =.7; on the Middle, e =1; on the right, e = 2.

For any conic, a line perpendicular to d and passing through F is an axis of symmetry. The ellipse and the hyperbola have an additional axis of symmetry, perpendicular to the first, so there is an alternate focus and directrix F' and d' obtained as the reflection of F and d with respect to this axis. (By contrast, the focus and directrix are uniquely defined for a parabola.)

The simplest analytic form for the ellipse and hyperbola is obtained when the two symmetry axes coincide with the coordinate axes. The ellipse in Figure has equation

$$\frac{x^2}{a^2}+\frac{y^2}{b^2}=1,$$

with $b < a$. The x-axis is the major axis, and the y-axis is the minor axis. These names are also applied to the segments determined on the axes by the ellipse, and to the lengths of these segments: $2a$ for the major axis and $2b$ for the minor. The vertices are the intersections of the major axis with the ellipse, and have coordinates $(a,0)$ and $(-a,0)$. The distance from the center to either focus is $\sqrt{a^2-b^2}$, and the sum of the distances from a point in the ellipse to the foci is $2a$. The latera recta (in the singular, latus rectum) are the chords perpendicular to the major axis and going through the foci; their length is $2b/a$.

The eccentricity is $\sqrt{a^2-b^2}/a$. All ellipses of the same eccentricity are similar; in other words, the shape of an ellipse depends only on the ratio b/a. The distance from the center to either directrix is $a^2/\sqrt{a^2-b^2}$.

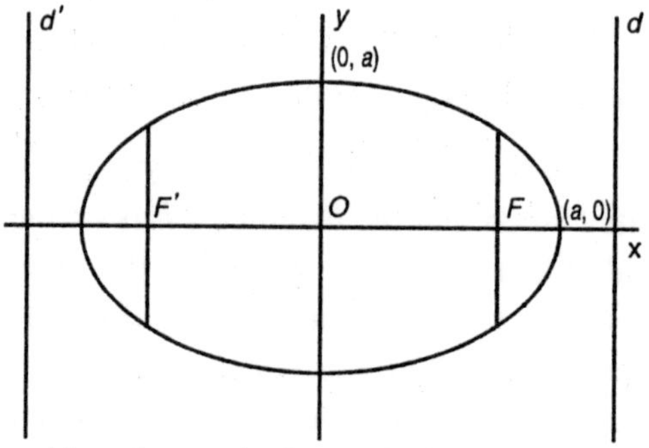

Fig: Left: Ellipse with major semiaxis a and minor semiaxis b. Here $b/a = 0.6$.

The hyperbola has equation

$$\frac{x^2}{a^2}-\frac{y^2}{b^2}=1.$$

The x-axis is the transverse axis, and the y-axis is the conjugate axis. The vertices are the intersections of the transverse axis with the ellipse, and have coordinates $(a,0)$ and $(-a,0)$. The segment thus determined, and its length $2a$, is also called the transverse axis, while the length $2b$ is also called the conjugate axis.

The distance from the center to either focus is $\sqrt{a^2+b^2}$, and the difference between the distances from a point in the hyperbola to the foci is $2a$. The latera recta are the chords perpendicular to the transverse axis and going through the foci; their length is $2b/a$. The eccentricity is $\sqrt{a^2+b^2}/a$. The distance from the center to either directrix is $a^2/\sqrt{a^2+b^2}$. The legs of the hyperbola approach the asymptotes, lines of slope $\pm b/a$ that cross at the center.

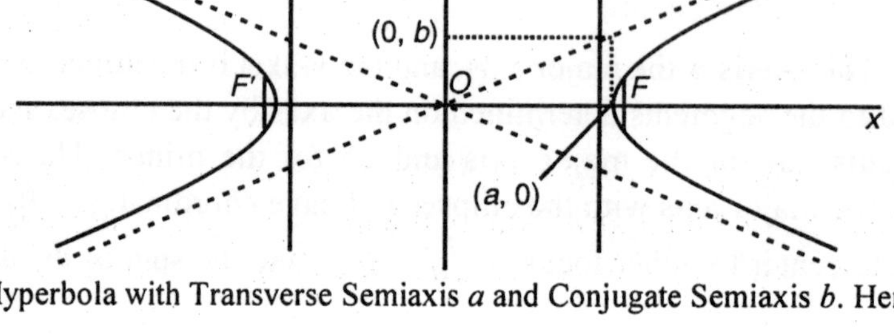

Fig. Left: Hyperbola with Transverse Semiaxis a and Conjugate Semiaxis b. Here $b/a = 0.4$.

All hyperbolas of the same eccentricity are similar; in other words, the shape of a hyperbola depends only on the ratio b/a. Unlike the case of the ellipse (where the major axis, containing the foci, is always longer than the minor axis), the two axes of a hyperbola can have arbitrary lengths. When they have the same length, so that $a = b$, the asymptotes are perpendicular, and $e = \sqrt{2}$, the hyperbola is called rectangular.

The simplest analytic form for the parabola is obtained when the symmetry axis coincides with one coordinate axis and the vertex (the intersection of the axis with the curve) is at the origin. The parabola on the right has equation

$$y^2 = 4ax,$$

where a is the distance from the vertex to the focus, or, which is the same, from the vertex to the directrix. The latus rectum is the chord perpendicular to the axis and going through the focus; its length is $4a$. All parabolas are similar: they can be made to look identical by scaling, translation and rotation.

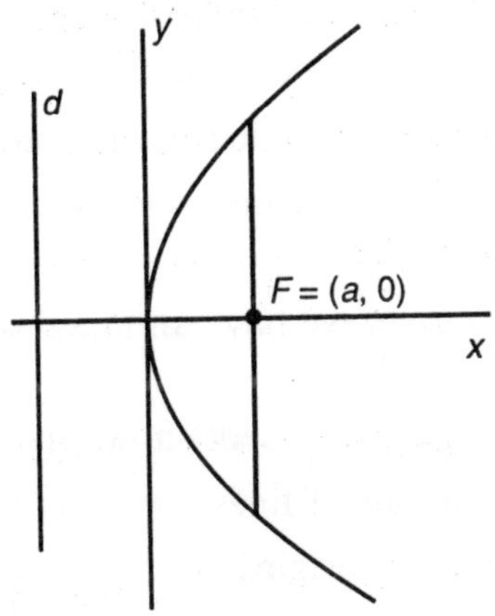

THE GENERAL QUADRATIC EQUATION

The analytic equation for a conic in arbitrary position is the following:

$$Ax^2 + By^2 + Cxy + Dx + Ey + F = 0,$$

where at least one of A, B, C is nonzero. To reduce this to one of the forms given previously, perform the following steps (note that the decisions are based on the most recent values of the coefficients, taken after all the transformations so far):

- If $C \neq 0$, simultaneously perform the substitutions $x \to \theta x + y$ and $y \to \theta y - x$, where

$$\theta = \sqrt{\left(\frac{B-A}{C}\right)^2 + 1} + \frac{B-A}{C}.$$

Now $C = 0$. (This step corresponds to rotating and scaling about the origin.)

- If $B = 0$, interchange x and y. Now $B \neq 0$.

- If $E \neq 0$, perform the substitution $y \to y$ -½ (E/B). (This corresponds to translating in the y direction.) Now $E = 0$.
- If $A = 0$:
- If $D \neq 0$, perform the substitution $x \to x$-(F/D) (translation in the x direction) and divide the equation by B to get (3) . The conic is a parabola.
- If $D = 0$, the equation gives a degenerate conic. If $F = 0$, we have the line $y = 0$ with multiplicity two. If $F < 0$ we have two parallel lines $y = \pm\sqrt{F/B}$. If $F > 0$ we have two imaginary lines; the equation has no solution over the real numbers.
- If $A \neq 0$:
- If $D \neq 0$, perform the substitution $x \to x$ -½ (D/A). Now $D = 0$. (This corresponds to translating in the x direction.)
- If $F \neq 0$, divide the equation by F to get a form with $F = 1$.
- If A and B have opposite signs, the conic is a hyperbola; to get to (2) , interchange x and y if necessary so that A is positive, then make $a = 1/\sqrt{A}$ and $b = 1/\sqrt{B}$.
- If A and B are both positive, the conic is an ellipse; to get to (1) , interchange x and y if necessary so that $A \leq B$, then make $a = 1/\sqrt{A}$ and $b = 1/\sqrt{B}$ .The circle is the particular case $a = b$.
- If A and B are both negative, we have an imaginary ellipse; the equation has no solution over the real numbers.
- If $F = 0$, the equation again represents a degenerate conic: when A and B have different signs we have a pair of lines $y = \pm\sqrt{-B/A}\ x$, and when they have the same sign we get a point (the origin).

We work out an example for clarity. Suppose the original equation is

$$4x^2 + y^2 - 4xy + 3x - 4y + 1 = 0$$

In step 1 we apply the substitutions $x \to 2x + y$ and $y \to 2y - x$. This gives $25x^2 + 10x - 5y + 1 = 0$. Next we interchange x and y (step 2) and get $25y + 10y - 5x + 1 = 0$. Replacing y by $y - \frac{1}{5}$ in step 3 we get $25y - 5x = 0$. Finally, in step $4a$ we divide the equation by 25, thus giving it the form (3) with $a = \frac{1}{20}$. We have reduced the conic to a parabola with vertex at the origin and focus at $(\frac{1}{20}, 0)$. To locate the features of the original curve, we work our way back along the chain of substitutions :

$$\text{substitution} \qquad y - y - \frac{1}{5} \quad \begin{matrix} x \to y \\ y \to x \end{matrix} \quad \begin{matrix} x \to 2x + y \\ y \to 2y - x \end{matrix}$$

vertex	$(0, 0)$	$\left(0, -\frac{1}{5}\right)$.	$\left(-\frac{1}{5}, 0\right)$	$\left(-\frac{2}{5}, -\frac{1}{5}\right)$
focus	$\left(\frac{1}{20}, 0\right)$	$\left(\frac{1}{20}, -\frac{1}{5}\right)$	$\left(-\frac{1}{5}, -\frac{1}{20}\right)$	$\left(-\frac{7}{20}, -\frac{8}{20}\right)$

We conclude that the original curve (3) is a parabola with vertex $\left(-\frac{2}{5}, -\frac{1}{5}\right)$ and focus $\left(-\frac{7}{20}, -\frac{8}{20}\right)$.

If one just wants to know the type of the conic defined by (1) , an alternative analysis consists in forming the quantities

$$\Delta = \begin{vmatrix} A & \frac{1}{2}C & \frac{1}{2}D \\ \frac{1}{2}C & B & \frac{1}{2}E \\ \frac{1}{2}D & \frac{1}{2}E & F \end{vmatrix}, \quad f = \begin{vmatrix} A & \frac{1}{2}C \\ \frac{1}{2}C & B \end{vmatrix}, \quad 1 = A + B,$$

$$K = \begin{vmatrix} A & \frac{1}{2}D \\ \frac{1}{2}D & F \end{vmatrix} + \begin{vmatrix} B & \frac{1}{2}E \\ \frac{1}{2}E & F \end{vmatrix},$$

and finding the appropriate case in the following table, where an entry in parentheses indicates that the equation has no solution over the real numbers:

Δ	J	Δ/l	K	Type of Conic
$\neq 0$	< 0			Hyperbola
$\neq 0$	0			Parabola
$\neq 0$	> 0	< 0		Elllipse
$\neq 0$	> 0	> 0		(Imaginary ellipse)
0	< 0			Intersecting lines
0	> 0			Point
0	0	< 0		Distinct parallel lines
0	0	> 0		(Imaginary parallel lines)
0	0	0		Coincident lines

For the central conics (the ellipse, the point, and the hyperbola), the center (x^0, y^0) is the solution of the system of equations

$$2Ax + Cy + D = 0,$$
$$Cx + 2By + E = 0,$$

and the axes have slope q and $-1/q$, where q is given by (2) .

ADDITIONAL PROPERTIES OF ELLIPSES

Let C be the ellipse with equation $x/a + y/b = 1$, with $a > b$, and let $F, F' = (\pm\sqrt{a^2 - b^2}, 0)$ be its foci.

- A parametric representation for C is given by

$$(a \cos\theta, b \sin\theta).$$

The area of the shaded sector below is

$$\tfrac{1}{2} ab\,\theta = \tfrac{1}{2} ab \arccos(x/a).$$

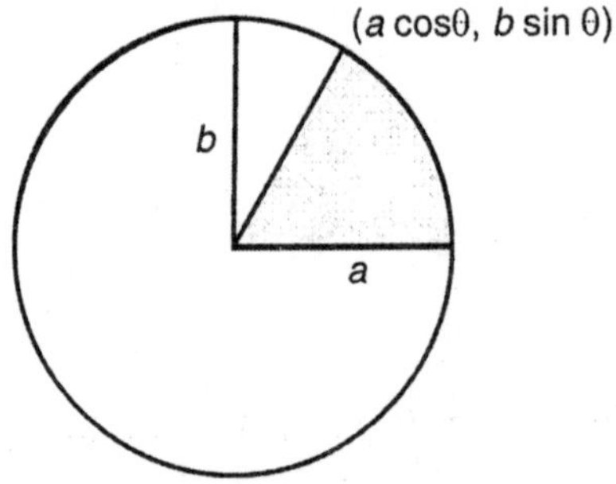

The length of the arc from $(a,0)$ to the point $(a \cos\theta, b \sin\theta)$ is given by $a(E(pi/2,e) - E(pi/2 - theta, e))$, where E is an elliptic integral. Setting $\theta = 2\pi$ we get

$$\text{area } C = \pi ab,$$

$$\text{perimeter } C = 4a\, E(pi/2,e).$$

- A rational parametric representation for C is given by

$$\left(a\frac{1-t^2}{1+t^2}, \frac{2bt}{1+t^2} \right).$$

- The polar equation for C in the usual polar coordinate system is

$$r = \frac{ab}{\sqrt{a^2 \sin\theta + b^2 \cos^2\theta}}.$$

With respect to a coordinate system with origin at a focus, the equation is

$$r = \frac{l}{1 \pm e\cos\theta},$$

where $l = b^2/a$ is half the latus rectum. (Use the + sign for the focus with positive x-coordinate and the - sign for the other.)

- Let P be any point of C. The sum of the distances PF and PF' is constant, and equal to $2a$.

- Let P be any point of C. Then the rays PF and PF' make the same angle with the tangent to C at P. Thus any light ray originating at F and reflected in the ellipse will also go through F'.
- Let T be any line tangent to C. The product of the distances from F and F' to T is constant, and equals b .
- Lahire's theorem: Let D and D' be fixed lines in the plane, and consider a third, moving, line on which three points P, P' and P$''$ are marked. If we constrain P to lie in D and P' to lie in D', then P'' describes an ellipse.

ADDITIONAL PROPERTIES OF HYPERBOLAS

Let C be the hyperbola with equation $x^2/a^2 - y^2/b^2 = 1$, and let $F, F' = (\pm\sqrt{a^2+b^2}, 0)$ be its foci. The conjugate hyperbola of C is the hyperbola C with equation $-x^2/a^2 + y^2/b^2 = 1$. It has the same asymptotes as C, the same axes (transverse and conjugate axes being interchanged), and its eccentricity e' is related to that of C by $1/e'^2 + 1/e^2 = 1$.

- A parametric representation for C is (a sec q , b tan q).

  A different parametric representation, which gives one branch only, is

  $$(a \cosh \theta,\ b \sinh \theta):$$

  The area of the shaded sector above is

  $$\frac{1}{2}ab\theta = \frac{1}{2}ab \text{ arccosh } (x/a) = \frac{1}{2}ab \log \frac{x+\sqrt{x^2-a^2}}{a}.$$

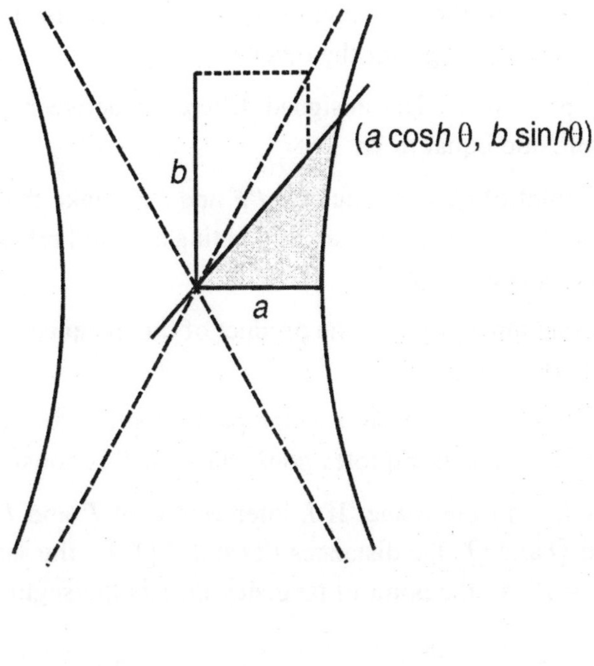

The length of the arc from $(a,0)$ to the point

$$(a \cosh \theta, b \sinh \theta)$$

is given by the elliptic integral

$$a \int_0^\theta \sqrt{e^2 \cosh^2 \phi - 1}\, d\phi = -bi\, E\left(\theta i, \frac{ea}{b}\right) = a \int_1^x \sqrt{\frac{e^2 \xi^2 - a^2}{\xi^2 - a^2}}\, d\xi,$$

where e is the eccentricity, $i = \sqrt{-1}$, and $x = a \cosh \theta$.

- A rational parametric representation for C is given by $\left(a\frac{1+t^2}{1-t^2}, \frac{2bt}{1-t^2}\right)$.

- The polar equation for C in the usual polar coordinate system is

$$r = \frac{ab}{\sqrt{a^2 \sin^2 \theta - b^2 \cos^2 \theta}}.$$

With respect to a system with origin at a focus the equation is

$$r = \frac{l}{1 \pm e \cos \theta},$$

where $l = b^2/a$ is half the latus rectum. (Use the - sign for the focus with positive x-coordinate and the + sign for the other.)

- Let P be any point of C. The unsigned difference between the distances PF and PF' is constant, and equal to $2a$.
- Let P be any point of C. Then the rays PF and PF' make the same angle with the tangent to C at P. Thus any light ray originating at F and reflected in the hyperbola will appear to emanate from F'.
- Let T be any line tangent to C. The product of the distances from F and F' to T is constant, and equals b^2.
- Let P be any point of C. The area of the parallelogram formed by the asymptotes and the parallels to the asymptotes going through P is constant, and equals ½ ab.
- Let L be any line in the plane. If L intersects C at P and P', and intersects the asymptotes at Q and Q', the distances PQ and $P'Q'$ are the same. If L is tangent to C we have $P = P'$, so the point of tangency bisects the segment QQ'.

ADDITIONAL PROPERTIES OF PARABOLAS

Let C be the parabola with equation $y^2 = 4ax$, and let $F = (a,0)$ be its focus.

- Let $P = (x,y)$ and $P' = (x',y')$ be points on C. The area bounded by the chord PP' and the corresponding arc of the parabola is

$$\frac{|y'-y|^2}{24a}.$$

It equals four-thirds of the area of the triangle PQP', where Q is the point on C whose tangent is parallel to the chord PP' (formula due to Archimedes). 11

- The length of the arc from (0,0) to the point (x,y) is

$$\frac{y}{4}\sqrt{4+\frac{y^2}{a^2}}+a\operatorname{arcsin}h\left(\frac{y}{2a}\right)=\frac{y}{4}\sqrt{4+\frac{y^2}{a^2}}+a\log\frac{y+\sqrt{y^2+4a^2}}{2a}.$$

- The polar equation for C in the usual polar coordinate system is

$$r=\frac{4a\cos\theta}{\sin^2\theta}.$$

With respect to a coordinate system with origin at F the equation is

$$r=\frac{l}{1-\cos\theta},$$

where $l = 2a$ is half the latus rectum.

- Let P be any point of C. Then the ray PF and the horizontal line through P make the same angle with the tangent to C at P. Thus light rays parallel to the axis and reflected in the parabola converge onto F (principle of the parabolic reflector).

Chapter 10

Hyperbolic Geometry

In mathematics, hyperbolic geometry is a non-Euclidean geometry, meaning that the parallel postulate of Euclidean geometry is replaced. The parallel postulate in Euclidean geometry states, in two dimensional space, for any given line l and point P not on l, there is exactly one line through P that does not intersect l; i.e., that is parallel to l. In hyperbolic geometry there are at least two distinct lines through P which do not intersect l, so the parallel postulate is false. Models have been constructed within Euclidean geometry that obey the axioms of hyperbolic geometry, thus proving that the parallel postulate is independent of the other postulates of Euclid.

Since there is no precise hyperbolic analogue to Euclidean parallel lines, the hyperbolic use of *parallel* and related terms varies among writers. In this article, the two limiting lines are called *asymptotic* and lines sharing a common perpendicular are called *ultraparallel*; the simple word *parallel* may apply to both.

THE HYPERBOLIC AXIOM AND ITS CONSEQUENCES[MH]

What is *hyperbolic geometry*? It is an example of one of many non-euclidean geometries. It is often called *Bolyai-Lobachevskiian geometry* after two of its *discovers* János Bolyai and Nikolai Ivanovich Lobachevsky. Bolyai first announced his discoveries in a 26 page appendix to a book by his father, the *Tentamen*, in 1831. You can read about what happened to him and his work in the text. Another of the great mathematicians who seems to have preceeded Bolyai in his work is Carl Fredrich Gauss.

He seems to have done some work in the area dating from 1792, but never published it. The first to publish a complete account of non-euclidean geometry was Lobachevsky in 1829. It was first published in Russian and was not widely read. In 1840 he published a treatise in German. We shall add one more axiom to the list of axioms for Neutral Geometry.

HYPERBOLIC AXIOM

In hyperbolic geometry there exists a line l and a point P not on l such that at least two distinct lines parallel to l pass through P.

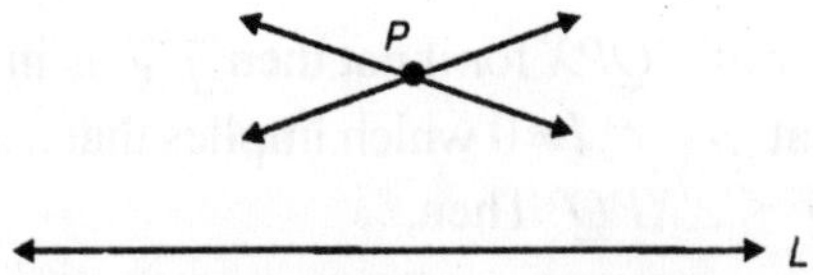

We shall denote the set of all points in the plane by H^2 and call this the *hyperbolic plane*.

Lemma: *There exists a triangle whose angle sum is less than* 180°

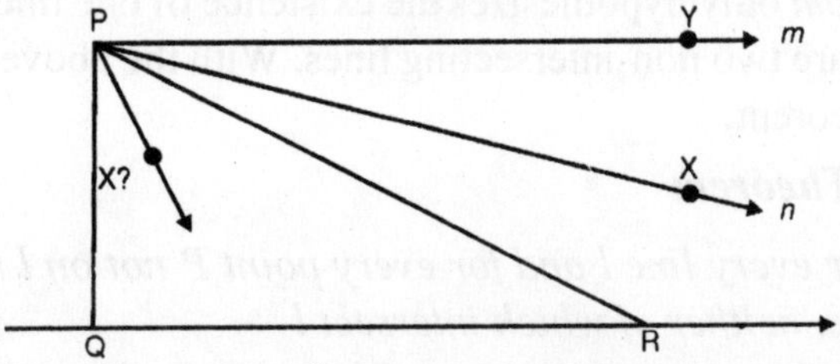

Proof: Let l be a line and P a point not on l such that two parallels to l pass through P. We can construct one of these parallels as previously done using perpendiculars. Let Q be the foot of the perpendicular to l through P. Let m be the perpendicular to $\overleftrightarrow{PQ}$ through P. Then m and l are non-intersecting. Let n be another line through P which does not intersect l. This line exists by the *Hyperbolic Axiom*. Let $\overrightarrow{PX}$ be a ray of n lying between $\overrightarrow{PQ}$ and a ray $\overrightarrow{PY}$ of m.

Claim: There is a point $R \in l$ on the same side of $\overleftrightarrow{PQ}$ as X and Y so that $\angle QRP < \angle XPY$.

PROOF OF CLAIM. The idea is to construct a sequence of angles

$\angle QR_1 P, \angle QR_2 P, ..., \angle AR_n P, ...$

So that $\angle QR_{j+1} P° < \frac{1}{2} \angle QRj P°$ We will then apply Archimedes Axiom for real numbers to complete the proof.

There is a point $R_1 \in l$ so that $QR_1 \cong PQ$ Then ÄQR_1P is isosceles and $\angle QR_1P° \leq 45°$ Also, there is a point $R2 \in l$ so that $F * R_1 * R_2$ and $R_1R_2 \cong PR_1$ Then ÄPR_1R_2 is isosceles and $\angle R_1PR_2 \cong \angle QR_2P$ Since $\angle QR_1P$ is exterior to ÄPR_1R_2 it follows that.

$$\angle R_1PR°_2 + \angle QR_2P° \leq \angle QR_1P°,$$

So then $\angle QR_2P° \leq 22\frac{1}{2}°$ Continuing with this construction, we find a point $R_n \in l$ so that $Q * R_{n-1} * R_n$ and

$$\angle QR_nP° \leq \left(\frac{45}{2^n}\right)°.$$

Applying the Archimedean axiom we see that for any positive real number, for example $\angle XPY°$ there is a point $R \in l$ so that R is on the same side of $\overleftrightarrow{PQ}$ as X and Y and $\angle QRP° < \angle XPY°$ Thus, we have proved our claim.

Now, $\overrightarrow{PR}$ lies in the interior of $\angle QPX$ for if not then $\overrightarrow{PX}$ is in the interior of $\angle QRP$ By the *Crossbar Theorem* it follows that $\overrightarrow{PX} \cap l \neq \theta$ which implies that n and l are not non-intersecting --a contradiction. Thus, $\angle RPQ° < \angle XPQ°$ Then,

$$\angle RPQ° + \angle QRP° < \angle XPQ° + \angle QRP° < \angle XPQ° + \angle XPY° = 90°.$$

Therefore, $\angle P° + \angle Q° + \angle R° < 180°$ and *defect* (ÄPQR) > 0.

The *Hyperbolic Axiom* only hypothesizes the existence of one line and one point not on that line for which there are two non-intersecting lines. With the above theorem we can now prove a much stronger theorem.

Universal Hyperbolic Theorem

Theorem: *In* H^2 *for every line l and for every point P not on l there pass through P at least two distinct lines, neither of which intersect l*

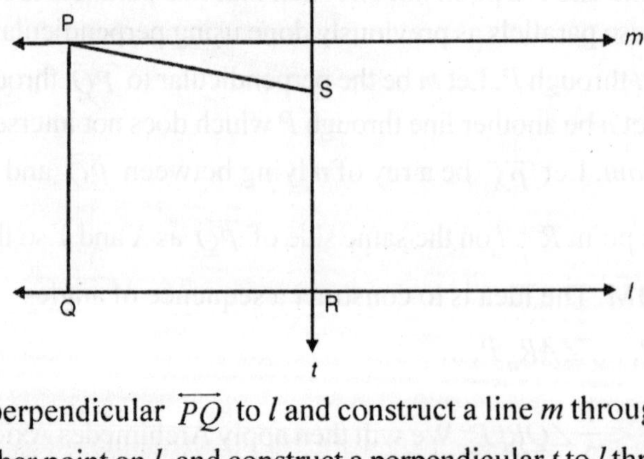

Proof: Drop a perpendicular $\overleftrightarrow{PQ}$ to l and construct a line m through P perpendicular to $\overleftrightarrow{PQ}$. Let R be any other point on l, and construct a perpendicular t to l through R. Now, let S be the foot of the perpendicular to t through P. Now, $\overleftrightarrow{PS}$ does not intersect l since both are perpendicular to t. At the same time $\overleftrightarrow{PS} \neq m$ Assume that $S \in m$ then $PQRS$ is a rectangle. By Theorem, if one rectangle exists all triangles have defect 0. We have a contradiction to Lemma. Thus, $\overleftrightarrow{PS} \neq m$, and we are done.

ANGLE SUMS

We have just proven the following theorem.

Theorem: *In* H^2 *rectangles do not exist and all triangles have angle sum less than* 180°.

This tells us that in hyperbolic geometry the defect of any triangle is a *positive* real number. We shall see that it is a very important quantity in hyperbolic geometry.

Corollary: *In* H^2 *all convex quadrilaterals have angle sum less than* 360°

Proof: Given any quadrilateral *ABCD* Take the diagonal *AC* and consider triangles Ä*ABC* and Ä*ACD* By Theoremboth of these triangles have angle sum less than 180° The assumption that ŒABCD is convex implies that $\angle BAC^\circ + \angle CAD^\circ = \angle BAD^\circ$ and $\angle ACB^\circ + \angle ACD^\circ = \angle BCD^\circ$ By adding all six angles we have that the angle sum of ŒABCD is less that 360°

SIMILAR TRIANGLES

In euclidean geometry we are used to having two triangles similar if their angles are congruent. It is obvious that we can construct two non-congruent, yet similar, triangles. In fact John Wallis attempted to prove the Parallel Postulate of Euclid by adding another postulate.

WALLIS' POSTULATE

Given any triangle ÄABC and given any segment DE. There exists a triangle ÄDEF having DE as one of its sides that is similar to ÄABC.

However, it is proven that Wallis' Postulate is equivalent to Euclid's Parallel Postulate. Thus, we know that the negation of Wallis' Postulate must hold in hyperbolic geometry. That is, under certain circumstances similar triangles do not exist. We can prove a much stronger statement.

AAA Criterion

Theorem: *In* H^2 *if* $\angle A \cong \angle D$, $\angle B \cong \angle E$ *and* $\angle C \cong \angle F$ *then ÄABC ≅ ÄDEF That is, if two triangles are similar, then they are congruent.*

Proof: Assume that $\Delta ABC \not\cong \Delta DEF$ Then there can be no corresponding sides which are congruent, else by *ASA* the triangles would be congruent. Consider the triples $\{AB, BC, AC\}$ and $\{DE, EF, DF\}$ One of these triples contains at least two segments that are larger than the corresponding segments of the other. Without loss of generality, we can assume that

$$AB > DE \text{ and } AC > DF.$$

There exist points $B' \in AB$ and $C' \in AC$ such that $AB' \cong DE$ and $AC' \cong DF$ Then by *SAS* $\Delta AB'C' \cong \Delta DEF$. Then $\angle AB'C' \cong \angle E \cong \angle B$ and $\angle AC'B' \cong \angle F \cong \angle C$ Thus, by the *Alterate Interior Angles Theorem* we can show that $\overleftrightarrow{B'C'}$ does not intersect $\overleftrightarrow{BC}$ Thus, it follows that ŒBB'C'C is convex. Using the fact that $\angle AB'C'$ and $\angle BB'C'$ are supplementary, as are $\angle AC'B'$ and $\angle CC'B'$ the angle sum of ŒBB'CC' is 360°, contradicting Corollary.

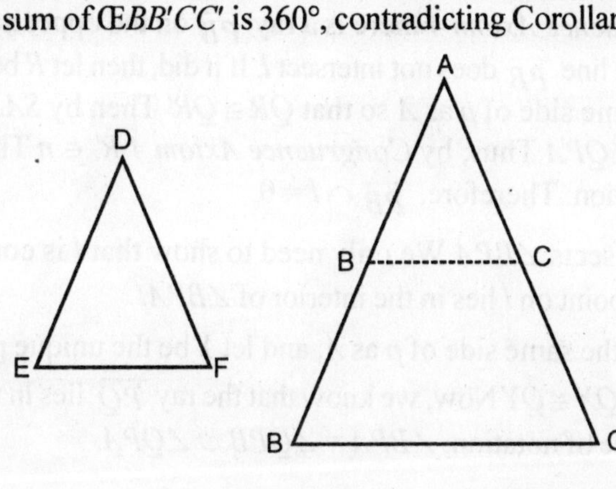

As a consequence of Theorem we shall see that in hyperbolic geometry a segment can be determined with the aid of an angle. For example, an angle of an equilateral triangle determines the length of a side uniquely. Thus in hyperbolic geometry there is an *absolute unit of length* as there is in elliptic geometry.

CLASSIFICATION OF PARALLELS

FAN ANGLES

A line l subdivides an angle if it passes through the vertex of the angle and intersects the interior of the angle. Let P be an arbitrary point in the hyperbolic plane. The pencil of lines through P, denoted $P(P)$ is the set of all lines coincident with P.

Theorem: *If $P \notin l$ then there exists an angle with vertex at P whose interior contains l and whose bisector is perpendicular to l.*

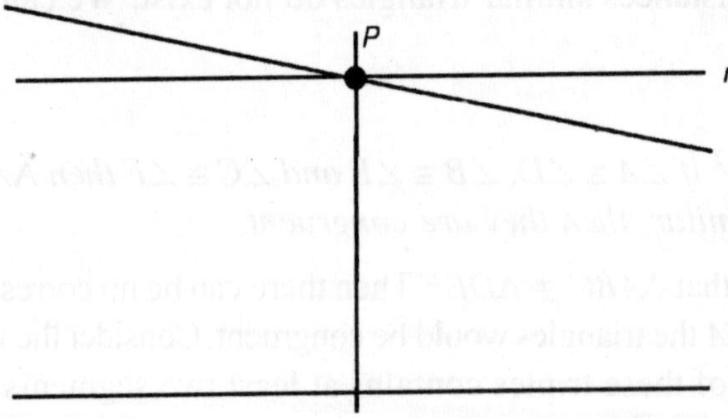

Proof: In the pencil $P(P)$ let p be the perpendicular to l and let $m \in P(p)$ be perpendicular to p at P. Thus, we know that $m \cap l = \theta$ By the *Universal Hyperbolic Theorem* there is another line $n \in p(P)$ which does not intersect l. Since n is distinct from m, n is not perpendicular to p. Thus, one pair of the vertical angles formed by p and n is a pair of acute angles. Thus, there exists a ray $\overrightarrow{PA} \subset n$ so that

where Q is the foot of P in l.

Now, by *Congruence Axiom 4* there is a ray $\overrightarrow{PB}$ on the opposite side of p from A so that $\angle QPB \cong \angle QPA$. The line $\overrightarrow{PB}$ does not intersect l. If it did, then let R be the point of intersection. Choose R' on the same side of p as A so that $QR \cong QR'$ Then by *SAS* ÄPRQ ≅ ÄPR'Q Thus, $\angle QPR \cong \angle QPR \cong \angle QPA$ Thus, by *Congruence Axiom 4* $R' \in n$ This implies that $n \cap l \neq \theta$ which is a contradiction. Therefore, $\overrightarrow{PB} \cap l = \theta$.

Clearly, $\overrightarrow{PQ}$ bisects $\angle BPA$ We only need to show that l is contained in the interior of $\angle BPA$ or that every point on l lies in the interior of $\angle BPA$.

Let $X \in l$ be on the same side of p as A, and let Y be the unique point on l on the opposite side of p as X so that $QX \cong QY$ Now, we know that the ray $\overrightarrow{PQ}$ lies in the interior of $\angle APB$ and, in fact, using an abuse of notation, $\angle BPA = \angle QPB \cup \angle QPA$.

Claim: $\overrightarrow{QX}$ lies in the interior of $\angle QPA$ Likewise, $\overrightarrow{QY}$ lies in the interior of $\angle QPB$.

To prove this claim, first note that since X is on the same side of $P = \overleftrightarrow{PQ}$ as A, every point of $\overrightarrow{QX}$ lies on the same side of p as A. Thus, we are left only to show that if $R \in \overrightarrow{QX}$ then R is on the same side of $n = \overleftrightarrow{PA}$ as Q. This must be the case, for if not we would have that $n \cap l \neq \theta$, a contradiction. Thus, the above claim is true.

We then have that

$$l = \overrightarrow{QY} \cup \{Q\} \cup \overrightarrow{QX},$$

which from above must lie in the interior of $\angle APB$

Corollary: *Every line in p(P) that intersects l subdivides the angle ∠APB.*

We have almost proven the following result.

Theorem: *If $P \notin l$ then there is a unique angle with the following properties:*

- *The angle contains l in its interior;*
- *The bisector of this angle is perpendicular to l;*
- *Every line passing through P that subdivides the angle intersects l*

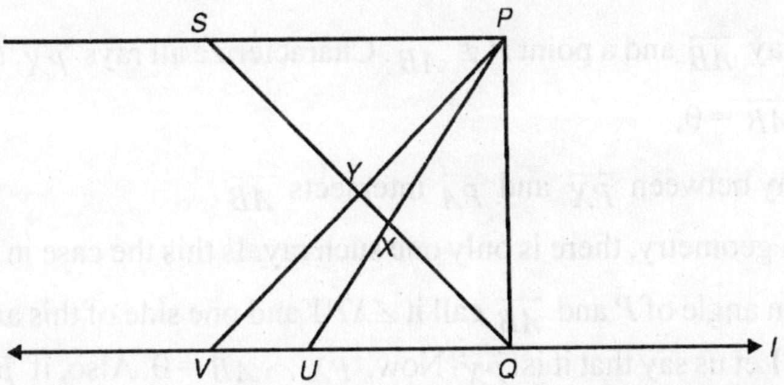

Proof: To prove that this angle exists, we need to construct the sides of the angle. We already know that there are angles with vertex at P that contain the line l. We want to prove much more in this theorem. In some sense we are proving that there is a smallest such angle.

Let S be a point on the line m which is the perpendicular at P to the perpendicular through P to l. Consider the line $\overleftrightarrow{SQ}$ Let Ó$_1$ be the set of all points T on SQ, such that $\overrightarrow{PT}$ intersects l together with all points on the ray opposite to $\overrightarrow{QS}$ Let Ó$_2$ = $\overleftrightarrow{SQ}$ \ Ó$_1$ By the *Crossbar Theorem* if $T \in SQ$ and $T \in$ Ó$_1$ then the entire subsegment $TQ \subset$ Ó$_1$ Hence, (Ó$_1$, Ó$_2$) is a Dedekind cut. By Dedekind's Axiom there is a unique point X on $\overleftrightarrow{SQ}$ such that for $P_1, P_2 \in \overleftrightarrow{SQ}, P_1 * X * P_2$ if and only if $X \neq P_1$ and $X \neq P_2$, $P_1 \in$ Ó$_1$ and $P_2 \in$ Ó$_2$.

By the definition of Ó$_1$ and Ó$_2$ rays below $\overrightarrow{PX}$ all intersect l and rays above $\overrightarrow{PX}$ do not. Now, we wish to prove that $\overrightarrow{PX} \cap l = \theta$.

Assume on the contrary that $\overrightarrow{PX} \cap l = \{U\}$ Choose a point V on l so that $V * U * Q$ Since V and U are on the same side of $\overleftrightarrow{SQ}$, V and P are on opposite sides, so that VP meets SQ in a point Y. By Propositionwe have that $Y * X * Q$. Thus, $Y \in Ó_2$ This is impossible, for if $Y \in Ó_2$, then $\overrightarrow{PY} \cap l = \theta$ Thus, $\overrightarrow{PX}$ does not meet

Doing the same construction on the opposite side of $\overrightarrow{PQ}$ at P gives us a similar ray $\overrightarrow{PX'}$ with similar properties. We need to show that $\angle QPX \cong \angle QPX'$ Assume not, then we may assume that $\angle QPX < \angle QPX'$ By Congruence Axiom 4 there is a ray $\overrightarrow{PR'}$ between $\overrightarrow{PX'}$ and $\overrightarrow{PQ}$ such that $\angle QPR' \cong \angle QPX$. By the properties of $\overrightarrow{PX'}$, $\overrightarrow{PR'}$ must intersect l. Let R be the point on l on the opposite of $\overrightarrow{PQ}$ from R' so that $QR \cong QR'$ Then by SAS $ÄQPR' \cong ÄQPR$ This implies that $\angle QPR \cong \angle QPR' \cong \angle QPX$. By Congruence Axiom 4 we have to have that $R \in \overrightarrow{PX}$ or $\overrightarrow{PX} \in l \neq \theta$, a contradiction. Thus, $\angle QPX \cong \angle QPX'$.

Checking the properties, clearly $\overrightarrow{PQ}$ bisects the angle, the angle contains l and by the definition of the ray $\overrightarrow{PX}$ any line that subdivides the angle intersects l.

For $P \notin l$ this angle at P is called the fan angle of P and l, and will be denoted by $\angle(P, l)$.

LIMITING PARALLEL RAYS

Consider a ray $\overrightarrow{AB}$ and a point $P \notin \overrightarrow{AB}$. Characterize all rays $\overrightarrow{PX}$ satisfying:

- $\overrightarrow{PX} \cap \overrightarrow{AB} = \theta$,
- Every ray between $\overrightarrow{PX}$ and $\overrightarrow{PA}$ intersects $\overrightarrow{AB}$.

In euclidean geometry, there is only one such ray. Is this the case in H^2 ?

There is a fan angle of P and $\overrightarrow{AB}$ call it $\angle XPY$ and one side of this angle lies on the same side of $\overleftrightarrow{AP}$ as B. Let us say that it is $\overrightarrow{PX}$ Now, $\overrightarrow{PX} \cap \overrightarrow{AB} = \theta$. Also, if $\overrightarrow{PR}$ lies between $\overrightarrow{PX}$ and $\overrightarrow{PA}$ then $\overrightarrow{PR}$ subdivides *angle* XPY. Thus, $\overrightarrow{PR} \cap \overleftrightarrow{AB} \neq \theta$, but due to its relative position with respect to $\overrightarrow{PA}$ and $\overrightarrow{PX}$ it must, in fact, intersect $\overrightarrow{AB}$ Thus, the side of the fan angle satisfies the two aforementioned conditions. Do any others?

Let $\overrightarrow{PT}$ be different from $\overrightarrow{PX}$.

If $\overrightarrow{PT}$ is on the line $\overleftrightarrow{PA}$ then $\overrightarrow{PT}$ nintersects $\overrightarrow{AB}$ and the ray $\overrightarrow{PT}$ does not have property

If $\overrightarrow{PT}$ lies on the same side of $\overleftrightarrow{PA}$ as Y, let $\overrightarrow{PU}$ be between $\overrightarrow{PA}$ and $\overrightarrow{PT}$ Then the line $\overleftrightarrow{PA}$ separates $\overrightarrow{PT}$ and $\overrightarrow{PU}$ from $\overrightarrow{AB}$ which means that $\overrightarrow{PU} \cap \overrightarrow{AB} = \theta$ so that $\overrightarrow{PT}$ does not satisfy property

If $\overrightarrow{PT}$ lies on the same side of $\overleftrightarrow{PA}$ as X, then one of $\overrightarrow{PT}$ and $\overrightarrow{PX}$ is between the other and $\overrightarrow{PA}$

- *If $\overrightarrow{PT}$ is between $\overrightarrow{PX}$ and $\overrightarrow{PA}$ then $\overrightarrow{PT}$ must intersect $\overrightarrow{AB}$ by the properties of the fan angle. Thus, $\overrightarrow{PT}$ would not satisfy property (1).*
- *If $\overrightarrow{PX}$ is between $\overrightarrow{PA}$ and $\overrightarrow{PT}$ then $\overrightarrow{PT}$ cannot satisfy property (2), for $\overrightarrow{PX} \cap \overrightarrow{AB} = \theta$.*

Thus, $\overrightarrow{PX}$ is the only ray at P with these two properties. This gives rise to the following definition.

Definition: A ray $\overrightarrow{PQ}$ is parallel to a ray $\overrightarrow{AB}$ if

- $\overleftrightarrow{PQ} \cap \overleftrightarrow{AB} = \theta$,
- Every ray between $\overrightarrow{FQ}$ and $\overrightarrow{PA}$ intersects $\overrightarrow{AB}$.

We have just shown that this ray is unique. In this case the rays $\overrightarrow{PQ}$ and $\overrightarrow{AB}$ are called limiting parallel rays and we say that $\overrightarrow{PQ}$ is *limiting parallel* to $\overrightarrow{AB}$.

QUESTION: If $\overrightarrow{PQ}$ is limiting parallel to $\overrightarrow{AB}$ is $\overrightarrow{AB}$ is limiting parallel to $\overrightarrow{PQ}$?

We need to following lemma for the proof of the main theorem. I leave the proof to you as an exercise.

Lemma: *If $\angle PAB$ is acute, then the foot of P in the line $\overleftrightarrow{AB}$ lies in the ray $\overrightarrow{AB}$ and is different from A.*

I offer the following theorems without proof.

Theorem: *If $\overrightarrow{PQ}$ is limiting parallel to $\overrightarrow{AB}$ then $\overrightarrow{AB}$ lies in the interior of the angle $\angle APQ$.*

Theorem: *Let $P \notin l$ and let $A * B * C$ in l. Then $\overrightarrow{PQ}$ is limiting parallel to $\overrightarrow{AB}$ if and only if $\overrightarrow{PQ}$ is limiting parallel to $\overrightarrow{BC}$.*

Theorem: *Let $A * B$ C on l $\overrightarrow{AB}$ is limiting parallel to $\overrightarrow{DE}$ if and only if $\overrightarrow{BC}$ is limiting parallel to $\overrightarrow{DE}$.*

Theorem: *If $\overrightarrow{PQ}$ is limiting parallel to $\overrightarrow{AB}$ then $\overrightarrow{AB}$ is limiting parallel to $\overrightarrow{PQ}$.*

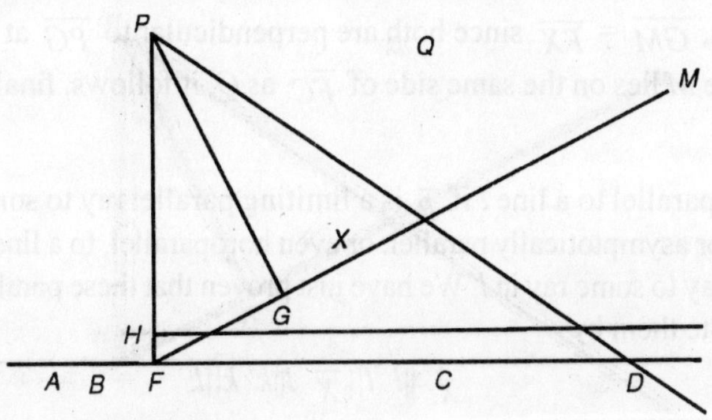

Proof: Let F be the foot of P in $\overrightarrow{AB} = l$ Let $C \in l$ so that $\overrightarrow{FC}$ is like directed to $\overrightarrow{AB}$ This means that either $\overrightarrow{AB} \subset \overrightarrow{FC}$ or $\overrightarrow{FC} \subset \overrightarrow{AB}$ It follows from Theorem that $\overrightarrow{PQ}$ is limiting parallel to $\overrightarrow{FC}$ so that C lies on the same side of $\overleftrightarrow{PF}$ as does Q and $\overrightarrow{FC} \cap \overrightarrow{PQ} = \theta$ Let $\overrightarrow{FX}$ lie in the interior of $\angle PFC$

Claim: $\overrightarrow{FX} \cap \overrightarrow{PQ} \neq \theta$.

From this it follows that $\overrightarrow{FC}$ is limiting parallel to $\overrightarrow{PQ}$ Applying Theorem again, we have that $\overrightarrow{AB}$ is limiting parallel to $\overrightarrow{PQ}$ Thus, we need to establish this claim to finish the proof.

Since $\angle PFX° < \angle PFC° = 90°$, $\angle PFX$ is acute. Thus, by Lemmathe foot of P in $\overleftrightarrow{FX}$ must lie in the ray $\overrightarrow{FX}$ Label this point G. It follows that G lies on the same side of $\overleftrightarrow{PF}$ as C.

If $G \in \overleftrightarrow{PQ}$ then it must lie in $\overrightarrow{PQ}$. It follows that $\overrightarrow{FX} \cap \overrightarrow{PQ} \neq \theta$, and we are done.

If G lies on the opposite side of $\overleftrightarrow{PQ}$ from F, then again $\overrightarrow{FX}$ intersects $\overrightarrow{PQ}$ and we are done.

Assume then, as the final case, that G lies on the same side of $\overleftrightarrow{PQ}$ as F. Then, this puts G in the interior of $\angle QPF$ Let á° = $\angle Q° PG$ We have that á° < $\angle QPF°$ By Congruence Axiom 4 there is a unique ray $\overrightarrow{PZ}$ in the interior of $\angle QPF$ so that

$$\angle FPZ° = á° = \angle QPG°$$

Since $\overrightarrow{PQ}$ is limiting parallel to $\overrightarrow{FC}$, $\overrightarrow{PZ}$ must intersect in some point D. From Lemma $G \neq F$ so that $PG<PF$ by an Exercise (the hypotenuse of a right triangle is longer than either leg). Thus, there is a point $H \in PF$ so that $PH \cong PG$.

Let $\overleftrightarrow{HY}$ be perpendicular to $\overleftrightarrow{PF}$ at H. Since $\overleftrightarrow{AB}$ is perpendicular to $\overleftrightarrow{PF}$ it follows that $\overleftrightarrow{AB} \cap \overleftrightarrow{HY} = \theta$ Now, $\overleftrightarrow{HY}$ does intersect one side of the triangle ÄPFD By *Pasch's Theorem* it must intersect a second side. That side must be PD. Let the point of intersection be E. Now, $E \neq P$ and $E \neq D$.

On $\overrightarrow{PQ}$ there is a unique point M so that $PM \cong PE$ Recalling that $PH \cong PG$ and $\angle HPE = \angle FPZ \cong \angle QPG$ we then have that Ä$PHE \cong$ ÄPGM by *SAS*. Thus, $\angle PGM \cong \angle PHE$ and is a right angle. Thus, $\overleftrightarrow{GM} = \overleftrightarrow{FX}$ since both are perpendicular to $\overleftrightarrow{PG}$ at G. Therefore, $\overleftrightarrow{FX} \cap \overrightarrow{PQ} = \{M\}$ Since M lies on the same side of $\overleftrightarrow{FC}$ as Q, it follows, finally, that $M \in \overrightarrow{FX}$ and $\overrightarrow{FX} \in \overrightarrow{PQ} \neq \theta$.

A ray $\overrightarrow{X}$ is parallel to a line l if $\overrightarrow{R}$ is a limiting parallel ray to some ray in l. A line k is limiting parallel or asymptotically parallel, or even horoparallel, to a line l. if some ray in k is a limiting parallel ray to some ray in l. We have just proven that these parallelisms are symmetric and we may denote them by

$$\overrightarrow{X} \parallel l \quad l \parallel \overrightarrow{X} \quad l \parallel k \quad k \parallel l.$$

If $k = \overleftrightarrow{AB}$, $l = \overleftrightarrow{CD}$ and $\overrightarrow{AB} \parallel \overrightarrow{CD}$, then $\overrightarrow{AB}$ is said to be parallel to l in the direction of $\overrightarrow{CD}$ on l. Furthermore, k and are said to be parallel in the direction of $\overrightarrow{AB}$ on k and in the direction of $\overrightarrow{CD}$ on l.

Theorem: *If $P \notin l$ then there are exactly two lines through P that are limiting parallel to l. Each contains an arm of the fan angle $\angle(P, l)$ and they are limiting parallel to l in opposite directions.*

As we have mentioned several times already, there is no simple transitivity of parallelism in hyperbolic geometry. There is a weak form of transitivity.

Weak Transitivity of Parallels

Theorem: *Two lines parallel to a third in the same direction on the third are parallel to each other.*

HYPERPARALLEL LINES

Let k and l be lines in H^2 and let A_n $n = 1, 2, \ldots$ be points on k. Let A_n' be the foot of A_n in l. We say that the points $\{A_n\}$ are equidistant from l. if $A_iA_i' \cong A_jA_j'$ for all choices of i and j.

Theorem: *If l and l′ are non-intersecting lines in $H^{2,}$ then any set of points on l equidistant from l′ has at most two elements.*

Proof: Assume not; *i.e.*, assume that $A, B, C \in l$ are equidistant from l' Thus, the quadrilaterals Œ$AA'B'B$, $AA'CC$, and are all Saccheri quadrilaterals. Thus, $\angle A'AB \cong \angle B'BA$, $\angle AAC \cong \angle C'CA$, and $\angle B'BC \cong C'CB$. Stringing these together, we have

$$\angle BBA \cong \angle AAB \cong \angle C'CA \cong \angle B'BC.$$

Since $\angle B'BA$ and $\angle B'BC$ are supplementary, they are right angles. Thus all three quadrilaterals are rectangles, a contradiction.

This theorem states that at most two points at a time on l can be equidistant from l' It does not put a limit on the number of pairs of such points, though. We may have pairs (A,B) and (C,D) so that $AA' \cong BB'$ and $CC' \cong DD'$ but it is not possible that $AA' \cong CC'$ or any of the other possibilities. It may also occur that the number of pairs of such points will be zero. There is no guarantee that such a pair will exist. This will help us distinguish types of parallel lines.

Theorem: *Let l and l′ be non-intersecting lines for which there is a pair of points A, $B \in l$ which are equidistant from l′. Then, l and l′ have a common perpendicular segment which is the shortest segment between l and l′.*

Proof: $AA'B'B$ is a Saccheri quadrilateral. Its altitude is perpendicular to both l and l' by Theorem. Also, the quadrilateral formed by this altitude and any other segment from a point $X \in l$ and its foot $X' \in l'$ is a Lambert quadrilateral. By Theorem the altitude is less than XX' and hence is the shortest segment between l and l'.

***Theorem:** In H^2 if lines l and l′ have a common perpendicular segment MM’, then they are non-intersecting and MM’ is unique. Moreover, if A, B ∈ l so that M is the midpoint of AB then A and B are equidistant from l′.*

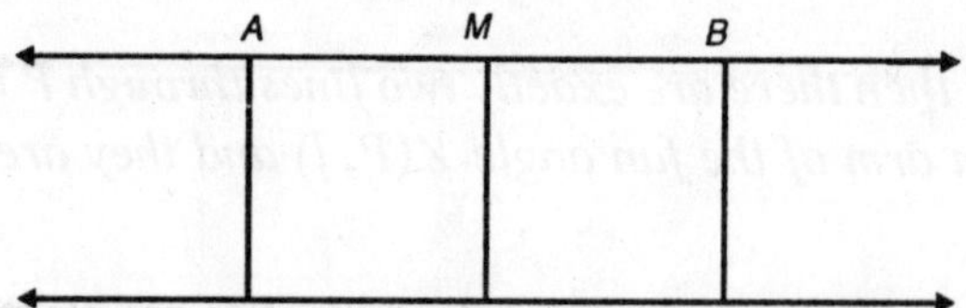

Proof: We already know that the lines do not intersect from Corollary to the *Alternate Interior Angles Theorem*. If *PP*’ were another common perpendicular then Œ*MM*′ *P*′*P* would be a rectangle, which cannot exist. Thus, *MM*′ is unique.

Let *A*, *B* ∈ *l* be so that *M* is the midpoint of *AB*. We have that *AM* ≅ *BM* and ∠*AMM*′ ≅ ∠*BMM*′ so by *SAS* Ä*AMM*′ ≅ Ä*BMM*′ Thus, *AM*′ ≅ *BM*′ and ∠*AMM* ≅ ∠*BM*′*M* By *Angle Subtraction* ∠*A*′*M*′*A* ≅ ∠*B*′*M*′*B*. Then by *AAS* Ä*AA*′*M*′ ≅ Ä*BB*′*M*′ and *AA*′ ≅ *BB*′.

***Lemma:** Let MM’ be the common perpendicular to l andl′ Let A, B ∈ l so that M * A * B, then AA’<BB’.*

If *k* and *l* are non-intersecting lines which admit a common perpendicular then they are said to be hyperparallel. This is denoted by *k*)(*l*. They are sometimes called *divergently parallel lines*.

Recall that for *P* ∉ *l* there is a unique fan angle ∠(*P*, *l*) for *P* and *l*. Let *F* be the foot of *P* in *l* and let *X* and *Y* be on opposite arms of the fan angle. ∠*FPX* ≅ ∠*FPY* The angle is called the angle of parallelism at *P* with respect to *l* and its angle measure is denoted by ∏ (*PF*)° We know that its measure must be less than 90°.

***Theorem:** If l is any line in* H² *and* 0 < á < 90 *then there is a point P ∉ l so that* ∏ (*PF*)° = á°.

CLASSIFICATION OF PARALLELS

We have defined two types of parallel lines--hyperparallel and horoparallel. Are these the only types of parallel lines in hyperbolic geometry? Are there lines that are both hyperparallel and limiting parallel? These are the questions that we wish to answer in this section.

***Theorem:** Let $\overrightarrow{PX}$ be the limiting parallel ray to l through P and let Q and X’ be the feet of P and X, respectively, in l. Then PQ > XX’.*

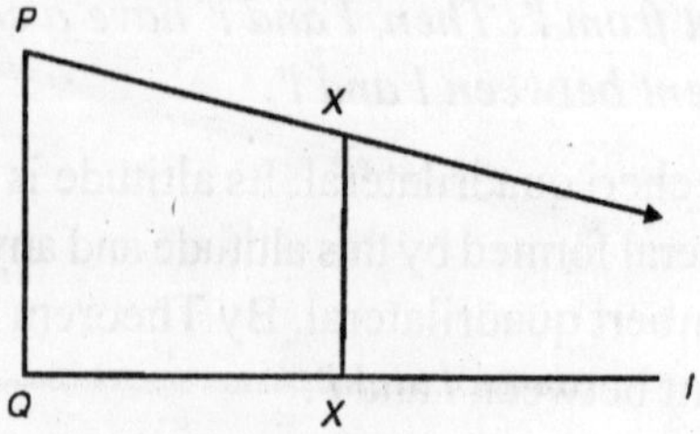

Proof: If $P * X * Y$ then $\overrightarrow{XY} \parallel l$ by Theorem. Thus, $\angle X'XY$ is the angle of parallelism for X and l. Thus, it is acute. Therefore, $\angle X'XP$ is obtuse. This implies that $\angle QPX < \angle X'XP$ and therefore $PQ > XX'$.

This shows that the distance form X to l decreases as the distance from P to X increases along the limiting parallel ray. In fact, one can show that the distance from X to l approaches 0. Note that we have seen that for hyperparallel lines the further one gets from the common perpendicular, the greater the distance between the lines. This gives one the impression that these types of parallel lines should be distinct.

Corollary: *If $l \| k$ in a given direction, then l and k have no common perpendicular.*

Proof: Let $P \in l$ so that $\overrightarrow{PX} \| k$ and assume that there is a point $M \in l$ so that

- *$P * M * X$, and*
- *MM' is a common perpendicular to l and k.*

From Theorem $PQ > MM' > XX'$. By Theorem $MM' < XX'$, a contradiction. Thus, l and k have no common perpendicular.

This tells us that if two lines are limiting parallel, then they are not hyperparallel. We now need to show that if two lines are non-intersecting and are not limiting parallel, then they must be hyperparallel. Having done this we will then know that there are only two types of parallel lines.

Theorem: *Given lines m and n so that*

- *$m \cap n = \theta$, and*
- *m does not contain a limiting parallel ray to n in either direction.*

Then $m)(n$.

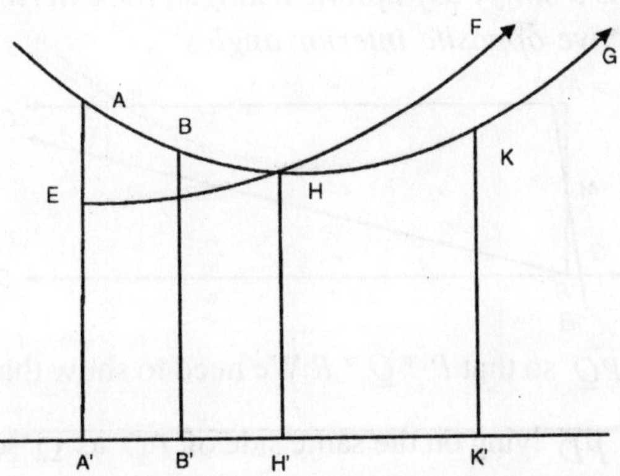

Proof: We need to find two points $H, K \in n$ that are equidistant from m; *i.e.*, $HH' \cong KK'$ If this is true, then by Theorem the perpendicular bisector of HK is the common perpendicular to m and n. By Theorem this common perpendicular is unique. It follows then that $m)(n$.

(i) The search for H and K.

Let $A, B \in n$ with feet $A', B' \in m$ If $AA' \cong BB'$ we are done. Assume that these segments are not congruent. Then, we may assume that $AA' > BB'$, since one must be greater than the other. There is a point $E \in AA'$ so that $A'E \cong BB'$.

There is a unique ray $\overrightarrow{EF}$ on the same side of $\overleftrightarrow{AA'}$ as B so that $\angle A'EF \cong \angle B'BG$ with $A * B * G$.

Claim: $\overrightarrow{EF} \cap \overrightarrow{AG} \neq \theta$ (Proof will follow.)

Call this point of intersection H. The there is a unique point $K \in \overrightarrow{BG}$ so that $BK \cong EH$ Then, by *SAS* Ä$EHA' \cong BKB'$ Thus, $A'HB'K'$ We also have that $\angle EA'H \cong \angle BB'K$ so that by *Angle Subtraction* it follows that $\angle HA'H' \cong \angle KB'K'$ Since HH' and KK' are perpendicular to m, we have by *Hypotenuse-Angle* that Ä$HA'H' \cong$ Ä$KB'K'$ Therefore, $HH' \cong KK'$ and H and K are equidistant from m.

Thus, we are finished once we have proved the claim.

It turns out that the claim is the hardest item to prove in the whole theorem. It takes us through some strange territory.

PROOF OF CLAIM AND ASYMPTOTIC TRIANGLES[H]

If $\overrightarrow{PB}$ and $\overrightarrow{QD}$ are limiting parallel rays, the convention is to *pretend* that they pass through an ideal point Ω. We then denote the rays by $P\Omega$ and $Q\Omega$ The figure consisting of these rays and the segment PQ is called a singly asymptotic triangle. You can have doubly asymptotic triangles, with 2 ideal points, and trebly asymptotic triangles, with 3 ideal points.

Exterior Angle Theorem

Lemma: *If* Ä$PQ\Omega$ *is a singly asymptotic triangle, the exterior angles at P and Q are greater than their respective opposite interior angles.*

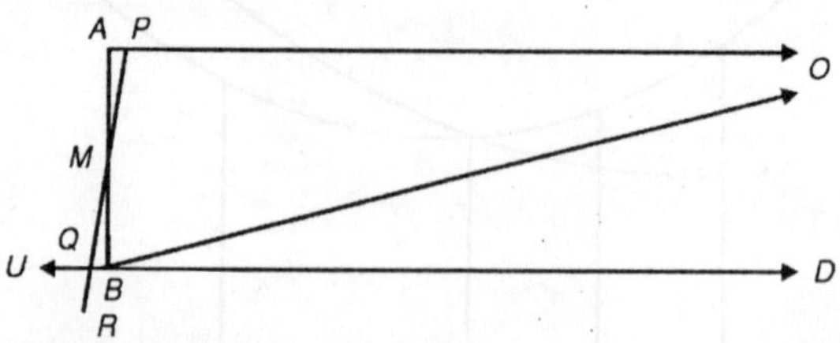

Proof: Choose $R \in \overline{PQ}$ so that $P * Q * R$ We need to show that $\angle RQ\Omega > \angle QP\Omega$.

There is a unique ray $\overrightarrow{PD}$ lying on the same side of $\overleftrightarrow{PQ}$ as Ω so that $\angle RQD \cong \angle QP\Omega$. If $U \in \overleftrightarrow{QD}$ is so that $U * Q * D$ then $\angle UQP \cong \angle RQD \cong \angle QP\Omega$ Thus, we have that the alternate interior angles are congruent. Let M be the midpoint of PQ and let A and B be the feet of M in $\overleftrightarrow{Q\Omega}$ and $\overleftrightarrow{QD}$ respectively. By *AAS* Ä$APM \cong$ ÄBQM so that $AM \cong BM$ and $\angle AMP \cong \angle BMQ$ Thus, A, M, and B are collinear. Thus, $\overleftrightarrow{Q\Omega}$ and $\overleftrightarrow{QD}$ have a common perpendicular.

Thus, $\overrightarrow{Q\Omega})(\overrightarrow{QD}$ Thus $\overrightarrow{QD}$ lies between $\overrightarrow{Q\Omega}$ and $\overrightarrow{QR}$ from which it follows that $\angle RQ\Omega > \angle QP\Omega$.

Lemma: *If in the singly asymptotic triangles* ÄABΩ *and* ÄCDΘ *we have* $\angle BA\Omega \cong \angle DC\Theta$ *then*

$$\angle AB\Omega \cong \angle CD\Theta \text{ if and only if } AB \cong CD.$$

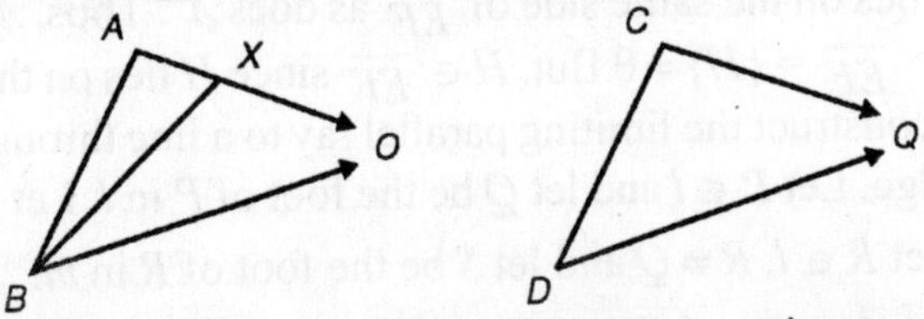

Proof: First, let us assume that $AB \cong CD$ but $\angle AB\Omega \not\cong \angle CD\Theta$ We may suppose that $\angle AB\Omega > \angle CD\Theta$ Then, there is a unique ray $\overrightarrow{BW}$ between $\overrightarrow{B\Omega}$ and $\overrightarrow{BA}$ so that $\angle ABW \cong \angle CD\Theta$ Since $\overrightarrow{A\Omega}$ and $\overrightarrow{B\Omega}$ are limiting parallel, $\overrightarrow{BW}$ must intersect $\overrightarrow{A\Omega}$ in a point X. There is a point $Y \in \overrightarrow{C\Theta}$ so that $AX \cong CY$ Then by *SAS* Ä$ABX \cong$ ÄCDY Therefore, $\angle CDY \cong \angle ABX \cong \angle CD\Theta$ Then, $\overrightarrow{C\Theta}$ and $\overrightarrow{D\Theta}$ are not limiting parallel, a contradiction.

Now assume that $\angle AB\Omega \cong \angle CD\Theta$ and $AB \not\cong CD$ We may then assume that $AB > CD$. There is a unique point $P \in AB$ so that $BP \cong CD$ Let $\overrightarrow{P\Omega}$ be the limiting parallel ray to $\overrightarrow{A\Omega}$ from P. Thus, $\overrightarrow{P\Omega}$ is also parallel to $\overrightarrow{B\Omega}$ The first part of this proof now implies $\angle BP\Omega \cong \angle DC\Theta \cong \angle BA\Omega$ But $\angle BP\Omega$ is exterior to Ä$AP\Omega$ by the *Exterior Angle Theorem* $\angle BP\Omega > \angle BA\Omega$ a contradiction.

Corollary: *$PQ \cong RS$ if and only if $\prod(PQ)^\circ = \prod(RS)^\circ$.*

We are now in a position to show that $\overrightarrow{EF} \cap \overrightarrow{AG} \neq \theta$.

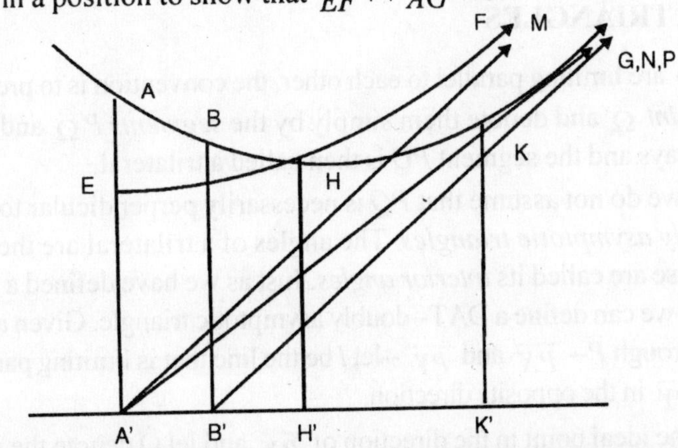

Let $\overrightarrow{A'M} \parallel \overrightarrow{EF}$, $\overrightarrow{A'N} \parallel \overrightarrow{AG}$, and $\overrightarrow{B'P} \parallel \overrightarrow{BG}$ Recall that $EA' \cong BB'$ and $\angle A'EF \cong \angle B'BG$ By Lemma $\angle EA'M \cong \angle BB'P$ Choose a point $L \in m$ so that $A' * B' * L$.

Since m and n were not limiting parallel lines, we have that $\overrightarrow{B'L} \neq \overrightarrow{B'P}$ and $\overrightarrow{A'L} \neq \overrightarrow{A'N}$

Now, $\angle MA'L \cong \angle PB'L$ by *Angle Subtraction*. Since $\overrightarrow{A'N} \parallel \overrightarrow{AG}$, $\overrightarrow{A'N} \parallel \overrightarrow{BG}$ and $\overrightarrow{B'P} \parallel \overrightarrow{BG}$ we must have that $\overrightarrow{A'N} \parallel \overrightarrow{B'P}$ Thus, $\angle PB'L$ is exterior to the singly asymptotic triangle $\ddot{A}A'B'\Omega$ Therefore, $\angle NA'L < \angle PB'L$ This means that $\overrightarrow{A'N}$ lies between $\overrightarrow{A'M}$ and $\overrightarrow{A'L}$ which implies that $\overrightarrow{A'M}$ lies between $\overrightarrow{A'N}$ and $\overrightarrow{A'A}$ But since $\overrightarrow{A'N} \parallel \overrightarrow{AG}$, we have that

$$\overrightarrow{A'M} \cap \overrightarrow{AG} = \{J\}.$$

Since $J \in \overrightarrow{A'M}$ it lies on the same side of $\overleftrightarrow{EF}$ as does A'. Thus, A and J are on opposite sides of $\overleftrightarrow{EF}$ Thus, $AJ \cap \overleftrightarrow{EF} = \{H\} \neq \theta$ But, $H \in \overrightarrow{EF}$ since H lies on the same side of $\overleftrightarrow{AA'}$ as does J. It is possible to construct the limiting parallel ray to a line through a given point with a compass and straightedge. Let $P \notin l$ and let Q be the foot of P in l. Let m be perpendicular to $\overleftrightarrow{PQ}$ at P so that $m)(l$ Let $R \in l, R \neq Q$ and let S be the foot of R in m.

Then $PQRS$ is a Lambert quadrilateral and $PS<QR$. $QR<PR$ since PR is a hypotenuse. Take a circle centered at P of radius QR. S is inside this circle and R is outside this circle. There is a unique point $X \in RS$ where RS intersects the circle. One can show that the ray $\overrightarrow{PX}$ is limiting parallel to l!

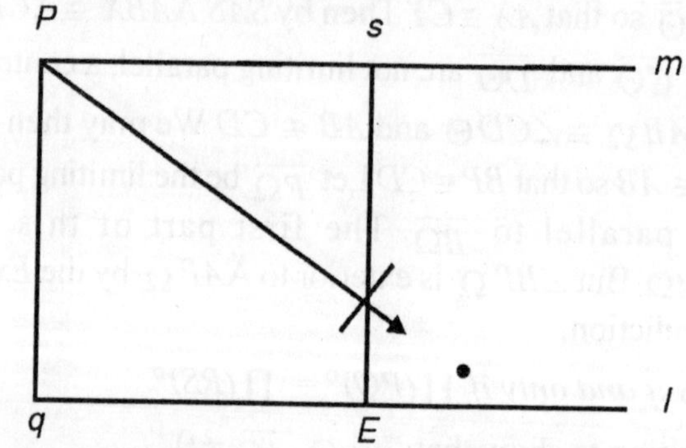

STRANGE NEW TRIANGLES

If $\overrightarrow{PB}$ and $\overrightarrow{QD}$ are limiting parallel to each other, the convention is to pretend that they go through an *ideal point* Ω and denote them simply by the *segments* $P\Omega$ and $Q\Omega$ The figure consisting of these rays and the segment PQ is then called a trilateral.

Note that here we do not assume that PQ is necessarily perpendicular to $Q\Omega$ Trilaterals are also called *singly asymptotic triangles*. The angles of a trilateral are the angles $\angle \Omega PQ$ and $\angle \Omega QP$ and these are called its *interior angles*. Just as we have defined a trilateral--singly asymptotic triangle--we can define a DAT--doubly asymptotic triangle. Given a point P and two non-opposite rays through P-- $\overrightarrow{PX}$ and $\overrightarrow{PY}$ --let l be the line that is limiting parallel to $\overrightarrow{PX}$ and limiting parallel to $\overrightarrow{PY}$ in the opposite direction.

Let Ω denote the ideal point in the direction of $\overrightarrow{PX}$ and let Ó denote the ideal point in the direction of $\overrightarrow{PY}$ Then we can identify the line l with the *segment* ΩÓ and the *segments* $P\Omega$, PÓ and ΩÓ form a doubly asymptotic triangle. The angle $\angle\Omega P$Ó is the fan angle of l and P.

Many of the properties of trilaterals are analogous to those of triangles.

Crossbar Theorem for Trilaterals

Lemma: *A line which subdivides an angle of a trilateral intersects the opposite side.*

Pasch's Theorem for Trilaterals

Lemma: *A line which intersects a side of the trilateral ÄAB*Ω *but does not pass through a vertex will intersect another side, provided the line is not limiting parallel to either A*Ω *or B*Ω *at* Ω.

Lemma: *The sum of the interior angles of a trilateral is less than* 180°.

Exterior Angle Theorem

Lemma: *If ÄPQ*Ω *is a trilateral, the exterior angles at P and Q are greater than their respective opposite interior angles.*

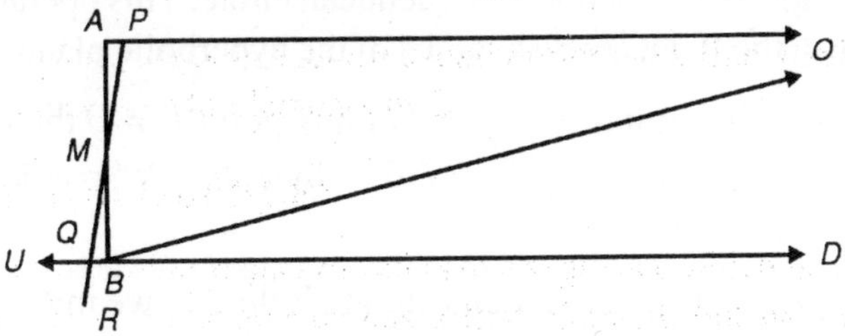

Proof: Choose $R \in \overrightarrow{PQ}$ so that $P * Q * R$ We need to show that $\angle RQ\Omega > \angle QP\Omega$.

There is a unique ray $\overrightarrow{PD}$ lying on the same side of $\overleftrightarrow{PQ}$ as Ω so that $\angle RQD \cong \angle QP\Omega$. If $U \in \overleftrightarrow{QD}$ is so that $U * Q * D$ then $\angle UQP \cong \angle RQD \cong \angle QP\Omega$ Thus, we have that the alternate interior angles are congruent. Let M be the midpoint of PQ and let A and B be the feet of M in $\overrightarrow{Q\Omega}$ and $\overrightarrow{QD}$ respectively. By *AAS* ÄAPM ≅ ÄBQM so that $AM \cong BM$ and $\angle AMP \cong \angle BMQ$ Thus, A, M, and B are collinear. Thus, $\overleftrightarrow{Q\Omega}$ and $\overleftrightarrow{QD}$ have a common perpendicular. Thus, $\overleftrightarrow{Q\Omega})(\overleftrightarrow{QD}$ Thus $\overrightarrow{QD}$ lies between $\overrightarrow{Q\Omega}$ and $\overrightarrow{QR}$ from which it follows that $\angle RQ\Omega > \angle QP\Omega$.

Lemma: *If in the singly asymptotic triangles ÄAB*Ω *and ÄCD*Θ *we have* $\angle BA\Omega \cong \angle DC\Theta$ *then*

$$\angle AB\Omega \cong \angle CD\Theta \text{ if and only if } AB \cong CD.$$

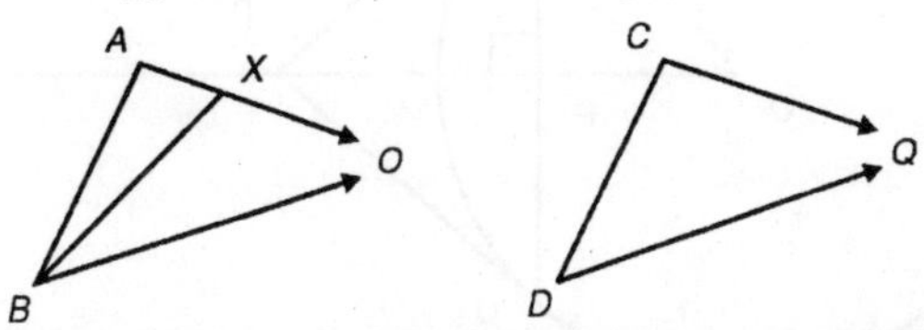

Proof: First, let us assume that $AB \cong CD$ but $\angle AB\Omega \not\cong \angle CD\Theta$ We may suppose that $\angle AB\Omega > \angle CD\Theta$ Then, there is a unique ray $\overrightarrow{BW}$ between $\overrightarrow{B\Omega}$ and $\overrightarrow{BA}$ so that $\angle ABW \cong \angle CD\Theta$ Since $\overrightarrow{A\Omega}$ and $\overrightarrow{B\Omega}$ are limiting parallel, $\overrightarrow{BW}$ must intersect $\overrightarrow{A\Omega}$ in a point

X. There is a point $Y \in \overrightarrow{C\Theta}$ so that $AX \cong CY$ Then by *SAS* ÄABX ≅ ÄCDY Therefore, $\angle CDY \cong \angle ABX \cong \angle CD\Theta$ Then, $\overrightarrow{C\Theta}$ and $\overrightarrow{D\Theta}$ are not limiting parallel, a contradiction.

Now assume that $\angle AB\Omega \cong \angle CD\Theta$ and $AB \not\cong CD$ We may then assume that $AB>CD$. There is a unique point $P \in AB$ so that $BP \cong CD$ Let $\overrightarrow{P\Omega}$ be the limiting parallel ray to $\overrightarrow{A\Omega}$ from *P*. Thus, $\overrightarrow{P\Omega}$ is also parallel to $\overrightarrow{B\Omega}$ The first part of this proof now implies $\angle BP\Omega \cong \angle DC\Theta \cong \angle BA\Omega$ But $\angle BP\Omega$ is exterior to ÄAPΩ by the *Exterior Angle Theorem* $\angle BP\Omega > \angle BA\Omega$ a contradiction.

INVERSION IN EUCLIDEAN CIRCLES

In order to define congruence in the Poincaré model and verify the axioms of congruence, we must study the operation of inversion in a Euclidean circle. This operation is the process of reflecting across a line in the Poincaré disk model of the hyperbolic plane.

Let ã be a circle of radius *r* and center *O*. For any point $P \neq O$ the inverse *P*' of *P* with respect to the circle ã is the unique point $P' \in \overrightarrow{OP}$ so that $OP.\ OP' = r^2$ In this case ã is called the circle of inversion.

These are Euclidean geometry theorems about circles, so we may use the facts that we know about Euclidean geometry.

Proposition: *For P, P' and ã as above,*

1. $P = P'$ *if and only if* $P \in$ ã.
2. *If* $P \in$ int ã *then* $P' \in$ int ã *and vice versa.*
3. $(P')' = P$.

Proposition: *Let* $P \in$ int ã *and let TU be the chord through P perpendicular to* $\overrightarrow{OP}$ *Then P '= P(TU), the pole of TU, i.e., the intersection of the tangents to ã at T and U.*

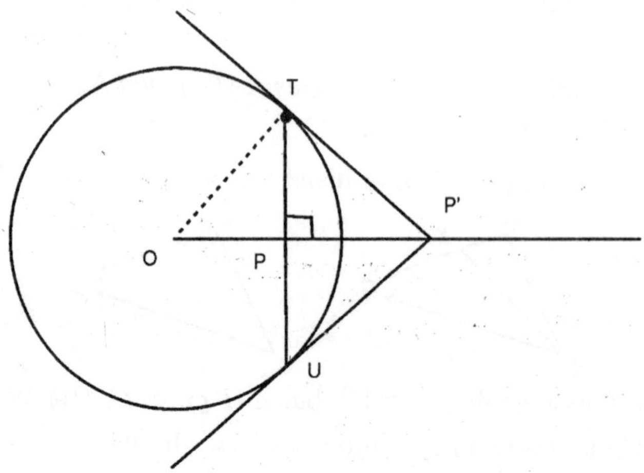

Proof: Suppose the tangent to ã at *T* cuts $\overrightarrow{OP}$ at the point *P*'. The right triangle Ä*OPT* is similar to right triangle *OTP*' (since they have $\angle TOP$ in common and the angle sum is 180°). Hence corresponding sides are proportional. Since $l(OT) = r$ we get that

$$\frac{l(OP)}{r} = \frac{r}{l(OP')}$$

which shows that P' is the inverse to P. Reflect across the line $\overleftrightarrow{OP}$ and we see that the tangent to ã at *U* also passes through P', so that P' is the pole of *TU*.

Proposition: *If P is outside of ã let Q be the midpoint of OP. Let be the circle of radius QP centered at Q. Then*

- ó $\cap$ ã = {*T*, *U*}.
- $\overleftrightarrow{PT}$ *and* $\overleftrightarrow{PU}$ *are tangent to* ã.
- $P' = TU \cap OP$.

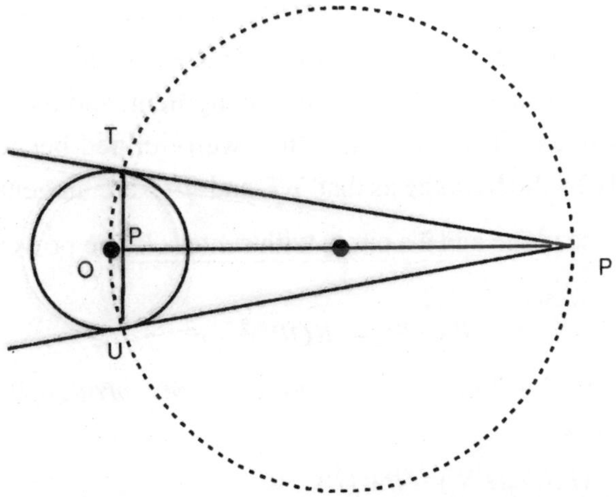

Proof: By the circular continuity principle, ó and ã do intersect in two points *T* and *U*. Since $\angle OTP$ and $\angle OUP$ are inscribed in semicircles of ó they are right angles. Therefore $\overrightarrow{PT}$ and $\overrightarrow{PU}$ are tangent to ã If *TU* intersects *OP* in a point *P*', then *P* is the inverse of P', by the previous proposition. Thus, *P*' is the inverse of *P* in ã.

The next proposition shows how to construct the Poincaré line joining two ideal points--the line of enclosure. Its proof shows that $\angle OP'T$ in the previous figure is indeed a right angle, as we needed.

Proposition: *Let T and U be points on* ã *that are not contained on a the same diameter, and let P be the pole of TU. Then*

- $PT \cong PU$,
- $\angle PTU \cong \angle PUT$,

- $\overrightarrow{OP} \perp \overleftrightarrow{TU}$, *and*
- *The circle Ý with center P and radius PT cuts ã orthogonally at T and U.*

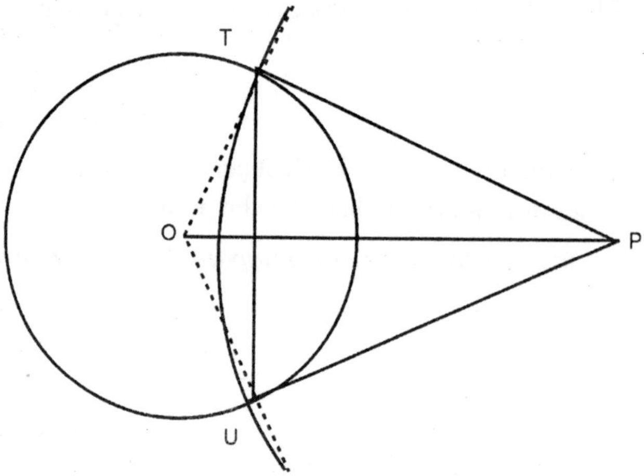

Proof: By definition of pole, $\angle OTP$ and $\angle OUP$ are right angles, so by the Hypotenuse-Leg criterion, Ä$OTP \cong$ ÄOUP Therefore, $PT \cong PU$ and $\angle OPT \cong \angle OPU$ The base angles $\angle PTU$ and $\angle PUT$ of the isosceles triangle ÄTPU are thus congruent, and the angle bisector $\overrightarrow{PQ}$ is perpendicular to the base TU. The circle Ý is then well-defined because $PU \cong PT$ and Ý intersects ã orthogonally by the hypothesis that $\overrightarrow{PT}$ and $\overrightarrow{PU}$ are tangent to ã.

Let P be a point in the plane and ã a circle with center O. The power of P with respect to ã is defined to be

$$Pw\,(P) = (l(OP))^2 - r^2.$$

Proposition: *Assume $P \notin$ ã and assume that two lines through P intersect ã in points $\{R_1, R_2\}$ and $\{S_1, S_2\}$. Then*

- $(PR_1).\ (PR_2) = (PS_1).\ (PS_2) = Pw\ (P)$.
- *If one of these lines through P is tangent to ã at T, then $Pw(P) = PT^2$.*

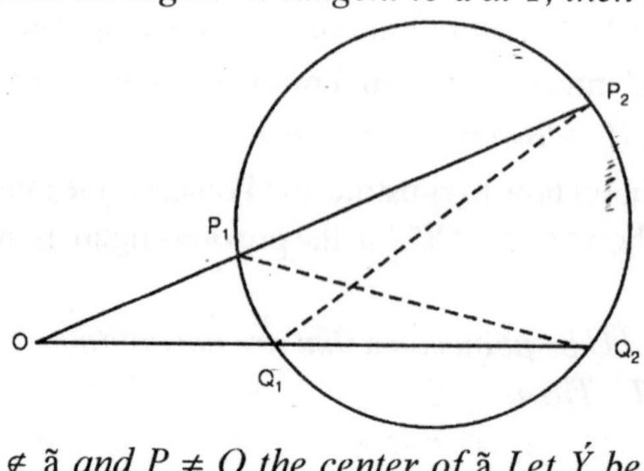

Proposition: *Let $P \notin$ ã and $P \neq O$ the center of ã Let Ý be a circle through P. ä is orthogonal to ã if and only if Ý passes through P', the inverse of P with respect to ã.*

Proposition: *Let ã have radius r, ä have radius t and let P be the center of ä. ä is orthogonal to ã if and only if Pw(P) =* t^2 *where the power is computed with respect to ã.*

Let O be a point and $k > 0$. The dilation with center O and ratio k is a mapping of the Euclidean plane that fixes O and maps every point $P \neq O$ to a unique $P^* \in \overrightarrow{OP}$ such that $OP^* = kOP$ Call this map *Do,k.*

Proposition: *Let ä be a circle with center* $C \neq O$ *and radius s. Do, á maps ä to a circle ä with center* C^* *= Do,á(C) and radius ás If Q ∈ ä the tangent to ä* at* Q* *is parallel to the tangent to ä at Q.*

Proof: Choose rectangular coordinates so that O is the origin. Then the dilation is given by the mapping $(x, y) \mapsto$ (áx, áy). The image of the line have equation $ax + by = c$ is the line having equation áx + by = ác so the image is parallel to the original line. In particular, $\overleftrightarrow{CQ}$ is parallel to $\overleftrightarrow{C^*Q^*}$ and their perpendiculars at Q and Q^* respectively, are also parallel. If *ä* has equation $(x - h)^2 + (y - k)^2 = s^2$ then Ý* has equation (x – á*h*)^2 + (y – á*k*)^2 = (á*s*)^2.

Proposition: *Let ã be the circle of radius r centered at O and let Ý be the circle of radius s centered at C. Assume O lies outside ä; let p = Pw(O) with respect to ä and let á =* r^2/p *The image ä' of ä under inversion in ã is the circle of radius ás whose center is* C^* *= Do,* $_{á}$*(C). If P ∈ ä and P' is the inverse of P with respect to ã then the tangent t' to ä at P' is the reflection across the perpendicular bisector of PP' of* the tangent to *ä* at P.

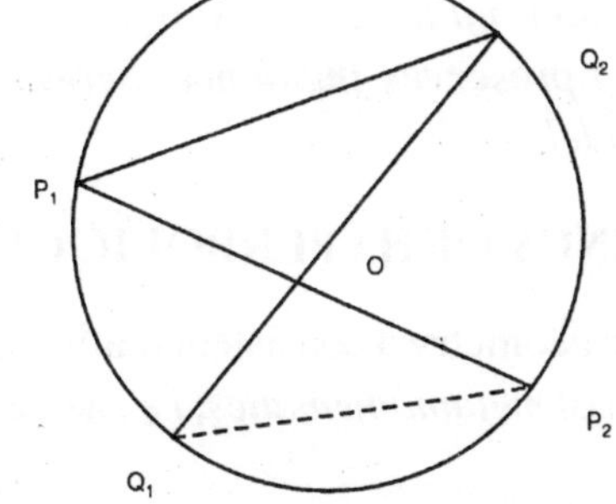

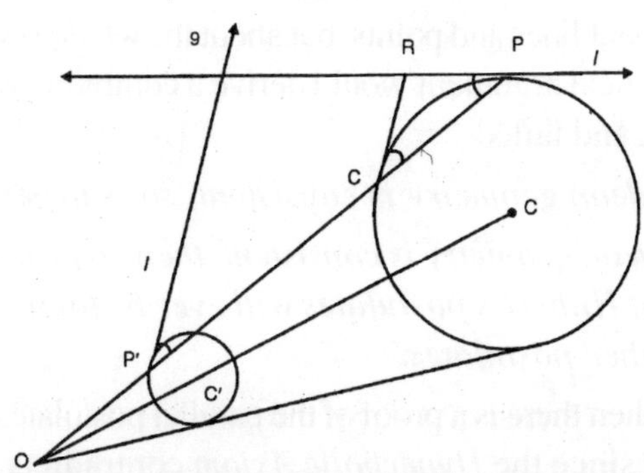

Corollary: *ä is orthogonal to ã if and only if Y is mapped to itself by inversion in ã.*

Proposition: *Let l be a line so that O ∉ l The image of l under inversion by ã is a punctured circle with missing point O. The diameter through O of the completed circle ä is perpendicular to l.*

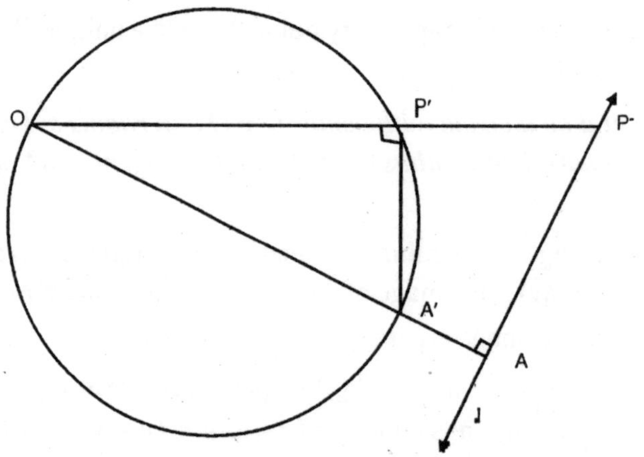

Proposition: *Let ä be a circle passing through the center O of ã The image of ä minus O under inversion in ã is a line l so that O ∉ l and l is parallel to the tangent to ä at O.*

Proposition: *A directed angle of intersection of two circles is preserved in magnitude by an inversion. The same applies to the angle of intersection of a circle and a line or the intersection of two lines.*

Proposition: *Let Ý be orthogonal to Ã Inversion in ä maps Ã onto Ã and the interior of Ã onto itself. Inversion in ä preserves incidence, betweenness and congruence in the sense of the Poincaré disk model.*

CONSISTENCY OF HYPERBOLIC GEOMETRY

Is this system of hyperbolic geometry a consistent mathematical system? Does it lead to a contradiction? This is a question of *metamathematics*; *i.e.*, a question outside of a mathematical system about the system itself.

The question is not about lines and points, but about the *whole system* itself. If it is inconsistent, then an ordinary mathematical argument would derive a contradiction. This is what Saccheri and Lambert attempted to do, and failed.

Theorem: *If Euclidean geometry is consistent, so is hyperbolic geometry.*

Corollary: *If Euclidean geometry is consistent, then no proof or disproof of the parallel postulate from the rest of Hilbert's postulates will ever be found, i.e., the parallel postulate is independent of the other postulates.*

Proof: Assume not, then there is a proof of the parallel postulate. This implies that hyperbolic geometry is inconsistent since the *Hyperbolic Axiom* contradicts a proven result. Our above theorem asserts that hyperbolic geometry is as consistent as Euclidean geometry.

Thus, Euclidean geometry is inconsistent. This contradiction implies that no proof o the parallel postulate exists.

Note also that the hypothesis that euclidean geometry is consistent implies that do disproof exists either.

You should recognize that had mathematicians succeeded in proving *Euclid's Fifth Postulate* from the other axioms with the intention of making Euclidean geometry more secure and elegant, they would have completely destroyed Euclidean geometry as a consistent body of thought.

What we will do now, in order to *see* that hyperbolic geometry is as consistent as Euclidean geometry, is to construct a model of hyperbolic geometry. We will look at three different models for H^2

THE BELTRAMI-KLEIN MODEL

This model is often referred to as the *Klein model*, because of the extensive work done by the German mathematician Felix Klein in geometry with this model. Fix, once and for all, a circle $\tilde{A}$ in E^2 with center O and radius OR. Let

$$\text{int } \tilde{A} = \{X \in E^2 \mid OX < OR\}.$$

Identify the points of H^2 with the points in int $\tilde{A}$.

A chord of $\tilde{A}$ is the Euclidean segment AB joining two points $A, B \in \tilde{A}$. $AB \cap \text{int } \tilde{A} = (A, B)$ is an open chord. These open chords represent the lines of the hyperbolic plane. To say that P lies on (A,B) means $P \in \overleftrightarrow{AB}$, the Euclidean line, and $A * P * B$.

It is easy to see that this model satisfies the *Hyperbolic Axiom*.

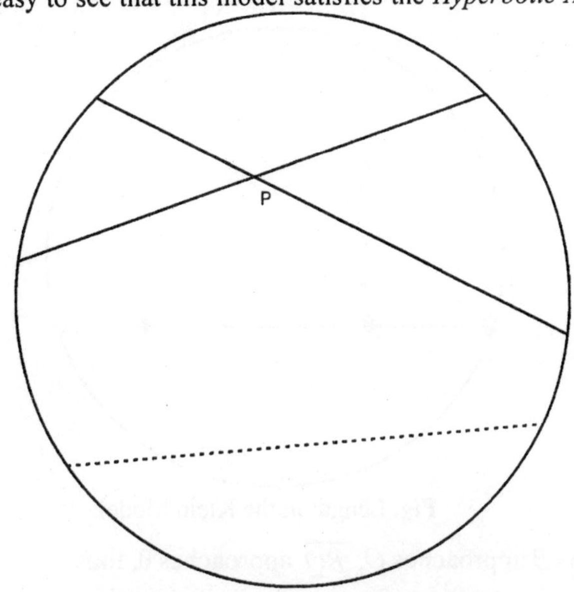

Fig. Parallel Lines in the Klein Model

Here, $k \cap \tilde{A} = \{A, B\}$ The lines $\overrightarrow{PA}$ and $\overrightarrow{PB}$ are two lines that do not intersect k in H^2 as is the line l The difference is that $\overrightarrow{PA}$ and $\overrightarrow{PB}$ are limiting parallel to k in different directions, while l and k are hyperparallel, admitting a common perpendicular. Recall that the points of $\tilde{A}$ do not belong to the hyperbolic plane. They are called ideal points of H^2 The points outside of $\tilde{A}$ are called ultraideal points.

By saying that the lines k and l admit a common perpendicular raises the question about how one defines congruence of segments and angles. It is not obvious, for it would seem that lines must be of finite extent, no line measuring more than twice the diameter of $\tilde{A}$. If this were the case, we would not have a model for hyperbolic geometry, as the congruence axioms, Archimedes' axiom, and Dedekind's axiom definitely would not hold. We must define a different method for measuring the length of segments and the measure of angles in this model.

I will at this point only mention the method for measuring the length of segments and the definition of right angles. The remaining measurement will follow from work that we do later, and will be noted then.

Let $A, B \in H^2$ and let $P, Q \in \tilde{A}$ denote the endpoints of the chord through A and B. Let $\overline{AB}$ denote the Euclidean distance from A to B, or the length of the segment AB. Define the *Klein distance*

$$dk\,(A, B) = \frac{1}{2}\left|\log\,(AB, PQ)\right|$$

$$= \frac{1}{2}\left|\log\left(\frac{\overline{AP}.\,\overline{BQ}}{\overline{BP}.\,\overline{AQ}}\right)\right|$$

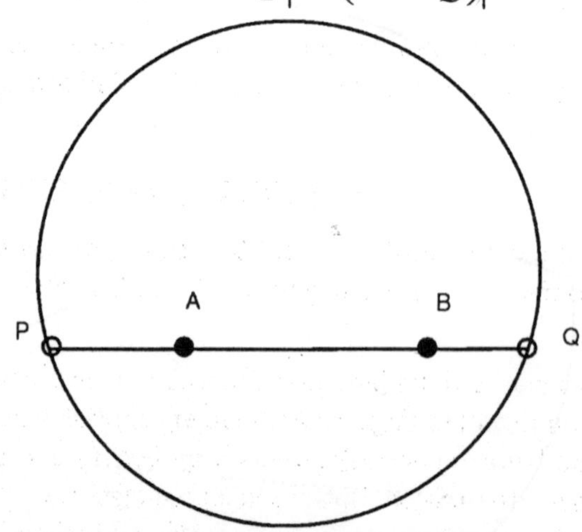

Fig. Length in the Klein Model

Note then that as B approaches Q, $\overline{BQ}$ approaches 0, thus

$$\lim_{B \to Q} d_k\,(A, B) = +\infty$$

This allows us to see then that we can find a segment of any length on any ray (Axiom C-1). Also, it will follow that Dedekind's and Archimedes' axioms are valid.

Our technique for measuring angles will be introduced at a later time. It depends on the model for hyperbolic geometry due to Poincaré. It is not the way in which angles are measured in Euclidean geometry. For this reason the Klein model is not a conformal model of geometry. We can talk about right angles, without too much difficulty. Let l and m be two lines in the Klein model of the hyperbolic plane, or *K-lines*. $l \perp m$ in the Klein sense if,

- l or m is a diameter and l and m are perpendicular in the Euclidean sense, or
- If neither l or m is a diameter, do the following: Let t_1 and t_2 be the tangents to Ã at the ends of l. Since l is not a diameter, $t_1 \cap t_2 \neq \theta$ Put $P(l) = t_1 \cap t_2$ $P(l)$ is called the pole of l. m is perpendicular to l in the Klein sense if and only if the Euclidean line extending m passes through $P(l)$.

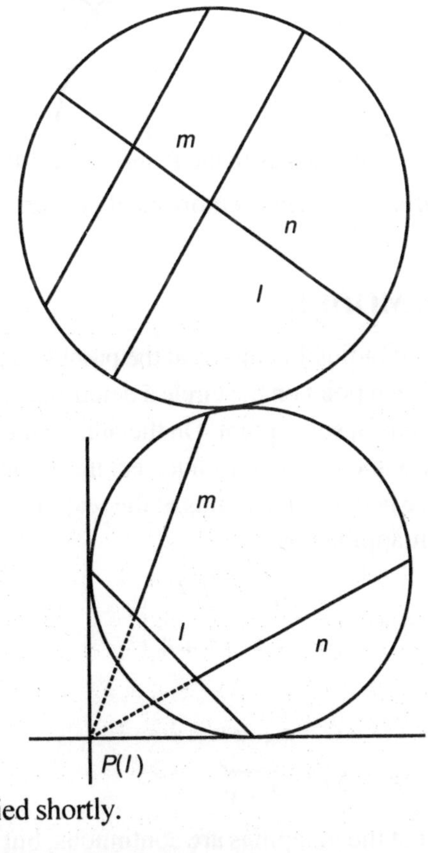

All of this will be verified shortly.

THE POINCARÉ HALF-PLANE MODEL

In this model take H^2 to be the set of all points in the coordinate Euclidean plane that satisfy $H^2 = \{(x, y) \in E^2 \mid y > 0\}$

We interpret lines in this model to be one of the following:

- The intersection of a Euclidean vertical line with H^2 or
- The intersection of a Euclidean circle, whose center lies on the x-axis, with H^2.

Once again, our concept of *congruence* must change. As before, we must define a new measure for segments. However, the manner of measuring angles is exactly the method that is used in Euclidean geometry. To measure the angle formed by two of these lines, we simply measure the Euclidean angle formed by the tangent lines to the two Poincaré lines.

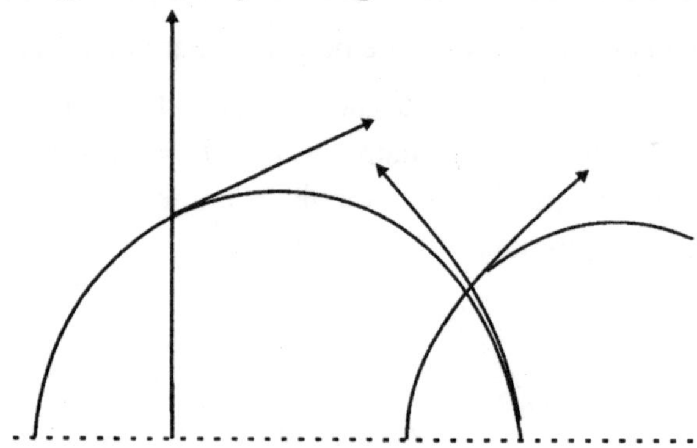

Fig. Measuring angles in the Poincaré Half-Plane Model

The method of measuring segments is more easily described in the next model, which is derived from this model.

THE POINCARÉ DISK MODEL

Let ã denote the circle of radius 1 centered at the point (0,1) in the real plane. Let N denote the point (0,2). Then N and each point on the circle ã determine a line which has non-zero slope. It must then intersect the x-axis in some point. On the other hand, given any point on the x-axis, it and the point N determine a line which must intersect the circle ã in some point. Thus, there is a one-to-one correspondence between the points of the real line (here, the x-axis) and the points of the circle ã, less N. The mappings are

$$f : \mathbb{R} \to \tilde{a} \setminus N \text{ given by } f(r) = \left(\frac{-4r}{r^2 + 4}, \frac{2r^2}{r^2 + 4} \right);$$

And

$$g : \tilde{a} \setminus N \to \mathbb{R} \text{ given by } g(x, y) = \frac{2r^2}{2 - y}.$$

Not only can you see that the mappings are continuous, but are differentiable as well.

What this means to us is that we may add a *point at infinity* to the real line and we get a circle, in some sense. Think of taking the Poincaré half-plane model and adding a point at infinity to the real axis. We will then have the interior of a circle sitting in the real plane. The vertical lines will correspond to diameters of the circle and the circles in the half-plane model will correspond to circles which are *orthogonal* to the boundary circle.

To make the setting more precise, let $H^2 = \text{int } \tilde{A}$. A *p*-line in this model is either

- An open diameter of Ã, or
- An open arc of a circle orthogonal to Ã. A circle is orthogonal to Ã if at each point of intersection of Ý and Ã the radii of Ã and Ý through that point are perpendicular. $\acute{Y} \cap \text{int } \tilde{A} = m$ an open arc which represents a *p*-line in H^2.

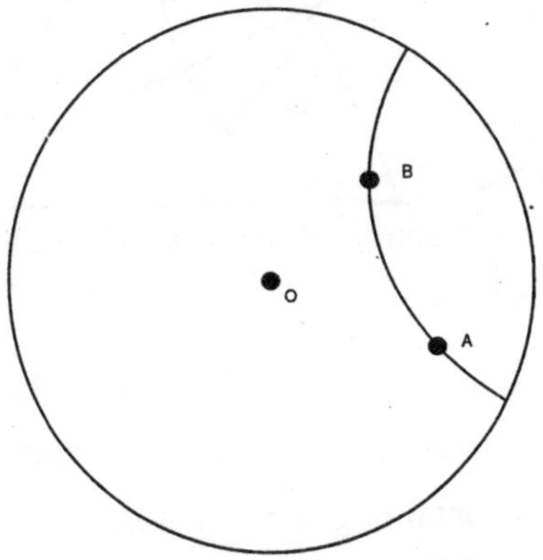

Fig. A *p*-line in the Poincaré disk model

A point in H^2 lies on a *p*-line if and only if it lies on it in the Euclidean sense. Betweenness has the same interpretation, though if *A*, *B*, and *C* are on an open arc from Ý with center *P* then $A * B * C$ if and only if $\overrightarrow{PB}$ is between $\overrightarrow{PA}$ and $\overrightarrow{PC}$.

If $A, B \in H^2$ and if *P* and *Q* are the ends of the *p*-line through *A* and *B*, then we define the Poincaré distance from *A* to *B* to be

$$d_p(A, B) = |\log(AB, PQ)| = \left|\log \frac{\overline{AP}.\overline{BQ}}{\overline{AQ}.\overline{BP}}\right|.$$

As with the above model, this Poincaré model is conformal. If two directed circular arcs meet at *A* then the measure of the angle they form is the measure of the angle between their tangent rays at *A*.

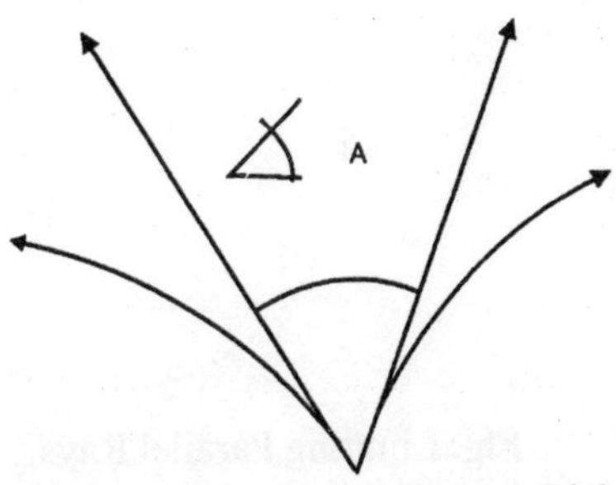

Fig. Angle Measure in the Poincaré Disk Model

Question: IF $A, B \in H^2$ and A and b do not lie on a diameter of $\tilde{A}$, how do you construct the p-line through A and B?

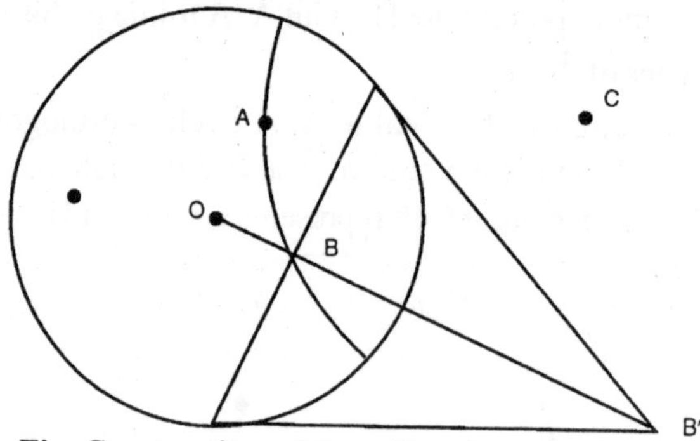

Fig. Construction of the p-line through A and B

We will verify these steps later.

- Construct the perpendicular to $\overrightarrow{OA}$ at A, intersecting $\tilde{A}$ in P and Q.
- Construct the tangents to $\tilde{A}$ at P and Q. These tangents intersect at a point A'.
- Construct the circle, δ containing A, A', and B.

δ is orthogonal to $\tilde{A}$ and $m = \delta \cap int\,\tilde{A}$ is the p-line through A and B.

Let us compare the two disk models by looking at certain familiar objects in each of them.

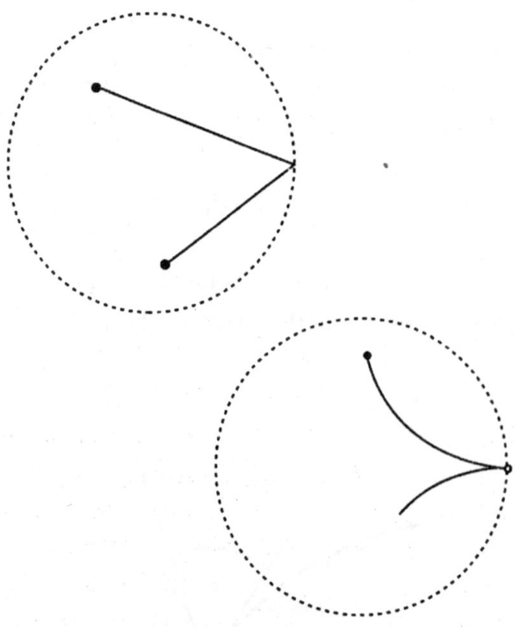

Fig. Limiting Parallel Rays

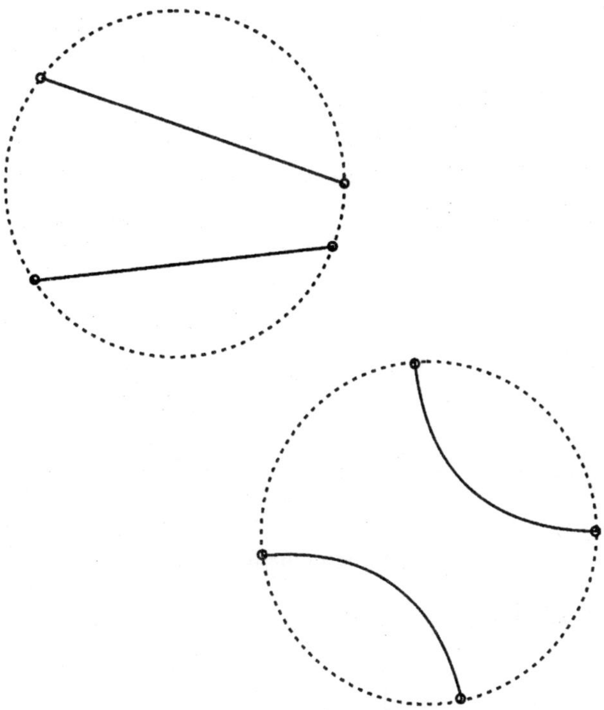

Fig. Hyperparallel Lines

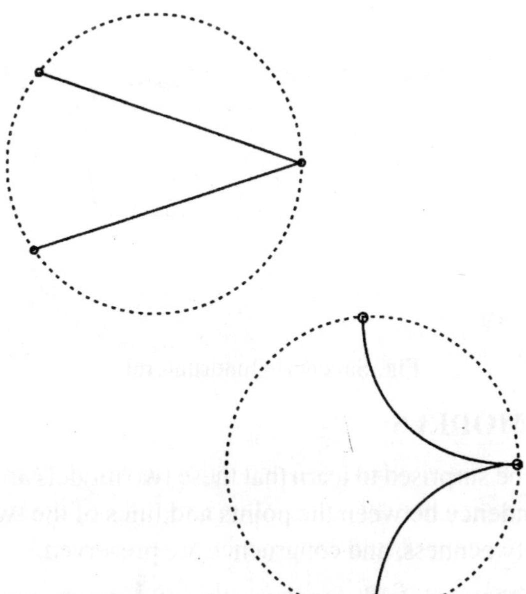

Fig. Horoparallel Lines

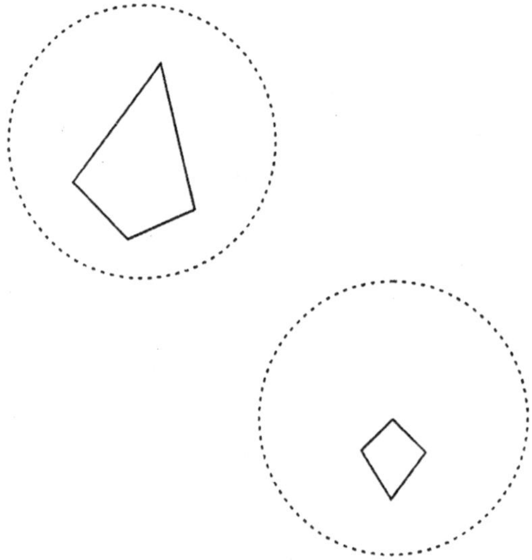

Fig. Lambert Quadrilateral

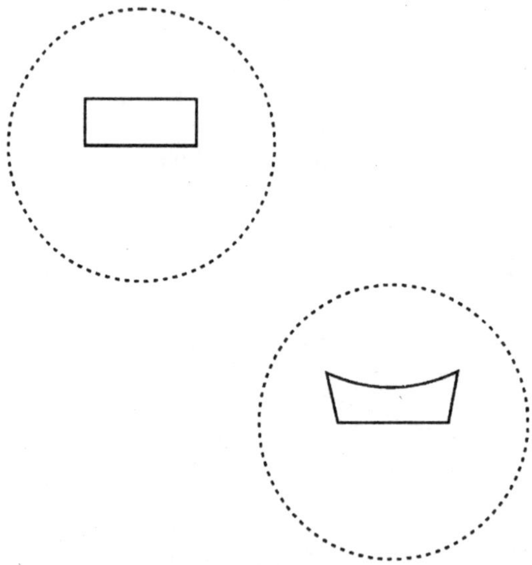

Fig. Saccheri Quadrilateral

ISOMORPHISM OF MODELS

You will, doubtless, be surprised to learn that these two models are *isomorphic;*, *i.e.*, there is a one-to-one correspondence between the points and lines of the two models such that the relations of incidence, betweenness, and congruence are preserved.

Let S^2 be a sphere of the same radius as the radius of Ã, tangent to *int* Ã at the origin. Use *orthogonal projection* of the Klein model in Ã upward onto the lower hemisphere of the sphere and then stereographic projection from the north pole back to the plane. The image will be the Poincaré model.

Let $(x, y) \in int\ \tilde{A}$ Orthogonal projection maps (x,y) onto the bottom hemisphere of $x^2 + y^2 + (z - a)^2 = a^2$ by

$$(x, y) \mapsto (x, y, a - \sqrt{a^2 - x^2 - y^2}).$$

Stereographic projection of S^2 onto the coordinate plane is the analogous map to that described for mapping the punctured circle onto the line. A point $(x, y, z) \in S^2$ is sent to the point on the xy-plane at which the line through $(0,0,2a)$ and (x,y,z) intersects the plane. We can parameterize the line by:

$$l(t) = \langle 0, 0, 2a \rangle + t\langle x, y, z - 2a \rangle$$

$$= \langle tx, ty, 2a + t(z - 2a) \rangle$$

$l(t)$ passes through the xy-plane when

$$2a + t(z - 2a) = 0$$

$$t = \frac{2a}{2a - z}$$

Thus, stereographic projection sends (x,y,z) to $\left(\frac{2ax}{2a - z}, \frac{2ay}{2a - z}\right)$.

We need a shrinking map to shrink the image back into $\tilde{A}$. The equator of the sphere (x,y,a) maps to $(2x,2y) = (x',y')$. Thus, $(x')^2 + (y')^2 = 4r^2$ Shrinking back into $\tilde{A}$ it is only necessary to divide by 2.

Combining the three maps, we have a mapping from the Klein model to the Poincaré model given by

$$(x, y) \mapsto \left(\frac{ax}{a + \sqrt{a^2 - x^2 - y^2}}, \frac{ay}{a + \sqrt{a^2 - x^2 - y^2}}\right).$$

It is easy to check that this is a continuous and differentiable map from *int* $\tilde{A}$ to itself. It sends *k*-lines to *p*-lines and is an isomorphism of geometries.

CONSTRUCTIONS IN THE POINCARÉ MODEL

To construct the Poincaré line joining two ideal points, or the so-called line of enclosure, we have the following result.

Lamma: *Let T, U* $\in \tilde{A}$, *not on a diameter, and Let P = P(TU), the pole of TU. Then*

- $PT \cong PU$,
- $\angle PTU \cong \angle PUT$,

- $\overleftrightarrow{OP} \perp \overleftrightarrow{TU}$,
- *The circle δ with center P and radius $PU \cong PT$ is orthogonal to ã at T and U.*

The previous description for the construction of the unique line between two ordinary points follows from the fifth item above. We have already discussed the measure of segments, as well.

How do you copy a given angle at a given point, A, on a given ray, $\overrightarrow{AB}$?

- If A is the center of Ã then the ray $\overrightarrow{AB}$ is a diameter of Ã and the angle is copied as in Euclidean geometry, using another diameter of Ã.
- If $A \neq O$ the center of Ã, then we must find the unique circle δ through A orthogonal to Ã and tangent to the given Euclidean line l which forms the Euclidean angle with the tangent to $\overrightarrow{AB}$ at A.
- For δ orthogonal to Ã, by the above facts, δ must pass through $A'_{\tilde{A}}$ the inverse of A with respect to Ã. The center, C, of δ lies on the perpendicular bisector of AA'. Let m be this bisector.
- For δ to be tangent to l at A, we must have $\overleftrightarrow{CA}$ perpendicular to l Thus, C lies on the unique perpendicular to l at A, say $n.\{C\} = m \cap n$.

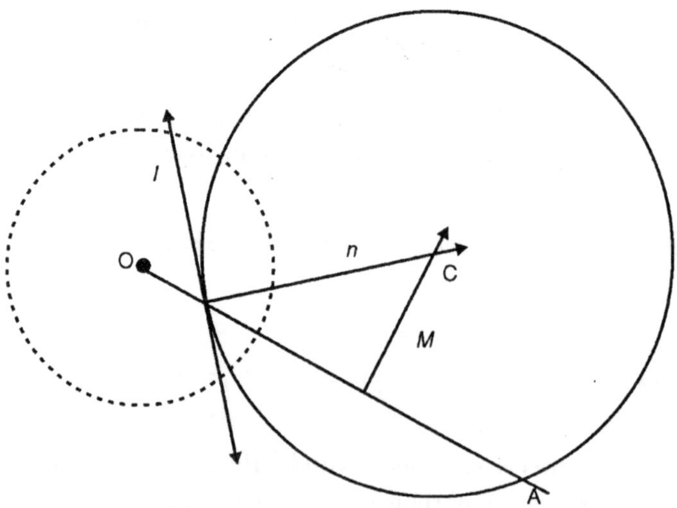

Lemma: *If $d_p(O, B) = x$ then*

$$d(O, B) = r\frac{e^x - 1}{e^x + 1},$$

Where r is the radius of Ã and e is the number so that $\log e = 1$.

Proof: If P and Q are the ends of the diameter through OB then

$$x = \log(OB, PQ)$$

$$e^x = \frac{OP.BQ}{OQ.BP}$$

$$= \frac{BQ}{BP} = \frac{r+OB}{r-OB}$$

$$OB = r\frac{e^x+1}{e^x-1}$$

Which is what was to be proven.

This will give us enough information to prove that all of the axioms of neutral geometry are satisfied, except for the *SAS* axiom. It requires a bit of work. Assume that we are given two *p*-triangles Ä*ABC* and Ä*XYZ* inside Ã so that $\angle A \cong \angle X$, $AC \cong XY$ and $AB \cong XY$ We need to show that these triangles are congruent in the Poincaré model.

First, we show that we may assume that $A = X = O$, where O is the center of Ã. Let δ be the circle through A and B orthogonal to Ã and let be the circle through A and C orthogonal to Ã. Then, since both circles are orthogonal to Ã it follows that δ ∩ ó = $\{A, A'\}$.

Let $s^2 = (AA')(A'O)$ and let be the circle of radius s centered at A'. Then

$$AA' = A'O - AO, \text{ so}$$

$$s^2 = (A'O)^2 - (A'O)(AO)$$

$$= (A'O)^2 - r^2, \text{ by the definition of inverse}$$

Thus, $r^2 + s^2 = (A'O)^2$ so that r is "perpendicular" to s by the converse of the Pythagorean theorem. This means then that Ã and c are orthogonal circles.

By the construction of this circle c O is the inverse of A in c By the last item above, inversion in c maps Ä*ABC* onto a *p*-congruent triangle Ä*OB'C'* Likewise, Ä*XYZ* can be mapped by inversion onto a *p*-congruent triangle Ä*OY'Z'*.

By Lemma and the SAS axiom for Euclidean geometry, we have that $d(O,B) = d(O,Y)$, $d(O,C) = d(O,Z)$ and $\angle BOC \cong \angle YOZ$ Thus, the Euclidean triangles, Ä*OBC* and Ä*OYZ* are congruent by *SAS*.

Therefore, there is a rotation about the center of Ã and possibly a reflection through a diameter which sends Ä*OBC* to Ä*OYZ* This will map Ã onto itself and the orthogonal circle through B and C to the orthogonal circle through Y and Z. This preserves the *p*-length, and hence the triangles are *p*-congruent.

Theorem: *Two triangles in the Poincaré model of hyperbolic geometry are p-congruent if and only if they can be mapped onto each other by a succession of inversions in circles orthogonal to Ã and/or reflections in diameters of Ã.*

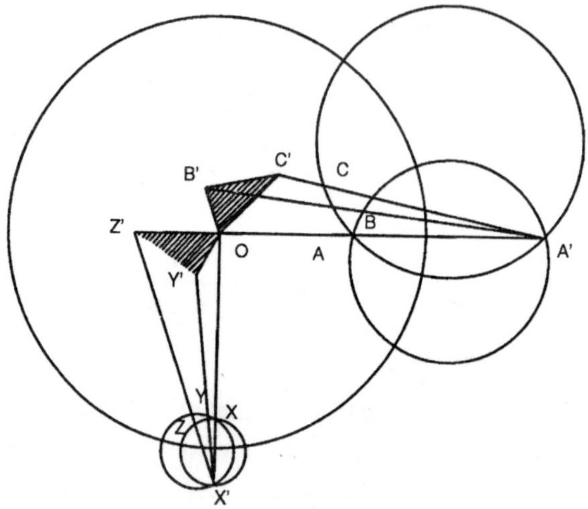

Fig. SAS in the Poincaré Model

At this point, I would rather think of measuring angles in radian measure, not degree measure. In this case we know that the sum of the angles of any triangle in hyperbolic geometry is less than π. Also, let $\prod(d)$ denote the radian measure of the angle of parallelism corresponding to the hyperbolic distance d. We can define the standard trigonometric functions, not as before--using right triangles--but in a standard way. Define

$$\sin x = \sum_{n=0}^{\infty}(-1)^n \frac{x^{2n+1}}{(2n+1)!}$$

$$\cos x = \sum_{n=0}^{\infty}(-1)^n \frac{x^{2n}}{(2n)!}$$

$$\text{ten } x = \frac{\sin x}{\cos x}.$$

In this way we have avoided the problem of the lack of similarity in triangles, the premise upon which all of real Euclidean trigonometry is based. What we have done is to define these functions *analytically*, in terms of a power series expansion. These functions are defined for all real numbers x and satisfy the usual properties of the trigonometric functions.

Theorem: *In the Poincaré model of hyperbolic geometry the angle of parallelism satisfies the equation*

$$e^{-d} = \tan\left(\frac{\Pi(d)}{2}\right).$$

Proof: By the definition of the angle of parallelism, $d = d_p(P, Q)$ for some point P to its foot Q in some p-line l Now, $\prod(d)$ is half of the radian measure of the fan angle at P, or is the radian measure of ÐQPÙ where $\overrightarrow{P\Omega}$ is the limiting parallel ray to l through P.

We may choose l to be a diameter of Ã and $Q = O$, the center of Ã, so that P lies on a diameter of Ã perpendicular to l.

The limiting parallel ray through P is the arc of a circle δ so that

- δ is orthogonal to Ã,
- l is tangent to δ at Ù.

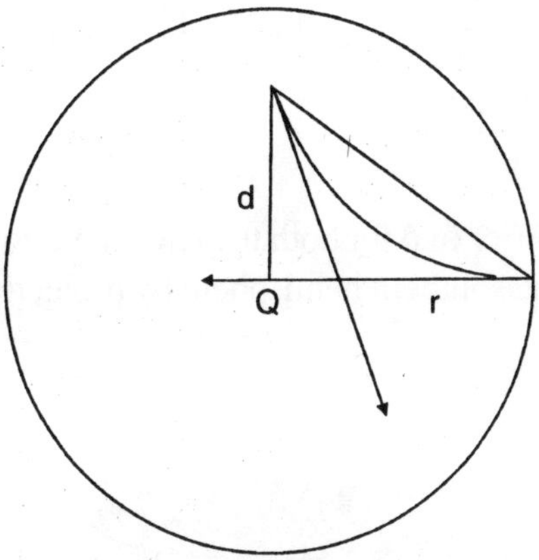

Fig. Angle of Parallelism

The tangent line to δ at P must meet l at a point R inside Ã. This point R is the pole of PÙ from δ By Lemma∠QPÙ = ∠QÙP = â radians. Let us denote Đ(d) = á Then in ÄPQÙ,

$$\text{á} + 2\text{â} = \frac{\pi}{2} \text{ or}$$

$$\text{â} = \frac{\pi}{4} - \frac{\alpha}{2}.$$

Now, $d(P, Q) = r \tan \text{â} = r \text{ ten} \left(\frac{\pi}{4} - \frac{\alpha}{2}\right)$ Applying Lemma we have

$$e^d = \frac{r + d(P,Q)}{r - d(P,Q)} = \frac{1 + \tan\beta}{1 - \tan\beta}.$$

Using the identity $\tan\left(\frac{\pi}{4} - \frac{\alpha}{2}\right) = \frac{1 + \tan\alpha/2}{1 - \tan\alpha/2}$ it follows that

Simplifying this it becomes $e^d = \dfrac{1}{1 - \tan\alpha/2}$.

$$= e^{-d} = \tan\left(\frac{\Pi(d)}{2}\right).$$

Also, we can write this as $\angle(d) = 2 \arctan(e^{-d})$

RETURN TO THE KLEIN MODEL

Now, we have that the Poincaré model is a model for hyperbolic geometry. Since the Klein model is isomorphic to the Poincaré model, it too is a model of the hyperbolic plane, but that does not explain why we do the things the way that we do.

We shall use another isomorphism to study the Klein model. Let Ó denote the unit sphere in Euclidean three-space,

$$Ó = \{(x, y, z) \mid x^2 + y^2 + z^2 = 1\}$$

And let ó denote the equatorial circle

$$ó = \{(x, y, 0) \mid x^2 + y^2 = 1\}$$

We shall consider the interior of ó for both the Klein and Poincaré models. Let $N = (0,0,1)$ and map the interior of onto the southern hemisphere by projection from N.

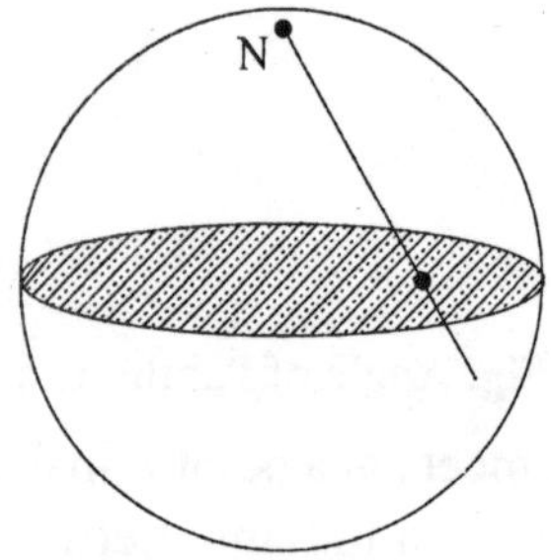

Fig. Central Projection from N

$$f(x, y, 0) = \left\langle \frac{2x}{1+x^2+y^2}, \frac{2y}{1+x^2+y^2}, \frac{x^2+y^2-1}{1+x^2+y^2} \right\rangle.$$

Now, apply the projection up to the xy-plane,

$$\pi(x, y, z) = (x, y, 0).$$

The composition of these two maps is the map from the interior of ó to itself given by:

$$F(x, y, 0) = \left\langle \frac{2x}{1+x^2+y^2}, \frac{2y}{1+x^2+y^2} 0 \right\rangle.$$

A common way of studying the circle is to recall that the Euclidean plane can represent the complex line by the point (a,b) represents the complex number $a + bi$. The circle is then the set $ó = \{z \in \mathbb{C} \mid z\bar{z} = 1\}$ Defining the modulus of z by $|z| = z\bar{z}$ we can rewrite the above function F by

$$F(z) = \frac{2z}{1+|z|^2}.$$

Clearly, F maps a diameter of onto itself, but fixes only the origin and the endpoints of the diameter. In fact, as F maps the interior of onto itself, it fixes only the origin and the points on ó

Consider a p-line in ó say ä --a circle orthogonal to ó. ä $\cap$ ó = $\{P, Q\}$ What is F (ä) ?

Claim: If $A \in$ ä $\cap$ int ó and PQ is the chord of ä in *int ó* then $F(A) = PQ \cap \overrightarrow{OA}$.

Proof: Let $C = (a,b)$ be the center of ä. Then we found P and Q by taking the circle with CO as diameter and intersecting it with ó. It s center is $\left(\frac{\alpha}{2}, \frac{b}{2}\right)$ and its radius is $\frac{1}{2}\sqrt{a^2 + b^2}$. Applying standard analytic geometry we find that its equation is $x^2 -$ á$^x + y^2 - by = 0$ Intersecting this with ó ($x^2 + y^2 = 1$) gives

$$ax + by = 1$$

for the line joining P and Q.

Now CQ is perpendicular to OQ since ó and ä are orthogonal circles. Thus, $OC^2 = OQ^2 + CQ^2$ or $r^2 = CQ^2 =$ á$^2 + b^2 - 1$ where r is the radius of ä Then ä has equation

$$x^2 + y^2 = 2ax + 2by - 1.$$

Let A = (á, â) $\in$ ä $\cap$ *int* ó Put $F(A) = (u,v)$. Solving we find $u = \frac{\alpha}{a\alpha + b\beta}$ and $u = \frac{\beta}{a\alpha + b\beta}$ Thus, $au + bv = 1$ which implies that $(u, v) \in PQ$.

We can now justify the definitions we made earlier in the Klein model.

- $AB \cong_k CD$ if and only if $F^1(AB) \cong_p F^{-1}(CD)$.
- $\angle ABC \cong_k \angle DEF$ if and only if F–1 $(\angle ABC) \cong F^{-1}(\angle DEF)$.

PERPENDICULARITY

Now, we see that $l \perp m$ if and only if $F^{-1}(l) \perp F^{-1}(m)$ in the Poincaré model.

- If l and m are both diameters, then $F^{-1}(l)$ and $F^{-1}(m)$ are both diameters. In fact, $F^{-1}(l) = l$ and $F^{-1}(m) = m$ Hence, the usual Euclidean meaning applies.
- Assume that only l is a diameter. Then $F^{-1}(l) = l$. $F^{-1}(m)$ is a circular arc perpendicular to l Thus, the tangent to $F^{-1}(m)$ is perpendicular to l and l passes through the center of $F^{-1}(m)$. m is the chord of F–1 (m). Since l passes through this center, l is the perpendicular bisector of m in the Euclidean sense. Therefore, l is perpendicular to m in our previous definition. Conversely, if $l \perp_k m$ then l is the perpendicular bisector of m. Thus, l passes through the center of the circle and l is perpendicular to the circular arc $F^{-1}(m)$.
- Assume that neither l nor m is a diameter. $F^{-1}(l)$ and $F^{-1}(m)$ are arcs of circles ë and ì orthogonal to ó.

- Suppose ë ⊥ ì in the Poincaré model. By one of the facts, the center of ë is $P(l)$ and that of ì is $P(m)$. We need to show that m passes through $P(l)$ then we will have that $m \perp_k l$.

  Let m Ç ó = $\{P, Q\}$ Since ë and ì are orthogonal inversion in ë interchanges P and Q, since ë and ì are also orthogonal. Thus, $P = Q$' in ì and the Euclidean line through P and Q--namely m--passes through the center of ë which is $P(l)$.
- Suppose m passes through $P(l)$ Then $P = Q$' in ë so that ë and ì are orthogonal.

AREA IN HYPERBOLIC GEOMETRY

A *polygonal region* is a plane figure which can be expressed as the union of a finite number of triangular regions, in such a way that if two of the triangular regions intersect, their intersection is an edge or a vertex of each of them.

Let R be a polygonal region. A *triangulation* of R is a finite collection,

$K = \{T_1, T_2, ..., T_n\}$

Of triangular regions T_i such that

- The T_i 's intersect only at edges and vertices, and
- Their union is R.

Let R_1 and R_2 be polygonal regions. Suppose that they have triangulations

$$K_1 = \{T_1, T_2, ..., T_n\},$$
$$K_2 = \{T'_1 T'_2, ..., T'_n\},$$

Such that for each i we have $T_i \cong T'_i$ Then we say that R_1 and R_2 are *equivalent by finite decomposition*, and we write $R_1 \equiv R_2$.

Let ÝT = *defect* (Ä(T)) for any triangular region T.

Theorem: *If* K_1 *and* K_2 *are triangulations of the same polygonal region R, then* Ý$K1$ = ÝK_2

Theorem: *If two Saccheri quadrilaterals have the congruent summits and equal defects, then their summit angles are congruent, in which case, the two Saccheri quadrilaterals are congruent.*

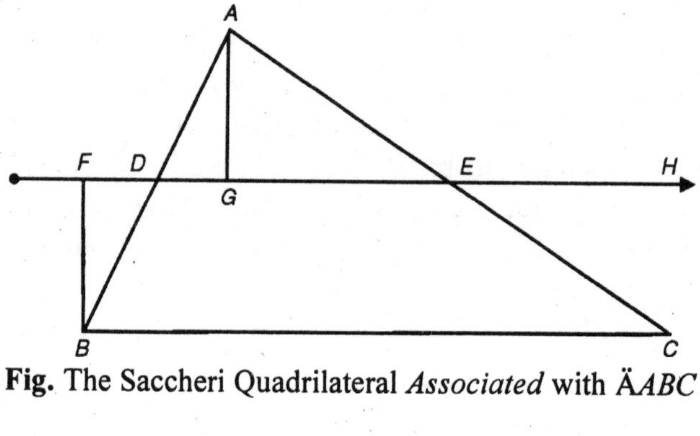

Fig. The Saccheri Quadrilateral *Associated* with ÄABC

Given Ä*ABC* with *BC* considered as the base. Let *D* and *E* be the midpoints of *AB* and *AC*; let *F*, *G*, and *H* be the feet of the perpendiculars from *B*, *A*, and *C*, respectively, to $l = \overleftrightarrow{DE}$ As you have proven in the homework, *HCBF* is a Saccheri quadrilateral. It is known as the quadrilateral *associated* withÄ*ABC* It depends on the choice of the base, but it should be clear which base we mean.

Theorem: *Every triangular region is equivalent by finite decomposition to its associated quadrilateral region.*

Theorem: *Every triangular region has the same defect as its associated quadrilateral region.*

Theorem: *If ÄABC and ÄDEF have the same defect and a pair of congruent sides, then the two triangular regions are equivalent by finite decomposition.*

Bolyai's Theorem in the Hyperbolic Plane

Theorem: *If* T_1 *and* T_2 *are triangular regions, and* ÝT_1 = ÝT_2, *then* $T_1 \equiv T_2$.

Theorem: *If ÝÄABC* > ÝÄ*DEF*, *then there is a point P between A and C such that* ÝÄ*ABP* = *ÝÄDEF*.

REQUIREMENTS FOR AN AREA FUNCTION

Now, how do we define the area of a polygonal region in the hyperbolic plane? Should we define the area in the same way that we do in the Euclidean plane? If so, what are the minimum requirements for an area function of the Euclidean plane? Minimally it should satisfy the following: Let *R* be the set of all polygonal regions in $H^{2.}$ An area function should be a function

á : $R \to R$

Such that

- á$R > 0$ for every R;
- if R_1 and R_2 intersect only in edges and vertices, then

  $$á(R_1 \cup R_2) = áR_1 + áR_2;$$
- if T_1 and T_2 are triangular regions with the same base and altitude, then áT_1 = áT_2.

  If there is such a function á, then by (2) and (3) it will satisfy
- If $R_1 \equiv R_2$, then áR_1 = áR_2 because congruent triangles have the same bases and altitudes.

  In Euclidean geometry we can show that this area function is unique and it must satisfy the formula á$T = \frac{1}{2}bh$ for each triangular region, *T*, where *b* is the length of the base and *h* is the length of the altitude.

Theorem: *There is no such function* á : $R \to \mathbb{R}$ *satisfying* (1), (2), (3), *and, hence,* (4).

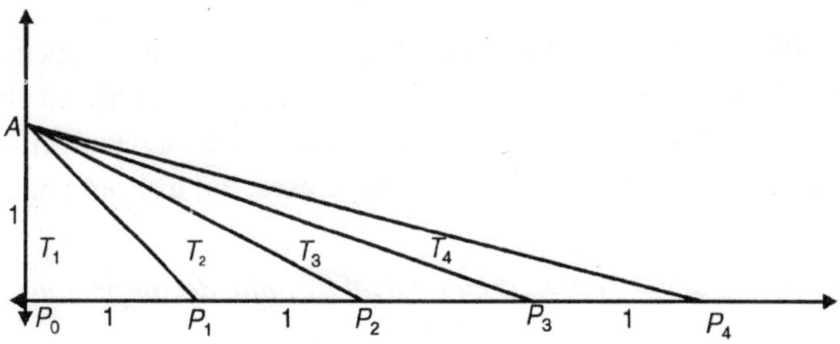

Proof: Consider the right angle $\angle AP_0P_1$, with $AP_0 = PP_0 = 1$ For each n, let P_n be the point of $\overrightarrow{P_0P_1}$ such that $P_0P_n = n$. This gives a sequence of triangles

$$ÄAP_0P_1, ÄAP_1P_2, ...,$$

And a corresponding sequence of triangular regions

$$T_1, T_2, ...$$

By condition (3) all the regions T_i have the same "area" á$T_i = A$.

Now consider the defects of these triangles and let d_i = ÝT_i. For each n

$d_1 + d_2 + ... + d_n$ = äÄ$AP_0P_n < 180°$

Since the partial sums $d_1 + d_2 + ... + d_n$ are bounded, we have that the infinite series,

$$\sum_{n=0}^{\infty} d_n$$

Is convergent. Therefore,

$$\lim_{n\to\infty} d_n = 0$$

Hence $d_n < d_1$ for some n.

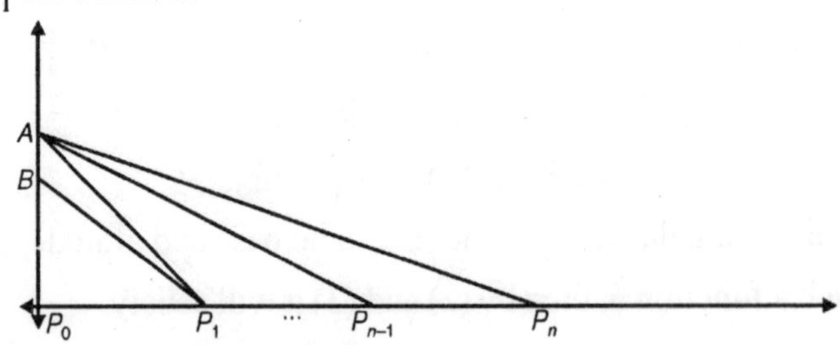

By Theorem there is a point, B, between A and P_0 such that

$$äÄBP_0P_1 = äÄAP_{n-1}\, P_n = d_n$$

Therefore by Bolyai's Theorem the regions T and T_n determined by these triangles are equivalent by finite decomposition. By condition (4) this means that áT = áT_n. But

$$áT_n = áT_1 = A.$$

Therefore, $$áT = áT_1.$$

Because $\acute{a}\ddot{A}ABP_1 > 0$ and

$$\acute{a}T_1 = \acute{a}T + \acute{a}\ddot{A}ABP_1$$

This must be impossible.

THE UNIQUENESS OF HYPERBOLIC AREA THEORY

Any reasonable area function á should have the following properties:

1. $\acute{a}R > 0$ for every R;
2. If R_1 and R_2 intersect only in edges and vertices, then
$$\acute{a}(R_1 \cup R_2) = \acute{a}R_1 + \acute{a}R_2;$$
3. If $R_1 \equiv R_2$ then $\acute{a}R_1 = \acute{a}R_2$.

There is such a function--the defect. Recall *defect* ($\ddot{A}ABC$) = $\ddot{a}\ddot{A}ABC$. We know that satisfies (1), (2) and (3). In fact, we have the following theorem:

Theorem: *Let*

$$\acute{a} : R \rightarrow \mathbb{R}$$

Be an area function satisfying (1), (2), *and* (3). *Then there is a positive constant* $k>0$ *such that*

$$\acute{a}\, R = \frac{\pi}{180} K^2 \delta R$$

for every $R \in R$..

We are not able to prove this at this time. The proof is within your grasp, but we do not have time to do it.

Corollary: *In* H^2 *the area of any triangle is at most* $\pi\, k^2$,

There is no finite triangle whose area equals the maximal value $\pi\, k^2$, although you can approach this area as closely as you wish (and achieve it with a trebly asymptotic triangle). J. Bolyai proved that you can construct a *circle* of area $\pi\, k^2$ and a regular 4-sided polygon with a 45° angle that also has this area.

ANGLE OF PARALLELISM

In the Poincaré model of hyperbolic geometry, we found a formula relating the angle measurement of the *angle of parallelism* and the distance to which it is associated. This formula is a special case of the general formula discovered independently by Bolyai and Lobachevsky.

Formula of Bolyai-Lobachevsky

Theorem:

$$\tan \frac{\alpha}{2} = e^{-d/k},$$

Where k is the proportionality constant from Theorem.

We have proven this formula in the Poincaré model where $k = 1$.

HYPERCYCLES AND HOROCYCLES

There is a curve peculiar to hyperbolic geometry, called the *horocycle*.Consider two limiting parallel lines, *l* and *m*, with a common direction, say Ù Let *P* be a point on one of these lines *P* ∈ *l*. If there exists a point *Q* ∈ *m* such that the *trilateral*, Ä*PQ*Ù, has the property that

∠*PQ*Ù ≅ ∠*QP*Ù

Then we say that *Q corresponds* to *P*. If the trilateral Ä*PQ*Ù has the above property we shall say that it is *equiangular*. Note that it is obvious from the definition that if *Q* corresponds to *P*, then *P* corresponds to *Q*. The points *P* and *Q* are called a pair of *corresponding points*.

Theorem: *If points P and Q lie on two limiting parallel lines in the direction of the ideal point,* Ù, *they are corresponding points on these lines if and only if the perpendicular bisector of PQ is limiting parallel to the lines in the direction of* Ù.

Theorem: *Given any two limiting parallel lines, there exists a line each of whose points is equidistant from them. The line is limiting parallel to them in their common direction.*

Proof: Let *l* and *m* be limiting parallel lines with common direction Ù. Let *A* ∈ *l* and *B* ∈ *m* The bisector of ∠*BA*Ù in the trilateral Ä*AB*Ù meets side *B*Ù in a point *X* and the bisector of Ä*AB*Ù meets side *AX* of the triangle Ä*ABX* in a point *C*. Thus the bisectors of the angles of the trilateral Ä*AB*Ù meet in a point *C*. Drop perpendiculars from *C* to each of *l* and *m*, say *P* and *Q*, respectively. By *Hypothesis-Angle* Ä*CAP* ≅ Ä*CAM* (*M* is the midpoint of *AB*) and Ä*CBQ* ≅ Ä*CBM* Thus, *CP* ≅ *CM* ≅ *CQ* Thus, by *SAS* for trilaterals, we have that

Ä*CP*Ù ≅ Ä*CQ*Ù

And thus the angles at *C* are congruent. Now, consider the line $\overleftrightarrow{C\Omega}$ and let *F* be any point on it other than *C*. By *SAS* we have Ä*CPF* ≅ Ä*CQF* If *S* and *T* are the feet of *F l* in and *m*, then we get that Ä*PSF* ≅ Ä*QTF* and *FS* ≅ Ä*FT*. Thus, every point on the line $\overleftrightarrow{C\Omega}$ is equidistant from *l* and *m*.

This line is called the *equidistant line*.

Theorem: *Given any point on one of two limiting parallel lines, there is a unique point on the other which corresponds to it.*

Theorem: *If three points P, Q, and R lie on three parallels in the same direction so that P and Q are corresponding points on their parallels and Q and R are corresponding points on theirs, then P, Q, and R are noncollinear.*

Theorem: *If three points P, Q, and R lie on three parallels in the same direction so that P and Q are corresponding points on their parallels and Q and R are corresponding points on theirs, then P and R are corresponding points on their parallels.*

Consider any line *l*, any point *P* ∈ *l* and an ideal point in one direction of *l*, say Ù. On each line parallel to *l* in the direction Ù there is a unique point *Q* that corresponds to *P*. The set consisting of *P* and all such points *Q* is called a horocycle, or, more precisely, the horocycle determined by *l*, *P*, and Ù The lines parallel to *l* in the direction Ù together with *l*, are called the *radii* of the horocycle. Since *l* may be denoted by $\overleftrightarrow{P\Omega}$ we may regard the horocycle as determined simply by *P* and Ù, and hence call it the *horocycle through P with direction* Ù, or in symbols, the horocycle (*P*, Ù).

All the points of this horocycle are mutually corresponding points by Theorem, so the horocycle is equally well determined by any one of them and Ù. In other words if *Q* is any point of horocycle (*P*, Ù) other than *P*, then horocycle (*Q*, Ù) is the same as horocycle (*P*, Ù) If, however, *P*' is any point of *l* other than *P*, then horocycle (*P*′, Ù) is different from horocycle (*P*, Ù) even though they have the same direction and the same radii. Such horocycles, having the same direction and the same radii, are called *codirectional horocycles*.

There are analogies between horocycles and circles. We will mention a few.

Lemma: *There is a unique horocycle with a given direction which passes through a given point. (There is a unique circle with a given center which passes through a given point.)*

Lemma: *Two codirectional horocycles have no common point. (Two concentric circles have no common point.)*

Lemma: *A unique radius is associated with each point of a horocycle. (A unique radius is associated with each point of a circle.)*

A *tangent to a horocycle* at a point on the horocycle is defined to be the line through the point which is perpendicular to the radius associated with the point.

No line can meet a horocycle in more than two points. This is a consequence of the fact that no three points of a horocycle are collinear inasmuch as it is a set of mutually corresponding points.

Theorem: *The tangent at any point A of a horocycle meets the horocycle only in A. Every other line through A except the radius meets the horocycle in one further point B. If á is the acute angle between this line and the radius, then d(A,B) is twice the segment which corresponds to á as angle of parallelism.*

Proof: Let *t* be the tangent to the horocycle at *A* and let Ù be the direction of the horocycle. If *t* met the horocycle in another point *B*, we would have a trilateral with two right angles, since *A* and *B* are corresponding points. In fact the entire horocycle, except for *A*, lies on the same side of *t*, namely, the side containing the ray $\overrightarrow{A\Omega}$.

Let *k* be any line through *A* other than the tangent or radius. We need to show that *k* meets the horocycle in some other point.

Let á be the acute angle between *k* and the ray $\overrightarrow{A\Omega}$ Let *C* be the point of *k*, on the side of *t* containing the horocycle, such that *AC* is a segment corresponding to as angle of parallelism.

(RECALL: $e^{-d} = \tan$ (á/2). The line perpendicular to k at C is then parallel to $\overleftrightarrow{A\Omega}$ in the direction Ù. Let B be the point of k such that C is the midpoint of AB. The trilaterals ÄACÙ and ÄBCÙ are congruent. Hence $\angle CB$Ù = á B corresponds to A, and $B \in$ (A, Ù).

A *chord* of a horocycle is a segment joining two points of the horocycle.

Theorem: *The line which bisects a chord of a horocycle at right angles is a radius of the horocycle.*

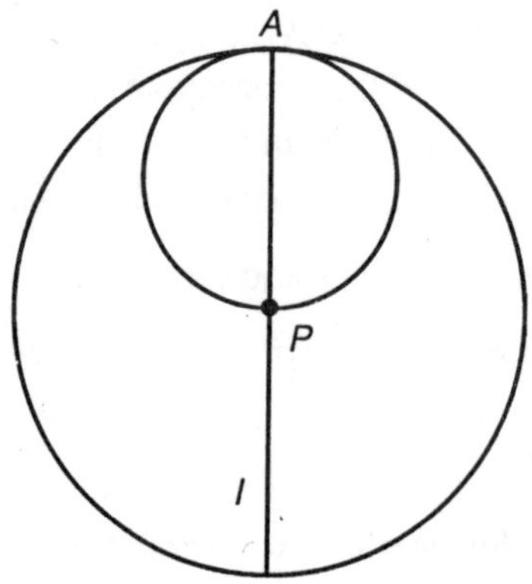

Fig. A Horocycle in the Poincaré Model

We can visualize a horocycle in the Poincaré model as follows.

Let l be the diameter of the euclidean circle Ã whose interior represents the hyperbolic plane, and let O be the center of Ã It is a fact that the *hyperbolic circle* with hyperbolic center P is represented by a euclidean circle whose euclidean center R lies between P and A.

As P recedes from A towards the ideal point Ù, R is pulled up to the euclidean midpoint of ÙA, so that the horocycle (A, Ù) is a euclidean circle tangent to Ã at Ù and tangent to l at A. It can be shown that all horocycles are represented in the Poincaré model by euclidean circles inside Ã and tangent to Ã. Moreover, all the Poincaré lines passing through the ideal point Ù are orthogonal to the horocycle (A, Ù).

In the Poincaré model two horocycles tangent to Ã at Ù are said to be *concentric*. It is by studying the ratio of corresponding arcs on concentric horocycles that the Bolyai-Lobachevsky formula can be derived.

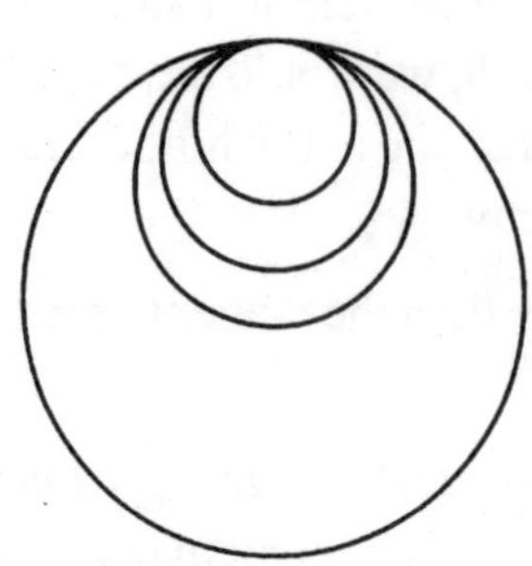

Fig. Concentric Horocycles in the Poincaré Model

Another curve found specifically in the hyperbolic plane and nowhere else is the *equidistant curve*, or *hypercycle*. Given a line l and a point P not on l consider the set of all points Q on one side of l so that the perpendicular distance from Q to l is the same as the perpendicular distance from P to l

The line l is called the *axis*, or *base line*, and the common length of the perpendicular segments is called the *distance*. The perpendicular segments defining the hypercycle are called its *radii*. The following statements about hypercycles are analogous to statements about regular euclidean circles.

- Hypercycles with equal distances are congruent, those with unequal distances are not. (Circles with equal radii are congruent, those with unequal radii are not.)
- A line cannot cut a hypercycle in more than two points.
- If a line cuts a hypercycle in one point, it will cut it in a second unless it is tangent to the curve or parallel to it base line.
- A tangent line to a hypercycle is defined to be the line perpendicular to the radius at that point. Since the tangent line and the base line have a common perpendicular, they must be hyperparallel. This perpendicular segment is the shortest distance between the two lines. Thus, each point on the *tangent line* must be at a greater perpendicular distance from the base line than the corresponding point on the hypercycle. Thus, the hypercycle can intersect the hypercycle in only one point.
- A line perpendicular to a chord of a hypercycle at its midpoint is a radius and it bisects the arc subtended by the chord.
- Two hypercycles intersect in at most two points.
- No three points of a hypercycle are collinear.

In the Poincaré model let P and Q be the ideal end points of l It can be shown that the hypercycle to l through P is represented by the arc of the euclidean circle passing through A, B, and P. This curve is orthogonal to all Poincaré lines perpendicular to the line l

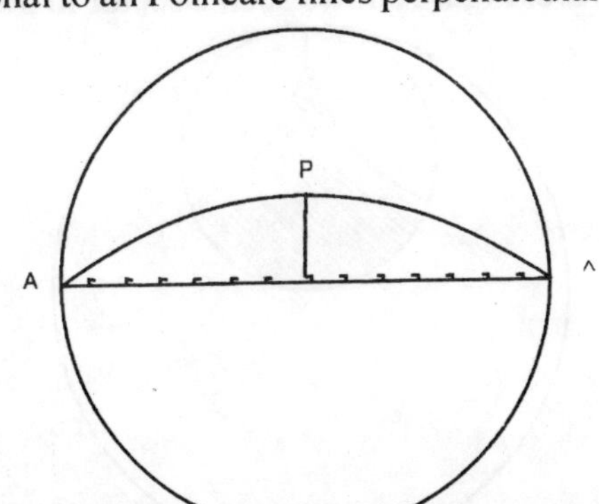

Fig. A hypercycle in the Poincaré Model

In the Poincaré model a euclidean circle represents:

- A *hyperbolic circle* if it is entirely inside Ã;
- A *horocycle* if it is inside Ã except for one point where it is tangent to Ã;
- An *hypercycle* if it cuts Ã nonorthogonally in two points;
- A *hyperbolic line* if it cuts Ã orthogonally.

It follows that in the hyperbolic plane three non-collinear points lie either on a circle, a horocycle, or a hypercycle accordingly, as the perpendicular bisectors of the triangle are *concurrent* in an ordinary point, an ideal point, or an ultraideal point.

THE PSEUDOSPHERE

Some of the difficulties of the different plane models for the hyperbolic plane are readily apparent. The Klein model is not conformal--angles are not preserved--we cannot use a Euclidean protractor--and it distorts distances. The Poincaré disk model, while conformal, distorts distances--we cannot use a Euclidean ruler. People have searched for another model which is conformal and that represents hyperbolic lengths faithfully by Euclidean lengths. Such a model would be called *isometric*. We are able to construct a model of elliptic geometry on the surface of the sphere in 3-space. Is there a surface in 3-space on which we can isometrically model hyperbolic geometry? If so, the lines of the hyperbolic plane would then be represented by *geodesics*∠ on the surface. Since there seems to be some feeling that hyperbolic lines cannot be straight (*i.e.*, modelled on Euclidean lines), we might expect this surface to be curved in Euclidean 3-space.

A theorem first proved by David Hilbert states that it is impossible to embed the entire hyperbolic plane isometrically as a surface in Euclidean space. On the other hand it *is* possible to embed the Euclidean plane isometrically in hyperbolic space, as the surface of the horosphere.

It is possible to embed a portion of the hyperbolic plane isometrically in Euclidean space, the portion called a *horocyclic sector*, bounded by an arc of a horocycle and the two diameters of the horocycle cutting off this arc. Figure shows such a sector in the Poincaré model.

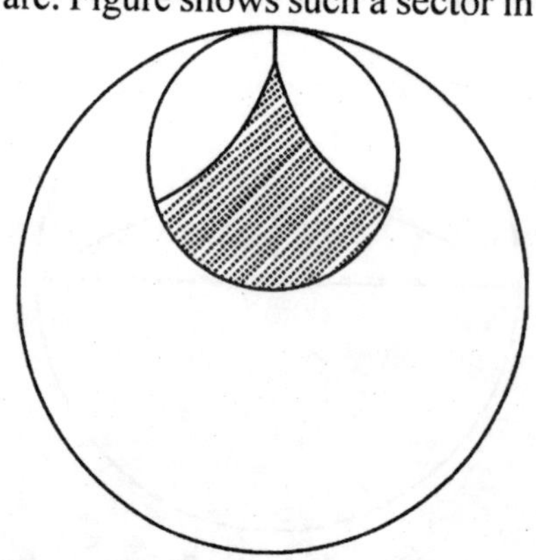

Fig. Horocyclic Sector in the Poincaré Model

We must embed this sector isometrically onto a surface in Euclidean 3-space. The surface that represents this region is called a *pseudosphere*. It is obtained by rotating a curved known as the *tractrix* around its asymptote. The tractrix is the curve characterized by the condition that the length of the segment of the tangent line to the curve from the curve to the y-axis is constant. It has the following equation for a given constant a:

$$x = a \ln\left(\frac{a+\sqrt{a^2 - y^2}}{y}\right) - \sqrt{a^2 - y^2}.$$

And has graph shown in Figure.

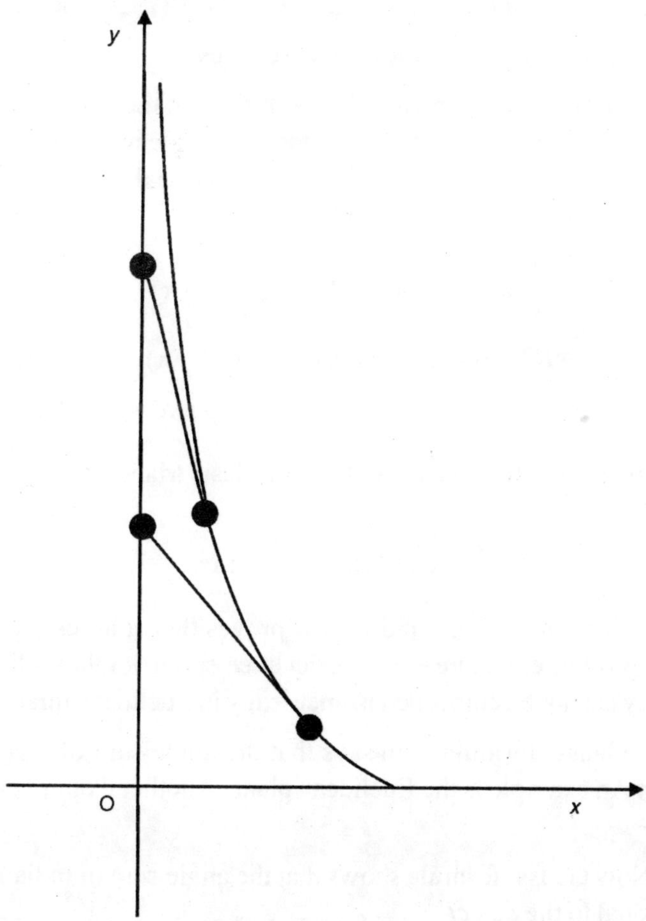

Fig. Tractrix Curve

The representation on the pseudosphere gives us the opportunity to give some geometric meaning to the fundamental constant k in the theorem on area. Gauss discovered a way in which we can measure the *curvature* of a surface. Intuitively, it is a number K which measures how much a surface bends. The sphere of radius R, for example, has constant curvature $1/R^2$

This makes sense as you think about, because the larger a sphere is the less it bends. In general the curvature will change from point to point, or region to region, on a surface. It should be zero at points where the surface is flat and it should be large where the surface bends sharply. Surfaces where the curvature is the same at all points are called *surfaces of constant curvature*. The sphere is such an example.

For a surface of constant curvature K, Gauss found a fundamental formula which relates the curvature, area, and angular measure. Take a geodesic triangle, ÄABC --a triangle where the sides are geodesic segments. Let $(\angle A)'$ denote the radian measure of angle A. Gauss showed that

$$K \times \text{arca } (\text{Ä}ABC) = (\angle A)' + (\angle B)' + (\angle C)' - \pi.$$

What does this mean? There are three possible cases.

Case 1. Let K be positive. This means that both sides of the above equation are positive. This means that the angle sum in radians of a geodesic triangle is greater than π(or in degrees, greater than 180°), and the area of the triangle is proportional to the *excess*. On a sphere of radius R, then we know that the area of a geodesic triangle is

$$\text{area } (\text{Ä}ABC) = \frac{1}{K} \times ((\angle A)' + (\angle B)' + (\angle C)' - \pi)$$
$$= R2 \times ((\angle A)' + (\angle B)' + (\angle C)' - \pi)$$

If we have a geodesic triangle on the surface of the sphere with three right angles, then the excess of this triangle is $\frac{\pi}{2}$. Thus, the area of this geodesic triangle is

$$\text{area } (\text{Ä}ABC) = \frac{1}{2} \pi R^2.$$

A theorem by Liebmann, Hopf, and Rinow proves that spheres are the only complete surfaces of constant positive curvature in Euclidean three-space. It follows from this information that elliptic geometry cannot be embedded isometrically in Euclidean three-space.

Case 2. $K = 0$. Gauss' formula then says that the angle sum of a geodesic triangle, in radians, is exactly π An example is the Euclidean plane. Another example is an infinitely long cylinder.

Case 3. $K<0$. Now Gauss' formula shows that the angle sum in radians is less than π and the area is proportional to the *defect*.

The pseudosphere is a surface of constant negative curvature. Since the pseudosphere represents a portion of the hyperbolic plane isometrically, we can compare Gauss' formula with the formula for the area in the hyperbolic plane. The comparison gives $K = 1/k^2$ Thus, $-1/k^2$ *is the curvature of the hyperbolic plane*.

Of interest, let $r = ik$ where $i = \sqrt{-1}$ Then $K = 1/r^2$ so that the hyperbolic plane can be described as a *sphere of imaginary radius* $r = ik$, as was noted by the mathematician Lambert. This description is very reasonable, for all the formulas of hyperbolic trigonometry can be obtained from the formulas known for spherical trigonometry by replacing r with ik.

Finally, notice that as k gets very large, the curvature K approaches zero, and the geometry of the surface resembles more and more the geometry of the Euclidean plane. It is in this sense that Euclidean geometry is a *limiting case* of hyperbolic geometry.

HYPÈRBOLIC TRIGONOMETRY

Trigonometry is the study of the relationships among sides and angles of a triangle. In Euclidean geometry we use similar triangles to define the trigonometric functions—but the theory of similar triangles in not valid in hyperbolic geometry. Thus, we have defined them in terms of their power series expansion for any real number.

Define two hyperbolic trigonometric functions

$$\sin h\, x = \frac{e^x - e^{-x}}{2}$$

$$\cos h\, x = \frac{e^x + e^{-x}}{2}$$

Since $e^x = \sum_{n=0}^{\infty} \frac{x^n}{n!}$ converges for all $x \in \mathbb{R}$ the power series expansions of the hyperbolic trigonometric functions are

$$\sin h\, x = \sum_{n=0}^{\infty} \frac{x^{2n+1}}{(2n+1)!}$$

$$\cos h\, x = \sum_{n=0}^{\infty} \frac{x^{2n}}{(2n)!}$$

And converge for all real x. In fact, using complex analysis and letting $i = \sqrt{-2}$, we can easily see that

$$\sinh x = -i \sin(ix) = i \sin\left(\frac{x}{i}\right)$$

$$\cosh x = \cos(ix) = \cos\left(\frac{x}{i}\right)$$

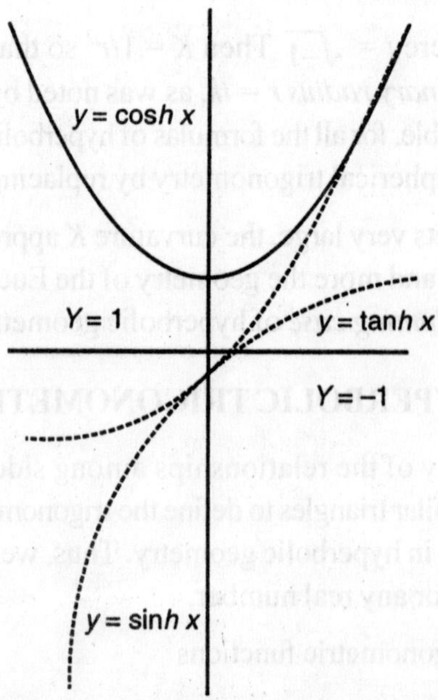

Fig. Graphs of Hyperbolic Functions

We get to choose our unit of measurement in H^2 -- not a particular model now. We would like to choose a unit so that k=1. It can be shown that this is analogous to the choice of a unit of angle measurement so that a right angle has *radian* measure $\frac{\pi}{2}$. By measuring in radians the theorem on area becomes

á(Ä*ABC*) = *defect* (Ä*ABC*),

And the Fundamental Formula becomes

$$\prod(x) = 2 \arctan^{e-x}.$$

A straightforward calculation using double angle formulas for the circular functions gives the following formulas:

$$\sin \prod(x) = \text{sech}\, x$$

$$\cos \prod(x) = \text{tenh}\, x$$

$$\tan \prod(x) = \text{esch}\, x$$

For example, to derive the first equation:

$$\sin \prod(x) = \sin(2 \arctan e^{-x})$$

$$= 2 \sin(\text{arctane}\ e{-}x) \cos(\arctan e{-}x)$$

$$= 2\left(\frac{e^{-x}}{1+e^{-2x}}\right) = 2\left(\frac{1}{e^{x}+e^{-x}}\right)$$

$$= \frac{1}{\cosh x} = \text{sech } x.$$

This function $\Pi\left(0, \frac{\pi}{2}\right) \rightarrow H^2$ provides a connection between the hyperbolic and circular functions.

Given ÄABC, let $a = d_p(B, C)$, $b = d_p(A, C)$, and $c = d_p(A, B)$ First, we will derive some formulæ of hyperbolic geometry from the Poincaré model. Choose our circle Ã to have radius 1.

Let O denote the center of Ã and let $B \in H^2$ Let $x = dp(O, B)$ be the Poincaré distance from O to B and let $t = d(O,B)$ be the Euclidean distance from O to B. From Lemma we have

$$ex = \frac{1+t}{1-t}.$$

We then get that

$$\sinh x = \frac{2t}{1-t^2}$$

$$\cosh x = \frac{1+t^2}{1-t^2}$$

$$\tanh x = \frac{2t}{1+t^2} = F(t)$$

Where F is the isomorphism of the Poincaré model onto the Klein model described in the previous chapter.

Theorem: *Given any right triangle ÄABC in the hyperbolic plane and k=1 with the right angle (having measure π/2 then*

$$\sin A = \frac{\sinh a}{\sinh c} \quad \cos A = \frac{\tanh b}{\tanh c}$$

$$\cosh C = \cosh a \cosh = \cot A \cot B$$

$$\cosh a = \frac{\cos A}{\sin B}$$

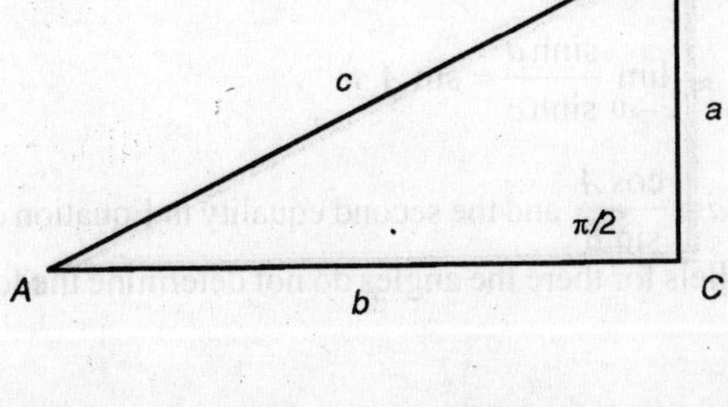

Before we prove these equations, compare them with the formulæ for a right triangle in Euclidean geometry.

- Equation $\cosh a = \dfrac{\cos A}{\sin B}$ is the hyperbolic analogue of the Pythagorean theorem.

$$\cosh c = \cosh a \cosh b$$

$$1+\frac{c^2}{2!}+\frac{c^4}{4!}+\ldots = \left(1+\frac{a^2}{2!}+\frac{a^4}{4!}+\ldots\right)\left(1+\frac{b^2}{2!}+\frac{b^4}{4!}+\ldots\right)$$

$$= 1+\left(\frac{a^2}{2!}+\frac{b^4}{2!}\right)+\ldots$$

Thus, if ÄABC is sufficiently small so that higher powers of a, b, and c are negligible, then

$$1+\frac{c^2}{2} \approx 1+\frac{1}{2}(a^2+b^2)$$

$$c^2 \approx a^2+b^2$$

- Equation $\cosh a = \dfrac{\cos A}{\sin B}$ says that for ΔABC sufficiently small $\sin A \approx \dfrac{a}{c}$ and $\sin B \approx \dfrac{b}{c}$. How close are these approximations? Consider right triangles $\angle A$ with fixed and let c approach 0. Since $a<c$ $a \to 0$.

$$\frac{1}{\sinh x} \approx \frac{1}{c+\frac{c^3}{3!}} = \frac{1}{c}\frac{1}{1+u} = \frac{1}{c}(1-u+u^2-u^3+\ldots)$$

where $\lim\limits_{c\to 0} u = 0$. Thus,

$$\text{Thus, } \frac{\sinh a}{\sinh c} = \frac{a}{c}\left(1+\frac{a^2}{3!}+\frac{a^4}{5!}+\ldots\right)(1-u+u^2-u^3+\ldots)$$

$$\lim_{c\to 0}\frac{a}{c} \approx \lim_{c\to 0}\frac{\sinh a}{\sinh c} = \sin A.$$

- Equation $\cosh a = \dfrac{\cos A}{\sin B}$ and the second equality in Equation $\cosh a = \dfrac{\cos A}{\sin B}$ have no euclidean parallels for there the angles do not determine the lengths of the sides.

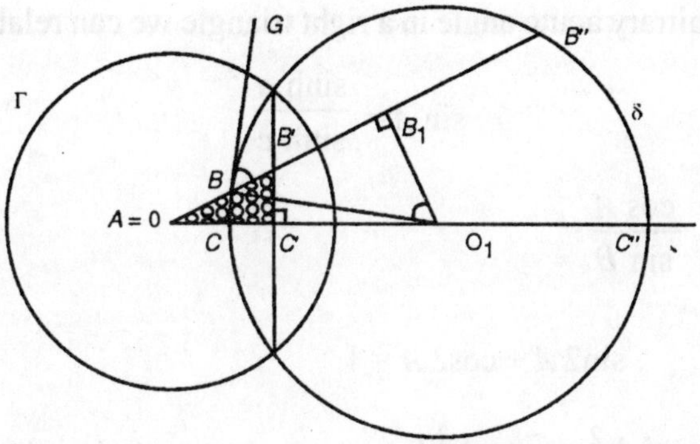

Proof: We may assume that $A = O$ is the center of Γ Let $B' = F(B)$ and $C'=F(C)$. Then

$$\cos A = \frac{AC'}{AB'}$$

From our above observation we have

$$\tanh b = F(d_p(A, C)) = AC'$$
$$\tanh c = F(d_p(A, B)) = AB'$$

The best way to see this is to think of t as $t = t + 0i$, so that t sits on the x-axis and is sent to a point on the x-axis. The point to which it is sent must be at a distance $F(t)$ from the origin.

It is then easy to see that

$$\cos A = \frac{\tanh b}{\tanh c}.$$

Now, let $B'' = \overrightarrow{OB} \cap$ Ý, $B'' \neq B$ By our previous results on the inversion of circles, B'' is the inverse of B in Ã Thus,

$$BB'' = OB'' - OB = \frac{1}{t} - t = \frac{1-t^2}{t} = \frac{2}{\sinh c}.$$

Likewise, $CC'' = \dfrac{2}{\sinh b}$.

Let O_1 be the center of Ý Let B_1 be the foot of O_1 on $\overleftrightarrow{BB''}$.$O_1B \cong O_1B''$ so that B_1 is the midpoint of BB''. Thus, if $\overrightarrow{BG}$ is tangent to Ý at B, $\angle GBO_1 = \angle BB_1O_1 = \pi/2$ Since they are opposite angles, $\angle B \cong \angle GBB_1$ Thus, $\angle B = \angle GBO_1 - \angle B_1BO_1 = \frac{\pi}{2} - \angle B_1BO_1 = \angle BO_1B_1$ Then

$$\sin B = \frac{BB_1}{O_1B} = \frac{BB''}{CC''} = \frac{\sinh b}{\sinh c}.$$

Since $\angle B$ is an arbitrary acute angle in a right triangle we can relabel to get that

$$\sin A = \frac{\sinh a}{\sinh c}.$$

To prove $\cosh a = \dfrac{\cos A}{\sin B}$

$$\sin 2A + \cos 2A = 1$$

$$\frac{\sinh^2 a}{\sinh^2 c} + \frac{\tanh^2 b}{\tanh^2 c} = 1$$

$$1 + \sinh^2 a + \cosh^2 c \tanh^2 b = \sinh^2 c$$

$$1 + \sinh^2 a + \cosh^2 c \left(\frac{\sinh^2 b}{\cosh^2 b}\right) = 1 + \sinh^2 c$$

$$\cosh^2 a \cosh^2 b + \cosh^2 c \sinh^2 b = \cosh^2 b \cosh^2 c$$

$$\cosh^2 c(\cosh^2 b - \sin^2 b) = \cosh^2 a \cosh^2 b$$

$$\cosh^2 c = \cosh^2 a \cosh^2 b$$

$$\cosh c = \cosh a \cosh b$$

Then,

$$\frac{\cos A}{\sin B} = \frac{\tanh b}{\tanh c}.\frac{\sin c}{\sin b} = \frac{\cosh c}{\cosh b} = \cosh a.$$

$$\cot A \cot B = \frac{\cos A \cos B}{\sin B \sin A} = \cosh a \cosh b.$$

follows from what we have done above.

Theorem: *For any triangle ÄABC in the hyperbolic plane, with k =1*

$$\cosh c = \cosh a \cosh b - \sinh a \sinh b \cos C$$

Hyperbolic Law of Cosines

$$\frac{\sin A}{\sinh a} = \frac{\sin B}{\sinh b} = \frac{\sin C}{\sinh c}.$$

Hyperbolic Law of Cosines

$$\cosh c = \frac{\cos A \cos B + \cos C}{\sin A \sin B}$$

Let us take a look at a specific example. Consider equilateral triangles. In Euclidean geometry all are similar, since they all must have angles measuring 60°. If this were true in hyperbolic geometry, they would have to be congruent by AAA. What then are the angles in an equilateral triangle of differing sides?

Table: Hyperbolic Equilateral Triangles

Sides	*Radians*	*Degrees*
10	0.0135	0.77
5	0.1633	9.35
2.5	0.5359	30.71
1.5	0.7930	45.43
1	0.9188	52.64
.5	1.0122	57.99
.1	1.0458	59.92

Other interesting examples would be how the angles in a right isosceles triangle vary with the sides and what triangle is analogous to the standard 30-60-90 triangle in Euclidean geometry.

CIRCUMFERENCE AND AREA OF A CIRCLE

Theorem: *The circumference C of a circle of radius r is*

$$C = 2\pi \sinh r.$$

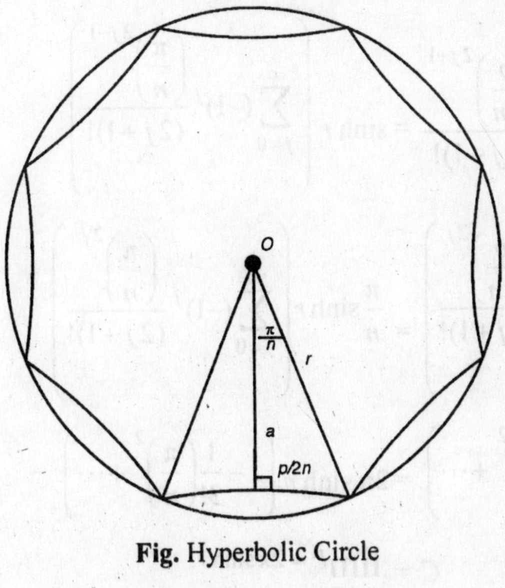

Fig. Hyperbolic Circle

In Euclidean geometry $C = \lim_{n\to\infty} P_n$ where P_n is the perimeter of the regular n-gon inscribed in the circle.

$$Pn = 2nr \sin\frac{\pi}{n} = 2nr\left[\frac{\pi}{n} - \frac{1}{3!}\left(\frac{\pi}{n}\right)^3 + \frac{1}{5!}\left(\frac{\pi}{n}\right)^5 - \ldots\right]$$

$$= 2\pi r - 2\pi r\left[\frac{1}{3!}\left(\frac{\pi}{n}\right)^2 + \frac{1}{5!}\left(\frac{\pi}{n}\right)^4 - \ldots\right]$$

Thus, $C = \lim_{n\to\infty} P_n$.

In hyperbolic geometry we can still compute the perimeter and compute the limit, but we will use Theorem to compute the perimeter.

The proof is nothing but the following computation.

$$\sin\frac{\pi}{2n} = \frac{\sinh\frac{p}{2n}}{\sinh r}$$

$$\sinh\frac{p}{2n} = \sinh r \sin\frac{\pi}{n}$$

$$\sum_{j=0}^{\infty}\frac{\left(\frac{p}{2n}\right)^{2j+1}}{(2j+1)!} = \sinh r\left(\sum_{j=0}^{\infty}(-1)^j\frac{\left(\frac{\pi}{n}\right)^{2j+1}}{(2j+1)!}\right)$$

$$\frac{p}{2n}\left(\sum_{j=0}^{\infty}\frac{\left(\frac{p}{2n}\right)^{2j}}{(2j+1)!}\right) = \frac{\pi}{n}\sinh r\left(\sum_{j=0}^{\infty}(-1)^j\frac{\left(\frac{\pi}{n}\right)^{2j}}{(2j+1)!}\right)$$

$$p\left(1 + \frac{1}{3!}\left(\frac{p}{2n}\right)^2 + \cdots\right) = 2\pi \sinh r\left(1 - \frac{1}{3!}\left(\frac{\pi}{n}\right)^2 + \cdots\right)$$

$$C = \lim_{n\to\infty} P = 2\pi \sinh r.$$

Let K be the area of ÄABC so that according to our choices, with k=1 and angles measured in radians, $K = \pi - \angle A - \angle B - \angle C$ Let ÄABC have a right angle at C, then $K = \frac{\pi}{2} - A - B$.

Theorem: *tan K/2 = tanh a/2 tanh b/2*

Once again the proof is a computation.

$$\tanh^2 \frac{a}{2} \tanh^2 \frac{b}{2} = \frac{\cosh a - 1}{\cos a + 1} . \frac{\cosh b - 1}{\cos b + 1}$$

$$= \frac{\cos A - \sin B}{\cos A + \sin B} . \frac{\cos B - \sin A}{\cos B + \sin A}$$

$$= \frac{1 - \sin(A+B)\cos(A-B)}{1 + \sin(A+B)\cos(A-B)}$$

$$= \frac{1 - \cos K}{1 + \cos K} = \tan^2 \frac{K}{2} .$$

Using this and our limiting approach we can now compute the area of a circle.

Theorem: *The area, A, of a circle of radius r is*

$$A = 4\pi \sinh^2 \frac{\pi}{2}$$

Proof: We do this just as before. If K_n is the area of the inscribed regular n-gon, then $A = \lim_{n\to\infty} K_n$. In the right triangle in Figure let K, a, and p denote K_n, a_n and P_n The area of the right triangle is $K_n/2_n$.

$$\tan \frac{K}{4n} = \tanh \frac{a}{2} \tanh \frac{p}{4n} 1$$

$$4n \tanh \frac{K}{4n} = (\tanh \frac{a}{2}) 4n \tanh \frac{p}{4n}$$

Now,

$$4n \tan \frac{K}{4n} = 4n \left[\frac{K}{4n} + \frac{1}{3}\left(\frac{K}{4n}\right)^3 + \frac{2}{15}\left(\frac{K}{4n}\right)^5 + \cdots \right]$$

$$= K \left[1 + \frac{1}{3}\left(\frac{K}{4n}\right)^2 + \frac{2}{15}\left(\frac{K}{4n}\right)^4 + \cdots \right]$$

and

$$4n \tanh \frac{p}{4n} = 4n \left[\frac{p}{4n} - \frac{1}{3}\left(\frac{p}{4n}\right)^3 + \frac{2}{15}\left(\frac{p}{4n}\right)^5 + \cdots \right]$$

$$= p\left[1+\frac{1}{3}\left(\frac{p}{4n}\right)^2-\frac{2}{15}\left(\frac{p}{4n}\right)^4+\cdots\right]$$

Thus, we find that

$$\lim_{n\to\infty} 4n\tanh\frac{p}{4n} = \lim_{n\to\infty} K = A$$

$$\lim_{n\to\infty} 4n\tanh\frac{p}{4n} = \lim_{n\to\infty} P = C$$

$$\lim_{n\to\infty} a = r$$

Putting all of this together we have that

$$A = C\tanh\frac{r}{2}$$

$$= 2\pi\sinh r\tanh\frac{r}{2}$$

$$= 2\pi\sinh r\frac{\cosh r-1}{\sinh r}$$

$$= 2\pi(\cosh r - 1)$$

$$= 4\pi\sinh^2\frac{r}{2}$$

Thus, we have computed the area of a circle.

HYPERBOLIC ANALYTIC GEOMETRY

MORE ON QUADRILATERALS

Now we need to consider a Saccheri quadrilateral which has base b, sides each with length a, and summit with length c. We showed that $c>a$, but we would like to know

- How much bigger?
- How are the relative sizes related to the lengths of the sides?

Theorem: For a Saccheri quadrilateral

$$\sinh\frac{c}{2} = (\cosh a)\cdot\left(\sinh\frac{b}{2}\right)$$

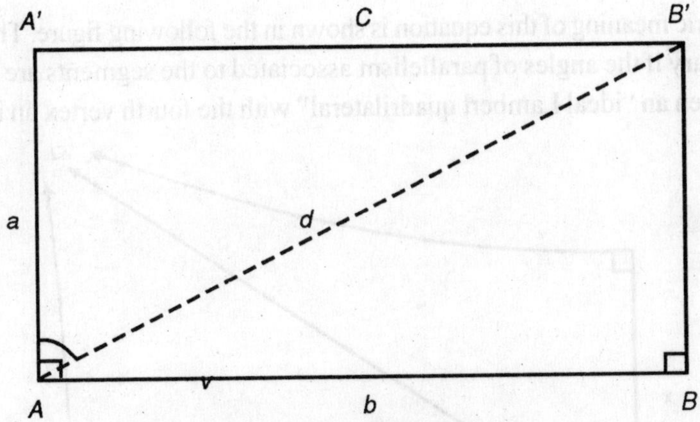

Fig. Saccheri Quadrilateral

Proof: Applying the Hyperbolic Law of Cosines from Theorem, we have

From Theoremwe know that

$$\cos\theta) = \sin\left(\frac{\pi}{2} - \theta\right) = \frac{\sinh a}{\sinh d}$$

$$\cosh d = \cosh a \cosh b$$

Using these in Equationwe eliminate the variable d and have

$$\cosh c = \cosh^2 a \cosh b - \sinh^2 a$$

$$= \cosh^2 a(\cosh b - 1) + 1$$

Now, we need to apply the identity

$$2\sinh^2\left(\frac{x}{2}\right) = \cosh x - 1,$$

and we have the formula.

Corollary: *Given a Lambert quadrilateral, if c is the length of a side adjacent to the acute angle, a is the length of the other side adjacent to the acute angle, and b is the length of the opposite side, then*

Two segments are said to be complementary segments if their lengths x and x^* are related by the equation

$$\Pi(x) + \Pi(x^*) = \frac{x}{2}$$

The geometric meaning of this equation is shown in the following figure. These lengths then are complementary if the angles of parallelism associated to the segments are complementary angles. This is then an "ideal Lambert quadrilateral" with the fourth vertex an ideal point Ù.

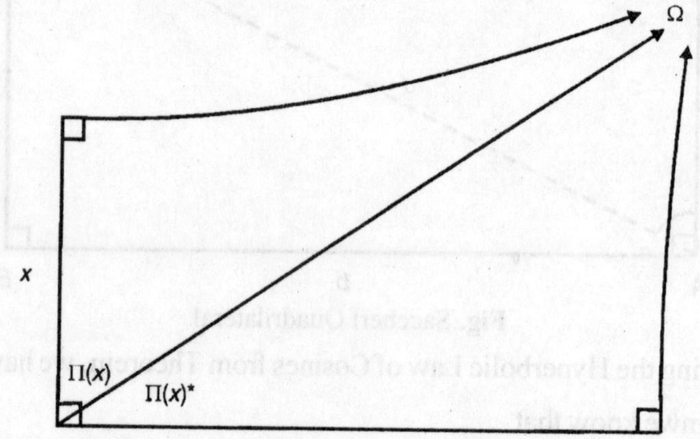

Fig. Complementary Segments

If we apply the earlier formulas for the anglê of parallelism to these segments, we get

$$\sinh x^* = \operatorname{csch} x$$

$$\cosh x^* = \coth x$$

$$\tanh x^* = \operatorname{scch} x$$

$$\tanh \frac{x^*}{2} = e^{-x}.$$

Engel's Theorem

Theorem: There is a right triangle with sides and angles as shown in Figure below if and only if there is a Lambert quadrilateral with sides as shown in Figure. Note that PQ is a complementary segment to the segment whose angle of parallelism is $\angle A$.

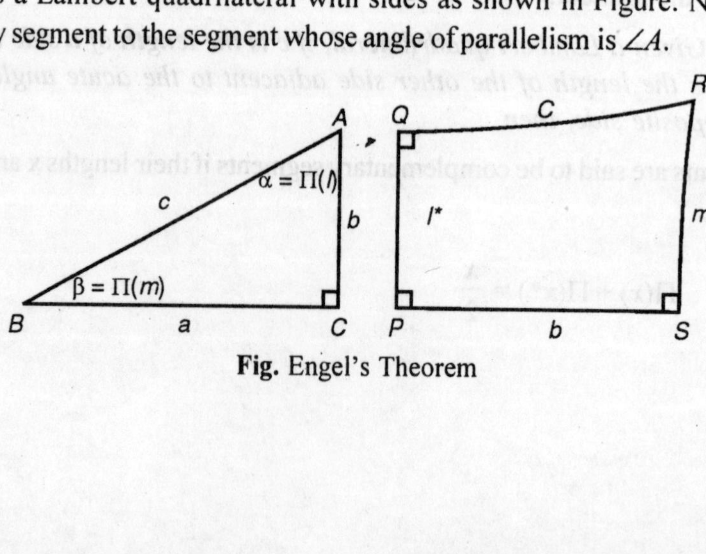

Fig. Engel's Theorem

COORDINATE GEOMETRY IN THE HYPERBOLIC PLANE

In the hyperbolic plane choose a point O for the origin and choose two perpendicular lines through $O - \overleftrightarrow{OX}$ and $\overleftrightarrow{OY}$. In our models—both the Klein and Poincaré—we will use the Euclidean center of our defining circle for this point O. We need to fix coordinate systems on each of these two perpendicular lines. By this we need to choose a positive and a negative direction on each line and a unit segment for each. There are other coordinate systems that can be used, but this is standard. We will call these the u-axis and the v-axis. For any point let $P \in H^2$ U and V be the feet of P on these axes, and let u and v be the respective coordinates of U and V. Then the quadrilateral $UOVP$ is a Lambert quadrilateral. If we label the length of UP as w and that of VP as z, then by the Corollary we have

$$\tanh w = \tanh v - \cosh u$$

$$\tanh z = \tanh u - \cosh v$$

Let $r = d_n(OP)$ be the hyperbolic distance from O to P and let θ be a real number so tha $-\pi < \theta < \pi$. Then

$$\tanh u = \cos\theta - \tanh r$$

$$\tanh v = \tanh\theta - \tanh r$$

We also set

$$x = \tanh u, \quad y = \tanh v$$

$$T = \cosh u, \cosh w \quad X = xT, Y = yT.$$

The ordered pair $\left\{\overleftrightarrow{OX}, \overleftrightarrow{OY}\right\}$ is called a frame with axes $\overleftrightarrow{OX}$ and $\overleftrightarrow{OY}$. With respect to this frame, we say the point P has

- Axial coordinates (u,v),
- Polar coordinates (r, θ),
- Lobachevsky coordinates (u,w),
- Beltrami coordinates (x,y),
- Weierstrauss coordinates (T,X,Y).

If a point has Beltrami coordinates (x,y) and $t = 1 + \sqrt{1 - x^2 - y^2}$ put

$$p = x/t \quad q = y/t$$

then (p,q) are the Poincaré coordinates of the point.

Every point has a unique ordered pair of Lobachevsky coordinates, and, conversely, every ordered pair of real numbers is tha pair of Lobachevsky coordinates for some unique point. In Lobachevsky coordinates

- For $a \neq 0$, $u = a$ is the equation of a line;

- For $a \neq 0$, $w = a$ is the equation of a hypercycle;
- $e^{-u} = \tanh w$ is an equation of the line in the first quadrant that is horoparallel to both axes.
- $e^{u} = \cosh w$ is an equation of the horocycle with radius

Thus, a line does not have a linear equation in Lobachevsky coordinates, and a linear equation does not necessarily describe a line.

Every point has a unique ordered pair of axial coordinates. However, not every ordered pair of real numbers is a pair of axial coordinates.

Let U and V be points on the axes with $V \neq 0$. Now the perpendiculars at U and V do not have to intersect. It is easy to see that they might be horoparallel or hyperparallel, especially by looking in the Poincaré model. If the two lines are limiting parallel (horoparallel) then that would make the segments OU and OV complementary segments. It can be shown then that these perpendiculars to the axes at U and V will intersect if and only if $|u| < |v|^*$. It then can be shown that (u,v) are the axial coordinates of a point if and only if $\tanh^2 u + \tanh^2 v < 1$.

***Lemma:** With respect to a given frame*

- Every point has a unique ordered pair of Beltrami coordinates, and (x, y) is an ordered pair of Beltrami coordinates if and only if $x^2 + y^2 < 1$
- If the point P_1 has Beltrami coordinates (x_1, y_2) and point has P_2 Beltrami coordinates (x_2, y_2) then the distance $d_h(P_1, P_2) = P_1P_2$ is given by the following formulae:

$$\cosh P_1 P_2 = \frac{1 - x_1x_2 - y_1y_2}{\sqrt{1 - x_1^2 - y_1^2}\,\sqrt{1 - x_2^2 - y_2^2}}$$

$$\tanh P_1 P_2 = \frac{\sqrt{(x_2 - x_1)^2 + (y_2 - y_1)^2 - (x_1y_2 - x_2y_1)^2}}{1 - x_1x_2 - y_1y_2}$$

- $Ax + By + C = 0$ is an equation of a line in Beltrami coordinates if and only if $A^2 + B^2 > C^2$, and every line has such an equation.
- Given an angle $\angle PQR$ and given that the Beltrami coordinates of P are (x_1, y_1) of Q are (x_2, y_2) and of R are (x_3, y_3), then the cosine of this angle is given by

$\cos(\angle PQR) =$

$$\frac{(x_2 - x_1)(x_3 - x_1) + (y_2 - y_1)(y_3 - y_1) - (x_2y_1 - x_1y_2)(x_3y_1 - x_1y_3)}{\sqrt{(x_2 - x_1)^2 + (y_2 - y_1)^2 - (x_1y_2 - x_2y_1)^2}\,\sqrt{(x_3 - x_1)^2 + (y_3 - y_1)^2 - (x_1y_3 - x_3y_1)^2}}$$

- If $Ax + By + C = 0$ and $Dx + Ey + F = 0$ are equations of two intersecting line in Beltrami coordinates and is the angle formed by their intersection, then

$$\cos\theta = \pm \frac{AD + BE - CE}{\sqrt{A^2 + B^2 - C^2}\sqrt{D^2 + E^2 - F^2}}.$$

In particular the lines are perpendicular if and only if $AD + BE = CF$.

- If (x_1, y_1) and (x_2, y_2) are the Beltrami coordinates of two distinct points, let $t_1 = \sqrt{1 - x_1^2 - y_1^2}$ and $t_2 = \sqrt{1 - x_2^2 - y_2^2}$. Then the midpoint of the segment joining the two points has Beltrami coordinates

$$\left(\frac{x_1t_2 + x_2t_1}{t_1 + t_2}, \frac{y_1t_2 + y_2t_1}{t_1 + t_2}\right)$$

and the perpendicular bisector of the two points has an equation

$$(x_1 t_2 - x_2 t_1)x + (y_1 t_2 - y_2 t_1)y + (t_1 - t_2) = 0.$$

- If $A_1x + B_1y + C_1 = 0$ and $A_2x + B_2y + C_2 = 0$ are equations of lines in Beltrami coordinates and if $A_1B_2 = A_2B_1$, then the two lines are hyperparallel.
- Every cycle has an equation in Beltrami coordinates that is of the form

$$\sqrt{1 - x^2 - y^2} = ax + by + c.$$

  - The cycle is a circle if and only if $-1 < a^2 + b^2 - a^2 < 0$ and $c > 0$.
  - The cycle is a horocycle if and only if $a^2 + b^2 - c^2$ and $c > 0$.
  - The cycle is a hypercycle if and only if $a^2 + b^2 - c^2 < 0$.

In Poincaré coordinates (p,q)

$$C((p^2 + q^2) + 2Ap + 2Bq + C = 0$$

is an equation of a line if and only if $A^2 + B^2 > C^2$, and every line has such an equation.

$$C(u^2 + v^2) + 2Au + 2Bv + C = 0$$

